Free Student Aid.

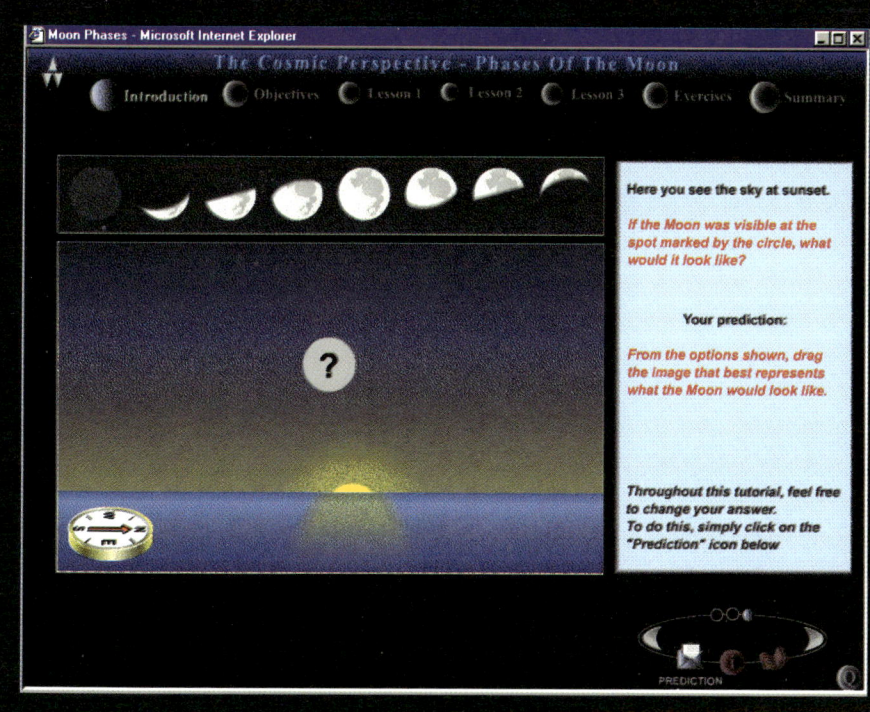

Log on.

Tune in.

Succeed.

To help you succeed in astronomy, your professor has arranged for you to enjoy access to great media resources, including planetarium software, Voyager: Skygazer, College Edition CD-ROM and The Astronomy Place web site. You'll find that these resources that accompany your textbook will enhance your course materials.

Here's your personal ticket to success:

How to log on to www.astronomyplace.com:

1. Go to www.astronomyplace.com
2. Click Bennett, *The Cosmic Perspective, Second Edition* book cover.
3. Click "Register Here."
4. Enter your pre-assigned access code exactly as it appears below and click "Submit."
5. Complete online registration form to create your own personal user ID and password.
6. Once your user ID and password are confirmed by email, go back to www.astronomyplace.com, type in your new user ID and password, and click "Enter."

Your Access Code is:

USTS-GAUSS-AZOLE-TIGON-ROSSI-POSSE

Detach this card and keep it handy. It's your ticket to valuable information.

Got technical questions?

For technical support, please visit www.aw.com/techsupport, send an email to online.support@pearsoned.com (for web site questions) or send an email to media.support@pearsoned.com (for CD-ROM questions) with a detailed description of your computer system and the technical problem. You can also call our tech support hotline at 1-800-677-6337 Monday-Friday, 8 a.m. to 5 p.m. CST.

What your system needs to use these media resources:

WINDOWS
250 MHz Intel Pentium processor or greater
Windows 95, 98 NT4, 2000
32 MB RAM installed, 64 MB preferred
800 X 600 screen resolution
4x CD-ROM drive
Browser: Internet Explorer 5.0 or Netscape Communicator 4.7
Plug-Ins: Shockwave Player 8, Flash Player 5, QuickTime 4.1

MACINTOSH
233 MHz G3 or higher
OS 8.1 or higher
32 MB RAM available, 64 MB preferred
800 x 600 screen resolution, thousands of colors
4x CD-ROM Drive
Browser: Internet Explorer 5.0 or Netscape Communicator 4.7
Plug-Ins: Shockwave Player 8, Flash Player 5, QuickTime 4.1

Important: Please read the License Agreement, located on the launch screen before using The Astronomy Place web site or Voyager: Skygazer, College Edition CD-ROM. By using the web site or CD-ROM, you indicate that you have read, understood and accepted the terms of this agreement.

0-8053-8556-8

Your User ID

_____

Your Password

_____

The Solar System

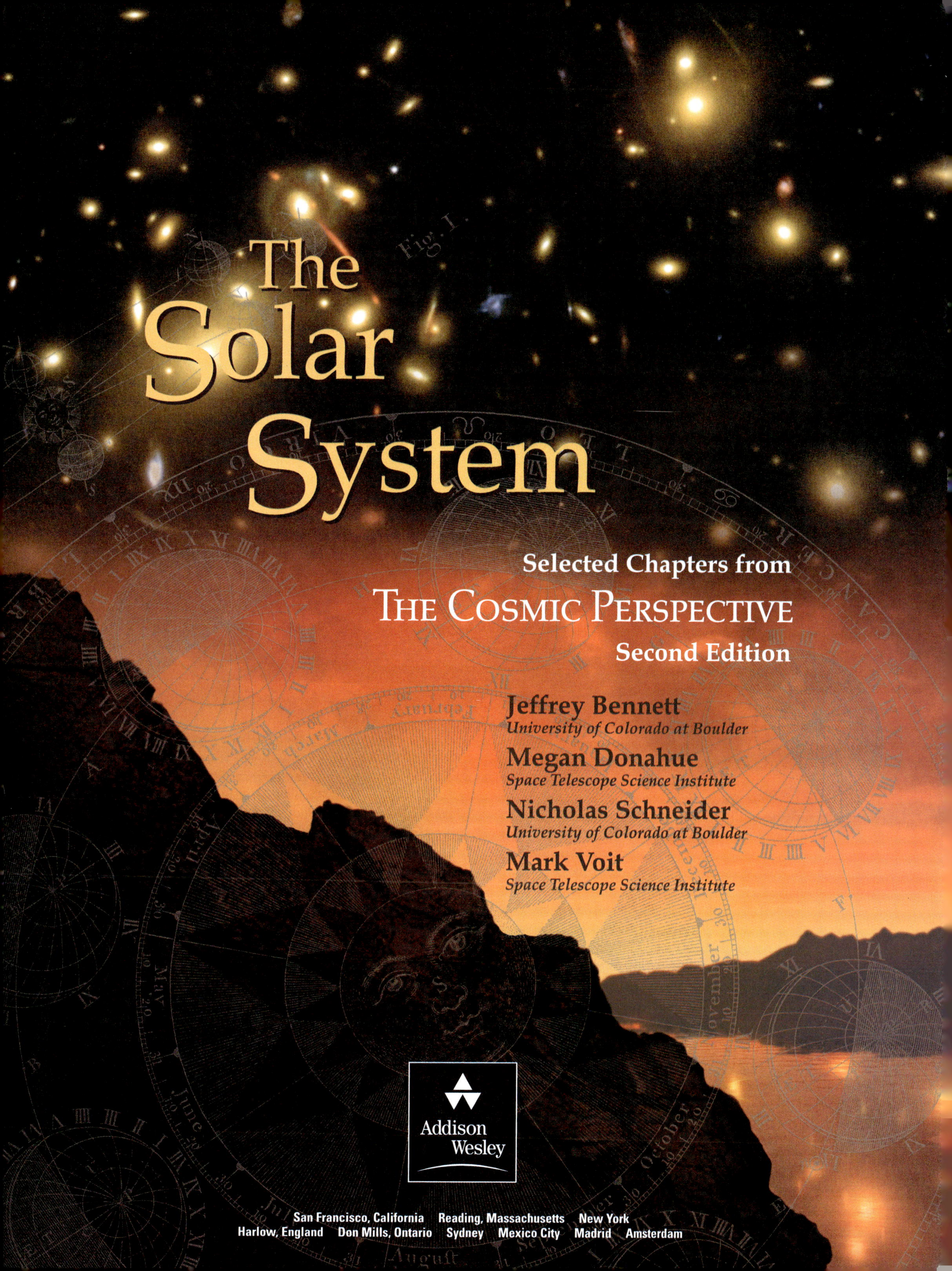

The Solar System

Selected Chapters from
THE COSMIC PERSPECTIVE
Second Edition

Jeffrey Bennett
University of Colorado at Boulder

Megan Donahue
Space Telescope Science Institute

Nicholas Schneider
University of Colorado at Boulder

Mark Voit
Space Telescope Science Institute

Addison Wesley

San Francisco, California   Reading, Massachusetts   New York
Harlow, England   Don Mills, Ontario   Sydney   Mexico City   Madrid   Amsterdam

Acquisitions Editor: *Adam Black*
Executive Editor: *Ben Roberts*
Market Developer: *Chalon Bridges*
Marketing Manager: *Christy Lawrence*
Publishing Associate: *Liana Allday*
Developmental Editor: *Laura Maria Bonazzoli*
Production Coordination: *Joan Marsh*
Production: *Mary Douglas, Rogue Valley Publications*
Photo Research: *Myrna Engler*
Graphic Artists: *Joe Bergeron, John Goshorn/Techarts, Blakeley Kim, Emiko-Rose Koike/fiVth.com, Quade Paul/fiVth.com*
Copyeditor: *Mary Roybal*
Text and Cover Designer: *Blakeley Kim, Andrew Ogus*
Partial Cover Photo: *Quade Paul/fiVth.com; PhotoDisk; and Hubble Space Telescope*
Composition: *Thompson Type/Alma Bell, Lori Shranko, Janice Adamski*
Cover Printer: *Coral Graphics*
Printer and Binder: *Von Hoffmann Press*

For permission to use copyrighted material, grateful acknowledgment is made to the copyright holders on pp. C-1–C-3, which are hereby made part of this copyright page.

Copyright © 2002 Pearson Education, Inc., publishing as Addison Wesley, 1301 Sansome St., San Francisco, CA 94111. All rights reserved. Manufactured in the United States of America. This publication is protected by copyright and permission should be obtained from the publisher prior to any prohibited reproduction, storage in a retrieval system, or transmission in any form or by any means, electronic, mechanical, photocopying, recording, or likewise. To obtain permission(s) to use material from this work, please submit a written request to Pearson Education, Inc., Permission Department, 1900 E. Lake Ave., Glenview, IL 60025. For information regarding permissions, call 847/486-2635.

Many of the designations used by manufacturers and sellers to distinguish their products are claimed as trademarks. Where those designations appear in this book, and the publisher was aware of a trademark claim, the designations have been printed in initial caps or all caps.

ISBN: 0-8053-8554-1

1 2 3 4 5 6 7 8 9 10—VHP—04 03 02 01

*We shall not cease from exploration
And the end of all our exploring
Will be to arrive where we started
And know the place for the first time.*
    T. S. ELIOT

Dedication

TO ALL WHO HAVE EVER WONDERED about the mysteries of the universe. We hope this book will answer some of your questions—and that it will also raise new questions in your mind that will keep you curious and interested in the ongoing human adventure of astronomy.

And, especially, to the members of the "baby boom" that has occurred among the authors and editors during the writing of this book: Michaela, Emily, Rachel, Sebastian, Elizabeth, Nathan, Grant, Georgia, Brooke, and Brian. The study of the universe begins at birth, and we hope that you will grow up in a world with far less poverty, hatred, and war so that all people will have the opportunity to contemplate the mysteries of the universe into which they are born.

BRIEF CONTENTS

(The chapters included in this volume are printed in bold type.)

PART I

DEVELOPING PERSPECTIVE

1 Our Place in the Universe 1
2 Discovering the Universe for Yourself 39
S1 Celestial Timekeeping and Navigation 65

PART II

KEY CONCEPTS FOR ASTRONOMY

3 The Science of Astronomy 91
4 A Universe of Matter and Energy 111
5 Universal Motion: From Copernicus to Newton 125
6 Light: The Cosmic Messenger 153
7 Telescopes and Spacecraft 171

PART III

LEARNING FROM OTHER WORLDS

8 Formation of the Solar System 197
9 Planetary Geology: Earth and the Other Terrestrial Worlds 225
10 Planetary Atmospheres: Earth and the Other Terrestrial Worlds 259
11 Jovian Planet Systems 287
12 Remnants of Rock and Ice: Asteroids, Comets, and Pluto 323
13 Planet Earth and Its Lessons on Life in the Universe 351

PART IV

A DEEPER LOOK AT NATURE

S2 Space and Time 383
S3 Spacetime and Gravity 405
S4 Building Blocks of the Universe 429

PART V

STELLAR ALCHEMY

14 Our Star 449
15 Stars 471
16 Star Stuff 493
17 The Bizarre Stellar Graveyard 519

PART VI

GALAXIES AND BEYOND

18 Our Galaxy 541
19 Galaxies: From Here to the Horizon 567
20 Galaxy Evolution 589
21 Dark Matter and the Fate of the Universe 613
22 The Beginning of Time 633
S5 Interstellar Travel and the Search for Extraterrestrial Civilizations 655

APPENDIXES

Appendix A Useful Numbers A-2
Appendix B Useful Formulas A-3
Appendix C A Few Mathematical Skills A-4
Appendix D The Periodic Table of the Elements A-13
Appendix E Planetary Data A-15
Appendix F Stellar Data A-18
Appendix G Galaxy Data A-20
Appendix H Selected Astronomical Web Sites A-23
Appendix I The 88 Constellations A-26
Appendix J Star Charts A-29

Glossary G-1
Acknowledgments C-1
Index I-1

DETAILED CONTENTS

PART I

DEVELOPING PERSPECTIVE

1 Our Place in the Universe 1

1.1 A Modern View of the Universe 4
1.2 The Scale of the Universe 10
Virtual Tour of the Universe 12
1.3 Spaceship Earth 27
1.4 The Human Adventure of Astronomy 34

The Big Picture 35
Review Questions 36
Discussion Questions 36
Problems 36
Web Projects 37

Basic Astronomical Definitions 4
*Mathematical Insight 1.1:
  How Far Is a Light-Year? 8*
Common Misconceptions: Light-Years 9
*Common Misconceptions:
  Confusing Very Different Things 25*
*Common Misconceptions:
  The Cause of Seasons 29*
Common Misconceptions: Sun Signs 30

2 Discovering the Universe for Yourself 39

2.1 Patterns in the Sky 40
2.2 The Circling Sky 42
2.3 The Moon, Our Constant Companion 50
2.4 The Ancient Mystery of the Planets 58

The Big Picture 63
Review Questions 63
Discussion Questions 63
Problems 63
Web Projects 64

*Common Misconceptions:
  Stars in the Daytime 41*
Common Misconceptions: A Flat Earth 44
*Common Misconceptions:
  What Makes the North Star Special? 47*
Common Misconceptions: High Noon 50
*Common Misconceptions:
  Moon in the Daytime 51*
*Common Misconceptions:
  Moon on the Horizon 52*
*Common Misconceptions:
  The "Dark Side" of the Moon 53*
*Thinking About . . .
  The Moon and Human Behavior 57*
Thinking About . . . Aristarchus 60
Thinking About . . . And Yet It Moves 62

S1 Celestial Timekeeping and Navigation 65

S1.1 Astronomical Time Periods 66
S1.2 Daily Timekeeping 69
S1.3 The Calendar 72
S1.4 Locations in the Sky 73
S1.5 Understanding Local Skies 78
S1.6 Principles of Celestial Navigation 84

The Big Picture 87
Review Questions 88
Discussion Questions 88
Problems 88
Web Projects 89

Thinking About . . .
   Solar Days and the Analemma 71
Mathematical Insight S1.1: Time by the Stars 76
Common Misconceptions: Compass Directions 85

PART II

KEY CONCEPTS FOR ASTRONOMY

3 The Science of Astronomy 91

3.1 Everyday Science 92
3.2 Ancient Observations 92
3.3 The Modern Lineage 100
3.4 Modern Science and the Scientific Method 103
3.5 Astronomy Today 106
3.6 Astrology 106

The Big Picture 109
Review Questions 109
Discussion Questions 109
Problems 110
Web Projects 110

Mathematical Insight 3.1: The Metonic Cycle 96
Thinking About . . .
   Eratosthenes Measures the Earth 101
Thinking About . . .
   Logic and the Scientific Method 104
Common Misconceptions:
   Eggs on the Equinox 105

4 A Universe of Matter and Energy 111

4.1 Matter and Energy in Everyday Life 112
4.2 A Scientific View of Energy 113
4.3 The Material World 117
4.4 Energy in Atoms 120

The Big Picture 122
Review Questions 122
Discussion Questions 123
Problems 123
Web Projects 124

Mathematical Insight 4.1: Temperature Scales 113
Mathematical Insight 4.2: Density 115
Mathematical Insight 4.3: Mass-Energy 116
Common Misconceptions:
   The Illusion of Solidity 118
Common Misconceptions:
   One Phase at a Time? 119
Common Misconceptions: Orbiting Electrons? 122

5 Universal Motion: From Copernicus to Newton 125

5.1 Describing Motion: Examples from Daily Life 126
5.2 Understanding Motion: Newton's Laws 129
5.3 Planetary Motion: The Copernican Revolution 133
5.4 The Force of Gravity 139
5.5 Tides 142
5.6 Orbital Energy and Escape Velocity 145
5.7 The Acceleration of Gravity 147

The Big Picture 148
Review Questions 149
Discussion Questions 149
Problems 149
Web Projects 151

Common Misconceptions:
  No Gravity in Space? 128
Mathematical Insight 5.1:
  Units of Force, Mass, and Weight 130
Common Misconceptions:
  What Makes a Rocket Launch? 131
Thinking About . . . Aristotle 138
Mathematical Insight 5.2:
  Using Newton's Version of Kepler's Third Law 142
Common Misconceptions:
  The Origin of Tides 143
Mathematical Insight 5.3:
  Calculating the Escape Velocity 146
Mathematical Insight 5.4:
  The Acceleration of Gravity 147

6 Light: The Cosmic Messenger 153

6.1 Light in Everyday Life 154
6.2 Properties of Light 155
6.3 The Many Forms of Light 158
6.4 Light and Matter 159
6.5 The Doppler Shift 165

The Big Picture 168
Review Questions 168
Discussion Questions 169
Problems 169
Web Projects 170

Mathematical Insight 6.1:
  Wavelength, Frequency, and Energy 157
Common Misconceptions:
  Is Radiation Dangerous? 158
Movie Madness: Lois Lane's Underwear 159
Common Misconceptions:
  You Can't Hear Radio Waves and You Can't See X Rays 159
Mathematical Insight 6.2:
  Laws of Thermal Radiation 162
Mathematical Insight 6.3: The Doppler Shift 167

7 Telescopes and Spacecraft 171

7.1 Eyes and Cameras: Everyday Light Sensors 172
7.2 Telescopes: Giant Eyes 175
7.3 Uses of Telescopes 180
7.4 Atmospheric Effects on Observations 182
7.5 Telescopes Across the Spectrum 184
7.6 Spacecraft 189

The Big Picture 193
Review Questions 193
Discussion Questions 193
Problems 194
Web Projects 195

Mathematical Insight 7.1: Angular Separation 174
Common Misconceptions:
  Magnification and Telescopes 177
Mathematical Insight 7.2:
  The Diffraction Limit 180
Common Misconceptions:
  Twinkle, Twinkle Little Star 183
Common Misconceptions:
  Closer to the Stars? 186

PART III

LEARNING FROM OTHER WORLDS

8 Formation of the Solar System 197

8.1 Comparative Planetology 198
8.2 The Origin of the Solar System: Four Challenges 198
8.3 The Nebular Theory of Solar System Formation 203
8.4 Building the Planets 206
8.5 Leftover Planetesimals 212
8.6 The Age of the Solar System 215
8.7 Other Planetary Systems 218

The Big Picture 221
Review Questions 222
Discussion Questions 222
Problems 223
Web Projects 224

Common Misconceptions:
  Solar Gravity and the Density of Planets 208
Mathematical Insight 8.1: Radioactive Decay 216

9 Planetary Geology: Earth and the Other Terrestrial Worlds 225

9.1 Comparative Planetary Geology 226
9.2 Inside the Terrestrial Worlds 227
9.3 How Interiors Work 232
9.4 Shaping Planetary Surfaces 235
9.5 A Geological Tour of the Terrestrial Worlds 246

The Big Picture 255
Review Questions 256
Discussion Questions 256
Problems 257
Web Projects 258

Thinking About . . . Seismic Waves 231
Common Misconceptions: Pressure and Temperature 233
Mathematical Insight 9.1: The Surface Area–to–Volume Ratio 234

10 Planetary Atmospheres: Earth and the Other Terrestrial Worlds 259

10.1 Planetary Atmospheres 260
10.2 Atmospheric Structure 263
10.3 Magnetospheres and the Solar Wind 270
10.4 Weather and Climate 270
10.5 Atmospheric Origins and Evolution 278
10.6 History of the Terrestrial Atmospheres 282

The Big Picture 284
Review Questions 284
Discussion Questions 285
Problems 285
Web Projects 286

Mathematical Insight 10.1: "No Greenhouse" Temperatures 266
Common Misconceptions: Higher Altitudes Are Always Colder 269
Common Misconceptions: The Greenhouse Effect Is Bad 270
Thinking . . . Weather and Chaos 272
Common Misconceptions: Do Toilets Flush Backward in the Southern Hemisphere? 274
Mathematical Insight 10.2: Thermal Escape from an Atmosphere 280

11 Jovian Planet Systems 287

11.1 The Jovian Worlds: A Different Kind of Planet 288
11.2 Jovian Planet Interiors 291
11.3 Jovian Planet Atmospheres 293
11.4 Jovian Planet Magnetospheres 300
11.5 A Wealth of Worlds: Satellites of Ice and Rock 302
11.6 Jovian Planet Rings 316

The Big Picture 320
Review Questions 321
Discussion Questions 321
Problems 321
Web Projects 322

12 Remnants of Rock and Ice: Asteroids, Comets, and Pluto 323

12.1 Remnants from Birth 324
12.2 Asteroids 324
12.3 Meteorites 330
12.4 Comets 333
12.5 Pluto: Lone Dog or Part of a Pack? 338
12.6 Cosmic Collisions: Small Bodies Versus the Planets 341

The Big Picture 348
Review Questions 349
Discussion Questions 349
Problems 350
Web Projects 350

13 Planet Earth and Its Lessons on Life in the Universe 351

13.1 How Is Earth Different? 352
13.2 Our Unique Geology 352
13.3 Our Unique Atmosphere 362
13.4 Life 365
13.5 Lessons from the Solar System 371
13.6 Lessons from Earth: Life in the Solar System and Beyond 374

The Big Picture 378
Review Questions 378
Discussion Questions 379
Problems 379
Web Projects 380

Common Misconceptions: Ozone—Good or Bad? 364

APPENDIXES

Appendix A Useful Numbers A-2

Appendix B Useful Formulas A-3

Appendix C A Few Mathematical Skills A-4

Appendix D The Periodic Table of the Elements A-13

Appendix E Planetary Data A-15

Appendix F Stellar Data A-18

Appendix G Galaxy Data A-20

Appendix H Selected Astronomical Web Sites A-23

Appendix I The 88 Constellations A-26

Appendix J Star Charts A-29

Glossary G-1

Acknowledgments C-1

Index I-1

PREFACE

We humans have gazed into the sky for countless generations, wondering how our lives are connected to the Sun, Moon, planets, and stars that adorn the heavens. Today, through the science of astronomy, we know that these connections go far deeper than our ancestors ever imagined. This book tells the story of modern astronomy and the new perspective—*The Cosmic Perspective*—with which it allows us to view ourselves and our planet. It is written for anyone who is curious about the universe, but it is designed primarily as a textbook for college students not planning to major in mathematics or science.

Approach

This book grew out of our experience teaching astronomy both to college students and the general public over the past 20 years. During this time, a flood of new discoveries fueled a revolution in our understanding of the cosmos, but the basic organization and approach of most astronomy textbooks remained unchanged. We felt the time had come to rethink how to organize and teach the major concepts in astronomy. This book is the result. Several major innovations in organization set this book apart.

A "Big Picture" Context Throughout the Book.

Many traditional textbooks begin with the sky as viewed by the ancients and gradually work through to the expanding universe and modern ideas of cosmology at the very end of the book. The potential advantage to this traditional organization is that, if students fully absorb it, they recreate for themselves the thought patterns and paradigm shifts that led from ancient superstitions about the sky to modern understanding of the universe. Unfortunately, this potential advantage is very difficult to realize in practice, because it means that students must learn a sequence of successive paradigms, why each was rejected, and why the next one followed, before finally arriving at what we now consider the correct understanding. That is, students must essentially recreate the thinking produced over thousands of years by some of the greatest minds in history, all with just a few hours of study per week over one or two course terms. Not only is such a process unrealistic for most students, but it also tends to lead to confusion among the various paradigms, thereby preventing students from understanding either the modern view or how we arrived at it. We have therefore chosen a different approach in which we begin the book with a broad overview of modern scientific understanding of the cosmos, and then use the rest of the book to help students understand how and why we have reached this understanding and what gaps remain in our knowledge. We believe that, for the vast majority of students, this approach enables them to understand the scientific process much better than they would with the historical approach, because knowing the end goal helps them focus their studies. We maintain this "big picture" context throughout the book by always telling students where we are going when we introduce new topics, and by reviewing the context at the end of each chapter (in the end-of-chapter sections called *The Big Picture*).

Stronger Emphasis on the Scientific Process.

Nearly all astronomy textbooks emphasize the scientific process, but we believe that our approach allows us to go beyond what is possible with a more traditional, historical approach to astronomy. In particular, by starting with the context of modern astronomy, we are able to focus clearly on how science works (Chapter 3) and on key physical principles that apply universally (Chapters 4 through 6). For example, we use the concept of conservation of energy over and over throughout the text to show how it helps explain everything from whether planets are geologically active to the processes that govern the lives of stars.

Fully Integrated History.

Most introductory astronomy textbooks lump the historical development of astronomy, from the Greeks through the Copernican revolution, into a single segment (for example, one or two chapters) of text. While the chronological flow of this approach certainly has aesthetic appeal, we have found that it also tends to encourage students to memorize the history for the test and then quickly forget it afterward. We therefore believe it is far more effective to integrate historical development throughout the narrative so that each historical topic appears in the context of how it helped lead to particular modern topics. This is precisely the approach to history that helped make Carl Sagan's *Cosmos* series so popular and informative, and we have found that it works equally well in the college classroom. Moreover, because it spreads the historical topics throughout the text, it actually allows us to have *more* coverage of the history of astronomy than can comfortably fit in most other astronomy texts. (Also, we have taken extraordinary care to make sure that our historical discussions are based on the most current understanding of history; to help ensure accuracy, noted historian of science Owen Gingerich reviewed our historical material.)

True Comparative Planetology. We offer a true comparative planetology approach, in which the discussion emphasizes general principles that apply to planetary geology and planetary atmospheres as opposed to simply describing one planet at a time. This approach has several crucial advantages over the more common, planet-by-planet approach. For example, by focusing on planetary processes rather than simply on morphological features, we strengthen students' grasp of how a few key physical principles apply throughout the universe, thereby furthering our goal of helping students understand the scientific process. In addition, long after the course is over, students are much more likely to retain the understanding of physical processes built in this way than they are of any particular planetary features that they memorize through a planet-by-planet approach. Perhaps most important, the comparative planetology approach helps students see the relevance of planetary science to their own lives by enabling them to gain a far deeper appreciation of our own unique world.

An Evolutionary Approach to Galaxies and Dark Matter. Over the past few decades, a revolution has occurred in our understanding of galactic evolution. For example, whereas quasars were once considered a distinct class of objects, the evidence now strongly contradicts this idea; instead, it now appears that quasars represent an unusually active stage of life in the centers of some young galaxies. Despite this revolution in scientific understanding, most introductory textbooks still present galaxies in the order that made sense when early galaxy classification schemes were first developed a half-century ago. We have taken an approach that teaches about galaxies in a way that closely parallels how nearly all textbooks (including ours) teach about stars: we begin with our own galaxy as a paradigm (Chapter 18), then discuss the variety of galaxies and how we determine key parameters such as distances (Chapter 19), and then discuss how we've arrived at the current state of knowledge regarding galaxy evolution (Chapter 20). Because dark matter plays such an important role in this evolutionary understanding, we then devote an entire chapter (Chapter 21) to its study.

An Integrated Approach to Cosmology. Ultimately, one of the primary goals of astronomy is to understand our cosmic origins, which means understanding cosmology. Most introductory textbooks present cosmology only in a final chapter. We integrate the study of cosmology throughout much of the text. Students first encounter modern cosmological ideas, including the notion of the Big Bang and the idea of an expanding universe, as part of our big picture overview in Chapter 1. Our entire presentation of galaxies is steeped in cosmology, since the study of galactic evolution cannot be effectively separated from cosmological ideas. Thus, we do not have a single cosmology chapter in the traditional sense, but instead we have a chapter devoted to the Big Bang theory (Chapter 22) that builds upon the understanding students develop throughout the text.

Supplementary Chapters on Relativity and Quantum Mechanics. It is impossible to teach a class in astronomy without receiving questions such as: Why can't we go faster than the speed of light? What is the universe expanding into? What is a black hole? The common thread to these questions is that the answers, or possible answers, can be understood only through relativity. We believe that the ideas of relativity can be made accessible to anyone and have therefore included two chapters (Chapters S2 and S3) covering basic ideas of special and general relativity. We have also included a chapter on quantum mechanics and its astronomical implications (Chapter S4). These three chapters will enable students to understand stars, galaxies, and cosmology in much greater depth than they could otherwise. And, from our teaching experience, students will find the topics of relativity and quantum mechanics to be among their favorite topics in the course. However, recognizing that many instructors will not have the time or inclination to cover these chapters, we have made them supplemental chapters—covering them will enhance student understanding of the chapters that follow, but they are not prerequisite to the remaining chapters.

Pedagogical Features

Besides the main narrative, the book includes a number of pedagogical features designed to enhance student learning:

End-of-Chapter Questions, Problems, and Projects. The following four sets of questions appear at the end of each chapter:

- *Review Questions.* The Review Questions replace a standard chapter summary. Like a summary, they review the important concepts from the chapter. However, they are more effective as a study tool than a standard summary because they require students to think rather than enabling mindless use of a highlight pen. While it is possible to assign review questions as homework, these questions generally can be answered with little more than a quick re-reading of the relevant sections of the text.

- *Discussion Questions.* These questions are meant to be particularly thought provoking and generally do not have objective answers. As such, they are ideal for in-class discussion.

- *Problems.* Each chapter includes several problems of varying levels of difficulty that are designed to be assigned as written homework. Problems are generally arranged from easier to harder, and to the extent possible follow the order of presentation in the chapter. Problems with an asterisk require mathematical manipulation, generally following methods described in Mathematical Insight boxes.

- *Web Projects.* Each chapter ends with one or more Web Projects, which are designed for independent research. The projects typically ask students to learn more about a topic of relevance to the chapter, such as a current or planned mission. Students can find useful links for all the Web Projects at The Astronomy Place (www.astronomyplace.com).

Time Out to Think. This feature gives students the opportunity to reflect on important new concepts. It also serves as an excellent starting point for classroom discussions. Answers or discussion points for all the Time Out to Think questions can be found in the Instructor's Guide.

Common Misconceptions. These boxes address and correct popularly held but incorrect ideas related to the chapter material. In a few cases in which the boxes focus on misconceptions perpetrated by television or movies, the boxes carry the title "Movie Madness."

Mathematical Insights. These boxes contain most of the mathematics used in the book, and they can be covered or skipped depending on the level of mathematics that you wish to include in your course.

Thinking About. These boxes contain supplementary discussion of topics related to the chapter material but not prerequisite to the continuing discussion.

The Big Picture. This end-of-chapter feature helps students put what they've learned into the context of the overall goal of gaining a new perspective on ourselves and our planet.

Cross-References. When we discuss a concept that is covered in greater detail elsewhere in the book, we include a cross-reference in brackets. For example, [Section 5.2] means that the concept being discussed is covered in greater detail in Section 5.2.

Glossary. A detailed glossary makes it easy for students to look up unfamiliar terms.

Appendices. The appendices include a number of useful references, including summaries of key constants (Appendix A), key formulas (Appendix B), and a brief review of key mathematical skills such as working with scientific notation and operating with units (Appendix C).

Star Charts. The list of constellations and the star charts help interested students learn their way around the night sky.

Alternate Versions of the Book

The Cosmic Perspective is available in four versions tailored to particular course types:

- *The Cosmic Perspective, Second Edition* (full version, ISBN 0-8053-8044-2)
  This is the most complete version of the book, containing sufficient material for astronomy courses ranging in length from one quarter to a full year.

- *The Solar System: The Cosmic Perspective, Second Edition, Volume 1* (ISBN 0-8053-8544-1)
  This version contains Chapters 1–13 and Chapter S1 of the full version of the book. It is designed for courses that focus on the solar system rather than on stars, galaxies, and cosmology.

- *Stars, Galaxies, and Cosmology: The Cosmic Perspective, Second Edition, Volume 2* (ISBN 0-8053-8557-6)
  This version contains Chapters 1, 4–7, 14–22, and S2–S5 of the full version of the book. It is designed for courses that focus on stars, galaxies, and cosmology.

- *The Cosmic Perspective, Brief Edition* (ISBN 0-201-69734-3)
  This is a condensed version of the full book, designed primarily for use in courses that provide an overview of all of astronomy but at a lower level of depth than offered in the full version of the book.

Supplements and Resources

The Cosmic Perspective is much more than just a textbook; it is a complete package of resources designed to help instructors and students. Here, we've broken the resources into two categories, the first designed to help students directly and the second designed to help instructors prepare course materials and lectures.

Resources To Help Students Learn

The following resources are designed to reinforce the content knowledge students will learn from the textbook.

The Astronomy Place (www.astronomyplace.com). This Web site is designed specifically to accompany *The Cosmic Perspective* textbooks. Among its constantly growing list of features, instructors and students will find:

- A complete set of student study resources for each chapter in the textbook; the resources include chapter summaries, additional information on Common Misconceptions and Mathematical Insights, self-assessment quizzes, and chapter-specific links (including links for the end-of-chapter Web Projects). In addition, all Time Out To Think questions and all end-of-chapter Problems are digitized for online assignments. The entire textbook is available on-line in XML and can be browsed by topic.

- A suite of interactive tutorials that build understanding through a carefully constructed, step-by-step process based on the type of interaction and discussion that typically would occur during office hours.

- Ten 5-minute animated and narrated mini-movies that provide engaging summaries of key topics.

- Activity worksheets to be used with the *SkyGazer, College Edition* software packaged with the textbook.

- On-line versions of the textbook glossary and appendices.

- Updates on the latest astronomy news.

Carl Sagan's Cosmos Series. The newly released *Best of Cosmos* (ISBN: 0-8053-8571-1) and complete, revised, enhanced, and updated *Cosmos* series (Video ISBN: 0-8053-8570-3, DVD ISBN: 0-8053-8572-X) are available free to qualified adopters of *The Cosmic Perspective.*

SkyGazer, College Edition. Based on *Voyager III*, one of the world's most popular planetarium programs, *SkyGazer, College Edition* makes it easy for students to learn constellations and explore the wonders of the sky through interactive exercises. The *SkyGazer* CD is packaged free with all new copies of the textbook. It can also be ordered separately (CD-ROM ISBN: 0-8053-8022-1).

The Astronomy Tutor Center. This center provides one-to-one tutoring by qualified college instructors in any of four ways—phone, fax, email, and the Internet—during afternoon/evening hours and on weekends. The tutor center instructors will answer questions and provide help with examples, exercises, and other content found in *The Cosmic Perspective.* Students who register to use the astronomy tutor center can receive as much tutoring as they wish for the duration of their course. Registration is free with purchase of a new textbook or can be purchased separately; see www.aw.com/tutorcenter for more information.

Course Preparation Resources

Instructor's Resource Guide. (ISBN: 0-8053-8559-2) Includes an overview of the philosophy of the text, discussion of strategies for teaching astronomy, suggestions on preparing an astronomy course, sample syllabi for courses of different length and emphasis, a reference guide for integrating the *Cosmos* video series and the Astronomy Place Web tutorials and movies with this textbook, a chapter-by-chapter commentary with notes and teaching hints, answers or discussion points for all the Time Out to Think questions in the book, and solutions to end-of-chapter problems.

The Astronomy Place Instructor Resources (www.astronomyplace.com). The Astronomy Place has numerous resources designed specifically for instructors, including:

- A digitized version of the Instructor's Resource Guide (not including solutions to problems).

- Syllabus Manager, an easy to use on-line course management system. It can be used to build and maintain one or more syllabi on the Web. The system also allows you to receive and grade student assignments on-line.

- Customization resources that allow you to use the content of The Astronomy Place in WebCT, BlackBoard, or CourseCompass. If you are interested in these customized options, please contact your local Addison Wesley sales representative.

Printed Test Bank. (ISBN: 0-8053-8545-2) Includes over 1500 multiple-choice, true/false, and free-response questions for each chapter.

TestGen EQ CD-ROM. (ISBN: 0-8053-8045-0) Provides identical questions from the Printed Test Bank on one CD-ROM (Windows/Mac compatible). Allows the instructor to view and edit questions, add questions, and create and print different versions of tests. A built-in question editor creates graphs, imports graphics, and inserts text.

Transparency Acetates. (ISBN: 0-8053-8546-0) A selection of over 200 figures from the text printed on full-color transparency acetates.

Addison Wesley Science Digital Library. (ISBN: 0-8053-8569-X) This cross-platform CD-ROM includes illustrations, tables, and star charts from the book and many animations/video clips from the Web site that can be customized for lecture presentation.

Acknowledgments

A textbook may carry author names, but it is the result of hard work by a long list of committed individuals. We could not possibly list everyone who has helped, but we would like to call attention to a few people who have played particularly important roles. First, we thank our editors and friends at Addison Wesley Longman who have stuck with us through thick and thin, including Ben Roberts, Adam Black, Robin Heyden, Sami Iwata, Bill Poole, Linda Davis, Joan Marsh, Stacy Treco, Christy Lawrence, Nancy Gee, Liana Allday, and Tony Asaro. Special thanks to our production team, especially Mary Douglas, Myrna Engler, and Mary Roybal; our art team, led by Blakeley Kim and Emiko-Rose Koike; and our Web team, led by Claire Masson, Jim Dove, and Ian Shakeshaft.

We've also been fortunate to have an outstanding group of reviewers whose extensive comments and suggestions helped us shape the book. We thank all those who have reviewed drafts of the book in various stages, including:

Christopher M. Anderson, University of Wisconsin
Peter S. Anderson, Oakland Community College
John Beaver, University of Wisconsin at Fox Valley
Priscilla J. Benson, Wellesley College
Eric Carlson, Wake Forest University
Supriya Chakrabarti, Boston University
Dipak Chowdhury, Indiana University–Purdue University at Fort Wayne
Robert Egler, North Carolina State University at Raleigh
Robert A. Fesen, Dartmouth College
Sidney Freudenstein, Metropolitan State College of Denver
Richard Gelderman, Western Kentucky University
Richard Gray, Appalachian State University
David Griffiths, Oregon State University
David Grinspoon, University of Colorado
Jim Hamm, Big Bend Community College
Charles Hartley, Hartwick College
Joe Heafner, Catawba Valley Community College
Richard Ignace, University of Iowa
Bruce Jakosky, University of Colorado
Kurtis Koll, Cameron University
Kristine Larsen, Central Connecticut State University
Larry Lebofsky, University of Arizona
David M. Lind, Florida State University
Michael LoPresto, Henry Ford Community College
William R. Luebke, Modesto Junior College
Marie Machacek, Massachusetts Institute of Technology
Steven Majewski, University of Virginia
Phil Matheson, Salt Lake Community College
Barry Metz, Delaware County Community College
John P. Oliver, University of Florida
Jorge Piekarewicz, Florida State University
Harrison B. Prosper, Florida State University
Christina Reeves-Shull, Richland College
John Safko, University of South Carolina
James A. Scarborough, Delta State University
Joslyn Schoemer, Denver Museum of Nature and Science
James Schombert, University of Oregon
Gregory Seab, University of New Orleans
Mark H. Slovak, Louisiana State University
Dale Smith, Bowling Green State University
John Spencer, Lowell Observatory
Darryl Stanford, City College of San Francisco
John Stolar, West Chester University
Jack Sulentic, University of Alabama
C. Sean Sutton, Mount Holyoke College
Beverley A. P. Taylor, Miami University
Donald M. Terndrup, Ohio State University
David Trott, Metro State College
Darryl Walke, Rariton Valley Community College
Fred Walter, State University of New York
Mark Whittle, University of Virginia
Jonathan Williams, University of Florida
J. Wayne Wooten, Pensacola Junior College
Arthur Young, San Diego State University
Dennis Zaritsky, University of California, Santa Cruz

Historical Accuracy Reviewer—Owen Gingerich, Harvard–Smithsonian

In addition, we thank the following colleagues who helped us clarify technical points or checked the accuracy of technical discussions in the book:

Thomas Ayres, University of Colorado
Cecilia Barnbaum, Valdosta State University
Rick Binzel, Massachusetts Institute of Technology
Howard Bond, Space Telescope Science Institute
Humberto Campins, University of Florida
Robin Canup, Southwest Research Institute
Mark Dickinson, Space Telescope Science Institute
Jim Dove, Metropolitan State College of Denver
Harry Ferguson, Space Telescope Science Institute
Andrew Hamilton, University of Colorado
Todd Henry, Georgia State University
Dave Jewitt, University of Hawaii
Hal Levison, Southwest Research Institute
Mario Livio, Space Telescope Science Institute
Mark Marley, New Mexico State University
Kevin McLin, University of Colorado, Boulder
Rachel Osten, University of Colorado, Boulder
Bob Pappalardo, Brown University
Michael Shara, American Museum of Natural History
Glen Stewart, University of Colorado
John Stolar, West Chester University
Dave Tholen, University of Hawaii
Nick Thomas, MPI/Lindau (Germany)
John Weiss, University of Colorado, Boulder
Don Yeomans, Jet Propulsion Laboratory

Finally, we thank the many people who have greatly influenced our outlook on education and our perspective on the universe over the years, including Tom Ayres, Fran Bagenal, Forrest Boley, Robert A. Brown, George Dulk, Erica Ellingson, Katy Garmany, Jeff Goldstein, David Grinspoon, Don Hunten, Catherine McCord, Dick McCray, Dee Mook, Cheri Morrow, Charlie Pellerin, Carl Sagan, Mike Shull, John Spencer, and John Stocke.

Jeff Bennett
Megan Donahue
Nick Schneider
Mark Voit

HOW TO SUCCEED IN YOUR ASTRONOMY COURSE

Most readers of this book are enrolled in a college course in introductory astronomy. If you are one of these readers, we offer you the following hints to help you succeed in your astronomy course.

Using This Book

Before we address general strategies for studying, here are a few guidelines that will help you use *this* book most effectively.

- Read assigned material twice.
  - Make your first pass *before* the material is covered in class. Use this pass to get a "feel" for all the material and to identify concepts that you may want to ask about in class.
  - Read the material for the second time shortly after it is covered in class. This will help solidify your understanding and allow you to make notes that will help you study for exams later.

- Take advantage of the features that will help you study.
  - Always read the *Time Out to Think* features, and use them to help you absorb the material and to identify areas where you may have questions.
  - Use the *cross-references* indicated in brackets in the text and the *glossary* at the back of the book to find more information about terms or concepts that you don't recall.
  - Go to The Astronomy Place at **www.astronomyplace.com** to find additional study aids, including tutorials and quizzes with answers.

- It's your book, so don't be afraid to make notes in it that will help you study later.
  - Don't highlight—underline! Using a pen or pencil to underline material requires greater care than highlighting and therefore helps keep you alert as you study. And be selective in your underlining—for purposes of studying later, it won't help if you underlined everything.
  - There's plenty of "white space" in the margins and elsewhere, so use it to make notes as you read. Your own notes will later be very valuable when you are doing homework or studying for exams.

- After you complete the reading, and again when you study for exams, make sure you can answer the *Review Questions* at the end of each chapter. If you are having difficulty with a review question, reread the relevant portions of the chapter until the answer becomes clear.

Budgeting Your Time

One of the easiest ways to ensure success in any college course is to make sure you budget enough time for studying. A general rule of thumb for college classes is that you should expect to study about 2 to 3 hours per week *outside* of class for each unit of credit. For example, based on this rule of thumb, a student taking 15 credit hours should expect to spend 30 to 45 hours each week studying outside of class. Combined with time in class, this works out to a total of 45 to 60 hours spent on academic work—not much more than the time a typical job requires, and you get to choose your own hours. Of course, if you are working while you attend school, you will need to budget your time carefully.

As a rough guideline, your studying time in astronomy might be divided as shown in the table at the top of p. xix. If you find that you are spending

If Your Course Is:	Time for Reading the Assigned Text (per week)	Time for Homework Assignments (per week)	Time for Review and Test Preparation (average per week)	Total Study Time (per week)
3 credits	2 to 4 hours	2 to 3 hours	2 hours	6 to 9 hours
4 credits	3 to 5 hours	2 to 4 hours	3 hours	8 to 12 hours
5 credits	3 to 5 hours	3 to 6 hours	4 hours	10 to 15 hours

fewer hours than these guidelines suggest, you can probably improve your grade by studying more. If you are spending more hours than these guidelines suggest, you may be studying inefficiently; in that case, you should talk to your instructor about how to study more effectively.

General Strategies for Studying

- Don't miss class. Listening to lectures and participating in discussions is much more effective than reading someone else's notes. Active participation will help you retain what you are learning.

- Budget your time effectively. An hour or two each day is more effective, and far less painful, than studying all night before homework is due or before exams.

- If a concept gives you trouble, do additional reading or studying beyond what has been assigned. And if you still have trouble, ask for help: You surely can find friends, colleagues, or teachers who will be glad to help you learn.

- Working together with friends can be valuable in helping you understand difficult concepts. However, be sure that you learn *with* your friends and do not become dependent on them.

- Be sure that any work you turn in is of *collegiate quality*: neat and easy to read, well organized, and demonstrating mastery of the subject matter. Although it takes extra effort to make your work look this good, the effort will help you solidify your learning and is also good practice for the expectations that future professors and employers will have.

Preparing for Exams

- Study the review questions, and rework problems and other assignments; try additional questions to be sure you understand the concepts. Study your performance on assignments, quizzes, or exams from earlier in the term.

- Try the quizzes available on the text Web site at **www.astronomyplace.com**

- Study your notes from lectures and discussions. Pay attention to what your instructor expects you to know for an exam.

- Reread the relevant sections in the textbook, paying special attention to notes you have made on the pages.

- Study individually *before* joining a study group with friends. Study groups are effective only if every individual comes prepared to contribute.

- Don't stay up too late before an exam. Don't eat a big meal within an hour of the exam (thinking is more difficult when blood is being diverted to the digestive system).

- Try to relax before and during the exam. If you have studied effectively, you are capable of doing well. Staying relaxed will help you think clearly.

ABOUT THE AUTHORS

Jeffrey Bennett received a B.A. in biophysics from the University of California at San Diego in 1981 and a Ph.D. in astrophysics from the University of Colorado in 1987. His thesis research focused on Sun-like stars, but he now specializes in mathematics and science education. He has extensive teaching experience at the elementary and secondary levels and has taught more than fifty college courses in astronomy, physics, mathematics, and education. He led the development of the Colorado Scale Model Solar System (pictured here) and proposed and served as a co–principal investigator on the Voyage scale model solar system for the National Mall. Besides his astronomy textbooks, he has written textbooks in mathematics (Bennett and Briggs, *Using and Understanding Mathematics,* Addison Wesley, 2002) and statistics (Bennett, Briggs, & Triola, *Statistical Reasoning for Everyday Life,* Addison Wesley, 2001), a popular book (*On the Cosmic Horizon,* Addison Wesley, 2001), and is currently working on children's books. When not working, he enjoys participating in masters swimming and hiking the trails of Boulder, Colorado, with his family.

Megan Donahue is an astronomer at the Space Telescope Science Institute in Baltimore, Maryland. After growing up in rural Nebraska, she obtained an S.B. in physics from the Massachusetts Institute of Technology in 1985 and a Ph.D. in astrophysics from the University of Colorado in 1990. Her thesis on intergalactic gas and clusters of galaxies won the Robert J. Trumpler Award (1993). She continued her research as a Carnegie Fellow at the Observatories of the Carnegie Institution in Pasadena, California, and later was an Institute Fellow at the Space Telescope Science Institute. She is an active observer, using ground-based telescopes, the Hubble Space Telescope, and orbiting X-ray telescopes. Her research focuses on questions of galaxy evolution, the nature of intergalactic space, large-scale structure formation, and dark matter and the fate of the universe. She married Mark Voit while in graduate school and they are the parents of two children, Michaela and Sebastian.

Nicholas Schneider is an associate professor in the Department of Astrophysical and Planetary Sciences at the University of Colorado and a researcher in the Laboratory for Atmospheric and Space Physics. He received his B.A. in physics and astronomy from Dartmouth College in 1979 and his Ph.D. in planetary science from the University of Arizona in 1988. In 1991, he received the National Science Foundation's Presidential Young Investigator Award. His research interests include planetary atmospheres and planetary astronomy, with a focus on the odd case of Jupiter's moon Io. He enjoys teaching at all levels and is active in efforts to improve undergraduate astronomy education. Off the job, he enjoys exploring the outdoors with his family and figuring out how things work.

Mark Voit is an astronomer in the Office of Public Outreach at the Space Telescope Science Institute. He earned his A.B. in physics at Princeton University in 1983 and his Ph.D. in astrophysics at the University of Colorado in 1990. He continued his studies at the California Institute of Technology, where he was Research Fellow in theoretical astrophysics. NASA then awarded him a Hubble Fellowship, under which he conducted research at the Johns Hopkins University. His research interests range from interstellar processes in our own galaxy to the clustering of galaxies in the early universe. Occasionally he escapes to the outdoors, where he and his wife, Megan Donahue, enjoy running, hiking, orienteering, and playing with their children. Mark is also author of the popular book *Hubble Space Telescope: New Views of the Universe*.

PART I
Developing Perspective

How vast those Orbs must be, and how inconsiderable this Earth, the Theatre upon which all our mighty Designs, all our Navigations, and all our Wars are transacted, is when compared to them. A very fit consideration, and matter of Reflection, for those Kings and Princes who sacrifice the Lives of so many People, only to flatter their Ambition in being Masters of some pitiful corner of this small Spot.

CHRISTIAAN HUYGENS,
DUTCH ASTRONOMER AND SCHOLAR (C. 1690)

CHAPTER 1
Our Place in the Universe

Far from city lights on a clear night, you can gaze upward at a sky filled with stars. If you lie back and watch for a few hours, you will observe the stars marching steadily across the sky. Confronted by the seemingly infinite heavens, you might wonder how the Earth and the universe came to be. With these thoughts, you will be sharing an experience common to humans around the world and in thousands of generations past.

Remarkably, modern science offers answers to many fundamental questions about the universe and our place within it. We now know the basic content and scale of the universe. We know the age of the Earth and the approximate age of the universe. And, although much remains to be discovered, we are rapidly learning how the simple constituents of the early universe developed into the incredible diversity of life on Earth.

In this first chapter, we will survey the content and history of the universe, the scale of the universe, and the motions of the Earth in our universe. We'll thereby develop a "big picture" perspective on our place in the universe that will provide a base on which we can build a deeper understanding in the rest of the book. You may wish to begin by studying the two paintings that appear in the first few pages of the chapter, summarizing (respectively) our cosmic address and cosmic origins.

FIGURE 1.1 Our place in the universe. This painting illustrates our cosmic address: The Earth is one of nine planets in our solar system; our solar system is one among more than 100 billion star systems in the Milky Way Galaxy; the Milky Way is one of the two largest of about 30 galaxies in the Local Group; the Local Group lies near the outskirts of the Local Supercluster; and the Local Supercluster fades into the background painting of structure throughout the universe.

1.1 A Modern View of the Universe

If you observe the sky carefully, you can see why most of our ancestors believed that the heavens revolved about the Earth. The Sun, Moon, planets, and stars appear to circle around our sky each day, and we cannot feel the constant motion of the Earth as it rotates on its axis and orbits the Sun. Thus, it seems quite natural to assume that we live in an Earth-centered, or *geocentric,* universe.

Nevertheless, we now know that the Earth is a planet orbiting a rather average star in a vast cosmos. (In astronomy, the term *cosmos* is synonymous with *universe.*) The historical path to this knowledge was long and complex, involving the dedicated intellectual efforts of thousands of individuals. In later chapters, we'll encounter many of these individuals and explore how their discoveries changed human understanding of the universe. We'll see that many ancient beliefs made a lot of sense and changed only when people were confronted by strong evidence to the contrary. We'll also see how the process of science has enabled us to acquire this evidence and thereby discover that we are connected to the stars in ways our ancestors never imagined.

First, however, it's useful to have at least a general picture of the universe as we know it today. This big picture will make it easier for you to understand the historical development of astronomy, the evidence for our modern ideas, and the mysteries that remain. Let's begin by examining what modern astronomy has to say about our cosmic location and origins.

Our Cosmic Address

Take a look at Figure 1.1 on the preceding pages. Going counterclockwise from Earth, this painting illustrates the basic levels of structure that describe what we might call our "cosmic address."

Earth is a planet, by which we mean that it is a reasonably large object that orbits a star—our Sun. Our **solar system** consists of the Sun and all the objects that orbit it: nine planets (including Earth) and their moons, the chunks of rock that we call asteroids, the balls of ice that we call comets, and countless tiny particles of interplanetary dust.

Our Sun is a star, just like the stars we see in our night sky. The Sun and all the stars we can see with the naked eye make up only a small part of a huge, disk-shaped collection of stars called the **Milky Way Galaxy**. A galaxy is a great island of stars in space, containing from a few hundred million to a trillion or more stars. The Milky Way Galaxy is relatively large, containing more than 100 billion stars. Our solar system is located a little over halfway from the galactic center to the edge of the galactic disk. Many of the stars in the Milky Way are grouped together in

---

Basic Astronomical Definitions

Star A large, glowing ball of gas that generates energy through nuclear fusion in its core. The term *star* is sometimes applied to objects that are in the process of becoming true stars (e.g., protostars) and to the remains of stars that have died (e.g., neutron stars).

Planet An object that orbits a star and that, while much smaller than a star, is relatively large in size. There is no "official" minimum size for a planet, but the nine planets in our solar system are all at least 2,000 kilometers in diameter. Planets may be rocky, icy, or gaseous in composition, and they shine primarily by reflecting light from their star.

Moon (or satellite) An object that orbits a planet. The term *satellite* is also used more generally to refer to any object orbiting another object.

Asteroid A relatively small, rocky object that orbits a star. Asteroids are sometimes called *minor planets* because they are similar to planets but smaller.

Comet A relatively small, icy object that orbits a star.

Star system One or more stars and any planets or other material that orbits them. The term *solar system* refers to our own star system (*solar* means "of the Sun"). Many star systems contain two or more stars that orbit each other; two-star systems are called *binary star systems.*

Star cluster A group of stars (ranging in number from a few hundred to a few million) that are closely associated in space. In general, all the stars in a star cluster were born at roughly the same time from the same cloud of interstellar gas.

Galaxy A great island of stars in space, containing from a few hundred million to a trillion or more stars.

Galaxy cluster A collection of galaxies held together by gravity. Small clusters are generally called *groups* (such as our Local Group), with the term *cluster* reserved for collections of hundreds or thousands of galaxies.

Supercluster A giant structure encompassing many groups and clusters of galaxies.

Universe (or cosmos) The sum total of all matter and energy. By current understanding, only part of the universe is observable to us even in principle; we call this our *observable universe.*

star clusters; on a clear night, you can see many beautiful star clusters with a pair of binoculars.

Many galaxies congregate in groups; groups that contain more than a few dozen galaxies are called **galaxy clusters**. Our Milky Way belongs to a group of 30 or so galaxies called the **Local Group**.

On the largest scale, the universe has a frothlike appearance in which galaxies and galaxy clusters are loosely arranged in giant chains and sheets. In some places the galaxies and galaxy clusters are more tightly packed than in others; we call these giant structures **superclusters**. The supercluster to which our Local Group belongs is called, not surprisingly, the **Local Supercluster**. Between these vast structures lie huge voids containing few, if any, galaxies.

Finally, the **universe** is the sum total of all matter and energy. That is, it encompasses the superclusters and voids, and everything within them. As you progress through the text, you will learn much more about the different levels of structure in the universe. To keep the hierarchy straight, you might imagine how a faraway friend would address a postcard to someone on Earth (Figure 1.2).

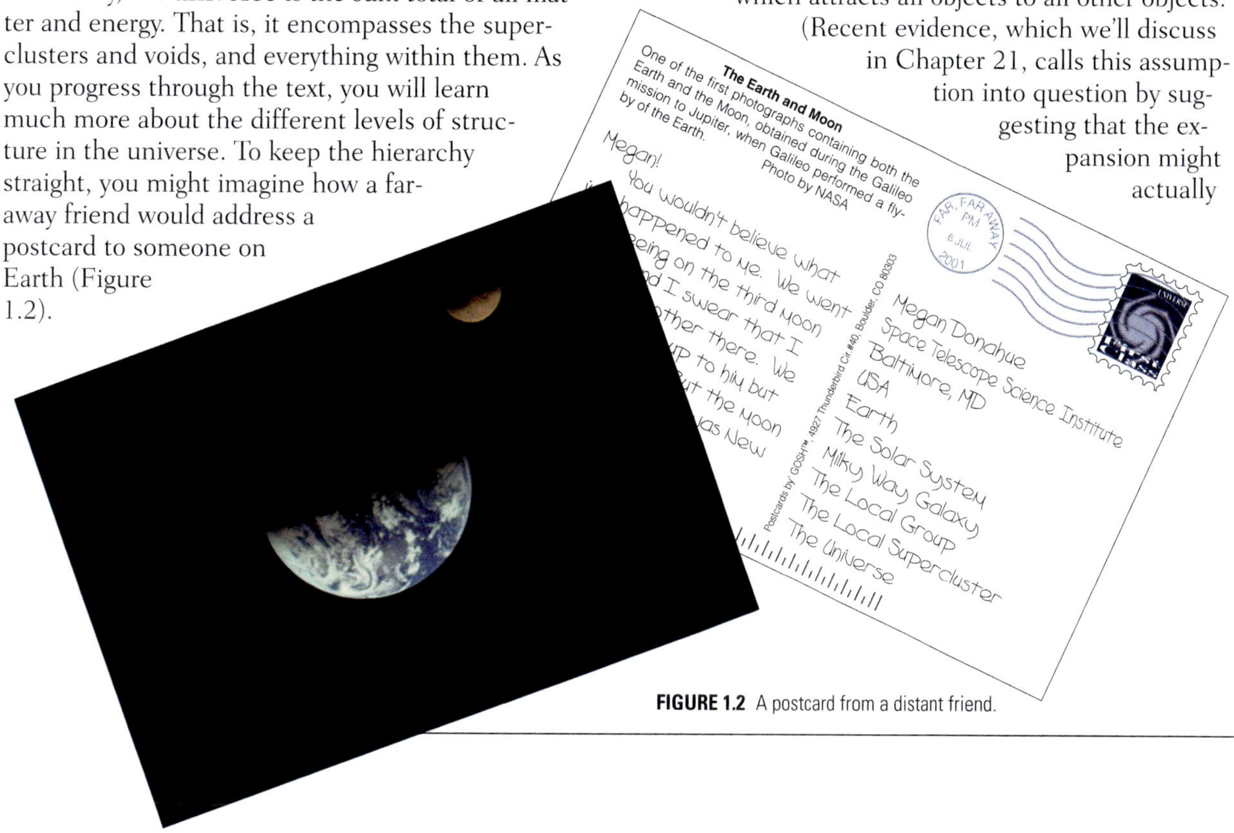

quick overview of the scientific story of creation, as summarized in Figure 1.3.

As we'll discuss shortly, telescopic observations of distant galaxies show that the entire universe is *expanding*. That is, average distances between galaxies are increasing with time. If the universe is expanding, everything must have been closer together in the past. From the observed rate of expansion, astronomers estimate that the expansion started somewhere between 12 billion and 16 billion years ago. Astronomers call this beginning the **Big Bang**.

The universe has continued to expand ever since the Big Bang, but not always at the same rate. Until about a decade ago, astronomers assumed that the expansion rate slowed with time because of gravity, which attracts all objects to all other objects. (Recent evidence, which we'll discuss in Chapter 21, calls this assumption into question by suggesting that the expansion might actually

FIGURE 1.2 A postcard from a distant friend.

TIME OUT TO THINK *Some people think that our tiny physical size in the vast universe makes us insignificant. Others think that our ability to learn about the wonders of the universe gives us significance despite our small size. What do you think?*

Our Cosmic Origins

How did we come to be? Much of the rest of this text discusses the scientific evidence concerning our cosmic origins, and we'll see that humans are relative newcomers in an old universe. For now, let's look at a

be accelerating). In addition, gravity attracts matter to other matter, and in structures such as galaxies and clusters of galaxies gravity has won out against the overall expansion. That is, while the universe as a whole continues to expand, individual galaxies and their contents do *not* expand.

Most galaxies, including our own Milky Way, probably formed within a few billion years after the Big Bang. Within galaxies, gravity drives the collapse of clouds of gas and dust, forming stars and planets. Stars have limited lifetimes, and when they die they may spew much of their content back into interstellar space. This material can then become part of new clouds of gas, which gravity collapses into new star systems. Thus, galaxies function as cosmic recycling

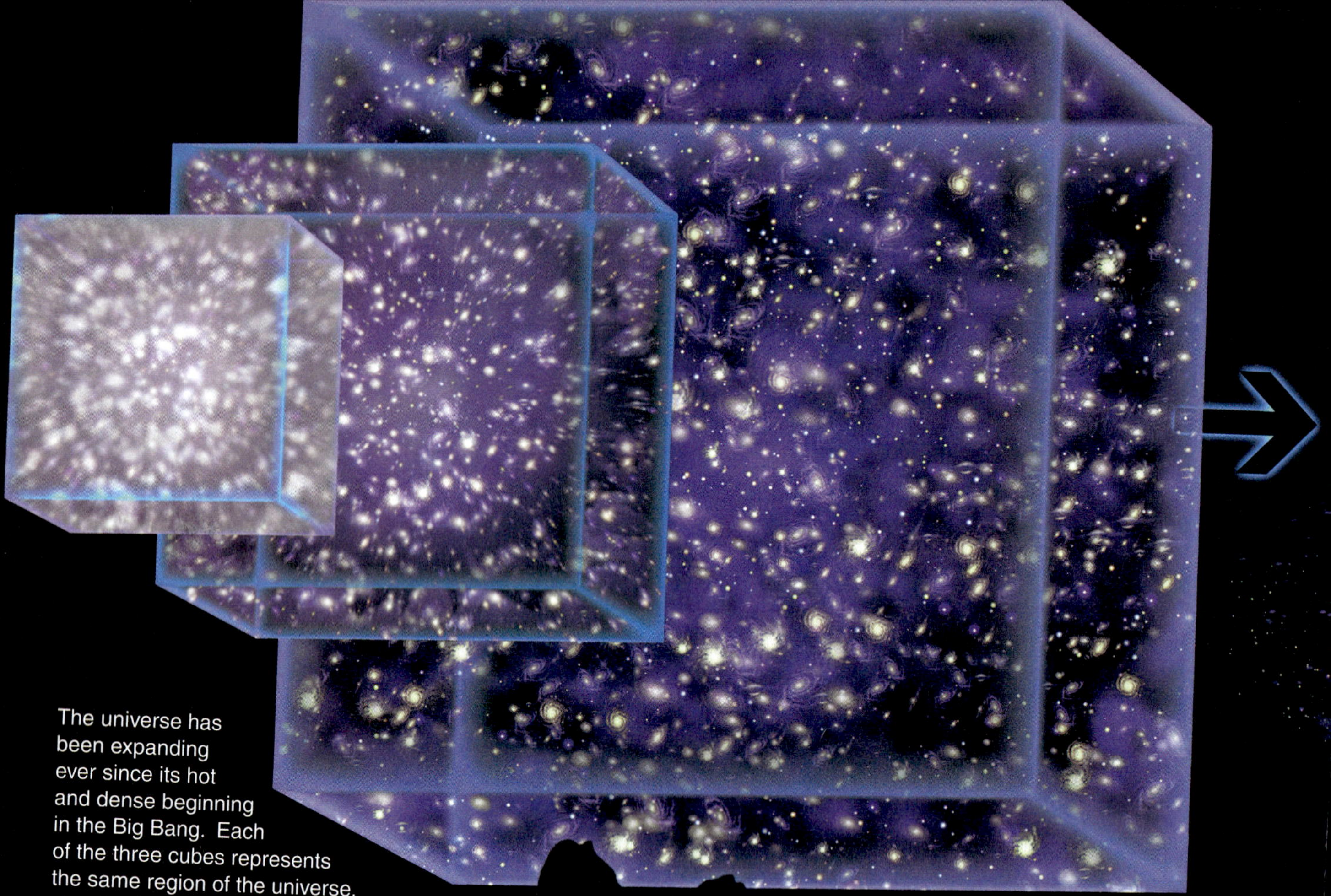

The universe has been expanding ever since its hot and dense beginning in the Big Bang. Each of the three cubes represents the same region of the universe, showing how the region expands with time.

FIGURE 1.3 Our cosmic origins: All the matter and energy in the universe was created in the Big Bang. This sequence of paintings shows the progression of that matter and energy from the Big Bang to human life. Note that the elements from which we are made were produced in stars that shined long ago, and these elements formed the Earth, thanks to the recycling role played by our galaxy.

The Earth was built with elements produced in stars that lived and died in the Milky Way before our solar system formed.

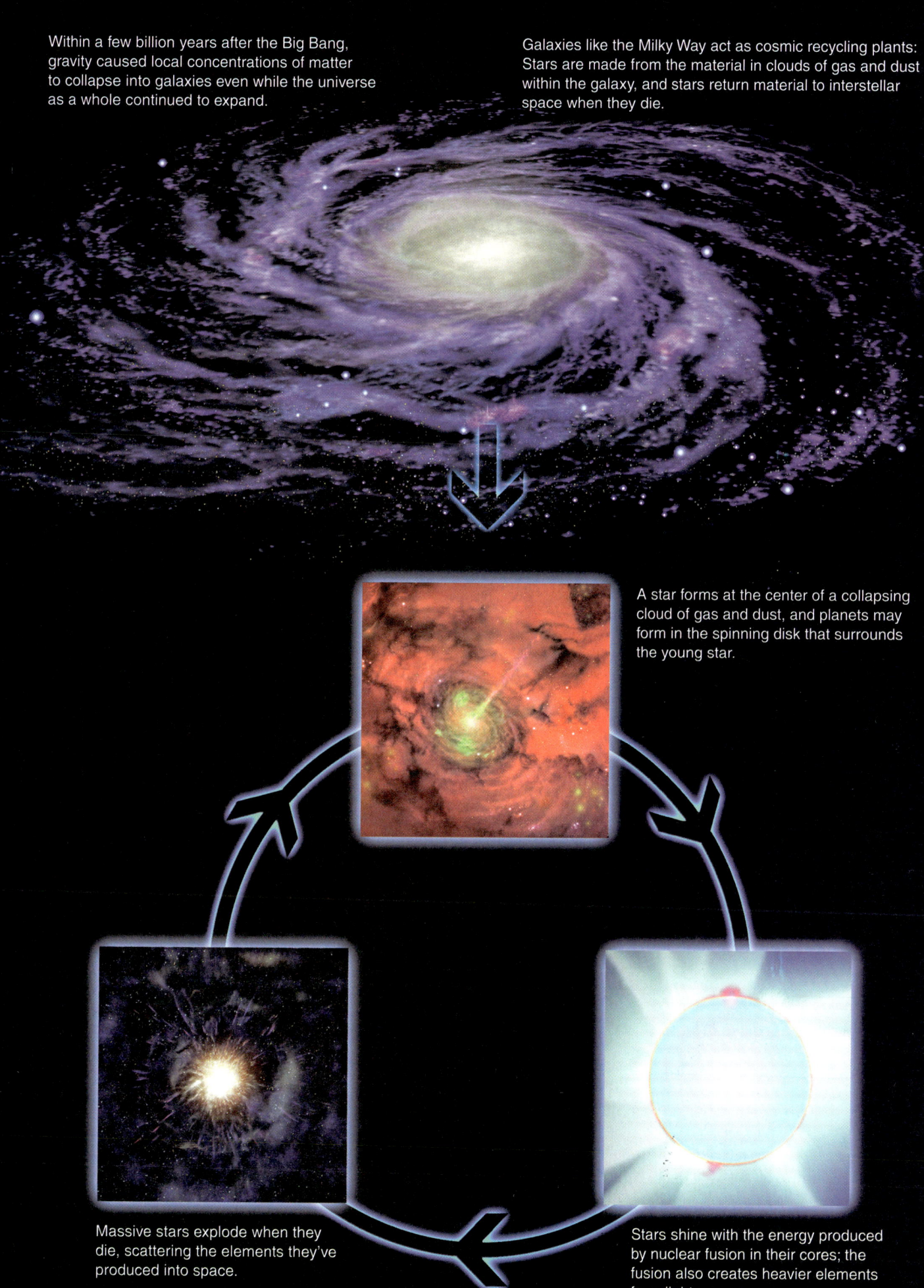

plants, recycling material expelled from dying stars into subsequent generations of stars and planets.

Not only is material recycled, but its chemical nature also changes with time. All the matter in the universe was created in the Big Bang, but the early universe contained only the simplest chemical elements: hydrogen and helium (and trace amounts of lithium). All other elements were manufactured by stars through **nuclear fusion**, in which lightweight elements fuse (i.e., their nuclei join together) to form heavier elements. The energy released by nuclear fusion, which occurs deep in stellar cores, is what makes stars shine. For most of their lives stars shine by fusing hydrogen into helium, but near the ends of their lives the more massive stars generate energy with advanced fusion reactions that produce elements such as carbon, oxygen, nitrogen, and iron. Many of these massive stars die in titanic explosions that release these heavy elements into space, where they mix with other gas and dust in the galaxy to be incorporated into later generations of star systems.

The processes of heavy-element production and cosmic recycling had already been taking place for several billion years by the time our solar system formed, about 4.6 billion years ago. The cloud that gave birth to our solar system was about 98% hydrogen and helium; the other 2% contained all the other chemical elements. The small rocky planets of our solar system, including the Earth, were made from a small part of this 2%. We do not know exactly how the elements on the Earth's surface developed into the first forms of life, but we do know that life was already flourishing on Earth more than 3.5 billion years ago. Biological evolution took over once life arose, leading to the great diversity of life on Earth today.

In summary, all the material from which we and the Earth are made (except hydrogen and most helium) was created inside stars that died before the birth of our Sun. We are intimately connected to the stars because we are products of stars. In the words of astronomer Carl Sagan (1934–96), we are "star stuff."

Images of Time

We study the universe by studying light from distant stars and galaxies. Light travels extremely fast by earthly standards: The speed of light is 300,000 kilometers per second, a speed at which it would be possible to circle the Earth nearly eight times in just 1 second. Nevertheless, even light takes a substantial amount of time to travel the vast distances in space. For example, light takes about 1 second to reach the Earth from the Moon and about 8 minutes to reach the Earth from the Sun. Light from the stars takes many years to reach us, so we measure distances to the stars in units called **light-years**. One light-year is the distance that light can travel in 1 year—about 10 trillion kilometers, or 6 trillion miles. Note that a light-year is a unit of *distance*, not of time.

The brightest star in the night sky, Sirius, is about 8 light-years from our solar system, which means that it takes light from Sirius about 8 years to reach us. Thus, when we look at Sirius, we see light that left the star about 8 years ago. The Orion Nebula, a star-forming region visible to the naked eye as a small, cloudy patch in the sword of the constellation Orion, lies about 1,500 light-years from Earth. Thus, we see the Orion Nebula as it looked about 1,500 years ago—about the time of the fall of the Roman Empire. If any major events have occurred in the Orion Nebula since that time, we cannot yet know about them because the light from these events would not yet have reached us.

Because it takes time for light to travel through space,

the farther away we look in distance, the further back we look in time.

If we look at a galaxy that lies 10 million light-years away, we see it as it was 10 million years ago. If we observe a distant cluster of galaxies that lies 1 billion light-years away, we see the cluster as it was 1 billion years ago.

Mathematical Insight 1.1   How Far Is a Light-Year?

It's easy to calculate the distance represented by a light-year if you recall that

$$\text{distance} = \text{speed} \times \text{time}$$

For example, if you travel at a speed of 50 kilometers per hour for 2 hours, you will travel 100 kilometers. A light-year is the distance covered by light, traveling at a speed of 300,000 kilometers per second, in a time of 1 year. In the process of multiplying the speed and the time, you must convert the year to seconds in order to arrive at a final answer in units of kilometers.

$$1 \text{ light-year} = (\text{speed of light}) \times (1 \text{ yr})$$
$$= \left(300{,}000 \, \frac{\text{km}}{\text{s}}\right) \times \left(1 \, \text{yr} \times \frac{365 \, \text{days}}{1 \, \text{yr}}\right.$$
$$\left. \times \frac{24 \, \text{hr}}{1 \, \text{day}} \times \frac{60 \, \text{min}}{1 \, \text{hr}} \times \frac{60 \, \text{s}}{1 \, \text{min}}\right)$$
$$= 9{,}460{,}000{,}000{,}000 \text{ km}$$

That is, "1 light-year" is just an easy way of saying "9.46 trillion kilometers" or "almost 10 trillion kilometers."

FIGURE 1.4 M31, the Great Galaxy in Andromeda, is about 2.5 million light-years away, so this photo captures light that traveled through space for 2.5 million years to reach us. Because the galaxy is 100,000 light-years in diameter, the photo also captures 100,000 years of time in M31: We see the galaxy's near side as it looked 100,000 years later than the time at which we see the far side.

Ultimately, the speed of light limits the portion of the universe that we can see. For example, if the universe is 12 billion years old, then light from galaxies more than 12 billion light-years away would not have had time to reach us. We would say that the **observable universe** extends 12 billion light-years in all directions from Earth. (As we'll discuss in Chapter 19, this explanation is somewhat oversimplified, but it captures the basic point that our *observable* universe may not include the *entire* universe.)

It is amazing to realize that any "snapshot" of a distant galaxy or cluster of galaxies is a picture of both space and time. For example, the Great Galaxy in Andromeda, also known as M31, lies about 2.5 million light-years from Earth. Figure 1.4, therefore, is a picture of how M31 looked about 2.5 million years ago, when early humans were first walking the Earth. Moreover, the diameter of M31 is about 100,000 light-years, so light from the far side of the galaxy required 100,000 years more to reach us than light from the near side. Thus, the picture of M31 shows

Common Misconceptions: Light-Years

A recent advertisement illustrated a common misconception by claiming "It will be light-years before anyone builds a better product." This advertisement makes no sense, because light-years are a unit of *distance*, not a unit of time. If you are unsure whether the term *light-years* is being used correctly, try testing the statement by remembering that 1 light-year is approximately 10 trillion kilometers, or 6 trillion miles. The advertisement then reads "It will be 6 trillion miles before anyone builds a better product," which clearly does not make sense.

FIGURE 1.5 This artist's rendering shows the locations of the Sun and inner planets in the *Voyage* scale model solar system on the National Mall in Washington, DC (opening Fall 2001). Voyage was created for the National Mall and other locations around the world by the Challenger Center for Space Science Education, the Smithsonian Institution, and NASA.

100,000 years of time. This single photograph captured light that left the near side of the galaxy some 100,000 years later than the light it captured from the far side. When we study the universe, it is impossible to separate space and time.

1.2 The Scale of the Universe

We've given some numbers in our description of the size and age of the universe, but these numbers probably have little meaning for you—after all, they are literally astronomical. In this section, we will try to give meaning to incredible cosmic distances and times.

A Walking Tour of the Solar System

One of the best ways to develop perspective on cosmic sizes and distances is to start with a scale model of our solar system. Let's imagine a walk through a model with a scale of 1 to 10 billion, such as the Voyage scale model solar system on the National Mall in Washington, DC (Figure 1.5); that is, diameters and distances in the model are *one tenbillionth* (10^{-10}) of actual diameters and distances in the solar system (Table 1.1). Figure 1.6 shows the sizes of the Sun and the planets on this scale.

Pages 12–23 take you on a virtual walk through a model solar system. Each page represents one stop, starting from the Sun and continuing to each of the planets. We also include stops for asteroids and comets. After you complete the 12-page tour, be sure that you have noticed the following key ideas about the scale of the solar system:

- The Sun is roughly the size of a grapefruit, while the planets range in size from dust-speck-size Pluto to marble-size Jupiter. Earth is about the size of a pinhead.

- The inner planets (Mercury, Venus, Earth, Mars) are all located within a couple of dozen steps of the Sun, with the Earth located 15 meters from the Sun. The outer planets (Jupiter, Saturn, Uranus, Neptune, Pluto) are spread much farther apart, with Pluto located about 600 meters (just over $\frac{1}{3}$ mile) from the Sun. Walking the full length of the model would take about 10 minutes, not including stops.

- Perhaps the most striking feature of the solar system is its *emptiness*. Although our model shows the planets laid out in a straight line from the Sun, a better model would show their orbits extending all the way around the Sun. Such a model would require an area measuring well over a kilometer (0.6 mile) on a side, equivalent to more than 300 football fields! Aside from the grapefruit-size Sun at the center, all we would find in the rest of this area would be the nine planets and their moons, plus scattered asteroids and comets—no object on this scale is larger than marble-size Jupiter, and most are far smaller than pinhead-size Earth.

Table 1.1 Solar System Sizes and Distances, 1-to-10-Billion Scale

Object	Real Diameter	Real Distance from Sun (average)	Model Diameter	Model Distance from Sun
Sun	1,392,500 km	—	139 mm = 13.9 cm	—
Mercury	4,880 km	57.9 million km	0.5 mm	6 m
Venus	12,100 km	108.2 million km	1.2 mm	11 m
Earth	12,760 km	149.6 million km	1.3 mm	15 m
Mars	6,790 km	227.9 million km	0.7 mm	23 m
Jupiter	143,000 km	778.3 million km	14.3 mm	78 m
Saturn	120,000 km	1,427 million km	12.0 mm	143 m
Uranus	52,000 km	2,870 million km	5.2 mm	287 m
Neptune	48,400 km	4,497 million km	4.8 mm	450 m
Pluto	2,260 km	5,900 million km	0.2 mm	590 m

If you are having difficulty visualizing the model, it's easy to make your own model. Use a grapefruit or similar-size object to represent the Sun and a pinhead to represent the Earth. In a long hallway or outdoors, place your Sun on a chair or table and place your tiny Earth about 15 meters from your Sun. Once you have these objects in place, you may find it easier to visualize the sizes and locations for the rest of the planets (using the information in Table 1.1).

TIME OUT TO THINK *Hold your pinhead-size model Earth in your hand and look toward your grapefruit-size model Sun 15 meters away. Aside from the tiny Earth that you hold, there is nowhere else in the solar system—and nowhere we yet know of in the universe—where humans can survive outside the artificial environment of a spacecraft or spacesuit. Does holding the model Earth affect your perspective on human existence in any way? Does it affect your perspective on the Earth? Explain.*

(continued on p. 24, following the 12-page virtual tour)

FIGURE 1.6 The sizes of the Sun and the planets on the 1-to-10-billion scale.

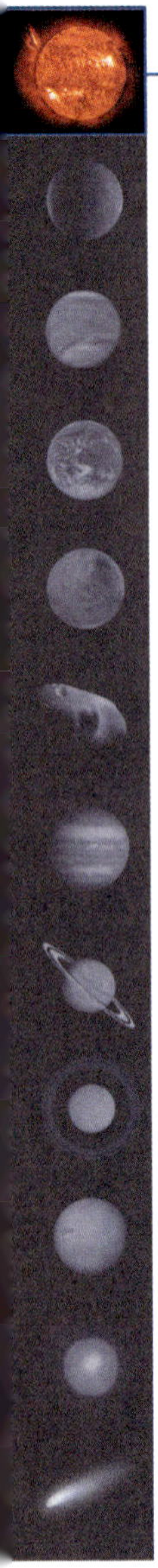

The Sun

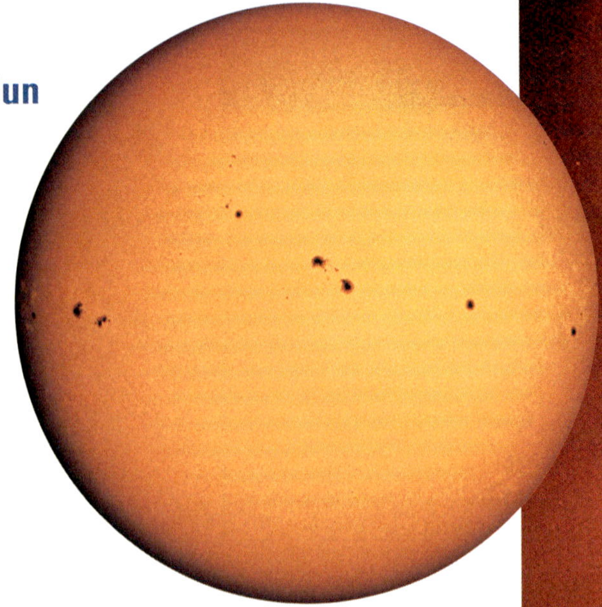

A visible light photograph of the Sun's surface. The dark splotches are sunspots—each large enough to swallow several Earths.

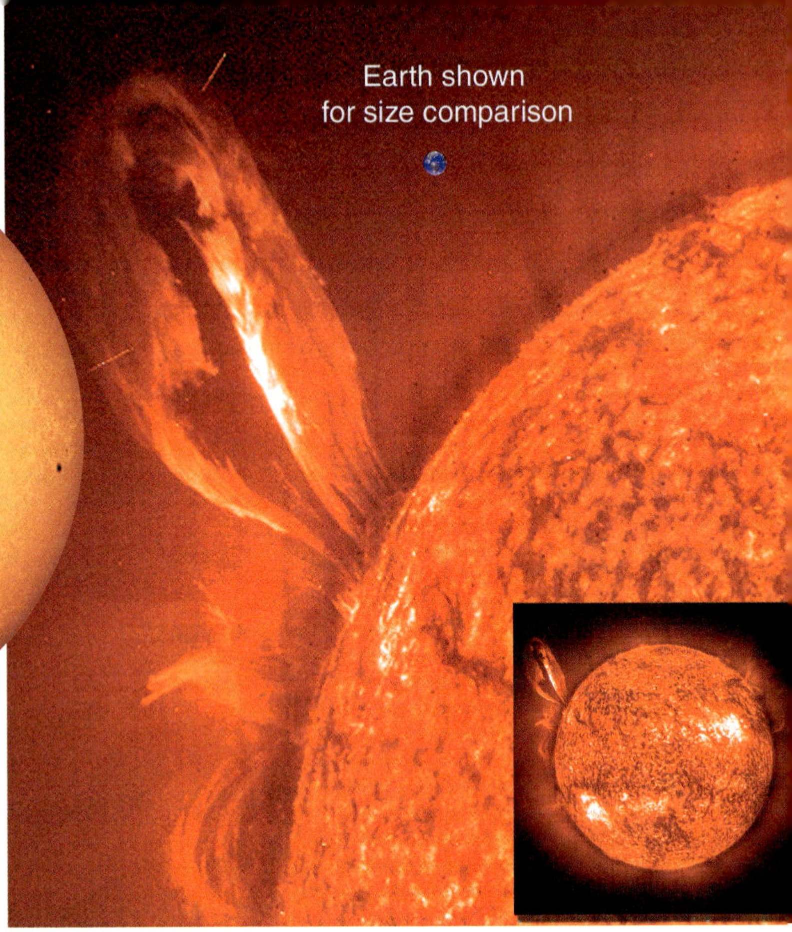

This photograph, from the *SOHO* spacecraft, shows a huge streamer of hot gas on the Sun; the image of the Earth was added for size comparison. The photograph was taken with a camera sensitive to ultraviolet light from the sun; the features shown would have been nearly invisible to our eyes.

We start our tour at the Sun, the central object of our solar system, which is about the size of a grapefruit in our model. The Sun is by far the largest object in our solar system and the only object easily visble anywhere within the model. In fact, the Sun contains about 99.9% of the solar system's total mass; that is, it outweighs everything else in the solar system combined by a factor of one thousand.

Although the Sun's surface looks solid in photographs, it is a roiling sea of hot (about 6000°C, or 10,000°F) hydrogen and helium gas. The surface is speckled with sun spots that appear dark in photographs only because they are slightly cooler than their surroundings. Solar storms sometimes send streamers of hot gas soaring far above the surface and can disrupt radio communications on Earth and disable orbiting satellites.

The Sun is gaseous throughout. If you could plunge beneath the Sun's surface, you'd find ever higher temperatures as you went deeper. You'd find the ultimate source of the Sun's energy deep in it's core, where the temperatures are so high that the Sun becomes a natural nuclear-fusion power plant. Each second, fusion transforms about 600 million tons of the Sun's hydrogen into 596 million tons of helium; the "missing" 4 million tons becomes energy in accord with Einstein's famous equation $E = mc^2$, which we will discuss in Chapter 4. Despite losing 4 million tons of mass each second, the Sun contains so much hydrogen that it has shone steadily for almost 5 billion years already and will continue to shine for some 5 billion years to come.

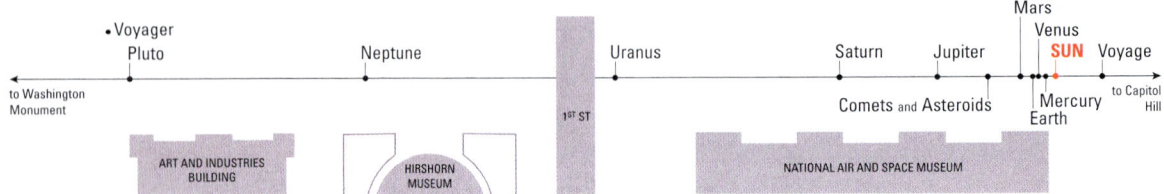

This map shows the Sun's location in the Voyage scale model solar system on the National Mall in Washington, DC. The Sun itself is about the size of a grapefruit on this scale.

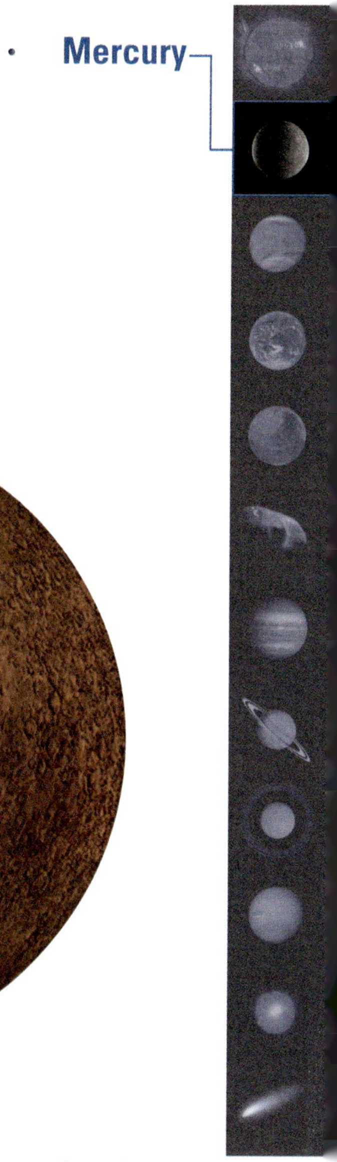

Mercury

This image shows what it would look like to be orbiting a few hundred kilometers above Mercury's surface with your back toward the Sun. Among the stars, you can see the Earth and the Moon as the blue speck and its tiny companion. The view was created using imagery from NASA's *Mariner 10* spacecraft but with computer manipulation to provide color and the orbital viewpoint. The inset (right) is a composite photograph of the full disk of Mercury by *Mariner 10*; the blank strip at the upper right was not photographed. (Image above from the Voyage scale model solar system, developed by Challenger Center for Space Science Education, the Smithsonian Institution, and NASA. Image created by ARC Science Simulations © 2001.)

The innermost planet, Mercury, lies just a few steps from the model Sun. Mercury is the smallest planet except for Pluto, and in our model it is only about as big as the period at the end of this sentence. Mercury is a desolate, cratered world with no active volcanoes, no earthquakes, no wind, no rain, and no life. Because there is virtually no air to scatter sunlight or color the sky, you could see stars even in the daytime if you stood on Mercury with your back toward the Sun. You wouldn't want to stay long, however, because the ground on Mercury's day side is nearly as hot as hot coals (about 425°C). Nighttime would not be much more comfortable: With no atmosphere to retain heat during the long nights (which last about 3 months), temperatures plummet below −150°C (about −240°F)—far colder than Antartica in winter.

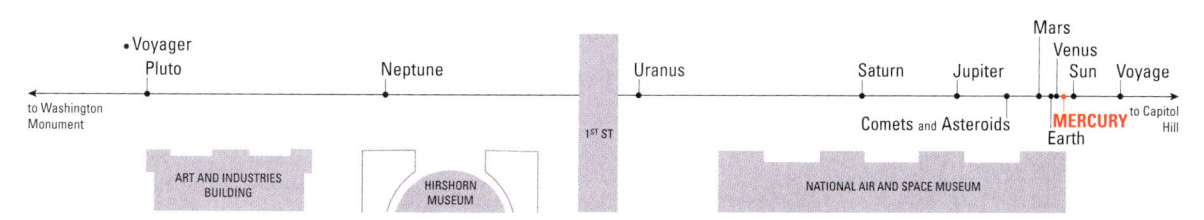

This map shows Mercury's location in the Voyage scale model solar system on the National Mall in Washington, DC. The dot at the top of the page (next to title) shows Mercury's size on the scale.

13

Venus

This image shows a realistic picture of the surface of Venus. The surface topography is based on actual data from NASA's *Magellan* spacecraft, and the colors represent what scientists believe our eyes would see on Venus. The inset (left) shows the full disk of Venus as photographed by NASA's *Pioneer Venus Orbiter*. This photograph was taken with cameras sensitive to ultraviolet light; with visible light, cloud features cannot be distinguished from the general haze. (Image above from the Voyage scale model solar system, developed by Challenger Center for Space Science Education, the Smithsonian Institution, and NASA. Image by David P. Anderson, Southern Methodist University © 2001.)

We find the second planet from the Sun, Venus, just a few steps beyond Mercury in our model. Because Venus and the Earth are nearly identical in size—both pinheads in the model solar system—Venus is sometimes called our sister planet.

Venus is covered in dense clouds, and because it is not much closer to the Sun than the Earth science fiction writers once speculated that Venus might be a lush, tropical paradise. We now know that an extreme *greenhouse effect* bakes its surface to an incredible 450°C (about 850°F) and traps heat so effectively that nighttime offers no relief—day and night, Venus is hotter than a pizza oven. All the while, the thick atmosphere bears down on the surface with a pressure equivalent to that nearly a kilometer (0.6 mile) beneath the ocean surface on Earth. Besides the crushing pressure and searing temperature, a visitor to Venus would feel the corrosive effects of sulfuric acid and other toxic chemicals in its atmosphere. Far from being a beautiful sister planet to Earth, Venus resembles a traditional view of hell.

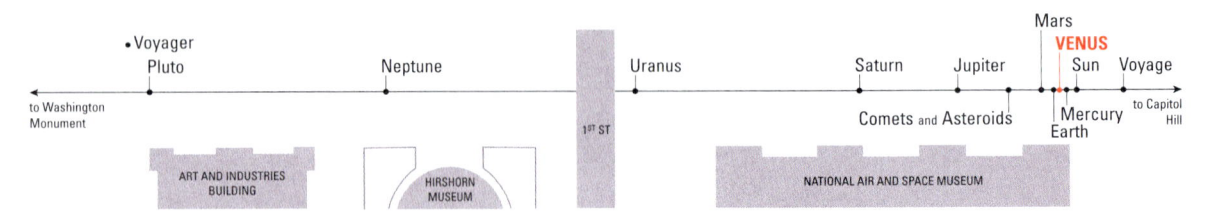

This map shows Venus's location in the Voyage scale model solar system on the National Mall in Washington, DC. The dot at the top of the page (next to title) shows Venus's size on the scale.

Earth

This image, computer generated from satellite data, shows the striking contrast between the daylight and nighttime hemispheres of Earth. The day side reveals little evidence of human presence, but at night our presence is revealed by the lights of human activity (mostly from cities as well as from agricultural, oil, and gas fires). (From the Voyage scale model solar system, developed by Challenger Center for Space Science Education, the Smithsonian Institution, and NASA. Image created by ARC Science Simulations © 2001.)

This famous photograph shows astronaut Buzz Aldrin standing on the Moon in July 1969 during the Apollo 11 mission. The reflection in his visor shows Neil Armstrong and the lunar module (spacecraft), along with some scientific equipment. Aldrin and Armstrong were the first of 12 humans who walked on the Moon; Armstrong took the first step, saying, "That's one small step for [a] man, one giant leap for mankind."

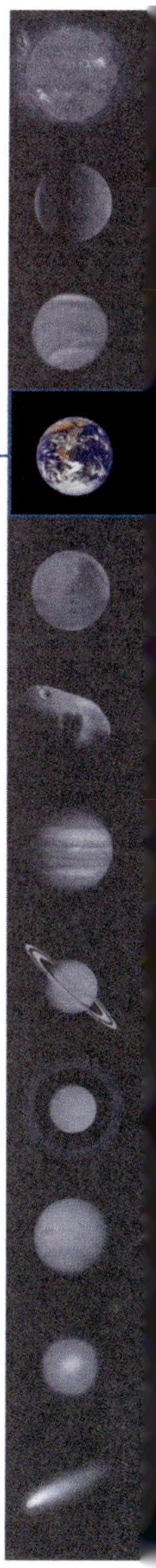

Beyond Venus and only about 15 meters from the model Sun, we find our home planet. Its tiny size in a big solar system reminds us that Earth is a rare and precious oasis of life. It is the only planet with oxygen for us to breathe and ozone to shield us from deadly solar radiation. Only on Earth does water flow freely to nurture life. And temperatures are ideal for life because Earth's atmosphere contains just enough carbon dioxide and water vapor to maintain a moderate greenhouse effect.

Despite its small size, Earth is striking in it beauty. Blue oceans cover nearly three-fourths of the surface, broken by the continental land masses and scattered islands. The polar caps are white with snow and ice, and white clouds are scattered above the surface. At night, the glow of artificial lights clearly reveals the presence of an intelligent civilization.

Earth is the first planet on our tour with a moon. On our model scale, the Moon is barely visible speck that orbits the Earth at a distance of about 4 centimeters. Thus, your thumb could easily cover both the Earth and the Moon. The Moon is cratered and desolate, much like Mercury. though it has its own interesting history and geology. It is also the only world besides Earth on which humans have ever stepped. From July 1969 through December 1972, NASA's Apollo program successfully landed 12 people (six crews of two) on the lunar surface.

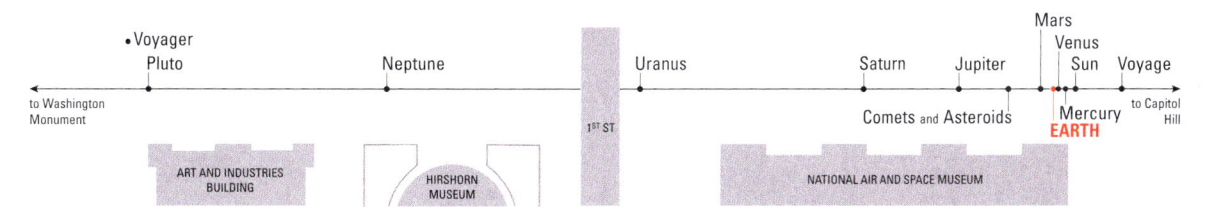

This map shows Earth's location in the Voyage scale model solar system on the National Mall in Washington, DC. The dot at the top of the page (next to title) shows Earth's size on the scale.

Mars

A few more steps take us to the model Mars, which is about half the size (diameter) of Earth. Mars has two tiny moons, Phobos and Deimos, but they are so small as to be microscopic on our 1-to-10-billion scale.

Mars is a world of wonders, with extinct volcanoes that dwarf the largest mountains on Earth, a great canyon that runs nearly one-fifth of the way around the planet, and polar caps made of frozen carbon dioxide ("dry ice") and water ice. Although Mars is frozen today, the presence of dried-up riverbeds and rock-strewn floodplains offers clear evidence that Mars was warm and wet sometime in the distant past. Thus, Mars may once have been hospitable for life, though its wet era probably ended at least 3 billion years ago.

Mars looks almost Earth-like in photographs taken by spacecraft on its surface, but you wouldn't want to visit without a space suit. The air pressure is far less than that on top of Mount Everest, the temperature is usually well below freezing, the trace amounts of oxygen would not be nearly enough to breathe, and the lack of atmospheric ozone would leave you exposed to deadly ultraviolet radiation from the Sun.

Mars is the most studied planet besides Earth. More than a dozen spacecraft have flown past, orbited, or landed on Mars, and plans are in the works for many more. We may even send humans to Mars within our lifetime. Overturning rocks in ancient riverbeds or chipping away at ice in the polar caps, explorers will search for fossil evidence of past life—and perhaps even find a few places where microbes survive today.

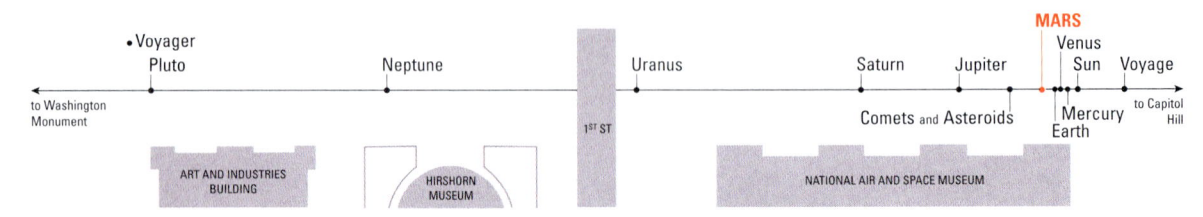

The surface of Mars, photographed from NASA's *Pathfinder* lander in 1997; part of the lander is visible in the foreground. The little rover called *Sojourner* is in the upper right, studying a rock that scientists dubbed Yogi. The inset (right) shows the full disk of Mars; the "gash" going horizontally across the center is the giant canyon known as Valles Marineris.

This map shows Mars's location in the Voyage scale model solar system on the National Mall in Washington, DC. The dot at the top of the page (next to title) shows Mars's size on the scale.

The Asteroid Belt

The asteroid Eros, photographed by the *NEAR* spacecraft that orbited it during 2000. At the end of its mission in February 2001, NASA engineers guided *NEAR* to a soft landing on Eros.

As we pass by Mars, we notice one of the most important characteristics of our solar system. In contrast to the few steps separating the four inner planets in the model solar system, the walk from Mars to Jupiter covers more than 50 meters—over half the length of a football field. As we look ahead, we see that increasingly large distances separate the remaining planets. Thus, the solar system has two major regions: the **inner solar system**, where Mercury, Venus, Earth, and Mars orbit the Sun relatively closely, and the **outer solar system**, where the remaining planets are widely separated. In between we find the **asteroid belt**, where thousands of asteroids orbit the Sun.

Asteroids are chunks of rock and metal left over from the formation of our solar system—bits and pieces that never became part of a moon or a planet. The largest asteroid, Ceres, is about 1,000 kilometers (600 miles) across. But most asteroids are far smaller and would be microscopic on our 1-to-10-billion scale. Despite their large number, the total mass of the asteroids is less than that of any of the inner planets. The average distance between asteroids is millions of kilometers—so the real asteroid belt does not at all resemble the crowded rock fields depicted in many science fiction films.

Not all asteroids are confined to the asteroid belt, and those that orbit elsewhere in the solar system may collide with moons or planets; impact craters are the scars of such collisions. (Impact craters may also result from comet collisions.) Asteroid impacts may have played an important role in the development of life on Earth. For example, as we'll discuss in Chapter 12, an asteroid impact may have been responsible for the extinction of the dinosaurs some 65 million years ago. Smaller pieces of asteroids fall to the Earth quite often; we call these fragments *meteorites*.

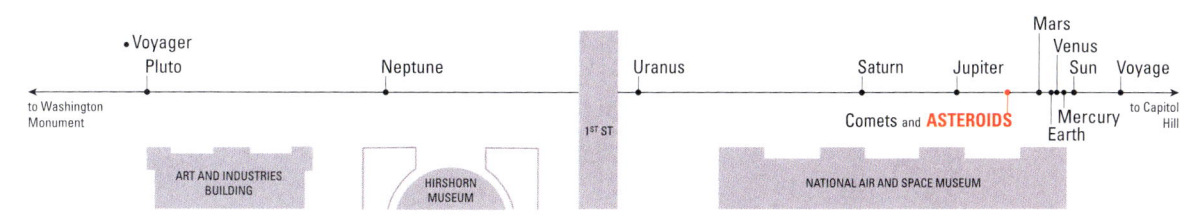

This map shows the Asteroid Belt's location in the Voyage scale model solar system on the National Mall in Washington, DC. Individual asteroids are too small to see on this scale.

Jupiter

This image shows a realistic view of what it would look like to be orbiting near Jupiter's moon Io as Jupiter comes into view displaying its Great Red Spot. The extraordinarily dark rings discovered in the Voyager missions are exaggerated to make them visible. This computer visualization was created using data from both NASA's Voyager and Galileo missions. (From the Voyage scale model solar system, developed by Challenger Center for Space Science Education, the Smithsonian Institution, and NASA. Image created by ARC Science Simulations © 2001.)

The model Jupiter is the size of a marble, making it a giant in comparison to the planets of the inner solar system. Indeed, Jupiter is so different from the planets of the inner solar system that we must create an entirely new mental image of the term *planet*. Its mass is more than 300 times that of the Earth, and its volume is more than 1,000 times that of the Earth. Its most famous feature—a long-lived storm called the Great Red Spot—is itself large enough to swallow two or three Earths. Like the Sun, Jupiter is made primarily of hydrogen and helium and has no solid surface. If you plunged deep into Jupiter, you would be crushed by the increasing gas pressure long before you ever reached its core.

Jupiter reigns over at least 28 moons and a thin set of rings (too faint to be seen in most photographs). The four largest moons—Io, Europa, Ganymede, and Callisto (often called the *Galilean moons* because they were discovered by Galileo)—are fascinating worlds in their own right and are easily visible on the scale of the model solar system. Io is the most volcanically active place in the solar system. Europa's icy crust probably hides a subsurface ocean of liquid water, making Europa a promising place to search for life. Indeed, as we'll discuss in Chapter 11, the moons of Jupiter and the other outer planets are at least as interesting as the planets themselves.

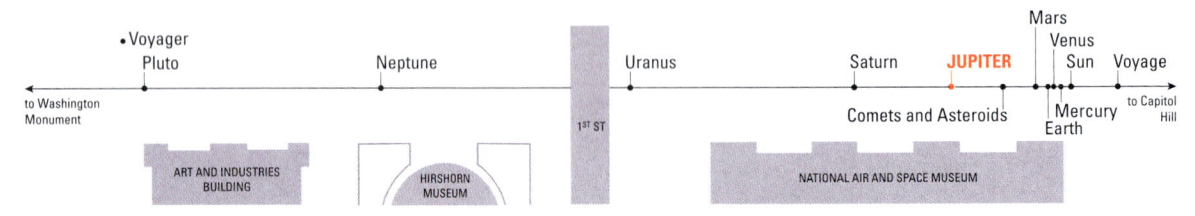

This map shows Jupiters's location in the Voyage scale model solar system on the National Mall in Washington, DC. The image at the top of the page (next to title) shows Jupiters's size on the scale.

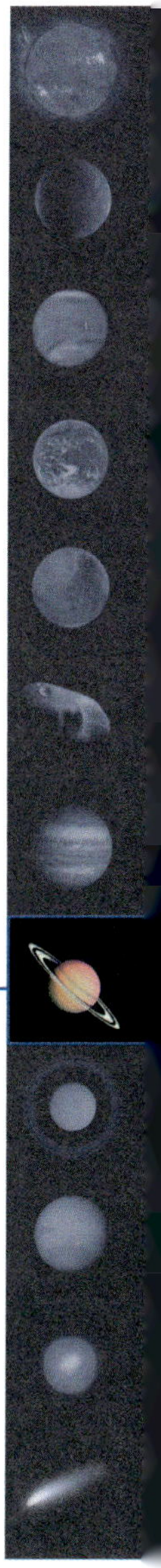

This computer simulation, built upon imagery from the *Voyager 1* mission, recreates the striking view from the spacecraft at somewhat higher detail. We see the shadow of the rings on Saturn's sunlit face, and the rings become lost in Saturn's shadow on the night side. (From the Voyage scale model solar system, developed by Challenger Center for Space Science Education, the Smithsonian Institution, and NASA. Image created by ARC Science Simulations © 2001.)

The distance from Jupiter to Saturn is about 65 meters in our model, or nearly three-fourths the length of a football field; notice how much longer this walk takes than the few steps between the planets of the inner solar system. Saturn is only slightly smaller than Jupiter in size but is considerably less massive (about one-third Jupiter's mass) because it is less dense. Like Jupiter, Saturn is made mostly of hydrogen and helium and has no solid surface.

Saturn is famous for its spectacular rings. Although all four of the giant outer planets have rings, only Saturn's can be seen easily through a small telescope on Earth. The rings may look solid from a distance, but this appearance is deceiving. If you could wander into the rings, you'd find yourself surrounded by countless individual particles of rock and ice, ranging in size from dust grains to city blocks. Each particle orbits Saturn like a tiny moon.

At least 30 moons orbit Saturn. Most are far too small to be visible in our model solar system (though they are still considerably larger than the ring particles). But Saturn's largest moon, Titan, is bigger than the planet Mercury and is blanketed by a thick atmosphere. On Titan's surface, you'd find an atmospheric pressure slightly greater than that on Earth, and you would inhale air with roughly the same nitrogen content as air on Earth. However, you'd need to bring your own oxygen, and you'd certainly want a warm spacesuit to protect yourself against frigid outside temperatures. NASA's *Cassini* spacecraft is currently on the way to Saturn; it will drop a probe to Titan's surface after it arrives in 2004.

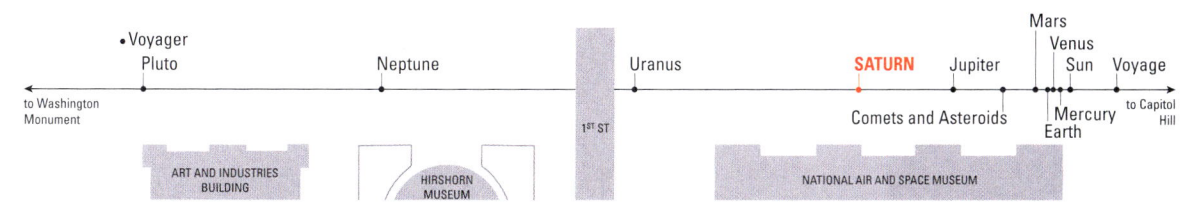

This map shows the Saturn's location in the Voyage scale model solar system on the National Mall in Washington, DC. The image at the top of the page (next to title) shows Saturn's size on this scale.

19

Uranus

To go from Saturn to Uranus in our model, we must walk as far as we've walked in our entire tour so far, again illustrating the vast distances that separate planets in the outer solar system. Uranus is much smaller than either Jupiter or Saturn, but it is still much larger than Earth. It is made largely of hydrogen, helium, and hydrogen compounds such as methane (CH_4); the latter gives Uranus its pale blue-green color. Like the other giants of the outer solar system, it lacks a solid surface. At least 21 moons orbit Uranus, along with a set of rings similar to those of Saturn but much darker and more difficult to see.

The entire Uranus system—planet, rings, and moon orbits—is tipped on its side compared to the rest of the planets. This unusual orientation may be the result of a cataclysmic collision suffered by Uranus as it was forming some 4.6 billion years ago. It also is responsible for the most extreme pattern of seasons on any planet. If you lived on a platform floating in Uranus's atmosphere near the north pole, you'd have continuous daylight for half of each orbit, or 42 years. Then, after a very gradual sunset, you'd be plunged into a 42-year-long night.

Uranus was the first "discovered" planet—all the planets closer to the Sun are easily visible to the naked eye and thus were known to all ancient cultures. English astronomer William Herschel discovered Uranus in 1781. He originally suggested naming the planet *Georgium Sidus,* Latin for "George's star," in honor of his patron, King George III. Fortunately, the idea of "Planet George" never caught on. Instead, many eighteenth- and nineteenth-century astronomers referred to the new planet as "Herschel." The modern name Uranus, after the mythological father of Saturn, was first suggested by one of Herschel's contemporaries, astronomer Johann Bode, and was generally accepted by the mid-nineteenth century.

This image shows a view from a vantage point high above Uranus's moon Ariel. Several other moons are visible in the image, and the thin vertical line is the ring system (actually too dark to see from this vantage point). Computer simulation based upon data from NASA's *Voyager 2* mission. (From the Voyage scale model solar system, developed by Challenger Center for Space Science Education, the Smithsonian Institution, and NASA. Image created by ARC Science Simulations © 2001.)

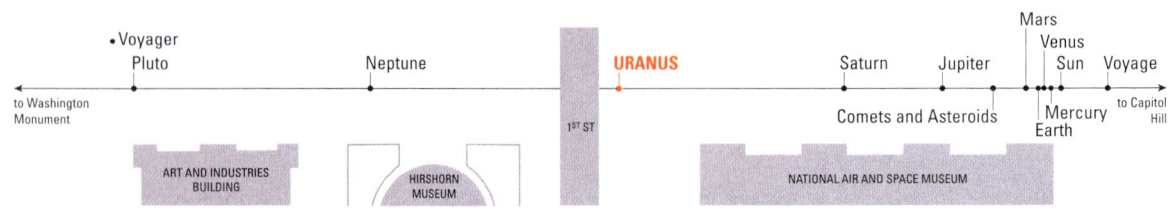

This map shows Uranus's location in the Voyage scale model solar system on the National Mall in Washington, DC. The image at the top of the page (next to title) shows Uranus's size on the scale.

Neptune

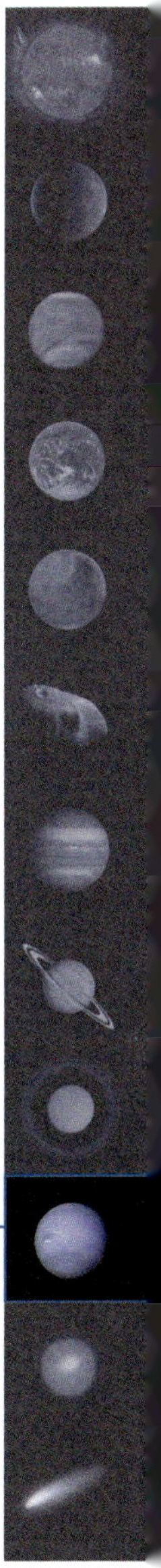

This image shows what it would look like to be orbiting Neptune's moon Triton as Neptune itself comes into view. The dark rings are exaggerated to make them visible in this computer simulation using data from NASA's *Voyager 2* mission. (From the Voyage scale model solar system, developed by Challenger Center for Space Science Education, the Smithsonian Institution, and NASA. Image created by ARC Science Simulations © 2001.)

From Uranus, we must cross a distance equivalent to more than one and a half football fields to reach Neptune in our model. Neptune looks nearly like a twin of Uranus, with very similar size and composition, although it is more strikingly blue. Neptune has rings and at least eight moons. Its largest moon, Triton, is unusual both in having a "backward" orbit around Neptune (orbiting in a direction opposite to Neptune's rotation) and in having active sites that spew plumes of gas into the sky. The backward orbit leads astronomers to believe that Triton originally may not have been a moon but instead was captured by Neptune sometime after it formed.

Neptune's discovery was a triumph for the theory of gravity, developed by Isaac Newton in the late 1600s. By the mid-nineteenth century, careful observations of Uranus had shown its orbit to be slightly inconsistent with that predicted by Newton's theory of gravity. In the early 1840s, Englishman John Couch Adams suggested that the inconsistency could be explained by a previously unseen "eighth planet" orbiting the Sun beyond Uranus. By making calculations based on Newton's theory, he even predicted the location of the planet and urged a telescopic search. Unfortunately, Adams was a student at the time and was unable to convince British astronomers to carry out the search.

In the summer of 1846, French astronomer Urbain Leverrier made similar calculations independently. He sent a letter to Johann Galle, of the Berlin Observatory, suggesting a search for the eighth planet. On the night of September 23, 1846, Galle pointed his telescope to the position suggested by Leverrier. There, within 1° of its predicted position, he saw the planet Neptune. Hence, Neptune's discovery truly was made by mathematics and physics and was merely confirmed with a telescope.

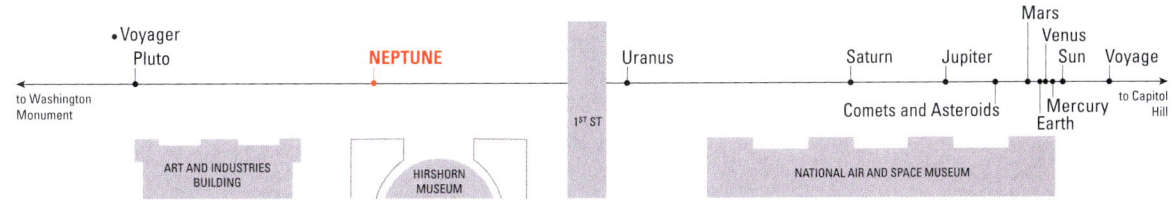

This map shows Neptune's location in the Voyage scale model solar system on the National Mall in Washington, DC. The image at the top of the page (next to title) shows Neptune's size on this scale.

21

Pluto

Pluto and Charon, as photographed by the Hubble Space Telescope. Because no spacecraft has been to Pluto, we do not yet have clear pictures of this tiny planet or its moon.

At its average distance from the Sun, the model Pluto lies another 140 meters beyond Neptune. The grapefruit-size Sun, more than a half kilometer away, appears tiny from here. It is easy to imagine that the world of Pluto must be cold and dark.

One look at the model Pluto, along with its moon Charon, shows it to be out of character with the rest of the planets. Pluto is neither large and gaseous like the other planets of the outer solar system nor rocky like the planets of the inner solar system. Instead, it is very small—the smallest planet by far—and is made mostly of ices. Pluto's orbit is also unusual. Whereas all the other planets orbit the Sun along nearly circular paths and in nearly the same plane, Pluto's orbit is highly elongated and substantially inclined relative to the orbits of the other planets. Pluto actually comes closer to the Sun than Neptune for 20 years of each 284-year orbit. The last such period ended in 1999, so the next won't begin until 2263.

Although Pluto's characteristics make it a "misfit" among the planets, they give it much in common with the only other inhabitants of the outskirts of the solar system: the balls of ice and dust we call *comets*. In the past decade or so, astronomers have discovered many other Pluto-like objects, though none quite as large as Pluto. Indeed, it now seems likely that Pluto is merely the largest (or one of the largest) among thousands of "giant comets" that orbit the Sun beyond Neptune. Some astronomers have gone so far as to suggest demoting Pluto from its status as a planet, but most favor keeping its popular title as the ninth planet—a title it has held since its discovery in 1930 by Clyde Tombaugh.

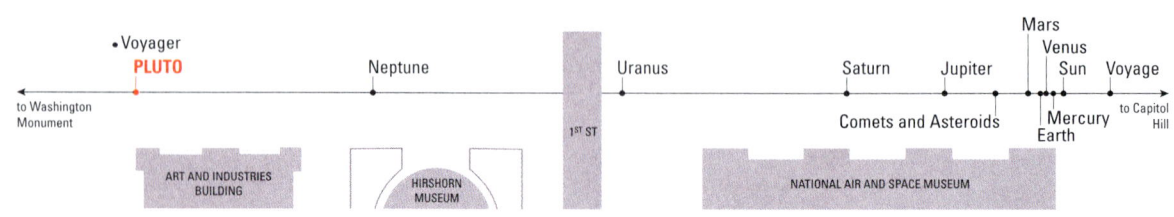

This map shows Pluto's location in the Voyage scale model solar system on the National Mall in Washington, DC. The dot at the top of the page (next to title) shows Pluto's size on the scale.

Realm of the Comets

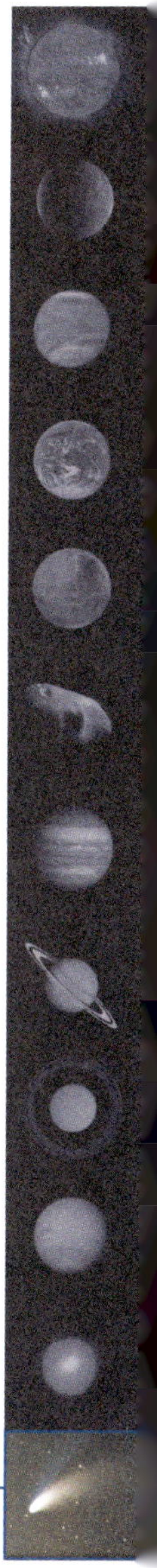

Comet Hale–Bopp, photographed over Boulder, Colorado, during its appearance in 1997.

After Pluto's discovery, many people continued searching the skies in hopes of discovering a tenth planet, sometimes called "Planet X." Given the sensitivity with which our telescopes now scan the skies, it is unlikely that another large planet orbits our Sun (though other Pluto-size objects are possible). Thus, Pluto's orbit marks the end of the realm of the planets. However, the most numerous objects in the solar system still lie ahead: comets.

You are probably familiar with the occasional appearance of a comet in the inner solar system, where the heat of the Sun evaporates some of its ice and it grows a long, beautiful tail. But with very few exceptions, nearly all comets spend nearly their entire lifetime in the extreme outer reaches of the solar system. Our present technology does not allow us to detect comets when they are very far from the Sun. However, to account for the frequency with which we see comets in the inner solar system, astronomers have calculated that there must be something like a trillion (10^{12}) comets inhabiting the outskirts of our solar system.

As we will discuss in Chapter 12, the realm of the comets probably consists of two vast regions of space. The region containing the Pluto-like "giant comets"—called the *Kuiper belt*—begins near the orbit of Neptune and probably continues out to several times this distance. The comets in this region have orbits that lie close to the plane of planetary orbits and go around the Sun in the same direction as the planets. The second and much larger region—called the *Oort cloud*—may extend more than one-fourth of the way to the nearest stars. Comets in this region have orbits that are inclined at all angles to the plane of planetary orbits. Thus, the Oort cloud would look roughly spherical in shape if we could see it. But it is so vast that, even if it has a trillion comets, each comet would typically be separated from the next by more than a billion kilometers.

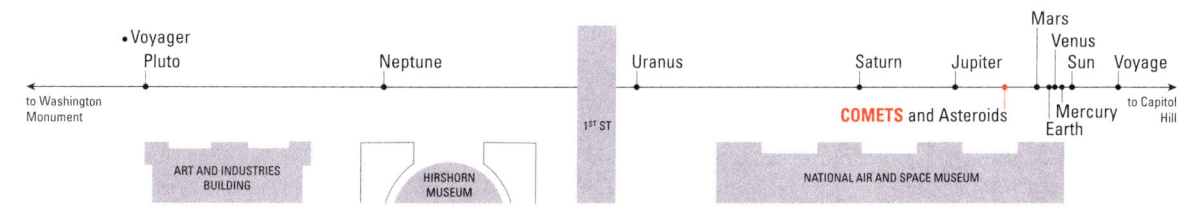

Comets and asteroids are discussed together in the Voyage scale model solar system on the National Mall in Washington, DC. Most comets actually reside well beyond the orbit of Pluto and are too small to see on this scale.

Onward to the Stars

The nearest star system to our own, called Alpha Centauri, contains three stars (Figure 1.7). The largest is about the same size as our Sun—a grapefruit on the 1-to-10-billion scale. After completing the less than 1 kilometer walk from the Sun to Pluto, how much farther would we have to walk to reach Alpha Centauri?

TIME OUT TO THINK *Before reading further, make a guess. Look at the grapefruit-size model Sun that you placed nearby. Where do you think you'll find the next grapefruit-size star? Within a few blocks? Within your city? Farther away?*

To answer this question, recall that a light-year is about 10 trillion kilometers, which becomes 1,000 kilometers on the 1-to-10-billion scale (10 trillion ÷ 10 billion = 1,000). Alpha Centauri's real distance, about 4.4 light-years, therefore becomes about 4,400 kilometers (2,700 miles) on this scale—roughly equivalent to the distance from New York to Los Angeles. In other words, looking at the largest star of Alpha Centauri from the Earth is equivalent to looking from New York at a very bright grapefruit in Los Angeles (neglecting the problems introduced by the curvature of the Earth). All other stars in our night sky are even farther away. No wonder the stars appear as mere points of light in the sky even when viewed through powerful telescopes.

Now, consider the difficulty of seeing *planets* orbiting other stars. Imagine looking from New York and trying to see a pinhead-size model Earth orbiting a grapefruit-size model Sun in Los Angeles. You probably won't be surprised to learn that we have not yet discovered Earth-size planets around other stars. Indeed, the bigger surprise may be that we *have* discovered dozens of extra-solar planets (planets around other stars), though most are closer in size to Jupiter than to Earth [Section 8.7]. With planet-detecting technology improving rapidly, we will probably know whether Earth-size planets orbit nearby stars within a couple of decades.

Although telescopic technology allows us to learn much about the stars, travel to the stars is another story. Consider the *Voyager 2* spacecraft. Launched in 1977, *Voyager 2* flew by Jupiter in 1979, Saturn in 1981, Uranus in 1986, and Neptune in 1989. (*Voyager*'s trajectory did not take it near Pluto.) *Voyager 2* is now bound for the stars at a speed near 50,000 kilometers per hour—about as fast as anything ever built by humans. But even at this speed, *Voyager 2* would take about 100,000 years to reach Alpha Centauri if it were headed in that direction, which it's not. Clearly, convenient interstellar travel remains well beyond our present technology. (We discuss prospects for interstellar travel in Chapter S5.)

FIGURE 1.7 Stars of the constellation Centaurus, including Alpha Centauri, which is visible only from tropical and southern latitudes. Stars are so far away that they should appear as mere points of light; photographs make them seem to have a measurable size only because of the limitations of cameras (brighter stars appear larger than dimmer stars because they are overexposed). In fact, despite being separated by about the distance between the Sun and Uranus, the two largest stars in the Alpha Centauri system look like a single point of light in this photograph. (The nearest of the three stars, Proxima Centauri, is too dim to see with the naked eye.)

The Milky Way Galaxy

The vast separation between our solar system and Alpha Centauri is typical of the separations among star systems here in the outskirts of the Milky Way Galaxy. Thus, the 1-to-10-billion scale is useless for modeling even just a few dozen of the nearest stars, because they could not all be spaced properly on the surface of the Earth. Visualizing the entire galaxy requires a new scale.

Let's further reduce our solar system scale by a factor of 1 billion (making it a scale of 1 to 10^{19}). On this new scale, each light-year becomes 1 millimeter. For example, the 4.4-light-year separation between the Sun and Alpha Centauri becomes just 4.4 millimeters on this scale—smaller than the width of your little finger. Aside from the fact that it makes the stars themselves become microscopic (about the size of individual atoms), this new scale makes it easy to visualize the Milky Way Galaxy. Its 100,000-light-year diameter becomes 100 meters on this scale, or about the length of a football field.

TIME OUT TO THINK *Visualize (or go to) a football field with our scale model of the galaxy centered over midfield. Standing on the 20-yard line puts you at about the correct location for our solar system. Hold your thumb and forefinger about 4 millimeters apart to represent the separation between the Sun and Alpha Centauri, and then remember what you learned about the separation between stars from the 1-to-10-billion solar system scale. How do these scale models affect your perspective on our place in the universe?*

Another way to put the galaxy into perspective is to consider its number of stars—more than 100 billion. Imagine that tonight you are having difficulty falling asleep (perhaps because you are contemplating the scale of the universe). Instead of counting sheep, you decide to count stars. If you are able to count about one star each second, on average, how long would it take you to count 100 billion stars in the Milky Way? Clearly, the answer is 100 billion (10^{11}) seconds. But how long is that? Amazingly, 100 billion seconds turns out to be more than 3,000 years. (You can confirm this by dividing 100 billion by the number of seconds in 1 year.) Thus, you would need thousands of years just to *count* the stars in the Milky Way Galaxy, and this assumes you never take a break—no sleeping, no eating, and absolutely no dying!

Beyond the Milky Way

As incredible as the scale of our galaxy may seem, the Milky Way is only one of at least 100 billion galaxies in the observable universe. Just as it would take thousands of years to count the stars in the Milky Way, it would also take thousands of years to count all the galaxies. Think for a moment about the total number of stars in all these galaxies. Assuming an average of 100 billion stars in each of these 100 billion galaxies, the total number of stars in the observable universe is roughly

$$100 \text{ billion} \times 100 \text{ billion} = 10^{11} \times 10^{11} = 10^{22}$$

This number, which we write out as

$$10,000,000,000,000,000,000,000$$

is so large that it does not even have a common name. How big is it?

Visit a beach. Run your hands through the fine-grained sand. Try to imagine counting every one of the tiny grains of sand as they slip through your fingers (Figure 1.8). Then imagine counting every grain of sand on the beach where you are sitting. Next think about counting *all* the grains of dry sand on *all* the beaches everywhere on Earth. The number you would count would be less than the number of stars in the observable universe.

The Scale of Time

Now that we have developed some perspective on the scale of space, we can do the same for the scale of time. Imagine the entire history of the universe, from the Big Bang to the present, compressed into a single year. We can represent this history with a *cosmic calendar*, on which the Big Bang takes place at the first instant of January 1 and the present day is just before the stroke of midnight on December 31 (Figure 1.9). If we assume that the universe is about 12 billion years old (recall that we think the age is between 10 billion and 16 billion years), then each month on the cosmic calendar represents about 1 billion years.

Common Misconceptions: Confusing Very Different Things

Most people are familiar with the terms *solar system* and *galaxy,* but many people sometimes mix them up. Notice how incredibly different our solar system is from our galaxy. The solar system is a *single* star system consisting of our Sun and the various objects that orbit it, including Earth and eight other planets. The galaxy is a collection of some 100 billion star systems—so many that it would take thousands of years just to count them. Thus, confusing the terms *solar system* and *galaxy* represents a mistake by a factor of 100 billion—a fairly big mistake!

Jan. 1
The Big Bang.

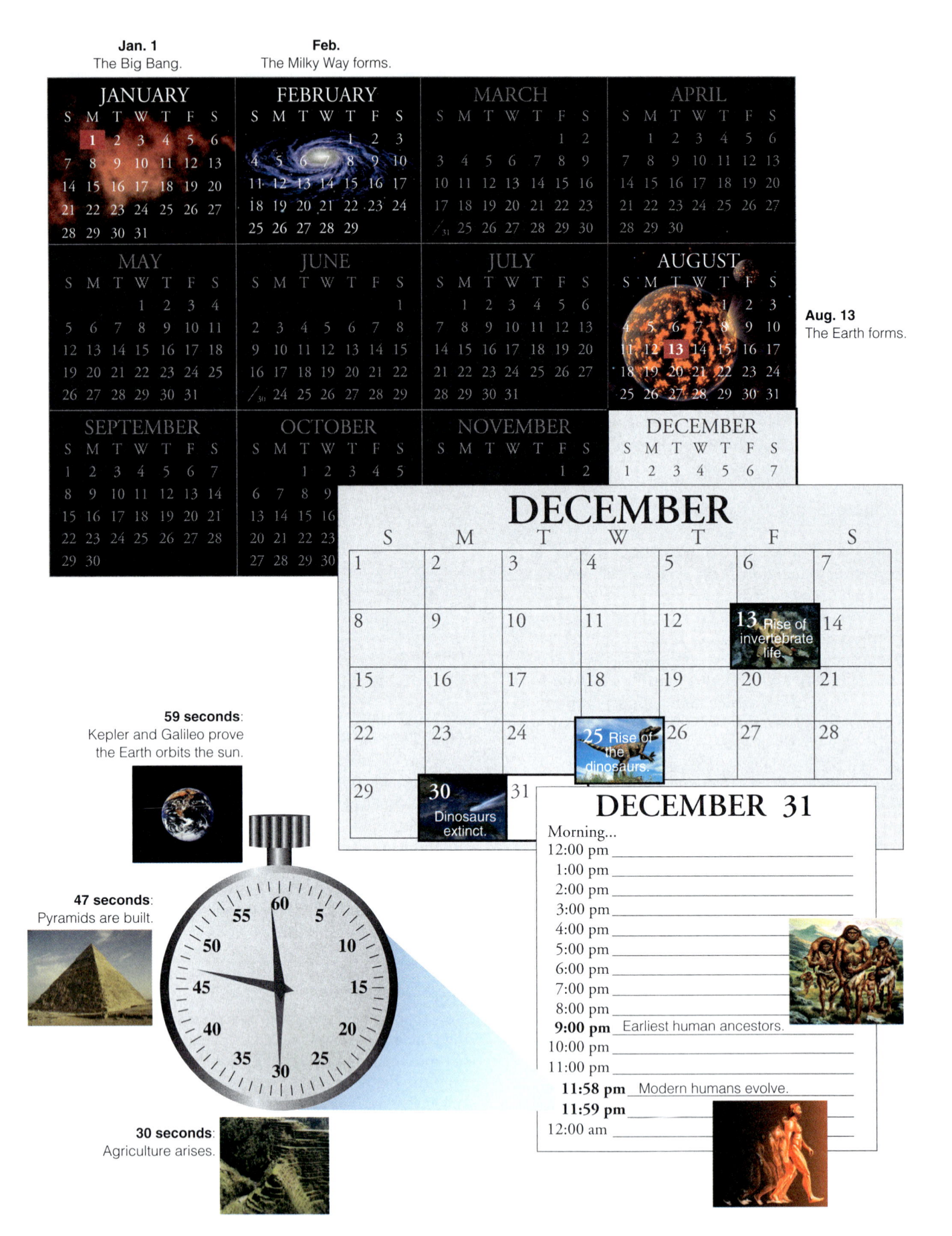

Feb.
The Milky Way forms.

Aug. 13
The Earth forms.

59 seconds: Kepler and Galileo prove the Earth orbits the sun.

47 seconds: Pyramids are built.

30 seconds: Agriculture arises.

DECEMBER 31
Morning...
12:00 pm _____
1:00 pm _____
2:00 pm _____
3:00 pm _____
4:00 pm _____
5:00 pm _____
6:00 pm _____
7:00 pm _____
8:00 pm _____
9:00 pm Earliest human ancestors.
10:00 pm _____
11:00 pm _____
11:58 pm Modern humans evolve.
11:59 pm _____
12:00 am _____

FIGURE 1.9 The cosmic calendar compresses the history of the universe into 1 year; this version assumes that the universe is 12 billion years old, so that each month represents about 1 billion years. It is only within the last few seconds of the last day that human civilization has taken shape. (This version of the cosmic calendar is adapted from a version created by Carl Sagan.)

26 PART I DEVELOPING PERSPECTIVE

FIGURE 1.8 The number of stars in the universe is larger than the total number of grains of sand on all the beaches on Earth.

On this scale, the Milky Way Galaxy may have formed sometime in February. Many generations of stars lived and died in the subsequent cosmic months, enriching the galaxy with the heavier elements from which we and the Earth are made. Not until about August 13, which represents a time 4.6 billion years ago, did our solar system form from a cloud of gas and dust in the Milky Way.

No one knows exactly when the earliest life arose on Earth, but it certainly was within the first billion years, or by mid-September on the cosmic calendar. For most of the Earth's history, living organisms remained relatively primitive and microscopic in size. Larger invertebrate life arose only about 600 million years ago, or about December 13 on the cosmic calendar. The age of the dinosaurs began shortly after midnight on Christmas (December 25) on the cosmic calendar. The dinosaurs vanished abruptly around 1:00 A.M. on December 30 (65 million years ago in real time), probably due to the impact of an asteroid or comet [Section 12.6].

With the dinosaurs gone, small furry mammals inherited the Earth. Some 60 million years later, or around 9:00 P.M. on December 31 of the cosmic calendar, the earliest hominids (human ancestors) walked upright. Most of the major events of human history took place within the final seconds of the final minute of the final day on the cosmic calendar. Agriculture arose about 30 seconds ago. The Egyptians built the pyramids about 13 seconds ago. It was only about 1 second ago on the cosmic calendar that Kepler and Galileo first proved that the Earth orbits the Sun. The average-age college student was born less than 0.1 second ago, around 11:59:59.9 P.M.

on December 31. On this scale of cosmic time, the human species is the youngest of infants, and a human lifetime is a mere blink of an eye.

TIME OUT TO THINK *Suppose the universe is 16 billion years old rather than the 12 billion years assumed in our cosmic calendar. How would this change the cosmic calendar? Would the changes affect the general point that human existence has been brief on the scale of cosmic time?*

1.3 Spaceship Earth

Now that we have discussed the scale of space and of time, the next step in our "big picture" overview is understanding motion in the universe. Wherever you are as you read this book, you probably have the feeling that you're "just sitting here." Nothing could be further from the truth. In fact, you are being spun in circles as the Earth rotates, you are racing around the Sun in the Earth's orbit, and you are careening through the cosmos in the Milky Way galaxy. In the words of noted inventor and philosopher R. Buckminster Fuller (1895–1983), you are a traveler on *spaceship Earth*.

Rotation and Revolution

The Earth **rotates** on its axis once each day. (The axis is a line joining the North and South Poles and passing through the center of the Earth.) As a result, you are whirling around the Earth's axis at a speed of 1,000 kilometers per hour (600 miles per hour) or more—faster than most airplanes travel (Figure 1.10). It is this rotation that makes the Sun and the stars appear to make their daily circles around the sky. You see sunrise when the rotating Earth first allows you to glimpse the Sun in the east, noon when your location faces directly toward the Sun, and sunset just as your location rotates to the Earth's night side where the Sun is hidden from view. (Because the Earth orbits the Sun at the same time that it rotates, our 24-hour day is slightly longer—by about 4 minutes—than the Earth's actual rotation period [Section S1.1].)

The Earth **revolves** around the Sun once each year, following an orbit that is not quite circular (Figure 1.11). The *average* distance of the Earth from the Sun is called an **astronomical unit**, or **AU** (more technically, an astronomical unit is the *semimajor axis* of the Earth's orbit):

$$1 \text{ AU} \approx 150 \text{ million km (93 million miles)}$$

The Earth travels slightly faster in its orbit when it is nearer to the Sun and slightly slower when it is farther from the Sun. But at all times the Earth is carrying you around the Sun at a speed in excess of 100,000 kilometers per hour (60,000 miles per hour).

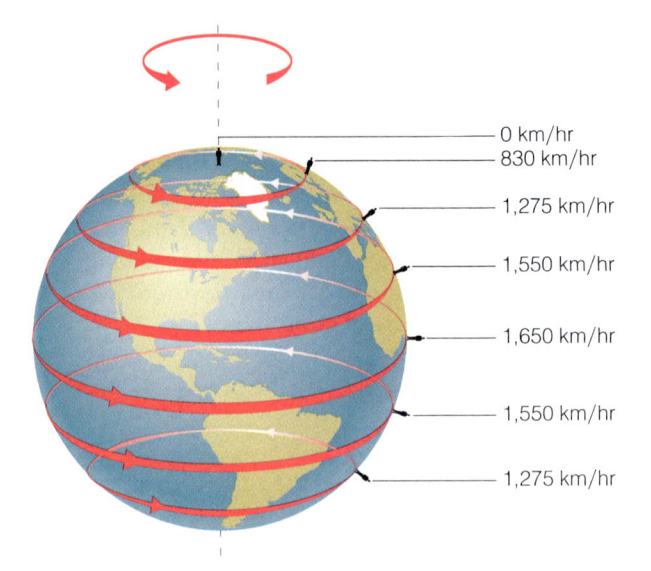

FIGURE 1.10 As the Earth rotates, your speed around the Earth's axis depends on your latitude. Unless you live at very high latitude, your speed is over 1,000 km/hr. Notice that the Earth rotates counterclockwise as viewed from above the North Pole, so you are always rotating from west to east (which is why the Sun rises in the east and sets in the west).

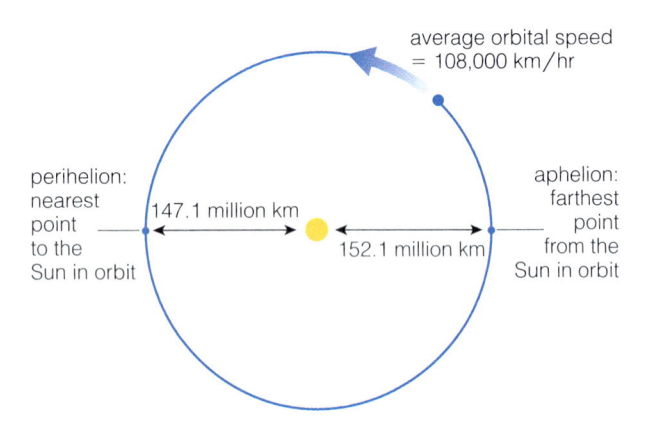

FIGURE 1.11 Basic properties of the Earth's orbit. The orbit is not quite circular, so the Earth–Sun distance varies between 147.1 million and 152.1 million kilometers. The average distance, which defines 1 astronomical unit (AU), is 149.6 million kilometers.

FIGURE 1.12 The Earth's axis is tilted by $23\frac{1}{2}°$ from a line *perpendicular* to the ecliptic plane. Note that, as viewed from above the North Pole, the Earth rotates *and* revolves counterclockwise.

The plane of the Earth's orbit around the Sun is called the **ecliptic plane** (Figure 1.12). The Earth's rotation axis happens to be tilted by $23\frac{1}{2}°$ from a line perpendicular to the ecliptic plane, pointing nearly in the direction of Polaris, the North Star. Viewed from above the North Pole, the Earth both rotates and revolves counterclockwise.

Seasons The combination of the Earth's rotation and its revolution around the Sun explains why we have seasons. Because the Earth's rotation axis remains pointed in the same direction (toward Polaris) throughout the year, its orientation *relative to the Sun* changes as it orbits the Sun, as shown in Figure 1.13. The familiar seasonal changes—longer and warmer days in summer, and shorter and cooler days in winter—arise because of changes in how directly the Sun's rays strike a particular location on Earth. When the rays are fairly direct, the Sun's path through the sky is long and high and the days are warm. (By analogy, think of how you sunburn more easily on parts of your body that face the Sun more directly.) When the Sun's rays strike less directly, the Sun's path is short and low through the sky and the days are cool. Notice that the Sun's rays strike the Northern Hemisphere more directly on the side of the orbit near point 2 in Figure 1.13, while they strike the Southern Hemisphere more directly on the side of the orbit near point 4. That is why the seasons are opposite in the Northern and Southern Hemispheres.

The four labeled points in Figure 1.13 have special names related to the seasons. The Earth is at point 1 on about March 21 each year, which represents the **spring equinox** (or *vernal equinox*). Both hemispheres receive equal amounts of sunlight at this time; it is the beginning of spring for the Northern Hemisphere but the beginning of fall for the Southern Hemisphere. The Earth reaches point 2, which represents the **summer solstice**, around June 21. This is the day on which the Northern Hemisphere receives its most direct sunlight and has the

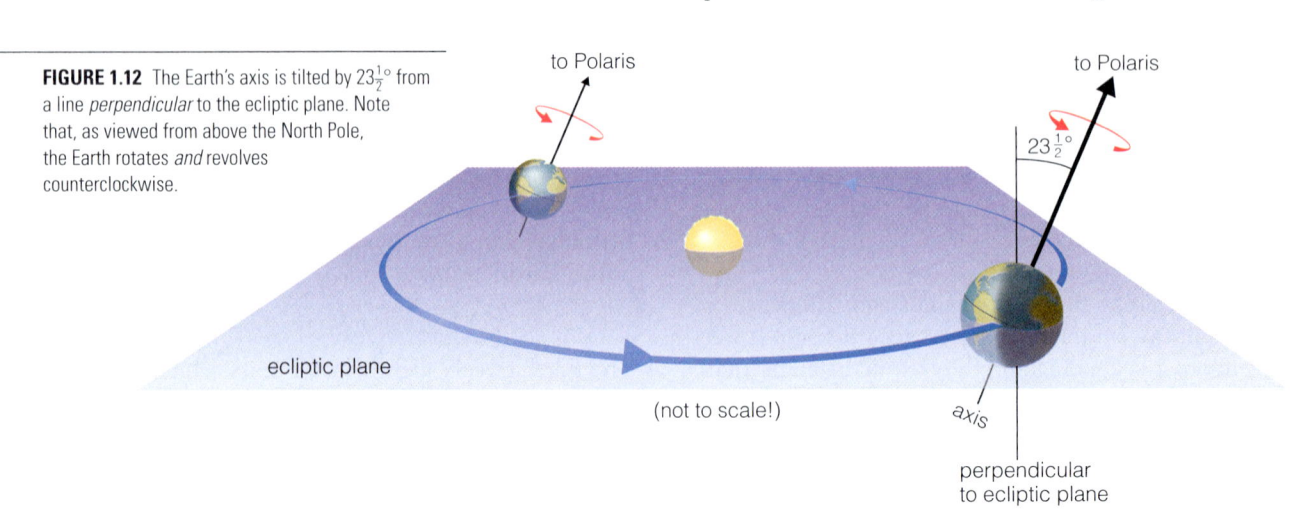

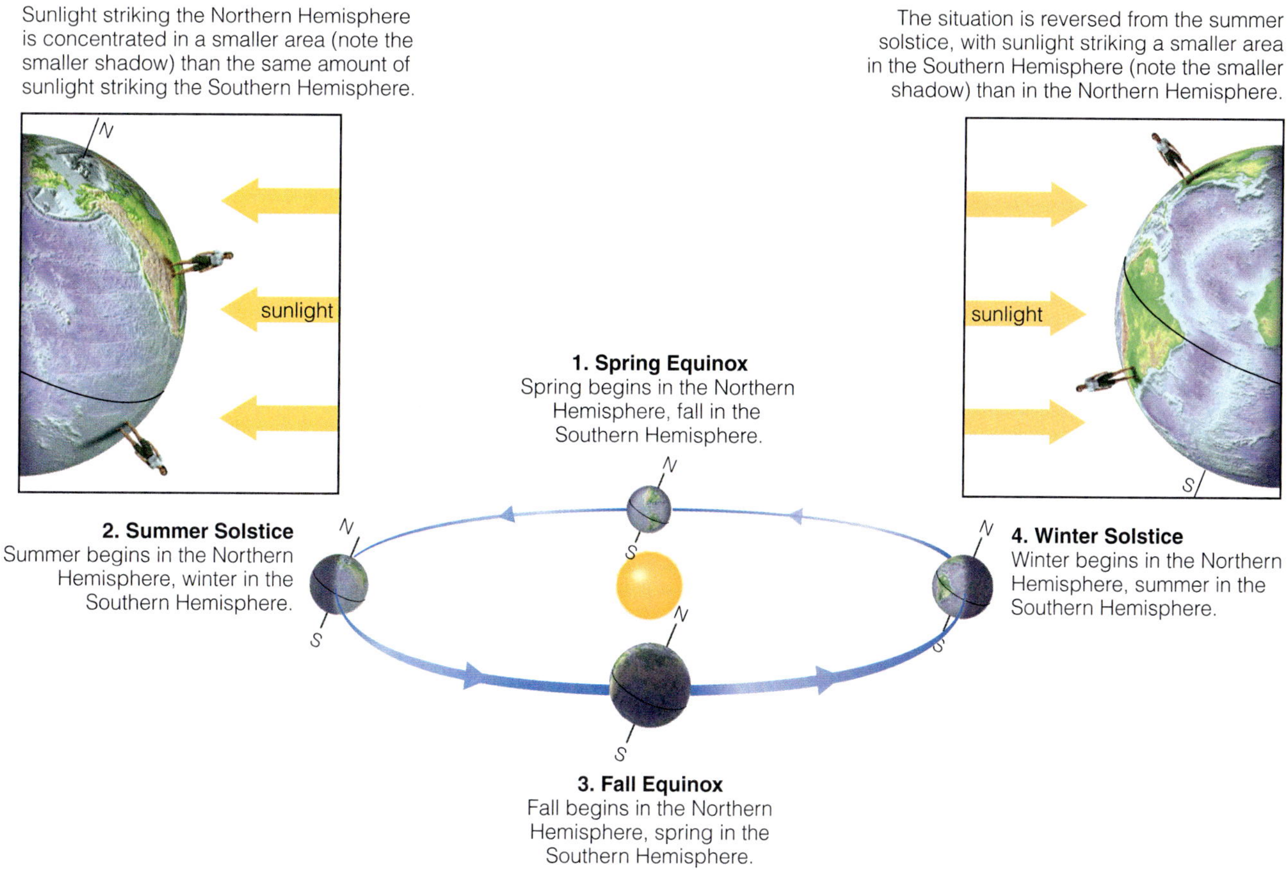

Sunlight striking the Northern Hemisphere is concentrated in a smaller area (note the smaller shadow) than the same amount of sunlight striking the Southern Hemisphere.

The situation is reversed from the summer solstice, with sunlight striking a smaller area in the Southern Hemisphere (note the smaller shadow) than in the Northern Hemisphere.

1. Spring Equinox
Spring begins in the Northern Hemisphere, fall in the Southern Hemisphere.

2. Summer Solstice
Summer begins in the Northern Hemisphere, winter in the Southern Hemisphere.

4. Winter Solstice
Winter begins in the Northern Hemisphere, summer in the Southern Hemisphere.

3. Fall Equinox
Fall begins in the Northern Hemisphere, spring in the Southern Hemisphere.

FIGURE 1.13 Seasons occur because, even though the Earth's axis remains pointed toward Polaris throughout the year, the orientation of the axis *relative to the Sun* changes as the Earth orbits the Sun. Around the time of the summer solstice, the Northern Hemisphere has summer because it is tipped toward the Sun, and the Southern Hemisphere has winter because it is tipped away from the Sun. The situation is reversed around the time of the winter solstice, when the Northern Hemisphere has winter and the Southern Hemisphere has summer. At the equinoxes, both hemispheres receive equal amounts of light. (The solstices and equinoxes were named by people in the Northern Hemisphere, so the names of the four points reflect the Northern Hemisphere seasons.)

longest period of daylight of any day of the year; it is usually considered the first day of summer. Note that the Southern Hemisphere receives its least direct sunlight and has its shortest daylight period on the summer solstice, so this is the first day of winter for the Southern Hemisphere. Point 3 represents the **fall equinox** (or *autumnal equinox*), which occurs around September 21; both hemispheres again receive the same amount of sunlight, but it is the beginning of fall in the Northern Hemisphere and the beginning of spring in the Southern Hemisphere. Finally, point 4 represents the **winter solstice**, which occurs around December 21. The situation here is the reverse of that on the summer solstice; it is usually considered the first day of winter for the Northern Hemisphere and the first day of summer for the Southern Hemisphere.

TIME OUT TO THINK *After studying Figure 1.13, explain why regions near the Earth's equator do not experience seasons in the same way as mid-latitudes such as North America, Europe, and Australia. When is the Sun's path in the sky highest for equatorial regions?*

Common Misconceptions: The Cause of Seasons

When asked the cause of the seasons, many people mistakenly answer that they are caused by variations in Earth's distance from the Sun. By knowing that the Northern and Southern Hemispheres experience opposite seasons, you'll realize that Earth's varying distance from the Sun *cannot* be the cause of the seasons; if it were, both hemispheres would have summer at the same time. Although Earth's distance from the Sun *does* vary slightly over the course of a year (Earth is closest to the Sun in January and farthest away in July), this factor is greatly overwhelmed by the way the tilt of the rotation axis causes the Northern and Southern Hemispheres to alternately receive more or less direct sunlight. Earth's varying distance from the Sun has no noticeable effect on our seasons. (This is not necessarily the case for other planets; seasons on Mars, for example, *are* affected by its varying distance from the Sun.)

Precession Although the Earth's axis will remain pointed toward Polaris throughout our lifetimes, it has not always been so and will not always remain so. The reason is a gradual motion called **precession**. Like the axis of a spinning top, the Earth's axis also sweeps out a circle (precesses)—but much more slowly (Figure 1.14). Note that the axis tilt remains close to $23\frac{1}{2}°$ throughout the cycle; only the axis *orientation* changes. Each cycle of the Earth's precession takes about 26,000 years. Today, the axis points toward Polaris, which is what makes it our North Star. In about 13,000 years, the axis will point nearly in the direction of the star Vega, making *it* our North Star. During most of the precession cycle, the axis does not point very near any bright star.

Precession also changes the locations in the Earth's orbit at which the equinoxes and solstices occur. For example, the spring equinox today occurs when the Sun appears in the direction of the constellation Pisces, but 2,000 years ago it occurred when the Sun appeared in Aries. (That is why the spring equinox is sometimes called "the first point of Aries.") In about 600 years, precession will move the spring equinox into the constellation Aquarius, marking what astrologers call "the age of Aquarius." (Astrologers disagree among themselves as to the starting date of this "age," with claims ranging from as early as the late-1700s to as late as the mid-2700s.)

Traveling in the Milky Way Galaxy

Besides moving within our solar system, we are traveling with our solar system on a journey through the Milky Way Galaxy. For example, we are moving relative to nearby stars at an average speed of about 70,000 kilometers per hour (40,000 miles per hour)—about three times as fast as the Space Shuttle orbits the Earth. In fact, all stars are in constant motion, and it is as valid to say that the stars are moving relative to us as it is to say that we are moving relative to them.

If nearby stars move at such high speeds, why don't we see them racing around the sky? The answer lies in their vast distances from us. You've probably noticed that a distant airplane appears to move through your sky more slowly than one flying close overhead. If we extend this idea to the stars, we find that even at speeds of 70,000 kilometers per hour

---

Common Misconceptions: Sun Signs

You probably know your astrological "sign." When astrology began a few thousand years ago, your sign was supposed to represent the constellation in which the Sun appeared on your birth date. However, for most people, this is no longer the case. For example, if your birthday is the spring equinox, March 21, a newspaper horoscope will show that your sign is Aries, but the Sun appears in Pisces on that date. In fact, your astrological sign generally corresponds to the constellation in which the Sun *would have appeared* on your birth date if you had lived about 2,000 years ago. Why? Because the astrological signs are based on the positions of the Sun among the stars as described by the Greek scientist Ptolemy in his book *Tetrabiblios*, which was written in about A.D. 150 [Section 3.6].

FIGURE 1.14 Precession affects the orientation, but not the tilt, of Earth's axis. (Precession is caused by gravity's effect on a tilted axis. The Earth precesses in the opposite sense of the top [clockwise instead of counterclockwise] because the Sun's gravity tries to make the Earth's axis more vertical. In contrast, the Earth's gravity tries to pull a top's axis over, making it more horizontal.)

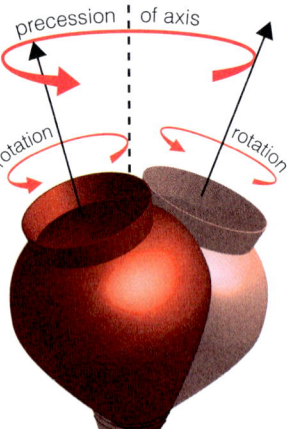

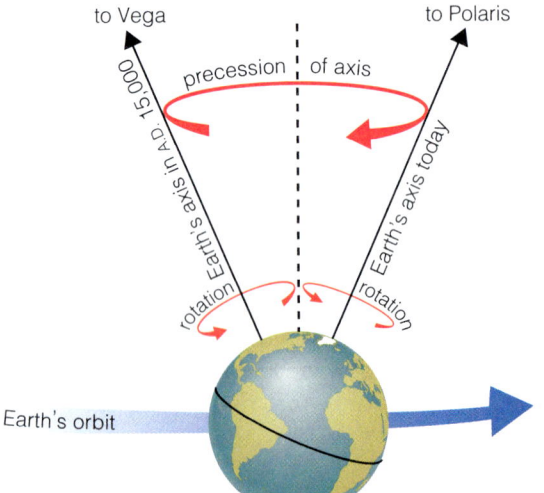

a A spinning top wobbles, or *precesses*, more slowly than it spins.

b The Earth's axis also precesses. Each precession cycle takes about 26,000 years. The axis tilt remains about the same ($23\frac{1}{2}°$) throughout the cycle, but the changing orientation of the axis means that Polaris is only a temporary North Star.

FIGURE 1.15 As shown in this painting, the local solar neighborhood is only a tiny portion of the Milky Way Galaxy. The stars in the local solar neighborhood actually move quite fast relative to our solar system, but the enormous distances between stars make this motion barely detectable on human time scales. If you could watch a time-lapse movie made over a time period of millions of years, you would see the stars in the local solar neighborhood racing around in seemingly random directions.

The box represents stars and their motions in the local solar neighborhood.

stellar motions would be noticeable to the naked eye only if we watched for thousands of years. That is why the patterns in the constellations seem to remain fixed. Nevertheless, in 10,000 years the constellations will be noticeably different from those we see today. In 500,000 years they will be unrecognizable. If you could watch a time-lapse movie made over millions of years, you would see the stars of our *local solar neighborhood* (the Sun and nearby stars) racing around in seemingly random directions (Figure 1.15).

TIME OUT TO THINK *Despite the chaos of motion in the local solar neighborhood over millions of years, collisions between star systems are extremely rare. Explain why. (Hint: Consider the sizes of star systems, such as the solar system, relative to the distances between them.)*

If you look closely at leaves floating in a stream, their motions relative to one another might appear random, just like the motions of stars in the local solar neighborhood. As you widen your view, you see that all the leaves are being carried in the same general direction by the downstream current. In the same way, if you widened your view beyond the local solar neighborhood, you would see that all the stars are moving in a simple and general way: The entire Milky Way Galaxy is rotating. Stars at different distances from the galactic center take different amounts of time to complete an orbit. Our solar system, located about 28,000 light-years from the galactic center, completes one orbit of the galaxy in about 230 million years (Figure 1.16). Even if you could watch from outside our galaxy, this motion would be unnoticeable to your naked eye. However, if you calculate the speed of our solar system as we orbit the center of the galaxy, you will find that it is close to 1 million kilometers per hour (600,000 miles per hour).

It is worth noting that the galactic rotation reveals one of the greatest mysteries in science—one that we will study in depth later in the book. The speeds at which stars orbit the galactic center depend on the strength of gravity, and the strength of gravity depends on how mass is distributed throughout the galaxy. Thus, careful study of the galactic rotation allows us to determine the distribution of mass in the galaxy. Remarkably, we find that the stars in the disk of the galaxy represent only the "tip of the iceberg" compared to the mass of the entire galaxy (Figure 1.17). That is, most of the mass of the galaxy is located outside the visible disk, in the galaxy's *halo*. We don't know the nature of this mass because we have

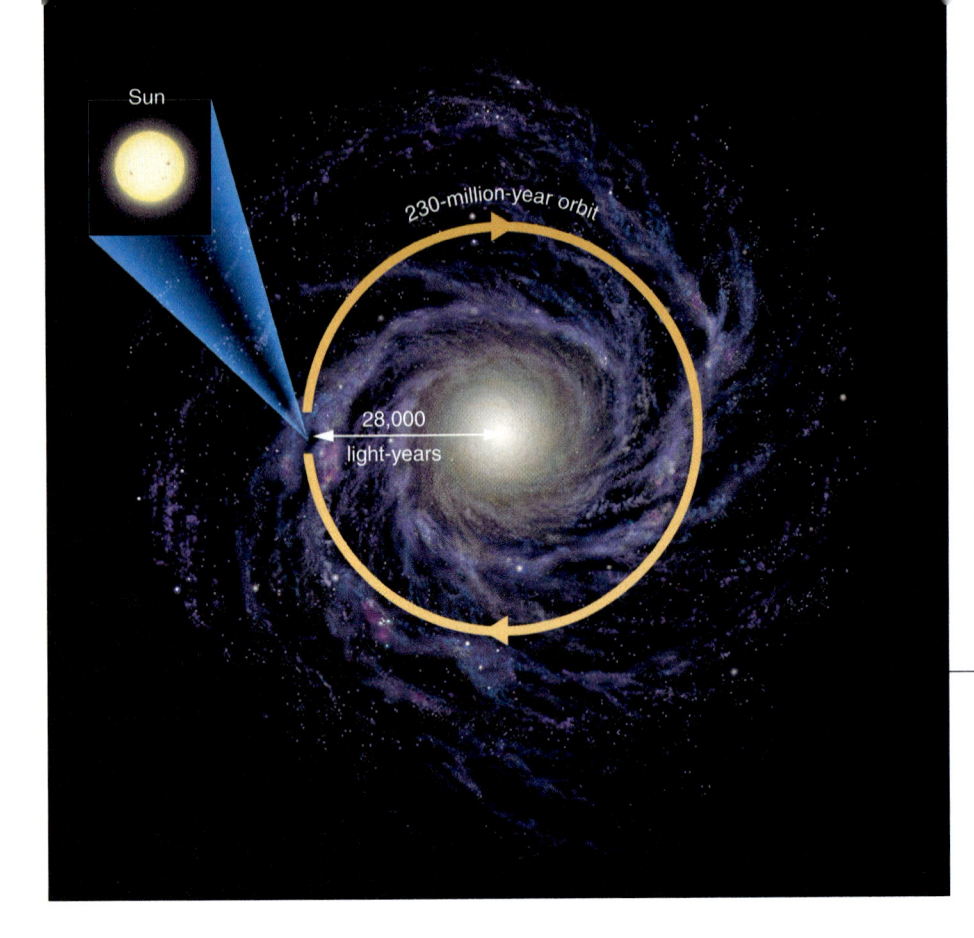

FIGURE 1.16 This painting shows how the entire Milky Way Galaxy rotates (in a direction that would tend to wind up the spiral arms). Our Sun and solar system are located about 28,000 light-years from the galactic center. At this distance, each orbit around the galactic center takes about 230 million years.

not detected any light coming from it, and we therefore call it **dark matter**. Studies of other galaxies show that they also are made mostly of dark matter. In fact, most of the mass in the universe appears to be made of this mysterious dark matter, but we do not yet know what it is.

The Expanding Universe

The billions of galaxies in the universe also move relative to one another. Within the Local Group (see Figure 1.1), the motions are about as we might expect. Some of the galaxies move toward the Milky Way Galaxy, some move away, and some move in more complex ways. (For example, two small galaxies, known as the Large and Small Magellanic Clouds, apparently orbit the Milky Way.) Again, the speeds are enormous by earthly standards—the Milky Way is moving toward the Great Galaxy in Andromeda (M31) at about 300,000 kilometers per hour (180,000 miles per hour)—but are unnoticeable to our eyes. We needn't worry about a collision any time soon. Even if the Milky Way and Andromeda Galaxies are approaching each other head-on (which they might not be), it will be nearly 10 billion years before the collision begins.

When we look outside the Local Group, however, we find two astonishing facts that were first discovered in the 1920s by Edwin Hubble, for whom the Hubble Space Telescope was named:

1. Virtually every galaxy outside the Local Group is moving *away* from us.
2. The more distant the galaxy, the faster it appears to be racing away from us.

Upon first becoming aware of these two facts, you might be tempted to conclude that our Local Group (which is held together by gravity) suffers a cosmic case of chicken pox. However, there is a much more natural explanation: *The entire universe is expanding.* We'll save details about this expansion for later in the book (Chapter 19), but you can understand the basic idea by thinking about a raisin cake baking in an oven.

Imagine that you make a raisin cake in which the distance between adjacent raisins is 1 centimeter. You place the cake in the oven, where it expands as it bakes. After 1 hour, you remove the cake, which has expanded so that the distance between adjacent raisins is 3 centimeters (Figure 1.18).

Pick any raisin (it doesn't matter which one), call it the Local Raisin, and identify it in the pictures of the cake both before and after baking; Figure 1.18 shows one possible choice for the Local Raisin, with several other nearby raisins labeled. Before baking,

FIGURE 1.17 This painting shows an edge-on view of the Milky Way Galaxy. Most visible stars reside within the galaxy's thin *disk,* which runs horizontally across the page in this figure. But careful study of galactic rotation shows that most of the mass lies in the galactic *halo*—a large, spherical region that surrounds and encompasses the disk. Because this mass emits no light that we have detected, we call it dark matter. (The dark matter may extend quite far out; see Figure 21.1.)

Raisin 1 is 1 centimeter away from the Local Raisin, Raisin 2 is 2 centimeters away, and Raisin 3 is 3 centimeters away. After baking, Raisin 1 is 3 centimeters away from the Local Raisin, Raisin 2 is 6 centimeters away, and Raisin 3 is 9 centimeters away. Nothing should seem surprising so far, because the new distances simply reflect the fact that the cake has expanded uniformly everywhere.

Now, suppose you live *inside* the Local Raisin. From your vantage point, the other raisins appear to move away from you as the cake expands. For example, Raisin 1 is 1 centimeter away before baking and 3 centimeters away after baking. It therefore appears to move 2 centimeters during the hour, so you'll see it moving away from you at a speed of 2 centimeters per hour. Raisin 2 is 2 centimeters away before baking and 6 centimeters away afterward; thus, it appears to move 4 centimeters during the hour of baking, giving it a speed of 4 centimeters per hour away from you. The following table lists the distances of other raisins before and after baking, along with their speeds as seen from the Local Raisin.

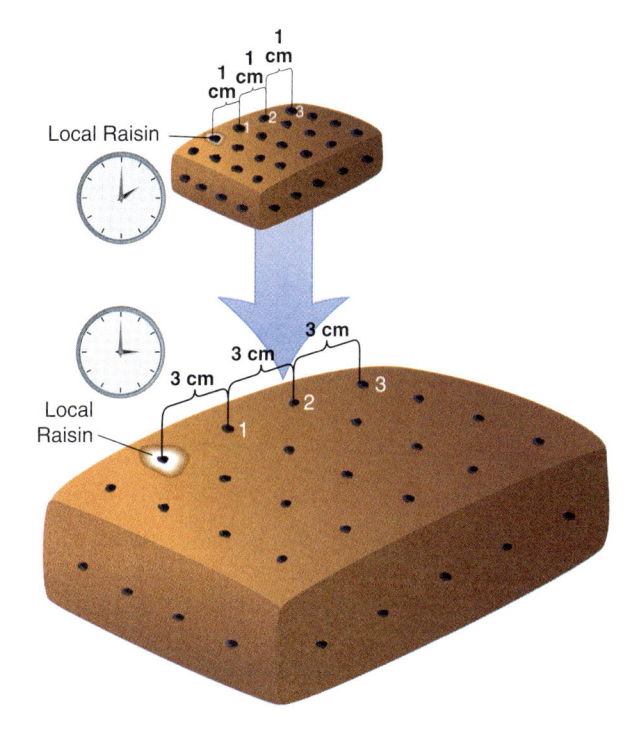

FIGURE 1.18 An expanding raisin cake illustrates basic principles of the expansion of the universe.

Distances and Speeds As Seen from the Local Raisin

Raisin Number	Distance Before Baking	Distance After Baking (1 hour later)	Speed
1	1 cm	3 cm	2 cm/hr
2	2 cm	6 cm	4 cm/hr
3	3 cm	9 cm	6 cm/hr
4	4 cm	12 cm	8 cm/hr
⋮	⋮	⋮	⋮
10	10 cm	30 cm	20 cm/hr
⋮	⋮	⋮	⋮

Note that, although the entire cake expanded uniformly, from the vantage point of the Local Raisin each subsequent raisin moved away at increasingly faster speeds. Also, because selection of the Local Raisin was arbitrary, you would find the same results no matter which raisin you chose to represent your Local Raisin.

If you now imagine the Local Raisin to represent our Local Group of galaxies and the other raisins to represent more distant galaxies or clusters of galaxies, you have a basic picture of the expansion of the universe. That is, *space* itself is growing, and more distant galaxies move away from us faster because they are carried along with this expansion like raisins in an expanding cake. And, just as the raisins themselves do not expand, the individual galaxies and clusters of galaxies do not expand (because they are bound together by gravity). The effects of expansion would appear basically the same viewed from any place in the universe.

Summary of Motion

Let's summarize the motions we have covered. We spin around the Earth's axis as we revolve around the Sun. Our solar system moves among the stars of the local solar neighborhood, while this entire neighborhood orbits the center of the Milky Way Galaxy. Our galaxy, in turn, moves among the other galaxies of the Local Group, while the Local Group is carried along with the overall expansion of the universe. Table 1.2 lists the motions and their associated speeds. Spaceship Earth is indeed carrying us on a remarkable journey!

1.4 The Human Adventure of Astronomy

In a relatively few pages, we've laid out a fairly complete overview of our modern scientific view of the universe. But our goal in this book is not for you simply to be able to recite this view. Rather it is to help you understand the evidence supporting this view and the extraordinary story of how it developed. In many ways, this book is devoted to this goal.

Astronomy is a human adventure in the sense that it affects virtually everyone—even those who have

Table 1.2 The Motions of Spaceship Earth

Motion	Typical Speed
rotation	1,000 km/hr or more around axis
revolution	100,000 km/hr around Sun
motion within local solar neighborhood	70,000 km/hr relative to nearby stars
rotation of the Milky Way Galaxy	1,000,000 km/hr around galactic center
motion within Local Group	300,000 km/hr toward Andromeda Galaxy
universal expansion	more distant galaxies moving away faster, with the most distant moving at speeds close to the speed of light

never looked at the sky—because the development of astronomy has been so deeply intertwined with the development of civilization as a whole. Revolutions in astronomy have gone hand in hand with the revolutions in science and technology that have shaped modern life. Witness the repercussions of the Copernican revolution, which changed our view of the Earth from the center of the universe to just one planet orbiting the Sun. This revolution, which we will discuss further in Chapter 5, began when Copernicus published his idea of a Sun-centered solar system in 1543. Three subsequent figures—Tycho Brahe, Johannes Kepler, and Galileo—provided the key evidence that eventually led to wide acceptance of the Copernican idea, and the revolution culminated with Isaac Newton's uncovering of the laws of motion and gravity. Newton's work, in turn, became the foundation of physics that helped fuel the industrial revolution.

More recently, the development of space travel and the computer revolution have helped fuel tremendous progress in astronomy. We've learned a lot about our solar system by sending probes to the planets, and many of our most powerful observatories, including the Hubble Space Telescope, reside in space. On the ground, computer design and control have led to tremendous growth in the size and power of telescopes, particularly in the past decade.

Many of these efforts, along with the achievements they spawned, have led to profound social change. The most famous example involved Galileo, whom the Vatican put under house arrest in 1633 for his claims that the Earth orbits the Sun. Although the Church soon recognized that Galileo was right, he was formally vindicated only with a statement by Pope John Paul II in 1992. In the meantime, his case spurred great debate in religious circles and had a profound influence on both theological and scientific thinking.

FIGURE 1.19 This fanciful image, which is also the cover of this book, imagines hikers looking at a sky filled with galaxies; the galaxies are actually part of a photograph taken by the Hubble Space Telescope (see Figure 21.11). Although such a view could not really exist anywhere in the universe, the image shows how our minds can connect the deepest mysteries of the cosmos to our lives here on Earth. Indeed, a primary goal of this book is to help you understand the extraordinary story of how we have reached our current scientific understanding of the universe and to appreciate the many mysteries that still remain.

As you progress through this book and learn about astronomical discovery, try to keep in mind the context of the human adventure. You will then be learning not just about a science, but about one of the great forces that have helped shape our modern world. This context will also lead you to think about how the many astronomical mysteries that remain—such as the makeup of dark matter, the events of the first instant of the Big Bang, and the question of life beyond Earth—may influence our future. What would it mean to us if we were ever to learn the complete story of our cosmic origins? How would our view of the Earth be changed if we came to learn that Earth-like planets are common or exceedingly rare? Only time may answer these questions, but the chapters ahead give you the foundation you need to understand how we changed from primitive people looking at patterns in the night sky to a civilization capable of asking deep questions about our existence.

THE BIG PICTURE

In this first chapter, we developed a broad overview of our place in the universe. It is not yet necessary for you to remember the details—everything presented in this chapter will be covered in greater depth later in the book. However, you should understand enough so that all the following "big picture" ideas are clear:

- The Earth is not the center of the universe but instead is a planet orbiting a rather ordinary star in the Milky Way Galaxy. The Milky Way Galaxy, in turn, is one of billions of galaxies in our observable universe.

- We are "star stuff." The atoms from which we are made began as hydrogen and helium in the Big Bang and were later fused into heavier elements in massive stars. When these stars died, they released these atoms into space, where our galaxy recycled them. Our solar system formed from such recycled matter some 4.6 billion years ago.

- Cosmic distances are literally astronomical, but we can put them in perspective with the aid of scale models and other scaling techniques. When you think about these enormous scales, don't forget that every star is a sun and every planet is a unique world.

- We are latecomers on the scale of cosmic time. The universe was more than halfway through its history by the time our solar system formed, and then it took billions of years more before humans arrived on the scene.

- All of us are being carried through the cosmos on spaceship Earth. Although we cannot feel this motion in our everyday lives, the associated speeds are surprisingly high. Learning about the motions of spaceship Earth not only gives us a new perspective on the cosmos, but also helps us understand its nature and history.

- Throughout history, the development of astronomy has gone hand in hand with social and technological development. In this sense, astronomy touches everyone, making it a human adventure to be enjoyed by all.

Review Questions

1. What do we mean by a *geocentric* universe? In broad terms, contrast a geocentric universe with our modern view of the universe.

2. Describe the major levels of structure in the universe.

3. What do we mean when we say that the universe is *expanding*? Why does an expanding universe suggest a beginning in what we call the *Big Bang*?

4. Summarize our cosmic origins and explain what Carl Sagan meant when he said that we are "star stuff."

5. Explain the statement *The farther away we look in distance, the further back we look in time.*

6. What do we mean by the *observable universe*?

7. Describe key features of our solar system as they would appear on a scale of 1 to 10 billion.

8. How do the distances to the stars compare to distances within our solar system? Give a few examples to put the comparison in perspective.

9. Describe at least two ways to put the scale of the Milky Way Galaxy in perspective.

10. How many galaxies are in the observable universe? How many stars are in the observable universe? Put these numbers in perspective.

11. Imagine describing the cosmic calendar to friends. In your own words, give them a feel for how the human race fits into the scale of time.

12. Briefly summarize the major motions of the Earth through the universe.

13. What is an *astronomical unit*? What is the *ecliptic plane*?

14. Describe the cause of the Earth's seasons, and explain why the seasons are opposite in the Northern and Southern Hemispheres.

15. What is *precession*, and how does it affect the appearance of our sky?

16. Briefly describe the motion of the Sun relative to stars in the local solar neighborhood and to the entire Milky Way Galaxy. Given the relatively high speeds of stars relative to us, why don't we notice them racing around the sky?

17. Describe the raisin cake model of the universe, and explain how it shows that a uniform expansion would lead us to see more distant galaxies moving away from us at higher speeds.

18. Briefly describe how astronomy is interwoven with social and technological change.

Discussion Questions

1. *Vast Orbs.* The chapter-opening quotation from Christiaan Huygens suggests that humans might be less inclined to wage war if everyone appreciated the Earth's place in the universe. Do you agree? Defend your opinion.

2. *Infant Species.* In the last few tenths of a second before midnight on December 31 of the cosmic calendar, we have developed an incredible civilization and learned a great deal about the universe, but we also have developed technology through which we could destroy ourselves. The midnight bell is striking, and the choice for the future is ours. How far into the next cosmic year do you think our civilization will survive? Defend your opinion.

3. *A Human Adventure.* How important do you think astronomical discoveries have been to our social development? Defend your opinion with examples drawn from your knowledge of history.

Problems

(Quantitative problems are marked with an asterisk.)

Sensible Statements? For **problems 1–8**, decide whether the statement is sensible and explain why it is or is not.

Example: I walked east from our base camp at the North Pole.

Solution: The statement does not make sense because *east* has no meaning at the North Pole—all directions are south from the North Pole.

1. The universe is between 10 billion and 16 billion years old.

2. The universe is between 10 billion and 16 billion light-years old.

3. It will take me light-years to complete this homework assignment!

4. Someday we may build spaceships capable of traveling at a speed of 1 light-minute per hour.

5. Astronomers recently discovered a moon that does not orbit a planet.

6. NASA plans soon to launch a spaceship that will leave the Milky Way Galaxy to take a photograph of the galaxy from the outside.

7. The observable universe is the same size today as it was a few billion years ago.

8. Photographs of distant galaxies show them as they were when they were much younger than they are today.

9. *No Axis Tilt.* Suppose the Earth's axis had no tilt. Would we still have seasons? Why or why not?

10. *Tour Report.* A friend asks you the following questions concerning your tour of a scale-model solar system. Answer each question in one paragraph.
    a. Is the Sun really much bigger than the Earth?
    b. Is it true that the Sun uses nuclear energy?
    c. Would it be much harder to send humans to Mars than to the Moon?
    d. In elementary school, I heard that Neptune is farther from the Sun than Pluto. Is this true?
    e. I read that Pluto is not really a planet. What's the story?
    f. Why didn't they have any stars besides the Sun in the scale model?
    g. What was the most interesting thing you learned during your tour?

11. *Raisin Cake Universe.* Suppose that all the raisins in a cake are 1 centimeter apart before baking and after baking they are 4 centimeters apart.
    a. Draw diagrams to represent the cake before and after baking.
    b. Identify one raisin as the Local Raisin on your diagrams. Construct a table showing the distances and speeds of other raisins as seen from the Local Raisin.
    c. Briefly explain how your expanding cake is similar to the expansion of the universe.

12. *Scaling the Local Group of Galaxies.* Both the Milky Way Galaxy and the Great Galaxy in Andromeda (M31) have a diameter of about 100,000 light-years. The distance between the two galaxies is about 2.5 million light-years.
    a. Using a scale on which 1 centimeter represents 100,000 light-years, draw a sketch showing both galaxies and the distance between them to scale.
    b. How does the separation between galaxies compare to the separation between stars? Based on your answer, discuss the likelihood of galactic collisions in comparison to the likelihood of stellar collisions.

*13. *Distances by Light.* Just as a light-year is the distance that light can travel in 1 year, we define a light-second as the distance that light can travel in 1 second, a light-minute as the distance that light can travel in 1 minute, and so on. Following the method of Mathematical Insight 1.1, calculate the distance in kilometers represented by each of the following: 1 light-second; 1 light-minute; 1 light-hour; 1 light-day.

*14. *Driving to the Planets (and Stars).* Imagine that you could drive your car in space at a constant speed of 100 km/hr (62 mi/hr). (In reality, the law of gravity would make driving through space at a constant speed all but impossible.)
    a. How long would it take to circle the Earth in low orbit? (*Hint:* Use the Earth's circumference of approximately 40,000 km.)
    b. Suppose you started driving from the Sun. How long would it take to reach Earth? How long would it take to reach Pluto? Use the data in Table 1.1.
    c. How long would it take to drive the 4.4 light-years to Alpha Centauri? (*Hint:* You'll need to convert the distance to kilometers.)
    d. The *Voyager 2* spacecraft is traveling at about 50,000 km/hr. At this speed, how long would it take to reach Alpha Centauri (if it were headed in the right direction, which it is not)?

*15. *Speed Calculations.* Calculate each of the following speeds in both kilometers per hour and miles per hour. In each case, assume a circular orbit; recall that the formula for the circumference of a circle is $2 \times \pi \times$ radius. (*Hint:* Divide the distance traveled in the circular orbit by the time it takes to complete one orbit.)
    a. The speed of the Earth going around the Sun; use the average Earth–Sun distance of 149.6 million km.
    b. The speed of our solar system around the center of our galaxy; assume that we are located 28,000 light-years from the center and that each orbit takes 230 million years.

Web Projects

Find useful links for Web projects on the text Web site.

1. *Astronomy on the Web.* The Web contains a vast amount of astronomical information. Starting from the links on the textbook Web site, spend at least an hour exploring astronomy on the Web. Write two or three paragraphs summarizing what you learned from your Web surfing. What was your favorite astronomical Web site, and why?

2. *Voyage.* Visit the Web site for the Voyage scale-model solar system on the National Mall. Write a one-page summary of what you learn from your virtual tour.

3. *NASA Missions.* Visit the NASA Web site to learn about upcoming missions that concern astronomy. Write a one-page summary of the mission you feel is most likely to give us new astronomical information during the time you are enrolled in your astronomy course.

We had the sky, up there, all speckled with stars, and we used to lay on our backs and look up at them, and discuss about whether they was made, or only just happened.

MARK TWAIN, *HUCKLEBERRY FINN*

CHAPTER 2

Discovering the Universe for Yourself

It is an exciting time in the history of astronomy. A new generation of telescopes is probing the depths of the universe. Increasingly sophisticated space probes are collecting new data about the planets and other objects in our solar system. Rapid advances in computing technology allow scientists to analyze the vast amount of new data and to model the processes that occur in planets, stars, galaxies, and the universe.

One of the goals of this book is to help *you* share in the ongoing adventure of astronomical discovery. One of the best ways to become a part of this adventure is to discover the universe for yourself by doing what other humans have done for thousands of generations: Go outside, observe the sky around you, and contemplate the awe-inspiring universe of which you are a part. In this chapter, we'll discuss a few key ideas that will help you understand what you see in the sky.

FIGURE 2.1 This photograph shows stars in the constellation Orion. Stars are so far away that they have no visible size; brighter stars appear larger in photographs only because they are overexposed. The crosses on bright stars are an artifact of the telescope used to take the photograph.

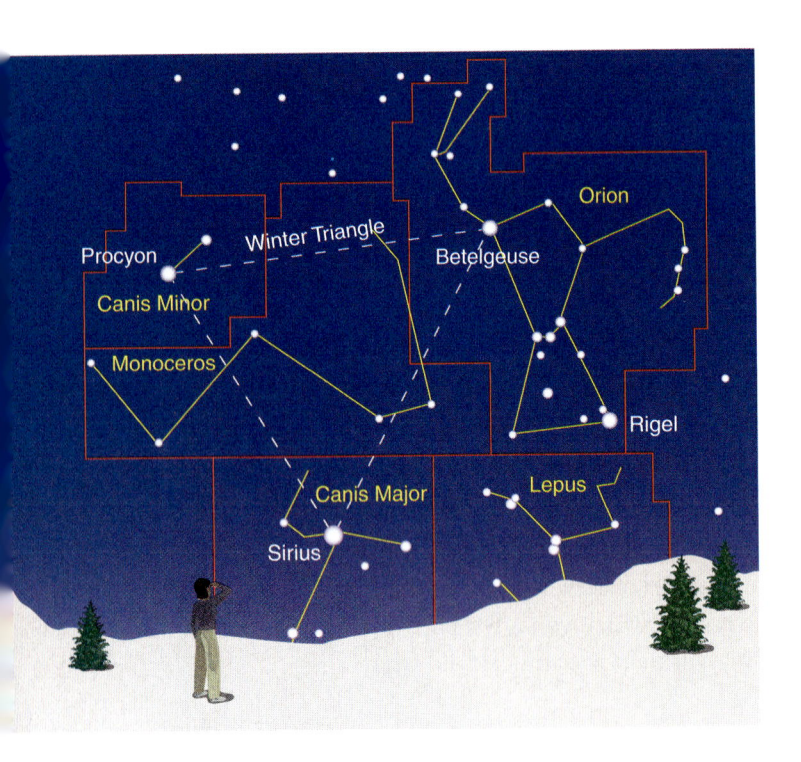

FIGURE 2.2 Red lines mark official borders of several constellations near Orion. Yellow lines connect recognizable patterns of stars within constellations. Sirius, Procyon, and Betelgeuse form a pattern spanning several constellations and called the Winter Triangle; it is easy to find on clear winter evenings.

2.1 Patterns in the Sky

Shortly after sunset, as daylight fades to darkness, the sky appears to fill slowly with stars. On clear, moonless nights far from city lights, as many as 2,000–3,000 stars may be visible to your naked eye. As you look at the stars, your mind might group them into many different patterns. If you observe the sky night after night or year after year, you will recognize the same patterns of stars.

People of nearly every culture gave names to patterns in the sky. The pattern that the Greeks named Orion, the hunter (Figure 2.1), was seen by the ancient Chinese as a supreme warrior called *Shen*. Hindus in ancient India also saw a warrior, called *Skanda*, who as the general of a great celestial army rode a peacock. The three stars of Orion's belt were seen as three fishermen in a canoe by Aborigines of northern Australia. As seen from southern California, these three stars climb almost straight up into the sky as they rise in the east, which may explain why the Chemehuevi Indians of the California desert saw them as a line of three sure-footed mountain sheep.[1] These are but a few of the many names, each accompanied by a rich folklore, that have been given to the pattern of stars we call Orion.

Constellations

The patterns of stars seen in the sky are usually called constellations. Astronomers, however, use the term **constellation** to refer to a specific *region* of the sky. Any place in the sky belongs to some constellation; familiar patterns of stars merely help locate particular constellations. For example, the constellation Orion includes all the stars in the familiar pattern of the hunter, along with the region of the sky in which these stars are found (Figure 2.2).

The official borders of the constellations were set in 1928 by members of the International Astronomical Union (IAU), an association of astronomers from around the world. The IAU divided the sky into 88 constellations whose borders correspond roughly to the star patterns recognized by Europeans. Thus, despite the wide variety of names given to patterns of stars by different cultures, the "official" names of constellations visible from the Northern Hemisphere can be traced back to the ancient Greeks and to other cultures of southern Europe, the Middle East, and northern Africa. No one knows exactly when these constellations were first named, although some names probably go back at least 5,000 years. The official names of the constellations visible from the Southern Hemisphere are primarily those given by seventeenth-century European explorers.

[1]These and other constellation stories are found in E. C. Krupp, *Beyond the Blue Horizon* (Oxford University Press, 1991).

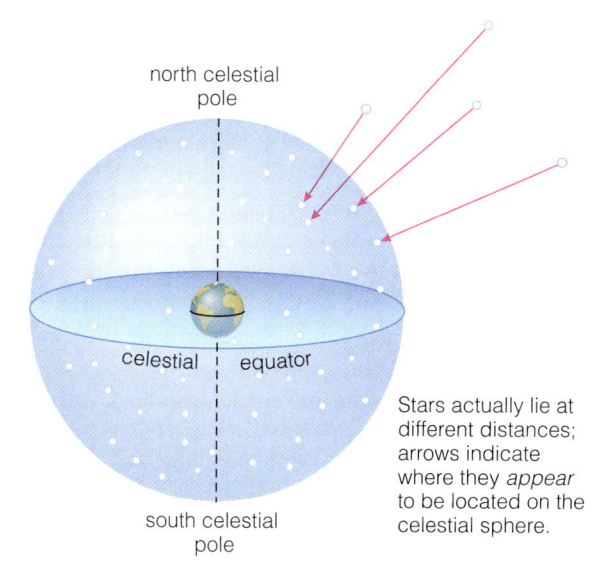

FIGURE 2.3 Our lack of depth perception when we look into space creates the illusion that the Earth is surrounded by a *celestial sphere*. However, stars that appear very close together in our sky may actually lie at very different distances from Earth.

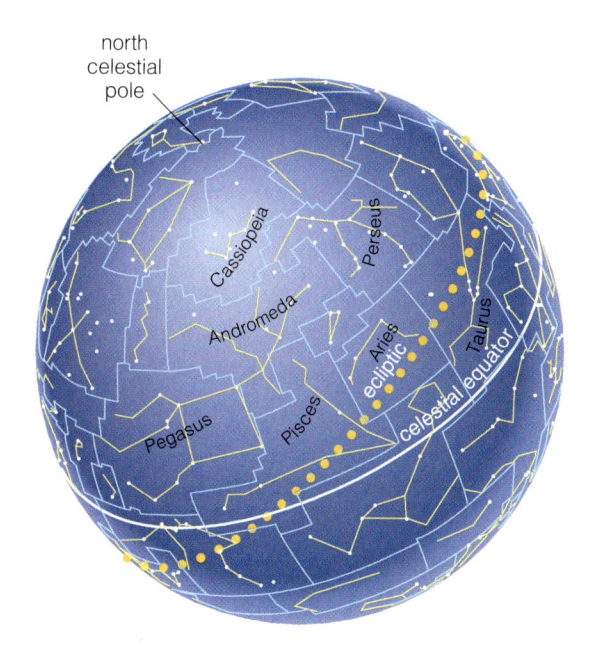

FIGURE 2.4 A model of the celestial sphere shows the patterns of the stars, the borders of the 88 official constellations, the ecliptic, and the celestial equator and poles. Because the celestial sphere shows the view from Earth, we imagine the Earth to reside in the center of the sphere. Thus, when we look into the sky, we see patterns of stars as they would appear from *inside* this imaginary sphere (which means the patterns appear left-right reversed compared to what we see when we look at the model from the outside).

Learning your way around the constellations is no more difficult than learning your way around your neighborhood, and recognizing the patterns of just 20–40 constellations is enough to make the entire sky seem familiar. The best way to learn the constellations is to go out and view them, guided by the help of a few visits to a planetarium and the star charts in the back of this book. The *Skygazer* software that comes with this book can also help you learn constellations.

The Celestial Sphere

The stars in a particular constellation may appear to lie close to one another, but this is an illusion. All stars in a particular constellation lie in similar directions from Earth but not necessarily at similar distances (Figure 2.3). However, because we lack depth perception when we look into space, it looks like all the stars lie at the same distance. The ancient Greeks mistook this illusion for reality, imagining the Earth to be surrounded by a great **celestial sphere**.

Today, the concept of a celestial sphere is still useful for our learning about the sky, even though we know that the Earth does not really lie in the center of a giant ball. We give names to special locations on the imaginary celestial sphere. The point directly above the Earth's North Pole is called the **north celestial pole** (NCP), and the point directly above the Earth's South Pole is called the **south celestial pole**. The **celestial equator** represents an extension of the Earth's equator into space (see Figure 2.3).

The stars form the fixed patterns of the constellations on the celestial sphere, while the Sun, Moon, and planets slowly wander among the stars. As we'll discuss shortly, the apparent motion of the Moon and the planets is fairly complex. The Sun, however, appears to slowly circle the celestial sphere on a simple annual path called the **ecliptic**; this apparent path is the projection of the ecliptic plane (see Figure 1.12) into space. A model of the celestial sphere typically shows the patterns of the stars, the borders of the 88 official constellations, the ecliptic, and the celestial equator and poles (Figure 2.4).

Common Misconceptions: Stars in the Daytime

Because we don't see stars in the daytime, some people believe that the stars vanish in the daytime and "come out" at night. In fact, the stars are always present. The only reason you cannot see stars in the daytime is that their dim light is overwhelmed by the bright daytime sky. You *can* see bright stars in the daytime with the aid of a telescope, and you may see stars in the daytime if you are fortunate enough to observe a total eclipse of the Sun. Astronauts also see stars in the daytime—above the Earth's atmosphere, where no air is present to scatter sunlight through the sky, the Sun is a bright disk against a dark sky filled with stars. (However, because the Sun is so bright, astronauts must block its light and allow their eyes to adapt to darkness if they wish to *see* the stars.)

FIGURE 2.5 A "fish-eye" photograph of the Milky Way in the Australian sky. Near the upper left, Comet Hale–Bopp is visible in this 1997 photo.

The Milky Way

As your eyes adapt to darkness at a dark site, you'll begin to see the whitish band of light called the *Milky Way*. It is from this band of light that our Milky Way Galaxy gets its name. You can see only part of the Milky Way at any particular time, but it stretches all the way around the celestial sphere. If you look carefully, you will notice that the Milky Way varies in width and has dark fissures running through it. The widest and brightest parts of the Milky Way are most easily seen from the Southern Hemisphere (Figure 2.5), which probably explains why the Aborigines of Australia gave names to the patterns they saw within the Milky Way in the same way that other cultures gave names to patterns of stars.

The band of light called the Milky Way bears an important relationship to the Milky Way Galaxy: *It traces the galactic plane as it appears from our location in the outskirts of the galaxy* (Figure 2.6). Recall that the Milky Way Galaxy is shaped like a thin pancake with a bulge in the middle (see Figures 1.16 and 1.17). We view the universe from our location more than halfway outward from the center of this "pancake." When we look in any direction *within* the plane of the galaxy, we see countless stars, along with interstellar gas and dust. These stars and glowing clouds of gas form the band of light we call the Milky Way. The dark fissures appear in regions where particularly dense interstellar clouds obscure our view of stars behind them. The central bulge of the galaxy makes the Milky Way wider in the direction of the galactic center, which is the direction of the constellation Sagittarius in our sky.

We see fewer stars when we look in directions pointing *away* from our location within the galactic plane, and there is relatively little gas and dust to obscure our view of more distant objects. Thus, we have a clear view to the far reaches of the universe, limited only by what our eyes or instruments allow. If you are fortunate enough to have a very dark sky, you may see a fuzzy patch in the constellation Andromeda (Figure 2.7). Although this patch may look like nothing more than a small cloud, you are actually seeing the Great Galaxy in Andromeda (M31)—some 2.5 million light-years away.

TIME OUT TO THINK *Contemplate the fact that light from M31 journeyed through space for some 2.5 million years to reach you and is the combined light of more than 100 billion stars. Do you think it possible that among some of those stars there are students like yourself gazing outward in amazement at the Milky Way Galaxy?*

2.2 The Circling Sky

If you spend a few hours out under a starry sky, you'll see stars (and the Moon and planets) rising and setting much like the Sun. In reality, it is we who are moving, not the stars. The Sun and stars appear to rise in the east and set in the west because the Earth rotates in the opposite direction, from west to east. In fact, the Earth's rotation makes the entire celestial sphere appear to rotate around us each day. If you could view the celestial sphere from outside, the daily circles of the stars would appear very simple (Figure 2.8).

However, because we live on the Earth, we see only *half* the celestial sphere at any one moment; the ground beneath us blocks our view of the other half. The particular half that we see depends on the time, the date, and our location on Earth. In this section, we'll discuss how to make sense of the sky as seen from Earth.

FIGURE 2.6 Artist's conception of the Milky Way Galaxy from afar, showing how the galaxy's structure affects our view from Earth. When we look *into* the galactic plane in any direction, our view is blocked by stars, gas, and dust. Thus, we see the galactic plane as the band of light we call the Milky Way, stretching a full 360° around our sky (i.e., around the celestial sphere). We have a clear view to the distant universe only when we look away from the galactic plane.

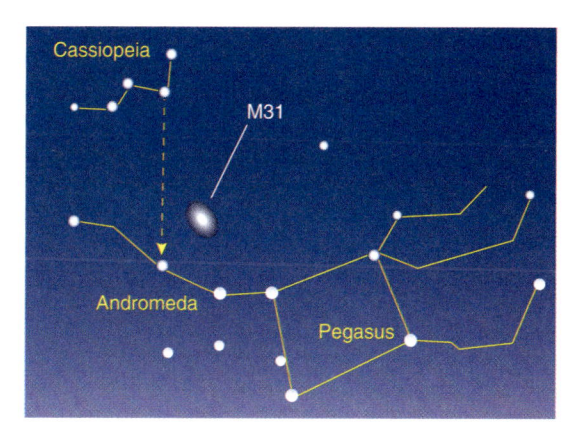

FIGURE 2.7 Location of M31 in the night sky.

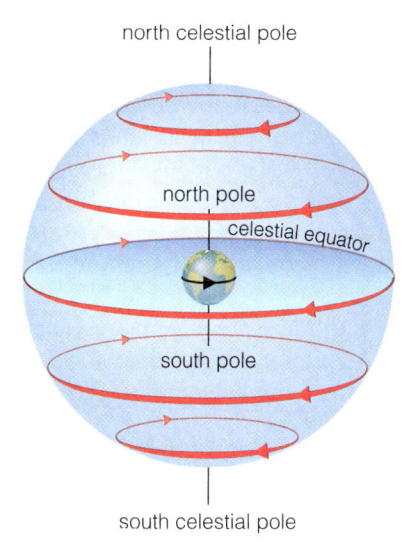

FIGURE 2.8 The Earth rotates from west to east (black arrows), making the celestial sphere *appear* to rotate around us from east to west. Viewed from outside, the stars (and the Sun, Moon, and planets) therefore appear to make simple daily circles around us. The red circles represent the apparent daily paths of a few selected stars.

CHAPTER 2 DISCOVERING THE UNIVERSE FOR YOURSELF 43

The Dome of the Sky

Picture yourself standing in a flat, open field. The sky appears to take the shape of a dome, making it easy to understand why people of many ancient cultures believed we lived on a flat Earth lying under a great dome that encompassed the world. Today, we use the appearance of a dome to define the **local sky**—the sky as seen from wherever you happen to be standing (Figure 2.9). The boundary between Earth and sky is what we call the **horizon**. The point directly overhead is your **zenith**. Your **meridian** is an imaginary half-circle stretching from your horizon due south, through your zenith, to your horizon due north.

You can pinpoint the position of any object in your local sky by stating its **direction** along your horizon (more technically measured as *azimuth*[2]) and its **altitude** above your horizon. For example, Figure 2.9 shows a person pointing to a star located in a southeasterly direction at an altitude of 60°. (The zenith has an altitude of 90° but no direction, since it is straight overhead.)

Because of our lack of depth perception in the sky, we cannot tell the true sizes of objects or the true distances between objects just by looking at them. For example, the Sun and the Moon look about the same size in our sky, but the Sun is actually about 400 times larger than the Moon in diameter. We can, however, measure *angles* in the sky. The **angular size** of an object like the Sun or the Moon is the angle it appears to span in your field of view. The **angular distance** between a pair of objects is the angle that appears to separate them. For example, Figure 2.10 shows that the angular size of the Moon is about $\frac{1}{2}°$ (about the same as that of the Sun), while the angular distance between the "pointer stars" at the end of the Big Dipper's bowl is about 5°. You can use your outstretched hand to make rough estimates of angles in the sky (Figure 2.10c).

For more precise astronomical measurements, we subdivide each degree into 60 **arcminutes** and subdivide each arcminute into 60 **arcseconds**. Thus, there are 60 arcseconds in 1 arcminute, 60 arcminutes in 1°, and 360° in a full circle. We abbreviate arcminutes with the symbol ′ and arcseconds with the symbol ″. For example, we read 35°27′15″ as "35 degrees, 27 arcminutes, 15 arcseconds."

Variation with Latitude

If you stay in one place, you'll see the same set of stars following the same paths through your sky from one year to the next. But if you travel north or south,

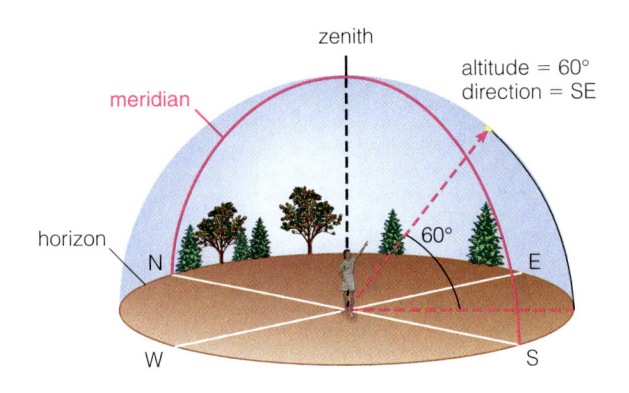

FIGURE 2.9 Definitions in your local sky.

you'll notice somewhat different sets of stars moving through the sky on somewhat different paths. This observation convinced ancient scientists that the sky must be more than a simple dome.

According to historical records, the idea of the sky as a celestial sphere was first proposed by the Greek scientist Anaximander (c. 610–547 B.C.). Based on what he learned about how the sky changes with travel north or south, Anaximander also concluded that the Earth must not be flat. However, because *east–west* travel does *not* change the stars visible in the sky, he did not realize that the Earth is round and instead suggested that it might be a cylinder curved in the north–south direction. By about 500 B.C., the famous mathematician Pythagoras suggested that the Earth is a sphere. More than a century later, Aristotle (384–322 B.C.) cited observations of the Earth's curved shadow on the Moon during lunar eclipses as evidence for a spherical Earth.

Common Misconceptions: A Flat Earth

A widespread myth holds that Columbus proved the Earth to be round rather than flat. In fact, knowledge of the round Earth predated Columbus by nearly 2,000 years. However, it is probably true that most people in Columbus's day believed the Earth to be flat, largely because of the poor state of education. The vast majority of the public was illiterate and unaware of the scholarly evidence for a spherical Earth. Interestingly, Columbus's primary argument with other scholars concerned the distance from Europe to Asia going westward—and it was Columbus who was wrong. He underestimated the true distance and as a result was woefully unprepared for the voyage to Asia that he thought he was undertaking. Indeed, his voyages would almost certainly have ended in disaster had it not been for the presence of the Americas, which offered a safe landing well to the east of Asia.

---

[2]Azimuth is usually measured clockwise around the horizon from due north. By this definition, the azimuth of due north is 0°, due east is 90°, due south is 180°, and due west is 270°.

a The angular size of the Moon is about $\frac{1}{2}°$.

b The angular distance between the pointer stars of the Big Dipper is about 5°.

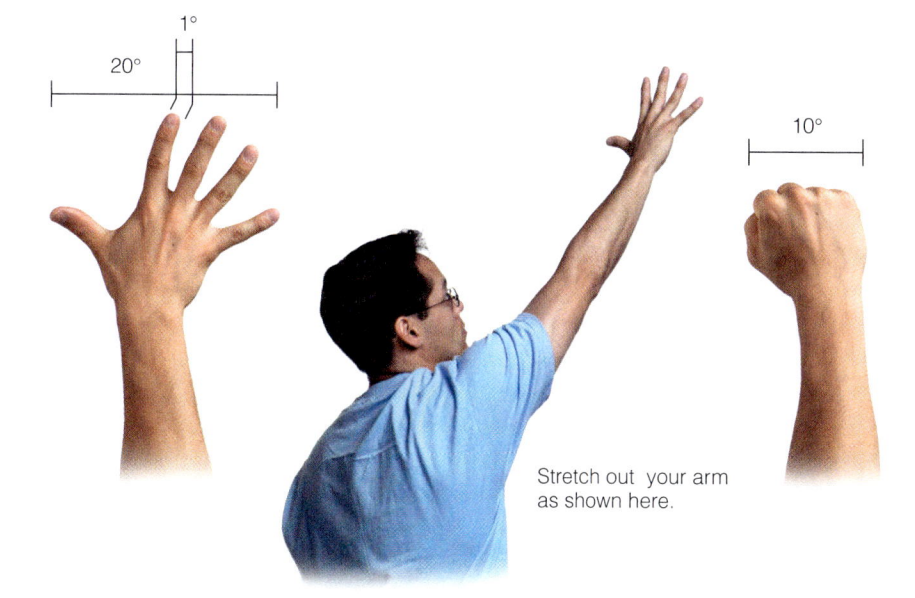

c You can estimate angular sizes or distances with your outstretched hand.

Stretch out your arm as shown here.

FIGURE 2.10 We measure *angular sizes* or *angular distances*, rather than actual sizes or distances, when we look at objects in the sky.

To understand why the sky changes with north–south travel, we must first review how we locate points on the Earth (Figure 2.11a). **Latitude** measures positions north or south; it is defined to be 0° at the equator, so the North Pole and the South Pole have latitude 90°N and 90°S, respectively. Note that "lines of latitude" actually are circles running parallel to the equator. **Longitude** measures east–west position, so "lines of longitude" are semicircles extending from the North Pole to the South Pole. The line of longitude passing through Greenwich, England, is defined to be longitude 0° (Figure 2.11b). This line is sometimes called the **prime meridian**; the decision to denote the line of longitude passing through Greenwich as the prime meridian was made by international treaty in 1884. Stating a latitude and a longitude pinpoints a location. For example, Figure 2.11a shows that Rome lies at about 42°N latitude and 12°E longitude and that Miami lies at about 26°N latitude and 80°W longitude.

If we combine the idea of the sky's dome-shaped appearance when seen from the ground with the idea that stars make simple daily circles around the celestial sphere, we can see why the apparent paths of stars depend on *latitude*. Your latitude determines your orientation on Earth relative to the celestial sphere; "down" is toward the center of the Earth, and "up" (toward your zenith) is away from the center of the Earth. Figure 2.12 shows sample orientations relative to the celestial sphere for mid-latitudes in each hemisphere. If you study the diagrams carefully, you'll discover the following key features:

■ In the Northern Hemisphere, stars relatively near the north celestial pole remain constantly above the horizon. We say that such stars are **circumpolar**. They never rise or set but instead make daily counterclockwise circles around the north celestial pole. In the Southern Hemisphere,

CHAPTER 2 DISCOVERING THE UNIVERSE FOR YOURSELF   45

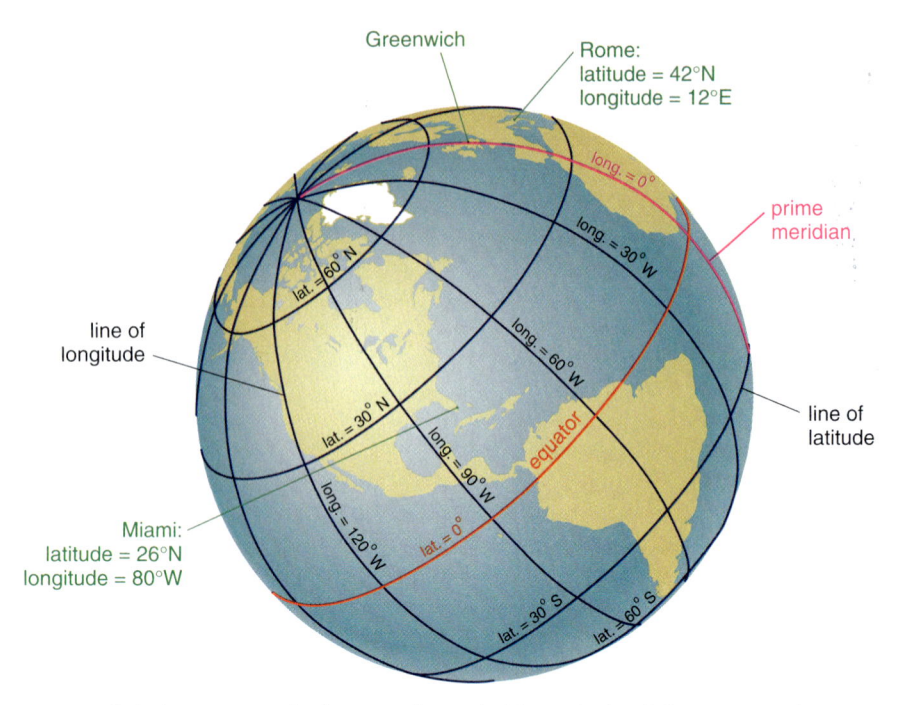

FIGURE 2.11 We can locate any place on Earth's surface by its latitude and longitude.

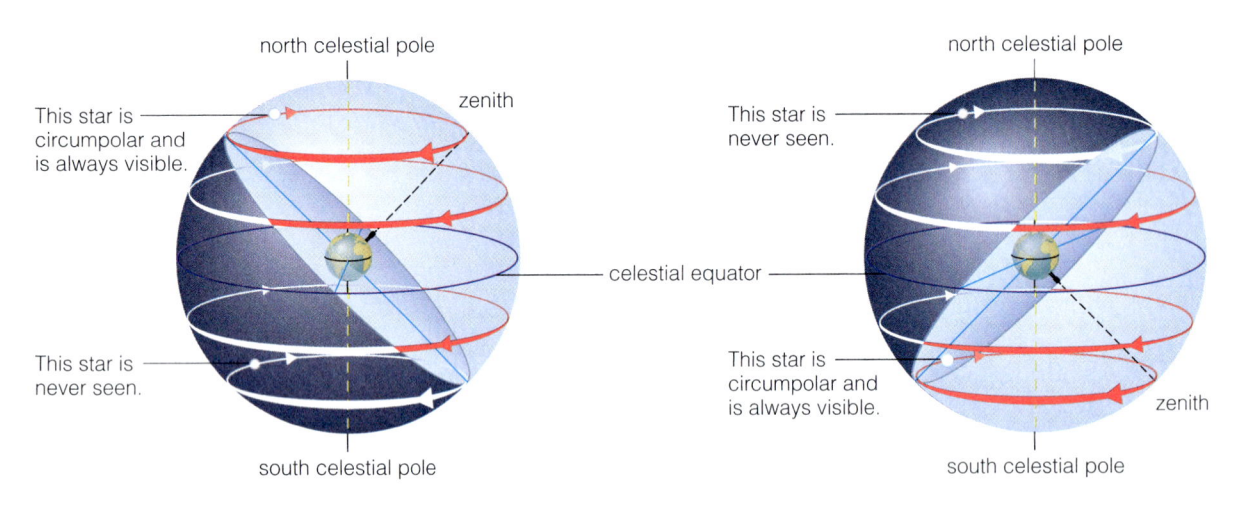

a Latitude measures angular distance north or south of the equator. Longitude measures angular distance east or west of the prime meridian, which passes through Greenwich, England.

b The entrance to the Old Royal Greenwich Observatory, near London. The line emerging from the door marks the prime meridian.

a Northern Hemisphere

b Southern Hemisphere

FIGURE 2.12 Sample local skies for (**a**) the Northern Hemisphere and (**b**) the Southern Hemisphere. Compare to Figure 2.8; note that your latitude determines your orientation relative to the celestial sphere, and the daily rotation determines the paths of stars through your local sky. To view the local skies more clearly, rotate the page so the zenith points up.

circumpolar stars make daily clockwise circles around the south celestial pole. Figure 2.13 shows a beautiful photograph of the daily circles of stars.

- In the Northern Hemisphere, stars relatively near the south celestial pole remain constantly below the horizon and are never seen. In the Southern Hemisphere, stars near the north celestial pole are never visible. Thus, different sets of constellations are visible in northern and southern skies.

Note that the set of constellations changes only with latitude, not with longitude.

- All other stars (those that are neither circumpolar nor never visible) rise daily in the east and set daily in the west. Stars located north of the celestial equator on the celestial sphere rise north of due east and set north of due west, stars located on the celestial equator rise due east and set

46   PART I   DEVELOPING PERSPECTIVE

FIGURE 2.13 This time-exposure photograph, taken at Arches National Park in Utah, shows how the Earth's rotation causes stars to trace daily circles around the sky. The north celestial pole lies at the center of the circles. Over the course of a full day, circumpolar stars trace complete circles, and stars that rise in the east and set in the west trace partial circles. Here we see only a portion of the full daily paths, because the time exposure lasted only a couple of hours.

due west, and stars located south of the celestial equator rise south of due east and set south of due west.

- If you study the geometry of the diagrams carefully, you'll see that in the Northern Hemisphere the altitude of the north celestial pole in your sky is equal to your latitude. For example, if the north celestial pole appears in your sky 40° above your horizon, your latitude is 40°N. Similarly, in the Southern Hemisphere the altitude of the south celestial pole in your sky is equal to your latitude.

The last feature listed above is very useful for navigation, since it allows you to determine your latitude just by finding the celestial pole in your sky. Finding the north celestial pole is fairly easy, because it lies very close to the star Polaris (Figure 2.14a). In the Southern Hemisphere, you can find the south celestial pole with the aid of the Southern Cross (Figure 2.14b). We'll discuss celestial navigation and how the sky varies with latitude in more detail in Chapter S1.

TIME OUT TO THINK *Answer the following questions for your latitude: Where is the north (or south) celestial pole in your sky? Where should you*

Common Misconceptions: What Makes the North Star Special?

Most people are aware that the North Star, Polaris, is a special star. Contrary to a relatively common belief, however, it is *not* the brightest star in the sky; more than 50 other stars are either considerably brighter or comparable in brightness. Polaris is special because it so closely marks the direction of due north and because its altitude in your sky is nearly equal to your latitude on Earth, making it very useful in navigation in the Northern Hemisphere.

CHAPTER 2 DISCOVERING THE UNIVERSE FOR YOURSELF

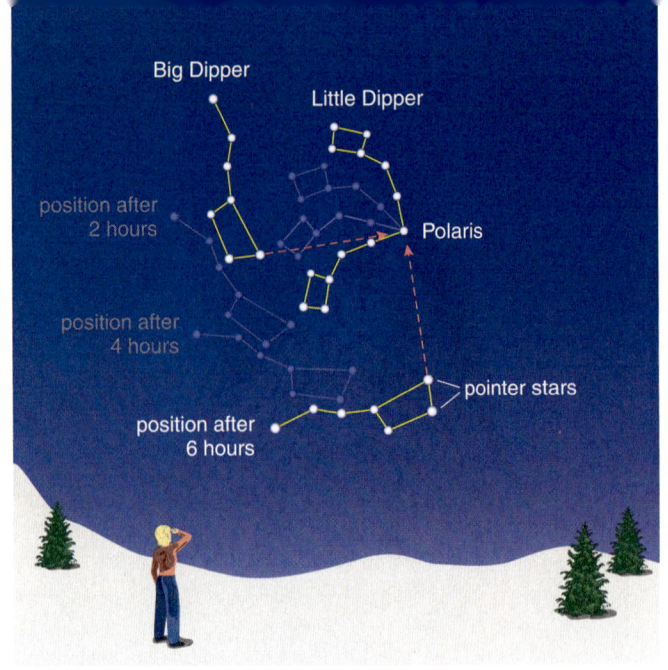

a In the Northern Hemisphere, the pointer stars of the Big Dipper point to Polaris, which lies within 1° of the north celestial pole. Note that the sky appears to turn *counterclockwise* around the north celestial pole.

b In the Southern Hemisphere, the Southern Cross points to the south celestial pole, which is not marked by any bright star. The sky appears to turn *clockwise* around the south celestial pole.

FIGURE 2.14 The altitude of the celestial pole in your sky is equal to your latitude.

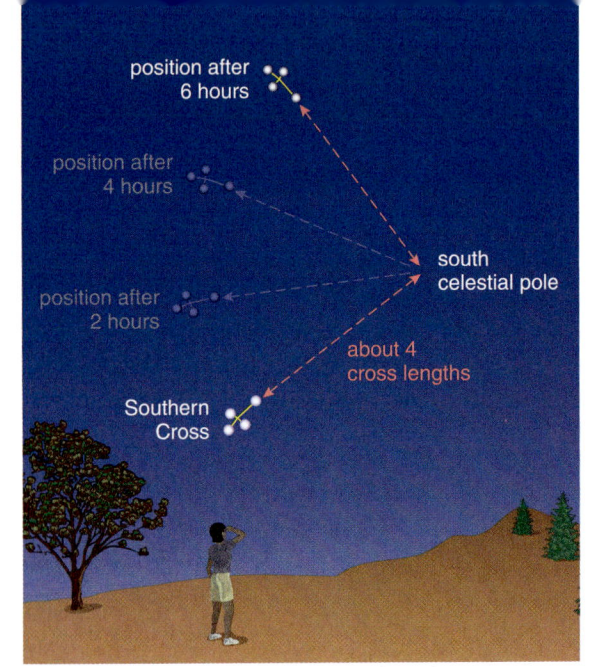

look to see circumpolar stars? What portion of the celestial sphere is never visible in your sky?

Seasonal Changes in the Sky

The basic patterns of motion in the sky remain the same from one day to the next: The Sun, Moon, planets, and stars trace daily circles around the sky. However, if you observe the sky night after night, you will notice changes that cannot be seen over a single night. One pattern that becomes obvious after a few months is that the constellations visible at a particular time change with the seasons. For example, Orion is prominent in the evening sky in February, but by September it is visible only if you look shortly before dawn.

The nighttime constellations change with the seasons because of the Earth's orbit around the Sun. As shown in Figure 2.15, at different times of year we view the Sun from different places in our orbit. As a result, the Sun appears to move steadily *eastward* along a circular path through the constellations each year; we've already identified this path as the *ecliptic* (see Figure 2.4). The constellations along the ecliptic are called the constellations of the **zodiac**. (Tradition places 12 constellations along the zodiac, but the official borders include a wide swath of a thirteenth constellation, Ophiuchus.) The Sun's apparent location along the ecliptic determines which constellations you see at night. In late August, for example, the Sun appears to be in the constellation Leo, so you won't see the stars of Leo at all—they're above your horizon only during the daytime. Aquarius, however, will be visible on your meridian at midnight because it is opposite Leo on the celestial sphere. Six months later, in February, it is Aquarius that cannot be seen and Leo that is on the meridian at midnight.

TIME OUT TO THINK *Based on Figure 2.15 and today's date, where among the zodiacal constellations does the Sun currently appear? What zodiacal constellation will be on your meridian at midnight? What zodiacal constellation will you see in the west shortly after sunset? Go outside at night to confirm your answers.*

The Sun's path through the daytime sky also changes with the seasons. Recall that the Earth's orientation *relative to the Sun* changes during the year because of the $23\frac{1}{2}°$ tilt of the Earth's axis (see Figure 1.13). This is also why the ecliptic crosses the celestial equator at angles of $23\frac{1}{2}°$ on the celestial sphere (see Figure 2.4). When the Sun appears north of the celestial equator, the Sun's daily path is long and high for the Northern Hemisphere (making the long, hot days of summer) but short and low for the Southern Hemisphere (making the short, cool days of winter). The situation for the two hemispheres is reversed when the Sun appears south of the celestial equator. At any particular location, you can notice this annual change by observing the Sun's position in your sky at the same time each day (Figure 2.16).

48 PART I DEVELOPING PERSPECTIVE

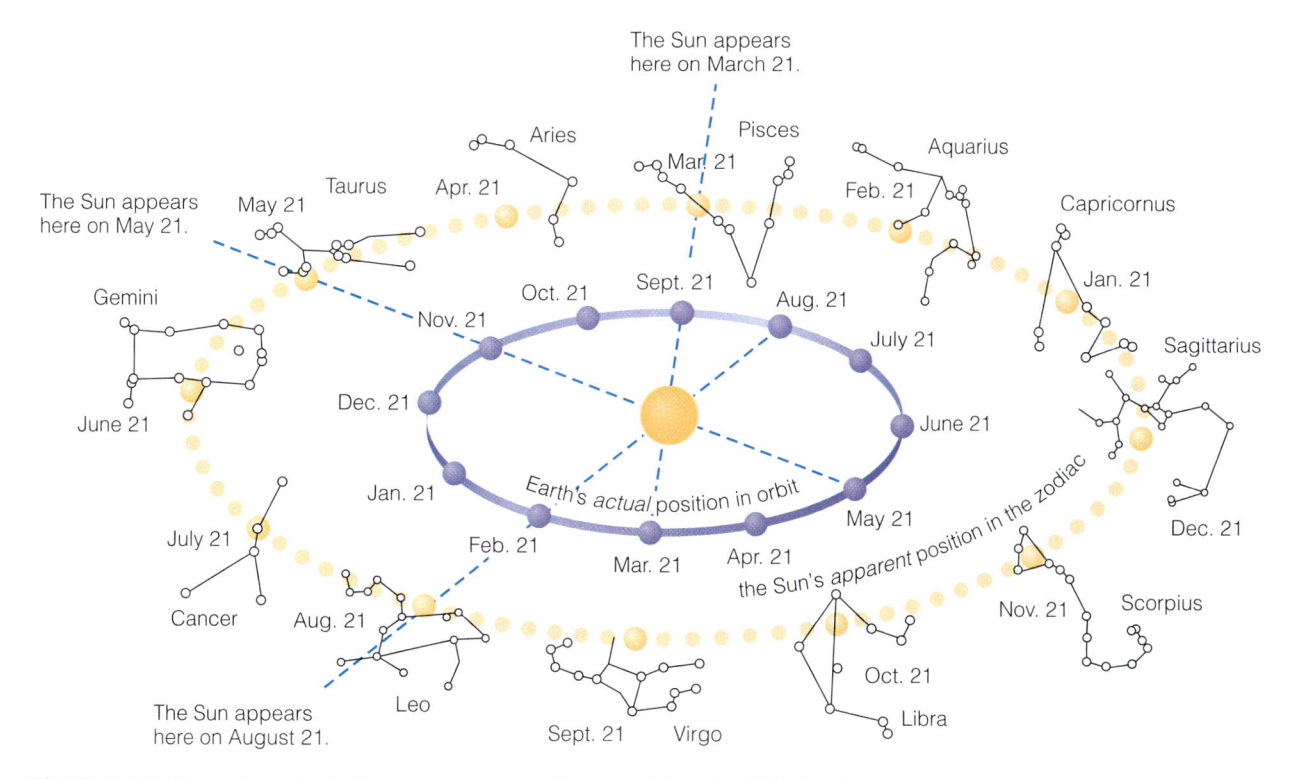

FIGURE 2.15 This diagram shows why the Sun appears to move steadily eastward along the ecliptic, through the constellations of the zodiac. As the Earth orbits the Sun, we see the Sun against the background of different zodiac constellations at different times of year. For example, on May 21, the Sun appears to be in Taurus, and on August 21, the Sun appears to be in Leo. Note that the zodiac constellations for particular dates are *not* the astrological sun signs (see "Common Misconceptions: Sun Signs" in Section 1.3).

FIGURE 2.16 This composite photograph shows images of the Sun, always at the same time of day (mean solar time), snapped at 10-day intervals over an entire year. The three bright streaks show the path of the Sun's rise on three particular dates. This photograph looks east, so north is to the left and south is to the right; it was taken in the Northern Hemisphere. Notice that the Sun's altitude varies considerably: It is high in the summer and low in the winter. The sunrise position also changes: The Sun rises north of due east in the summer and south of due east in the winter. We'll discuss the reasons for the "figure 8" (called an *analemma*) in Chapter S1.

CHAPTER 2 DISCOVERING THE UNIVERSE FOR YOURSELF 49

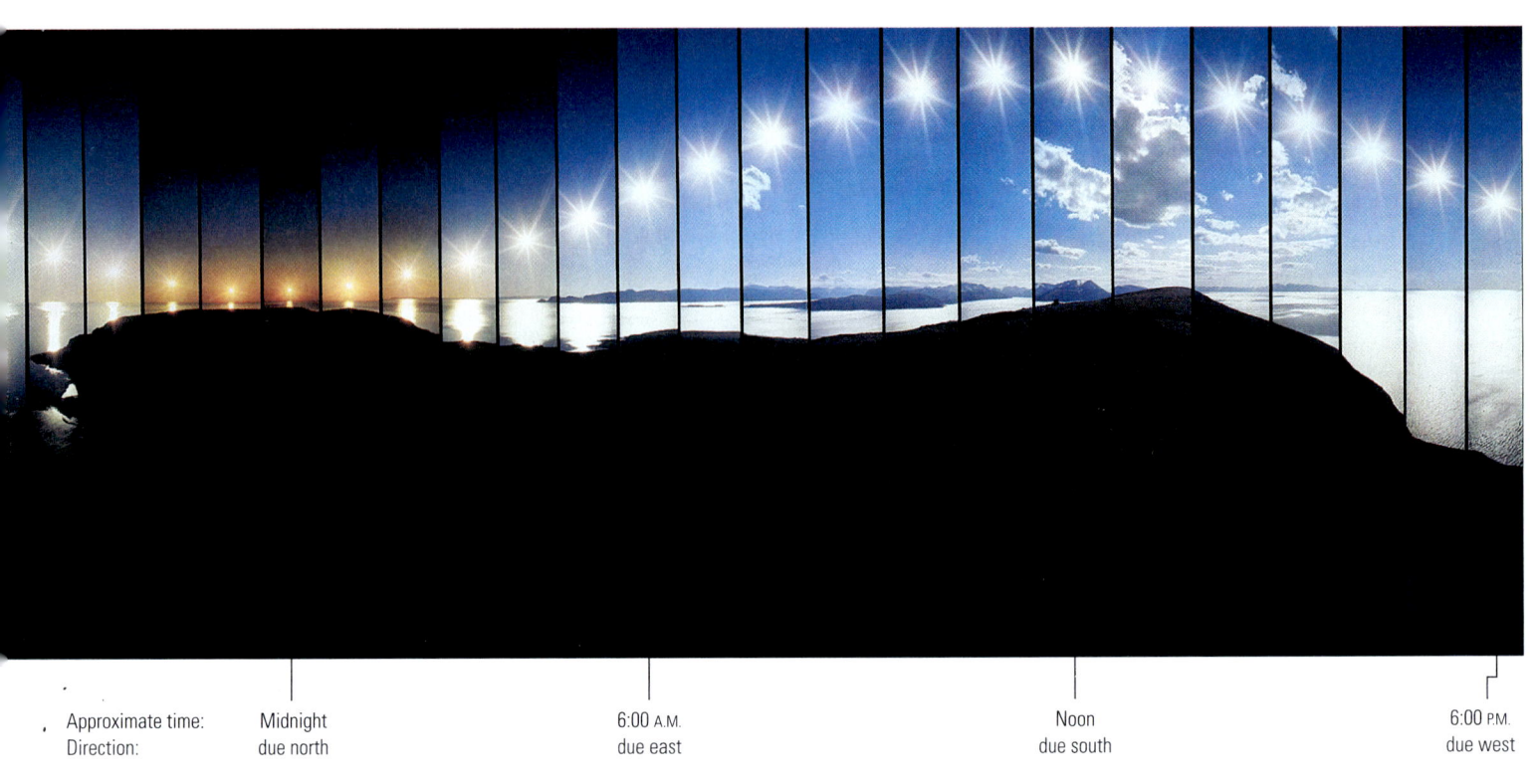

Approximate time:	Midnight	6:00 A.M.	Noon	6:00 P.M.
Direction:	due north	due east	due south	due west

FIGURE 2.17 This sequence of photos shows the progression of the Sun around the horizon on the summer solstice at the Arctic Circle. Note that the Sun does not set but instead skims the northern horizon at midnight. It then gradually rises higher, reaching its highest point at noon, when it appears due south.

Moreover, just as the daily path of any star through your sky depends on your latitude, so does the daily path of the Sun at any time of year. During the summer, the Sun's daily path is longer at higher latitudes than at latitudes nearer the equator. The Sun's daily path is correspondingly shorter at higher latitudes in the winter. That is why higher latitudes have longer summer days and longer winter nights. In fact, the Sun becomes circumpolar at very high latitudes (within the *Arctic* or *Antarctic Circle*) in the summer. The Sun never sets during these summer days, giving these latitudes the name *land of the midnight Sun* (Figure 2.17). Of course, the name "land of noon darkness" is more appropriate in the winter, when the Sun never rises above the horizon at these high latitudes.

Common Misconceptions: High Noon

When is the Sun directly overhead in your sky? If you ask this question of friends and acquaintances, you'll probably find that many people answer "at noon." It's true that the Sun reaches its *highest* point each day when it crosses the meridian (hence the term "high noon," though the meridian crossing is rarely at precisely 12:00 [Section S1.3]), but unless you live in the Tropics (between latitudes 23.5°S and 23.5°N), the Sun is *never* directly overhead. In fact, any time you can see the Sun as you walk around, you can be sure it is *not* at your zenith; unless you are lying down, seeing objects at the zenith requires tilting your head back into a very uncomfortable position. (See Chapter S1 to learn how to determine the Sun's maximum altitude in your sky.)

2.3 The Moon, Our Constant Companion

Like all objects on the celestial sphere, each day the Moon rises in the east and sets in the west. Like the Sun, the Moon also moves gradually eastward through the constellations of the zodiac. However, it takes the Moon only about a month to make a complete circuit around the celestial sphere. If you carefully observe the Moon's position relative to bright stars over just a few hours, you can notice the Moon's drift among the constellations.

As the Moon moves through the sky, both its appearance and the time at which it rises and sets change with the cycle of **lunar phases**. Each com-

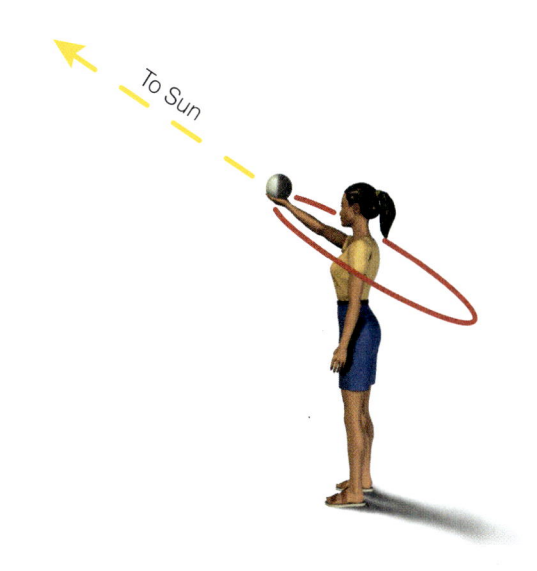

FIGURE 2.18 The phases demonstration. Your head represents the Earth and the ball represents the Moon. Hold the ball toward the Sun at arm's length. Then turn in a circle, always keeping the ball at arm's length and always looking directly at it. Be sure to swing the ball in a circle that carries it from being directly toward the Sun (when you are facing the Sun) to directly opposite the Sun (when you are facing away from the Sun). As you turn, you should see the ball go through phases just like the phases of the Moon.

plete cycle from one new moon to the next takes about $29\frac{1}{2}$ days—hence the origin of the word *month* (think of "moonth"). The easiest way to understand the lunar phases is with a simple demonstration. Use a small ball to represent the Moon while your head represents the Earth. If it's daytime and the Sun is shining, take your ball outside and watch how you see phases as you move the ball around your head (Figure 2.18); if it's dark or cloudy, you can place a flashlight a few meters away to represent the Sun. As you hold your ball at various places in its "orbit" around your head, you'll observe that phases result from just two basic facts:

1. At any particular time, half of the ball faces the Sun (or flashlight) and therefore is bright, while the other half faces away from the Sun and therefore is dark.

2. As you look at the ball, you see some combination of its bright and dark faces. This combination is the phase of the ball.

We see lunar phases for the same reason. Half of the Moon is always illuminated by the Sun, but the amount of this illuminated half that we see from Earth depends on the Moon's position in its orbit (Figure 2.19). The time of day during which the Moon is visible also depends on its phase. For example, full moon occurs when the Moon is opposite the Sun in the sky, so the full moon rises around sunset, reaches the meridian at midnight, and sets around sunrise.

**Common Misconceptions:
Moon in the Daytime**

In traditions and stories, night is so closely associated with the Moon that many people mistakenly believe that the Moon is visible only in the nighttime. In fact, the Moon is above the horizon as often in the daytime as at night, though it is easily visible only when its light is not drowned out by sunlight. For example, a first-quarter moon is easy to spot in the late afternoon as it rises through the eastern sky, and a third-quarter moon is visible in the morning as it heads toward the western horizon (see rise and set times in Figure 2.19). A related misconception appears in illustrations that show a star in the dark portion of the crescent moon (diagram below): A star in the dark portion appears to be in front of the Moon, which is impossible because the Moon is much closer to us than is any star.

This view, though common in art, can never occur because the star would have to be in front of the Moon.

CHAPTER 2 DISCOVERING THE UNIVERSE FOR YOURSELF 51

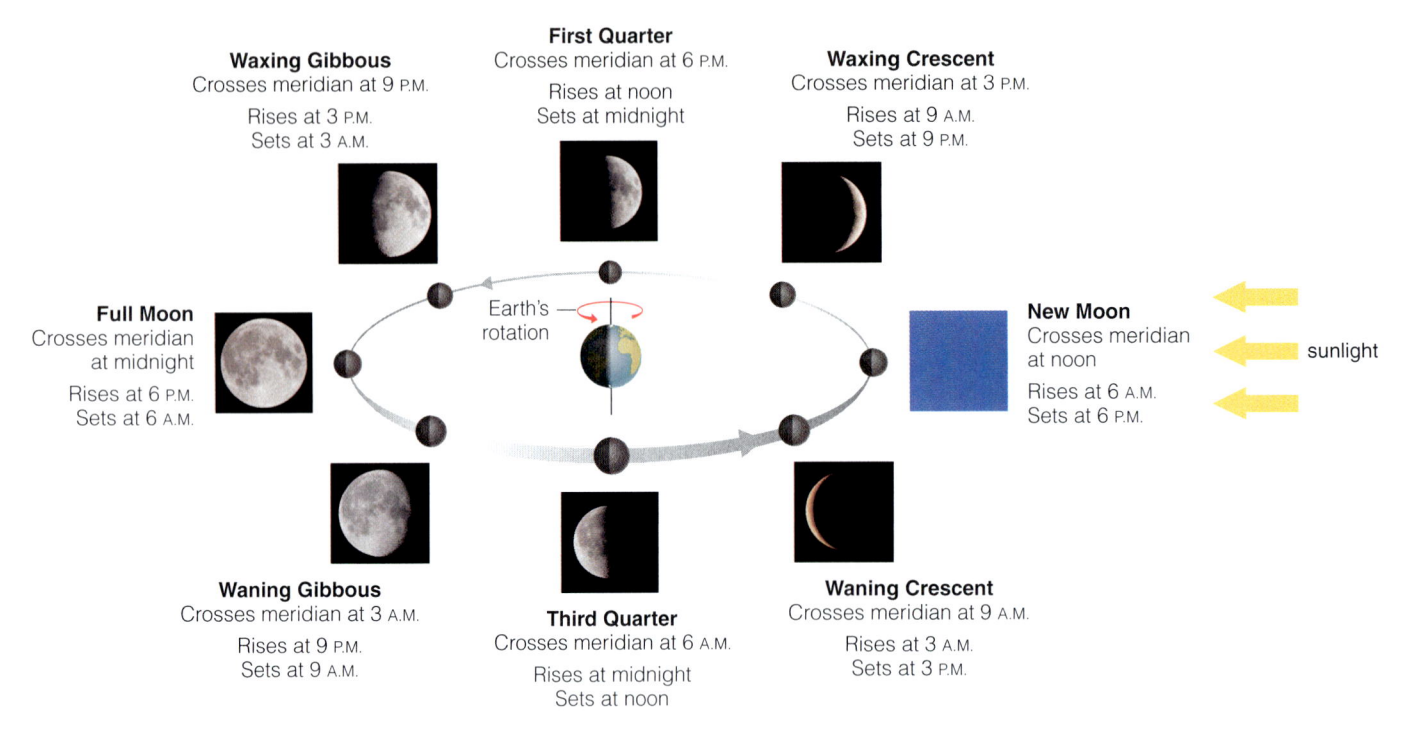

Waxing Gibbous
Crosses meridian at 9 P.M.
Rises at 3 P.M.
Sets at 3 A.M.

First Quarter
Crosses meridian at 6 P.M.
Rises at noon
Sets at midnight

Waxing Crescent
Crosses meridian at 3 P.M.
Rises at 9 A.M.
Sets at 9 P.M.

Full Moon
Crosses meridian at midnight
Rises at 6 P.M.
Sets at 6 A.M.

Earth's rotation

New Moon
Crosses meridian at noon
Rises at 6 A.M.
Sets at 6 P.M.

sunlight

Waning Gibbous
Crosses meridian at 3 A.M.
Rises at 9 P.M.
Sets at 9 A.M.

Third Quarter
Crosses meridian at 6 A.M.
Rises at midnight
Sets at noon

Waning Crescent
Crosses meridian at 9 A.M.
Rises at 3 A.M.
Sets at 3 P.M.

FIGURE 2.19 The phases of the Moon depend on its location relative to the Sun in its orbit. To understand rise and set times, imagine yourself standing on the rotating Earth and looking at the Moon in a particular phase. (The times given are approximate; precise times depend on several factors including your latitude and the time of year. The photo for the new moon shows only a blue sky, because a new moon is above the horizon in daylight but cannot be seen because it is very close to the Sun in the sky.)

TIME OUT TO THINK *Suppose you go outside in the morning and notice that the visible face of the Moon is half light and half dark. Is this a first-quarter or third-quarter moon? How do you know?* (Hint: *Study Figure 2.19.*)

Common Misconceptions: Moon on the Horizon

You've probably noticed that the full moon appears far larger when it is near the horizon than when it is high in your sky. However, this appearance is an illusion. If you measure the angular size of the full moon on a particular night, you'll find it is about the same whether it is near the horizon or high in the sky. (It actually appears slightly larger when overhead than when on the horizon, because you are viewing it from a position on Earth that is closer to the Moon by the radius of the Earth.) In fact, the Moon's angular size in the sky depends only on its distance from the Earth. Although this distance varies over the course of the Moon's monthly orbit, it does not change enough to cause a noticeable effect on a single night. (You can eliminate the illusion by viewing the Moon upside down between your legs when it is on the horizon.)

Although we see many *phases* of the Moon, we do not see many *faces*. In fact, from the Earth we always see (nearly) the same face of the Moon.[3] This tells us that the Moon must rotate once on its axis in the same time that it makes a single orbit of the Earth, as you can observe with another simple demonstration. Place a ball on a table to represent the Earth, while you represent the Moon. Start by facing the ball. If you do not rotate as you walk around the ball, you'll be looking away from it by the time you are halfway around your orbit (Figure 2.20a). The only way you can face the ball at all times is by completing exactly one rotation while you complete one orbit (Figure 2.20b). (We'll learn why the Moon's periods of rotation and orbit are the same in Chapter 5.)

The Moon's appearance is affected to a lesser degree by its varying distance from Earth. Just as the Earth's orbit around the Sun is not a perfect circle (see Figure 1.11), the Moon's orbit around the Earth is not perfectly circular. As a result, the Moon's distance from Earth varies during its orbit (from a minimum of about 356,000 kilometers to a maximum of

---

[3]Because the Moon's orbital speed varies while its rotation rate is steady, the Moon's visible face appears to wobble slightly back and forth as it orbits the Earth. This effect, called *libration*, allows us to see a total of about 59 percent of the Moon's surface by looking at different times of the month, even though we see only 50 percent of the Moon at any single time.

a If you do not rotate while walking around the model, you will not always face it.

b You will face the model at all times only if you rotate exactly once during each orbit.

FIGURE 2.20 The fact that we always see the same face of the Moon means that the Moon must rotate once in the same amount of time that it takes to orbit the Earth once. You can see why by imagining that you are the Moon and walking around a model of the Earth.

about 407,000 kilometers, with an average distance of 380,000 kilometers). When the Moon happens to be closer to the Earth, it appears larger in angular size; when it is farther from Earth, it appears smaller in angular size.

The View from the Moon

A good way to solidify your understanding of the lunar phases is to imagine that you live on the side of the Moon that faces the Earth. Look again at Figure 2.19. Note that at new moon you would be facing the day side of the Earth. Thus, you would see *full earth* when people on Earth see new moon. Similarly, at full moon you would be facing the night side of the Earth. Thus, you would see *new earth* when people on Earth see full moon. In general, you'd always see the Earth in a phase opposite the phase of the Moon that people on Earth see. Moreover, because the Moon always shows nearly the same face to Earth, the Earth would appear to hang nearly stationary in your sky. In addition, because the Moon takes about a month to rotate, your "day" would last about a month; that is, you'd have about 2 weeks of daylight followed by about 2 weeks of darkness as you watched the Earth hanging in your sky and going through its cycle of phases.

Thinking about the view from the Moon clarifies another interesting feature of the lunar phases: The dark portion of the lunar face is not *totally* dark. Imagine that you are standing on the Moon when it is in a crescent phase. Because it's nearly new moon as seen from Earth, you would see nearly full earth in your sky. Just as we can see at night by the light of the Moon, the light of the Earth would illuminate your night moonscape. (In fact, because the Earth is much larger than the Moon, the full earth is much bigger and brighter in the lunar sky than the full moon is in Earth's sky.) This faint light illuminating the "dark" portion of the Moon's face is often called the *ashen light* or *earthshine;* it is the light that enables us to see the outline of the full face of the Moon even when the Moon is not full.

Eclipses

Look once more at Figure 2.19. If this figure told the whole story of the lunar phases, a new moon would always block our view of the Sun, and the Earth would always prevent sunlight from reaching a full moon. More precisely, this figure makes it look as if the Moon's shadow should fall on Earth during new moon and that Earth's shadow should fall on the Moon during full moon. Any time one astronomical

---

**Common Misconceptions:
The "Dark Side" of the Moon**

The term *dark side of the Moon* really should be used to mean the side facing away from the Sun. Unfortunately, *dark side* traditionally meant what would better be called the *far side*—the hemisphere that never can be seen from the Earth. Many people still refer to the far side as the "dark side," even though this side is not necessarily dark. For example, during new moon the far side faces the Sun and hence is completely sunlit; the only time the far side actually is completely dark is at full moon, when it faces away from both the Sun and the Earth.

FIGURE 2.21 This painting represents the ecliptic plane as the surface of a pond. The Moon's orbit is slightly tilted to the ecliptic plane. Thus, in this illustration, the Moon spends half of each orbit above the pond surface and half below the surface; the points at which the orbit crosses the surface represent the *nodes* of the Moon's orbit. Eclipses occur only when the Moon both is at a node (splashing through the pond surface) *and* has a phase of either new moon or full moon—as is the case with the lower left and top right orbits shown. At all other times, new moons and full moons occur above or below the ecliptic plane, so no eclipse is possible.

---

object casts a shadow on another, we say that an **eclipse** is occurring. Thus, Figure 2.19 makes it look as if we should have an eclipse with every new moon and every full moon—but we don't.

The missing piece of the story in Figure 2.19 is that the Moon's orbit is inclined to the ecliptic plane by about 5°. An easy way to visualize this inclination is to imagine the ecliptic plane as the surface of a pond, as shown in Figure 2.21. Because of the inclination of its orbit, the Moon spends most of its time either above or below this surface. It crosses *through* this surface only twice during each orbit: once coming out and once going back in. The two points in each orbit at which the Moon crosses the surface are called the **nodes** of the Moon's orbit.

Figure 2.21 shows the position of the Moon's orbit at several different times of year. Note that the nodes are aligned the same way in each case (diagonally on the page). As a result, the nodes lie in a straight line with the Earth and the Sun only about twice each year. (For reasons we'll discuss shortly, it is not *exactly* twice each year.) Because an eclipse can occur only when the Earth, Moon, and Sun lie along a straight line, two conditions must be met simultaneously for an eclipse to occur:

1. The nodes of the Moon's orbit must be nearly aligned with the Earth and the Sun.

2. The phase of the Moon must be either new or full.

There are two basic types of eclipse. A **lunar eclipse** occurs when the Moon passes through the Earth's shadow and therefore can occur only at *full moon*. A **solar eclipse** occurs when the Moon's

54 PART I DEVELOPING PERSPECTIVE

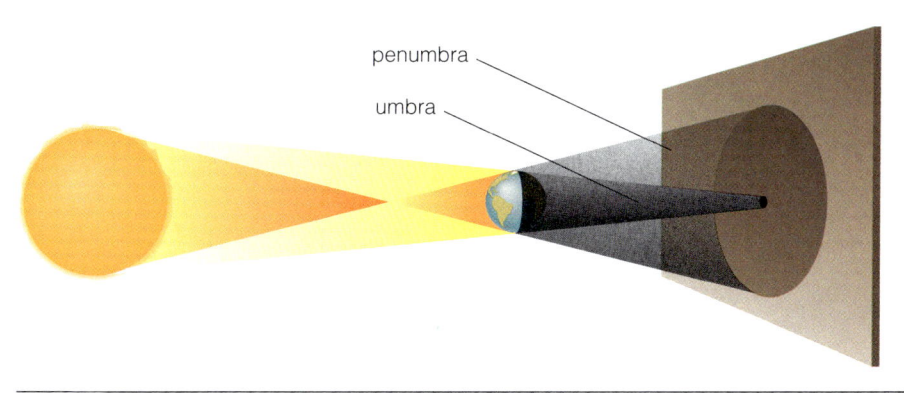

FIGURE 2.22 The shadow cast by an object in sunlight. Sunlight is fully blocked in the umbra and partially blocked in the penumbra.

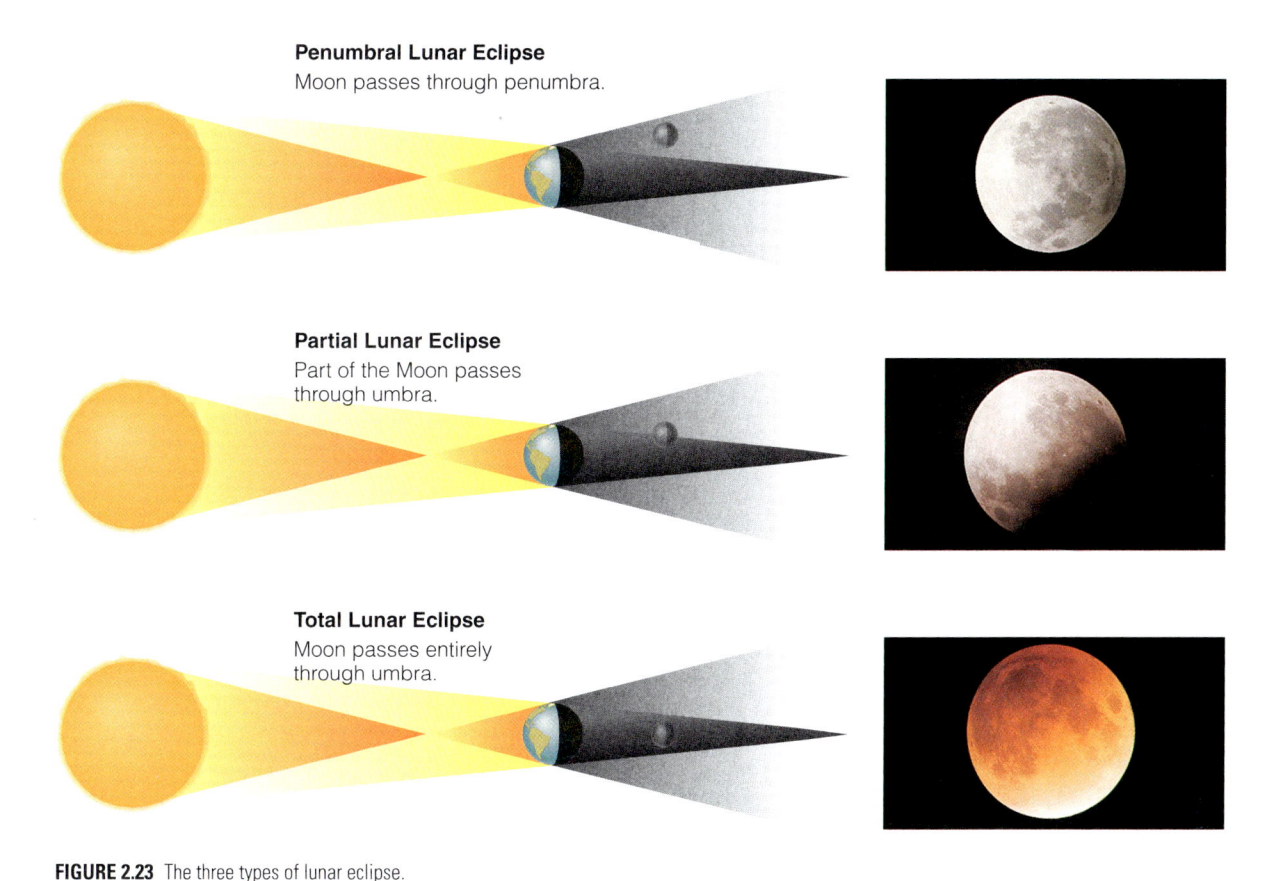

Penumbral Lunar Eclipse
Moon passes through penumbra.

Partial Lunar Eclipse
Part of the Moon passes through umbra.

Total Lunar Eclipse
Moon passes entirely through umbra.

FIGURE 2.23 The three types of lunar eclipse.

shadow falls on the Earth and therefore can occur only at *new moon*. But the full story of eclipse types is more complex, because the shadow of the Moon or the Earth consists of two distinct regions: a central **umbra** where sunlight is completely blocked and a surrounding **penumbra** where sunlight is only partially blocked (Figure 2.22). Thus, an umbral shadow is totally dark, while a penumbral shadow is only slightly darker than no shadow.

Lunar Eclipses A lunar eclipse begins at the moment that the Moon's orbit first carries it into the Earth's penumbra. After that, we will see one of three types of lunar eclipse (Figure 2.23). If the Sun, the Earth, and the Moon are nearly perfectly aligned, the Moon will pass through the Earth's umbra, and we will see a **total lunar eclipse**. If the alignment is somewhat less perfect, only part of the full moon will pass through the umbra (with the rest in the penumbra), and we will see a **partial lunar eclipse**. If the Moon passes *only* through the Earth's penumbra, we will see a **penumbral lunar eclipse**.

Penumbral eclipses are the most common type of lunar eclipse, but they are difficult to notice because the full moon darkens only slightly. Partial lunar

CHAPTER 2 DISCOVERING THE UNIVERSE FOR YOURSELF 55

A **total solar eclipse** occurs in this region.

A **partial solar eclipse** occurs in the lighter area surrounding the area of totality.

If the Moon's umbral shadow does not reach the Earth, an **annular eclipse** occurs in this region.

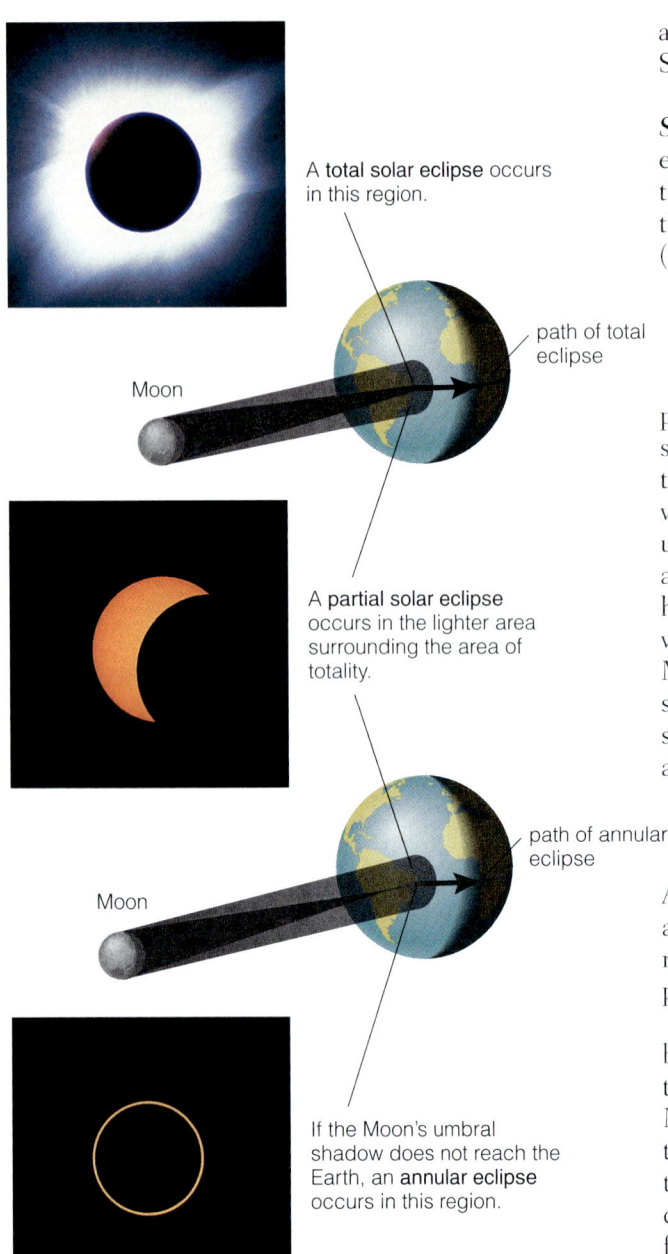

FIGURE 2.24 The three types of solar eclipse. (The photographs on the left are, from top to bottom, a total solar eclipse, a partial solar eclipse, and an annular eclipse.)

atmosphere bends some of the red light from the Sun around the Earth and toward the Moon.

Solar Eclipses We can also see three types of solar eclipse (Figure 2.24). If a solar eclipse occurs when the Moon is relatively close to the Earth in its orbit, the Moon's umbra touches a small area of the Earth (no more than about 270 kilometers in diameter). Anyone within this area will see a **total solar eclipse**. Surrounding this region of totality is a much larger area (typically about 7,000 kilometers in diameter) that falls within the Moon's penumbral shadow. Anyone within this region will see a **partial solar eclipse**, in which only part of the Sun is blocked from view. If the eclipse occurs when the Moon is relatively far from the Earth, the umbra may not reach the Earth at all. In that case, anyone in the small region of the Earth directly behind the umbra will see an **annular eclipse**, in which a ring of sunlight surrounds the disk of the Moon. (Again, anyone in the surrounding penumbral shadow will see a partial solar eclipse.) During any solar eclipse, the combination of the Earth's rotation and the orbital motion of the Moon causes the circular umbral and penumbral shadows to race across the face of the Earth at a typical speed of about 1,700 kilometers per hour (relative to the ground). As a result, the umbral (or annular) shadow traces a narrow path across the Earth, and totality (or annularity) never lasts more than a few minutes in any particular place.

A total solar eclipse is a spectacular sight. It begins when the disk of the Moon first appears to touch the Sun. Over the next couple of hours, the Moon appears to take a larger and larger "bite" out of the Sun. As totality approaches, the sky darkens and temperatures fall. Birds head back to their nests, and crickets begin their nighttime chirping. During the few minutes of totality, the Moon completely blocks the normally visible disk of the Sun, allowing the faint *corona* to be seen (Figure 2.25). The surrounding sky takes on a twilight glow, and planets and bright stars become visible in the daytime. As totality ends, the Sun slowly emerges from behind the Moon over the next couple of hours. However, because your eyes have adapted to the darkness, totality appears to end far more abruptly than it began.

Predicting Eclipses Few phenomena have so inspired and humbled humans throughout the ages as eclipses. For many cultures, eclipses were mystical events associated with fate or the gods, and countless stories and legends surround eclipses. One legend holds that the Greek philosopher Thales (c. 624–546 B.C.) successfully predicted the year (but presumably not the precise time) that a total eclipse of the Sun would be visible in the area where

eclipses are easier to see because the Earth's umbral shadow always clearly darkens part of the Moon's face. Note that the Earth's umbra always casts a curved shadow on the Moon, demonstrating that the Earth is round. A total lunar eclipse is particularly spectacular, because the Moon becomes dark and eerily red during **totality** (the time during which the Moon is entirely engulfed in the umbra)—dark because it is in shadow, and red because the Earth's

THINKING ABOUT . . .

The Moon and Human Behavior

From myths of werewolves to stories of romance under the full moon, human culture is filled with claims that our behavior is influenced by the phase of the Moon. Can we say anything scientific about such claims?

The Moon clearly has important influences on Earth. The most noticeable influence is on the tides, for which the Moon is primarily responsible [Section 5.5]. However, tidal forces are felt only by relatively large objects, such as Earth, and we can rule out the possibility that tidal forces due to the Moon affect small objects such as people.

If a physical force from the Moon cannot affect human behavior, could we be influenced in other ways? Certainly, anyone who lives near the oceans is influenced by the rising and falling of the tides. For example, fishermen and boaters must follow the tides. Thus, although the Moon is not directly influencing their behavior, it is doing so indirectly through its effect on the oceans.

Many physiological patterns in many species appear to follow the lunar phases, and the human menstrual cycle is so close in length to a lunar month that it is difficult to believe the similarity is mere coincidence. Nevertheless, aside from the obvious physical cycles and the influence of tides on people who live near the oceans, claims that the lunar phase affects human behavior are difficult to verify scientifically. For example, although it is possible that the full moon brings out certain behaviors, it may also simply be that some behaviors are easier to exhibit when the sky is bright. Although a beautiful full moon may bring out your desire to walk on the beach under the moonlight, there is no scientific evidence that the full moon would affect you the same way if you lived in a cave and couldn't see it.

he lived, which is now part of Turkey. Coincidentally, the eclipse occurred as two opposing armies (the Medes and the Lydians) were massing for battle. The eclipse so frightened the armies that they put down their weapons, signed a treaty, and returned home. Because modern research shows that the only eclipse visible in that part of the world at about that time occurred on May 28, 585 B.C., we know the precise date on which the treaty was signed—the first historical event that can be dated precisely.

Much of the mystery of eclipses probably stems from the fact that they are relatively difficult to predict. Look again at Figure 2.21. The two periods each year when the nodes of the Moon's orbit are nearly aligned with the Sun are called **eclipse seasons**. Each eclipse season lasts a few weeks, so some type of lunar eclipse occurs during each eclipse season's full moon, and some type of solar eclipse occurs during its new moon. If Figure 2.21 told the whole story, eclipse seasons would occur every 6 months,

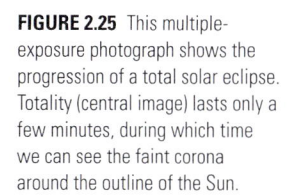

FIGURE 2.25 This multiple-exposure photograph shows the progression of a total solar eclipse. Totality (central image) lasts only a few minutes, during which time we can see the faint corona around the outline of the Sun.

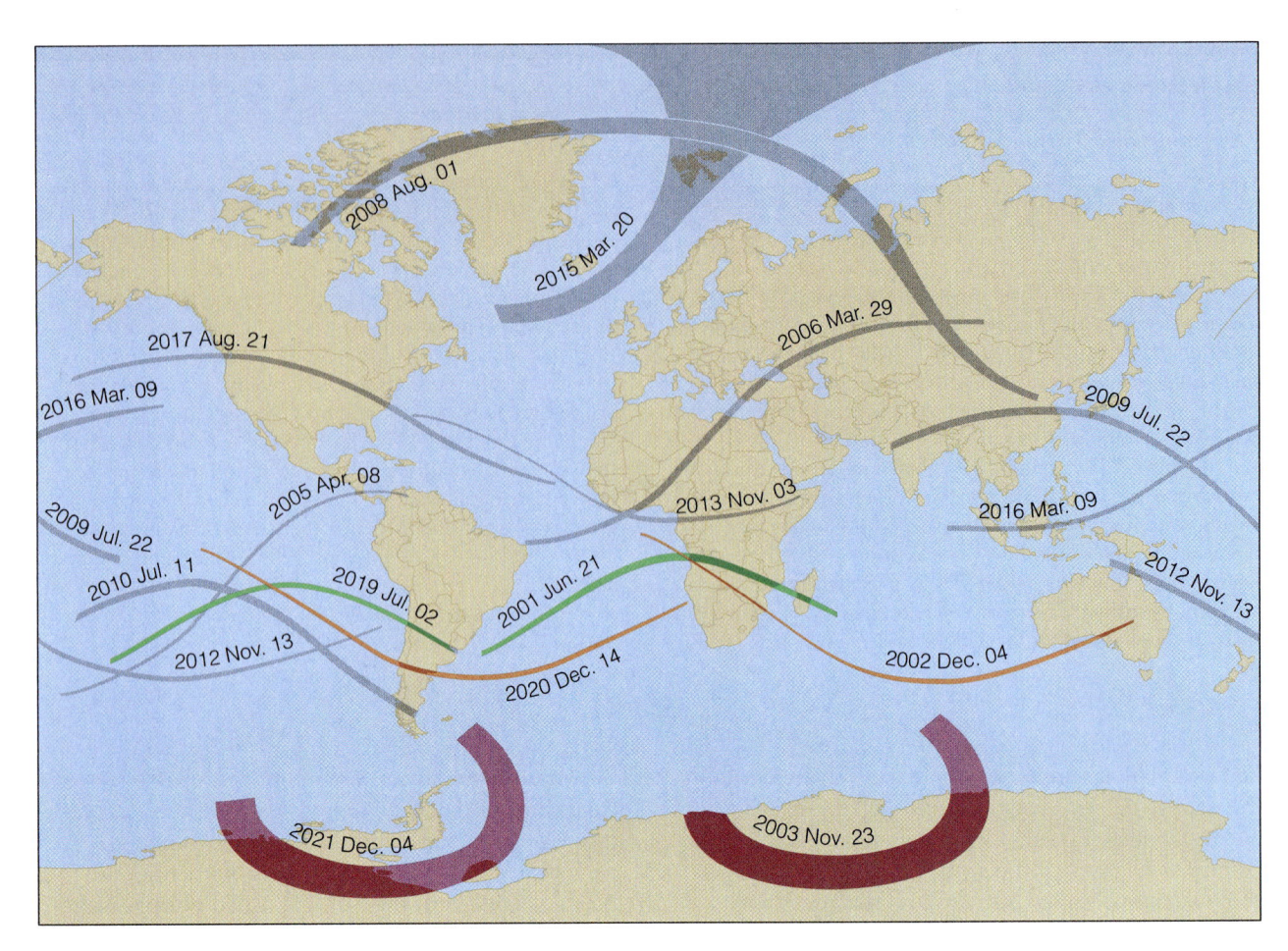

FIGURE 2.26 This map shows the paths of totality for solar eclipses from 2001–2022. Paths of the same color represent eclipses occurring in successive saros cycles, separated by 18 years 11 days. For example, the 2019 eclipse occurs 18 years 11 days after the 2001 eclipse (both shown in green).

and predicting eclipses would be easy. For example, if eclipse seasons always occurred in January and July, eclipses would occur only on the dates of new and full moons in those months. But Figure 2.21 does not show one important fact about the Moon's orbit: The nodes slowly precess around the orbit. As a result, eclipse seasons actually occur slightly less than 6 months apart (about 173 days apart) and therefore do not recur in the same months year after year.

The combination of the changing dates of eclipse seasons and the $29\frac{1}{2}$-day cycle of lunar phases makes eclipses recur in a cycle of about 18 years $11\frac{1}{3}$ days. For example, if a solar eclipse were to occur today, another would occur 18 years $11\frac{1}{3}$ days from now. This roughly 18-year cycle is called the **saros cycle**.

Astronomers in many ancient cultures identified the saros cycle and therefore could predict *when* eclipses would occur. However, the saros cycle does not account for all the complications involved in predicting eclipses. If a solar eclipse occurred today, the one that would occur 18 years $11\frac{1}{3}$ days from now would not be visible from the same places on the Earth and might not be of the same type (e.g., one might be total and the other only partial). As a result, no ancient culture achieved the ability to predict eclipses in every detail. Today, eclipses can be predicted because we know the precise details of the orbits of the Earth and the Moon; many astronomical software packages can do the necessary calculations. Table 2.1 lists upcoming total lunar eclipses, and Figure 2.26 shows paths of totality for upcoming total solar eclipses. It is well worth your while to plan a trip to see a total solar eclipse.

2.4 The Ancient Mystery of the Planets

Five planets are easy to find with the naked eye: Mercury, Venus, Mars, Jupiter, and Saturn. Mercury can be seen only infrequently, and then only just

Table 2.1 Total Lunar Eclipses 2002–2010

May 16, 2003
November 9, 2003
May 4, 2004
October 28, 2004
March 3, 2007
August 28, 2007
February 21, 2008
December 21, 2010

after sunset or before sunrise because it is so close to the Sun. Venus often shines brightly in the early evening in the west or before dawn in the east; if you see a very bright "star" in the early evening or early morning, it probably is Venus. Jupiter, when it is visible at night, is the brightest object in the sky besides the Moon and Venus. Mars is recognizable by its red color, but be careful not to confuse it with a bright red star. Saturn is easy to see with the naked eye, but because many stars are just as bright as Saturn, it helps to know where to look. (Also, planets tend not to twinkle as much as stars.) Sometimes several planets may appear close together in the sky, offering a particularly beautiful sight (Figure 2.27).

Like the Sun and the Moon, the planets appear to move slowly through the constellations of the zodiac. (The word *planet* comes from the Greek for "wandering star.") However, although the Sun and the Moon always appear to move eastward relative to the stars, the planets occasionally reverse course and appear to move *westward* through the zodiac. A period during which a planet appears to move westward relative to the stars is called a period of **apparent retrograde motion** (*retrograde* means "backward"). Figure 2.28 shows a period of apparent retrograde motion for Jupiter.

FIGURE 2.27 This photograph shows Mercury, Venus, Jupiter, and Saturn appearing close together in the sky on February 23, 1999.

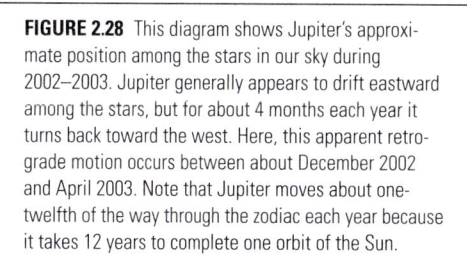

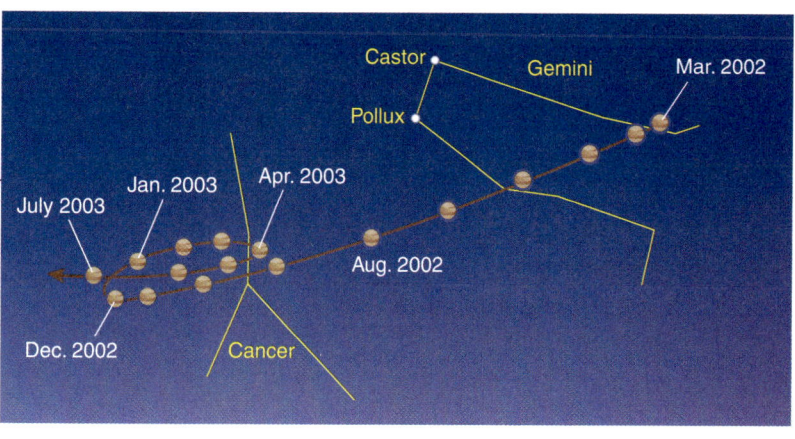

FIGURE 2.28 This diagram shows Jupiter's approximate position among the stars in our sky during 2002–2003. Jupiter generally appears to drift eastward among the stars, but for about 4 months each year it turns back toward the west. Here, this apparent retrograde motion occurs between about December 2002 and April 2003. Note that Jupiter moves about one-twelfth of the way through the zodiac each year because it takes 12 years to complete one orbit of the Sun.

Dots represent Jupiter's approximate position at 1-month intervals.
(Jupiter not to scale.)

CHAPTER 2 DISCOVERING THE UNIVERSE FOR YOURSELF

THINKING ABOUT . . .

Aristarchus

Until the early 1600s, nearly everyone believed that the Earth was the center of the universe. Yet Aristarchus (c. 310–230 B.C.) argued otherwise almost 2,000 years earlier.

Aristarchus proposed a Sun-centered system in about 260 B.C. Little of Aristarchus's work survives to the present day, so we cannot know why he made this proposal. It may have been an attempt to find a more natural explanation for the apparent retrograde motion of the planets. To account for the lack of detectable stellar parallax, Aristarchus suggested that the stars were extremely far away.

He further strengthened his argument by estimating the sizes of the Moon and the Sun. By observing the shadow of the Earth on the Moon during a lunar eclipse, Aristarchus estimated the Moon's diameter to be about one-third of the Earth's diameter—only slightly higher than the actual value. He then used a geometric argument, based on measuring the angle between the Moon and the Sun at first- and third-quarter phases, to conclude that the Sun must be larger than the Earth. (His measurements were imprecise, so he estimated the Sun's diameter to be about 7 times the Earth's, rather than the correct value of about 100 times.) His conclusion that the Sun is larger than the Earth may have been another reason why he felt that the Earth should orbit the Sun, rather than vice versa.

Like most scientific work, Aristarchus's work built upon the work of others. In particular, Heracleides (c. 388–315 B.C.) had suggested that the Earth rotates; Aristarchus may have drawn on this idea to explain the apparent daily rotation of the stars in our sky. Heracleides also was the first to suggest that not all heavenly bodies circle the Earth; based on the fact that Mercury and Venus always are close to the Sun in the sky, he argued that these two planets must orbit the Sun. Thus, in suggesting that all the planets orbit the Sun, Aristarchus was extending the ideas of Heracleides and others before him. Alas, Aristarchus's arguments were not widely accepted in ancient times and were revived only with the work of Copernicus some 1,800 years later.

FIGURE 2.29 The retrograde motion demonstration. Watch how your friend usually appears to you to move forward against the background of the building in the distance but appears to move backward as you catch up to and pass him or her in your "orbit."

Ancient astronomers could easily "explain" the daily paths of the stars through the sky by imagining that the celestial sphere was real and that it really rotated around the Earth each day. But the apparent retrograde motion of the planets posed a far greater mystery: What could cause the planets sometimes to go backward? As we'll discuss in Chapter 3, the ancient Greeks came up with some very clever ways to explain the occasional backward motion of the planets. But these ideas never allowed them to predict planetary positions in the sky with great accuracy (though the accuracy was sufficient for their needs at that time), because they were wedded to the incorrect idea of an Earth-centered universe.

In contrast, apparent retrograde motion has a simple explanation in a Sun-centered solar system. You can demonstrate it for yourself with the help of a friend (Figure 2.29). Pick a spot in an open field to represent the Sun. You can represent the Earth, walking counterclockwise around the Sun, while your friend represents a more distant planet (e.g., Mars, Jupiter, or Saturn) by walking counterclockwise around the Sun at a greater distance. Your friend should walk more slowly than you because more distant planets orbit the Sun more slowly. As you walk, watch how your friend appears to move relative to buildings or trees in the distance. Although both of you always walk the same way around the Sun, your friend will *appear* to move backward against the background during the part of your "orbit" at which you catch up to and pass him or her. (To understand the apparent retrograde motions of Mercury and Venus, which are closer to the Sun than is the Earth, simply switch places with your friend and repeat the demonstration.)

The apparent retrograde motion demonstration applies directly to the planets. For example, because Mars takes about 2 years to orbit the Sun (actually 1.88 years), it covers about half its orbit during the 1 year in which Earth makes a complete orbit. If you trace lines of sight from Earth to Mars from different points in their orbits, you will see that the line of sight usually moves eastward relative to the stars but moves westward during the time when Earth is passing Mars in its orbit (Figure 2.30). Like your friend in the demonstration, Mars never actually changes direction; it only *appears* to change direction from our perspective on Earth.

If the apparent retrograde motion of the planets is so readily explained by recognizing that the Earth is a planet, why wasn't this idea accepted in ancient times? In fact, the idea that the Earth goes around the Sun was suggested as early as 260 B.C. by the Greek astronomer Aristarchus and likely was debated many times thereafter. In a book published in 1440, the German scholar Nicholas of Cusa wrote that the Earth goes around the Sun. (Interestingly, although Galileo was punished by the Church for promoting the same belief two centuries later, Nicholas was ordained a priest in the year his book was published and later was elevated to cardinal.) The desire for a simple explanation for apparent retrograde motion was a primary motivation of Copernicus when he revived Aristarchus's idea in the early 1500s. (Copernicus was aware of the claim by Aristarchus but probably was not aware of the book by Nicholas of Cusa.)

Nevertheless, the idea that the Earth goes around the Sun did not gain wide acceptance among scientists until the work of Kepler and Galileo in the early 1600s [Section 5.3]. Although there were many reasons for the historic reluctance to abandon the idea of an Earth-centered universe, perhaps the most prominent involved the inability of ancient peoples to detect something called **stellar parallax**.

Extend your arm and hold up one finger. If you keep your finger still and alternately close your left eye and right eye, your finger will appear to jump back and forth against the background. This apparent shifting, called *parallax*, occurs simply because your two eyes view your finger from opposite sides of your nose. Note that if you move your finger closer to your face, the parallax increases. In contrast, if you look at a distant tree or flagpole instead of your finger, you probably cannot detect any parallax by alternately closing your left eye and right eye. Thus, parallax depends on distance, with nearer objects exhibiting greater parallax than more distant objects.

If you now imagine that your two eyes represent the Earth at opposite sides of its orbit around the Sun and that your finger represents a relatively nearby star, you have the idea of stellar parallax. That is, because we view the stars from different places in our orbit at different times of year, nearby stars should *appear* to shift back and forth against distant stars in the background during the course of the year (Figure 2.31). The Greeks actually expected stellar parallax in a slightly different way, because they believed that all stars lay on the same celestial sphere: They

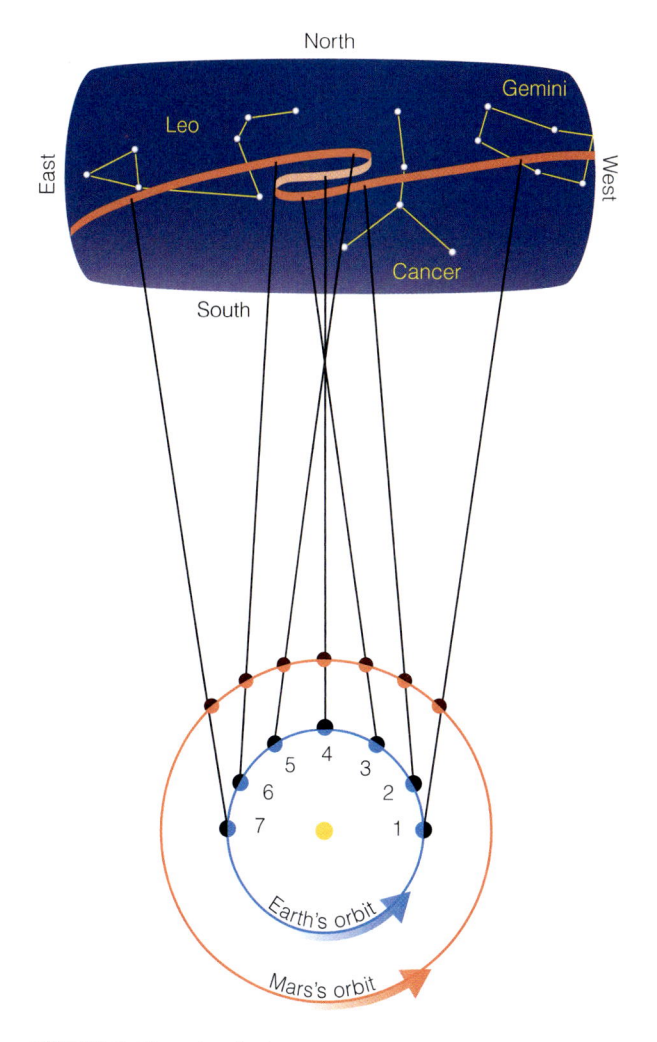

FIGURE 2.30 The explanation for apparent retrograde motion. Follow the lines of sight from Earth to Mars in numerical order. The period during which the lines of sight shift *westward* relative to the distant stars is the period during which we observe apparent retrograde motion for Mars.

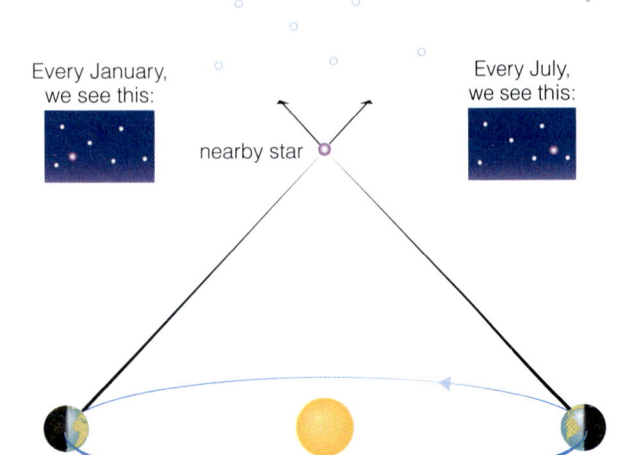

FIGURE 2.31 Stellar parallax is an apparent shift in the position of a nearby star as we look at it from different places in the Earth's orbit. This figure is greatly exaggerated; in reality, the amount of shift is far too small to detect with the naked eye.

assumed that (if Earth orbited the Sun) at different times of year we would be closer to different parts of the celestial sphere, which would change the angular separations of stars. However, no matter how hard they searched, ancient astronomers could find no sign of stellar parallax. They therefore concluded that one of the following must be true:

1. The Earth orbits the Sun but the stars are so far away that stellar parallax is undetectable to the naked eye.
2. There is no stellar parallax because the Earth doesn't move; it is the center of the universe.

Unfortunately, with notable exceptions such as Aristarchus, ancient astronomers rejected the correct answer (1) because they could not imagine that the stars could be *that* far away. Today, we can detect stellar parallax with the aid of telescopes, thereby providing direct proof that the Earth really does orbit the Sun. Careful measurements of stellar parallax also provide the most reliable means of measuring distances to nearby stars [Section 15.2].

TIME OUT TO THINK *How far apart are opposite sides of the Earth's orbit? How far away are the nearest stars? Describe the challenge of detecting stellar parallax. It may help to visualize the Earth's orbit and the distance to stars on the 1-to-10-billion scale used in Chapter 1.*

Thus, the ancient mystery of the planets drove much of the historical debate over the Earth's place in the universe. In many ways, the modern technological society that we take for granted today can be traced directly back to the scientific revolution that began because of the quest to explain the slow wandering of the planets among the stars in our sky.

THINKING ABOUT . . .

And Yet It Moves

On June 22, 1633, Galileo was brought before a Church inquisition in Rome and ordered to recant his heretical view that the Earth goes around the Sun rather than vice versa. Fearing for his life, Galileo recanted as ordered. However, legend has it that, as he rose from his knees following his renunciation, he whispered under his breath, *Eppur si muove*—Italian for "And yet it moves." (Given the likely consequences if church officials had heard him say this, most historians doubt the veracity of this legend.) The evidence supporting the idea that the Earth moves was quite strong by the mid-1600s, but it was still indirect. Today, we have much more direct proof.

French physicist Jean Foucault provided the first direct proof of *rotation* in 1851. Foucault built a large pendulum that he carefully started swinging. Because any pendulum tends to swing always in the same plane, the Earth's rotation made Foucault's pendulum appear to twist slowly in a circle. Today, *Foucault pendulums* are a popular attraction at many science centers and museums. A second direct proof that the Earth rotates is provided by the *Coriolis effect,* first described by French physicist Gustave Coriolis (1792–1843). The Coriolis effect, which would not happen if the Earth were not rotating, is responsible for things such as the swirling of hurricanes and the fact that missiles that travel great distances on the Earth deviate from straight-line paths [Section 10.3].

Stellar parallax provides direct proof that the Earth orbits the Sun, and it was first measured by German astronomer Friedrich Bessel in 1838. However, another direct proof that the Earth orbits the Sun actually was found about a century earlier by English astronomer James Bradley (1693–1762). To understand Bradley's proof, imagine that starlight is like rain, falling straight down. If you are standing still you should hold your umbrella straight over your head, but if you are walking through the rain you should tilt your umbrella forward, because your motion makes the rain appear to be coming down at an angle. Bradley discovered that observing light from stars requires that telescopes be slightly tilted in the direction of Earth's motion—just like the umbrella; this effect is called the *aberration of starlight.*

THE BIG PICTURE

In this chapter, we surveyed the phenomena of our sky. Keep the following "big picture" ideas in mind as you continue your study of astronomy:

- You can enhance your enjoyment of learning astronomy by spending time outside observing the sky. The more you learn about the appearance and apparent motions of the sky, the more you will appreciate what you can see in the universe.

- From our vantage point on Earth, it is convenient to imagine that we are at the center of a great celestial sphere—even though we really are on a planet orbiting a star in a vast universe. We can then understand what we see in the local sky by thinking about how the celestial sphere appears from our latitude.

- Most of the phenomena of the sky are relatively easy to observe and understand. But the more complex phenomena, particularly eclipses and apparent planetary motion, challenged our ancestors for thousands of years and helped drive the development of science and technology.

Review Questions

1. What is a *constellation?* How is a constellation related to a pattern of stars in the sky?

2. What is the *celestial sphere?* Describe the major features shown on a model of the celestial sphere.

3. What is the *ecliptic,* and how is it related to the ecliptic plane?

4. What is the *Milky Way* in our sky, and how is it related to the Milky Way Galaxy?

5. Why does the local sky look like a dome? Define *horizon, zenith,* and *meridian.* Describe how you can locate an object in the local sky by its altitude and its direction along the horizon.

6. Explain why we can measure only *angular* sizes and distances for objects in the sky. What are *arcminutes* and *arcseconds?*

7. Briefly describe how and why the sky varies with latitude.

8. Briefly describe how and why the sky changes with the seasons.

9. Describe the Moon's cycle of *phases* and explain why we see phases of the Moon.

10. Why don't we see an eclipse at every new and full moon? Describe the conditions that must be met to see a solar or lunar eclipse.

11. Why are eclipses so difficult to predict? Explain the importance of *eclipse seasons* and the *saros cycle* to eclipse prediction.

12. What is the *apparent retrograde motion* of the planets? Why was it difficult for ancient astronomers to explain but easy for us to explain?

13. What is *stellar parallax?* Describe the role it played in making ancient astronomers believe in an Earth-centered universe.

Discussion Questions

1. *Geocentric Language.* Many common phrases reflect the ancient Earth-centered view of our universe. For example, the phrase "the Sun rises each day" implies that the Sun is really moving over the Earth. In fact, the Sun only *appears* to rise as the rotation of the Earth carries us to a place where we can see the Sun in our sky. Identify other common phrases that imply an Earth-centered viewpoint.

2. *Flat Earth Society.* Believe it or not, there is an organization called the Flat Earth Society, whose members hold that the Earth is flat and that all indications to the contrary (such as pictures of the Earth from space) are fabrications made as part of a conspiracy to hide the truth from the public. Discuss the evidence for a round Earth and how you can check it for yourself. In light of the evidence, is it possible that the Flat Earth Society is correct? Defend your opinion.

Problems

Sensible Statements? For **problems 1–8**, decide whether the statement is sensible and explain why it is or is not. (For an example, see Chapter 1 problems.)

1. If you had a very fast spaceship, you could travel to the celestial sphere in about a month.

2. The constellation Orion didn't exist when my grandfather was a child.

3. When I looked into the dark fissure of the Milky Way with my binoculars, I saw what must have been a cluster of distant galaxies.

4. Last night the Moon was so big that it stretched for a mile across the sky.

5. I live in the United States, and during my first trip to Argentina I saw many constellations that I'd never seen before.

6. Last night I saw Jupiter right in the middle of the Big Dipper. (*Hint:* Is the Big Dipper part of the zodiac?)

7. Last night I saw Mars move westward through the sky in its apparent retrograde motion. (*Hint:* How long does it take to notice apparent retrograde motion?)

8. Although all the known stars appear to rise in the east and set in the west, we might someday discover a star that will appear to rise in the west and set in the east.

9. *View from Afar.* Describe how the Milky Way Galaxy would look in the sky of someone observing from a planet around a star in M31.

10. *Your View.*
    a. Find your latitude and longitude, and state the source of your information.
    b. Describe the altitude and direction in your sky at which the north or south celestial pole appears.
    c. Is Polaris a circumpolar star in your sky? Explain.
    d. Describe the path of the meridian in your sky.
    e. Describe the path of the celestial equator in your sky. (*Hint:* Study Figure 2.12.)

11. *View from the Moon.* Suppose you lived on the Moon, near the center of the face that we see from Earth.
    a. During the phase of full moon, what phase would you see for Earth? Would it be daylight or dark where you live?
    b. On Earth, we see the Moon rise and set in our sky each day. If you lived on the Moon, would you see the Earth rise and set? Why or why not? (*Hint:* Remember that the Moon always keeps the same face toward Earth.)
    c. What would you see when people on Earth were experiencing an eclipse? Answer for both solar and lunar eclipses.

12. *A Farther Moon.* Suppose the distance to the Moon was twice its actual value. Would it still be possible to have a total solar eclipse? An annular eclipse? A total lunar eclipse? Explain.

13. *A Smaller Earth.* Suppose the Earth was smaller in size. Would solar eclipses be any different? If so, how? What about lunar eclipses? Explain.

14. *Observing Planetary Motion.* Find out what planets are currently visible in your evening sky. At least once a week, observe the planets and draw a diagram showing the position of each visible planet relative to stars in a zodiac constellation. From week to week, note how the planets are moving relative to the stars. Can you see any of the apparently "erratic" features of planetary motion? Explain.

15. *A Connecticut Yankee.* Find the book *A Connecticut Yankee in King Arthur's Court,* by Mark Twain. Read the portion that deals with the Connecticut Yankee's prediction of an eclipse (or read the entire book). In a one- to two-page essay, summarize the episode and how it helps the Connecticut Yankee gain power.

Web Projects

Find useful links for Web projects on the text Web site.

1. *Sky Information.* Search the Web for sources of daily information about sky phenomena (such as lunar phases, times of sunrise and sunset, and unusual sky events). Identify and briefly describe your favorite source.

2. *Constellations.* Search the Web for information about the constellations and their mythology. Write a short report about one or more constellations.

3. *Upcoming Eclipse.* Find information about an upcoming solar or lunar eclipse that you might have a chance to witness. Write a short report about how you could best witness the eclipse, including any necessary travel to a viewing site, and what you can expect to see. Bonus: Describe how you could photograph the eclipse.

Socrates: Shall we make astronomy the next study? What do you say?
Glaucon: Certainly. A working knowledge of the seasons, months, and years is beneficial to everyone, to commanders as well as to farmers and sailors.
Socrates: You make me smile, Glaucon. You are so afraid that the public will accuse you of recommending unprofitable studies.

PLATO, REPUBLIC

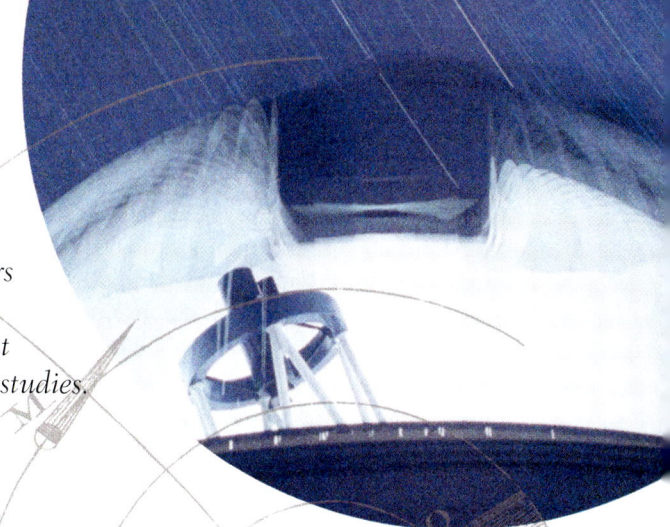

CHAPTER S1
Celestial Timekeeping and Navigation

In ancient times, the practical need for timekeeping and navigation was one of the primary reasons for the study of astronomy. The celestial origins of timekeeping and navigation are still evident. The time of day comes from the location of the Sun in the local sky, the month comes from the Moon's cycle of phases, and the year comes from the Sun's annual path along the ecliptic. With regard to navigation, we've already learned how the position of the north or south celestial pole can tell us latitude (see Section 2.2).

Today, we can tell the time by glancing at an inexpensive electronic watch and navigate with hand-held devices that receive signals from satellites of the global positioning system (GPS). But knowing the celestial basis of timekeeping and navigation can still be useful, particularly for understanding the rich history of astronomical discovery. In this chapter, we will explore the apparent motions of Sun, Moon, and planets in greater detail than we have previously, enabling us to study the principles of celestial timekeeping and navigation.

S1.1 Astronomical Time Periods

People began measuring time long before anyone knew that the Earth is round or that it is a planet that orbits the Sun. The only celestial motions that mattered were those that could be seen in the local sky. Many ancient cultures carefully tracked the motion of the Sun, Moon, planets, and stars through the sky. These ancient observations established astronomically based time periods such as the day, month, and year. Let's begin our study of timekeeping by examining several important astronomical time periods.

Solar Versus Sidereal Day

We often think of our 24-hour day as the Earth's rotation period, but that's not quite true. The Earth's rotation period is the time it takes the Earth to complete one full rotation. Because the Earth's daily rotation makes the celestial sphere appear to rotate around us (see Figure 2.8), we can measure the rotation period by timing how long it takes the celestial sphere to make one full turn through the local sky. For example, we could start a stopwatch at the moment when a particular star is on our meridian (the semicircle stretching from due south, through the zenith, to due north [Section 2.2]), then stop the watch the next day when the same star again is on the meridian (Figure S1.1a). Measured in this way, the Earth's rotation period is about 23 hours 56 minutes (more precisely, $23^h 56^m 4.09^s$)—or about 4 minutes short of 24 hours. This time period is called a **sidereal day**, because it is measured relative to the apparent motion of stars in the local sky; *sidereal* (pronounced sy-dear-ee-al) means "related to the stars."

Our 24-hour day, which we call a **solar day**, is based on the time it takes for the *Sun* to make one circuit around our local sky. We could measure this time period by starting the stopwatch at the moment when the Sun is on our meridian one day and stopping it when the Sun reaches the meridian the next day (Figure S1.1b). The solar day is indeed 24 hours on average, although it varies slightly over the course of a year (see the Thinking About box on p. 71).

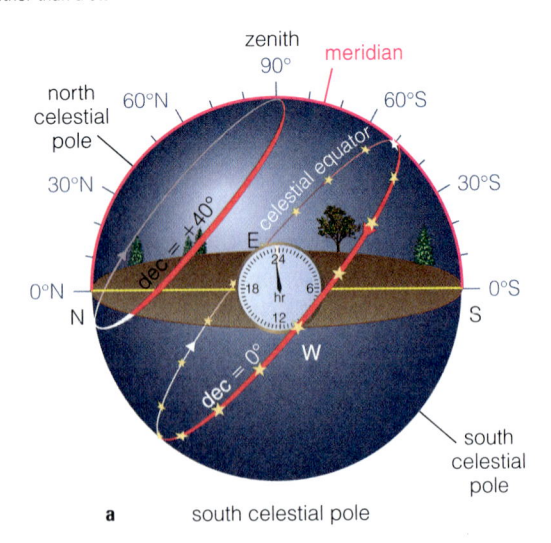

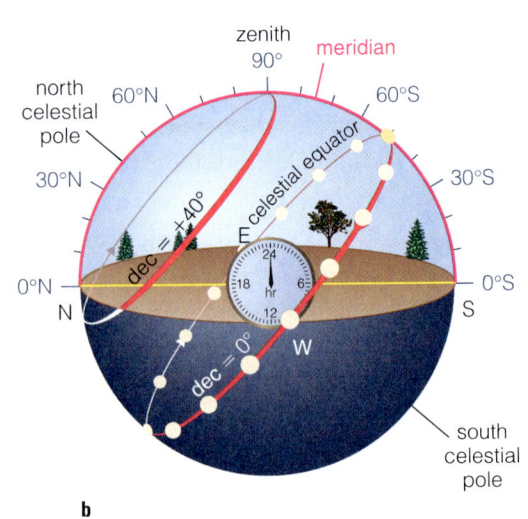

A simple demonstration shows why a solar day is slightly longer than a sidereal day. Set an object on a table to represent the Sun, and stand a few steps away from the object to represent the Earth. Point at the Sun and imagine that you also happen to be pointing toward some distant star. If you rotate (counterclockwise) while standing in place, you'll again be pointing at the Sun and the star after one full rotation (Figure S1.2a). But because the Earth orbits the Sun at the same time that it rotates, you can make the demonstration more realistic by taking a couple of steps around the Sun (counterclockwise) while you are rotating (Figure S1.2b). After one full rotation, you will again be pointing in the direction of the distant star; hence, this represents a sidereal day. However, because of your orbital motion, you'll need to rotate slightly more than once to be pointing again at the Sun; it is this "extra" bit of rotation that makes a solar day longer than a sidereal day.

The only problem with this demonstration is that it exaggerates the Earth's daily orbital motion. In reality, the Earth moves about 1° per day around its orbit (because it makes a full 360° orbit in 1 year, or about 365 days). Thus, if you recall that a single rotation means rotating 360°, the Earth must actually rotate about 361° with each solar day (Figure S1.2c).

FIGURE S1.1 (a) A sidereal day is the time it takes any star to make a circuit of the local sky, that is, the time from its meridian crossing one day to its meridian crossing the next day. (b) A solar day is measured similarly, but by timing the Sun rather than a star.

a One full rotation represents a sidereal day.

b While "orbiting" the Sun, one rotation still returns you to pointing at a distant star, but you need slightly more than one full rotation to return to pointing at the Sun.

c The Earth travels about 1° per day around its orbit, so a solar day requires about 361° of rotation.

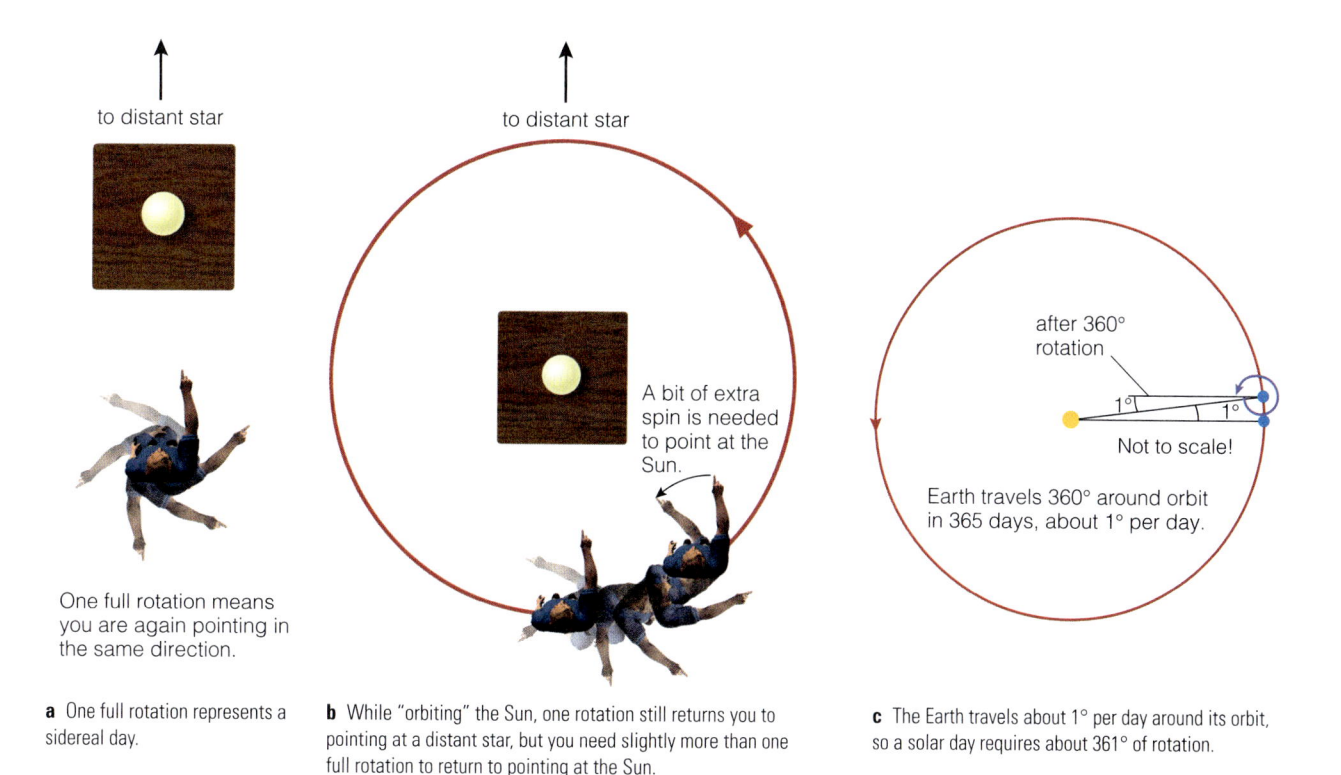

FIGURE S1.2 A demonstration showing why a solar day is slightly longer than a sidereal day.

The extra 1° rotation accounts for the extra 4 minutes by which the solar day is longer than the sidereal day. (To see why it takes 4 minutes to rotate 1°, note that the Earth rotates 360° in about 23 hours 56 minutes, or 1,436 minutes. Thus, 1° of rotation takes $\frac{1}{360} \times 1{,}436$ minutes ≈ 4 minutes.)

Synodic Versus Sidereal Month

As we discussed in Chapter 2, our month comes from the Moon's $29\frac{1}{2}$-day cycle of phases. More technically, the $29\frac{1}{2}$-day period required for each cycle of phases is called a **synodic month**. The word *synodic* comes from the Latin *synod*, which means "meeting," so a synodic month gets its name because the Sun and the Moon "meet" in the sky with every new moon.

Just as a solar day is not the Earth's true rotation period, a synodic month is not the Moon's true orbital period: the Earth's motion around the Sun means that the Moon must complete more than one full orbit of the Earth from one new moon to the next (Figure S1.3). The Moon's true orbital period, or **sidereal month**, is only about $27\frac{1}{3}$ days. Like the sidereal day, the sidereal month gets its name because it describes how long it takes the Moon to complete an orbit relative to the positions of distant stars.

Tropical Versus Sidereal Year

A year is related to the Earth's orbital period, but again there are two slightly different definitions for the length of the year. The time it takes for the Earth to complete one orbit relative to the stars is called a **sidereal year**. But our calendar is based on the cycle of the seasons, which we can measure from the time of the spring equinox one year to the spring equinox the next year. This time period, called a **tropical year**, is about 20 minutes shorter than the sidereal year.

FIGURE S1.3 The Moon completes one 360° orbit in about $27\frac{1}{3}$ days (a sidereal month), but the time from new moon to new moon is about $29\frac{1}{2}$ days (a synodic month).

Earth travels about 30° per month around the Sun, so the Moon must orbit around Earth about 360° + 30° = 390° from new moon to new moon.

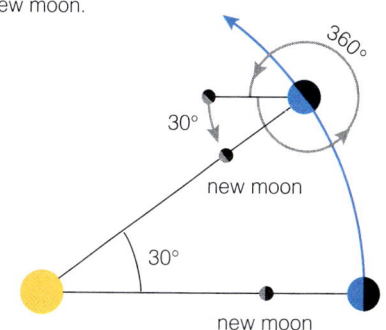

CHAPTER S1 CELESTIAL TIMEKEEPING AND NAVIGATION

(A 20-minute difference might not seem like much, but it would make a calendar based on the sidereal year get out of sync with the seasons by 1 day every 72 years—a difference that would add up over centuries.)

The difference between the sidereal year and the tropical year arises from the Earth's 26,000-year cycle of axis precession [Section 1.3]. Recall that precession not only changes the orientation of the axis in space, but also changes the locations in the Earth's orbit at which the seasons occur. Each year, the location of the equinoxes and solstices among the stars shifts about $\frac{1}{26,000}$ of the way around the orbit. Note that $\frac{1}{26,000}$ of a year is about 20 minutes, which explains the 20-minute difference between the tropical and sidereal years.

Planetary Periods (Synodic Versus Sidereal)

Planetary periods are not used in our modern timekeeping, but some ancient cultures used them. For example, the Mayan calendar was based in part on the apparent motions of Venus. Today, understanding planetary periods can help us make sense of what we see in the sky.

A planet's **sidereal period** is the time it takes to orbit the Sun; as usual, it has the name *sidereal* because it is measured relative to distant stars. For example, Jupiter's sidereal period is about 12 years (more precisely, 11.86 years), so it takes about 12 years for Jupiter to make a complete circuit around the constellations of the zodiac. Thus, Jupiter appears to move through roughly one zodiac constellation each year; if Jupiter is in Gemini this year (as it is in 2002), it will be in Cancer next year and Leo the following year, returning to Gemini in 12 years.

A planet's **synodic period** is the time between being lined up with the Sun in our sky one time and the next similar alignment; again, the term *synodic* refers to the planet's "meeting" the Sun in the sky. Figure S1.4 shows that the situation is somewhat different for planets nearer the Sun than Earth (that is, Mercury and Venus) and planets farther away (all the rest of the planets).

Look first at the situation for the more distant planet in Figure S1.4. As seen from Earth, this planet will sometimes line up with the Sun in what we call a **conjunction**. At other special times, it will appear exactly opposite the Sun in our sky, which we call **opposition**. We cannot see the planet during conjunction with the Sun, because it is hidden by the Sun's glare and rises and sets with the Sun in our sky. At opposition, the planet moves through the sky like the full moon, rising at sunset, reaching the meridian at midnight, and setting at dawn. Note that the planet is closest to Earth at opposition and hence appears brightest at this time.

Figure S1.4 shows that planets *nearer* than Earth to the Sun have two conjunctions—an "inferior conjunction" between the Earth and the Sun and a "superior conjunction" when the planet appears behind the Sun as seen from Earth—rather than one conjunction and one opposition. The planet is hidden by the Sun's glare at both conjunctions. Mercury and Venus usually appear slightly above or below the Sun at inferior conjunction, because they have slightly different orbital planes than the Earth. Occasionally, however, these planets appear to move across the disk of the Sun during inferior conjunction, creating a *transit* (Figure S1.5). Mercury transits occur an average of a dozen times per century, with the next two coming on May 7, 2003, and November 8, 2006. Venus transits typically come in pairs separated by a century or more. Its next pair of transits will occur on June 8, 2004, and June 6, 2012.

The inner planets are best viewed when they are near their points of greatest **elongation**, that is, when they are farthest from the Sun in our sky. At its greatest eastern elongation, Venus appears about 46° east of the Sun in our sky, which means it shines brightly in the evening sky. Similarly, at its greatest western elongation, Venus appears about 46° west of the Sun in our sky, shining brightly in the predawn

FIGURE S1.4 This diagram shows important positions of planets relative to the Earth and the Sun. For a planet farther from the Sun than Earth (such as Mars, Jupiter, Saturn), conjunction is when it appears aligned with the Sun in the sky, and opposition is when it appears on our meridian at midnight. Planets nearer the Sun (Mercury and Venus) have two conjunctions and never get farther from the Sun in our sky than at their greatest elongations.

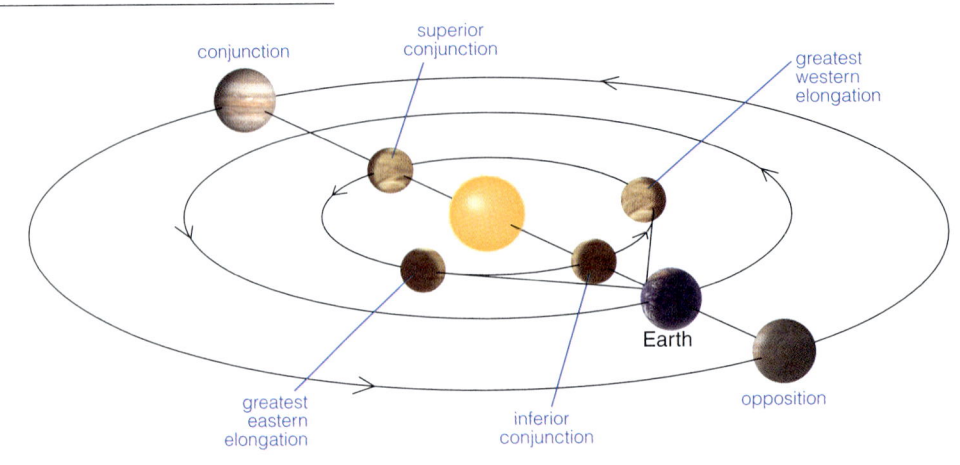

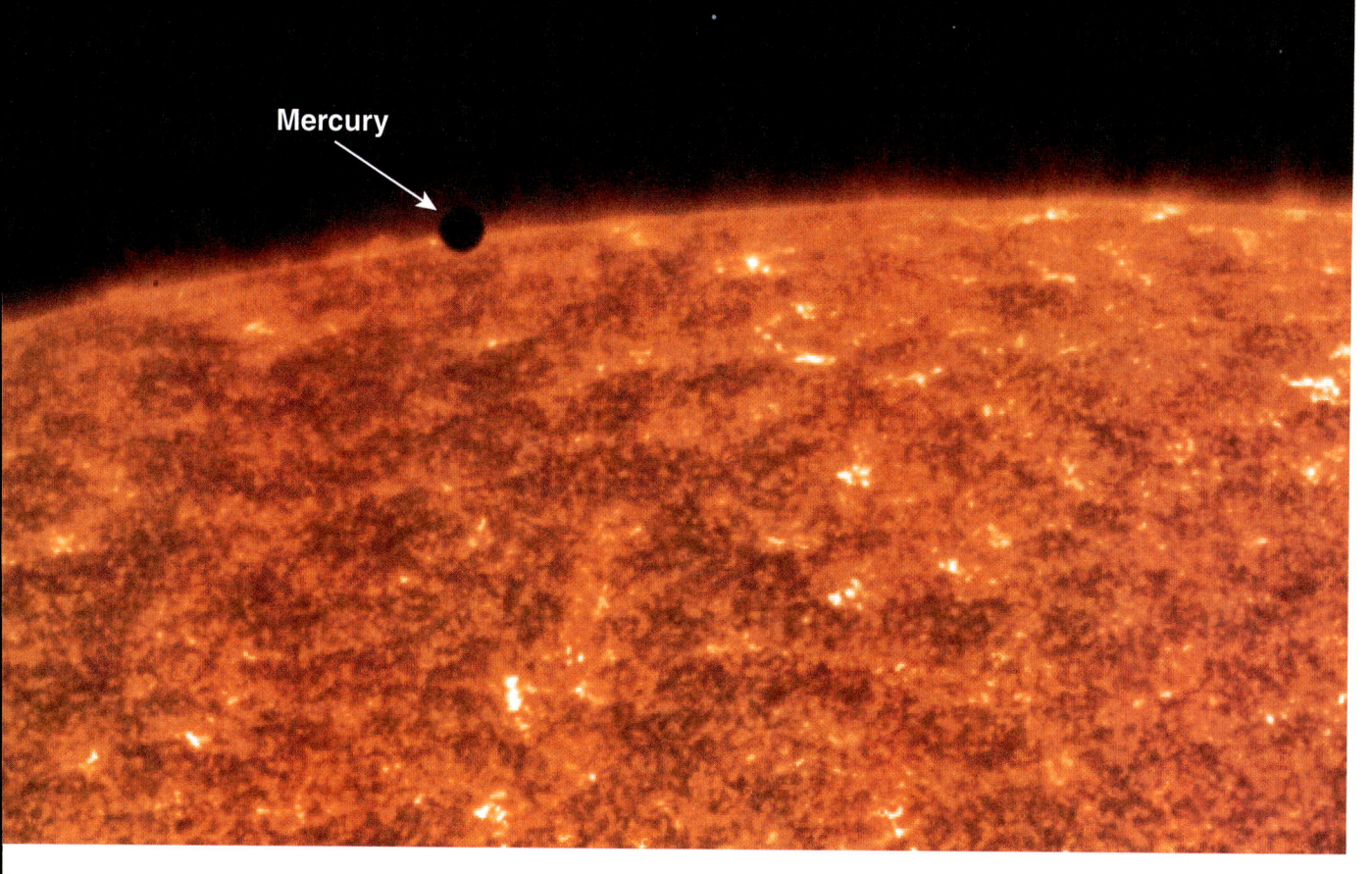

FIGURE S1.5 NASA's *TRACE* satellite captured this image of a Mercury transit on November 15, 1999. The photograph was taken with ultraviolet light, so the colors are not real; the structures seen with ultraviolet light are patches of hot gas just above the Sun's visible surface.

sky. In between the times when Venus appears in the morning sky and the times when it appears in the evening sky, Venus disappears from view for a few weeks with each conjunction. Mercury's pattern is similar, but because it is closer to the Sun it never appears more than about 28° from the Sun in our sky. (Mercury's angular distance from the Sun at greatest elongation varies from about 18° to 28°, depending on where it is in its fairly eccentric orbit.) This makes it difficult to see Mercury, because it is almost always obscured by the glare of the Sun.

TIME OUT TO THINK *Based on Figure S1.4, explain why neither Mercury nor Venus can ever be on the meridian in the midnight sky.*

Measuring a planet's synodic period is fairly easy, because it requires simply observing its position relative to the Sun in the sky. In contrast, we must calculate a planet's sidereal period from the geometry of planetary orbits. Copernicus was the first to perform these calculations, and they revealed a simple pattern in which more distant planets had longer sidereal periods. The simplicity of this pattern helped convince him that his idea of a Sun-centered solar system was correct. (We'll discuss Copernicus further in Chapter 5.)

S1.2 Daily Timekeeping

Now that we have discussed astronomical time periods, we can turn our attention to modern measures of time. Our clock is based on the 24-hour solar day. You are probably familiar with the basic principles. For example, we usually think of noon as the time when the Sun crosses the meridian and hence is highest in the local sky. However, several subtleties make this only an approximation. Let's investigate various ways of keeping daily time.

Apparent Solar Time

If we base time on the Sun's *actual* position in the local sky, as is the case when we use a sundial (Figure S1.6), we are measuring **apparent solar time**. Noon is the precise moment when the Sun is on the meridian and the sundial casts its shortest shadow. Before noon, when the Sun is on its way toward the meridian, the apparent solar time is *ante meridian,* or *a.m.* (*ante* means "before"). For example, if the Sun will reach the meridian 2 hours from now, the apparent solar time is 10 A.M. After noon, the apparent solar time is *post meridian,* or *p.m.* (*post* means "after"). If the Sun crossed the meridian 3 hours ago, the

FIGURE S1.6 A basic sundial consists of a stick, or *gnomon,* that casts a shadow, and a dial marked by numerals. Here, the shadow is on the Roman numeral III, indicating that the apparent solar time is 3:00 P.M. (The portion of the dial without numerals represents nighttime hours.) Because the Sun's path across the local sky depends on latitude, a particular sundial will be accurate only for a particular latitude.

apparent solar time is 3 P.M. Note that noon and midnight are *neither* a.m. nor p.m.; we must specify whether we mean 12:00 noon or 12:00 midnight. (Alternatively, we can use a 24-hour clock, in which case midnight is 0:00, noon is 12:00, 1 P.M. is 13:00, and so on.)

TIME OUT TO THINK *Is it daytime or nighttime at 12:01 A.M.? 12:01 P.M.? Explain.*

Mean Solar Time

Suppose you set your watch to read precisely 12:00 when a sundial reads noon today. If every solar day were precisely 24 hours, your watch would always remain synchronized with the sundial. However, while 24 hours is the *average* length of the solar day, the actual length of the solar day varies throughout the year. As a result, your watch will not remain perfectly synchronized with the sundial. For example, your watch is likely to read a few seconds before or after 12:00 when the sundial reads noon tomorrow, and over the course of the year your watch may read anywhere from 17 minutes before to 15 minutes after 12:00 when the sundial reads noon.

If you set your watch so it reads precisely 12:00 when the Sun is on the meridian on an "average" day, then your watch is set to read **mean solar time** (*mean* is another word for *average*). Mean solar time is more convenient than apparent solar time because, once set, a reliable mechanical or electronic clock can always tell you the mean solar time. In contrast, measuring apparent solar time requires a sundial, which is useless at night or when it is cloudy.

Note that, like apparent solar time, mean solar time is a *local* measure of time. That is, it varies with longitude because of the Earth's west-to-east rotation. You're probably familiar with this idea; for example, you probably know that clocks in New York are set 3 hours ahead of clocks in Los Angeles. When clocks are set precisely according to local mean solar time, they vary even over relatively short east–west distances. For example, mean solar clocks in central Los Angeles are about 2 minutes behind mean solar clocks in Pasadena, because Pasadena is slightly farther east.

Standard, Daylight, and Universal Time

Clocks reading mean solar time were common during the early history of the United States. However, by the late 1800s, widespread railroad travel made the use of mean solar time increasingly problematic. Some states had dozens of different "official" times, usually corresponding to mean solar time in dozens of different cities, and each railroad company made schedules according to its own "railroad time." The many time systems made it difficult for passengers to follow the scheduling of trains. On November 18, 1883, the railroad companies agreed to a new system that divided the United States into four time zones, setting all clocks within each zone to the same time. That was the birth of **standard time**, which divides the world today into time zones (Figure S1.8). Depending on where you live within a time zone, your standard time may vary by up to a half-hour or more from your mean solar time. (In principle, the standard time in a particular time zone is the mean solar time in the *center* of the time zone. In that case, local mean solar time would always be within a half-hour of standard time. However, time zones often have unusual shapes to conform to social, economic, and political realities, so larger variations between standard time and mean solar time sometimes occur.)

In most parts of the United States, clocks are set to standard time for only part of the year. Between the first Sunday in April and the last Sunday in October, most of the United States changes to **daylight**

THINKING ABOUT . . .

Solar Days and the Analemma

The average length of a solar day is 24 hours, but the precise length varies over the course of the year. Two effects contribute to this variation.

The first effect is due to the Earth's varying orbital speed. Recall that the Earth moves slightly faster when it is closer to the Sun in its orbit and slightly slower when it is farther from the Sun [Section 1.3]. Thus, the Earth moves slightly farther along its orbit each day when it is closer to the Sun, which means that the solar day requires more than the average amount of "extra" rotation (see Figure S1.2) during these periods—making these solar days longer than average. Similarly, the solar day requires less than the average amount of "extra" rotation when it is in the portion of its orbit farther from the Sun—making these solar days shorter than average.

The second effect is due to the tilt of the Earth's axis, which causes the ecliptic to be inclined by $23\frac{1}{2}°$ to the celestial equator on the celestial sphere. Because the length of a solar day depends on the Sun's *eastward* motion along the ecliptic, the inclination would cause solar days to vary in length even if the Earth's orbit were perfectly circular. To see why, suppose the Sun appeared to move exactly 1° per day along the ecliptic. Around the times of the solstices, this motion would be entirely eastward, making the solar day slightly longer than average. Around the times of the equinoxes, when the motion along the ecliptic has a significant northward or southward component, the solar day would be slightly shorter than average.

Together, the two effects make the actual length of solar days vary by up to about 25 seconds (either way) from the 24-hour average. Moreover, because the effects accumulate at particular times of year, the apparent solar time can differ by as much as 17 minutes from the mean solar time. The net result is often depicted visually by an **analemma**, which is printed on many globes (Figure S1.7a; Figure 2.16 shows a photographic version).

By using the horizontal scale on the analemma in Figure S1.7a, you can convert between mean and apparent solar time for any date. (The vertical scale shows the declination of the Sun, which is discussed in Section S1.5.) For example, the dashed line shows that on November 10 a mean solar clock is about 17 minutes "behind the Sun," or behind apparent solar time; that is, if the apparent solar time is 6:00 P.M., the mean solar time is only 5:43 P.M. The discrepancy between mean and apparent solar time is called the **equation of time**; it is often plotted as a graph, which gives the same results as reading from the analemma (Figure S1.7b).

The discrepancy between mean and apparent solar time also explains why the times of sunrise and sunset don't follow seasonal patterns perfectly. For example, the winter solstice around December 21 has the shortest daylight hours (in the Northern Hemisphere), but the earliest sunset occurs around December 7, when the Sun is still well "behind" mean solar time.

FIGURE S1.7 Two ways of illustrating discrepancies between mean and apparent solar time.

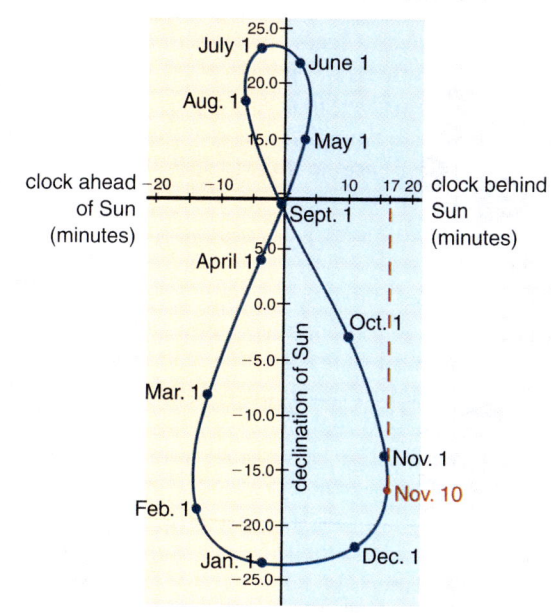

a The analemma shows the annual pattern of discrepancies between apparent and mean solar time. For example, the dashed line shows that on November 10, a mean solar clock reads 17 minutes behind (earlier than) apparent solar time.

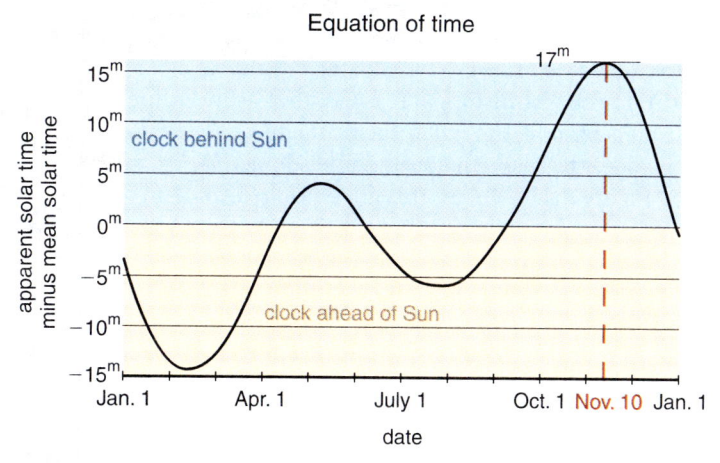

b The discrepancies can also be plotted on a graph as the equation of time.

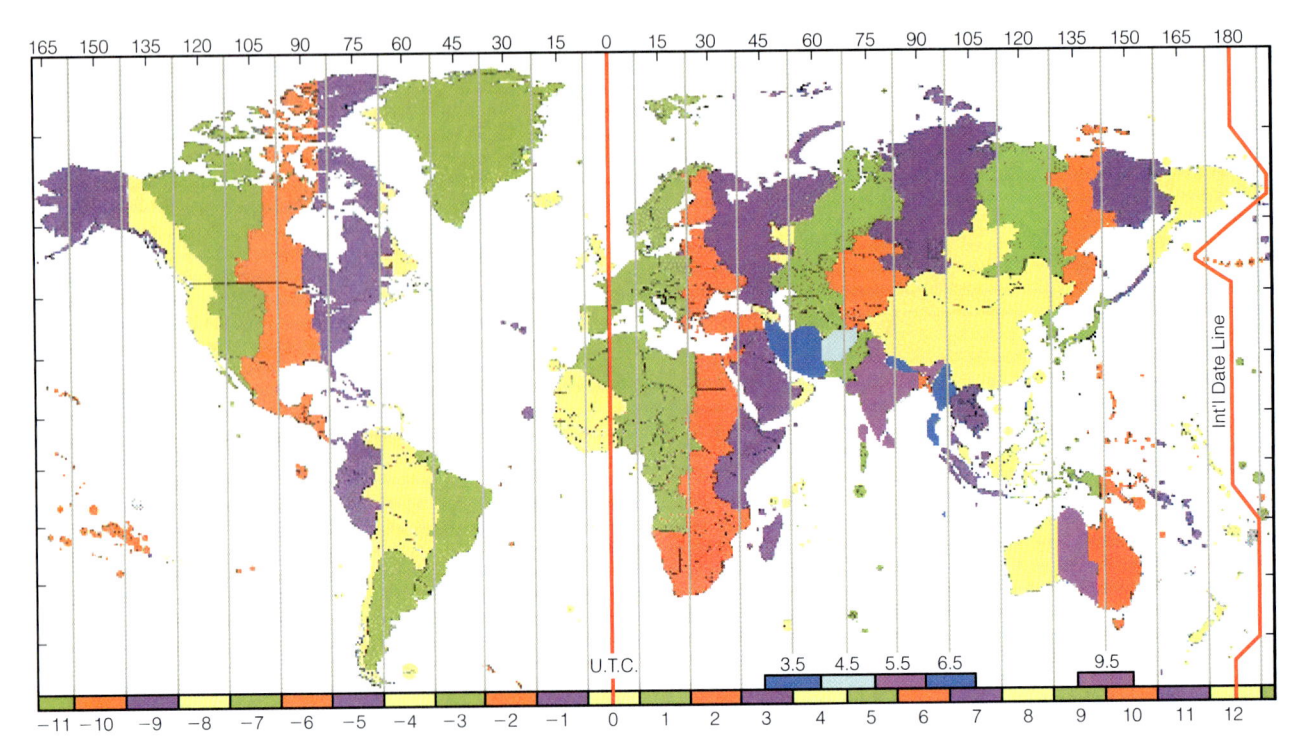

FIGURE S1.8 Time zones around the world. The numerical scale at the bottom shows hours ahead of (positive numbers) or behind (negative numbers) the time in Greenwich, England; the scale at the top is longitude. The vertical lines show standard time zones as they would be in the absence of political considerations. The color-coded regions show the actual time zones. Note, for example, that all of China uses the same standard time, even though the country is wide enough to span several time zones. Note also that a few countries use time zones centered on a half-hour, rather than an hour, relative to Greenwich time.

saving time, which is 1 hour ahead of standard time. Because of the 1-hour advance on daylight saving time, clocks read around 1 P.M. (rather than around noon) when the Sun is on the meridian.

As we will see, for purposes of navigation and astronomy it is useful to have a single time for the entire Earth. For historical reasons, this "world" time was chosen to be the mean solar time in Greenwich, England—the place that also defines longitude 0° (see Figure 2.11). Today, this *Greenwich mean time* (*GMT*) is often called **universal time** (**UT**). (Outside astronomy, it is often called universal coordinated time (UTC); many airlines and weather services call it "Zulu time," because Greenwich's time zone is designated Z and "zulu" is a common way of phonetically identifying the letter Z.)

S1.3 The Calendar

Our modern calendar is based on the length of the tropical year, which is the amount of time from one spring equinox to the next. The origins of our calendar go back to ancient Egypt: By 4200 B.C., the Egyptians were using a calendar that counted 365 days in a year.

Because the tropical year actually is closer to $365\frac{1}{4}$ days, the Egyptian calendar slowly drifted out of phase with the seasons by about 1 day every 4 years. For example, if the spring equinox occurred on March 21 one year, 4 years later it occurred on March 20, 4 years after that on March 19, and so on. Over many centuries, the spring equinox moved through many different months. To keep the seasons and the calendar in phase, in 46 B.C. Julius Caesar decreed the adoption of a new calendar. This **Julian calendar** introduced the concept of **leap year**: Every fourth year has 366 days, rather than 365, so that the average length of the calendar year is $365\frac{1}{4}$ days.

The Julian calendar originally had the spring equinox falling around March 24. If it had been perfectly synchronized with the tropical year, this calendar would have ensured that the spring equinox occurred on the same date every 4 years (that is, every leap-year cycle). It didn't work perfectly, however, because a tropical year is actually about 11 minutes short of $365\frac{1}{4}$ days. Thus, the moment of the spring equinox slowly advanced by an average of 11 minutes per year. By the late 1500s, the spring equinox was occurring on March 11.

In 1582, Pope Gregory XIII introduced a new calendar—the **Gregorian calendar**—designed to return the spring equinox to the same date after every

4-year cycle. The Gregorian calendar made two adjustments to the Julian calendar. First, Pope Gregory decreed that the day in 1582 following October 4 would be October 15. By eliminating the ten dates from October 5 through October 14, 1582, he pushed the date of the spring equinox in 1583 from March 11 to March 21. (He chose March 21 because it was the date of the spring equinox in A.D. 325, which was the time of the Council of Nicaea, the first ecumenical council of the Christian church.) Second, the Gregorian calendar added an exception to the rule of having leap year every 4 years: Leap year is skipped when a century changes (for example, in years 1700, 1800, 1900) *unless* the century year is divisible by 400. Thus, 2000 was a leap year because it is divisible by 400 (2000 ÷ 400 = 5), but 2100 will *not* be a leap year. These adjustments make the average length of the Gregorian calendar year almost exactly the same as the actual length of a tropical year, which ensures that the spring equinox will occur on March 21 every fourth year for thousands of years to come.

Today, the Gregorian calendar is used worldwide for international communication and commerce. (Many countries still use traditional calendars, such as the Chinese, Islamic, and Jewish calendars, for cultural purposes.) However, as you might guess, the Pope's decree was not immediately accepted in regions not bound to the Catholic Church. For example, the Gregorian calendar was not adopted in England or in the American colonies until 1752, and it was not adopted in China until 1912 or in Russia until 1919. In parts of the world that still used the Julian calendar, the dates October 5 through October 14 *did* occur in 1582, and the 10 days were removed from the calendar at some later time. Thus, a particular date, such as December 25, came at different times in different countries depending on whether a country had yet adopted the Gregorian calendar.

S1.4 Locations in the Sky

We are now ready to turn our attention from timekeeping to navigation. The goal of celestial navigation is to use the Sun and the stars to find our position on Earth. Before we can do that, we need to become more familiar with the apparent motions of the sky. In this section, we'll discuss how we locate the Sun and stars on the celestial sphere; this will enable us to understand motions in the local sky, the topic of the next section. Then, in the final section, we'll use these ideas to explain the principles of celestial navigation.

A Map of the Celestial Sphere

Recall that, for purposes of pinpointing objects in our sky, it's useful to think of the Earth as being in the center of a giant celestial sphere (see Figure 2.4). From our point of view on Earth, the celestial sphere appears to rotate around us each day (see Figure 2.8). We can use a model of the celestial sphere to locate stars or the Sun, much as we use a globe to locate places on Earth. The primary difference is that the celestial sphere models *apparent* positions in the sky, rather than true positions in space. As we discussed in Chapter 2, we can compare positions on the celestial sphere only by reference to the *angles* that separate them, not by actual distances. But aside from this important difference between a globe and the celestial sphere, we can use both as maps by identifying special locations (such as the equator and poles) and adding a system of coordinates (such as latitude and longitude).

We've already discussed the special locations we need for a map of the celestial sphere: the north and south celestial poles, the celestial equator, and the ecliptic. Figure S1.9 shows these locations on a schematic diagram; recall that the Earth is in the center because the celestial sphere represents the sky as we see it from Earth. The arrow along the ecliptic indicates the direction in which the Sun appears to move along it over the course of each year. It is much easier to visualize the celestial sphere if you make a model with a simple plastic ball. Use a felt-tip pen to mark the north and south celestial poles on your ball, then add the celestial equator and the ecliptic; note that the ecliptic crosses the celestial equator on

FIGURE S1.9 This schematic diagram of the celestial sphere, shown without stars, helps us create a map of the celestial sphere by showing the north and south celestial poles, the celestial equator, the ecliptic, and the equinoxes and solstices. As a study aid, you should use a plastic ball as a model of the celestial sphere, marking it with the same special locations.

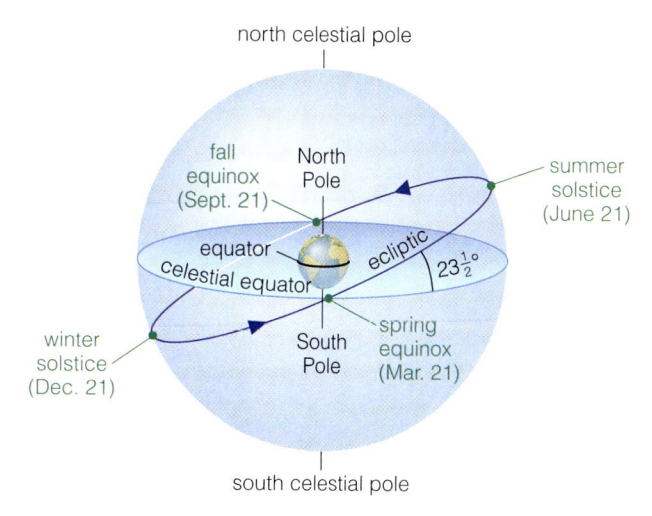

opposite sides of the celestial sphere at an angle of $23\frac{1}{2}°$ (because of the tilt of the Earth's axis).

Equinoxes and Solstices

Recall that the equinoxes and solstices are special moments in the year that help define the seasons [Section 1.3]. For example, the *spring equinox*, which occurs around March 21 each year, is the moment when spring begins for the Northern Hemisphere and fall begins for the Southern Hemisphere. These moments correspond to positions in the Earth's orbit (see Figure 1.13) and hence to apparent locations of the Sun along the ecliptic. As shown in Figure S1.9, the spring equinox occurs when the Sun is on the ecliptic at the point where it crosses from south of the celestial equator to north of the celestial equator; this point is also called the spring equinox. Thus, the term *spring equinox* has a dual meaning: It is the *moment* when spring begins and also the *point* on the ecliptic at which the Sun appears to be located at that moment.

Figure S1.9 also shows the points marking the summer solstice, fall equinox, and winter solstice; labels indicate the dates on which the Sun appears to be located at each point. Remember that the dates are approximate. Because of the leap-year cycle and the fact that a tropical year is not exactly $365\frac{1}{4}$ days, the precise moments of the equinoxes and solstices may come anywhere between about the 20th and 23rd of the indicated months. (For example, the spring equinox may occur anytime between March 20 and March 23.)

Although no bright stars mark the locations of the equinoxes or solstices among the constellations, you can find them with the aid of nearby bright stars (Figure S1.10). For example, the spring equinox is located in the constellation Pisces and can be found with the aid of the four bright stars in the Great Square of Pegasus. Of course, when the Sun is located at this point around March 21, we cannot see Pisces or Pegasus because they are close to the Sun in our daytime sky.

TIME OUT TO THINK *Using your plastic ball as a model of the celestial sphere (which you have already marked with the celestial poles, equator, and ecliptic), mark the locations and approximate dates of the equinoxes and solstices. Based on the dates for these points, approximately where along the ecliptic is the Sun on April 21? On November 21? How do you know?*

Celestial Coordinates

We can complete our map of the celestial sphere by adding a coordinate system similar to the coordinates that measure latitude and longitude on Earth. This

FIGURE S1.10 These diagrams show the locations among the constellations of (**a**) the spring equinox, (**b**) the summer solstice, (**c**) the fall equinox, and (**d**) the winter solstice. No bright stars mark any of these points, so you must find them by studying their positions relative to recognizable patterns. The time of day and night at which each point is above the horizon depends on the time of year.

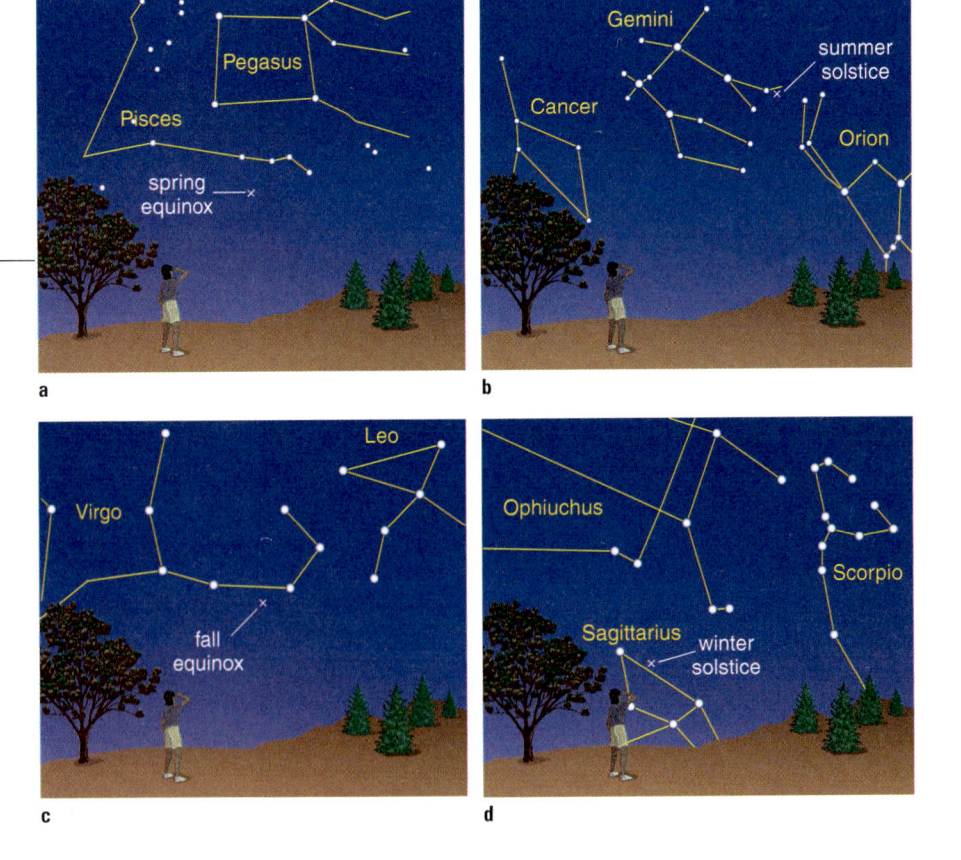

system will be the third coordinate system we've used in this book; Figure S1.11 reviews the three systems. Figure S1.11a shows the coordinates of *altitude* and *direction* (or *azimuth* [Section 2.2]) we use in the local sky. Figure S1.11b shows the coordinates of *latitude* and *longitude* we use on Earth's surface. Figure S1.11c shows the system of **celestial coordinates** we use to pinpoint locations on the celestial sphere; these coordinates are called **declination (dec)** and **right ascension (RA)**.

If you carefully compare Figures S1.11b and c, you'll see that declination on the celestial sphere is very similar to *latitude* on Earth:

- Just as lines of latitude are parallel to the Earth's equator, lines of declination are parallel to the celestial equator.

- Just as the Earth's equator has lat = 0°, the celestial equator has dec = 0°.

- Whereas latitudes are labeled north or south relative to the equator, declinations are labeled *positive* or *negative*. For example, the North Pole has lat = 90°N, but the north celestial pole has dec = +90; the South Pole has lat = 90°S, but the south celestial pole has dec = −90°.

The diagrams in Figures S1.11b and c also show that right ascension on the celestial sphere is very similar to *longitude* on Earth:

- Just as lines of longitude extend from the North Pole to the South Pole, lines of right ascension extend from the north celestial pole to the south celestial pole.

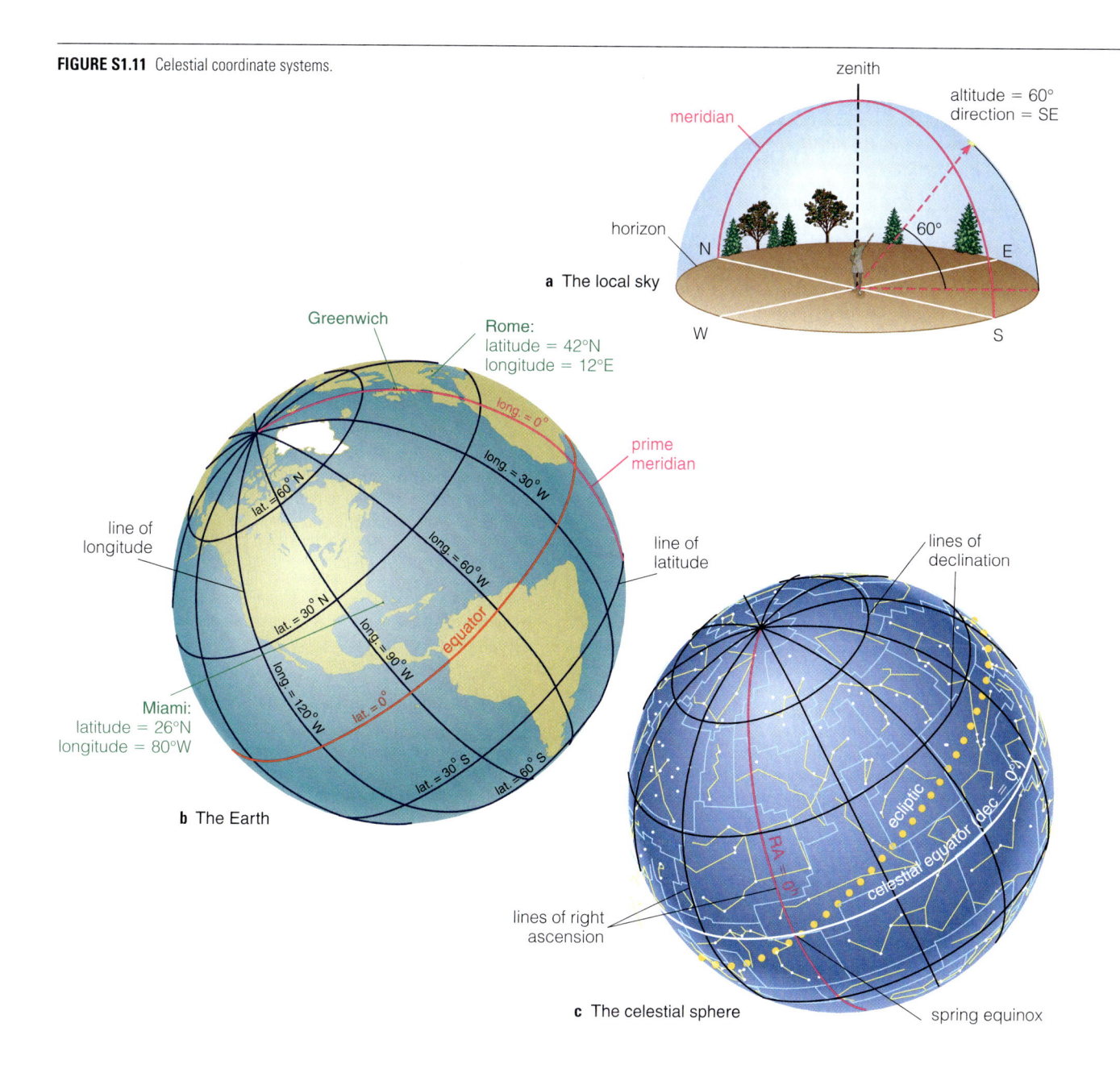

FIGURE S1.11 Celestial coordinate systems.

a The local sky

b The Earth

c The celestial sphere

CHAPTER S1 CELESTIAL TIMEKEEPING AND NAVIGATION 75

- Just as there is no natural starting point for longitude, there is no natural starting point for right ascension. By international treaty, longitude zero (the prime meridian) is the line of longitude that runs through Greenwich, England. By convention, right ascension zero is the line of right ascension that runs through the spring equinox.

- Whereas longitudes are measured in degrees east or west of Greenwich, right ascensions are measured in hours (and minutes and seconds) east of the spring equinox. A full 360° circle around the celestial equator goes through 24 hours of right ascension, so each hour of right ascension represents an angle of 360° ÷ 24 = 15°.

We can use celestial coordinates to describe the position of any object on the celestial sphere. For example, the bright star Vega has dec = +38°44′ and RA = $18^h 35^m$ (Figure S1.12). The positive declination tells us that Vega is 38°44′ *north* of the celestial equator. The right ascension tells us that Vega is 18 hours 35 minutes east of the spring equinox. Translating the right ascension from hours to angular degrees, we find that Vega is about 279° east of the spring equinox (because 18 hours represents 18 × 15° = 270° and 35 minutes represents $\frac{35}{60} \times 15 \approx 9°$).

TIME OUT TO THINK *On your plastic ball model of the celestial sphere, add a scale for right ascension along the celestial equator and also add a few circles of declination, such as declination 0°, ±30°, ±60°, and ±90°. Locate Vega on your model.*

Mathematical Insight S1.1   Time by the Stars

The clocks we use in daily life are set to solar time, ticking through 24 hours for each day of mean solar time. In astronomy, it is useful to have clocks that tell time by the stars, or **sidereal time**. Just as we define *solar time* according to the Sun's position relative to the meridian, *sidereal time* is based on the positions of stars relative to the meridian. We define the **hour angle (HA)** of any object on the celestial sphere to be the time since it last crossed the meridian. (For a circumpolar star, hour angle is measured from the *higher* of the two points at which it crosses the meridian each day.) For example:

- If a star is crossing the meridian now, its hour angle is 0^h.
- If a star crossed the meridian 3 hours ago, its hour angle is 3^h.
- If a star will cross the meridian 1 hour from now, its hour angle is -1^h or, equivalently, 23^h.

By convention, time by the stars is based on the hour angle of the spring equinox. That is, the **local sidereal time (LST)** is

$$LST = HA_{\text{spring equinox}}$$

For example, the local sidereal time is 00:00 when the spring equinox is *on* the meridian. Three hours later, when the spring equinox is 3 hours west of the meridian, the local sidereal time is 03:00.

Note that, because right ascension tells us how long after the spring equinox an object reaches the meridian, the local sidereal time is also equal to the right ascension (RA) of objects currently crossing your meridian. For example, if your local sidereal time is 04:30, stars with RA = $4^h 30^m$ are currently crossing your meridian. This idea leads to an important relationship between any object's current hour angle, the current local sidereal time, and the object's right ascension:

$$HA_{\text{object}} = LST - RA_{\text{object}}$$

This formula will make sense to you if you recognize that an object's right ascension tells us the time by which it trails the spring equinox on its daily trek through the sky. Because the local sidereal time tells us how long it has been since the spring equinox was on the meridian, the difference $LST - RA_{\text{object}}$ must tell us the position of the object relative to the meridian.

Sidereal time has one important subtlety: Because the stars (and the celestial sphere) appear to rotate around us in one sidereal day ($23^h 56^m$), sidereal clocks must tick through 24 hours of sidereal time in 23 hours 56 minutes of solar time. That is, a sidereal clock gains about 4 minutes per day over a solar clock. As a result, you cannot immediately infer the local sidereal time from the local solar time, or vice versa, without either doing some calculations or consulting an astronomical table. Of course, the easiest way to determine the local sidereal time is with a clock that ticks at the sidereal rate. Astronomical observatories always have sidereal clocks, and you can buy moderately priced telescopes that come with sidereal clocks.

Example 1: Suppose it is 9:00 P.M. on the spring equinox (March 21). What is the local sidereal time?

Solution: On the day of the spring equinox, the Sun is located at the point of the spring equinox in the sky. Thus, if the Sun is 9 hours past the meridian, so is the spring equinox. The local sidereal time is LST = 09:00.

Example 2: Suppose the local sidereal time is LST = 04:00. When will Vega cross your meridian?

Solution: Vega has RA = $18^h 35^m$. Thus, at LST = 04:00, Vega's hour angle is

$$HA_{\text{Vega}} = LST - RA_{\text{Vega}} = 4:00 - 18:35 = -14:35$$

Vega will cross your meridian in 14 hours 35 minutes, which also means it crossed your meridian 9 hours 25 minutes ago ($14^h 35^m + 9^h 25^m = 24^h$). (Note that these are intervals of sidereal time.)

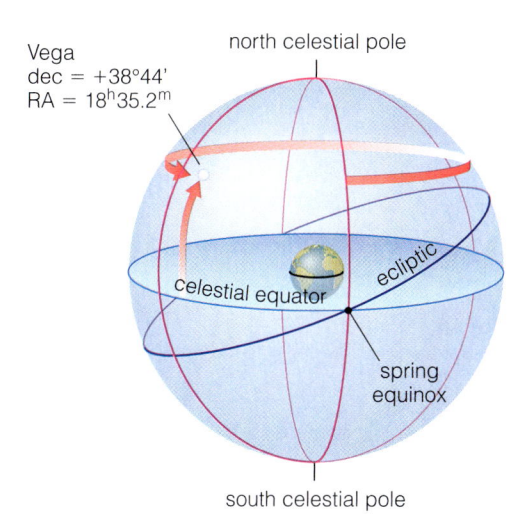

FIGURE S1.12 This diagram shows how we interpret the celestial coordinates of Vega. Its declination tells us that it is 38°44′ north of the celestial equator. We can interpret its right ascension in two ways: As an angle, it means it is about 279° (the angular equivalent of $18^h 35^m$) east of the vernal equinox; as a time, it means that Vega crosses the meridian about 18 hours 35 minutes after the spring equinox.

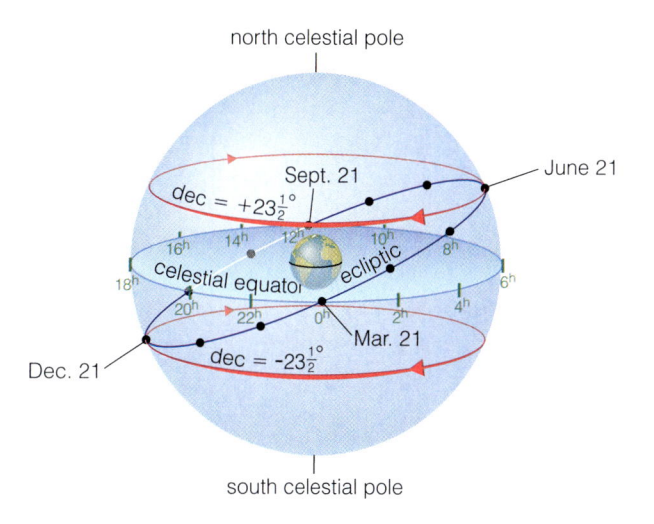

FIGURE S1.13 We can use this diagram of the celestial sphere to determine the Sun's right ascension and declination at monthly intervals.

You may be wondering why right ascension is measured in units of time. The answer is that time units are convenient for tracking the daily motion of objects through the local sky. All objects with a particular right ascension cross the meridian at the same time; for example, all stars with RA = 0^h cross the meridian at the same time that the spring equinox crosses the meridian. For any other object, the right ascension tells us when it crosses the meridian in hours *after* the spring equinox crosses the meridian. Thus, for example, Vega's right ascension, $18^h 35^m$, tells us that on any particular day it crosses the meridian about 18 hours 35 minutes after the spring equinox. (This is 18 hours 35 minutes of *sidereal time* later, which is not exactly the same as 18 hours 35 minutes of solar time; see Mathematical Insight S1.1.)

The celestial coordinates of stars are not quite constant; they change gradually with the Earth's 26,000-year precession cycle. The change occurs because celestial coordinates are tied to the celestial equator, which moves with precession relative to the constellations. (Precession does not affect the Earth's orbit, so it does not affect the location of the ecliptic among the constellations.) Thus, the celestial coordinates of stars change even while the stars themselves remain fixed in the patterns of the constellations. The coordinate changes are not noticeable to the eye on the time scale of a human lifetime, but precise astronomical work requires the updating of celestial coordinates almost constantly. Many astronomical software packages do these calculations automatically and can also give day-to-day celestial coordinates for the Sun, Moon, and planets as they wander among the constellations.

Celestial Coordinates of the Sun

Unlike the Moon and the planets, which wander among the constellations in complex ways that require detailed calculations, the Sun moves through the zodiac constellations in a simple way: It moves roughly 1° per day along the ecliptic. In a month, the Sun moves approximately one-twelfth of the way around the ecliptic, meaning that its right ascension changes by about 24 ÷ 12 = 2 hours per month. Figure S1.13 shows the ecliptic marked with the Sun's monthly position and a scale of celestial coordinates. From this figure, we can create a table of the Sun's month-by-month celestial coordinates. Table S1.1 starts from the spring equinox, when the Sun has declination 0° and right ascension 0^h.

Note that, unlike its steady advance in right ascension, the Sun's declination changes much more rapidly around the equinoxes than around the solstices. For example, the Sun's declination changes from −12° on February 21 to +12° on April 21, a change of 24° in just two months. In contrast, between May 21 and July 21, the declination varies only between +20° and +23$\frac{1}{2}$°. This explains why the daylight hours increase rapidly in spring and decrease rapidly in fall but stay long for a couple of months around the summer solstice and short for a couple of months around the winter solstice.

TIME OUT TO THINK *On your plastic ball model of the celestial sphere, add dots along the ecliptic to show the Sun's monthly positions. Use your model to estimate the Sun's celestial coordinates on your birthday.*

Table S1.1 The Sun's Approximate Celestial Coordinates at 1-Month Intervals

Approximate Date	RA	Dec	Approximate Date	RA	Dec
Mar. 21 (Spring equinox)	0 hr	0°	Sept. 21 (Fall equinox)	12 hr	0°
Apr. 21	2 hr	+12°	Oct. 21	14 hr	−12°
May 21	4 hr	+20°	Nov. 21	16 hr	−20°
June 21 (Summer solstice)	6 hr	$+23\frac{1}{2}°$	Dec. 21 (Winter solstice)	18 hr	$-23\frac{1}{2}°$
July 21	8 hr	+20°	Jan. 21	20 hr	−20°
Aug. 21	10 hr	+12°	Feb. 21	22 hr	−12°

S1.5 Understanding Local Skies

In Chapter 2, we briefly discussed how the daily circles of stars vary with latitude (see Figure 2.12). With our deeper understanding of the celestial sphere and celestial coordinates, we can now study local skies in more detail. We'll begin by focusing on star tracks through the local sky, which depend only on a star's declination. Then we'll discuss the daily path of the Sun, which varies with the Sun's declination and therefore with the time of year.

Star Tracks

The apparent daily rotation of the celestial sphere makes star tracks seem simple when viewed from the outside (see Figure 2.8). Local skies seem complex only because the ground always blocks our view of half of the celestial sphere. The half that is blocked depends on latitude. Let's first consider the local sky at the Earth's North Pole, the easiest case to understand, and then look at the local sky at other latitudes.

The North Pole Figure S1.14a shows the rotating celestial sphere and your orientation relative to it when you are standing at the North Pole. Your "up" points toward the north celestial pole, which therefore marks your zenith. The Earth blocks your view of anything south of the celestial equator, which therefore runs along your horizon. To make it easier for you to visualize the local sky, Figure S1.14b shows your horizon extending to the celestial sphere; note that the horizon is marked with directions and that all directions are south from the North Pole. (Because the meridian is defined as running from north to south in the local sky, there is no meridian at the North Pole.)

The daily circles of the stars keep them at constant altitudes above or below your horizon, and their

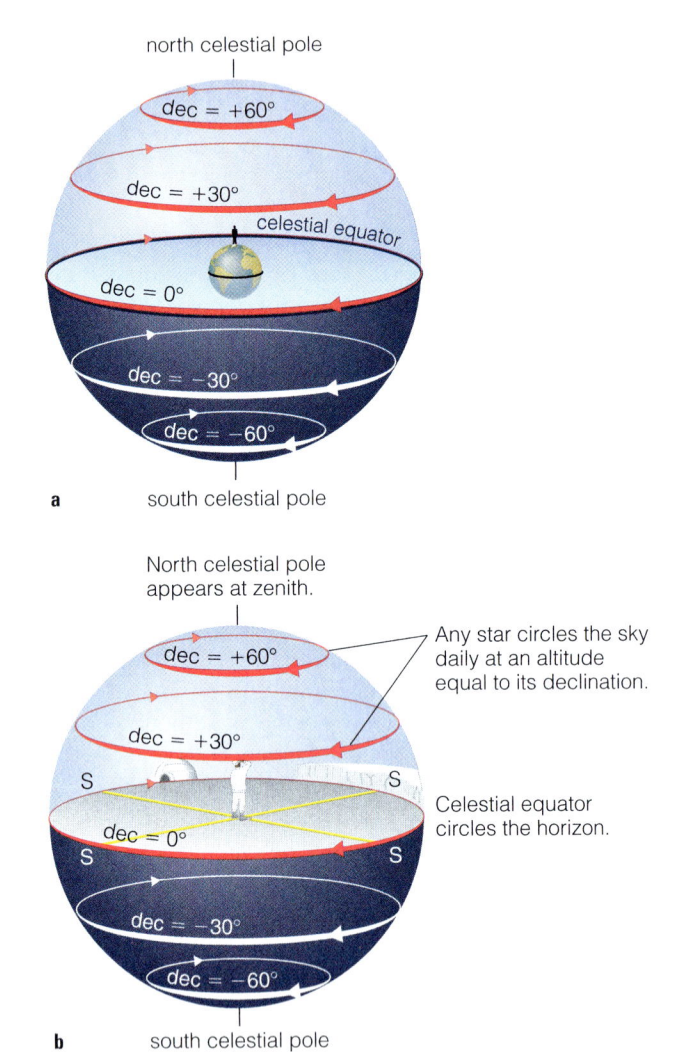

FIGURE S1.14 (a) The orientation of the local sky, relative to the celestial sphere, for the North Pole. (b) Extending the horizon to the celestial sphere makes it easier to visualize the local sky at the North Pole. (*Note:* To understand the extension, remember that the Earth should appear extremely small compared to the celestial sphere.)

altitudes are equal to their declinations. For example, a star with declination +60° circles the sky at an altitude of 60°, and a star with declination −30° remains 30° below your horizon at all times. As a result, all stars north of the celestial equator are circumpolar at the North Pole, never falling below the horizon. Similarly, stars south of the celestial equator never appear in the sky as seen from the North Pole.

Note that right ascension does not affect a star's path at all; it affects only the time of day and year at which a star is found in a particular direction along your horizon. If you are having difficulty visualizing the star paths, it may help you to watch star paths as you rotate your plastic ball model of the celestial sphere.

The Equator Next imagine that you are standing somewhere on the Earth's equator (lat = 0°), such as in Ecuador, in Kenya, or on the island of Borneo. Figure S1.15a shows that "up" points directly away from (perpendicular to) the Earth's rotation axis. Figure S1.15b shows the local sky more clearly by extending your horizon to the celestial sphere and rotating the diagram so your zenith is up. As everywhere except at the poles, the meridian extends from the horizon due south, through the zenith, to the horizon due north.

Look carefully at how the celestial sphere appears to rotate in the local sky. The north celestial pole remains stationary on your horizon due north; as we should expect, its altitude, 0°, is equal to the equator's latitude [Section 2.2]. Similarly, the south celestial pole remains stationary on your horizon due south. At any particular time, half of the celestial equator is visible, extending from the horizon due east, through the zenith, to the horizon due west; the other half lies below the horizon. As the equatorial sky appears to turn, all star paths rise straight out of the eastern horizon and set straight into the western horizon, with the following features:

- Stars with dec = 0° lie *on* the celestial equator and therefore rise due east, cross the meridian at the zenith, and set due west.

- Stars with dec > 0° rise north of due east, reach their highest point on the meridian in the north, and set north of due west. Their rise, set, and highest point depend on their declinations. For example, a star with dec = +30° rises 30° north of due east, crosses the meridian 30° to the north of the zenith—that is, at an *altitude* of 90° − 30° = 60° in the north—and sets 30° north of due west.

- Stars with dec < 0° rise south of due east, reach their highest point on the meridian in the south, and set south of due west. For example, a star with dec = −50° rises 50° south of due east, crosses the meridian 50° to the south of the zenith—that is, at an *altitude* of 90° − 50° = 40° in the south—and sets 50° south of due west.

Note that exactly half of any star's daily circle lies above the horizon. Thus, every star is above the horizon for exactly half of each sidereal day, or just under 12 hours, and below the horizon for the other half of the sidereal day.

TIME OUT TO THINK *Visualize the daily paths of stars as seen from the Earth's equator. Are any stars circumpolar? Are there stars that never rise above the horizon? Explain.*

FIGURE S1.15 (a) The orientation of the local sky, relative to the celestial sphere, for the Earth's equator. (b) Extending the horizon and rotating the diagram make it easier to visualize the local sky at the equator.

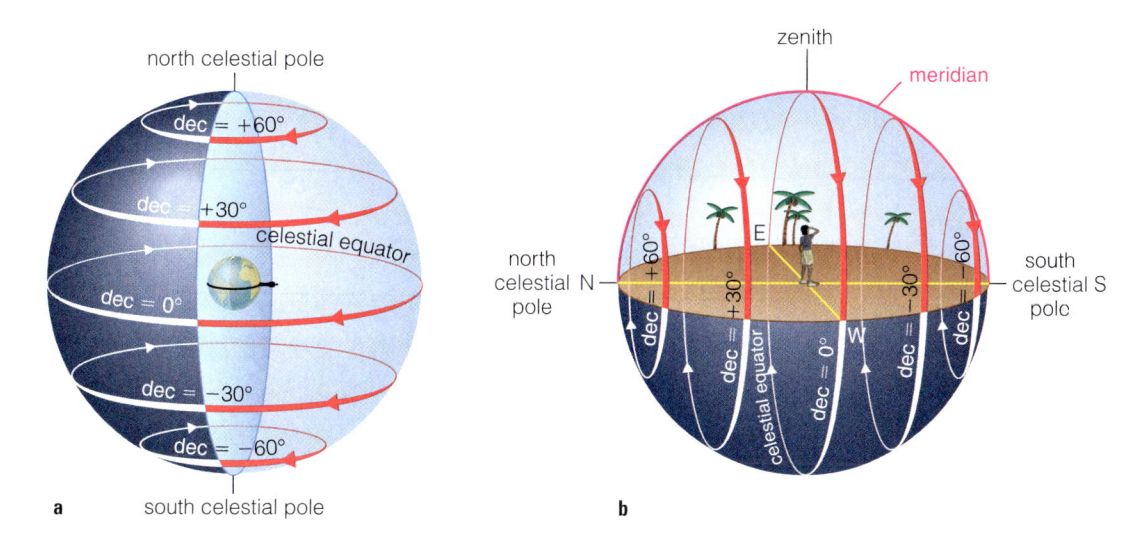

Other Latitudes We can use the same basic strategy to determine star tracks for other latitudes. Let's consider latitude 40°N, such as in Denver, Indianapolis, Philadelphia, or Beijing. First, as shown in Figure S1.16a, imagine standing at this latitude on a basic diagram of the rotating celestial sphere; note that "up" points to a location on the celestial sphere with declination +40°. To make it easier to visualize the local sky, we next extend the horizon and rotate the diagram so your zenith is up (Figure S1.16b).

As we would expect, the north celestial pole appears 40° above the horizon due north, since its altitude in the local sky is always equal to the latitude. Half of the celestial equator is visible; it extends from the horizon due east, to the meridian at an altitude of 50° in the south, to the horizon due west. By comparing this diagram to the local sky for the equator, you can probably notice a general rule for the celestial equator at any latitude: *Exactly half the celestial equator is always visible, extending from due east on the horizon to due west on the horizon and crossing the meridian at an altitude of 90° minus the latitude.* The celestial equator runs through the southern half of the sky at locations in the Northern Hemisphere and through the northern half of the sky at locations in the Southern Hemisphere. The diagram in Figure S1.16b also shows star tracks through the local sky for 40°N; it may be easier for you to visualize them if you rotate your plastic ball model of the celestial sphere while holding it in the orientation of Figure S1.16b. As you study the tracks, note the following key features:

- Stars with dec = 0° lie *on* the celestial equator and therefore follow the path of the celestial equator through the local sky. That is, for latitude 40°N, they rise due east, cross the meridian at 50° in the south, and set due west.

- Stars with dec > 0° follow paths parallel to the celestial equator but farther north. Thus, they rise north of due east, cross the meridian north of where the equator crosses it, and set north of due west. If they are within 40° of the north celestial pole on the celestial sphere (which means declinations greater than 90° − 40° = 50°), their entire circles are above the horizon, making them circumpolar. You can find the precise point at which a star crosses the meridian by adding its declination to the 50°S altitude at which the celestial equator crosses the meridian. For example, Figure S1.16b shows that a star with dec = +30° crosses the meridian at altitude 50° + 30° = 80° in the south and a star with dec = +60° crosses the meridian at altitude 70° in the north. (To calculate the latter result, note that the sum 50° + 60° = 110° goes 20° past the zenith altitude of 90°, making it equivalent to 90° − 20° = 70°.)

- Stars with dec < 0° follow paths parallel to the celestial equator but farther south. Thus, they rise south of due east, cross the meridian south of where the celestial equator crosses it, and set south of due west. If they are within 40° of the south celestial pole on the celestial sphere (which means declinations less than −90° + 40° =

FIGURE S1.16 (**a**) The orientation of the local sky, relative to the celestial sphere, for latitude 40°N. Because latitude is the angle to the Earth's equator, "up" points to the circle on the celestial sphere with declination +40°. (**b**) Extending the horizon and rotating the diagram so the zenith is up make it easier to visualize the local sky. The blue scale along the meridian shows altitudes and directions in the local sky.

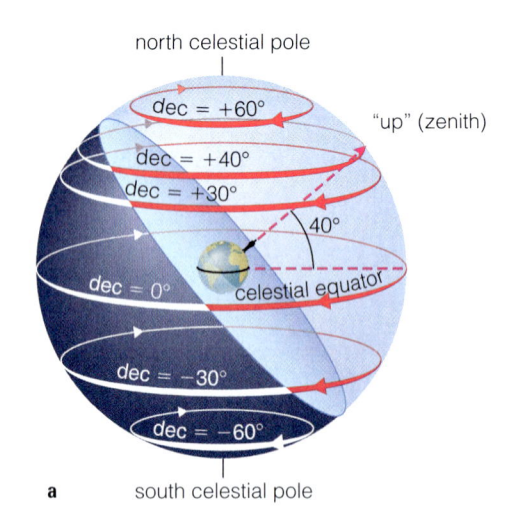

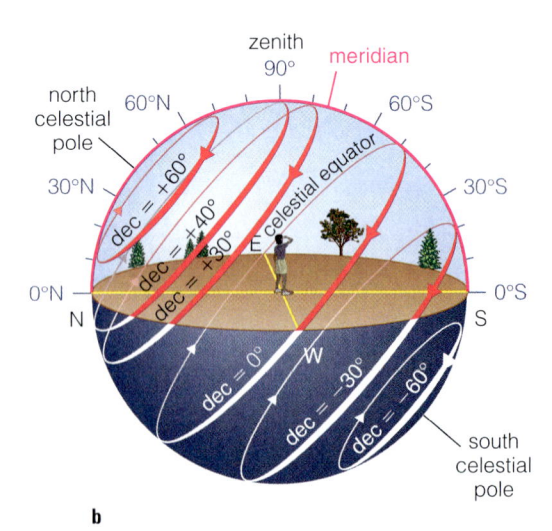

−50°), their entire circles are below the horizon and thus they are never visible.

Note that the fraction of any star's daily circle that is above the horizon—and hence the amount of time it is above the horizon each day—depends on its declination. Because exactly half the celestial equator is above the horizon, stars on the celestial equator (dec = 0°) are above the horizon for about 12 hours per day. Stars with positive declinations have more than half their daily circle above the horizon and hence are above the horizon for more than 12 hours each day (with the range extending to 24 hours a day for the circumpolar stars). Stars with negative declinations have less than half their daily circle above the horizon and hence are above the horizon for less than 12 hours each day (with the range going to zero for stars that are never above the horizon).

We can apply the same strategy we used in Figure S1.16 to find star paths for other latitudes. Figure S1.17 shows the process for latitude 30°S. Note that the south celestial pole is visible to the south and that the celestial equator passes through the northern half of the sky. If you study the diagram carefully, you can see how star tracks depend on declination.

TIME OUT TO THINK *Study Figure S1.17 for latitude 30°S. Describe the path of the celestial equator; explain how it obeys the 90° − latitude rule given above. Give a general description of how star tracks differ for stars with positive and negative declinations. What stars are circumpolar at this latitude?*

The Path of the Sun

Just as a star's path through the sky depends only on its declination, the Sun's path through the sky on any particular day depends only on its declination for that day. For example, because the Sun's declination is $+23\frac{1}{2}°$ on the summer solstice, the Sun's path through the local sky on June 21 is the same as that of any star with declination $+23\frac{1}{2}°$. Thus, as long as we know the Sun's declination for a particular day, we can find the Sun's path at any latitude with the same local sky diagrams we used to find star tracks.

Figure S1.18 shows the Sun's path on the equinoxes and solstices for latitude 40°N. On the equinoxes, when the Sun is on the celestial equator (dec = 0°), the Sun's path follows the celestial equator: It rises due east, crosses the meridian at altitude 50° in the south, and sets due west; like any object on the celestial equator, it is above the horizon for 12 hours. On the summer solstice, the Sun rises well north of due east,[1] reaches an altitude of $73\frac{1}{2}°$ when it crosses the meridian in the south, and sets well

---

[1] Calculating exactly how far north of due east the Sun rises is beyond the scope of this book, but astronomical software packages will tell you exactly where—and at what time—the Sun rises and sets along the horizon for any location and any date.

FIGURE S1.17 This pair of diagrams shows how we find star tracks for latitude 30°S. (**a**) The orientation of the local sky at latitude 30°S, relative to the celestial sphere; "up" points to the circle on the celestial sphere with declination −30°. (**b**) Extending the horizon and rotating the diagram so the zenith is up make it easier to visualize the local sky. Note that the south celestial pole is visible at altitude 30° in the south, while the celestial equator stretches across the northern half of the sky.

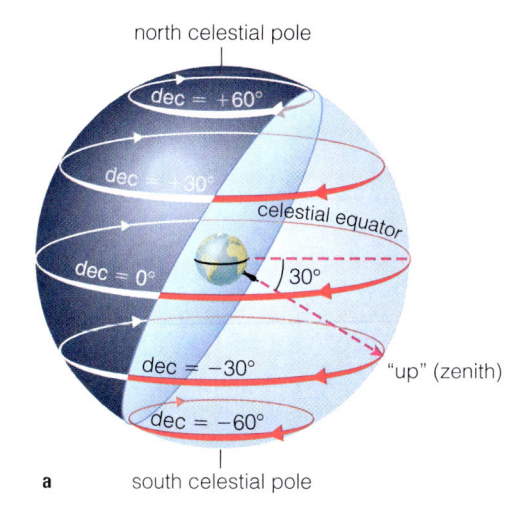

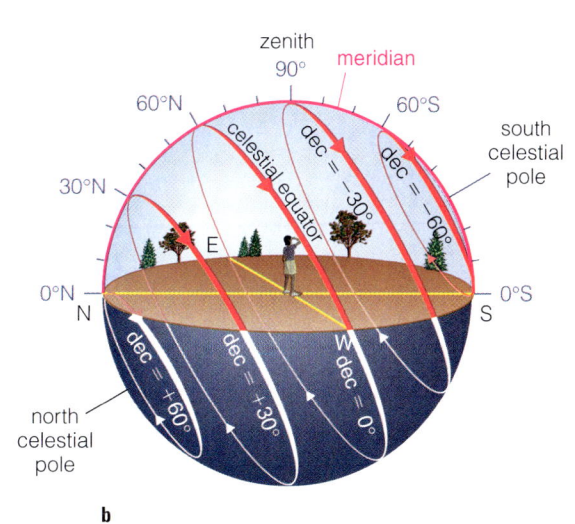

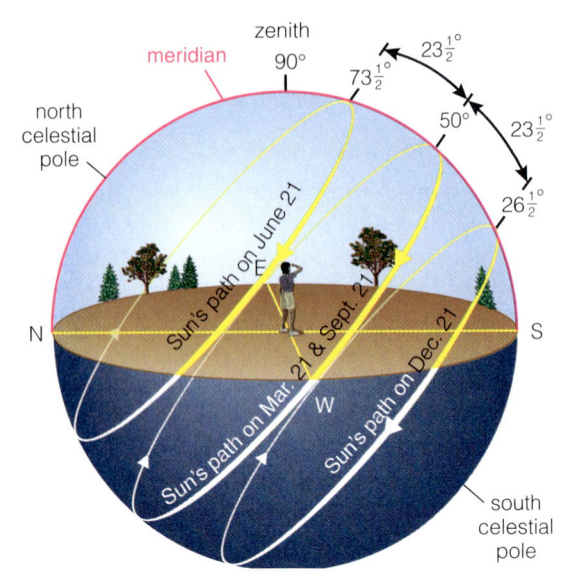

FIGURE S1.18 The Sun's daily paths for the equinoxes and solstices at latitude 40°N.

The North and South Poles Recall that the celestial equator circles the horizon at the North Pole. Because the Sun appears *on* the celestial equator on the day of the spring equinox, the Sun circles the north polar sky *on the horizon* on March 21 (Figure S1.20); it completes a full circle of the horizon in 24 hours (1 solar day). Over the next 3 months, the Sun continues to circle the horizon, circling at gradually higher altitudes as its declination increases. It reaches its highest point on the summer solstice, when its declination of $+23\frac{1}{2}°$ means that it circles the north polar sky at an altitude of $23\frac{1}{2}°$. After the summer solstice, the daily circles gradually fall lower over the next 3 months, reaching the horizon on the fall equinox. Then, because the Sun's declination is negative for the next 6 months (until the following spring equinox), it remains below the north polar horizon. Thus, the North Pole essentially has 6 months of daylight and 6 months of darkness, with an extended twilight that lasts a few weeks beyond the fall equinox and an extended dawn that begins a few weeks before the spring equinox.

The situation is the opposite at the South Pole. Here the Sun's daily circles slowly rise above the horizon on the fall equinox to a maximum altitude of $23\frac{1}{2}°$ on the *winter* solstice, then slowly fall back to the horizon on the spring equinox. Thus, the South Pole has the Sun above the horizon during the 6 months it is below the north polar horizon.

There's an important caveat to this discussion: Although we've correctly described the Sun's true position in the polar skies over the course of the year, two effects complicate what we actually see at the poles around the times of the equinoxes. First, the atmosphere bends light enough so that the Sun *appears* to be above the horizon even when it is actually below it; near the horizon, this bending makes the

north of due west; the daylight hours are long, because much more than half of the Sun's path is above the horizon. On the winter solstice, the Sun rises well south of due east, reaches an altitude of only $26\frac{1}{2}°$ when it crosses the meridian in the south, and sets well south of due west; the daylight hours are short, because much less than half of the Sun's path is above the horizon.

We could make a similar diagram to show the Sun's path on various dates for any latitude. However, the $23\frac{1}{2}°$ tilt of the Earth's axis makes the Sun's path particularly interesting at the special latitudes shown in Figure S1.19. Let's investigate these latitudes.

FIGURE S1.19 Special latitudes defined by the Sun's path through the sky.

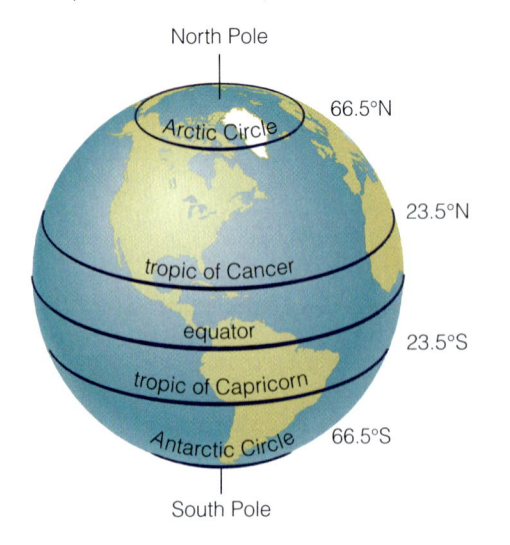

FIGURE S1.20 Daily paths of the Sun for the equinoxes and solstices at the North Pole.

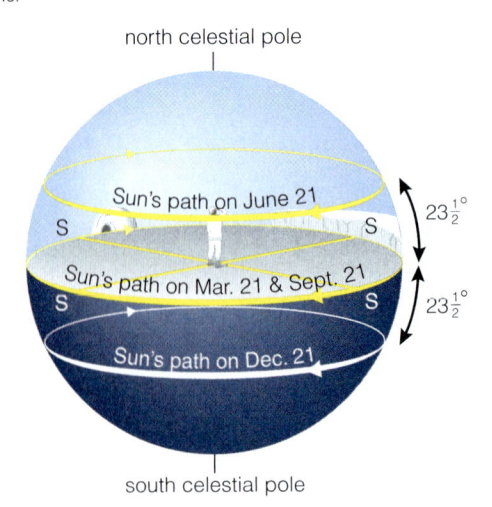

82　PART I　DEVELOPING PERSPECTIVE

Sun appear about 1° higher than it would in the absence of an atmosphere. Second, the Sun's angular size of about $\frac{1}{2}$° means that it does not fall below the horizon at a single moment but instead sets gradually. Together, these effects mean that the Sun appears above the polar horizons for slightly longer (by several days) than 6 months each year.

The Equator Recall that, at the equator, the celestial equator extends from the horizon due east, through the zenith, to the horizon due west. The Sun therefore follows this path on each equinox, reaching the zenith at local noon (Figure S1.21). Following the spring equinox, the Sun's increasing declination means that it follows a daily track that takes it gradually northward in the sky. It is farthest north on the summer solstice, when it rises $23\frac{1}{2}$° north of due east, crosses the meridian at altitude $66\frac{1}{2}$° in the north, and sets $23\frac{1}{2}$° north of due west. Over the next 6 months, it gradually tracks southward until the winter solstice, when its path is the mirror image (across the celestial equator) of its summer solstice path.

Like all objects in the equatorial sky, the Sun is always above the horizon for half a day and below it for half a day. Moreover, the Sun's track is highest in the sky on the equinoxes and lowest on the summer and winter solstices. Thus, equatorial regions, unlike regions at temperate northern and southern latitudes, do not have four seasons. Instead, equatorial regions generally have a rainy season and a dry season, with the weather determined by wind patterns on Earth. (The Sun's path in the equatorial sky also makes it rise and set perpendicular to the horizon, making for a more rapid dawn and a briefer twilight than at other latitudes.)

The Tropic Circles We've seen that, while the Sun reaches the zenith twice a year at the equator (on the spring and fall equinoxes), it never reaches the zenith at mid-latitudes (such as 40°N). The boundaries of the regions on Earth where the Sun sometimes reaches the zenith are the circles of latitude 23.5°N and 23.5°S. These latitude circles are called the **tropic of Cancer** and the **tropic of Capricorn**, respectively. (The region between these two circles is generally called the *tropics*.)

Figure S1.22 shows why the tropic of Cancer is special. The celestial equator extends from due east on the horizon to due west on the horizon, crossing the meridian in the south at an altitude of 90° − $23\frac{1}{2}$° (the latitude) = $66\frac{1}{2}$°; this is the path the Sun follows on the equinoxes (March 21st and September 21st). As a result, the Sun's path on the summer solstice, when it crosses the meridian $25\frac{1}{2}$° northward of the celestial equator, takes it to the zenith at local noon. Because the Sun has its maximum declination on the summer solstice, the tropic of Cancer marks the northernmost latitude at which the Sun ever reaches the zenith. Similarly, at the tropic of Capricorn the Sun reaches the zenith at local noon on the winter solstice, making this the southernmost latitude at which the Sun ever reaches the zenith. Between the two tropic circles, the Sun passes through the zenith twice a year (the dates vary with latitude).

Incidentally, the names of the tropics of Cancer and Capricorn come from the locations of the summer and winter solstices along the ecliptic—as they were about 2,000 years ago. Recall that the summer solstice is currently in the constellation Gemini and the winter solstice is in the constellation Sagittarius (see Figure S1.10). However, these positions gradually change with the Earth's precession cycle. Some 2,000 years ago, the summer and winter solstices

FIGURE S1.21 Daily paths of the Sun for the equinoxes and solstices at the equator.

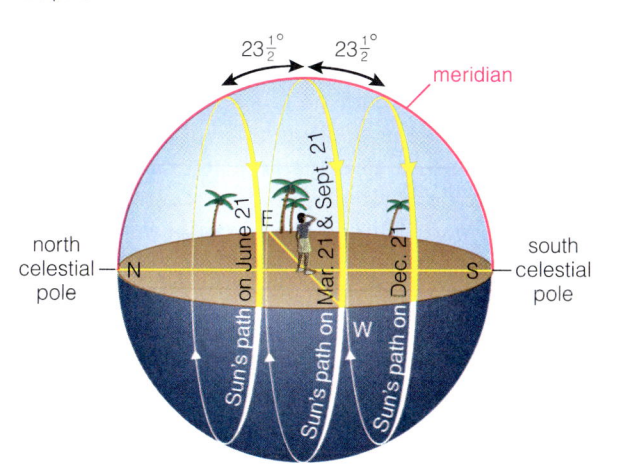

FIGURE S1.22 Daily paths of the Sun for the equinoxes and solstices at the tropic of Cancer.

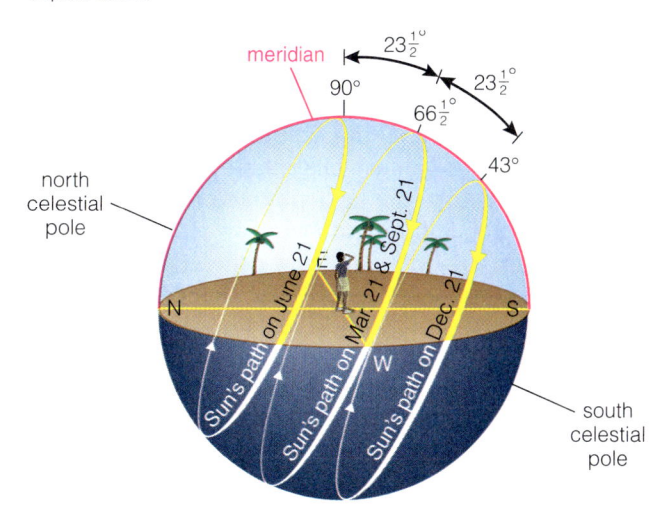

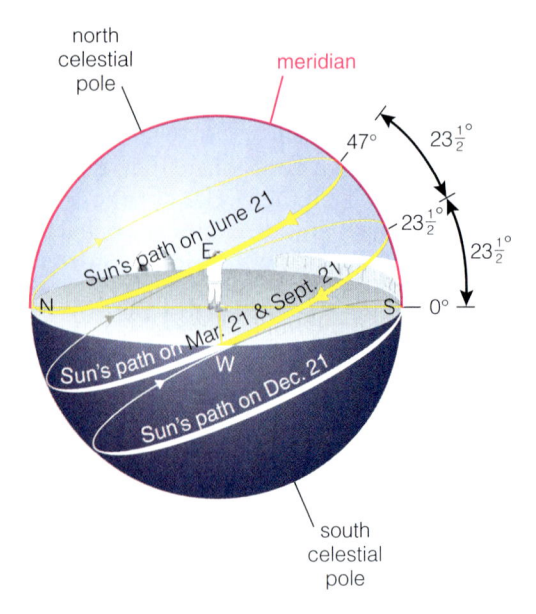

FIGURE S1.23 Daily paths of the Sun for the equinoxes and solstices at the Arctic Circle.

were located in Cancer and Capricorn, respectively, which is how the tropic circles got their names.

The Polar Circles At the equator, the Sun is above the horizon for 12 hours each day year-round. At latitudes progressively farther from the equator, the daily time that the Sun is above the horizon varies progressively more with the seasons. The special latitudes at which the Sun remains continuously above the horizon for a full day each year mark the polar circles: the **Arctic Circle** at latitude 66.5°N and the **Antarctic Circle** at latitude 66.5°S. Poleward of these circles, the length of continuous daylight (or darkness) increases beyond 24 hours, reaching the extreme of 6 months at the North and South Poles.

Figure S1.23 shows why the Arctic Circle is special. The celestial equator extends from due east on the horizon to due west on the horizon, crossing the meridian in the south at an altitude of $90° - 66\frac{1}{2}°$ (the latitude) $= 23\frac{1}{2}°$. As a result, the Sun's path is circumpolar on the summer solstice: It skims the northern horizon at midnight, rises through the eastern sky to a noon maximum altitude of 47° in the south, and then gradually falls through the western sky until it is back on the horizon at midnight (see the photograph of this path in Figure 2.17). At the Antarctic Circle, the Sun follows the same basic pattern on the winter solstice, except that it skims the horizon in the south and rises to a noon maximum altitude of 47° in the north.

However, as at the North and South Poles, what we actually see at the polar circles is slightly different from this idealization. Again, the bending of light by the Earth's atmosphere and the Sun's angular size of about $\frac{1}{2}°$ make the Sun *appear* to be above the horizon even when it is slightly below it. Thus, the Sun seems not to set for several days, rather than a single day, around the summer solstice at the Arctic Circle (or the winter solstice at the Antarctic Circle). Similarly, the Sun appears to peek above the horizon momentarily, rather than not at all, around the winter solstice at the Arctic Circle (or the summer solstice at the Antarctic Circle).

S1.6 Principles of Celestial Navigation

Imagine that you're on a ship at sea, far from any landmarks. How can you figure out where you are? It's easy, at least in principle, if you understand the apparent motions of the sky discussed in this chapter.

Latitude

Determining latitude is particularly easy if you can find the north or south celestial pole: Your latitude is equal to the altitude of the celestial pole in your sky. In the Northern Hemisphere at night, you can determine your approximate latitude by measuring the altitude of Polaris. Because Polaris has a declination within 1° of the north celestial pole, its altitude is within 1° of your latitude. For example, if Polaris has altitude 17°, your latitude is between 16°N and 18°N.

If you want to be more precise, you can determine your latitude from the altitude of *any* star as it crosses your meridian. For example, suppose Vega happens to be crossing your meridian at the moment and appears in your southern sky at altitude 78°44′. Because Vega has dec = +38°44′ (see Figure S1.12), it crosses your meridian 38°44′ north of the celestial equator. As shown in Figure S1.24a, you can conclude that the celestial equator crosses your meridian at an altitude of precisely 40° in the south. Your latitude must therefore be 50°N, because the celestial equator always crosses the meridian at an altitude of 90° minus the latitude; you know you are in the Northern Hemisphere because the celestial equator crosses the meridian in the south.

In the daytime, you can find your latitude from the Sun's altitude on your meridian, as long as you know the date and have a table that tells you the Sun's declination on that date. For example, suppose the date is March 21 and the Sun crosses your meridian at altitude 70° in the north (Figure S1.24b). Because the Sun has dec = 0° on March 21, you can conclude that the celestial equator also crosses your meridian in the north at altitude 70°. You must be in the Southern Hemisphere, because the celestial equator crosses the meridian in the north. From the rule that the celestial equator crosses the meridian at

84 PART I DEVELOPING PERSPECTIVE

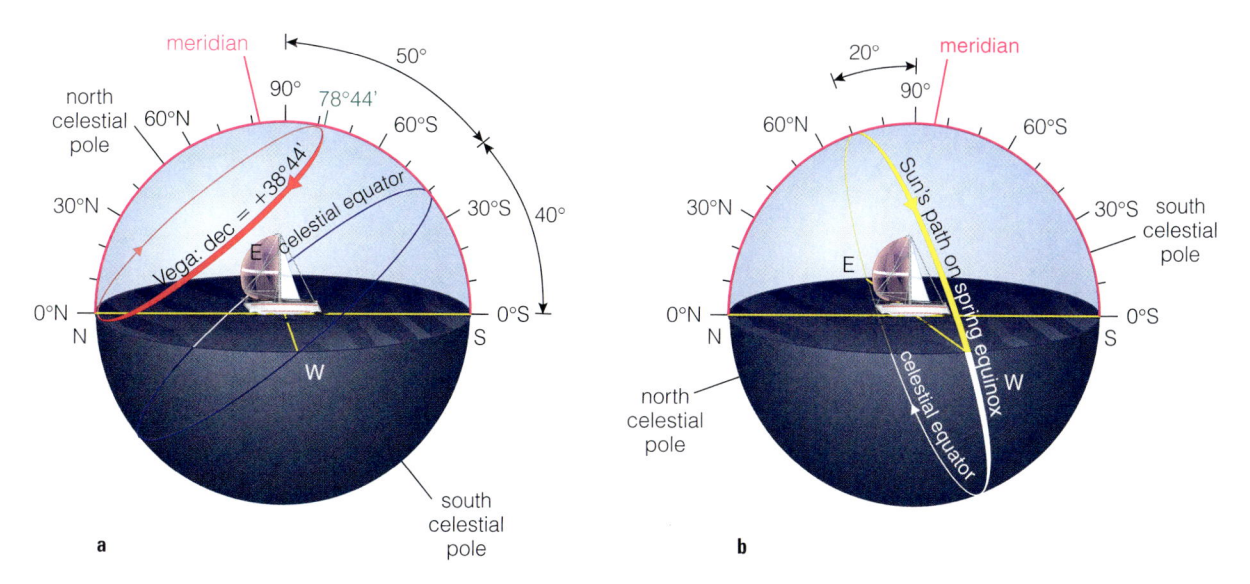

FIGURE S1.24 (a) Because Vega has dec = +38°44′, it crosses the meridian 38°44′ north of the celestial equator. From Vega's meridian crossing at altitude 78°44′ in the south, the celestial equator must cross the meridian at altitude 40° in the south. Thus, the latitude must be 50°N. (b) To determine latitude from the Sun's meridian crossing, you must know the Sun's declination, which you can determine from the date. The case shown is for the spring equinox, when the Sun's declination is 0° and hence follows the path of the celestial equator through the local sky. From the celestial equator's meridian crossing at 70° in the north, the latitude must be 20°S.

an altitude of 90° minus the latitude, you can conclude that you are at latitude 20°S.

Longitude

Determining longitude requires comparing the current positions of objects in your sky with their positions as seen from some known longitude. As a simple example, suppose you use a sundial to determine that the apparent solar time is 1:00 P.M. You immediately call a friend in England and learn that it is 3:00 P.M. in Greenwich. You now know that your local time is 2 hours earlier than the local time in Greenwich, which means you are 2 hours west of Greenwich. (An earlier time means you are *west* of Greenwich, because the Earth rotates from west to east.) Each hour corresponds to 15° of longitude, so "2 hours west of Greenwich" means longitude 30°W.

At night, you can find your longitude by comparing the positions of stars in your local sky and at some known longitude. For example, suppose Vega is on your meridian and a call to your friend reveals that it won't cross the meridian in Greenwich until 6 hours from now. In this case, your local time is 6 hours later than the local time in Greenwich. Thus, you are 6 hours east of Greenwich, or at longitude 90°E (because 6 × 15° = 90°).

Celestial Navigation in Practice

Although celestial navigation is easy in principle, at least three practical considerations make it more difficult in practice. First, finding either latitude or longitude requires a tool for measuring angles in the sky. One such device, called an *astrolabe*, was invented by the ancient Greeks and significantly improved by Islamic scholars during the Middle Ages. The astrolabe's faceplate (Figure S1.25a) could be used to tell time, because it consisted of a rotating star map and horizon plates for specific latitudes; today you can buy similar rotatable star maps, called *planispheres*.

Common Misconceptions: Compass Directions

Most people determine direction with the aid of a compass rather than the stars. However, to many people's surprise, a compass needle doesn't actually point to true geographic north. Instead, the compass needle responds to the Earth's magnetic field and points to *magnetic* north, which can be substantially different from true north. As a result, if you want to navigate precisely with a compass, you need a special map that shows local variations in the Earth's magnetic field. Such maps are available at most camping stores; however, they are not perfectly reliable because the magnetic field also varies with time. In general, celestial navigation is much more reliable than a compass for determining direction.

a The faceplate of an astrolabe; many astrolabes had sighting sticks on the back for measuring the positions of bright stars.

b A copper engraving of Italian explorer Amerigo Vespucci (for whom America was named) using an astrolabe to sight the Southern Cross. The engraving by Philip Galle, from the book *Nova Reperta*, was based on an original by Joannes Stradanus in the early 1580s.

c A woodcutting of Ptolemy holding a cross-staff (artist unknown).

d A sextant.

FIGURE S1.25 Navigational instruments

In addition, the astrolabe contained a sighting stick, on the back, that allowed users to measure the altitudes of bright stars in the sky; these measurements could be correlated against special markings under the faceplate (Figure S1.25b). Astrolabes were effective but difficult and expensive to make. As a result, medieval sailors often measured angles with a simple pair of calibrated perpendicular sticks, called a *cross-staff* or *Jacob's staff* (Figure S1.25c). A more modern device called a *sextant* allows much more precise angle determinations by incorporating a small telescope for sightings (Figure S1.25d). Sextants are still used for celestial navigation on many ships. If you want to practice celestial navigation yourself, you can buy an inexpensive plastic sextant at many science-oriented stores.

A second practical consideration is the need to know the celestial coordinates of stars and the Sun so that you can determine their paths through the local sky. At night, you can use a table listing the celestial coordinates of bright stars. In addition to knowing the celestial coordinates, you must either know the constellations and bright stars extremely well or carry star charts to help you identify them. For navigating by the Sun in the daytime, you'll need a table listing the Sun's celestial coordinates on each day of the year.

The third practical consideration is related to determining longitude: You need to know the current position of the Sun (or a particular star) in a known location, such as Greenwich, England. Although you could find this out by calling a friend who lives there, it's more practical to carry a clock set to universal time (that is, Greenwich mean time). In the daytime, the clock makes it easy to determine your longitude. If apparent solar time is 1:00 P.M. in your location and the clock tells you that it is 3:00 P.M. in Greenwich, then you are 2 hours west of Greenwich, or at longitude 30°W. The task is more difficult at night, because you must compare the position of a *star* in your sky to its current position in Greenwich. You can do this with the aid of detailed astronomical tables that allow you to determine the current position of any star in the Greenwich sky from the date and the universal time.

Historically, this third consideration created enormous problems for navigation. Before the invention of accurate clocks, sailors could easily determine their latitude but not their longitude. Indeed, most of the European voyages of discovery beginning in the 1400s relied on little more than guesswork about longitude, although some sailors learned complex mathematical techniques for estimating longitude through observations of the lunar phases. More accurate longitude determination, upon which the development of extensive ocean commerce and travel depended, required the invention of a clock that would remain accurate on a ship rocking in the ocean swells. By the early 1700s, solving this problem was considered so important that the British government offered a substantial monetary prize for the solution. The prize was claimed in 1761 by John Harrison, with a clock that lost only 5 seconds during a 9-week voyage to Jamaica.[2]

The Global Positioning System

In the past decade, a new type of celestial navigation has supplanted traditional methods. It involves finding positions relative to a set of satellites in Earth orbit—the satellites of the **global positioning system (GPS)**. In essence, the GPS satellites function like artificial stars; the satellite positions at any moment are known precisely from their orbital characteristics. The GPS currently involves about two dozen satellites orbiting the Earth at an altitude of 20,000 kilometers. Each satellite transmits a radio signal that can be received by a small radio receiver—rain or shine, day or night. GPS receivers have a built-in computer that calculates your precise position on Earth by comparing the signals received from several GPS satellites.

The United States originally built the GPS in the late 1970s to give its military troops an advantage over adversaries, and the system was designed so that secret information would be needed to use the system to its fullest. Thus, while civilians (and adversaries) could use the GPS to determine their location on Earth to within about 100 meters, U.S. military personnel could pinpoint their location to within 1 meter. However, civilian scientists eventually found ways to measure position with far more precision than the designers of the GPS had imagined possible, even without the secret information. For example, the GPS has been used by geologists to measure *millimeter*-scale changes in the Earth's crust. Today, the many applications of the GPS include automobile navigation systems as well as systems for helping airplanes land safely, guiding the blind around town, and helping lost hikers find their way.

With the rapid growth in the use of GPS navigation, the ancient practice of celestial navigation is in danger of becoming a lost art. Fortunately, many amateur clubs and societies are keeping the art of celestial navigation alive.

THE BIG PICTURE

In this chapter, we built upon concepts from Chapters 1 and 2 to form a more detailed understanding of celestial timekeeping and navigation. We also learned how to determine paths for the Sun and the stars in the local sky. As you look back at what you've learned, keep in mind the following "big picture" ideas:

- Our modern systems of timekeeping are rooted in the apparent motions of the Sun through the sky. Although it's easy to forget these roots when you look at a clock or a calendar, the sky was the only guide to time for most of human history.

- The term *celestial navigation* sounds a bit mysterious, but it involves simple principles that allow you to determine your location on Earth. Even if you're never lost at sea, you may find the basic techniques of celestial navigation useful to orient yourself at night (for example, on your next camping trip).

- If you understand the apparent motions of the sky discussed in this chapter and also learn the constellations and bright stars, you'll feel very much "at home" under the stars at night.

---

[2]The story of the difficulties surrounding the measurement of longitude at sea and how the problem was finally solved by Harrison is chronicled in Dava Sobel, *Longitude* (Walker and Company, 1995).

Review Questions

1. Briefly explain the difference between a *solar day* and a *sidereal day*.

2. Briefly explain the difference between a *synodic month* and a *sidereal* month.

3. Why is the *tropical year* slightly shorter than the *sidereal year*?

4. What is the difference between a planet's *sidereal period* and its *synodic period*? Explain the meaning of *conjunction, opposition,* and *greatest elongation* for planetary orbits viewed from Earth.

5. What is *apparent solar time*? Why is it different from *mean solar time*?

6. Why did it become necessary to create time zones in the late 1800s? Explain how local mean solar time is related to *standard time* and *daylight saving time*.

7. What is *universal time (UT)*? Why is it useful?

8. Describe the origins of the *Julian* and *Gregorian* calendars. What is the rule for leap years on the Gregorian calendar?

9. Explain what we mean when we say that the equinoxes and solstices are both *moments* in time and *points* on the celestial sphere.

10. What are *declination* and *right ascension*? How are these *celestial coordinates* similar to latitude and longitude on Earth? How are they different?

11. Briefly describe how the Sun's celestial coordinates change over the course of each year.

12. Suppose you are standing at the North Pole. Where is the celestial equator in your sky? Where is the north celestial pole? Describe the daily motion of the sky.

13. Repeat question 12 for the equator.

14. Repeat question 12 for latitude 40°N.

15. Describe the Sun's paths through the local sky for latitude 40°N on the equinoxes and on the solstices.

16. Repeat question 15 for the North Pole.

17. Repeat question 15 for the Earth's equator.

18. Repeat question 15 for the *tropic of Cancer*. What makes the tropic circles special?

19. Repeat question 15 for the *Arctic Circle*. What makes the polar circles special?

20. Briefly describe how you can find your latitude by finding the north or south celestial pole.

21. Briefly describe how you can find your latitude from the meridian altitude of any star or the Sun.

22. Briefly describe how you can find your longitude from the position of the Sun and a clock set to universal time.

Discussion Questions

1. *Northern Chauvinism.* Why is the solstice in June called the *summer solstice,* when it marks winter for places like Australia, New Zealand, and South Africa? Why is the writing on maps and globes usually oriented so that the Northern Hemisphere is at the top, even though there is no up or down in space? Discuss.

2. *When Do Seasons Begin?* It is usually said that the moments of the equinoxes and solstices mark the beginnings of the seasons. For example, it is said that the summer solstice marks the beginning of summer. Based on the annual progression of the Sun's celestial coordinates, do you think it is accurate to say that the equinoxes and solstices mark beginnings? Why or why not?

3. *Seasonal Weather.* The summer solstice occurs in June, but the months of July and August generally have warmer temperatures in the Northern Hemisphere. Similarly, the winter solstice occurs in December, but the months of January and February generally have cooler temperatures in the Northern Hemisphere. Why do you think this is the case? (*Hint:* Look at the distribution of land and oceans on a globe.)

Problems

Sensible Statements? For **problems 1–12**, decide whether the statement is sensible and explain why it is or is not. (*Hint:* For statements that involve coordinates—such as altitude, longitude, or declination—check whether the correct coordinates are used for the situation; for example, it is not sensible to describe a location on Earth by an altitude, since altitude makes sense only for positions in the local sky.)

1. Last night I saw Venus shining brightly on the meridian at midnight.

2. The apparent solar time was noon, but the Sun was just setting.

3. My mean solar clock said it was 2:00 P.M., but my friend who lives east of here had a mean solar clock that said it was 2:11 P.M.

4. When the standard time is 3:00 P.M. in Baltimore, it is 3:15 P.M. in Washington, D.C.

5. The Julian calendar differed from the Gregorian calendar because it was based on the sidereal year.

6. Last night around 8:00 P.M. I saw Jupiter at an altitude of 45° in the south.

7. The latitude of the stars in Orion's belt is about 5°N.

8. Today the Sun is at an altitude of 10° on the celestial sphere.

9. Los Angeles is west of New York by about 3 hours of right ascension.

10. The summer solstice is east of the vernal equinox by 6 hours of right ascension.

11. If it were being named today, the tropic of Cancer would probably be called the tropic of Gemini.

12. Even though my UT clock had stopped, I was able to find my longitude by measuring the altitudes of 14 different stars in my local sky.

13. *Opposite Rotation.* Suppose the Earth rotated in the opposite direction from its revolution; that is, suppose it rotated clockwise (as seen from above the North Pole) every 24 hours while revolving counterclockwise around the Sun each year. Would the solar day still be longer than the sidereal day? Explain.

14. *Fundamentals of Your Local Sky.* Answer each of the following for *your* latitude.
    a. Where is the north (or south) celestial pole in your sky?
    b. Describe the location of the meridian in your sky; be sure to specify its shape and at least three distinct points along it (such as the points at which it meets your horizon and its highest point).
    c. Describe the location of the celestial equator in your sky; be sure to specify its shape and at least three distinct points along it (such as the points at which it meets your horizon and crosses your meridian).
    d. Does the Sun ever appear at your zenith? If so, when? If not, why not?
    e. What is the range of declinations that makes a star circumpolar in your sky? Explain.
    f. What is the range of declinations for stars that you can never see in your sky? Explain.

15. *Sydney Sky.* Repeat problem 14 for the local sky in Sydney, Australia (latitude 34°S).

16. *Path of the Sun in Your Sky.* Describe the path of the Sun through your local sky for each of the following days.
    a. The spring and fall equinoxes.
    b. The summer solstice.
    c. The winter solstice.
    d. Today. (*Hint:* Estimate the right ascension and declination of the Sun for today's date by using the data in Table S1.1.)

17. *Sydney Sun.* Repeat problem 16 for the local sky in Sydney, Australia (latitude 34°S).

18. *Lost at Sea I.* During an upcoming vacation, you decide to take a solo boat trip. While contemplating the universe, you lose track of your location. Fortunately, you have some astronomical tables and instruments, as well as a UT clock. You thereby put together the following description of your situation:
    - It is the spring equinox.
    - The Sun is on your meridian at altitude 75° in the south.
    - The UT clock reads 22:00.
    a. What is your latitude? How do you know?
    b. What is your longitude? How do you know?
    c. Consult a map. Based on your position, where is the nearest land? Which way should you sail to reach it?

19. *Lost at Sea II.* Repeat problem 18, based on the following description of your situation:
    - It is the day of the summer solstice.
    - The Sun is on your meridian at altitude $67\frac{1}{2}°$ in the north.
    - The UT clock reads 06:00.

20. *Lost at Sea III.* Repeat problem 18, based on the following description of your situation:
    - Your local time is midnight.
    - Polaris appears at altitude 67° in the north.
    - The UT clock reads 01:00.

21. *Lost at Sea IV.* Repeat problem 18, based on the following description of your situation:
    - Your local time is 6 A.M.
    - From the position of the Southern Cross, you estimate that the south celestial pole is at altitude 33° in the south.
    - The UT clock reads 11:00.

*22. *Sidereal Time.*
    a. Suppose it is 4 P.M. on the spring equinox. What is the local sidereal time?
    b. Suppose the local sidereal time is 19:30. When will Vega cross your meridian?
    c. You observe a star that has an hour angle of +3 hours ($+3^h$) when the local sidereal time is 8:15. What is the star's right ascension?

Web Projects

Find useful links for Web projects on the text Web site.

1. *Sundials.* Although they are no longer necessary for timekeeping, sundials remain popular for their cultural and artistic value. Search the Web for pictures and information about interesting sundials around the world. Write a short report about at least three sundials that you find particularly interesting.

2. *The Analemma.* Learn more about the analemma and its uses from information available on the Web. Write a short report on your findings.

3. *Calendar History.* Investigate the history of the Julian or Gregorian calendar in greater detail. Write a short summary of some interesting aspect of the history you learn from your Web research. (For example, why did Julius Caesar allow one year to have 445 days? How did our months end up with 28, 30, or 31 days?)

4. *Global Positioning System.* Learn more about the global positioning system and its uses. Write a short report summarizing how you think the GPS will affect our lives over the next 10 years.

PART II
Key Concepts for Astronomy

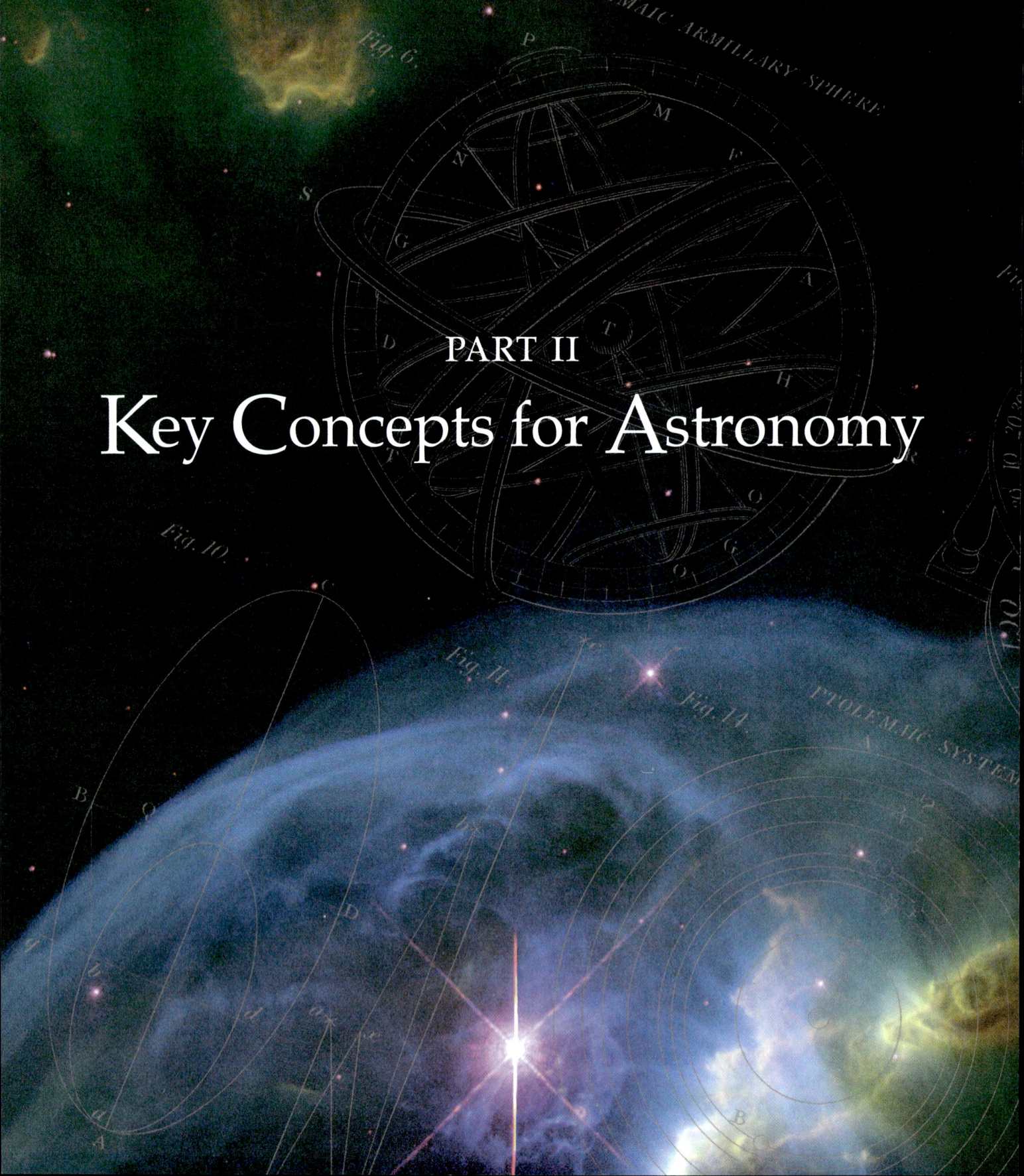

We especially need imagination in science.

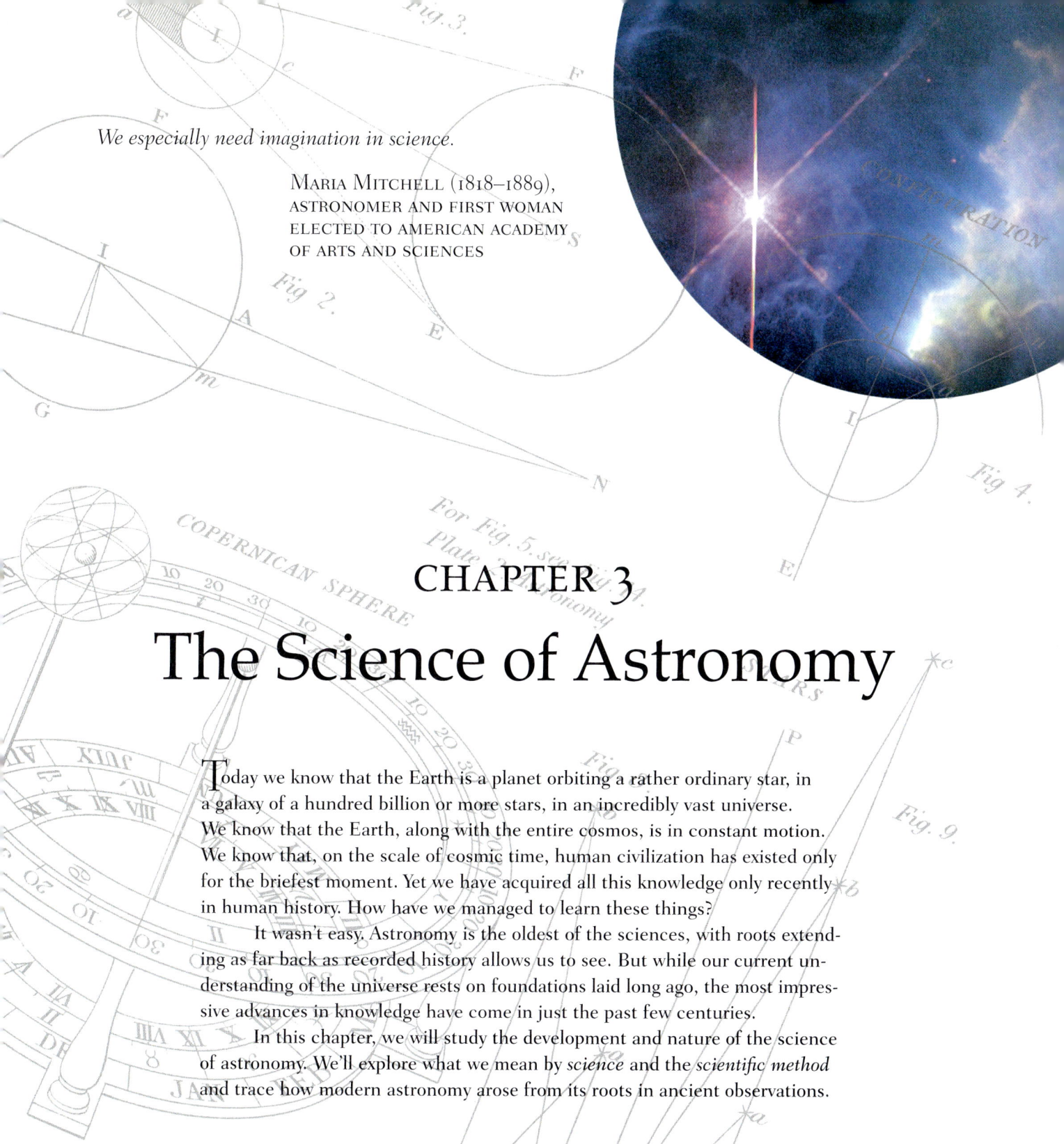

MARIA MITCHELL (1818–1889),
ASTRONOMER AND FIRST WOMAN
ELECTED TO AMERICAN ACADEMY
OF ARTS AND SCIENCES

CHAPTER 3

The Science of Astronomy

Today we know that the Earth is a planet orbiting a rather ordinary star, in a galaxy of a hundred billion or more stars, in an incredibly vast universe. We know that the Earth, along with the entire cosmos, is in constant motion. We know that, on the scale of cosmic time, human civilization has existed only for the briefest moment. Yet we have acquired all this knowledge only recently in human history. How have we managed to learn these things?

It wasn't easy. Astronomy is the oldest of the sciences, with roots extending as far back as recorded history allows us to see. But while our current understanding of the universe rests on foundations laid long ago, the most impressive advances in knowledge have come in just the past few centuries.

In this chapter, we will study the development and nature of the science of astronomy. We'll explore what we mean by *science* and the *scientific method* and trace how modern astronomy arose from its roots in ancient observations.

3.1 Everyday Science

A common stereotype holds that scientists are men and women in white lab coats who somehow think differently than other people. In reality, scientific thinking is a fundamental part of human nature.

Think about how a baby behaves. By about a year of age, she notices that objects fall to the ground when she drops them. She lets go of a ball; it falls. She pushes a plate of food from her high chair; it falls too. She continues to drop all kinds of objects, and they all plummet to Earth. Through powers of observation, the baby learns about the physical world: Things fall when they are unsupported. Eventually, she becomes so certain of this fact that, to her parents' delight, she no longer needs to test it continually.

One day somebody gives the baby a helium balloon. She releases it; to her surprise, it rises to the ceiling! Her rudimentary understanding of physics must be revised. She now knows that the principle "all things fall" does not represent the whole truth, although it still serves her quite well in most situations. It will be years before she learns enough about the atmosphere, the force of gravity, and the concept of density to understand *why* the balloon rises when most other objects fall. For now, she is delighted to observe something new and unexpected.

The baby's experience with falling objects and balloons exemplifies scientific thinking, which, in essence, is a way of learning about nature through careful observation and trial-and-error experiments. Rather than thinking differently than other people, modern scientists simply are trained to organize this everyday thinking in a way that makes it easier for them to share their discoveries and employ their collective wisdom. This organizational scheme is what we call the *scientific method*; it is the method by which humanity has acquired its present physical knowledge of the universe.

TIME OUT TO THINK *When was the last time you used trial and error to learn something? Describe a few cases where you have learned by trial and error in cooking, participating in sports, fixing something, or any other situation.*

Just as it takes years for a child to learn to communicate through language, art, or music, it took humanity a long time to develop the principles of the scientific method. In its modern form, the scientific method requires painstaking attention to detail, relentless testing of each piece of information to ensure its reliability, and a willingness to give up old beliefs that are not consistent with observed facts about the physical world. For professional scientists, the demands of the scientific method are the "hard work" part of the job. At heart, professional scientists are like the baby with the balloon, delighted by the unexpected and motivated by those rare moments when they—and all of us—learn something new about the universe.

3.2 Ancient Observations

We will discuss the modern scientific method shortly, but first we will explore how it arose from the observations of ancient peoples. Our exploration begins in central Africa, where people of many indigenous societies predict the weather with reasonable accuracy

FIGURE 3.1 This diagram shows how central Africans used the Moon to predict the weather. The graph depicts the annual rainfall pattern in central Nigeria, characterized by a wet season and a dry season. The Moon diagrams represent the orientation of a waxing crescent moon relative to the western horizon at different times of year. The angle of the crescent "horns" allows observers to determine the time of year and hence the expected rainfall. (The orientation is measured in degrees, where 0° means that the crescent horns are parallel to the horizon.)

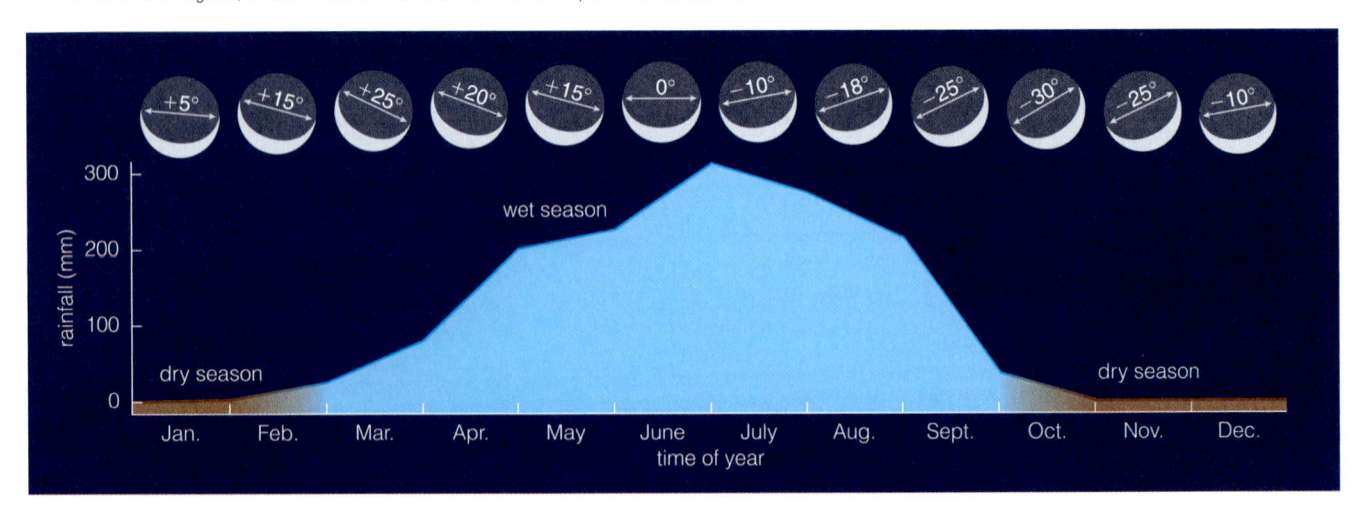

by making careful observations of the Moon. The Moon begins its monthly cycle as a crescent in the western sky just after sunset. Through long traditions of sky watching, central African societies learned that the orientation of the crescent "horns" relative to the horizon is closely tied to rainfall patterns (Figure 3.1).

No one knows when central Africans first developed the ability to predict weather using the lunar crescent. But the earliest-known written astronomical record comes from central Africa near the border of modern-day Congo and Uganda. It consists of an animal bone (known as the *Ishango* bone) etched with patterns that appear to be part of a lunar calendar, probably carved around 6500 B.C.

Why did ancient people bother to make such careful and detailed observations of the sky? In part, it was probably because they were just as curious and intelligent as we are today. In the daytime, they surely recognized the importance of the Sun to their lives. At night, without electric light, they were much more aware of the starry sky than we are today. Thus, it's not surprising that they paid attention to patterns of motion in the sky and developed ideas and stories to explain what they saw.

Astronomy played a practical role in ancient societies by enabling them to keep track of time and seasons, a crucial skill for people who depended on agriculture for survival. This ability may seem quaint today, when digital watches tell us the precise time and date, but it required considerable knowledge and skill in ancient times, when the only clocks and calendars were in the sky.

Modern measures of time still reflect their ancient astronomical roots. Our 24-hour day is the time it takes the Sun to circle our sky. The length of a month comes from the lunar cycle, and our calendar year is based on the cycle of the seasons. The days of the week are named after the seven naked-eye objects that appear to move among the constellations: the Sun, the Moon, and the five planets recognized in ancient times (Table 3.1).

FIGURE 3.2 This ancient Egyptian obelisk, which stands 83 feet tall and weighs 331 tons, resides in St. Peter's Square at the Vatican in Rome. It is one of 21 surviving obelisks from ancient Egypt, most of which are now scattered around the world. Shadows cast by the obelisks may have been used to tell time.

Determining the Time of Day

In the daytime, ancient peoples could tell time by observing the Sun's path through the sky. Many cultures probably used the shadows cast by sticks as simple sundials [Section S1.2]. The ancient Egyptians built huge obelisks, often inscribed or decorated in homage to the Sun, that probably also served as simple clocks (Figure 3.2).

At night, the Moon's position and phase give an indication of the time (see Figure 2.19). For example, a first-quarter moon sets around midnight, so it is not yet midnight if the first-quarter moon is still above the western horizon. The positions of the stars also indicate the time if you know the approximate date. For example, in December the constellation Orion rises around sunset, reaches the meridian around midnight, and sets around sunrise. Hence, if it is winter and Orion is setting, dawn must be approaching. Most ancient peoples probably were adept at estimating the time of night, although written evidence is sparse.

Our modern system of dividing the day into 24 hours arose in ancient Egypt some 4,000 years ago. The Egyptians divided the daylight into 12 equal parts,

Table 3.1 The Seven Days of the Week and the Astronomical Objects They Honor
The correspondence between objects and days is easy to see in French and Spanish. In English, the correspondence becomes clear when we look at the names of the objects used by the Teutonic tribes who lived in the region of modern-day Germany.

Object	Teutonic Name	English	French	Spanish
Sun	Sun	Sunday	dimanche	domingo
Moon	Moon	Monday	lundi	lunes
Mars	Tiw	Tuesday	mardi	martes
Mercury	Woden	Wednesday	mercredi	miércoles
Jupiter	Thor	Thursday	jeudi	jueves
Venus	Fria	Friday	vendredi	viernes
Saturn	Saturn	Saturday	samedi	sábado

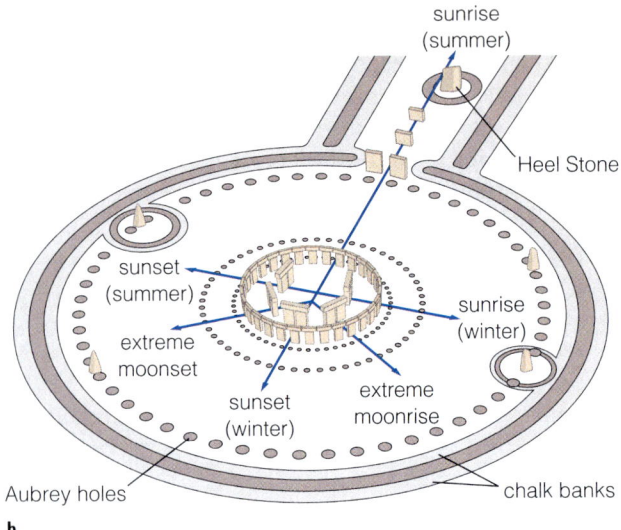

FIGURE 3.3 (**a**) Stonehenge today. (**b**) A sketch showing how archaeologists believe Stonehenge looked when construction was completed around 1550 B.C. Note, for example, that observers standing in the center would see the Sun rise directly over the Heel Stone on the summer solstice.

and we still break the 24-hour day into 12 hours each of a.m. and p.m.

The Egyptian "hours" were not a fixed amount of time because the amount of daylight varies during the year. For example, "summer hours" were longer than "winter hours," because one-twelfth of the daylight lasts longer in summer than in winter. Only much later in history did the hour become a fixed amount of time, subdivided into 60 equal minutes, each consisting of 60 equal seconds.

The Egyptians also divided the night into 12 equal parts, and early Egyptians used the stars to determine the time at night. Egyptian *star clocks,* often found painted on the coffin lids of Egyptian pharaohs, essentially cataloged where particular stars appear in the sky at particular times of night and particular times of year. By knowing the date from their calendar and observing the positions of particular stars in the sky, the Egyptians could use the star clocks to estimate the time of night.

By about 1500 B.C., Egyptians had abandoned star clocks in favor of clocks that measure time by the flow of water through an opening of a particular size, just as hourglasses measure time by the flow of sand through a narrow neck.[1] These *water clocks* had the advantage of being useful even when the sky was cloudy, and they eventually became the primary timekeeping instruments for many cultures, including the Greeks, Romans, and Chinese. The water clocks, in turn, were replaced by mechanical clocks in the 1600s and by electronic clocks in the twentieth century. Despite the availability of other types of clocks, sundials remained in use throughout ancient times and are still popular today both for their decorative value and as reminders that the Sun and stars once were our only guides to time.

Determining the Time of Year

Many cultures built structures to help them mark the seasons. One of the oldest standing human-made structures served such a purpose: Stonehenge in southern England, which was constructed in stages from about 2750 B.C. to about 1550 B.C. (Figure 3.3). Observers standing in its center see the Sun rise directly over the Heel Stone only on the summer solstice. Stonehenge also served as a social gathering place and probably as a religious site; no one knows whether its original purpose was social or astronomical. Perhaps it was built for both—in ancient times, social rituals and practical astronomy probably were deeply intertwined.

One of the most spectacular structures used to mark the seasons was the Templo Mayor in the Aztec city of Tenochtitlán, located on the site of modern-day Mexico City (Figure 3.4). Twin temples surmounted a flat-topped, 150-foot-high pyramid. From the location of a royal observer watching from the opposite side of the plaza, the Sun rose directly through the notch between the twin temples on the equinoxes. Like Stonehenge, the Templo Mayor served important social and religious functions in addition to its astronomical role. Before it was destroyed by the

---

[1] Hourglasses using sand were not invented until about the eighth century A.D., long after the advent of water clocks. Natural sand grains vary in size, so making accurate hourglasses required technology for getting uniform grains of sand.

FIGURE 3.4 This scale model shows the Templo Mayor and the surrounding plaza as they are thought to have looked before Aztec civilization was destroyed by the Conquistadors.

Conquistadors, other Spanish visitors reported stories of elaborate rituals, sometimes including human sacrifice, that took place at the Templo Mayor at times determined by astronomical observations.

Many ancient cultures aligned their buildings with the cardinal directions (north, south, east, and west), enabling them to mark the rising and setting of the Sun relative to the building orientation. Some even created monuments that had a single special astronomical purpose. Perhaps in a practical form of ancient art, someone among the ancient Anasazi people carved a spiral known as the Sun Dagger on a vertical cliff face near the top of a butte in Chaco Canyon, New Mexico (Figure 3.5). The Sun's rays form a dagger of sunlight that pierces the center of the carved spiral only once each year—at noon on the summer solstice.

FIGURE 3.5 (**a**) The Sun Dagger is an arrangement of rocks that shape the Sun's light into a dagger, with a spiral carved on a rock face behind them. The dagger pierces the center of the carved spiral each year at noon on the summer solstice. (**b**) The Sun Dagger is located on a vertical cliff face high on this butte in New Mexico.

Lunar Cycles

Many ancient civilizations paid particular attention to the lunar cycle, often using it as the basis for lunar calendars. The months on lunar calendars generally have either 29 or 30 days, chosen to make the average agree with the approximately $29\frac{1}{2}$-day lunar cycle. A 12-month lunar calendar has only 354 or 355 days, or about 11 days fewer than a calendar based on the tropical year. Such a calendar is still used in the Muslim religion, which is why the month-long fast of Ramadan (the ninth month) begins about 11 days earlier with each subsequent year.

Other lunar calendars take advantage of the fact that 19 years is almost precisely 235 lunar months. That is, every 19 years we get the same lunar phases on about the same dates. Because this fact was discovered in 432 B.C. by the Greek astronomer Meton, the 19-year period is called the **Metonic cycle**. A lunar calendar can be synchronized to the Metonic cycle by adding a thirteenth month to 7 of every 19 years (making exactly 235 months in each 19-year period), thereby ensuring that "new year" comes on approximately the same date every nineteenth year. The Jewish calendar follows the Metonic cycle, adding a thirteenth month in the third, sixth, eighth, eleventh, fourteenth, seventeenth, and nineteenth years of each cycle. This also explains why the date of Easter changes each year: The New Testament ties the date of Easter to the Jewish festival of Passover, which has its date set by the Jewish lunar calendar. In a slight modification of the original scheme, most Western Christians now celebrate Easter on *the first Sunday after the first full moon after March 21*; if the full moon falls on Sunday, Easter is the following Sunday. (Eastern Orthodox churches calculate the date of Easter differently, because they base the date on the Julian rather than the Gregorian calendar [Section S1.3].)

Some ancient cultures learned to predict eclipses by recognizing the 18-year saros cycle [Section 2.3]. In the Middle East, the ancient Babylonians achieved remarkable success in predicting eclipses more than 2,500 years ago. Perhaps the most successful eclipse predictions prior to modern times were made by the Mayans in Central America. The Mayan calendar featured a sacred cycle that almost certainly was related to eclipses. This Mayan cycle, called the *sacred round*, lasted 260 days—almost exactly $1\frac{1}{2}$ times the 173.32 days between successive eclipse seasons [Section 2.3]. Unfortunately, we know little more about the extent of Mayan knowledge, because the Spanish Conquistadors burned most Mayan writings.

The complexity of the Moon's orbit leads to other long-term patterns in the Moon's appearance. For example, the full moon rises at its most southerly point along the eastern horizon only once every 18.6 years, a phenomenon that may have been observed from the 4,000-year-old sacred stone circle at Callanish, Scotland (Figure 3.6).

Mathematical Insight 3.1 The Metonic Cycle

The Metonic cycle can be used to keep lunar calendars fairly closely synchronized with solar calendars, but only if the lunar calendar has 7 extra lunar months in every 19 years. To see why, recall that each cycle of lunar phases takes an average of 29.53 days (the synodic month [Section S1.1]). A 12-month lunar calendar has $12 \times 29.53 = 354.36$ days, or about 11 days less than our 365-day solar calendar. If the new year on a 12-month lunar calendar occurs this year on October 1, next year it will occur 11 days earlier, on September 20; the following year it will occur 11 days before that, on September 9; and so on. Without using the Metonic cycle, over many years the lunar new year would move all the way around the solar calendar.

The Metonic cycle is a period of 19 years on a solar calendar, which is almost precisely 235 months on a lunar calendar. To verify this fact, note that 19 solar years is equivalent to:

$$19 \text{ yr} \times \frac{365.25 \text{ days}}{1 \text{ yr}} = 6{,}939.75 \text{ days}$$

Similarly, 235 lunar months is:

$$235 \text{ months} \times \frac{29.53 \text{ days}}{1 \text{ month}} = 6{,}939.55 \text{ days}$$

Notice that the difference between 19 solar years and 235 lunar months is only about 0.2 day, or about 5 hours. This near equivalence means that the dates of lunar phases repeat with each Metonic cycle. For example, there was a new moon on October 1, 1998, and we will have a new moon on October 1, 2017. (In general, the dates may be off by 1 day because of the 5-hour "error" in the Metonic cycle and also because 19 years on our calendar may be either 6,939 days or 6,940 days, depending on the leap-year cycle.)

For a lunar calendar to remain roughly synchronized with a solar calendar, it must have exactly 235 months in each 19-year period. Because a 12-month lunar calendar has a total of only $19 \times 12 = 228$ months in 19 years, the lunar calendar needs 7 extra months to reach the required 235 months. One way to accomplish this is to have 7 years with a thirteenth lunar month in every 19-year cycle, as is the case for the Jewish calendar.

FIGURE 3.6 The Moon rising between two stones of the 4,000-year-old sacred stone circle at Callanish, Scotland (on the Isle of Lewis in the Scottish Hebrides). The full moon rises in this position only once every 18.6 years.

Observations of Planets and Stars

Many cultures also made careful observations of the planets and stars. Mayan observatories in Central America, such as the one still standing at Chichén Itzá (Figure 3.7), had windows strategically placed for observations of Venus.

More than 2,000 years ago, early Central American people marked the beginning of their year when the group of stars called the Pleiades first rose in the east after having been hidden from view for 40 days by the glare of the Sun. On the other side of the world, southern Greeks also used the Pleiades to time their planting and harvests, as described in this excerpt from an epic poem (*Works and Days*) composed by the Greek farmer Hesiod in about 800 B.C.:

> *When you notice the daughters of Atlas, the Pleiades, rising, start on your reaping, and on your sowing when they are setting. They are hidden from your view for a period of forty full days, both night and day, but then once again, as the year moves round, they reappear at the time for you to be sharpening your sickle.*

Observational aids could also mark the rising and setting of the stars. More than 800 lines, some stretching for miles, are etched in the dry desert sand of Peru between the Ingenio and Nazca Rivers. Many of these lines may simply have been well-traveled pathways, but others are aligned in directions that point to places where bright stars or the Sun rose at particular times of year. In addition to the many straight lines, the desert features many

FIGURE 3.7 The ruins of the Mayan observatory at Chichén Itzá.

CHAPTER 3 THE SCIENCE OF ASTRONOMY   97

large figures of animals, which may be representations of constellations made by the Incas who lived in the region (Figure 3.8).

TIME OUT TO THINK *The animal figures show up clearly only when seen from above. As a result, some UFO enthusiasts argue that the patterns must have been created by aliens. What do you think of this argument? Defend your opinion.*

The great Incan cities of South America clearly were built with astronomy in mind. Entire cities appear to have been designed so that particular arrangements of roads, buildings, or other human-made structures would point to places where bright stars rose or set, or where the Sun rose and set at particular times of year.[2]

Structures for astronomical observation also were popular in North America. Lodges built by the Pawnee people in Kansas featured strategically placed holes for observing the passage of constellations that figured prominently in their folklore. In the northern plains of the United States, Native American Medicine Wheels probably were designed for astronomical observations. The "spokes" of the Medicine Wheel at Big Horn, Wyoming, were aligned with the rising and setting of bright stars, as well as with the rising and setting of the Sun on the equinoxes and solstices (Figure 3.9). The 28 spokes of the Medicine Wheel probably relate to the month of the Native Americans, which they measured as 28 days (rather than 29 or 30 days) because they did not count the day of the new moon.

Perhaps the people most dependent on knowledge of the stars were the Polynesians, who lived and

FIGURE 3.8 Hundreds of lines and patterns are etched in the sand of the Nazca desert in Peru. This aerial photo shows a large etched figure of a hummingbird.

[2]For more details about the layout of Incan cities and other ancient astronomy discussed in this chapter, see Anthony F. Aveni, *Ancient Astronomers* (Smithsonian Books, St. Remy Press and Smithsonian Institution, 1993).

FIGURE 3.9 Aerial and ground views of the Big Horn Medicine Wheel in Wyoming. Note the 28 "spokes" radiating out from the center. These spokes probably relate to the month of the Native Americans.

98 PART II KEY CONCEPTS FOR ASTRONOMY

FIGURE 3.10 A traditional Polynesian navigational instrument.

traveled among the many islands of the mid- and South Pacific. Because the next island in a journey usually was too distant to be seen, poor navigation meant becoming lost at sea. As a result, the most esteemed position in Polynesian culture was that of the Navigator, a person who had acquired the detailed knowledge necessary to navigate great distances among the islands. The Navigators employed a combination of detailed knowledge of astronomy and equally impressive knowledge of the patterns of waves and swells around different islands (Figure 3.10). The stars provided their broad navigational sense, pointing them in the correct direction of their intended destination. As they neared a destination, the wave and swell patterns guided them to their precise landing point. The Navigator memorized all his skills and passed them to the next generation through a well-developed program for training future Navigators. Unfortunately, with the advent of modern navigational technology, many of the skills of the Navigators have been lost.

From Observation to Science

Before a structure such as Stonehenge could be built, careful observations had to be made and repeated over and over to ensure their validity. Careful, repeatable observations also underlie the modern scientific method. To this extent, elements of modern science were present in many early human cultures.

The degree to which scientific ideas developed in different societies depended on practical needs, social and political customs, and interactions with other cultures. Because all of these factors can change, it should not be surprising that different cultures were more scientifically or technologically advanced than others at different times in history. The ancient Chinese kept remarkably detailed records of astronomical observations beginning at least 5,000 years ago (Figure 3.11).

Yet, by the end of the 1600s, Chinese science and technology had fallen behind that of Europe. Some historians argue that a primary reason for this decline was that the Chinese tended to regard their science and technology as state secrets. This secrecy may have slowed Chinese scientific development

FIGURE 3.11 This photo shows a model of the celestial sphere and other instruments on the roof of the ancient astronomical observatory in Beijing. The observatory was built in the 1400s; the instruments shown here were built later and show the influence of Jesuit missionaries.

CHAPTER 3 THE SCIENCE OF ASTRONOMY 99

FIGURE 3.12 Map of the Middle East in ancient times. Mesopotamia was the ancient Greek name for the region between the Tigris and Euphrates Rivers, but modern historians use the name for the entire region today occupied by Iraq. Three major cultures thrived at various times in ancient Mesopotamia: Sumer, Babylonia, and Assyria.

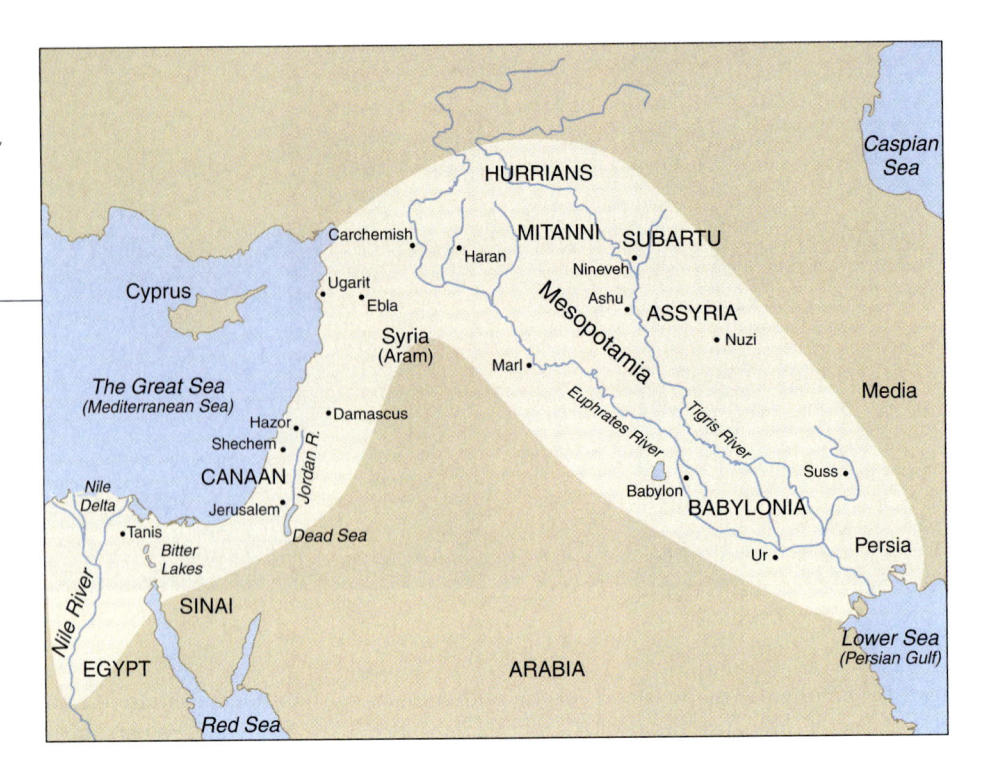

by preventing the broad-based collaborative science that fueled the European advance beginning in the Renaissance.

In Central America, the ancient Mayans also were ahead of their time in many ways. For example, their system of numbers and mathematics looks distinctly modern; their invention of the concept of zero preceded by some 500 years its introduction in the Eurasian world (by Hindu mathematicians, around A.D. 600). The Aztecs, Incas, Anasazi, and other ancient peoples of the Americas may have been quite advanced in many other areas as well, but few written records survive to tell the tale.

It appears that virtually all cultures employed scientific thinking to varying degrees. Had the circumstances of history been different, any one of these many cultures might have been the first to develop what we consider to be modern science. But, in the end, history takes only one of countless possible paths, and the path that led to modern science emerged from the ancient civilizations of the Mediterranean and the Middle East.

3.3 The Modern Lineage

By 3000 B.C., civilization was well established in two major regions of the Middle East: Egypt and Mesopotamia (Figure 3.12). Their geographical location placed these civilizations at a crossroads for travelers, merchants, and armies of Europe, Asia, and Africa. This mixing of cultures fostered creativity, and the broad interactions among peoples ensured that new ideas spread throughout the region. Over the next 2,500 years, numerous great cultures arose. For example, the ancient Egyptians built the Great Pyramids between 2700 and 2100 B.C., using their astronomical knowledge to orient the Pyramids with the cardinal directions. They also invented papyrus scrolls and ink-based writing. The Babylonians invented methods of writing on clay tablets and developed arithmetic to serve in commerce and later in astronomical calculations. Many more of our modern principles of commerce, law, religion, and science originated with the cultures of Egypt and Mesopotamia.

The development of principles of modern science accelerated with the rise of Greece as a power in the Middle East, beginning around 500 B.C. In 330 B.C., Alexander the Great led the expansion of the Greek empire throughout the Middle East, absorbing all the former empires of Egypt and Mesopotamia. Alexander had a keen interest in science and education, perhaps fueled by his association with Aristotle [Section 5.3], who was his personal tutor. He encouraged the pursuit of knowledge and respect for foreign cultures. On the Nile delta in Egypt, he founded the city of Alexandria, which he hoped would become a center of world culture.

Although Alexander died at the age of 35, his dream came true. His successors continued to build Alexandria, including the construction of a great

THINKING ABOUT...

Eratosthenes Measures the Earth

In a remarkable ancient feat, the Greek astronomer and geographer Eratosthenes (c. 276–196 B.C.) estimated the size of the Earth in about 240 B.C. He did it by comparing the altitude of the Sun on the summer solstice in the Egyptian cities of Syene (modern-day Aswan) and Alexandria.

Eratosthenes knew that the Sun passed directly overhead in Syene on the summer solstice (it lies on the tropic of Cancer [Section S1.5]). He also knew that in the city of Alexandria to the north the Sun came within only 7° of the zenith on the summer solstice. He therefore concluded that Alexandria must be 7° of latitude to the north of Syene (Figure 3.13). Because 7° is $\frac{7}{360}$ of a circle, he concluded that the north-south distance between Alexandria and Syene must be $\frac{7}{360}$ of the circumference of the Earth.

Eratosthenes estimated the north-south distance between Syene and Alexandria to be 5,000 stadia (the *stadium* was a Greek unit of distance). Thus, he concluded that

$$\frac{7}{360} \times \text{circumference of Earth} = 5{,}000 \text{ stadia}$$

from which he found the Earth's circumference to be about 250,000 stadia.

Today, we don't know exactly what distance a stadium meant to Eratosthenes; however, based on the actual sizes of Greek stadiums, it must have been about $\frac{1}{6}$ kilometer. Thus, Eratosthenes estimated the circumference of the Earth to be about $\frac{250{,}000}{6} = 42{,}000$ kilometers—remarkably close to the modern value of just over 40,000 kilometers.

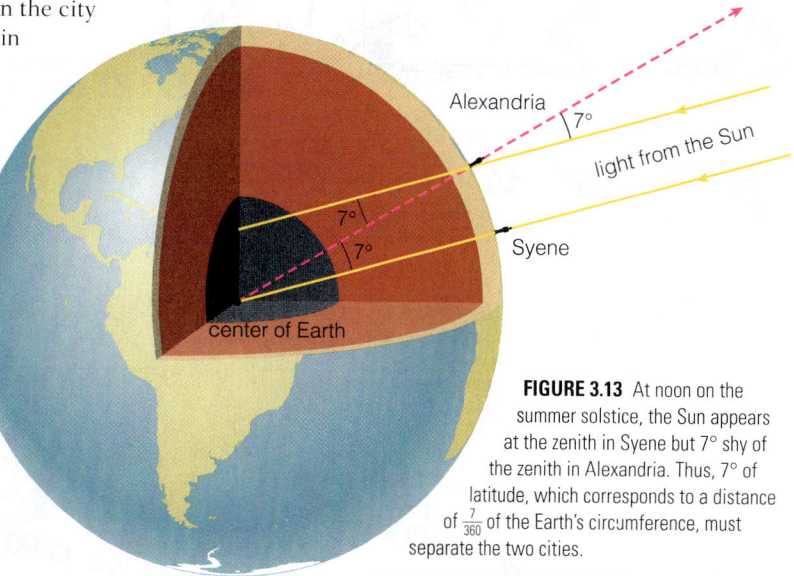

FIGURE 3.13 At noon on the summer solstice, the Sun appears at the zenith in Syene but 7° shy of the zenith in Alexandria. Thus, 7° of latitude, which corresponds to a distance of $\frac{7}{360}$ of the Earth's circumference, must separate the two cities.

---

library and research center around 300 B.C. (Figure 3.14). The Library of Alexandria was the world's preeminent center of research for the next 700 years. Its end is associated with the death of Hypatia, perhaps the most prominent female scholar of the ancient world. Hypatia was a resident scholar of the Library of Alexandria, director of the observatory in Alexandria, and one of the leading mathematicians and astronomers of her time. Unfortunately, she lived during a time of rising sentiment against free inquiry and was murdered by anti-intellectual mobs in A.D. 415. The final destruction of the Library of Alexandria followed not long after her death.

At its peak, the Library of Alexandria held more than a half million books, handwritten on papyrus scrolls. As the invention of the printing press was far in the future, most of the scrolls probably were original manuscripts or the single copies of original manuscripts. When the library was destroyed, most of its storehouse of ancient wisdom was lost forever.

TIME OUT TO THINK *How old is the school that you attend? Compare its age to the 700 years that the Library of Alexandria survived. Estimate the number of books you've read in your life. Compare this number to the half million books once housed in the Library of Alexandria.*

The Ptolemaic Model of the Universe

Perhaps the most important scientific idea developed by the ancient Greeks was that of creating **models** of nature. Just as a model airplane is a representation of a real airplane, a scientific model seeks to represent some aspect of nature. However, scientific models usually are conceptual models rather than miniature representations. The purpose of a scientific model is to explain and predict real phenomena, without any need to invoke myth, magic, or the supernatural. The ancient Greek idea of a celestial sphere is an example of a scientific model: The celestial sphere is the model, and it can be used to explain and predict the apparent motions of stars in our sky.

By combining the concept of modeling with advances in logic and mathematics, the ancient Greeks practiced a pursuit of knowledge very much like that of modern science. They used their models to make

FIGURE 3.14 These renderings show an artist's reconstruction, based on scholarly research, of how (**a**) the Great Hall and (**b**) a scroll room of the Library of Alexandria might have looked.

predictions, such as predicting where the planets should appear among the constellations of the night sky. If the predictions proved inaccurate, they changed and refined their models. In this way, the Greeks soon learned that the simple idea of a celestial sphere could not explain all the phenomena they saw in the sky (particularly the apparent retrograde motion of the planets). They therefore added spheres for the Sun, the Moon, and each of the planets and continued to change their ideas of how the spheres moved in an ongoing attempt to make the model agree with reality (Figure 3.15).

The culmination of this ancient modeling came with the work of Claudius Ptolemy (c. A.D. 100–170), pronounced *tol-e-mee*. Ptolemy's model of the universe is a tribute to human ingenuity in that it made reasonably accurate predictions despite being built on the flawed premise of an Earth-centered universe. Following a Greek tradition dating back to Plato (428–348 B.C.), Ptolemy imagined that all heavenly motions must proceed in perfect circles. To explain the apparent retrograde motion of planets, Ptolemy used an idea first suggested by Apollonius (c. 240–190 B.C.) and further developed by Hipparchus (c. 190–120 B.C.). This idea held that each planet moved along a small circle that, in turn, moved around a larger circle (Figure 3.16). As seen from Earth, this "circle upon circle" motion meant that a planet usually moved eastward relative to the stars but sometimes moved westward in apparent retrograde motion.

Ptolemy selected the sizes of the circles and the rates of motion of the planets along the circles

FIGURE 3.15 This diagram depicts the heavenly spheres, an ancient Greek model of the universe (c. 200 B.C.).

so that his model reproduced the apparent retrograde motion of all the known planets. He also incorporated more complex ideas into his model, such as allowing the Earth to be slightly off-center from the circles of a particular planet, so that it could forecast future planetary positions to within a few degrees—which was considered quite accurate at that time.

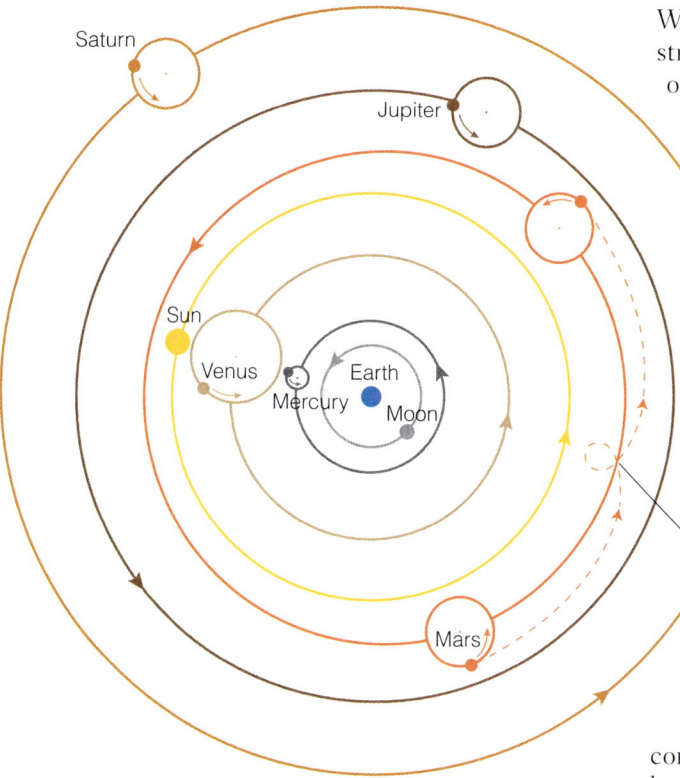

FIGURE 3.16 This diagram shows several key features of Ptolemy's model of the motions of the Sun, Moon, and planets around the Earth. The planets move around the small circles at the same time that these circles move around the Earth. By careful selection of the sizes of the circles and the rates at which planets moved, along with a few other adjustments, this model was used to predict planetary positions, including apparent retrograde motion, for 1,500 years. In order to show the circle-upon-circle idea clearly, this diagram exaggerates several features of the true Ptolemaic model; for example, for each retrograde loop, Mars moves much farther in the actual model than shown here.

Ptolemy's model remained the standard until the time of the Copernican revolution, some 1,400 years later.

The Islamic Role

Most of the great scholarly work of the ancient Greeks was lost with the fall of the Roman Empire and the destruction of the Library of Alexandria in the fifth century A.D. Much more would have been lost had it not been for the rise of a new center of intellectual achievement in Baghdad (present-day Iraq). While European civilization fell into the period of intellectual decline known as the Dark Ages, scholars of the new religion of Islam sought knowledge of mathematics and astronomy in hopes of better understanding the wisdom of Allah. During the eighth and ninth centuries A.D., scholars working in the Muslim empire centered in Baghdad translated and thereby saved many of the ancient Greek works.

Around A.D. 800, an Islamic leader named Al-Mamun (A.D. 786–833) established a "House of Wisdom" in Baghdad comparable in scope to the destroyed Library of Alexandria. Founded in a spirit of great openness and tolerance, the House of Wisdom employed Jews, Christians, and Muslims, all working together in scholarly pursuits. Using the translated Greek scientific manuscripts as building blocks, these scholars developed algebra and many new instruments and techniques for astronomical observation. Most of the official names of constellations and stars come from Arabic because of the work of the scholars at Baghdad. If you look at a star chart, you will see that the names of many bright stars begin with *al* (e.g., Aldebaran, Algol), which simply means "the" in Arabic. Ptolemy's book describing his astronomical system is also known by its Arabic name *Almagest*, from Arabic words meaning "the greatest compilation."

The Islamic world of the Middle Ages was in frequent contact with Hindu scholars from India, who in turn brought knowledge of ideas and discoveries from China. Hence, the intellectual center in Baghdad achieved a synthesis of the surviving work of the ancient Greeks and that of the Indians and the Chinese. The accumulated knowledge of the Arabs spread throughout the Byzantine Empire (the eastern part of the former Roman Empire). When the Byzantine capital of Constantinople (modern-day Istanbul) fell to the Turks in 1453, many Eastern scholars headed west to Europe, carrying with them the knowledge that helped ignite the European Renaissance.

3.4 Modern Science and the Scientific Method

As we have seen, many different cultures developed scientific modes of thinking, and the ancient Greeks developed principles of modeling very similar to those underlying the modern scientific method. But there are important differences between these ancient ideas and modern science. For example, Ptolemy's writings suggest that he designed his model to aid in calculations, with little concern for whether it reflected reality. While modern scientists also employ purely computational models when necessary, understanding the underlying reality of nature is an important goal of modern science.

The principles that govern modern science coalesced during the European Renaissance. Within 100 years after the fall of Constantinople, Copernicus began the process of overturning the Earth-centered,

THINKING ABOUT...

Logic and the Scientific Method

In science, we attempt to learn new knowledge through logical reasoning. The process of reasoning is carried out by constructing arguments in which we begin with a set of premises and try to draw appropriate conclusions. Note that this concept of a logical argument differs somewhat from the definition of *argument* in everyday life; in particular, a logical argument need not imply any animosity or dissension. There are two basic types of argument: *deductive* and *inductive*. Both are important in the scientific method.

The following is a simple example of a *deductive argument*:

> PREMISE: All planets orbit the Sun in ellipses with the Sun at one focus.
> PREMISE: Earth is a planet.
> CONCLUSION: Earth orbits the Sun in an ellipse with the Sun at one focus.

Note that, as long as the two premises are true, the conclusion must also be true; this is the essence of deduction. In addition, note that the first premise is a general statement that applies to all planets, whereas the conclusion is a specific statement that applies only to Earth. As this example suggests, deduction is often used to *deduce* a specific prediction from a more general theory. If the specific prediction proves to be false, then something must be wrong with the theoretical premises from which it was deduced. If it proves true, then we've acquired a piece of evidence in support of the theory.

Contrast the deductive argument above with the following example of an *inductive argument*:

> PREMISE: Birds fly up into the air but eventually come back down.
> PREMISE: People who jump into the air fall back down.
> PREMISE: Rocks thrown into the air come back down.
> PREMISE: Balls thrown into the air come back down.
> CONCLUSION: What goes up must come down.

Imagine that you lived in the time of ancient Greece, when the Earth was thought to be a realm distinct from the sky. Because each premise supports the conclusion, you might agree that this is a strong inductive argument and that its conclusion probably is true. However, no matter how many more examples you consider of objects that go up and come down, you could never *prove* that the conclusion is true—only that it seems likely to be true. A single counterexample can prove the conclusion to be false—for example, the fact that some rockets are launched into space and never return to Earth.

Inductive arguments *generalize* from specific facts to a broader model or theory and hence are the arguments used to build scientific theories. That is why theories can never be proved true beyond all doubt—they can only be shown to be consistent with ever larger bodies of evidence. Theories can be proved false, however, if they fail to account for observed or experimental facts.

Ptolemaic model, setting into motion a burst of scientific discovery that continues to this day. Before we can fully appreciate the Copernican revolution, which we'll discuss in Chapter 5, we must investigate a more basic question: What is modern science?

The word **science** comes from the Latin *scientia*, meaning "knowledge"; it is also the root of the words *conscience* (related to the idea of self-knowledge or self-awareness) and *omniscience* (the quality of being all-knowing). Today, science connotes a special kind of knowledge: the kind that makes predictions that can be confirmed by rigorous observations and experiments. In brief, the modern **scientific method** is an organized approach to explaining observed facts with a model of nature, subject to the constraint that any proposed model must be testable and the provision that the model must be modified or discarded if it fails these tests.

In its most idealized form, the scientific method begins with a set of observed facts. A fact is supposed to be a statement that is objectively true. For example, we consider it a fact that the Sun rises each morning, that the planet Mars appeared in a particular place in our sky last night, and that the Earth rotates. Facts are not always obvious, as illustrated by the case of the Earth's rotation—for most of human history, the Earth was assumed to be stationary at the center of the universe. In addition, our interpretations of facts often are based on beliefs about the world that others might not share. For example, when we say that the Sun rises each morning, we assume that it is the *same* Sun day after day—an idea that might not have been accepted by ancient Egyptians, whose mythology held that the Sun died with every sunset and was reborn with every sunrise. Nevertheless, facts are the raw material that scientific models seek to explain, so it is important that scientists agree on the facts. In the context of science, a fact must therefore be something that anyone can verify for himself or herself, at least in principle.

Once the facts have been collected, a model can be proposed to explain them. A useful model must also make predictions that can be tested through further observations or experiments. Ptolemy's model of the universe was useful because it predicted future locations of the Sun, Moon, and planets in the sky. However, although the Ptolemaic model remained in use for nearly 1,500 years, eventually it became clear that its predictions didn't quite match actual observations—a key reason why the Earth-centered model of the universe finally was discarded.

In summary, the idealized scientific method proceeds as follows:

- *Observation:* The scientific method begins with the collection of a set of observed facts.

- *Hypothesis:* A model is proposed to explain the observed facts and to make new predictions. A proposed model is often called a **hypothesis**, which essentially means an *educated guess.*

- *Further observations/experiments:* The model's predictions are tested through further observations or experiments. When a prediction is verified, we gain confidence that the model truly represents nature. When a prediction fails, we recognize that the model is flawed, and we therefore must refine or discard the model.

- *Theory:* A model must be continually challenged with new observations or experiments by many different scientists. A model achieves the status of a **scientific theory** only after a broad range of its predictions has been repeatedly verified. Note that, while we can have great confidence that a scientific theory truly represents nature, we can never prove a theory to be true *beyond all doubt.* Therefore, even well-established theories must be subject to continuing challenges through further observations and experiments.

In reality, scientific discoveries rarely are made by a process as mechanical as the idealized scientific method described here. For example, Johannes Kepler, who discovered the laws of planetary motion in the early 1600s, tested his model against observations that had been made previously, rather than verifying new predictions based on his model [Section 5.3]. Moreover, like most scientific work, Kepler's work involved intuition, collaboration with others, moments of insight, and luck. Nevertheless, with hindsight we can look back at Kepler's theory and see that other scientists eventually made plenty of observations to verify the planetary positions predicted by his model. In that sense, the scientific method represents an ideal prescription for judging objectively whether a proposed model of nature is close to the truth.

TIME OUT TO THINK *Look up* theory *in a dictionary. How does its definition differ from that given here for* scientific theory?

Science, Pseudoscience, and Nonscience

People often seek knowledge in ways that do not follow the basic tenets of the scientific method and therefore are not science. When we leave the realm of science, claims of knowledge become far more subjective and difficult to verify. Nevertheless, it can be useful to categorize such claims of knowledge as either *pseudoscience* or *nonscience.*

By **pseudoscience** we mean attempts to search for knowledge in ways that may at first seem scientific but do not adhere to the testing and verification requirements of the scientific method; the prefix *pseudo* means "false." For example, at the beginning of each year you can find tabloid newspapers offering predictions made by people who claim to be able to "see" the future. Because they make specific predictions, we can test their claims by checking whether their predictions come true, and numerous studies have shown that their predictions come true no more often than would be expected by pure chance. But these seers seem unconcerned with the results of such studies. The fact that they make testable claims but then ignore the results of the tests marks their claimed ability to see the future as pseudoscience.

By **nonscience** we mean knowledge sought through pure intuition, societal traditions, ancient scriptures, and other processes that make no claim

---

**Common Misconceptions:
Eggs on the Equinox**

A central principle of science is that you needn't take scientific claims on faith; in principle, at least, you can always test them for yourself. Consider the claim, repeated in news reports every year, that the spring equinox is the only day on which you can balance an egg on its end. Many people believe this claim, but you'll be immediately skeptical if you think about the nature of the spring equinox. The equinox is merely a point in the Earth's orbit at which sunlight strikes both hemispheres equally (see Figure 1.13), and it's difficult to see how sunlight could affect an attempt to balance eggs (especially if the eggs are indoors). More important, you can test this claim directly. It's not easy to balance an egg on its end, but with practice you'll find that you can do it on *any* day of the year, not just on the spring equinox. Not all scientific claims are so easy to test for yourself, but the basic lesson should be clear: Before you accept any scientific claim, you should demand at least a reasonable explanation of the evidence that backs it up.

to adhere to the scientific method. In nonscience, experimental testing is irrelevant because nonscientific beliefs are based on things such as faith, political conviction, and tradition. Because nonscience makes no pretense of following the scientific method, science can say nothing about the validity of nonscience.

TIME OUT TO THINK *Can the scientific method be applied in any way to questions of faith, prayer, emotion, or values? Defend your view.*

Is Science Objective?

The boundaries between science, nonscience, and pseudoscience are sometimes blurry. In particular, because science is practiced by human beings, individual scientists carry their personal biases and beliefs with them in their scientific work. These biases can influence how a scientist proposes or tests a model, and in some cases scientists have been known to cheat—either deliberately or subconsciously—to obtain the results they desire. For example, in the late nineteenth and early twentieth centuries, some astronomers claimed to see networks of "canals" in their blurry telescopic images of Mars, and they hypothesized that Mars was home to a dying civilization that used the canals to transport water to thirsty cities [Section 9.5]. Because no such canals actually exist, these astronomers apparently were allowing their beliefs to influence how they interpreted blurry images. It was, in essence, a form of cheating—even though it certainly was not intentional.

Sometimes, bias can show up even in the thinking of the scientific community as a whole. Thus, some valid ideas may not be considered by any scientist because the ideas fall too far outside the general patterns of thought, or the **paradigm**, of the time. Einstein's theory of relativity provides an example. Many other scientists had gleaned hints of this theory in the decades before Einstein but did not investigate them, at least in part because they seemed too outlandish.

The beauty of the scientific method is that it encourages continued testing by many people. Even if personal biases affect some results, tests by others eventually will uncover the mistakes. Similarly, even when a new idea falls outside the accepted paradigm, sufficient testing and verification of the idea eventually will force a change in the paradigm. Thus, although individual scientists rarely follow the scientific method in its idealized form, the collective action of many scientists over many years generally *does* follow the basic tenets of the scientific method. That is why, despite the biases of individual scientists, science as a whole is usually objective.

3.5 Astronomy Today

The modern scientific method has fueled a dramatic rise in human knowledge. It took humanity some 6,000 years to progress from the early lunar astronomy recorded in Africa to the ancient Greek idea of creating models of the universe, and another 2,000 years to finally recognize that the Earth orbits the Sun. In contrast, the past 400 years have seen discovery upon discovery, and we have come to know that the universe is filled with wonders far beyond the wildest imagination of our ancestors.

Today, the science of astronomy continues to develop at an ever-growing rate. New telescopes are fueling a renaissance in astronomy. In the past decade alone, we made the first discoveries of planets around other stars, witnessed the birth of galaxies near the edge of the observable universe, and discovered tantalizing hints of an accelerating universal expansion. If we measure information as a computer does, in *bytes* (where 1 byte represents the information content of one typed character and 1 megabyte represents the information in a 500-page book), then the information recorded at a single major observatory in a single night exceeds the total amount of astronomical information recorded by all ancient peoples combined.

The abundance of new data is disseminated worldwide via the Internet, and new data are collected at international observatories in many different countries and in space. The result is that, after a few centuries during which advances in astronomy were made primarily in Europe and the United States, astronomy is once again a worldwide endeavor. Just as the ancient civilizations of Egypt and Mesopotamia benefited from the cross-fertilization of ideas brought by people of many different cultures, astronomy today is benefiting from the great variety of perspectives of scientists from all over the world.

3.6 Astrology

Although the terms *astrology* and *astronomy* sound very similar, today they describe very different practices. In ancient times, however, astrology and astronomy often went hand in hand, and astrology played an important role in the historical development of astronomy.

In brief, the basic tenet of astrology is that human events are influenced by the apparent positions of the Sun, Moon, and planets among the stars in our sky. The origins of this idea are easy to understand. After all, there is no doubt that the position of the Sun in the sky influences our lives—it determines the seasons and hence the times of planting and harvesting, of warmth and cold, and of daylight and darkness. Similarly, the Moon determines the tides, and the cycle of lunar phases coincides with many biological cycles. Because the planets also appear to move among the stars, it seemed reasonable that planets also influence our lives, even if these influences were much more difficult to discover.

Ancient astrologers hoped that they might learn *how* the positions of the Sun, Moon, and planets influence our lives. They charted the skies, seeking correlations with events on Earth. For example, if an earthquake occurred when Saturn was entering the constellation Leo, might Saturn's position have been the cause of the earthquake? If the king became ill when Mars appeared in the constellation Gemini and the first-quarter moon appeared in Scorpio, might it mean another tragedy for the king when this particular alignment of the Moon and Mars next recurred? Surely, the ancient astrologers thought, the patterns of influence eventually would become clear. Thus, the astrologers hoped that they might someday learn to forecast human events with the same reliability with which astronomical observations of the Sun could forecast the coming of spring.

Astrology's Role in Astronomical History

Because forecasts of the seasons and forecasts of human events were imagined to be closely related, astrologers and astronomers usually were one and the same in the ancient world. For example, in addition to his books on astronomy, Ptolemy published a treatise on astrology called *Tetrabiblios* that remains the foundation for much of astrology today. Interestingly, Ptolemy himself recognized that astrology stood upon a far shakier foundation than astronomy. In the introduction to *Tetrabiblios*, Ptolemy compared astronomical and astrological predictions:

> [Astronomy], which is first both in order and effectiveness, is that whereby we apprehend the aspects of the movements of sun, moon, and stars in relation to each other and to the earth. . . . I shall now give an account of the second and less sufficient method [of prediction (astrology)] in a proper philosophical way, so that one whose aim is the truth might never compare its perceptions with the sureness of the first, unvarying science. . . .

Other ancient scientists probably likewise recognized that their astrological predictions were far less reliable than their astronomical ones. Nevertheless, if there was even a slight possibility that astrologers could forecast the future, no king or political leader would dare to be without one. Astrologers therefore held esteemed positions as political advisers in the ancient world and were provided with the resources they needed to continue charting the heavens and history. Much of the development of ancient astronomy was made possible through wealthy political leaders' support of astrology.

Ptolemy (c. A.D. 85–165)

Throughout the Middle Ages and into the Renaissance, many astronomers continued to practice astrology. For example, Kepler cast numerous **horoscopes**—the predictive charts of astrology—even as he was discovering the laws of planetary motion. However, given Kepler's later description of astrology as "the foolish stepdaughter of astronomy" and "a dreadful superstition," he may have cast the horoscopes solely as a source of much-needed income. Modern-day astrologers also claim Galileo as one of their own, in part for his having cast a horoscope for the Grand Duke of Tuscany. However, whereas his astronomical discoveries changed human history, the horoscope he cast was just plain wrong: It predicted a long and fruitful life for the duke, who died just a few weeks later.

The scientific triumph of Kepler and Galileo in showing the Earth to be a planet orbiting the Sun heralded the end of the linkage between astronomy and astrology. Nevertheless, astrology remains popular today. More people earn income by casting horoscopes than through astronomical research, and books and articles on astrology often outsell all but the most popular books on astronomy.

Scientific Tests of Astrology

Today, different astrologers follow different practices, which makes it difficult even to define *astrology*. Some astrologers no longer claim any ability to make testable predictions and therefore are practicing a form of *nonscience*—which modern science can say nothing about. However, for most astrologers the business of astrology is casting horoscopes. Horoscopes often are cast for individuals and either predict future events in the person's life or describe characteristics of the person's personality and life. If the horoscope predicts future events, it can be evaluated by whether the predictions come true. If it describes the person's personality and life, the description can be checked for accuracy.

A *scientific* test of astrology requires evaluating many horoscopes and comparing their accuracy to what would be expected by pure chance. For example, suppose a horoscope states that a person's best friend is female. Because that is true of roughly half the population in the United States, an astrologer who casts 100 such horoscopes would be expected by pure chance to be right about 50 times. Thus, we would be impressed with the predictive ability of the astrologer only if he or she were right much more often than 50 times out of 100. In hundreds of scientific tests, astrological predictions have never proved to be accurate by a substantially greater margin than expected from pure chance.[3] Similarly, in tests in which astrologers are asked to cast horoscopes for people they have never met, the horoscopes fail to correctly match personalities any more often than expected by chance. The verdict is clear: The methods of astrology are useless for predicting the past, the present, or the future.

What about newspaper horoscopes, which often appear to ring true? If you read them carefully, you will find that these horoscopes generally are so vague as to be untestable. For example, a horoscope that says "It is a good day to spend time with your friends" doesn't offer much for testing. Indeed, if you read the horoscopes for all 12 astrological signs, you'll probably find that several of them apply equally well to you.

TIME OUT TO THINK *Look in a local newspaper for today's weather forecast and for your horoscope. Contrast the nature of their predictions. By the end of the day, you will know if the weather forecast was accurate. Will you know whether your horoscope was accurate? Explain.*

Does It Make Sense?

In science, observations and experiments are the ultimate judge of any idea. No matter how outlandish an idea might appear, it cannot be dismissed if it successfully meets observational or experimental tests. The idea that the Earth rotates and orbits the Sun at one time seemed outlandish, yet today it is so strongly supported by the evidence that we consider it a fact. The idea that the positions of the Sun, Moon, and planets among the stars influence our lives might sound outlandish today, but if astrology were to make predictions that came true, adherence to the scientific method would force us to take it seriously. However, given that scientific tests of astrology have never found any evidence that its predictive methods work, it is worth looking back at its premises to see whether they make sense. Might there be a few kernels of wisdom buried within the lore of astrology?

Let's begin with one of the key premises of astrology: There is special meaning in the patterns of the stars in the constellations. This idea may have seemed quite reasonable in ancient times, when the stars were assumed to be fixed on an unchanging celestial sphere. But today we know that the patterns of the stars in the constellations are accidents of the moment; long ago the constellations did not look the same, and they will look still different far in the future [Section 1.3]. Moreover, the stars in a constellation don't necessarily have any *physical* association. Because stars vary in distance, two stars that appear on opposite sides of our sky might well be closer together than two stars in the same constellation. Thus, constellations are only *apparent* associations of stars, with no more physical reality than the water in a desert mirage.

Astrology also places great importance on the positions of the planets among the constellations. Again, this idea might have seemed quite reasonable in ancient times, when it was thought that the planets truly wandered among the stars. But today we know that the planets only *appear* to wander among the stars. In reality, the planets are in our own solar system, while the stars are vastly farther away. It is difficult to see how mere appearances could have profound effects on our lives.

Many other ideas at the heart of astrology are equally suspect. For example, most astrologers claim that a proper horoscope must account for the positions of *all* the planets. Does that mean that all horoscopes cast before the discovery of Pluto in 1930 were invalid? If so, why didn't astrologers notice that something was wrong with their horoscopes and thereby predict Pluto's existence? Given that several moons in the solar system are larger than Pluto, with two larger than Mercury, should astrologers also be tracking the positions of these moons? What about asteroids, which orbit the Sun like planets? What about planets orbiting other stars? Given seemingly unanswerable questions like these, there seems little hope that astrology will ever meet its ancient goal of being able to forecast human events.

---

[3] An excellent summary of scientific tests of astrology can be found in Roger B. Culver and Philip A. Ianna, *Astrology: True or False?* (Buffalo, New York: Prometheus Books, 1988).

THE BIG PICTURE

In this chapter, we focused on the scientific principles through which we have learned so much about the universe. Key "big picture" concepts from this chapter include the following:

- The basic ingredients of scientific thinking—careful observation and trial-and-error testing—are a part of everyone's experience. The modern scientific method simply provides a way of organizing this everyday thinking to facilitate the learning and sharing of new knowledge.

- Although knowledge about the universe is growing rapidly today, each new piece rests upon foundations of older discoveries. The foundations of astronomy reach far back into history and are intertwined with the general development of human culture and civilization.

- The concept of making models of nature lies at the heart of modern science. Models are created by generalizing from many specific facts. They then yield further predictions that allow the models to be tested. Models that withstand repeated testing rise to the status of scientific theories, while models that fail must be modified or discarded.

- Although astronomy and astrology once developed hand in hand, today they represent very different things. The predictions of astrology either fail rigorous testing or are so vague as to be untestable. Astronomy is a science and is the primary means by which humans learn about the physical universe.

Review Questions

1. In what ways is scientific thinking a fundamental part of human nature? What characteristics does the modern scientific method possess that go beyond this everyday type of thinking?

2. Explain how the people of central Africa predicted the weather by observing the Moon.

3. How are the names of the seven days of the week related to astronomical objects?

4. Briefly explain how the ancient Egyptians determined the time of day or night.

5. Briefly describe the astronomical uses of each of the following: Stonehenge, the Templo Mayor, the Sun Dagger, the Mayan observatory at Chichén Itzá, lines in the Nazca desert, Pawnee lodges, the Big Horn Medicine Wheel.

6. Explain why the Muslim fast of Ramadan occurs earlier with each subsequent year.

7. What is the *Metonic cycle?* How does the Jewish calendar follow the Metonic cycle? How does this influence the date of Easter?

8. What role did the Navigator play in Polynesian society?

9. What was the Library of Alexandria? What role did it play in the ancient empires of Greece and Rome?

10. What do we mean by a *model* of nature? Explain how the celestial sphere represents such a model.

11. Who was Ptolemy? Briefly describe how the Ptolemaic model of the universe explains apparent retrograde motion while preserving the idea of an Earth-centered universe.

12. Briefly describe the role played by Islamic scholars of the Middle Ages in the development of modern science.

13. Describe the basic process of the *scientific method.* What is the role of a *model* in this method? What do we mean by *facts* when discussing science?

14. Briefly describe how science differs from *pseudoscience* and *nonscience.*

15. Briefly explain how science as a whole can be objective even while individual scientists have personal biases and beliefs.

16. Briefly describe the international character of modern astronomy.

17. What is the basic tenet of *astrology?* Why might this idea have seemed quite reasonable in ancient times? What role did astrology play in the development of astronomy?

18. How do we make scientific tests of astrology? What have such tests found?

Discussion Questions

1. *The Impact of Science.* The modern world is filled with ideas, knowledge, and technology that developed through science and application of the scientific method. Discuss some of these things and how they affect our lives. Which of these impacts do you think are positive? Which are negative? Overall, do you think science has benefited the human race? Defend your opinion.

2. *The Importance of Ancient Astronomy.* Why was astronomy important to people in ancient times? Discuss both the practical importance of astronomy and the importance it may have had for religious or other traditions. Which do you think was more important in the development of ancient astronomy, its practical or its philosophical role? Defend your opinion.

3. *Secrecy and Science.* The text mentions that some historians believe that Chinese science and technology fell behind that of Europe because of the Chinese culture of secrecy. Do you agree that secrecy can hold back the advance of science? Why or why not? For the past 200 years, the United States has allowed a greater degree of free speech than most other countries in the world. How great a role do you think this has played in making the United States the world leader in science and technology? Defend your opinion.

4. *Lunar Cycles.* We have now discussed four distinct lunar cycles: the $29\frac{1}{2}$-day cycle of phases, the 18-year-$11\frac{1}{3}$-day saros cycle, the 19-year Metonic cycle, and the 18.6-year cycle over which a full moon rises at its most southerly place along the horizon. Discuss how each of these cycles can be observed and the role each cycle has played in human history.

5. *Astronomy and Astrology.* Why do you think astrology remains so popular around the world even though it has failed all scientific tests of its validity? Do you think the popularity of astrology has any positive or negative social consequences? Defend your opinions.

Problems

True Statements? For **problems 1–6**, decide whether the statements are true or false and clearly explain how you know.

1. If we defined hours as the ancient Egyptians did, we'd have the longest hours on the summer solstice and the shortest hours on the winter solstice.

2. The date of Christmas (December 25) is set each year according to a lunar calendar.

3. When navigating in the South Pacific, the Polynesians found their latitude with the aid of the pointer stars of the Big Dipper.

4. The Ptolemaic model reproduced apparent retrograde motion by having planets move sometimes counterclockwise and sometimes clockwise in their circles.

5. In science, saying that something is a theory means that it is really just a guess.

6. Ancient astronomers were convinced of the validity of astrology as a tool for predicting the future.

7. *Cultural Astronomy.* Choose a particular culture of interest to you, and research the astronomical knowledge and accomplishments of that culture. Write a two- to three-page summary of your findings.

8. *Astronomical Structures.* Choose an ancient astronomical structure of interest to you (e.g., Stonehenge, Nazca lines, Pawnee lodges) and research its history. Write a two- to three-page summary of your findings. If possible, also build a scale model of the structure or create detailed diagrams to illustrate how the structure was used.

9. *Venus and the Mayans.* The planet Venus apparently played a particularly important role in Mayan society. Research the evidence and write a one- to two-page summary of current knowledge about the role of Venus in Mayan society.

10. *Scientific Test of Astrology.* Find out about at least one scientific test that has been conducted to test the validity of astrology. Write a short summary of how the test was conducted and what conclusions were reached.

11. *Your Own Astrological Test.* Devise your own scientific test of astrology. Clearly define the methods you will use in your test and how you will evaluate the results. Then carry out the test.

Web Projects

Find useful links for Web projects on the text Web site.

1. *Easter.* Find out when Easter is celebrated by different sects of Christianity. Then research how and why different sects set different dates for Easter. Summarize your findings in a one- to two-page report.

2. *Greek Astronomers.* Many ancient Greek scientists had ideas that, in retrospect, seem well ahead of their time. Choose one or more of the following ancient Greek scientists, and learn enough about their work in science and astronomy to write a one- to two-page "scientific biography."

Thales	Anaximander	Pythagoras
Anaxagoras	Empedocles	Democritus
Meton	Plato	Eudoxus
Aristotle	Callipus	Aristarchus
Archimedes	Eratosthenes	Apollonius
Hipparchus	Seleucus	Ptolemy
Hypatia		

3. *The Ptolemaic Model.* This chapter gives only a very brief description of Ptolemy's model of the universe. Investigate the model in greater depth. Using diagrams and text as needed, give a two- to three-page description of the model.

The eternal mystery of the world is its comprehensibility. The fact that it is comprehensible is a miracle.

ALBERT EINSTEIN

CHAPTER 4
A Universe of Matter and Energy

In this and the next three chapters, we turn our attention to the scientific concepts that lie at the heart of modern astronomy. We begin by investigating the nature of matter and energy, the fundamental stuff from which the universe is made.

The history of the universe essentially is a story about the interplay between matter and energy since the beginning of time. Interactions between matter and energy govern everything from the creation of hydrogen and helium in the Big Bang to the processes that make Earth a habitable planet. Understanding the universe therefore depends on familiarity with how matter responds to the ebb and flow of energy.

The concepts of matter and energy presented in this chapter will enable you to understand most of the topics in this book. Some of the concepts and terminology may already be familiar to you. If not, don't worry. We will go over them again as they arise in various contexts, and you can refer back to this chapter as needed during the remainder of your studies.

4.1 Matter and Energy in Everyday Life

The meaning of **matter** is obvious to most people, at least on a practical level. Matter is simply material, such as rocks, water, or air. You can hold matter in your hand or put it in a box. The meaning of **energy** is not quite as obvious, although we certainly talk a lot about it. We pay energy bills to the power companies, we use energy from gasoline to run our cars, and we argue about whether nuclear energy is a sensible alternative to fossil fuels. On a personal level, we often talk about how energetic we feel on a particular day. But what *is* energy?

Table 4.1 Energy Comparisons

Item	Energy (joules)
Average daytime solar energy striking Earth, per m² per second	1.3×10^3
Energy released by metabolism of one average candy bar	1×10^6
Energy needed for 1 hour of walking (adult)	1×10^6
Kinetic energy of average car traveling at 60 mi/hr	1×10^6
Daily energy needs of average adult	1×10^7
Energy released by burning 1 liter of oil	1.2×10^7
Energy released by fission of 1 kg of uranium-235	5.6×10^{13}
Energy released by fusion of hydrogen in 1 liter of water	7×10^{13}
Energy released by 1-megaton H-bomb	5×10^{15}
Energy released by major earthquake (magnitude 8.0)	2.5×10^{16}
U.S. annual energy consumption	10^{20}
Annual energy generation of Sun	10^{34}
Energy released by supernova (explosion of a star)	10^{44}–10^{46}

Broadly speaking, energy is what makes matter move. For Americans, the most familiar way of measuring energy is in Calories, which we use to describe how much energy our bodies can draw from food. A typical adult uses about 2,500 Calories of energy each day. Among other things, this energy keeps our hearts beating and our lungs breathing, generates the heat that maintains our 37°C (98.6°F) body temperature, and allows us to walk and run.

Just as there are many different units for measuring height, such as inches, feet, and meters, there are many alternatives to Calories for measuring energy. If you look closely at an electric bill, you'll probably find that the power company charges you for electrical energy in units called *kilowatt-hours*. If you purchase a gas appliance, its energy requirements may be labeled in *British thermal units,* or *BTUs*. In science and internationally, the favored unit of energy is the **joule**, which is equivalent to $\frac{1}{4,184}$ of a Calorie. Thus, the 2,500 Calories used daily by a typical adult are equivalent to about 10 million joules. For purposes of comparison, some energies given in joules are listed in Table 4.1.

Although energy can always be measured in joules, it has many different forms. Already we have talked about food energy, electrical energy, the energy of a beating heart, the energy represented by our body temperature, and more. Fortunately, the many forms of energy can be grouped into three basic categories.

First, whenever matter is moving, it has energy of motion, or **kinetic energy** (*kinetic* comes from a Greek word meaning "motion"). Falling rocks, the moving blades on an electric mixer, a car driving down the highway, and the molecules moving in the air around us are all examples of objects with kinetic energy.

The second basic category of energy is **potential energy**, or energy being stored for later conversion into kinetic energy. A rock perched on a ledge has *gravitational* potential energy because it will fall if it slips off the edge. Gasoline contains *chemical* potential energy, which a car engine converts to the kinetic energy of the moving car. Power companies supply *electrical* potential energy, which we use to run dishwashers and other appliances.

The third basic category is energy carried by light, or **radiative energy** (the word *radiation* is often used as a synonym for *light*). Plants directly convert the radiative energy of sunlight into chemical potential energy through the process of *photosynthesis*. Radiative energy is fundamental to astronomy, because telescopes collect the radiative energy of light from distant stars.

TIME OUT TO THINK *We buy energy in many different forms. For example, we buy chemical potential energy in the form of food to fuel our bodies. Describe several other forms of energy that you commonly buy.*

Understanding how energy changes from one form to another helps us understand many common phenomena. For example, a diver standing on a 10-meter platform has gravitational potential energy owing to her height above the water and chemical potential energy stored in her body tissues. She uses the chemical potential energy to flex her muscles in such a way as to initiate her dive and then to help her execute graceful twists and spins (Figure 4.1). Meanwhile, her gravitational potential energy becomes kinetic energy of motion as she falls toward the water.

Following how energy changes from one form to another also helps us understand the universe. For example, the particles in a collapsing cloud of interstellar gas convert gravitational potential energy into kinetic energy as they fall inward, and their motion generates heat that can eventually ignite a star. But before we study astronomical phenomena, we need ways to describe energy quantitatively.

4.2 A Scientific View of Energy

In this section, we discuss a few ways to quantify kinetic and potential energy; we'll discuss radiative energy in Chapter 6, in which we study properties of light.

Kinetic Energy

We can calculate the kinetic energy of any moving object with a very simple formula:

$$\text{kinetic energy} = \tfrac{1}{2}mv^2$$

where *m* is the mass of the object and *v* is its speed (*v* for *velocity*). If we measure the mass in kilograms and the speed in meters per second, the resulting

FIGURE 4.1 Understanding energy can help us understand both the graceful movements of a diver and the story of the universe.

Mathematical Insight 4.1 Temperature Scales

Three temperature scales are commonly used today (Figure 4.2). In the United States, we usually use the **Fahrenheit** scale, defined so that water freezes at 32°F and boils at 212°F. Internationally, temperature is usually measured on the **Celsius** scale, which places the freezing point of water at 0°C and the boiling point at 100°C.

Scientists measure temperature on the **Kelvin** scale, which is the same as the Celsius scale except for its zero point. A temperature of 0 K is the coldest possible temperature, known as **absolute zero**; it is equivalent to −273.15°C. (The degree symbol ° is not used when writing temperatures on the Kelvin scale.) Thus, any particular temperature has a Kelvin value that is numerically 273.15 larger than its Celsius value. Using T to stand for temperature and a subscript to indicate the temperature scale, we can write the conversions between Celsius and Kelvin as follows:

$$T_{\text{Kelvin}} = T_{\text{Celsius}} + 273.15$$
$$T_{\text{Celsius}} = T_{\text{Kelvin}} - 273.15$$

To find the conversion between Fahrenheit and Celsius, note that the Fahrenheit scale has 180° (212°F − 32°F = 180°F) between the freezing and boiling points of water, whereas the Celsius scale has only 100° between these points. A temperature change of 1 Celsius degree therefore is equivalent to a temperature change of 1.8 Fahrenheit degrees. Furthermore, the freezing point of water is numerically 32 larger on the Fahrenheit scale than on the Celsius scale. Combining these two facts we find the conversions between Fahrenheit and Celsius:

$$T_{\text{Celsius}} = \frac{T_{\text{Fahrenheit}} - 32}{1.8}$$
$$T_{\text{Fahrenheit}} = 32 + (1.8 \times T_{\text{Celsius}})$$

Example: Convert human body temperature of 98.6°F into Celsius and Kelvin.

Solution: First, we convert 98.6°F to Celsius:

$$T_{\text{Celsius}} = \frac{T_{\text{Fahrenheit}} - 32}{1.8} = \frac{98.6 - 32}{1.8} = 37.0°C$$

Next, we convert Celsius to Kelvin:

$$T_{\text{Kelvin}} = T_{\text{Celsius}} + 273.15 = 37.0 + 273.15 = 310.15 \text{ K}$$

Thus, human body temperature of 98.6°F is 37.0°C or 310.15 K.

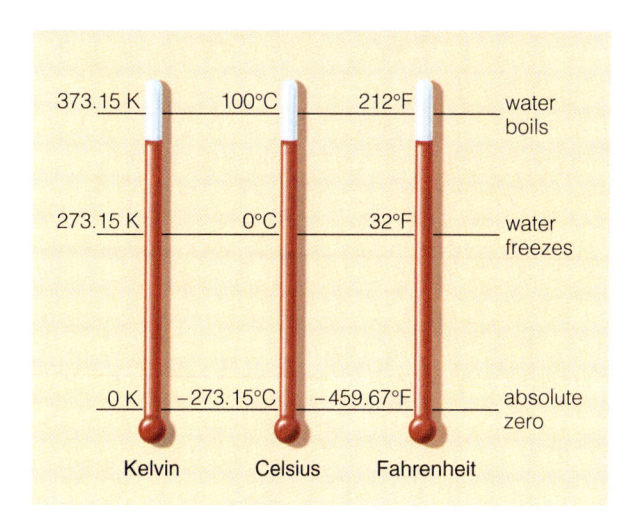

FIGURE 4.2 Three common temperature scales: Kelvin, Celsius, and Fahrenheit.

answer will be in joules. (That is, energy has units of a mass times a velocity squared, so 1 joule = 1 (kg × m²)/s².) The kinetic energy formula is easy to interpret. The *m* in the formula tells us that kinetic energy is proportional to mass: A 5-ton truck has 5 times the kinetic energy of a 1-ton car moving at the same speed. The v^2 tells us that kinetic energy increases with the *square* of the velocity: If you double your speed (e.g., from 30 km/hr to 60 km/hr), your kinetic energy becomes $2^2 = 4$ times greater.

Thermal Energy Suppose we want to know about the kinetic energy of the countless tiny particles (atoms and molecules) inside a rock or of the countless particles in the air or in a distant star. Each of these tiny particles has its own motion relative to surrounding particles, and these motions constantly change as the particles jostle one another. The result is that the particles inside a substance appear to move randomly: Any individual particle may be moving in any direction with any of a wide range of speeds. Despite the seemingly random motion of particles within a substance, it's easy to measure the *average* kinetic energy of the particles—**temperature** is a measure of this average. A higher temperature simply means that, on average, the particles have more kinetic energy and hence are moving faster (Figure 4.3). (Kinetic energy depends on both mass and speed, but for a particular set of particles, greater kinetic energy means higher speeds.) The speeds of particles within a substance can be surprisingly fast. For example, the air molecules around you move at typical speeds of about 500 meters per second (about 1,000 miles per hour).

The energy contained *within* a substance as measured by its temperature is often called **thermal energy**. Thus, thermal energy represents the collective kinetic energy of the many individual particles moving within a substance.

Temperature and Heat The concepts of *temperature* and *heat* are not quite the same. To understand the difference, imagine the following experiment (but don't try it!). Suppose you heat your oven to 500°F. Then you open the oven door, quickly thrust your arm inside (without touching anything), and immediately remove it. What will happen to your arm? Not much. Now suppose you boil a pot of water. Although the temperature of boiling water is only 212°F, you would be badly burned if you put your arm in the pot, even for an instant. Thus, the water in the pot transfers more thermal energy to your arm than does the air in the oven, even though it has a lower temperature. That is, your arm heats more quickly in the water. This is because thermal energy content depends on both the temperature and the total number of particles, which is much larger in a pot of water (Figure 4.4).

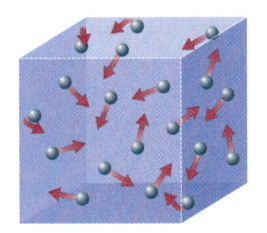

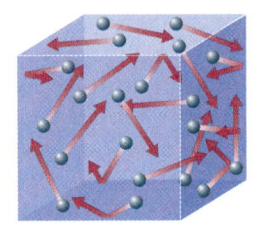

Longer arrows mean higher average speed.

FIGURE 4.3 The particles in the box on the right have a higher temperature because their average speeds are higher (assuming that both boxes contain particles of the same mass).

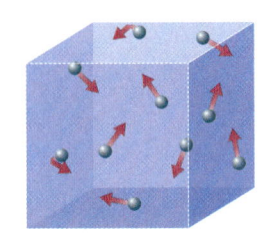

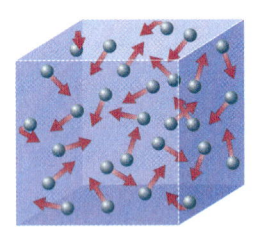

FIGURE 4.4 Both boxes have the same temperature, but the box on the right contains more *thermal energy* because it contains more particles.

Let's look at what is happening on the molecular level. If air or water is hotter than your body, molecules striking your skin transfer some of their thermal energy to your arm. The high temperature in a 500°F oven means that the air molecules strike your skin harder, on average, than the molecules in a 212°F pot of boiling water. However, because the *density* is so much higher in the pot of water, many more molecules strike your skin each second. Thus, while each individual molecular collision transfers a little less thermal energy in the boiling water than in the oven, the sheer number of collisions in the water transfers so much thermal energy that your skin burns rapidly.

TIME OUT TO THINK *In air or water that is colder than your body temperature, thermal energy transfers from you to the surrounding cold air or water. Use this fact to explain why falling into a 32°F (0°C) lake is much more dangerous than standing naked outside on a 32°F day.*

The environment in space provides another example of the difference between temperature and heat. Surprisingly, the temperature in low Earth orbit is several thousand degrees. However, astronauts working in Earth orbit (e.g., outside the Space Shuttle) tend to get very cold and therefore use heated space suits and gloves. The astronauts feel cold despite the high temperature because the extremely low density of space means that relatively few particles are available to transfer thermal energy to an astronaut. (You may wonder how the astronauts become cold given that the low density also means the astronauts cannot transfer much of their own thermal energy to the particles in space. It turns out that they lose their body heat by emitting *thermal radiation,* which we will discuss in Chapter 6.)

Potential Energy
Potential energy can be stored in many different forms and is not always easy to quantify. Fortunately, it is easy to describe two types of potential energy that we use frequently in astronomy.

Gravitational Potential Energy Gravitational potential energy is extremely important in astronomy. The conversion of gravitational potential energy into kinetic (or thermal) energy helps explain everything from the speed at which an object falls to the ground to the formation processes of stars and planets. The mathematical formula for gravitational potential energy can take a variety of forms, but in words the idea is simple: *The amount of gravitational potential energy released as an object falls depends on its mass, the strength of gravity, and the distance it falls.*

This statement explains the obvious fact that falling from a 10-story building hurts more than falling out of a chair. Your gravitational potential energy is much greater on top of the 10-story building than in your chair because you can fall much farther. Because your gravitational potential energy will be converted to kinetic energy as you fall, you'll have a lot more kinetic energy by the time you hit the ground after falling from the building than after falling from the chair, which means you'll hit the ground much harder.

Gravitational potential energy also helps us understand how the Sun became hot enough to sustain nuclear fusion. Before the Sun formed, its matter was contained in a large, cold, diffuse cloud of gas. Most of the individual gas particles were far from the center of this large cloud and therefore had considerable amounts of gravitational potential energy. As the cloud collapsed under its own gravity, the gravitational potential energy of these particles was converted to thermal energy, eventually making the center of the cloud hot enough to ignite nuclear fusion.

Mass-Energy Although matter and energy seem very different in daily life, they are intimately connected. Einstein showed that mass itself is a form of potential energy, often called **mass-energy**. The mass-energy of any piece of matter is given by the formula

$$E = mc^2$$

where E is the amount of potential energy, m is the mass of the object, and c is the speed of light. If the mass is measured in kilograms and the speed of light in meters per second, the resulting mass-energy has units of joules. Note that the speed of light is a large number ($c = 3 \times 10^8$ m/s) and the speed of light squared is much larger still ($c^2 = 9 \times 10^{16}$ m²/s²). Thus, Einstein's formula implies that a relatively small amount of mass represents a huge amount of mass-energy.

Mass-energy can be converted to other forms of energy, but noticeable amounts of mass become other forms of energy only under special but important circumstances. The process of nuclear fusion in the core of the Sun converts some of the Sun's mass into energy, ultimately generating the sunlight that sustains most life on Earth. On Earth, nuclear reactors and nuclear bombs also work in accord with Einstein's formula. In nuclear reactors, the splitting (fission) of elements such as uranium or plutonium converts some of the mass-energy of these materials into heat, which is then used to generate electrical

Mathematical Insight 4.2 Density

The term *density* usually refers to *mass density,* which quantifies how much mass is packed into each unit of volume. The more tightly matter is packed, the higher its density. In science, the most common unit of density is grams per cubic centimeter (a cubic centimeter is about the size of a sugar cube):

$$\text{density} = \frac{\text{mass (in g)}}{\text{volume (in cm}^3\text{)}}$$

For example, if a 30-gram rock has a volume of 10 cubic centimeters, we calculate its density as follows:

$$\text{density of rock} = \frac{\text{mass of rock}}{\text{volume of rock}} = \frac{30 \text{ g}}{10 \text{ cm}^3} = 3 \frac{\text{g}}{\text{cm}^3}$$

A useful guide for putting densities in perspective is the density of water, which is 1 gram per cubic centimeter; this isn't a coincidence—it's how the gram was defined. Rocks, like the one with a density of 3 grams per cubic centimeter, are denser than water and therefore sink in water. Wood floats in water because it is less dense than water.

The concept of density is sometimes applied to things other than mass. For example, an average of about 25,000 people live on each square kilometer of Manhattan, so we say that Manhattan has a *population density* of 25,000 people per square kilometer. As another example, 1 liter of oil releases about 12 million joules of energy when burned, so we say that the *chemical energy density* of oil is about 12 million joules per liter.

power. In an H-bomb, nuclear fusion similar to that in the Sun uses a small amount of the mass-energy in hydrogen to devastating effect. Incredibly, a 1-megaton H-bomb that could destroy a major city requires the conversion of only about 0.1 kilogram of mass (about 3 ounces) into energy (Figure 4.5).

Just as $E = mc^2$ tells us that mass can be converted into other forms of energy, it also tells us that energy can be transformed into mass. In *particle accelerators*, scientists accelerate subatomic particles to extremely high speeds. When these particles collide with one another or with a barrier, some of the energy released in the collision spontaneously turns into mass, appearing as a shower of subatomic particles. These showers of particles allow scientists to test theories about how matter behaves at extremely high temperatures such as those that prevailed in the universe during the first fraction of a second after the Big Bang. Among the most powerful particle accelerators in the world are Fermilab in Illinois, the Stanford Linear Accelerator in California, and CERN in Switzerland.

TIME OUT TO THINK *Einstein's formula $E = mc^2$ is probably the most famous physics equation of all time. Considering its role as described in the preceding paragraphs, do you think its fame is well deserved?*

Conservation of Energy

A fundamental principle in science is that, regardless of how we change the *form* of energy, the total *quantity* of energy never changes. This principle is called the **law of conservation of energy**. It has been carefully tested in many experiments, and it is a pillar upon which modern theories of the universe are built. Because of this law, the story of the universe is a story of the interplay of energy and matter: All actions in the universe involve exchanges of energy or the conversion of energy from one form to another.

For example, imagine that you've thrown a baseball so that it is moving and hence has kinetic energy. Where did this kinetic energy come from? The baseball got its kinetic energy from the motion of your

FIGURE 4.5 The energy from this H-bomb comes from converting only about 0.1 kg of mass into energy in accordance with $E = mc^2$.

Mathematical Insight 4.3   Mass-Energy

It's easy to calculate mass-energies with Einstein's formula $E = mc^2$. Once we calculate an energy, we can compare it to other known energies.

Example: Suppose a 1-kilogram rock were completely converted into energy. How much energy would it release? Compare this to the energy released by burning 1 liter of oil.

Solution: The total mass-energy of the rock is $E = mc^2$, where m is the 1-kg mass and $c = 3 \times 10^8$ m/s:

$$E = mc^2 = 1 \text{ kg} \times \left(3 \times 10^8 \frac{\text{m}}{\text{s}}\right)^2$$

$$= 1 \text{ kg} \times \left(9 \times 10^{16} \frac{\text{m}^2}{\text{s}^2}\right)$$

$$= 9 \times 10^{16} \frac{\text{kg} \times \text{m}^2}{\text{s}^2}$$

$$= 9 \times 10^{16} \text{ joules}$$

Burning 1 liter of oil releases 12 million joules (see Table 4.1). Dividing the mass-energy of the rock by the energy released by burning 1 liter of oil, we find:

$$\frac{9 \times 10^{16} \text{ joules}}{1.2 \times 10^7 \text{ joules}} = 7.5 \times 10^9$$

That is, if the rock could be converted completely to energy, its mass would supply as much energy as 7.5 billion liters of oil—roughly the amount of oil used by *all* cars in the United States in a week. Unfortunately, no technology available now or in the foreseeable future can release all the mass-energy of a rock.

arm as you threw it; that is, some of the kinetic energy of your moving arm was transferred to the baseball. Your arm, in turn, got its kinetic energy from the release of chemical potential energy stored in your muscle tissues. Your muscles got this energy from the chemical potential energy stored in the foods you ate. The energy stored in the foods came from sunlight, which plants convert into chemical potential energy through photosynthesis. The radiative energy of the Sun was generated through the process of nuclear fusion, which releases some of the mass-energy stored in the Sun's supply of hydrogen. Thus, the ultimate source of the energy of the moving baseball is the mass-energy stored in hydrogen—which was created in the Big Bang.

We have described where the baseball got its kinetic energy. Where will this energy go? As the baseball moves through the air, some of its energy is transferred to molecules in the air, generating heat or sound. If someone catches the baseball, its energy will cause his or her hand to recoil and perhaps will also generate some heat and sound. Ultimately, the energy of the moving baseball will be converted to a barely noticeable amount of heat (thermal energy) in the air, the ground, or a person's hand, making it extremely difficult to track. Nevertheless, the energy will never disappear. According to present understanding, the total energy content of the universe was determined in the Big Bang. It remains the same today and will stay the same forever into the future.

4.3 The Material World

Now that we have seen how energy animates the matter in the universe, it's time to consider matter itself in greater detail. You are familiar with two basic properties of matter on Earth from everyday experience. First, matter can exist in different **phases**: as a **solid**, such as ice or a rock; as a **liquid**, such as flowing water or oil; or as a **gas**, such as air. Second, even in a particular phase, matter exists in a great variety of different substances.

What is matter, and why does it have so many different forms? Let's follow the lead of the ancient Greek philosopher Democritus (c. 470–380 B.C.), who wondered what would happen if we broke a piece of matter, such as a rock, into ever smaller pieces. Democritus believed that the rock would eventually break into particles so small that nothing smaller could be possible. He called these particles *atoms,* a Greek term meaning "indivisible." (By modern definition, atoms are *not* indivisible because they are composed of smaller particles.) Building upon the beliefs of earlier Greek scientists, Democritus thought that all materials were composed of just four

Democritus

basic *elements:* fire, water, earth, and air. He proposed that the different properties of the elements could be explained by the physical characteristics of their atoms. Democritus suggested that atoms of water were smooth and round, so water flowed and had no fixed shape, while burns were painful because atoms of fire were thorny. He imagined atoms of earth to be rough and jagged, like pieces of a three-dimensional jigsaw puzzle, so that they could stick together to form a solid substance. He even explained the creation of the world with an idea that sounds uncannily modern, suggesting that the universe began as a chaotic mix of atoms that slowly clumped together to form the Earth.

Although Democritus was wrong about there being only four types of atoms and about their specific properties, his general idea was right. Today, we know that all ordinary matter is composed of **atoms** and that each different type of atom corresponds to a different chemical **element**. Among the most familiar chemical elements are hydrogen, helium, carbon, oxygen, silicon, iron, gold, silver, lead, and uranium. (Appendix D gives the periodic table of all the elements.)

The number of different material substances is far greater than the number of chemical elements because atoms can combine to form **molecules**.

Some molecules consist of two or more atoms of the same element. For example, we breathe O_2, oxygen molecules made of two oxygen atoms. Other molecules, such as water (H_2O) and sulfuric acid (H_2SO_4), are made up of atoms of two or more different elements; such molecules are called **compounds**. The chemical properties of a molecule are different from those of its individual atoms. For example, water behaves very differently than pure hydrogen or pure oxygen, even though each water molecule is composed of two hydrogen atoms and one oxygen atom, as indicated by the familiar symbol H_2O.

Common Misconceptions: The Illusion of Solidity

Bang your hand on a table. Although the table feels solid, it is made almost entirely of empty space! Nearly all the mass of the table is contained in the nuclei of its atoms. But the volume of an atom is more than a trillion times the volume of its nucleus, so relatively speaking the nuclei of adjacent atoms are nowhere near to touching one another. The solidity of the table comes about from a combination of electrical interactions between the charged particles in its atoms and the strange quantum laws governing the behavior of electrons. If we could somehow pack all the table's nuclei together, the table's mass would fit into a microscopic speck. Although *we* cannot pack matter together in this way, nature can and does—in *neutron stars,* which we will study in Chapter 17.

Atomic Structure

Atoms are incredibly small: Millions could fit end to end across the period at the end of this sentence, and the number in a single drop of water (10^{22} to 10^{23} atoms) may exceed the number of stars in the observable universe. Yet atoms are composed of even smaller particles. A small, dense **nucleus** lies at the center of an atom; atomic nuclei (plural of nucleus) are made of particles called **protons** and **neutrons**. The nucleus is surrounded by particles called **electrons**.

The properties of an atom depend mainly on the amount of **electrical charge** in its nucleus. Electrical charge is a fundamental physical property that is always conserved, just as energy is always conserved. Electrons carry a negative electrical charge of -1, protons carry a positive electrical charge of $+1$, and neutrons are electrically neutral. Oppositely charged particles attract one another, and similarly charged particles repel one another. The attraction between the positively charged protons in the nucleus and the negatively charged electrons that surround it is what holds an atom together. Ordinary atoms have identical numbers of electrons and protons, making them electrically neutral overall. (You may wonder why electrical repulsion doesn't cause the positively charged protons in a nucleus to fly apart from one another. It tries, but it is overcome by an even stronger force that holds nuclei together, called the *strong nuclear force,* which we will discuss in Chapter S4.)

Although electrons can be thought of as tiny particles, they are not quite like tiny grains of sand, and they don't really orbit the nucleus as planets orbit the Sun. Instead, the electrons in an atom are "smeared out," forming a kind of cloud that surrounds the nucleus and gives the atom its apparent size. The electrons aren't really cloudy; it's just that it is impossible to pinpoint their positions. An atomic nucleus is very tiny, even compared to the atom itself: If we imagine an atom on a scale on which its nucleus is the size of your fist, its electron cloud would be many miles wide. Nearly all the atom's mass resides in its nucleus, because protons and neutrons are each about 2,000 times more massive than an electron. Figure 4.6 shows the structure of a typical atom.

Each different chemical element contains a different number of protons in its nucleus called its **atomic number**. For example, a hydrogen nucleus contains just one proton, so its atomic number is 1; a helium nucleus contains two protons, so its atomic number is 2.

The *combined* number of protons and neutrons in an atom is called its **atomic mass** (or *atomic weight*). The atomic mass of ordinary hydrogen is 1 because its nucleus is just a single proton. Helium usually has two neutrons in addition to its two protons, giving it an atomic mass of 4. Carbon usually has six protons and six neutrons, giving it an atomic mass of 12.

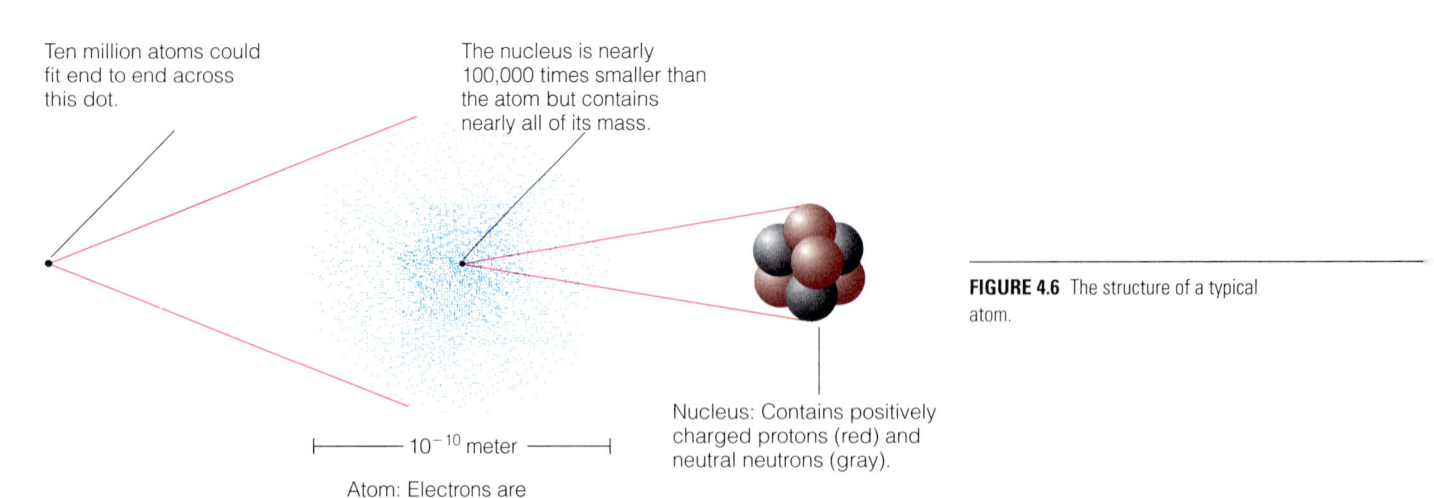

Ten million atoms could fit end to end across this dot.

The nucleus is nearly 100,000 times smaller than the atom but contains nearly all of its mass.

10^{-10} meter

Atom: Electrons are "smeared out" in a cloud around the nucleus.

Nucleus: Contains positively charged protons (red) and neutral neutrons (gray).

FIGURE 4.6 The structure of a typical atom.

Sometimes, the same element can have varying numbers of neutrons. For example, in rare cases a hydrogen nucleus contains a neutron in addition to its proton, making its atomic mass 2; hydrogen with atomic mass 2 is often called *deuterium*. The different forms of hydrogen are called **isotopes** of one another. A third isotope of hydrogen, tritium (^{3}H), contains two neutrons in addition to its one proton. Each isotope of an element has the same number of protons but a different number of neutrons.

We usually denote different isotopes by showing the atomic mass to the upper left of the chemical symbol. For example, most carbon atoms contain six neutrons in addition to their six protons, making their atomic mass 6 + 6 = 12; we write this isotope of carbon as ^{12}C and read it as carbon-12. The symbol ^{14}C (carbon-14) represents a rarer isotope of carbon with atomic mass 14 and hence eight neutrons rather than six. Figure 4.7 illustrates some atomic terminology.

TIME OUT TO THINK *The symbol ^{4}He represents helium with an atomic mass of 4; this is the most common form, containing two protons and two neutrons. What does the symbol ^{3}He represent?*

Phases of Matter

Everyday experience tells us that, depending on the temperature, the same substance exists in different phases. The main difference between phases is how tightly neighboring particles are bound together. As a substance is heated, the average kinetic energy of its particles increases, enabling the particles to break the bonds holding them to their neighbors. Each change in phase corresponds to the breaking of a different kind of bond. Phase changes occur in all substances; let's consider what happens when we heat water, starting from its solid phase, ice (Figure 4.8):

- Below 0°C (32°F), water molecules have a relatively low average kinetic energy, and each molecule is bound tightly to its neighbors, making the *solid* structure of ice.

- As the temperature increases (but remains below freezing), the rigid arrangement of the molecules in ice vibrates more and more. At 0°C, the molecules have enough energy to break the solid bonds of ice. The molecules can then move relatively freely among one another, allowing the water to flow as a *liquid*. Even in liquid water, a type of bond between adjacent molecules still keeps them close together.

- When a water molecule breaks free of all bonds with its neighbors, we call it a molecule of water vapor, which is a *gas*. Molecules in the gas phase move independently of other molecules. Even at temperatures at which water is a solid or liquid, a few molecules will have enough energy to enter the gas phase. We call the process **evaporation** when molecules escape from a liquid and **sublimation** when they escape from a solid. As temperatures rise, the rates of sublimation and

**Common Misconceptions:
One Phase at a Time?**

In daily life, we usually think of H$_2$O as being in the phase of either solid ice, liquid water, or water vapor, with the phase depending on the temperature. In reality, two or even all three phases can exist at the same time. In particular, some sublimation *always* occurs over solid ice, and some evaporation *always* occurs over liquid water. Thus, the phases of solid ice and liquid water never occur alone but instead occur in conjunction with water vapor. The *amount* of water vapor increases with the temperature, which is why more and more steam rises from a kettle as you heat it.

FIGURE 4.7 Terminology of atoms.

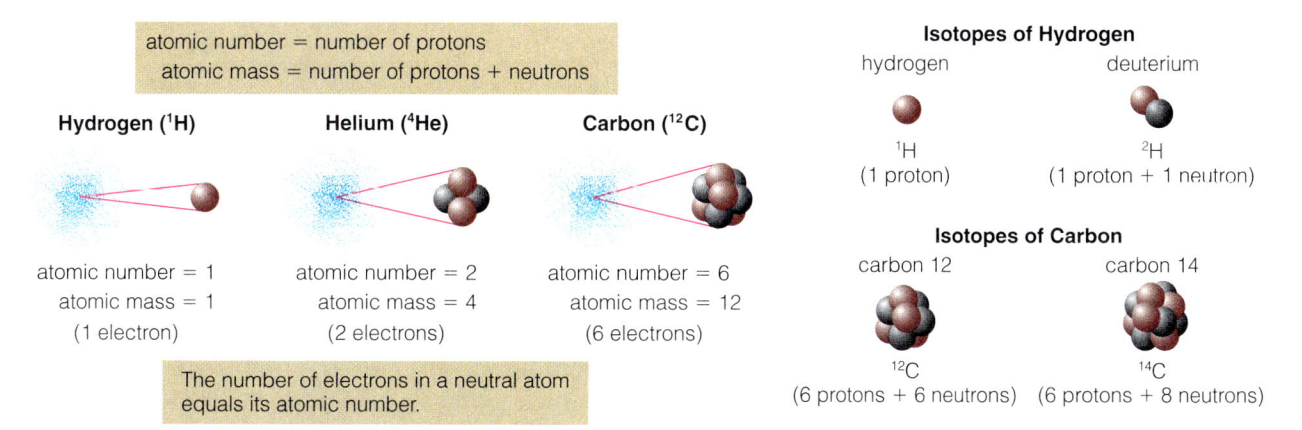

atomic number = number of protons
atomic mass = number of protons + neutrons

Hydrogen (^{1}H)
atomic number = 1
atomic mass = 1
(1 electron)

Helium (^{4}He)
atomic number = 2
atomic mass = 4
(2 electrons)

Carbon (^{12}C)
atomic number = 6
atomic mass = 12
(6 electrons)

The number of electrons in a neutral atom equals its atomic number.

Isotopes of Hydrogen

hydrogen
^{1}H
(1 proton)

deuterium
^{2}H
(1 proton + 1 neutron)

Isotopes of Carbon

carbon 12
^{12}C
(6 protons + 6 neutrons)

carbon 14
^{14}C
(6 protons + 8 neutrons)

evaporation increase, eventually changing all the solid and liquid water into water vapor.

What happens if we continue to raise the temperature of water vapor? As the temperature rises, the molecules move faster, making collisions among them more violent. These collisions eventually split the water molecules into their component atoms of hydrogen and oxygen. The process by which the bonds that hold the atoms of a molecule together are broken is called **molecular dissociation**. At still higher temperatures, collisions can break the bonds holding electrons around the nuclei of individual atoms, allowing the electrons to go free. The loss of one or more negatively charged electrons leaves a remaining atom with a net positive charge. Such charged atoms are called **ions**, and the process of stripping electrons from atoms is called **ionization**. Thus, at high temperatures, what once was water becomes a hot gas consisting of freely moving electrons and positively charged ions of hydrogen and oxygen. This type of hot gas, in which atoms have become ionized, is called a **plasma**, sometimes referred to as "the fourth phase of matter." Other chemical substances go through similar phase changes, but the temperatures at which phase changes occur depend on the types of atom or molecules involved.

Note that neutral hydrogen contains only one electron, which balances the single positive charge of the one proton in its nucleus. Thus, hydrogen can be ionized only once, and the remaining hydrogen ion, designated H$^+$, is simply a proton. Oxygen, with atomic number 8, has eight electrons when it is neutral, so it can be ionized multiple times. *Singly ionized* oxygen is missing one electron, so it has a charge of +1 and is designated O$^+$. *Doubly ionized* oxygen, or O^{++}, is missing two electrons; *triply ionized* oxygen, or O^{+3}, is missing three electrons; and so on. At extremely high temperatures, oxygen can be *fully ionized*, in which case all eight electrons are stripped away and the remaining ion has a charge of +8.

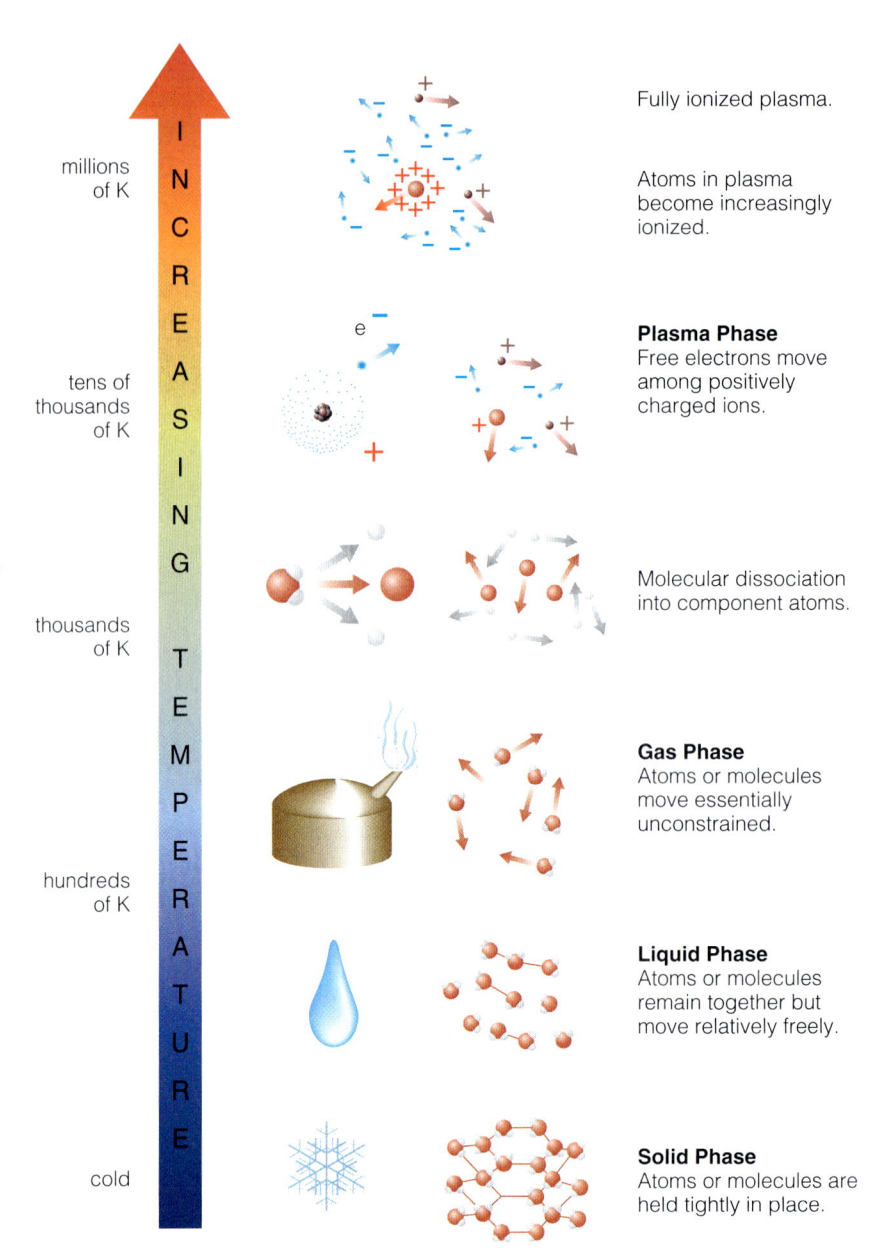

FIGURE 4.8 The general progression of phase changes.

4.4 Energy in Atoms

So far, we've seen two different ways in which atoms have energy. First, by virtue of their mass, they possess mass-energy in the amount mc^2. Second, they possess kinetic energy by virtue of their motion. Atoms also contain energy in a third way: as *electric potential energy* in the distribution of their electrons around their nuclei. The simplest case is that of hydrogen, which has only one electron. Remember that an electron tends to be "smeared out" into a cloud around the nucleus. When the electron is "smeared out" to the minimum extent that nature allows, the atom contains its smallest possible amount of electric potential energy; we say that the atom is in its **ground state** (Figure 4.9). If the electron somehow gains energy, then it becomes "smeared out" over a

ground state   excited state   ionization

FIGURE 4.9 In its ground state, an electron is "smeared out" to the minimum extent allowed by nature. Adding energy can raise the electron to an excited state that occupies a larger volume. Adding enough energy can ionize the atom.

greater volume, and we say that the atom is in an **excited state**. If the electron gains enough energy, it can escape the atom completely, in which case the atom has been *ionized*.

Perhaps the most surprising aspect of atoms was discovered in the 1910s, when scientists realized that electrons in atoms can have only *particular* energies (which correspond to particular sizes and shapes of the electron cloud). As a simple analogy, suppose you're washing windows on a building. If you use an adjustable platform to reach high windows, you can stop the platform at any height above the ground (Figure 4.10a). But if you use a ladder, you can stand only at *particular* heights—the heights of the rungs of the ladder—and not at any height in between (Figure 4.10b). The possible energies of electrons in atoms are like the possible heights on a ladder. Only a few particular energies are possible, and energies between these special few are not possible.

The possible energy levels of the electron in hydrogen are represented like the steps of a ladder in Figure 4.11. The ground state, or level 1, is the bottom rung of the ladder; its energy is labeled zero because the atom has no excess electrical potential energy to lose. Each subsequent rung of the ladder represents a possible excited state for the electron. Each level is labeled with the electron's energy above the ground state in units of **electron-volts**, or **eV** (1 eV = 1.60×10^{-19} joule). For example, the energy of an electron in energy level 2 is 10.2 eV greater than that of an electron in the ground state. That is, an electron must gain 10.2 eV of energy to "jump" from level 1 to level 2. Similarly, jumping from level 1 to level 3 requires gaining 12.1 eV of energy.

Note that, unlike a ladder built for climbing, the rungs on the electron's energy ladder are closer together near the top. The top itself represents the energy of ionization—if the electron gains this much energy, 13.6 eV above the ground state in the case of hydrogen, the electron breaks free from the atom. (Any excess energy beyond that needed for ionization becomes kinetic energy of the free-moving electron.)

Because energy is always conserved, an electron cannot jump to a higher energy level unless its atom

FIGURE 4.10 (**a**) A window washer on an adjustable platform can be at any height. (**b**) A window washer on a ladder can be at only the particular heights of the steps. Similarly, electrons in an atom can have only particular energy levels.

ionization level ——————— 13.6 eV
level 4 ——————— 12.8 eV
level 3 ——————— 12.1 eV

level 2 ——————— 10.2 eV

level 1 ——————— 0 eV
(ground state)

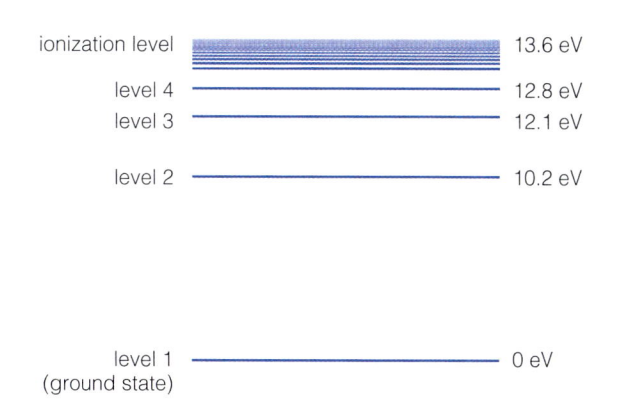

FIGURE 4.11 Energy levels for the electron in a hydrogen atom. There are many more closely spaced energy levels between level 4 and the ionization level.

gains the energy from somewhere else. Generally, the atom gains this energy either from the kinetic energy of another particle colliding with it or from the absorption of energy carried by light. Similarly, when an electron falls to a *lower* energy level, it either transfers its energy to another particle through a collision or emits light that carries the energy away. The key point is this: *Electron jumps can occur only with the particular amounts of energy representing differences between possible energy levels.*

The result is that electrons in atoms can absorb or emit only particular amounts of energy and not other amounts in between. For example, if you attempt to provide a hydrogen atom in the ground state with 11.1 eV of energy, the atom won't accept it because it is too high to boost the electron to level 2 but not high enough to boost it to level 3.

TIME OUT TO THINK *We will see in Chapter 6 that light comes in "pieces" called* photons *that carry specific amounts of energy. Can a hydrogen atom absorb a photon with 11.1 eV of energy? Why or why not? Can it absorb a photon with 10.2 eV of energy? Explain.*

If you think about it, the idea that electrons in atoms can jump only between particular energy levels is quite bizarre. It is as if you had a car that could go only particular speeds and not other speeds in between. How strange it would seem if your car suddenly jumped from 5 miles per hour to 20 miles per hour without gradually passing through a speed of 10 miles per hour! In scientific terminology, the electron's energy levels are said to be *quantized,* and the study of the energy levels of electrons (and other particles) is called *quantum mechanics.*

Electrons have quantized energy levels in all atoms, not just in hydrogen. Moreover, the allowed energy levels differ from element to element and even from one ion of an element to another ion of the same element. This fact holds the key to the study of distant objects in the universe. As we will see in Chapter 6 when we study light, the different energy levels of different elements allow light to carry "fingerprints" that can tell us the chemical composition of distant objects.

Common Misconceptions: Orbiting Electrons?

Most people have been taught to think of electrons in atoms as "orbiting" the nucleus like tiny planets orbiting a tiny sun. But this representation is simply not true. Electrons and other subatomic particles do not behave at all like baseballs, rocks, or planets. In fact, the behavior of subatomic particles is so strange that human minds may be incapable of visualizing it. Nevertheless, the *effects* of electrons are easy to observe, and one of the most important effects is that electrons give atoms their size. Thus, we say that electrons in atoms are "smeared out" into an "electron cloud" around the nucleus. This description is rather vague, but it is far more accurate than the misleading picture of electrons circling like tiny planets.

THE BIG PICTURE

In this chapter, we discussed the concepts of energy and matter in some detail. Key "big picture" ideas to draw from this chapter include the following:

- Energy and matter are the two basic ingredients of the universe. Therefore, understanding energy and matter, and the interplay between them, is crucial to understanding the universe.

- Energy is always conserved, and we can understand many processes in the universe by following how energy changes from one form to another in its interactions with matter. Don't forget that mass itself is a form of potential energy, called mass-energy.

- The strange laws of quantum mechanics govern the interactions of matter and energy on the atomic level. Electrons in atoms can have only particular energies and not energies in between, and each element has a different set of allowed energy levels.

Review Questions

1. Briefly describe and differentiate between *kinetic energy, potential energy,* and *radiative energy.*

2. What is the formula for the kinetic energy of an object? Based on this formula, explain why (a) a 4-ton truck moving at 100 km/hr has 4 times as much kinetic energy as a 1-ton car moving at 100 km/hr; (b) a 1-ton car moving at 100 km/hr has the *same* kinetic energy as a 4-ton truck moving at 50 km/hr.

3. What does *temperature* measure? How is it related to kinetic energy? What is *thermal energy?*

4. Briefly describe how a 500°F oven and a pot of boiling water illustrate the difference between *temperature* and *heat.* How is this difference important to astronauts in Earth orbit?

5. What is *gravitational potential energy?* How does an object's gravitational potential energy depend on its mass, the distance it has to fall, and the strength of gravity?

6. What do we mean by *mass-energy?* Explain the meaning of the formula $E = mc^2$ and what it has to do with the Sun, nuclear bombs, and particle accelerators.

7. What is the *law of conservation of energy?* Your body is using energy right now to keep you alive. Where does this energy come from? Where does it go?

8. Briefly define *atom, element,* and *molecule.* How was Democritus's idea of atoms similar to the modern concept of atoms? How was it different?

9. What is *electrical charge?* What type of electrical charge is carried by *protons, neutrons,* and *electrons?* Under what circumstances do electrical charges attract? Under what circumstances do they repel?

10. Briefly describe the structure of an atom. How big is an atom? How big is the *nucleus* in comparison to the entire atom?

11. Define and distinguish between *atomic number* and *atomic mass.* Under what conditions are two atoms different *isotopes* of the same element?

12. What do we mean by a *phase* of matter? Briefly describe how the idea of *bonds* between atoms (or molecules) explains the difference between the *solid, liquid,* and *gas* phases.

13. Briefly explain why a few atoms (or molecules) are always in gas phase around any solid or liquid. Then explain how *sublimation* and *evaporation* are similar and how they are different.

14. Explain why, at sufficiently high temperatures, molecules undergo *molecular dissociation* and atoms undergo *ionization.*

15. What is a *plasma?* Explain why, at very high temperatures, all matter is in this "fourth phase of matter."

16. Describe the three basic ways in which atoms can contain energy. In what way does an atom in an *excited state* contain more energy than an atom in the *ground state?*

17. How are the possible energy levels of electrons in atoms similar to the possible gravitational potential energies of a person on a ladder? How are they different?

Discussion Questions

1. *Knowledge of Mass-Energy.* Einstein's discovery that energy and mass are equivalent has led to technological developments both beneficial and dangerous. Discuss some of these developments. Overall, do you think the human race would be better or worse off if we had never discovered that mass is a form of energy? Defend your opinion.

2. *Perpetual Motion Machines.* Every so often, someone claims to have built a machine that can generate energy perpetually from nothing. Why isn't this possible according to the known laws of nature? Why do you think claims of perpetual motion machines sometimes receive substantial media attention?

3. *Indoor Pollution.* Given that sublimation and evaporation are very similar processes, why is sublimation generally much more difficult to notice? Discuss how sublimation, particularly from plastics and other human-made materials, can cause "indoor pollution."

4. *Democritus and the Path of History.* Besides his belief in atoms, Democritus held several other strikingly modern notions. For example, he maintained that the Moon was a world with mountains and valleys and that the Milky Way was composed of countless individual stars—ideas that weren't generally accepted until the time of Galileo, more than 2,000 years later. Unfortunately, we know of Democritus's work only secondhand because none of the 72 books he is said to have written survived the destruction of the Library of Alexandria. How do you think history might have been different if the work of Democritus had not been lost?

Problems

Sensible Statements? For **problems 1–4**, decide whether the statement is sensible and explain why it is or is not.

1. The sugar in my soda will provide my body with about a million joules of energy.

2. When I drive my car at 30 miles per hour, it has three times as much kinetic energy as it does at 10 miles per hour.

3. If you put an ice cube outside the Space Station, it would take a very long time to melt, even though the temperature in Earth orbit is several thousand degrees (Celsius).

4. Someday soon, scientists are likely to build an engine that produces more energy than it consumes.

5. *Gravitational Potential Energy.*
   a. Why does a bowling ball perched on a cliff ledge have more gravitational potential energy than a baseball perched on the same ledge?
   b. Why does a diver on a 10-meter platform have more gravitational potential energy than a diver on a 3-meter diving board?
   c. Why does a 100-kg satellite orbiting Jupiter have more gravitational potential energy than a 100-kg satellite orbiting the Earth, assuming both satellites orbit at the same distance from the planet centers?

6. *Einstein's Famous Formula.*
   a. What is the meaning of the formula $E = mc^2$? Be sure to define each variable.
   b. How does this formula explain the generation of energy by the Sun?
   c. How does this formula explain the destructive power of nuclear bombs?

7. *Atomic Terminology Practice.*
   a. The most common form of iron has 26 protons and 30 neutrons in its nucleus. State its atomic number, atomic mass, and number of electrons if it is electrically neutral.
   b. Consider the following three atoms: Atom 1 has 7 protons and 8 neutrons; atom 2 has 8 protons and 7 neutrons; atom 3 has 8 protons and 8 neutrons. Which two are *isotopes* of the same element?

c. Oxygen has atomic number 8. How many times must an oxygen atom be ionized to create an O^{+5} ion? How many electrons are in an O^{+5} ion?

d. Consider fluorine atoms with 9 protons and 10 neutrons. What are the atomic number and atomic mass of this fluorine? Suppose we could add a proton to this fluorine nucleus. Would the result still be fluorine? Explain. What if we added a neutron to the fluorine nucleus?

e. The most common isotope of gold has atomic number 79 and atomic mass 197. How many protons and neutrons does the gold nucleus contain? If it is electrically neutral, how many electrons does it have? If it is triply ionized, how many electrons does it have?

f. The most common isotope of uranium is ^{238}U, but the form used in nuclear bombs and nuclear power plants is ^{235}U. Given that uranium has atomic number 92, how many neutrons are in each of these two isotopes of uranium?

8. *The Fourth Phase of Matter.*
   a. Explain why nearly all the matter in the Sun is in the plasma phase.
   b. Based on your answer to part (a), explain why plasma is the most common phase of matter in the universe.
   c. Given that plasma is the most common phase of matter in the universe, why is it so rare on Earth?

9. *Energy Level Transitions.* The labeled transitions below represent an electron moving between energy levels in hydrogen. Answer each of the following questions and explain your answers.

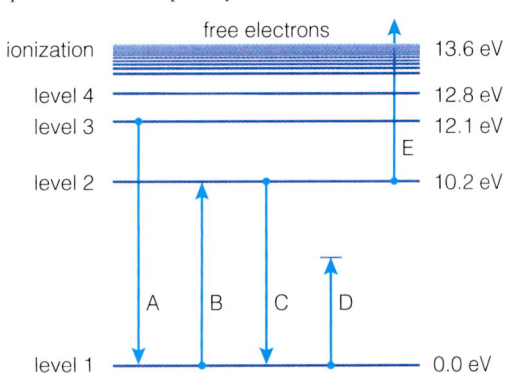

   a. Which transition could represent an electron that *gains* 10.2 eV of energy?
   b. Which transition represents an electron that *loses* 10.2 eV of energy?
   c. Which transition represents an electron that is breaking free of the atom?
   d. Which transition, as shown, is *not* possible?
   e. Describe the process taking place in transition A.

*10. *Energy Comparisons.* Use the data in Table 4.1 to answer each of the following questions.
   a. Compare the energy of a 1-megaton hydrogen bomb to the energy released by a major earthquake.
   b. If the United States obtained all its energy from oil, how much oil would be needed each year?
   c. Compare the Sun's annual energy output to the energy released by a supernova.

*11. *Moving Candy Bar.* Metabolizing a candy bar releases about 10^6 joules. How fast must the candy bar travel to have the same 10^6 joules in the form of kinetic energy? (Assume the candy bar's mass is 0.2 kg.) Is your answer faster or slower than you expected?

*12. *Calculating Densities.* Find the average density of the following objects in grams per cubic centimeter.
   a. A rock with volume 15 cm^3 and mass 50 g.
   b. Earth, with its mass of 6×10^{24} kg and radius of about 6,400 km. (*Hint:* The formula for the volume of a sphere is $\frac{4}{3} \times \pi \times$ radius^3.)
   c. The Sun, with its mass of 2×10^{30} kg and radius of about 700,000 km.

*13. *Spontaneous Human Combustion.* Suppose that, through a horrific act of an angry god (or a very powerful alien, if you prefer), all the mass in your body was suddenly converted into energy according to the formula $E = mc^2$. How much energy would be produced? Compare this to the energy released by a 1-megaton hydrogen bomb (see Table 4.1). What effect would your disappearance have on the surrounding region? (*Hint:* Multiply your mass in kilograms by c^2.)

*14. *Fusion Power.* No one has yet succeeded in creating a commercially viable way to produce energy through nuclear fusion. However, suppose we could build fusion power plants using the hydrogen in water as a fuel. Based on the data in Table 4.1, how much water would we need each minute in order to meet U.S. energy needs? Could such a reactor power the entire United States with the water flowing from your kitchen sink? Explain. (*Hint:* Use the annual U.S. energy consumption to find the energy consumption per minute; then divide by the energy yield from fusing 1 liter of water to figure out how many liters would be needed each minute.)

Web Projects

Find useful links for Web projects on the text Web site.

1. *Energy Comparisons.* Using information from the Energy Information Administration Web site, choose some aspect of U.S. or world energy use that interests you. Write a short report on this issue.

2. *Nuclear Power.* There are two basic ways to generate energy from atomic nuclei: through nuclear fission (splitting nuclei) and through nuclear fusion (combining nuclei). All current nuclear reactors are based on fission, but fusion would have many advantages if we could develop the technology. Research some of the advantages of fusion and some of the obstacles to developing fusion power. Do you think fusion power will be a reality in your lifetime? Explain.

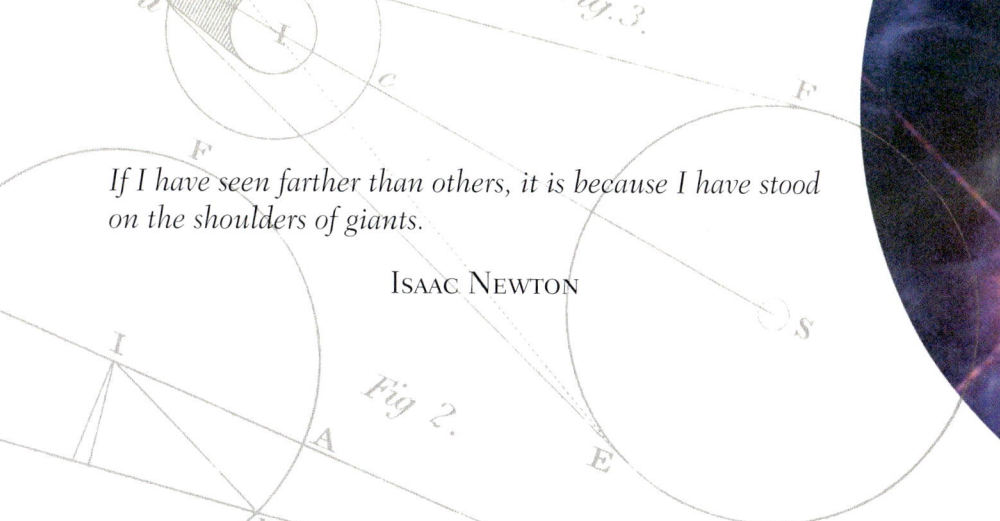

If I have seen farther than others, it is because I have stood on the shoulders of giants.

ISAAC NEWTON

CHAPTER 5
Universal Motion
From Copernicus to Newton

Everything in the universe is in constant motion, from the random meandering of molecules in the air to the large-scale drifting of galaxies in superclusters. Remarkably, just a few physical laws describe all this motion. The elucidation of these laws over the past several centuries is surely one of the greatest scientific triumphs of all time.

We have two primary goals in this chapter: understanding the laws that govern the motion of celestial objects and understanding how these laws were discovered. Achieving these goals requires that we first explore motion and the basic laws that govern it more generally. Once we develop this background, you'll be able to understand the extraordinary story of how the ancient belief in an Earth-centered universe was finally overthrown.

As in Chapter 4, much of the subject matter of this chapter may already be familiar to you. Again, don't worry if this is not the case. The material is not difficult, and studying it carefully will greatly enhance your understanding of the astronomy in the rest of the book as well as of many everyday phenomena.

5.1 Describing Motion: Examples from Daily Life

Think about what happens when you throw a ball to a dog: The dog runs and catches it. Now think about the complexity of this trick. The ball leaves your hand traveling in some particular direction with some particular amount of kinetic energy. As the ball rises, gravity converts some of its kinetic energy into potential energy, slowing the ball's rise until it reaches the top of its trajectory. Then gravity transforms the potential energy of the ball back into kinetic energy, bringing it back toward the ground. Meanwhile, the ball may lose some of its kinetic energy to air resistance or may be pushed by gusts of wind. Despite this complexity, the dog still catches the ball.

We humans can perform an even better trick: We have learned how to figure out where the ball will land even before throwing it, and we can perform this trick with extraordinary precision. Understanding how we perform this trick and applying it to problems of motion throughout the universe require understanding the laws that govern motion. We'll study these laws shortly, but first we need to discuss the scientific concepts used to describe motion.

We all have a great deal of experience with motion and natural intuition as to how motion works, so we begin our discussion with some familiar examples. Indeed, you probably are familiar with all the terms defined in this section, but their scientific definitions may differ subtly from those you use in casual conversation.

Speed, Velocity, and Acceleration

The concepts we use to determine the trajectory of a ball, a rocket, or a planet are familiar to you from driving a car. The speedometer indicates your **speed**, usually in units of both miles per hour (mi/hr) and kilometers per hour (km/hr); for example, 100 km/hr is a speed. Your **velocity** is your speed in a certain direction; "100 km/hr going due north" describes a velocity. It is possible to change your velocity without changing your speed, for example, by maintaining a steady 60 km/hr as you drive around a curve. Because your direction is changing as you round the curve, your *velocity* is also changing—even though your *speed* is constant.

Whenever your velocity is changing, you are experiencing **acceleration**. You are undoubtedly familiar with the term *acceleration* as it applies to increasing speed, such as when you accelerate away from a stop sign while driving your car. In science, we also say that you are accelerating when you slow down or turn. Slowing occurs when your acceleration is in a direction opposite to your motion; we say that your acceleration is negative, causing your velocity to decrease. Turning changes your direction, which means a change in velocity and thus involves acceleration even if your speed remains constant.

Note that you don't feel anything when you are traveling at *constant velocity*, which is why you don't feel any sensation of motion when you're traveling in an airplane on a smooth flight. In contrast, you can often feel acceleration: As you speed up in a car you feel yourself being pushed back into your seat, as you slow down you feel yourself being pulled forward from the seat, and as you drive around a curve you lean outward because of your acceleration.

The Acceleration of Gravity

One of the most important types of acceleration is that caused by gravity, which makes objects accelerate as they fall. In a famous (though probably apocryphal) experiment that involved dropping weights from the Leaning Tower of Pisa, Galileo demonstrated that gravity accelerates all objects by the same amount, regardless of their mass. This fact may be surprising because it seems to contradict everyday experience: A feather floats gently to the ground, while a rock plummets. However, this difference is caused by air resistance. If you dropped a feather and a rock on the Moon, where there is no air, both would fall at exactly the same rate.

TIME OUT TO THINK *Find a piece of paper and a small rock. Hold both at the same height, one in each hand, and let them go at the same instant. The rock, of course, hits the ground first. Next crumple the paper into a small ball and repeat the experiment. What happens? Explain how this experiment suggests that gravity accelerates all objects by the same amount.*

The acceleration of a falling object is called the **acceleration of gravity**, abbreviated g. On Earth, the acceleration of gravity causes falling objects to fall faster by 9.8 meters per second (m/s), or about 10 m/s, with each passing second. For example, suppose you drop a rock from a tall building. At the moment you let it go, its speed is 0 m/s. After 1 second, the rock will be falling downward at about 10 m/s. After 2 seconds, it will be falling at about 20 m/s. In the absence of air resistance, its speed will continue to increase by about 10 m/s each second until it hits the ground (Figure 5.1).

We therefore say that the acceleration of gravity is about 10 *meters per second per second*, or 10 *meters per second squared*, which we write as 10 m/s^2. (More precisely, $g = 9.8$ m/s^2.)

Momentum and Force

Imagine that you're innocently stopped in your car at a red light when a bug flying at a velocity of 30 km/hr

$t = 0$
$v = 0$

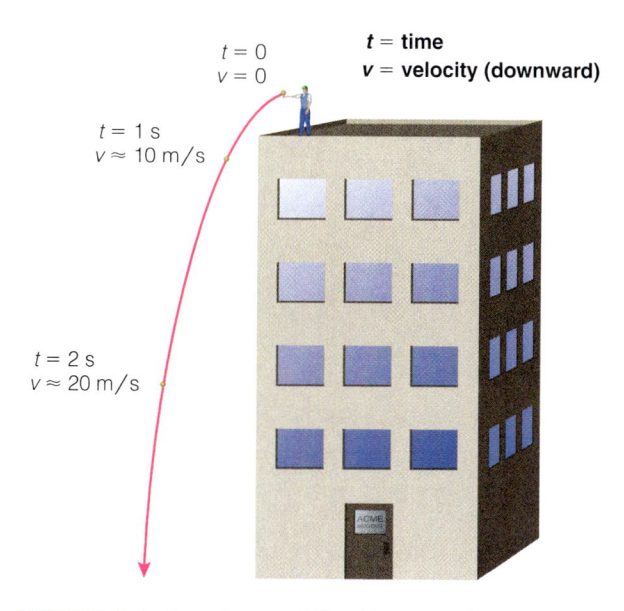

$t = 1$ s
$v \approx 10$ m/s

$t = 2$ s
$v \approx 20$ m/s

t = time
v = velocity (downward)

FIGURE 5.1 On Earth, gravity causes falling objects to accelerate downward at about 10 m/s^2. That is, a falling object's downward velocity increases by about 10 m/s with each passing second. (Gravity does not affect horizontal velocity.) More precisely, $g = 9.8$ m/s^2.

due south slams into your windshield. What will happen to your car? Not much, except perhaps a bit of a mess on your windshield. Next imagine that a 2-ton truck runs the red light and hits you head-on with the same velocity as the bug. Clearly, the truck will cause far more damage.

Scientifically, we say that the truck imparts a much larger jolt than the bug because it transfers more **momentum** to you. Momentum describes a combination of mass and velocity. We can describe the momentum of the truck before the collision as "2 tons moving due south at 30 km/hr," while the momentum of the bug is perhaps "1 gram moving due south at 30 km/hr." Mathematically, momentum is defined as mass × velocity.

In transferring some of its momentum to your car, the truck (or bug) exerts a **force** on your car. More generally, a force is anything that can cause a change in momentum. You are familiar with many types of force besides collisional force. For example, if you shift into neutral while driving along a flat stretch of road, the forces of air resistance and road friction will continually sap your car's momentum (transferring it to molecules in the air and the pavement), slowing your velocity until you come to a stop. You probably are also familiar with forces arising from gravity, electricity, and magnetism.

The mere presence of a force does not always cause a change in momentum. For example, if the engine works hard enough, a car can maintain constant velocity—and hence constant momentum—despite air resistance and road friction. In this case, the force generated by the engine to turn the wheels precisely offsets the forces of air resistance and road friction that act to slow the car, and we say that no **net force** is acting on the car.

As long as an object is not shedding (or gaining) mass, a change in momentum means a change in velocity (because momentum = mass × velocity), which means an acceleration. Hence, any net force will cause acceleration, and all accelerations must be caused by a force. That is why you feel forces when you accelerate in your car.

Mass and Weight

Up until now, we've been glossing over one key term: *mass*. Your **mass** refers to the amount of matter in your body, which is different from your *weight*. Imagine standing on a scale in an elevator (Figure 5.2). When the elevator is stationary or moving at constant velocity, the scale reads your "normal" weight. When the elevator is accelerating upward, the floor exerts an additional force that makes you feel heavier, and the scale verifies your greater apparent weight.[1] Note that your weight is greater than its "normal" value only while the elevator is *accelerating*, not while it is moving at constant velocity. When the elevator accelerates downward, the floor and the scale are dropping away, so your weight is reduced. Thus, your weight varies with the elevator's motion, but your mass remains the same. (You can verify these facts by taking a small bathroom scale with you on an elevator.)

More precisely, your (apparent) **weight** describes the *force* that acts on your mass, and it depends on the strength of gravity and other forces acting on you (such as the force due to the elevator's acceleration). Thus, while your mass is the same anywhere, your weight can vary. For example, your mass would be the same on the Earth and on the Moon, but you would weigh less on the Moon because of its weaker gravity.

Free-Fall, Weightlessness, and Orbit

If the cable breaks so that the elevator is in **free-fall**, the floor drops away at the same rate that you fall. You lose contact with the scale, so your apparent weight is zero and you feel **weightless**. In fact, you are in free-fall whenever nothing is *preventing* you from falling. For example, you are in free-fall when you jump off a chair or spring from a diving board or trampoline. Surprising as it may seem, you have therefore experienced weightlessness many times in your life and can experience it right now simply by jumping

---

[1]Many physics texts distinguish between *true weight* that is due only to the effects of gravity on mass and the *apparent weight* that a scale reads when other forces (such as in an accelerating elevator) also act. In this book, you may assume that "weight" refers to apparent weight, except when stated otherwise.

CHAPTER 5 UNIVERSAL MOTION: FROM COPERNICUS TO NEWTON

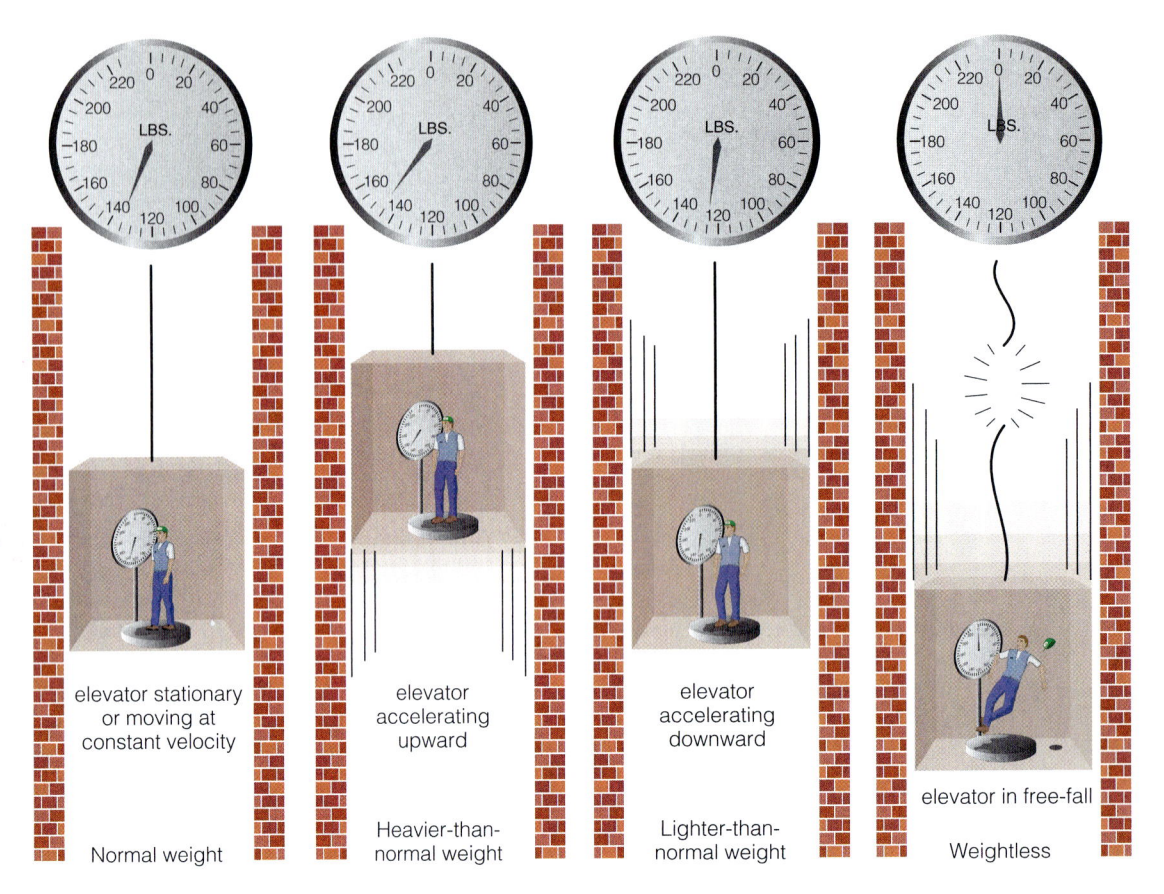

FIGURE 5.2 Mass is not the same as weight. The man's mass never changes, but his apparent weight does.

(From left to right: elevator stationary or moving at constant velocity — Normal weight; elevator accelerating upward — Heavier-than-normal weight; elevator accelerating downward — Lighter-than-normal weight; elevator in free-fall — Weightless.)

Common Misconceptions: No Gravity in Space?

Most people are familiar with pictures of astronauts floating weightlessly in Earth orbit. Unfortunately, because we usually associate weight with gravity, many people assume that the astronauts' weightlessness implies a lack of gravity in space. Actually, there's plenty of gravity in space; even at the distance of the Moon, the Earth's gravity is strong enough to hold the Moon in orbit. In fact, in the low-Earth orbit of the Space Station, the acceleration of gravity is scarcely less than it is on the Earth's surface. Why, then, are the astronauts weightless? Because the Space Station and all other orbiting objects are in a constant state of *free-fall,* and any time you are in free-fall you are weightless. Imagine being an astronaut. You'd have the sensation of free-fall—just as when falling from a diving board—the entire time you were in orbit. This constantly falling sensation makes most astronauts sick to their stomach when they first experience weightlessness. Fortunately, they quickly get used to the sensation, which allows them to work hard and enjoy the view.

off your chair. Of course, your weightlessness lasts for only the very short time until you hit the ground.

Astronauts are weightless for much longer periods because orbiting spacecraft are in a constant state of free-fall. To understand why, imagine a cannon that shoots a ball horizontally from a tall mountain. Once launched, the ball falls solely because of gravity and hence is in free-fall. Thus, the faster the cannon shoots the ball, the farther it goes (Figure 5.3). If the cannonball could go fast enough, its motion would keep it constantly "falling around" the Earth and it would never hit the ground (as long as we neglect air resistance). This is precisely what spacecraft such as the Space Shuttle and the Space Station do as they orbit the Earth. Their constant state of free-fall makes the spacecraft and everything in them weightless.

If an orbiting object travels at the **escape velocity** or faster, it can completely escape from the Earth, never to return. (It will still be weightless as long as it does not fire any engines, because it is still in free-fall—only now "falling" in response to the gravity of the Sun, planets, and other objects.) The escape velocity from the Earth's surface is about 40,000 km/hr (25,000 mi/hr). We'll discuss escape velocity in more detail in Section 5.6.

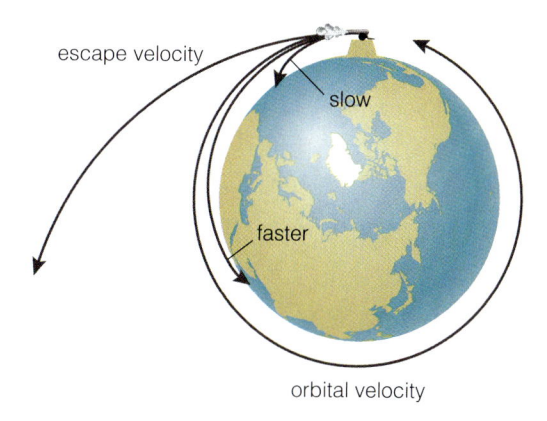

FIGURE 5.3 The faster the cannonball is shot, the farther it goes before hitting the ground. If it goes fast enough, it will continually "fall around," or orbit, the Earth. With an even faster speed, it may escape the Earth's gravity altogether.

TIME OUT TO THINK *In the* Hitchhiker's Guide to the Galaxy *books, author Douglas Adams says that the trick to flying is to throw yourself at the ground and miss. Although this phrase does not really explain flying, which involves lift from air, it does describe an orbit fairly well. Explain.*

5.2 Understanding Motion: Newton's Laws

The human "trick" of being able to figure out where a ball will land before it is thrown or to predict how other motions will unfold requires understanding precisely how forces affect objects in motion. At first, the complexity of motion in daily life might lead you to guess that the laws governing how forces affect motion are quite complex. For example, if you watch a piece of paper fall to the ground, you'll see it waft lazily back and forth in a seemingly unpredictable pattern. However, the complexity of this motion arises because the paper is affected by a variety of forces, including gravity and the changing forces caused by air currents. If you could analyze the forces individually, you'd find that each force affects the paper's motion in a simple, predictable way. Three simple laws, known as **Newton's laws of motion**, describe how forces affect motion. Let's investigate them, again using examples from everyday life. (Credit for discovering these laws really should go jointly to Galileo and Newton. They are called Newton's laws because he enumerated them in his book *Principia*.)

Newton's First Law of Motion

Newton's first law of motion deals with situations in which no net force is acting on an object. It states:

In the absence of a net force, an object moves with constant velocity.

Thus, objects at rest (velocity = 0) tend to remain at rest, and objects in motion tend to remain in motion with no change in either their speed or their direction.

The idea that an object at rest should remain at rest is rather obvious; after all, if a car is parked on a flat street, it won't suddenly start moving for no reason. But what if the car is traveling along a flat, straight road? Newton's first law says that the car should keep going forever *unless* a force acts on it. You know that the car eventually will come to a stop if you take your foot off the gas pedal, so we must conclude that some forces are stopping the car— in this case forces arising from friction and air resistance.[2] If the car were in space, and therefore unaffected by friction or air, it would keep moving forever (though gravity would eventually alter its speed and direction). That is why interplanetary spacecraft, once launched into space, need no fuel to keep going.

Although friction cannot be eliminated on Earth, it can sometimes be minimized enough so that Newton's first law becomes more evident. For example, friction is low on ice, so a single push-off allows an ice skater to glide for a long time. Popular arcade games like *air hockey* also minimize friction so that a puck can travel for a long time before friction finally stops it.

Newton's first law also explains why you don't feel any sensation of motion when you're traveling in an airplane on a smooth flight. As long as the plane is traveling at constant velocity, no net force is acting on it or on you. Therefore, you feel no different from how you would feel at rest. You can walk around the cabin, play catch with a person a few rows forward, or relax and go to sleep just as though you were "at rest" on the ground.

Newton's Second Law of Motion

Newton's second law of motion tells us what happens to an object when a net force *is* present. We have already said that a net force changes an object's momentum, accelerating it in the direction of the force. Newton's second law quantifies this relationship, which can be stated in two equivalent ways:

force = rate of change in momentum

force = mass × acceleration (or $F = ma$)

This law explains why you can throw a baseball farther than you can throw a shot-put. For both the

---

[2]Some readers might wonder why gravity does not help stop the car. A force can affect motion only if it is acting along the direction of motion. Because gravity acts downward, it cannot affect the motion of a car traveling along a flat (level) road.

CHAPTER 5 UNIVERSAL MOTION: FROM COPERNICUS TO NEWTON   129

FIGURE 5.4 (a) When you swing a ball on a string, the ball moves in a circle because the string exerts a force that pulls the ball inward. The acceleration must also be inward. (b) If the string breaks, the force disappears and the ball flies off in a straight line with velocity v.

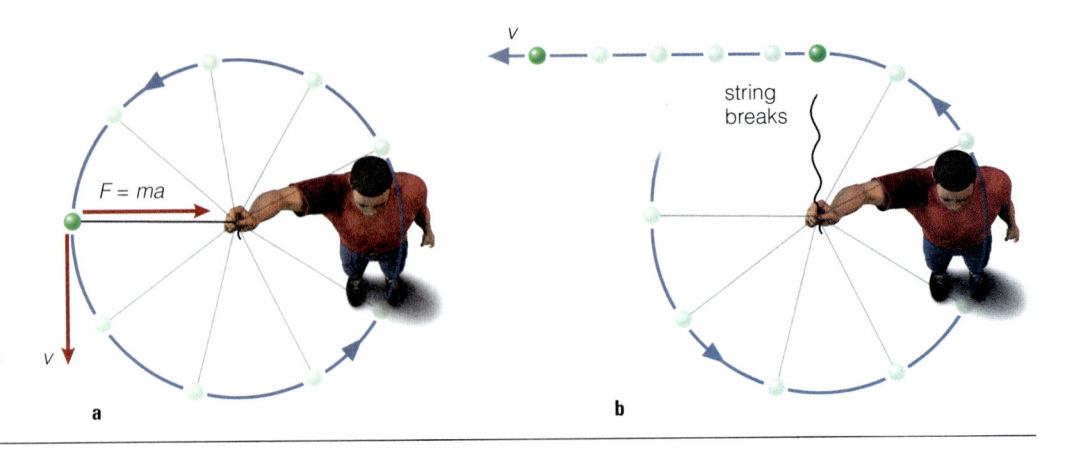

baseball and the shot-put, the force delivered by your arm equals the product of mass and acceleration. Because the mass of the shot-put is greater than that of the baseball, the same force from your arm gives the shot-put a smaller acceleration. Due to its smaller acceleration, the shot-put leaves your hand with less speed than the baseball and thus travels a shorter distance before hitting the ground.

We can also use Newton's second law of motion to understand acceleration around curves. Suppose you swing a ball on a string around your head (Figure 5.4a). The ball is accelerating even if it has a steady speed, because it is constantly changing direction. What makes it accelerate? From Newton's second law, the taut string must be applying a force to the ball. We can understand this force by considering what would happen if it disappeared, such as if the string were to break (Figure 5.4b). In that case, the ball would simply fly off in a straight line. Thus, when the string is intact, the force must be pulling the ball *inward* to keep it from flying off. Because acceleration must be in the same direction as the force, we conclude that the ball has an inward acceleration as it moves around the circle.

The same idea helps us understand the force on a car moving around a curve or a planet orbiting around the Sun. In the case of a car, the force comes from friction between the tires and the road. The tighter the curve or the faster the car is going, the greater the force needed to keep the car moving around it. If the force is not great enough, the car skids outward. For a planet moving around the Sun, the force pulling inward is gravity, as we'll discuss shortly. Thus, an orbiting planet is always accelerating toward the Sun. Indeed, it was Newton's discovery of the precise nature of this acceleration that helped him deduce the law of gravity.

Newton's Third Law of Motion

Think for a moment about standing still on the ground. The force of gravity acts downward on you,

---

Mathematical Insight 5.1 Units of Force, Mass, and Weight

Newton's second law, $F = ma$, shows that the units of force are equal to a unit of mass multiplied by a unit of acceleration. For example, if a mass of 1 kg accelerates at 10 m/s^2, the magnitude of the responsible force is

$$\text{force} = \text{mass} \times \text{acceleration}$$
$$= 1 \text{ kg} \times 10 \frac{\text{m}}{\text{s}^2} = 10 \frac{\text{kg} \times \text{m}}{\text{s}^2}$$
$$= 10 \text{ newtons}$$

The standard unit of force is the *kilogram-meter per second squared*, called the **newton** for short.

We can now further clarify the difference between mass and weight. When you stand on a scale, it records the downward force that you exert on it, which is equal and opposite to the upward force it exerts on you. If the scale were suddenly pulled out from under your feet, you would begin accelerating downward with the acceleration of gravity. Thus, when you are standing still, the scale must support you with a force equal to your mass times the acceleration of gravity. Your weight must also equal this force (but in an opposite direction):

$$\text{weight} = \text{mass} \times \text{acceleration of gravity}$$

Your apparent weight may differ from this value if forces besides gravity are affecting you.

Like any force, weight has units of mass times acceleration. Thus, although we commonly speak of weights in *kilograms*, this usage is not technically correct: Kilograms are a unit of mass, not of force. You may safely ignore this technicality as long as you are dealing with objects on the Earth that are not accelerating. In space or on other planets, the distinction between mass and weight is important and cannot be ignored.

FIGURE 5.5 Momentum conservation as demonstrated on a pool table.

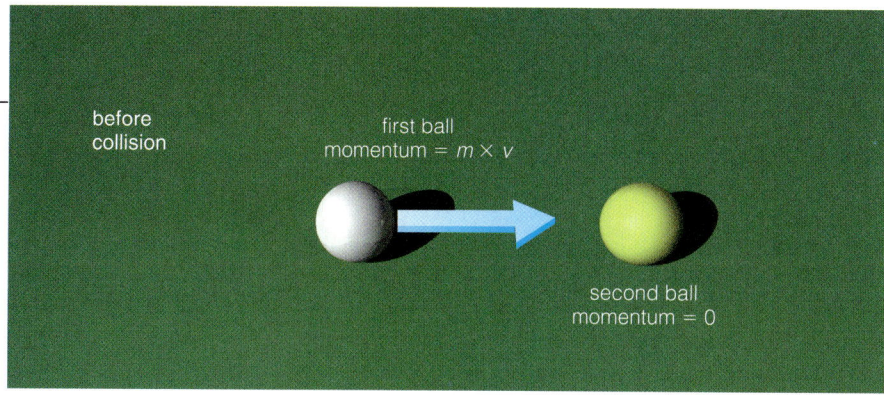

so if this force were acting alone Newton's second law would demand that you be accelerating downward. The fact that you are not falling means that the ground must be pushing back up on you with exactly the right amount of force to offset gravity. This fact is embodied in Newton's third law of motion:

For any force, there always is an equal and opposite reaction force.

According to this law, your body exerts a gravitational force on the Earth identical to the one the Earth exerts on you, except that it acts in the opposite direction. In this mutual pull, the ground just happens to be caught in the middle. Because other forces (between atoms and molecules in the Earth) keep the ground stationary with respect to the center of the Earth, the opposite, upward force is transmitted to you by the ground, holding you in place. Newton's third law also explains how rockets work: Engines generate an explosive force driving hot gas out the back, which creates an equal and opposite force propelling the rocket forward.

Conservation of Momentum

If you look more closely at Newton's laws, you will see that they all reflect aspects of a deeper principle: the *conservation of momentum.* Like the amount of energy, the total amount of momentum in the universe is conserved—that is, it does not change. Newton's first law says that an individual object's momentum will not change at all if the object is left alone. Newton's second law says that a force can change the object's momentum, but Newton's third law says that another equal and opposite force simultaneously changes some other object's momentum by a precisely opposite amount. Total momentum always remains unchanged.

Common Misconceptions: What Makes a Rocket Launch?

If you've ever watched a rocket launch, it's easy to see why many people believe that the rocket "pushes off" the ground. In fact, the ground has nothing to do with the rocket launch. The rocket takes off because of momentum conservation. Rocket engines are designed to expel hot gas with an enormous amount of momentum. To balance the explosive force driving gas out the back of the rocket, an equal and opposite force must propel the rocket forward, keeping the total momentum—gas plus rocket—unchanged. Thus, rockets can be launched horizontally as well as vertically, and a rocket can be "launched" in space (e.g., from a space station) with no need for any nearby solid ground.

On a pool table, momentum conservation is rather obvious. When one ball hits another ball "dead on," the first ball stops and the second one takes off with the speed of the first (Figure 5.5). Sometimes, momentum conservation is less apparent. When you jump into the air, how do you get your upward momentum? As your legs propel you skyward, they are actually pushing the Earth in the other direction,

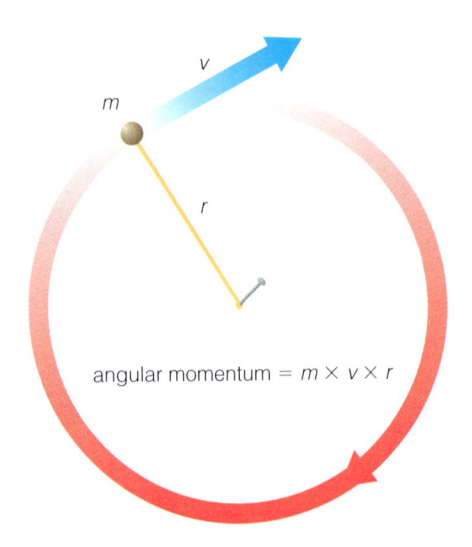

FIGURE 5.6 The angular momentum of an object moving in a circle is $m \times v \times r$.

or orbiting has angular momentum. We can write a simple formula for the angular momentum of an object moving in a circle:

$$\text{angular momentum} = m \times v \times r$$

where m is the object's mass, v is its speed around the circle, and r is the radius of the circle (Figure 5.6).

Just as momentum can be changed only by a force, the angular momentum of any object can be changed only by a twisting force, or **torque**. For example, opening a door requires rotating the door on its hinges. Making it rotate means giving it some angular momentum, which you can do by applying a torque (Figure 5.7). Note that the torque depends not only on how much force you use to push on the door, but also on *where* you push. The farther out from the hinges you push, the more torque you can apply and the easier it is to open the door.

TIME OUT TO THINK *Use the idea of torque to explain why it is easier to change a tire with a long wrench than with a short wrench.*

When no net torque is present, a law very similar to Newton's first law applies—the **law of conservation of angular momentum**:

In the absence of net torque (twisting force), the total angular momentum of a system remains constant.

A spinning ice skater illustrates this law. Because there is so little friction on ice, the ice skater essentially keeps a constant angular momentum. When she pulls in her extended arms, she effectively decreases her radius. Therefore, for the product $m \times v \times r$ to remain unchanged, her velocity of rotation must increase (Figure 5.8).

The law of conservation of angular momentum arises frequently in astronomy, because gravity is often the only significant force and gravity rarely acts

giving the Earth's momentum an equal and opposite kick. However, the Earth's huge mass renders its acceleration undetectable. During your brief flight, the gravitational forces between you and the Earth return your upward momentum to the Earth. Again, the total momentum remains the same at all times.

Conservation of Angular Momentum

In astronomy, a special kind of momentum is particularly important. Consider an ice skater spinning in place. She certainly has some kind of momentum, but it is a little different from the momentum we've discussed previously because she's not actually going anywhere. Technically, her "spinning momentum" is called **angular momentum**. (The term *angular* arises because each spin involves turning through an *angle* of 360°.) In fact, any object that is spinning

FIGURE 5.7 Opening a door requires applying a torque. Given the same amount of force, the torque on the door is greater if you push farther from the hinges (the door's rotation axis).

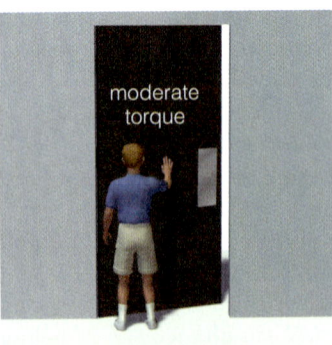

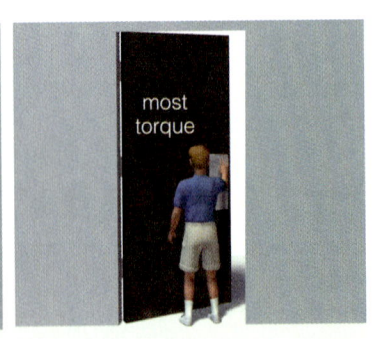

132 PART II KEY CONCEPTS FOR ASTRONOMY

FIGURE 5.8 A spinning skater conserves angular momentum.

In the product m × v × r, extended arms mean larger radius and smaller velocity of rotation.

Bringing in her arms decreases her radius and therefore increases her rotational velocity.

as a twisting force.[3] For example, angular momentum is generally conserved for rotating planets, planets orbiting a star, and rotating galaxies. Indeed, conservation of angular momentum explains why the Earth doesn't need any fuel to make it rotate: In the absence of any torque to stop the rotation, the Earth would keep rotating forever. (In Section 5.5, we'll see that Earth's rotation gradually slows, due to a torque caused by tides.)

5.3 Planetary Motion: The Copernican Revolution

Now that we have described motion in general terms, we can turn to the important issue of planetary motion in our solar system. Although planetary motion is among the simplest motions found in nature, it remained mysterious for most of human history because of its *apparent* complexity in our sky [Section 2.4]. The prevailing view of European scholars at the dawn of the Renaissance remained that of Ptolemy, whose complex model had planets spinning on circles upon circles around the Earth [Section 3.3]. The dramatic change in human understanding of planetary motion, one of the most remarkable stories in the history of science, began early in the sixteenth century, with the work of Nicholas Copernicus.

Nicholas Copernicus: The Revolution Begins

Copernicus was born in Toruń, Poland, on February 19, 1473. His family was wealthy, and he received a first-class education, studying mathematics, medicine, and law. He began studying astronomy in his late teens. By that time, tables of planetary motion based on the Ptolemaic model of the universe were noticeably inaccurate.[4] Copernicus concluded that planetary motion could be explained more simply in a Sun-centered solar system, as had been suggested by Aristarchus some 1,800 years earlier [Section 2.4], and he began developing a Sun-centered system for predicting planetary positions.

Copernicus was hesitant to publish his work for fear that his suggestion that the Earth moved would be considered absurd. Nevertheless, he discussed his system with other scholars and generated great interest in his work. At the urging of some of these scholars, including some high-ranking officials of the Catholic Church, he finally agreed to publish a book describing his system. The book, *De Revolutionibus Orbium Caelestium*, or "Concerning the Revolutions of the Heavenly Spheres," was published in 1543. Copernicus saw the first printed copy on the day he died—May 24, 1543.

In addition to its aesthetic advantages, the Sun-centered system of Copernicus allowed him to discover a mathematical relationship between a planet's true orbital period around the Sun and the time between successive appearances of the planet

---

[3]Gravity exerts a torque only when it acts on an object with a non-spherical mass distribution and a rotation axis tilted (i.e., not perpendicular) to the direction of the gravitational force; this torque causes the axis to precess. This type of torque, created by the gravitational attraction of the Sun and the Moon acting on the Earth's tilted rotation axis, is the cause of the Earth's 26,000-year precession cycle.

[4]The best set of tables, known as the *Alfonsine Tables*, was compiled in 1252 under the guidance of the Spanish monarch Alfonso X (1221–1284). Commenting on the tedious nature of the calculations required by the Ptolemaic model, Alfonso X supposedly said that if he had been present at the creation, he would have recommended a simpler design for the universe.

Copernicus (1473–1543)

at opposition in the sky [Section S1.1]. He was also able to use geometrical techniques to estimate the distances of the planets from the Sun in terms of the Earth–Sun distance (i.e., distances in astronomical units). However, the model published by Copernicus did not predict planetary positions substantially more accurately than did the old Ptolemaic model. Moreover, it remained complex because Copernicus held to the ancient Greek belief that all heavenly motions must follow perfect circles. Because the true orbits of the planets are *not* circles, Copernicus found it necessary to add circles upon circles to his system, just as in the Ptolemaic system. As a result, the Copernican system won relatively few converts in the 50 years after it was published. After all, why throw out thousands of years of tradition for a new system that predicted planetary motion equally poorly?

Tycho Brahe: The Greatest Naked-Eye Observer of All Time

Part of the difficulty faced by astronomers who sought to improve either the Ptolemaic or the Copernican system was a lack of quality data. The telescope had not yet been invented, and existing naked-eye observations were not very accurate. In the late 1500s, a Danish nobleman named Tycho Brahe (1546–1601) set about correcting this problem.

When Tycho was a young boy, his family discouraged his interest in astronomy. He therefore hid his passion, learning the constellations from a miniature model of a celestial sphere that he kept hidden. As he grew older, Tycho was often arrogant about both his noble birth and his learned abilities. At age 20, he fought a duel with another student over which of them was the better mathematician. Part of his nose was cut off, so he designed a replacement piece made of silver and gold.

In 1563, Tycho decided to observe a widely anticipated conjunction of Jupiter and Saturn. To his surprise, the conjunction occurred nearly two days later than Copernicus had predicted. He resolved to improve the state of astronomical prediction and set about compiling careful observations of stellar and planetary positions in the sky.

Tycho's fame grew after he observed what he called a *nova,* meaning "new star," in 1572 and proved that it was at a distance much farther away than the Moon. (He compared observations made by other astronomers at other locations on Earth to his own, proving that the nova had no observable parallax and must be more distant than the Moon.) Today, we know that Tycho saw a *supernova*—the explosion of a distant star. In 1577, Tycho observed a comet and proved that it too lay in the realm of the heavens; others, notably Aristotle, had argued that comets were phenomena of the Earth's atmosphere. King Frederick II of Denmark then decided to sponsor Tycho's ongoing work, providing him with money to build an unparalleled observatory for naked-eye observations. (After Frederick II died in 1588, Tycho moved to Prague, where his work was supported by German emperor Rudolf II.)

Over a period of three decades, Tycho and his assistants compiled naked-eye observations accurate to within less than 1 arcminute. Because the telescope was invented shortly after his death, Tycho's data remain the best set of naked-eye observations ever made.

Despite the quality of his observations, Tycho never succeeded in coming up with a satisfying explanation for planetary motion. He did, however, succeed in finding someone who could: In 1600, he hired a young German astronomer named Johannes Kepler (1571–1630). Kepler and Tycho had a strained relationship while Tycho was living.[5] But in 1601, as

---

[5] For a particularly moving version of the story of Tycho and Kepler, see episode 3 of Carl Sagan's *Cosmos* video series.

Tycho Brahe (1546–1601)

Tycho's naked-eye observatory

Kepler (1571–1630)

Tycho lay on his deathbed, he begged Kepler to find a system that would make sense of the observations so "that it may not appear I have lived in vain."

Kepler's Reformation: The Laws of Planetary Motion

Kepler was deeply religious and believed that understanding the geometry of the heavens would bring him closer to God. Kepler, like Copernicus, believed that Earth and the other planets traveled around the Sun in circular orbits, and he worked diligently to match circular motions to Tycho's data.

Kepler worked with particular intensity to find an orbit for Mars, which posed the greatest difficulties in matching the data to a circular orbit. After years of calculation, Kepler found a circular orbit that matched all of Tycho's observations of Mars' position along the ecliptic (east-west) to within 2 arcminutes. However, the same model did not correctly predict Mars' positions north or south of the ecliptic. Because Kepler sought a physically realistic orbit for Mars, he could not (as Ptolemy and Copernicus had done) tolerate one model for the east-west positions and another for the north-south positions. He attempted to find a unified model with a circular orbit, but in doing so he found that some of his predictions differed from Tycho's observations by as much as 8 arcminutes.

Kepler surely was tempted to ignore these discrepancies and attribute them to errors by Tycho. After all, 8 arcminutes is barely one-fourth the angular diameter of the full moon. But Kepler trusted Tycho's careful work, and the misses by 8 arcminutes finally led him to abandon the idea of circular orbits— and to find the correct solution to the ancient riddle of planetary motion. About this event, Kepler wrote:

> If I had believed that we could ignore these eight minutes [of arc], I would have patched up my hypothesis accordingly. But, since it was not permissible to ignore, those eight minutes pointed the road to a complete reformation in astronomy.

TIME OUT TO THINK *Some historians assert that Kepler's discovery represented the true birth of modern science because, for the first time, a scientist was willing to cast off long-held beliefs in a quest to match theory to observation. How did Kepler's work meet the criteria for the modern scientific method?*

Kepler summarized his discoveries with three simple laws that we now call **Kepler's laws of planetary motion**. He published the first two laws in 1610 and the third in 1618. (Be careful not to confuse Kepler's three laws of *planetary* motion with Newton's three laws that apply generally to *all* motion.)

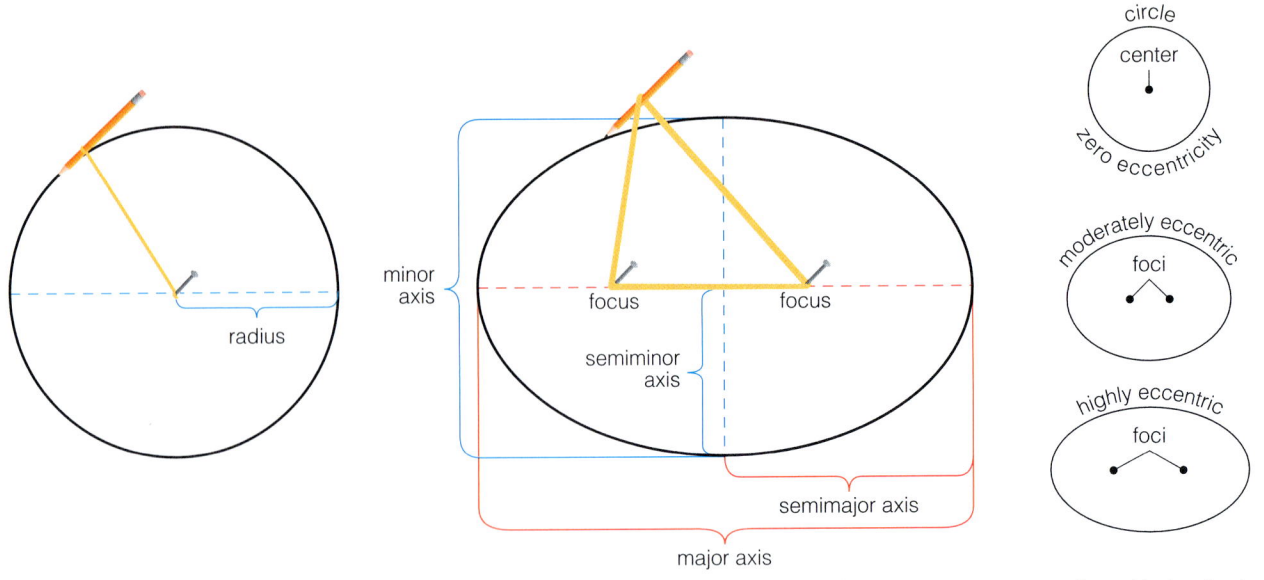

FIGURE 5.9 Basic characteristics of circles and ellipses.

a Drawing a circle with a string of fixed length.

b Drawing an ellipse with a string of fixed length.

c *Eccentricity* describes how much an ellipse deviates from a perfect circle.

Kepler's key discovery was that planetary orbits are not circles but instead are a special type of oval called an **ellipse**. You probably know how to draw a circle by putting a pencil on the end of a string, tacking the string to a board, and pulling the pencil around (Figure 5.9a). Drawing an ellipse is similar, except that you must stretch the string around *two* tacks (Figure 5.9b). The locations of the two tacks are called the **foci** (singular, **focus**) of the ellipse. By altering the distance between the two foci while keeping the same length of string, you can draw ellipses of varying **eccentricity**, a quantity that describes how much an ellipse deviates from a perfect circle (Figure 5.9c): A circle has zero eccentricity, and greater eccentricity means a more elongated ellipse. (Mathematically, eccentricity is the center-to-focus distance divided by the length of the semimajor axis.)

Kepler's first law states that *the orbit of each planet about the Sun is an ellipse with the Sun at one focus* (Figure 5.10). (There is nothing at the other focus.) This law tells us that a planet's distance from the Sun varies during its orbit: It is closest at the point called **perihelion**, and farthest at the point called **aphelion**. (*Helios* is Greek for the Sun, the prefix *peri* means "near," and the prefix *ap* (or *apo*) means "away." Thus, *perihelion* means "near the Sun" and *aphelion* means "away from the Sun.") The *average* of a planet's perihelion and aphelion distances is called its **semimajor axis**; we will refer to this simply as the planet's average distance from the Sun.

Kepler's second law states that *as a planet moves around its orbit, it sweeps out equal areas in equal times.*

As shown in Figure 5.11, this means that the planet moves a greater distance when it is near perihelion than it does in the same amount of time near aphelion; that is, the planet travels faster when it is nearer to the Sun and slower when it is farther from the Sun.

Kepler's third law describes how a planet's *orbital period* (the time it takes to complete one orbit of the Sun), measured in years, is related to its average distance from the Sun in astronomical units (1 AU ≈ 150 million km):

$$(\text{orbital period in years})^2 = (\text{average distance in AU})^3$$

This formula is often written more simply as $p^2 = a^3$, where p is the orbital period measured in years and a is the average distance measured in AU (Figure 5.12a). The formula also tells us that *more distant planets move at slower speeds* in their orbits about the Sun, which we can see by graphing planetary periods against average distances from the Sun (Figure 5.12b). For example, Saturn is slightly less than twice as far as Jupiter from the Sun but takes almost three times as long to orbit the Sun; thus, Saturn must be moving along its orbit at a slower average speed than Jupiter.

Note that a planet's period does not depend on the eccentricity of its orbit: All orbits with the same semimajor axis have the same period. Nor does the period depend on the mass of the planet: Any object located an average of 1 AU from the Sun would orbit the Sun in a year (as long as the object's mass was small compared to the Sun's mass).

Galileo: The Death of the Earth-Centered Universe

The success of Kepler's laws in matching Tycho's data provided strong evidence in favor of Copernicus's placement of the Sun, rather than the Earth, at the center of the solar system. Nevertheless, many scientists still voiced reasonable objections to the Copernican view. There were three basic objections, all rooted in the 2,000-year-old beliefs of Aristotle (384–322 B.C.) and other ancient Greeks. First, Aristotle had held that the Earth could not be moving because, if it were, objects such as birds, falling stones, and clouds would be left behind as the Earth moved along its way. Second, the idea of noncircular orbits contradicted the ancient Greek belief that the heavens—the realm of the Sun, Moon, planets, and stars—must be perfect and unchanging. The third objection was the ancient argument that stellar parallax ought to be detectable if the Earth orbits the Sun [Section 2.4]. Galileo (1564–1642), a contemporary and correspondent of Kepler, answered all three objections.

Galileo defused the first objection with experiments that almost single-handedly overturned the Aristotelian view of physics. His demonstration that gravity accelerates all objects by the same amount directly contradicted Aristotle's claim that heavier objects would fall to the ground faster. Aristotle also had taught that the natural tendency of any object was to come to rest. Galileo demonstrated that a moving object remains in motion *unless* a force acts to stop it, an idea now codified in Newton's first law of motion. Thus, he concluded that objects such as birds, falling stones, and clouds that are moving with the Earth should *stay* with the Earth unless some force knocks them away. This same idea explains why passengers in an airplane stay with the moving airplane even when they leave their seats.

Tycho's supernova and comet observations already had shown that the heavens could change, and Galileo shattered the idea of heavenly perfection after he built a telescope in late 1609. (Galileo did *not* invent the telescope; it was invented in 1608 by Hans Lippershey. However, Galileo took what was little more than a toy and turned it into a scientific instrument.) Through his telescope, Galileo saw sunspots on the Sun, which were considered "imperfections" at the time. He also used his telescope to prove that the Moon has mountains and valleys like the "imperfect" Earth by noticing the shadows cast near the dividing line between the light and dark portions of the lunar face (Figure 5.13). If the heavens were not perfect, then the idea of elliptical (rather than circular) orbits was not so objectionable.

The absence of observable stellar parallax had been of particular concern to Tycho. Based on his estimates of the distances of stars, Tycho believed that his naked-eye observations were sufficiently

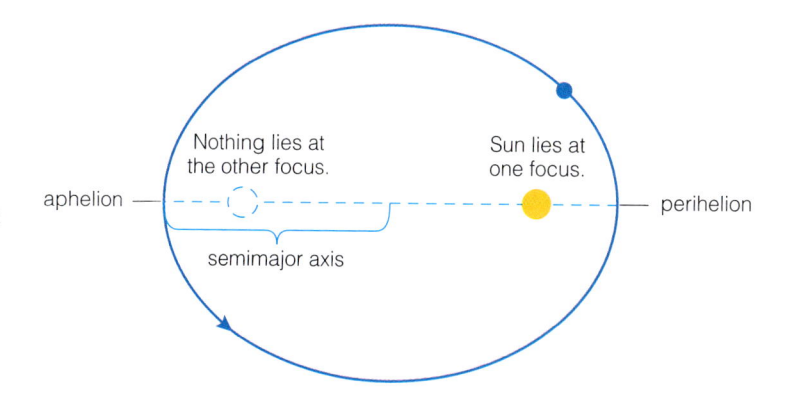

FIGURE 5.10 Kepler's first law: The orbit of each planet about the Sun is an ellipse with the Sun at one focus. (The eccentricity shown here is exaggerated compared to the actual eccentricities of the planets.)

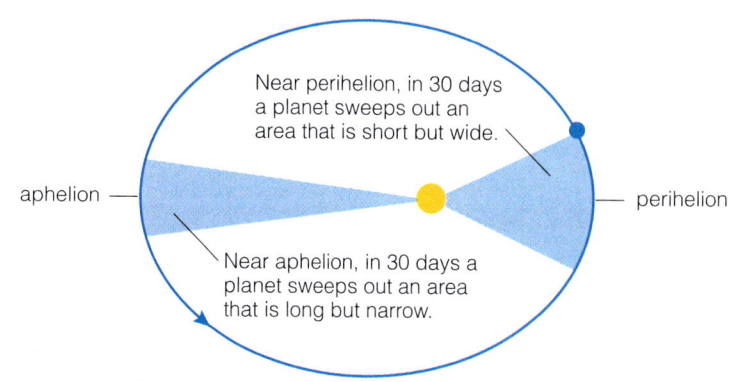

FIGURE 5.11 Kepler's second law: As a planet moves around its orbit, it sweeps out equal areas in equal times.

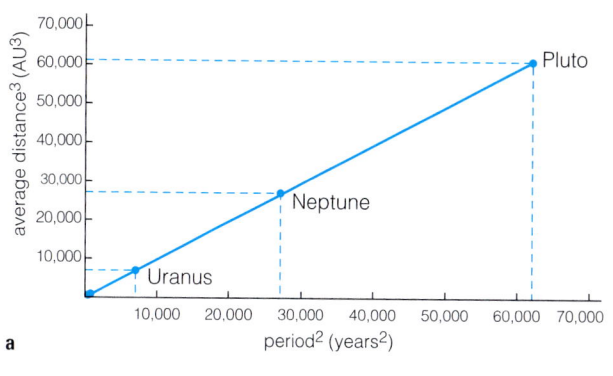

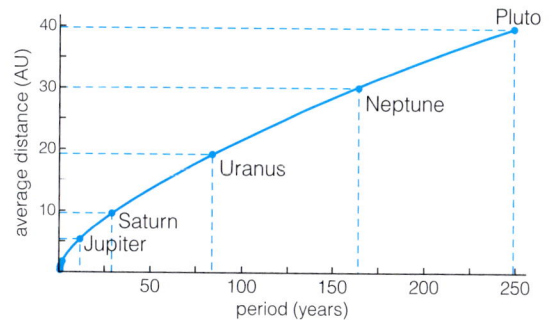

FIGURE 5.12 (a) Kepler's third law states that the period of a planet squared is equal to its average orbital distance cubed. (b) A plot of average distance against orbital period. We can use this plot to find that planets nearer the Sun orbit faster than more distant planets.

THINKING ABOUT . . .

Aristotle

Aristotle (384–322 B.C.) is among the best-known philosophers of the ancient world. Both his parents died when he was a child, and he was raised by a family friend. In his 20s and 30s, he studied under Plato (427–347 B.C.) at Plato's Academy. He later founded his own school, called the Lyceum, where he studied and lectured on virtually every subject. Historical records tell us that his lectures were collected and published in 150 volumes. About 50 of these volumes survive to the present day.

Many of Aristotle's scientific discoveries involved the nature of plants and animals. He studied more than 500 animal species in detail, including dissecting specimens of nearly 50 species, and came up with a strikingly modern classification system. For example, he was the first person to recognize that dolphins should be classified with land mammals rather than with fish. In mathematics, he is known for laying the foundations of mathematical logic. Unfortunately, he was far less successful in physics and astronomy, areas in which many of his claims turned out to be wrong.

Interestingly, Aristotle's philosophies were not particularly influential until many centuries after his death. His books were preserved and valued by Islamic scholars, but they were unknown in Europe until they were translated into Latin in the twelfth and thirteenth centuries. Aristotle achieved his near-reverential status only after St. Thomas Aquinas (1225–1274) integrated Aristotle's philosophy into Christian theology. In the ancient world, Aristotle's greatest influence came indirectly, through his role as the tutor of Alexander the Great [Section 3.3].

Galileo (1564–1642)

FIGURE 5.13 The shadows cast by mountains and crater rims near the dividing line between the light and dark portions of the lunar face prove that the Moon's surface is not perfectly smooth.

precise to detect stellar parallax if the Earth did in fact orbit the Sun.[6] (His planetary observations convinced him that the *planets* must orbit the Sun, so he advocated a model in which the Sun orbits the Earth while all other planets orbit the Sun. Few people took this model seriously, and Kepler's work soon made it a mere historical curiosity.) Refuting Tycho's argument required showing that the stars were more distant than Tycho had thought and therefore too distant for him to have observed stellar parallax. Although Galileo didn't actually prove this fact, he provided strong evidence in its favor. In particular, he saw with his telescope that the Milky Way resolved into countless individual stars, which helped

---

[6]Tycho based his estimates of stellar distances on what he *thought* were their angular sizes, but he was really measuring effects of atmospheric refraction.

138   PART II   KEY CONCEPTS FOR ASTRONOMY

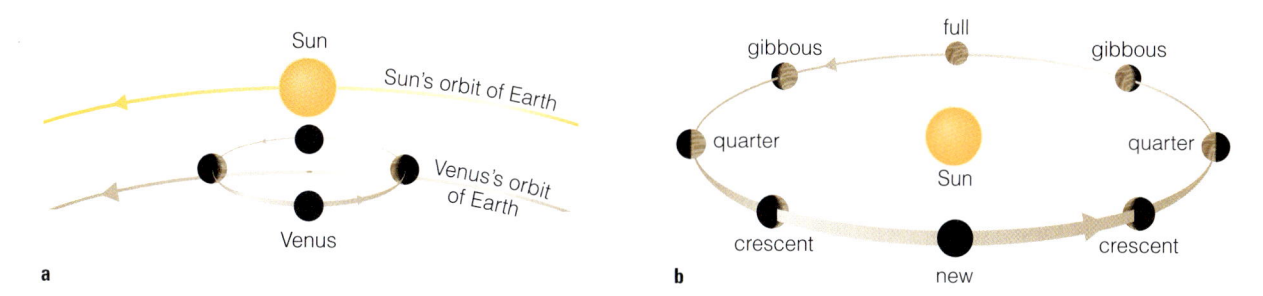

FIGURE 5.14 (**a**) In the Ptolemaic system, Venus follows a circle upon a circle that keeps it close to the Sun in our sky. Therefore, its phases would range only from new to crescent. (**b**) Galileo observed Venus go through a complete set of phases and therefore proved that it orbits the Sun.

him argue that the stars were far more numerous and more distant than Tycho had imagined.

The true death knell for an Earth-centered universe came with two of Galileo's earliest discoveries through the telescope. First, he observed four moons clearly orbiting Jupiter, *not* the Earth. (By itself, this observation still did not rule out a stationary, central Earth. However, it showed that moons can orbit a moving planet like Jupiter, which overcame some critics' complaints that the Moon could not stay with a moving Earth.) Soon thereafter, he observed that Venus goes through phases like the Moon, proving that Venus must orbit the Sun and not the Earth (Figure 5.14). With Earth clearly removed from its position at the center of the universe, the scientific debate turned to the question of whether Kepler's laws were the correct model for our solar system. The most convincing evidence came in 1631, when astronomers observed a transit of Mercury across the Sun's face [Section S1.1]; Kepler's laws had predicted the transit with overwhelmingly better success than any competing model.

At last, the Copernican revolution was complete. Thanks in large part to Galileo, by the mid-1600s most astronomers accepted Kepler's model of planetary motion. Note that it took roughly a century from the time Copernicus published his ideas in 1543 to the time that a Sun-centered system became widely accepted. The dramatic shift in thought that occurred during this period offers an excellent example of the process of science. Prior to the start of this revolution, Ptolemy's model seemed an adequate way of explaining planetary motions in the sky. But as more and better observations were gathered, it became apparent that Ptolemy's model did not agree with the data. Copernicus offered an alternative model that had the right geneal idea in displacing the Earth from the center of the universe, but he still did not make predictions that agreed sufficiently with observations. As a result, the idea of an Earth that orbits the Sun remained controversial until Kepler created a model that successfully matched the observations and Galileo defused the objections raised on other grounds. The agreement between theory and observation became so strong that the debate finally turned away from how planets move to the question of *why* Kepler's laws hold true.

5.4 The Force of Gravity

The Aristotelian view of the world, held in Europe as near-gospel truth, was in tatters. The Earth was not the center of the universe, and the laws of physics were not what Aristotle had believed. But the final blow was still to come. Aristotle had maintained that the heavens were totally distinct from Earth and that the physical laws on Earth could not be applied to understanding heavenly motion.

In 1666, Isaac Newton saw an apple fall to the ground. (The story of the apple may or may not be true, but Newton himself told the story.) He suddenly realized that the force that brought the apple to the ground and the force that held the Moon in orbit were the same. That insight brought the Earth and the heavens together in one *universe*. It also heralded the birth of the modern science of *astrophysics* (although the term wasn't coined until much later), in which physical laws discovered on Earth are applied to phenomena throughout the cosmos.

Newton (1642–1727)

Newton and the Universal Law of Gravitation

Newton was born prematurely in Lincolnshire, England, on Christmas Day in 1642. His father, a farmer who had never learned to read or write, died 3 months before he was born. Newton had a difficult childhood and showed few signs of unusual talent. He attended Trinity College at Cambridge, where he earned his keep by performing menial labor, such as cleaning the boots and bathrooms of wealthier students and waiting on their tables.

Shortly after he graduated, the plague hit Cambridge, and Newton returned home. It was there that he saw the apple fall and began his detailed investigations of gravity, light, optics, and mathematics. Over the next 20 years, Newton's work completely revolutionized mathematics and science. In addition to his work on motion and gravity, he conducted crucial experiments regarding the nature of light, built the first reflecting telescopes, and invented the branch of mathematics called *calculus*. The compendium of Newton's discoveries is so tremendous that it would take a complete book just to describe them, and many more books to describe their influence on civilization. When Newton died in 1727, at age 84, the English poet Alexander Pope composed the following epitaph:

> *Nature, and Nature's laws lay hid in the Night.*
> *God said, Let Newton be! and all was Light.*

In 1687, Newton published *Philosophiae Naturalis Principia Mathematica* ("Mathematical Principles of Natural Philosophy"), usually referred to as *Principia*. In this book, Newton stated his three laws of motion and another law, called the **universal law of gravitation**, that describes the force of gravity.

Three simple statements summarize the universal law of gravitation:

- Every mass attracts every other mass through the force called *gravity*.

- The force of attraction between any two objects is *directly proportional* to the product of their masses. For example, doubling the mass of *one* object doubles the force of gravity between the two objects.

- The force of attraction between two objects decreases with the *square* of the distance between their centers. That is, the force follows an **inverse square law** with distance. For example, doubling the distance between two objects weakens the force of gravity by a factor of 2^2, or 4.

TIME OUT TO THINK *How does the gravitational force between two objects change if the distance between them triples? If the distance between them drops in half?*

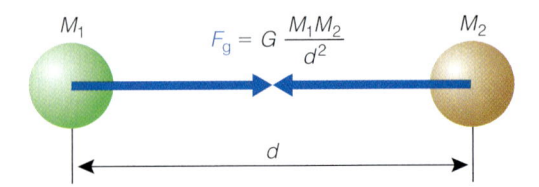

FIGURE 5.15 The universal law of gravitation.

Mathematically, Newton's law of universal gravitation is written:

$$F_g = G \frac{M_1 M_2}{d^2}$$

where F_g is the force of gravitational attraction, M_1 and M_2 are the masses of the two objects, and d is the distance between their *centers* (Figure 5.15). The symbol G is a constant called the **gravitational constant**. Its numerical value was not known to Newton but has since been measured by experiments to be $G = 6.67 \times 10^{-11}$ m³/(kg × s²).

The "Why" of Kepler's Laws, and More

For some 70 years after Kepler published his laws of planetary motion, the outstanding unsolved problem in science was what caused Kepler's laws to hold true. Kepler himself speculated that his laws might be explained by a force holding the planets in their orbits about the Sun. He incorrectly guessed that this force might be related to magnetism, an idea shared by Galileo. (This idea was first suggested by William Gilbert [1544–1603], an early believer in the Copernican system.) In *Principia*, Newton showed this force to be gravity.

Newton explained Kepler's laws by *solving* the law of universal gravitation together with the laws of motion. The solution is a bit like solving a pair of algebraic equations, but it requires the use of calculus. Newton found that the elliptical orbits with varying speeds described by Kepler's first two laws represent one possible solution. The easiest way to understand the varying speeds is to think of them in terms of conservation of angular momentum. A planet's orbital angular momentum is the product $m \times v \times r$, where m is the mass of the planet, v is its orbital speed, and r is its distance ("radius") from the Sun. To keep this product constant, the planet's orbital speed (v) must go up when its distance from the Sun (r) goes down, and vice versa.

Newton's work showed that Kepler's first two laws apply not just to planets but to *any* object going around another object under the force of gravity. (More technically, pairs of objects orbit each other in ellipses with their *center of mass* at one focus. If one object is much more massive than the other, the center of mass is very close to the massive object's center.) The orbits of satellites around the Earth, of moons around planets, of asteroids around the Sun, and of binary stars around each other are all ellipses in which orbital speeds vary so that angular momentum is conserved. Moreover, Newton found that elliptical orbits are not the only possible solution to the equations involving the law of gravity. Orbits in the shape of *parabolas* or *hyperbolas* are also allowed (Figure 5.16). Elliptical orbits are **bound orbits** because gravity creates a bond that makes one object go around and around the other. In contrast, parabolas and hyperbolas are **unbound orbits**, in which gravity governs the orbital path but cannot prevent the orbiting object from escaping. A comet with an unbound orbit comes in toward the Sun just once, looping around the Sun and never returning.

Kepler's third law also follows naturally from the law of gravity: The force is stronger nearer the Sun, so inner planets must move faster than outer planets to avoid falling toward the Sun. However, in solving the equations involving gravity, Newton showed that Kepler's statement $p^2 = a^3$ is only one special case of a more general law. Newton's generalization of Kepler's third law is:

$$p^2 = \frac{4\pi^2}{G(M_1 + M_2)} a^3$$

In this form, the law applies to any pair of orbiting objects with bound orbits, and it is not necessary to measure the period (p) in years and the average distance (a) in astronomical units; you can use any units, as long as you properly match them to the units you use for the gravitational constant, G. In addition, the general law includes the *masses* of the orbiting objects, which has important implications.

Suppose the two objects are the Sun and a planet. Because the Sun is much more massive than any planet, the sum $M_{Sun} + M_{planet}$ is pretty much just M_{Sun}. In that case, we can rewrite Newton's version of Kepler's third law as:

$$p^2_{planet} \approx \frac{4\pi^2}{G \times M_{Sun}} a^3_{planet}$$

We now find two remarkable results. First, note that the only characteristic of a planet that affects its orbital period is its average distance from the Sun. Generalizing this result, we find that the orbital period of any object orbiting a much more massive object depends only on its average distance. For example, the period of any satellite orbiting the Earth depends only on its average distance from the center of the Earth. All satellites orbiting 42,000 kilometers from the center of the Earth take 1 day to complete an orbit, and all satellites in the low-Earth orbit of the Space Shuttle (or Space Station) orbit the Earth in about 90 minutes. Thus, an astronaut on a space walk remains close to the Space Shuttle because the astronaut and the shuttle both orbit the Earth in the same amount of time (Figure 5.17).

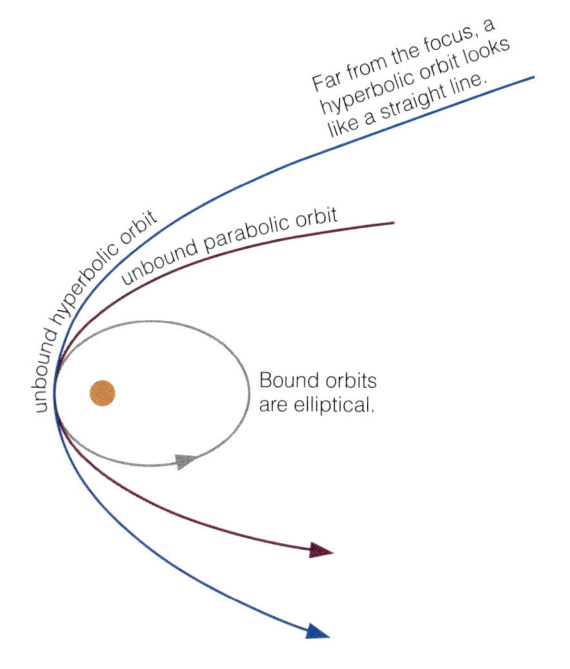

FIGURE 5.16 Orbits allowed by the law of gravity.

Second, by knowing the period p and the average distance a of any planet, we can calculate the mass of the Sun. More generally, we can calculate the mass of any massive object by measuring the period p and the average distance a of something that orbits it. By measuring the period and average distance of any one of Jupiter's moons, we can calculate the mass of Jupiter. By measuring the period and distance of a small star orbiting a more massive star, we can calculate (approximately) the larger star's mass. In fact, Newton's version of Kepler's third law provides

FIGURE 5.17 Newton's version of Kepler's third law shows that, when one object orbits a much more massive object, the orbital period depends only on its average orbital distance. Thus, the astronaut and the Space Shuttle share the same orbit despite their different masses—even as both orbit the Earth at a speed of some 25,000 km/hr.

the primary means by which we determine masses throughout the universe.

Newton's law of gravitation has applications that go far beyond explaining Kepler's laws. In the rest of this chapter, we'll explore three important concepts that we can understand with the help of the law of gravitation: tides, orbital energy and escape velocity, and the acceleration of gravity.

5.5 Tides

If you've spent time near an ocean, you're probably aware of the rising and falling of the tide twice each day. What causes the tides, and why are there two each day?

We can understand the basic idea by examining the gravitational attraction between the Earth and the Moon. Gravity attracts the Earth and the Moon toward each other (with the Moon staying in orbit as it "falls around" the Earth), but it affects different parts of the Earth slightly differently: Because the strength of gravity declines with distance, the side of the Earth facing the Moon feels a slightly stronger gravi-

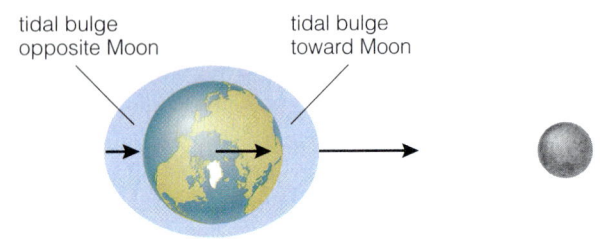

FIGURE 5.18 Tidal bulges face toward and away from the Moon because of the difference in the strength of the gravitational attraction in parts of the Earth at different distances from the Moon. (Arrows represent the strength and direction of the gravitational attraction toward the Moon.) There are two daily high tides as any location on Earth rotates through the two tidal bulges.

tational attraction than the side facing away from the Moon. As shown in Figure 5.18, this creates two tidal bulges, one facing the Moon and one opposite the Moon. The Earth's rotation carries your location through each of the two bulges each day, creating two high tides; low tides occur when your location is at the points halfway between the two tidal bulges.

Mathematical Insight 5.2 Using Newton's Version of Kepler's Third Law

The following examples show the remarkable power of Newton's version of Kepler's third law.

Example 1: Use the fact that the Earth orbits the Sun in 1 year at an average distance of 150 million km (1 AU) to calculate the mass of the Sun.

Solution: The Earth is much less massive than the Sun, so the sum of their masses is approximately the mass of the Sun alone; that is, $M_{Sun} + M_{Earth} \approx M_{Sun}$. We can therefore use Newton's version of Kepler's third law in the form:

$$(p_{Earth})^2 \approx \frac{4\pi^2}{G \times M_{Sun}} (a_{Earth})^3$$

We solve this equation for the mass of the Sun by multiplying both sides by M_{Sun} and dividing both sides by $(p_{Earth})^2$:

$$M_{Sun} \approx \frac{4\pi^2}{G} \frac{(a_{Earth})^3}{(p_{Earth})^2}$$

The Earth's orbital period is $p_{Earth} = 1$ year, which is the same as 3.15×10^7 seconds. Its average distance from the Sun is $a_{Earth} \approx 150$ million km, or 1.5×10^{11} m. Using these values and the experimentally measured value $G = 6.67 \times 10^{-11}$ m^3/(kg $\times$ s^2), we find:

$$M_{Sun} \approx \frac{4\pi^2}{G} \frac{(a_{Earth})^3}{(p_{Earth})^2}$$

$$= \frac{4\pi^2 (1.5 \times 10^{11} \text{ m})^3}{\left(6.67 \times 10^{-11} \frac{\text{m}^3}{\text{kg} \times \text{s}^2}\right)\left(3.15 \times 10^7 \text{ s}\right)^2}$$

$$= 2 \times 10^{30} \text{ kg}$$

The mass of the Sun is about 2×10^{30} kg. Simply by knowing the Earth's orbital period and distance from the Sun, and the gravitational constant G, we have used Newton's version of Kepler's third law to "weigh" the Sun.

Example 2. A *geosynchronous satellite* orbits the Earth in the same amount of time that Earth rotates: 1 sidereal day. If a geosynchronous satellite is also in an equatorial orbit, it is said to be *geostationary* because it remains fixed in the sky (i.e., it maintains a constant altitude and direction) as seen from the ground (see problem 11). Calculate the orbital distance of a geosynchronous satellite.

Solution: A satellite is much less massive than the Earth, so $M_{Earth} + M_{satellite} \approx M_{Earth}$ and we can use Newton's version of Kepler's third law in the form:

$$(p_{satellite})^2 \approx \frac{4\pi^2}{G \times M_{Earth}} (a_{satellite})^3$$

Because we want to know the satellite's distance, we solve for $a_{satellite}$ by multiplying both sides of the equation by $(G \times M_{Earth})$, dividing both sides by $4\pi^2$, and then taking the cube root of both sides:

$$a_{satellite} \approx \sqrt[3]{\frac{G \times M_{Earth}}{4\pi^2} (p_{satellite})^2}$$

We know that $p_{satellite} = 1$ sidereal day $\approx 86,164$ seconds. You should confirm that substituting this value and the mass of the Earth yields $a_{satellite} \approx 42,000$ km. Thus, a geosynchronous satellite orbits at a distance of 42,000 km above the *center* of the Earth.

142 PART II KEY CONCEPTS FOR ASTRONOMY

FIGURE 5.19 Photographs of high and low tide at the abbey at Mont-Saint-Michel, France, one of the world's most popular tourist destinations. Here the tide rushes in much faster than a person can swim. Before a causeway was built, the Mont was accessible by land only at low tide; at high tide, it became an island.

In a simple sense, the reason there are *two* daily high tides is that the oceans facing the Moon bulge because they are being pulled out from the Earth, while the oceans opposite the Moon bulge because the Earth is being pulled out from under them. However, a better way to look at tides is to recognize that the attraction toward the Moon gets progressively weaker with distance throughout the Earth. This varying attraction creates a "stretching force," or **tidal force**, that stretches the *entire Earth*—land and ocean—along the Earth–Moon line. The tidal bulges are more noticeable for the oceans than for the land only because liquids flow more readily than solids.

The two "daily" high tides actually come slightly more than 12 hours apart. Because the Moon orbits the Earth while the Earth rotates, the Moon is at its highest point (i.e., on the meridian) at any location about every 24 hours 50 minutes, rather than every 24 hours. Thus, the tidal cycle of two high tides and two low tides actually takes about 24 hours 50 minutes, with each high tide occurring about 12 hours 25 minutes after the previous one.

In addition, the height and timing of tides can vary considerably from place to place around the Earth, depending on factors such as latitude, the orientation of the coastline (e.g., north-facing or west-facing), and the depth and shape of any channel through which the rising tide must flow. For example, while the tide rises gradually in most locations, the incoming tide near the famous abbey on Mont-Saint-Michel, France, moves much faster than a person can swim (Figure 5.19). In centuries past, the Mont was an island twice a day at high tide but was connected to the mainland at low tide. (Today, a man-made land bridge connects the island to the mainland.) Many pilgrims drowned when they were caught unprepared by the inrushing tide. Another unusual tidal pattern occurs in coastal states along the northern shore of the Gulf of Mexico, where the topography and other factors combine to make only one noticeable high tide and low tide each day.

Common Misconceptions: The Origin of Tides

Many people believe that tides arise because the Moon pulls the Earth's oceans toward it. But if that were the whole story, there would be a bulge only on the side of Earth facing the Moon, and hence only one high tide each day. The correct explanation for tides must account for why the Earth has *two* tidal bulges. If you think about it, you'll realize there's only one simple way to explain this fact: The Earth must be stretching from its center in both directions (toward and away from the Moon). Once you see this, it becomes clear that tides must come from the difference between the force of gravity on one side of the Earth and that on the other, since a difference makes the Earth stretch. In fact, stretching due to tides affects many objects, not just the Earth. Many moons are stretched into oblong shapes by tidal forces caused by their parent planets, and mutual tidal forces stretch close binary stars into teardrop shapes. In regions where gravity is extremely strong, such as near a black hole, tides could even stretch spaceships or people [Section 17.4].

Spring and Neap Tides

The Sun also exerts a tidal force on the Earth, causing the Earth to stretch along the Sun–Earth line. You might at first guess that the Sun's tidal force would be substantial, since the Sun's mass is more than a million times the mass of the Moon. Indeed, the *gravitational* force between the Earth and the Sun is much greater than that between the Earth and the Moon, which is why the Earth orbits the Sun. However, the much greater distance to the Sun (than to the Moon) means that the *difference* in the Sun's pull on the near and far sides of the Earth is relatively small, and the overall tidal force caused by the Sun is only about one-third that caused by the Moon (Figure 5.20). When the tidal forces of the Sun and the Moon work together, as is the case at both new moon and full moon, we get the especially pronounced *spring tides* (so named because the water tends to "spring up" from the Earth). When the tidal forces of the Sun and the Moon oppose each other, as is the case at first- and third-quarter moon, we get the relatively small tides known as *neap tides*.

FIGURE 5.20 Tides also depend on tidal force from the Sun, which is about one-third as strong as that from the Moon. In these diagrams, yellow arrows represent tidal force due to the Sun, which causes the Earth to stretch along the Sun–Earth line, and black arrows represent tidal force due to the Moon, which causes the Earth to stretch along the Earth–Moon line.

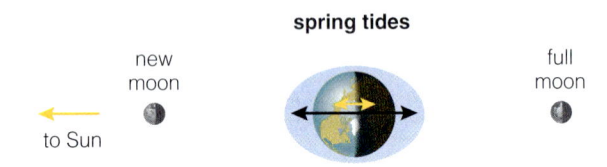

a At new moon and full moon, the tidal forces from both the Moon and the Sun stretch the Earth along the same line, leading to enhanced *spring tides*.

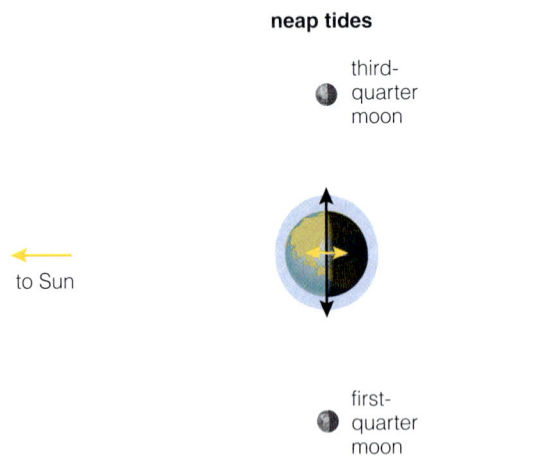

b At first- and third-quarter moon, the Sun's tidal force stretches the Earth along a line perpendicular to the Moon's tidal force. The tides still follow the Moon, because the Moon's tidal force is greater, but they are reduced in size, making *neap tides*.

TIME OUT TO THINK *Explain why any tidal effects (on Earth) caused by the other planets would be extremely small (in fact, so small as to be unnoticeable).*

Tidal Friction

So far, we have talked as if the Earth slides smoothly through the tidal bulges as it rotates. But because it is the Earth itself that stretches, this process involves some friction, called **tidal friction**. In essence, the friction arises because the tidal bulges try to stay on the Earth–Moon line, while the Earth's rotation tries to pull the bulges around with it. The resulting "compromise" puts the bulges just ahead of the Earth–Moon line at all times (Figure 5.21), ensuring that the Earth feels continuous friction as it rotates through the tidal bulges.

This tidal friction has two important effects. First, it causes the Earth's rotation to slow gradually, so that the length of a day gradually gets longer. Second, it makes the Moon move gradually farther from the Earth, because the tidal bulge is always slightly ahead of the Earth–Moon line and thus exerts a gravitational attraction that tends to pull the Moon slightly ahead in its orbit. This makes it harder for the Earth's overall gravity to hold on to the Moon, and as a result the Moon moves slightly farther from the Earth.

These two effects are barely noticeable on human time scales. For example, tidal friction increases the length of a day by only about 1 second every 50,000 years. (On shorter time scales, the length of the day fluctuates by up to a second or more per year due to slight changes in the Earth's internal mass distribution; this is why you may hear of "leap seconds" occasionally being added to or subtracted from the year.) But the effects add up over billions of years; early in the Earth's history, a day may have been only 5 or 6 hours long and the Moon may have been one-tenth or less its current distance from Earth. The changes in the Earth's rotation and the Moon's orbit also provide a remarkable example of conservation of angular momentum: The amount of angular momentum the Earth loses as its rotation slows is precisely the same as the amount the Moon gains through its growing orbit.

Synchronous Rotation

Tidal friction has had even more dramatic effects on the Moon. Recall that the Moon always shows (nearly) the same face to the Earth [Section 2.3]. This trait is called **synchronous rotation**, because it means that the Moon's rotation period and orbital period are the same (see Figure 2.20). Synchronous rotation may seem like an extraordinary coincidence, but it is a natural consequence of tidal friction.

The Moon probably once rotated much faster than it does today. But just as the Moon exerts a tidal

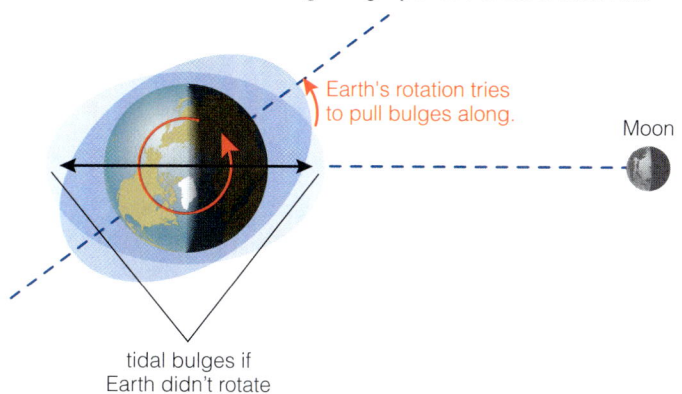

FIGURE 5.21 If the Earth always kept the same face to the Moon, the tidal bulges would stay fixed on the Earth–Moon line. But the Earth's rotation tries to pull them along with it. The result is that the bulges stay nearly fixed relative to the Moon but in a position slightly ahead of the Earth–Moon line. (The effect is exaggerated here for clarity.)

force on the Earth, the Earth exerts a tidal force on the Moon. In fact, because of its larger mass, the Earth exerts a greater tidal force on the Moon than vice versa. This tidal force stretches the Moon along the Earth–Moon line, creating two tidal bulges similar to those on Earth. As long as the Moon rotated through these bulges, the motion created tidal friction that slowed the Moon's rotation. But once the Moon's rotation slowed to the point at which the Moon and its bulges rotated at the same rate—that is, synchronously with the orbital period—there was no further source for tidal friction. The Moon has stayed in synchronous rotation ever since, with its two tidal bulges permanently fixed along the Earth–Moon line. (Interestingly, the Moon's tidal bulges cannot be seen outwardly because the Moon's diameter is not greater on this line; however, as we would expect, the Moon does have mass concentrations along the line of the tidal bulges.)

Tidal friction has led to synchronous rotation in many other cases. Most of the moons of the jovian planets rotate synchronously; for example, Jupiter's four large moons (Io, Europa, Ganymede, and Callisto) keep nearly the same face toward Jupiter at all times. Pluto and Charon *both* rotate synchronously: Like two dancers, they always keep the same face toward each other (Figure 5.22). If the Earth and the Moon were to stay together, tidal friction would eventually (in a few hundred billion years) create a similar situation: The Earth would always show the same face to the Moon.

Some moons and planets exhibit variations on synchronous rotation. For example, Mercury rotates exactly three times for every two orbits of the Sun (Figure 5.23). This pattern ensures that Mercury's tidal bulge always aligns with the Sun at perihelion, where the Sun exerts its strongest tidal force. As you study astronomy, you will encounter many more cases where tides and tidal friction play important roles.

5.6 Orbital Energy and Escape Velocity

Consider a satellite in an elliptical orbit around the Earth. Its gravitational potential energy is greatest when it is farthest from the Earth, and smallest

FIGURE 5.22 Pluto and Charon rotate synchronously with each other, so that each always shows the same face to the other. If you stood on Pluto, Charon would remain stationary in your sky, always showing the same face (but going through phases like the phases of our Moon). Similarly, if you stood on Charon, Pluto would remain stationary in your sky, always showing the same face (and going through phases). Tidal bulges are exaggerated in this figure.

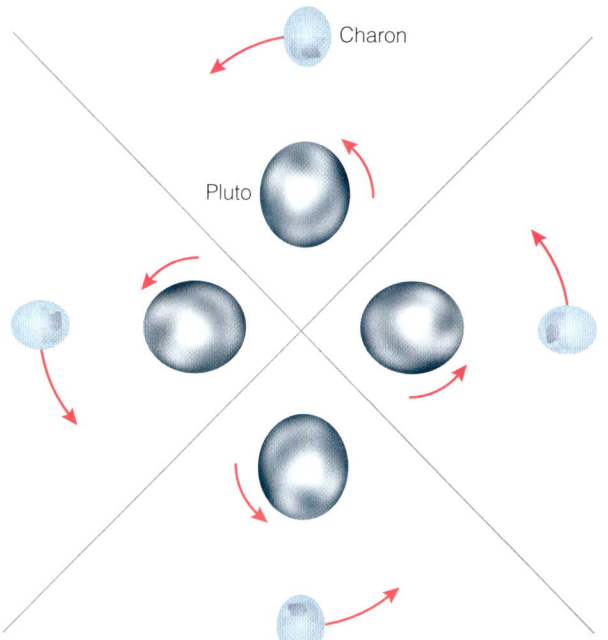

FIGURE 5.23 These "snapshots" of Mercury, in which Mercury's exaggerated shape represents its tidal bulges, show that it rotates exactly one and a half times per orbit, or three times for every two orbits. The rotation pattern ensures that the tidal bulges are always aligned with the Sun at perihelion, when the tidal forces are strongest. The orbital eccentricity is also exaggerated.

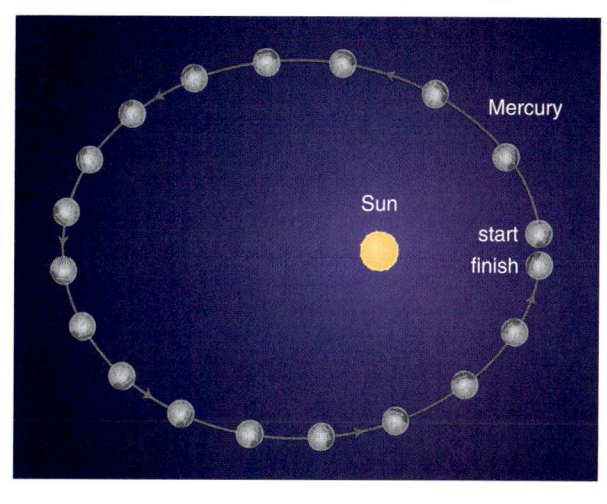

when it is nearest the Earth. Conversely, its kinetic energy is greatest when it is nearest the Earth and moving fastest in its orbit, and smallest when it is farthest from the Earth and moving slowest in its orbit. Throughout its orbit, its total *orbital energy*—the sum of its kinetic and gravitational potential energies—must be conserved.

Because any change in its orbit would mean a change in its total orbital energy, a satellite's orbit around the Earth cannot change if it is left completely undisturbed. If the satellite's orbit *does* change, it must somehow have gained or lost energy. For a satellite in low-Earth orbit, the Earth's thin upper atmosphere exerts a bit of drag that can cause the satellite to lose energy and eventually plummet back to Earth. The satellite's lost orbital energy is converted to thermal energy in the atmosphere, which is why a falling satellite usually burns up. Raising a satellite to a higher orbit requires that it gain energy by firing one of its rockets. The chemical potential energy of the rocket fuel is converted to gravitational potential energy as the satellite moves to a higher orbit.

Generalizing from the satellite example shows that conservation of energy has a very important implication for motion throughout the cosmos: *Orbits cannot change spontaneously.* For example, an asteroid or a comet passing near a planet cannot spontaneously be "sucked in" to crash on the planet. It can hit the planet only if its current orbit already intersects the planet's surface or if it somehow gains or loses orbital energy so that its new orbit intersects

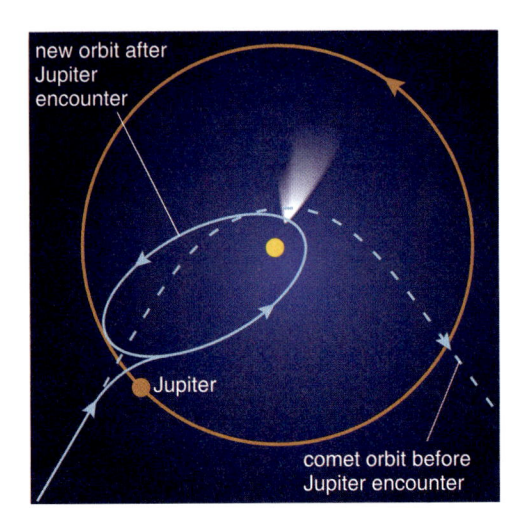

FIGURE 5.24 Depiction of a comet in an unbound orbit of the Sun that happens to pass near Jupiter. The comet loses orbital energy to Jupiter, thereby changing to a bound orbit around the sun.

the planet's surface. Of course, if the asteroid or comet *gains* energy, something else must *lose* exactly the same amount of energy, and vice versa.

One way two objects can exchange orbital energy is through a **gravitational encounter**, in which they pass near enough so that each can feel the effects of the other's gravity. For example, Figure 5.24 shows a gravitational encounter between Jupiter and a comet headed toward the Sun on an unbound orbit. The comet's close passage by Jupiter allows the

---

Mathematical Insight 5.3 Calculating the Escape Velocity

A simple formula allows us to calculate the escape velocity from any planet, moon, or star:

$$v_{escape} = \sqrt{\frac{2 \times G \times M}{R}}$$

where M is the object's mass, R is the starting distance above the object's center, and G is the gravitational constant. If you use this formula to calculate the escape velocity from an object's surface, replace R with the object's radius.

Example 1: Calculate the escape velocity from the Moon. Compare it to that from the Earth.

Solution: The mass and radius of the Moon are, respectively, $M = 7.4 \times 10^{22}$ kg and $R = 1.7 \times 10^6$ m. Plugging these numbers into the escape-velocity formula, we find:

$$v_{escape} = \sqrt{\frac{2 \times \left(6.67 \times 10^{-11} \frac{m^3}{kg \times s^2}\right) \times \left(7.4 \times 10^{22} \, kg\right)}{1.7 \times 10^6 \, m}}$$

$$\approx 2,400 \text{ m/s} = 2.4 \text{ km/s}$$

The escape velocity from the Moon is 2.4 km/s, or less than one-fourth the 11-km/s escape velocity from the Earth.

Example 2: Suppose a future space station orbits the Earth in geosynchronous orbit, which is 42,000 km above the center of the Earth (see Mathematical Insight 5.2). At what velocity must a spacecraft be launched from the station to escape the Earth? Is there any advantage to launching from the space station instead of from Earth's surface?

Solution: We use the escape-velocity formula with the mass of the Earth ($M_{Earth} = 6.0 \times 10^{24}$ kg) and the distance of the orbit above the center of the Earth ($R = 42,000$ km $= 4.2 \times 10^7$ m):

$$v_{escape} = \sqrt{\frac{2 \times \left(6.67 \times 10^{-11} \frac{m^3}{kg \times s^2}\right) \times \left(6.0 \times 10^{24} \, kg\right)}{4.2 \times 10^7 \, m}}$$

$$= 4,400 \text{ m/s} = 4.4 \text{ km/s}$$

The escape velocity from geosynchronous orbit is 4.4 km/s—considerably lower than the 11-km/s escape velocity from the Earth's surface. Thus, it requires substantially less fuel to launch the spacecraft from the space station than from the Earth. (In addition, the spacecraft would already have the orbital velocity of the space station.) Of course, this assumes that the space station is already in place and that the spacecraft is assembled at the space station.

comet and Jupiter to exchange energy: The comet loses orbital energy and changes to a bound, elliptical orbit; Jupiter must gain the energy the comet loses. However, because Jupiter is so much more massive than the comet, the effect on Jupiter is unnoticeable.

More generally, when two objects exchange orbital energy, we expect one to lose energy and fall to a lower orbit while the other gains energy and is thrown to a higher orbit. If an object gains enough energy, it may end up on an unbound orbit that allows it to *escape* from the gravitational influence of the object it is orbiting. For example, if we want to send a space probe to Mars, we must use a large rocket that gives the probe enough energy to achieve an unbound orbit (relative to Earth) and ultimately escape the Earth's gravitational influence. Although it would probably make more sense to say that the probe achieves "escape energy," we instead say that it achieves *escape velocity* (see Figure 5.3). For example, the escape velocity from the Earth's surface is about 40,000 km/hr, or 11 km/s, meaning that this is the *minimum* velocity required to escape the Earth's gravity if you start near the surface. The escape velocity does not depend on the mass of the escaping object; *any* object must travel at a velocity of 11 km/s to escape from the Earth, whether it is an individual atom or molecule escaping from the atmosphere, a spacecraft being launched into deep space, or a rock blasted into the sky by a large impact. But escape velocity does depend on whether you start from the surface or from someplace high above the surface. Because gravity weakens with distance, it takes less energy—and hence a lower escape velocity—to escape from a point high above the Earth than from the Earth's surface.

5.7 The Acceleration of Gravity

Throughout the remainder of the text, we will see many more applications of the universal law of gravitation. For the moment, let's look at just one more: Galileo's discovery that the acceleration of a falling object is independent of its mass.

Mathematical Insight 5.4 The Acceleration of Gravity

The text on page 148 shows that the acceleration of a falling rock near the surface of the Earth is:

$$a_{rock} = G \times \frac{M_{Earth}}{(R_{Earth})^2}$$

Because this formula applies to *any* falling object, it is the *acceleration of gravity*, g. Calculating g is easy. Simply look up the Earth's mass (6.0×10^{24} kg) and radius (6.4×10^6 m), and then "plug in":

$$g = G \times \frac{M_{Earth}}{(R_{Earth})^2}$$

$$= \left(6.67 \times 10^{-11} \frac{m^3}{kg \times s^2}\right) \times \frac{6.0 \times 10^{24} \text{ kg}}{(6.4 \times 10^6 \text{ m})^2}$$

$$= 9.8 \frac{m}{s^2}$$

We can find the acceleration of gravity on the surface of any other world by using the same formula, using the other world's mass and radius instead of Earth's.

Example 1: What is the acceleration of gravity on the surface of the Moon?

Solution: We use the Moon's mass (7.4×10^{22} kg) and radius (1.7×10^6 m) to find:

$$g_{Moon} = G \times \frac{M_{Moon}}{(R_{Moon})^2}$$

$$= \left(6.67 \times 10^{-11} \frac{m^3}{kg \times s^2}\right) \times \frac{7.4 \times 10^{22} \text{ kg}}{(1.7 \times 10^6 \text{ m})^2}$$

$$= 1.7 \frac{m}{s^2}$$

The acceleration of gravity on the Moon is 1.7 m/s², or about one-sixth that on the Earth. Thus, objects on the Moon weigh about one-sixth of what they would weigh on Earth. If you can lift a 50-kilogram barbell on Earth, you'll be able to lift a 300-kilogram barbell on the Moon.

Example 2: The Space Station orbits at an altitude roughly 300 kilometers above the Earth's surface. What is the acceleration of gravity at this altitude?

Solution: Because the Space Station is significantly above the Earth's surface, we cannot use the approximation $d \approx R_{Earth}$ that we used in the text. Instead, we must go back to Newton's second law, set the gravitational force on the Space Station equal to its mass times acceleration, and then solve for its acceleration:

$$G \times \frac{M_{Earth} \cancel{M_{station}}}{d^2} = \cancel{M_{station}} \times a_{station}$$

$$\Rightarrow a_{station} = G \times \frac{M_{Earth}}{d^2}$$

In this case, the distance d is the 6,400-km radius of the Earth *plus* the 300-km altitude of the station, or $d = 6{,}700$ km $= 6.7 \times 10^6$ m. Thus, the gravitational acceleration of the Space Station when orbiting the Earth is:

$$a_{station} = G \times \frac{M_{Earth}}{d^2}$$

$$= \left(6.67 \times 10^{-11} \frac{m^3}{kg \times s^2}\right) \times \frac{6.0 \times 10^{24} \text{ kg}}{(6.7 \times 10^6 \text{ m})^2}$$

$$= 8.9 \frac{m}{s^2}$$

The acceleration of gravity in low-Earth orbit is 8.9 m/s², or only slightly less than the 9.8 m/s² acceleration of gravity at the Earth's surface.

If you drop a rock, the force acting on the rock is the force of gravity. The two masses involved are the mass of the Earth and the mass of the rock, denoted M_{Earth} and M_{rock}, respectively. The distance between their *centers* is the distance from the *center of the Earth* to the center of the rock. If the rock isn't too far above the Earth's surface, this distance is approximately the radius of the Earth, R_{Earth} (about 6,400 km); that is, $d \approx R_{Earth}$. Thus, the force of gravity acting on the rock is

$$F_g = G \frac{M_{Earth} M_{rock}}{d^2} \approx G \frac{M_{Earth} M_{rock}}{(R_{Earth})^2}$$

According to Newton's second law of motion, this force is equal to the product of the mass and the acceleration of the rock; that is,

$$G \frac{M_{Earth} \cancel{M_{rock}}}{(R_{Earth})^2} = \cancel{M_{rock}} a_{rock}$$

Note that M_{rock} "cancels" because it appears on both sides of the equation (as a multiplier), giving Galileo's result that the acceleration of the rock—or of any falling object—does not depend on the object's mass.

The fact that objects of different mass fall with the same acceleration struck Newton as an astounding coincidence, even though his own equations showed it to be so. For the next 240 years, this seemingly odd coincidence remained just that—a coincidence—in the minds of scientists. However, in 1915 Einstein discovered that it is not a coincidence at all. Rather, it reveals something deeper about the nature of gravity and of the universe; the new insights were described by Einstein in his *general theory of relativity* (the topic of Chapter S3).

THE BIG PICTURE

We've covered a lot of ground in this chapter, from the scientific terminology of motion to the story of how universal motion was understood by Newton. Be sure you understand the following "big picture" ideas:

- Understanding the universe requires understanding motion. Although the terminology of the laws of motion may be new to you, the *concepts* are familiar from everyday experience. Think about these experiences so that you'll better understand the less familiar astronomical applications of the laws of motion.

- The Copernican revolution, which overthrew the ancient belief in an Earth-centered universe, did not occur instantaneously. It unfolded over a period of more than a century and involved careful observational, experimental, and theoretical work by many different people—especially Copernicus, Tycho Brahe, Kepler, and Galileo.

- Newton's discovery of the universal law of gravitation allowed him to explain *how* gravity holds planets in their orbits. Perhaps even more important, it showed that the same physical laws we observe on Earth apply throughout the universe. This universality of physics opens up the entire universe as a possible realm of human study.

Review Questions

1. How does *speed* differ from *velocity*? Give an example in which you can be traveling at constant speed but not at constant velocity.

2. What do we mean by *acceleration*? Explain the units of acceleration.

3. What is the *acceleration of gravity* on Earth? If you drop a rock from very high, how fast will it be falling after 4 seconds (neglecting air resistance)?

4. What is *momentum*? How can momentum be affected by a *force*? What do we mean when we say that momentum will be changed only by a *net force*?

5. What is the difference between *mass* and *weight*?

6. What is *free-fall*, and why does it make you *weightless*? Briefly describe why astronauts are weightless in the Space Station.

7. Why does a spaceship require a high speed to achieve orbit? What would happen if it were launched with a speed greater than the Earth's *escape velocity*?

8. State each of Newton's three laws of motion. For each law, give an example of its application.

9. What is *angular momentum*? What kind of force can cause a change in angular momentum? Explain.

10. What is the *law of conservation of angular momentum*? How can a skater use this principle to vary his or her rate of spin?

11. Was Copernicus the first person to suggest a Sun-centered solar system? What advantages did his model have over the Ptolemaic model? In what ways did it fail to improve on the Ptolemaic model?

12. How were Tycho Brahe's observations important to the development of modern astronomy?

13. State Kepler's three laws of motion, and explain the meaning of each law.

14. Describe how Galileo helped spur acceptance of Kepler's Sun-centered model of the solar system.

15. What is the *universal law of gravitation*? Summarize what this law says in words, and then state the law mathematically. Define each variable in the formula.

16. Describe how Newton's laws of motion and law of universal gravitation explain why each of Kepler's three laws is true.

17. What do we mean by *bound* and *unbound* orbits?

18. Why is Newton's version of Kepler's third law so important to our understanding of the universe?

19. Explain how the Moon creates tides on the Earth. How do tides vary with the phase of the Moon? Why?

20. What is *tidal friction*? Briefly describe how tidal friction has affected the Earth and the Moon.

21. What is *synchronous rotation*, and what causes it? Describe the synchronous rotation of Pluto and Charon. Describe the variation on synchronous rotation that applies to Mercury.

22. Explain why orbits cannot change spontaneously. How can atmospheric drag cause an orbit to change? How can a *gravitational encounter* cause an orbit to change?

Discussion Questions

1. *Kepler's Choice.* Casting aside the idea that orbits must be perfect circles meant going against deeply entrenched beliefs, and Kepler said that it shook his deep religious faith. Given that only two of Tycho's observations disagreed with a perfectly circular orbit—and only by 8 arcminutes—do you think that most other people would have made the choice Kepler made to abandon perfect circles? Have you ever performed an experiment that disagreed with theory? Which did you question, the theory or your experiment? Why?

2. *Aristotle and Modern English.* Aristotle believed that the Earth was made from the four elements fire, water, earth, and air, while the heavens were made from *ether* (literally, "upper air"). The literal meaning of *quintessence* is "fifth element," and the literal meaning of *ethereal* is "made of ether." Look up these words in the dictionary. Discuss how their modern meanings are related to Aristotle's ancient beliefs.

3. *Tidal Complications.* The ocean tides on Earth are much more complicated than they might at first seem from the simple physics that underlies tides. Discuss some of the factors that make the real tides so complicated and how these factors affect the tides. Some factors to consider: the distribution of land and oceans; the Moon's varying distance from Earth in its orbit; the fact that the Moon's orbital plane is not perfectly aligned with the ecliptic and neither the Moon's orbit nor the ecliptic is aligned with the Earth's equator.

Problems

Sensible Statements? For **problems 1–4**, decide whether the statement is sensible and explain why it is or is not.

1. If you could go shopping on the Moon to buy a pound of chocolate, you'd get a lot more chocolate than if you bought a pound on Earth.

2. Upon its publication in 1543, the Copernican model was immediately accepted by most scientists because its predictions of planetary positions were essentially perfect.

3. Newton's version of Kepler's third law allows us to calculate the mass of Saturn from orbital characteristics of its moon Titan.

4. If an asteroid passed by Earth at just the right distance, it would be captured by the Earth's gravity and become our second moon.

5. *Understanding Acceleration.*

   a. Some schools have an annual ritual that involves dropping a watermelon from a tall building. Suppose it takes 6 seconds for the watermelon to fall to the ground (which would mean it's dropped from about a 60-story building). If there were no air resistance so that the watermelon would fall with the acceleration of gravity, how fast would it be going when it hit the ground? Give your answer in m/s, km/hr, and mi/hr.

   b. As you sled down a steep, slick street, you accelerate at a rate of 4 m/s^2. How fast will you be going after 5 seconds? Give your answer in m/s, km/hr, and mi/hr.

   c. You are driving along the highway at a speed of 70 miles per hour when you slam on the brakes. If your acceleration is at an average rate of -20 miles per hour per second, how long will it take to come to a stop?

6. *Spinning Skater.* Suppose an ice skater wants to *start* spinning. Explain why she won't start spinning if she simply stomps her foot straight down on the ice. How should she push off on the ice to start spinning? Why? What should she do when she wants to stop spinning?

7. *New Comet.* Imagine that a new comet is discovered and studies of its motion indicate that it orbits the Sun with a period of 1,000 years. What is the comet's average distance (semimajor axis) from the Sun? (*Hint:* Use Kepler's third law in its original form.)

8. *The Gravitational Law.* Use the universal law of gravitation to answer each of the following questions.

   a. How does tripling the distance between two objects affect the gravitational force between them?

   b. Compare the gravitational force between the Earth and the Sun to that between Jupiter and the Sun. The mass of Jupiter is about 318 times the mass of the Earth.

   c. Suppose the Sun were magically replaced by a star with twice as much mass. What would happen to the gravitational force between the Earth and the Sun?

9. *Head-to-Foot Tides.* You and the Earth attract each other gravitationally, so you should also be subject to a tidal force resulting from the difference between the gravitational attraction felt by your feet and that felt by your head (at least when you are standing). Explain why you can't feel this tidal force.

10. *Eclipse Frequency in the Past.* Over billions of years, the Moon has gradually been moving farther from the Earth. Thus, the Moon used to be substantially nearer the Earth than it is today.

    a. How would the Moon's past angular size in our sky compare to its present angular size? Why?

    b. How would the length of a lunar month in the past compare to the length of a lunar month today? Why? (*Hint:* Think about Kepler's third law as it would apply to the Moon orbiting the Earth.)

    c. Based on your answers to parts (a) and (b), would eclipses (both solar and lunar) have been more or less common in the past? Why?

11. *Geostationary Orbit.* A satellite in geostationary orbit appears to remain stationary in the sky as seen from any particular location on Earth.

    a. Briefly explain why a geostationary satellite must orbit the Earth in 1 *sidereal* day, rather than 1 solar day.

    b. Communications satellites, such as those used for television broadcasts, are often placed in geostationary orbit. The transmissions from such satellites are received with satellite dishes, such as those that can be purchased for home use. In one or two paragraphs, explain why geostationary orbit is a convenient orbit for communications satellites.

12. *Elevator to Orbit.* Suppose that someday we build a giant elevator from the Earth's surface to geosynchronous orbit. The top of the elevator would then have the same orbital distance and period as any satellite in geosynchronous orbit.

    a. Suppose you were to drop an object out of the elevator at its top. Explain why the object would appear to float right next to the elevator rather than falling.

    b. Briefly explain why (not counting the huge costs for construction) the elevator would make it much cheaper and easier to put satellites in orbit or to launch spacecraft into deep space.

13. *Understanding Kepler's Third Law.* Use Newton's version of Kepler's third law to answer the following questions. (*Hint:* The calculations for this problem are so simple that you will not need a calculator.)

    a. Imagine another solar system, with a star of the same mass as the Sun. Suppose there is a planet in that solar system with a mass twice that of Earth orbiting at a distance of 1 AU from the star. What is the orbital period of this planet? Explain.

    b. Suppose a solar system has a star that is four times as massive as our Sun. If that solar system has a planet the same size as Earth orbiting at a distance of 1 AU, what is the orbital period of the planet? Explain.

*14. *Gees.* Acceleration is sometimes measured in *gees*, or multiples of the acceleration of gravity: 1 gee (1*g*) means 1 × *g*, or 9.8 m/s²; 2 gees (2*g*) means 2 × *g*, or 2 × 9.8 m/s² = 19.6 m/s²; and so on. Suppose you experience 6 gees of acceleration in a rocket.
   a. What is your acceleration in meters per second squared?
   b. You will feel a compression force from the acceleration. How does this force compare to your normal weight?
   c. Do you think you could survive this acceleration for long? Explain.

*15. *Measuring Masses.* Use Newton's version of Kepler's third law to answer each of the following questions.
   a. The Moon orbits the Earth in an average of 27.3 days at an average distance of 384,000 kilometers. Use these facts to determine the mass of the Earth. You may neglect the mass of the Moon and assume $M_{Earth} + M_{Moon} \approx M_{Earth}$. (The Moon's mass is about $\frac{1}{80}$ of Earth's.)
   b. Jupiter's moon Io orbits Jupiter every 42.5 hours at an average distance of 422,000 kilometers from the center of Jupiter. Calculate the mass of Jupiter. (Io's mass is very small compared to Jupiter's.)
   c. Calculate the orbital period of the Space Shuttle in an orbit 300 kilometers above the Earth's surface.
   d. Pluto's moon Charon orbits Pluto every 6.4 days with a semimajor axis of 19,700 kilometers. Calculate the *combined* mass of Pluto and Charon. Compare this combined mass to the mass of the Earth.

*16. *Weights on Other Worlds.* Calculate the acceleration of gravity on the surface of each of the following worlds. How much would *you* weigh, in pounds, on each of these worlds?
   a. Mars (mass = $0.11 M_{Earth}$, radius = $0.53 R_{Earth}$).
   b. Venus (mass = $0.82 M_{Earth}$, radius = $0.95 R_{Earth}$).
   c. Jupiter (mass = $317.8 M_{Earth}$, radius = $11.2 R_{Earth}$). Bonus: Given that Jupiter has no solid surface, how could you weigh yourself on Jupiter?
   d. Jupiter's moon Europa (mass = $0.008 M_{Earth}$, radius = $0.25 R_{Earth}$).
   e. Mars's moon Phobos (mass = 1.1×10^{16} kg, radius = 12 km).

*17. Calculate the escape velocity from each of the following. (Masses and radii are listed in problem 16.)
   a. The surface of Mars.
   b. The surface of Phobos.
   c. The cloud tops of Jupiter.
   d. Our solar system, starting from the Earth's orbit. (*Hint:* Most of the mass of our solar system is in the Sun; $M_{Sun} = 2.0 \times 10^{30}$ kg.)
   e. Our solar system, starting from Saturn's orbit.

Web Projects

Find useful links for Web projects on the text Web site.

1. *Players in the Copernican Revolution.* A number of interesting personalities played important roles in the Copernican revolution besides Copernicus, Tycho, Kepler, and Galileo. Research one or more of these people and write a short biography of their scientific lives and their contributions to the Copernican revolution. Among the people you might consider: Nicholas of Cusa (1401–1464), who argued for a Sun-centered solar system and believed that other stars might be circled by their own planets; Leonardo da Vinci (1452–1519), who made many contributions to science, engineering, and art and also believed the Earth *not* to be the center of the universe; Rheticus (1514–1574), a student of Copernicus who persuaded him to publish his work; William Gilbert (1540–1603), who studied the Earth's magnetism and influenced the thinking of Kepler; Giordano Bruno (1548–1600), who was burned at the stake for his beliefs in the Copernican system and the atomism of Democritus; and Francis Bacon (1561–1626), whose writings contributed to the acceptance of experimental methods in science.

2. *Tide Tables.* Find a tide table or tide chart for a beach town that you'd like to visit. Briefly explain how to read the table or chart, and discuss any differences between the actual tidal pattern and the idealized tidal pattern described in this chapter.

3. *Space Elevator.* Read more about space elevators (see problem 12) and how they might make it easier and cheaper to get to Earth orbit or beyond. Write a short report about the feasibility of building a space elevator, and briefly discuss the pros and cons of such a project.

May the warp be the white light of morning,
May the weft be the red light of evening,
May the fringes be the falling rain,
May the border be the standing rainbow.
Thus weave for us a garment of brightness.

SONG OF THE SKY LOOM
(NATIVE AMERICAN)

CHAPTER 6

Light
The Cosmic Messenger

Ancient observers could discern only the most basic features of the light that they saw—such as color and brightness. Over the past several hundred years, we have discovered that light carries far more information. Remarkably, analysis of light with special instruments can reveal the chemical composition of distant objects, their temperature, how fast they rotate, and much more.

It is fortunate that light can convey so much information. Our present spacecraft can reach only objects within our solar system, and except for an occasional meteorite falling from the sky the cosmos does not come to us. In contrast to the limited reach of spacecraft, light travels throughout the universe, carrying its treasury of information wherever it goes. Light, the cosmic messenger, brings the stories of distant objects to our home here on Earth.

FIGURE 6.1 A prism reveals that white light contains a spectrum of colors from red to violet.

6.1 Light in Everyday Life

Even without opening your eyes, it's clear that light is a form of energy, called *radiative energy* [Section 4.1]. Outside on a hot, sunny day, you can feel the radiative energy of sunlight being converted to thermal energy as it strikes your skin. On an economic level, electric companies charge you for the energy needed by your light bulbs (and other appliances). The rate at which a light bulb uses energy (converts electrical energy to light and heat) is usually printed on it—for example "100 watts." A watt is a unit of **power**, which describes the *rate* of energy use; 1 watt of power means that 1 joule of energy is being used each second:

$$1 \text{ watt} = 1 \text{ joule/s}$$

Thus, for every second that you leave a 100-watt light bulb turned on, you will have to pay the utility company for 100 joules of energy. Interestingly, the power requirement of an average human—about 10 million joules per day—is about the same as that of a 100-watt light bulb.

Another basic property of light is what our eyes perceive as *color*. You've probably seen a prism split light into a **spectrum** (plural, spectra), or rainbow of colors (Figure 6.1). You can also produce a spectrum with a **diffraction grating**—a piece of plastic or glass etched with many closely spaced lines. The colors in a spectrum are pure forms of the basic colors red, orange, yellow, green, blue, and violet. The wide variety of all possible colors comes from mixtures of these basic colors in varying proportions; *white* is simply what we see when the basic colors are mixed in roughly equal proportions. Your television takes advantage of this fact to simulate a huge range of colors by combining only three specific colors of red, green, and blue light.

TIME OUT TO THINK *If you have a magnifying glass handy, hold it close to your TV set to see the individual red, blue, and green dots. If you don't have a magnifying glass, try splashing a few droplets of water onto your TV screen (carefully!). What do you see? What are the drops of water doing?*

Energy carried by light can interact with matter in four general ways:

- **Emission**: When you turn on a lamp, electric current flowing through the filament of the light bulb heats it to a point at which it *emits* visible light.

- **Absorption**: If you place your hand near a lit light bulb, your hand *absorbs* some of the light, and this absorbed energy makes your hand warmer.

- **Transmission**: Some forms of matter, such as glass or air, *transmit* light; that is, they allow light to pass through them.

- **Reflection**: A mirror *reflects* light in a very specific way, similar to the way a rubber ball bounces off a hard surface, so that the direction of a reflected beam of light depends on the direction of the incident (incoming) beam of light (Figure 6.2a). Sometimes, reflection is more random, so that an incident beam of light is **scattered** in many different directions. The screen in a movie theater, for example, scatters a narrow beam of light from the projector into an array of beams that reach every member of the audience (Figure 6.2b).

Materials that transmit light are said to be **transparent**, and materials that absorb light are called **opaque**. Many materials are neither perfectly transparent nor perfectly opaque, and the technical term that describes *how much* light they transmit and absorb is **opacity**. For example, dark sunglasses have a higher opacity than clear glasses because they absorb more light.

Particular materials can affect different colors of light differently. For example, red glass transmits red light but absorbs other colors. A green lawn reflects green light but absorbs all other colors.

Now let's put all these ideas together and think about what happens when you walk into a dark room and turn on the light switch. The light bulb begins to emit white light, which is a mix of all the colors in the spectrum. Some of this light exits the room, transmitted through the windows. The rest of the light strikes the surfaces of objects inside the room, and each object's material properties determine the colors absorbed or reflected. The light coming from each object therefore carries an enormous amount of information about the object's location, shape and structure, and material makeup. You acquire this information when light enters your eyes, where it is absorbed by special cells (called *cones* and *rods*) that use the energy of the absorbed light to send signals to your brain. Your brain interprets the messages carried by the light, recognizing materials and objects in the process we call *vision*.

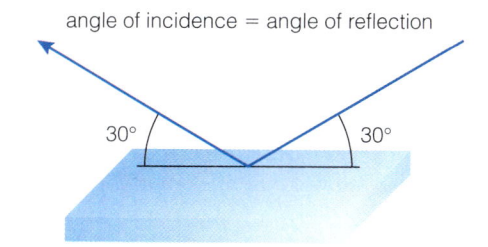

In fact, light carries even more information than your ordinary vision can recognize. Just as a microscope can reveal structure that is invisible to the naked eye, modern instruments can reveal otherwise invisible details in the spectrum of light. Learning to interpret these details is the key to unlocking the vast amount of information carried by light.

6.2 Properties of Light

Despite our familiarity with light, its nature remained a mystery for most of human history. The first real insights into the nature of light came with experiments performed by Isaac Newton in the 1660s. It was already well known that light passed through a prism separates into the rainbow of colors, but the most common belief held that the colors were a property of the prism rather than of the light itself. Newton dispelled this belief by placing a second prism in front of the light of just one color, such as red, from the first prism. He found that the color did not change any further, thereby proving that the colors were not a property of the prism but must be part of the white light itself.

Newton guessed that light, with all its colors, is made up of countless tiny particles. However, later experiments by other scientists demonstrated that light behaves like waves. Thus began one of the most important debates in scientific history: Is light a wave or a particle? We must address this question if we hope to understand the messages conveyed by light; first let's consider the difference between a wave and a particle.

Particles and Waves

Marbles, baseballs, and individual atoms are all examples of *particles*. A particle of matter can sit still or it can move from one place to another. If you throw

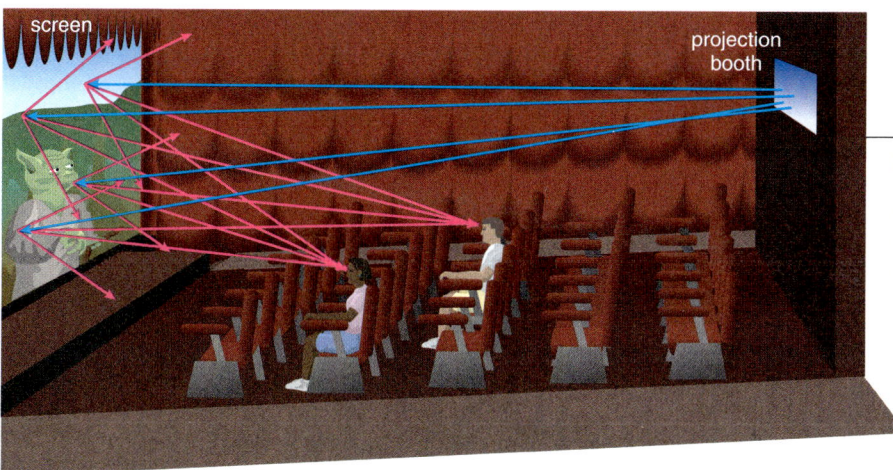

FIGURE 6.2 (**a**) A mirror reflects light along a path determined by the angle at which the light strikes the mirror. (**b**) A movie screen scatters light into an array of beams that reach every member of the audience.

a baseball at a wall, it moves from your hand to the wall. In contrast, imagine tossing a pebble into a pond (Figure 6.3). The ripples moving out from the place where the pebble lands are *waves*, consisting of *peaks* where the water is higher than average and *troughs* where the water is lower than average. If you watch as the waves pass by a floating leaf, you'll see the leaf rise up with the peak and drop down with the trough, but the leaf itself does *not* move across the pond's surface with the wave; this tells us that the water is moving up and down, but not outward. That is, the wave carries *energy* outward from the place where the pebble landed but does not carry matter along with it. In a sense, a particle is a *thing*, while a wave is a *pattern* revealed by its interaction with particles.

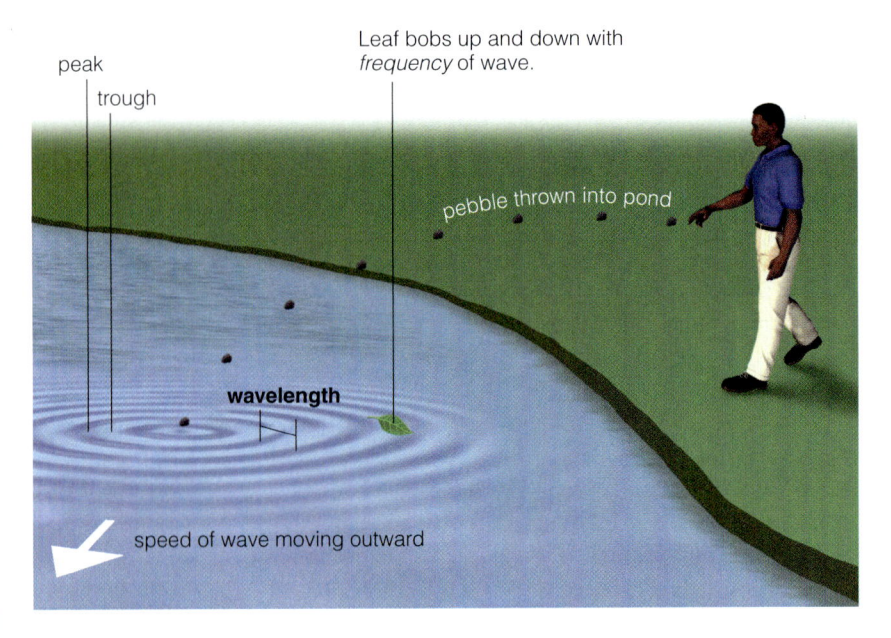

FIGURE 6.3 Tossing a pebble into a pond generates waves traveling outward.

TIME OUT TO THINK *Hold a piece of rope with one end in each hand. Make waves moving along the rope by shaking one end up and down. Watch the motion of the peaks and troughs. As a peak moves along the rope, does any material move with it? Explain.*

Three basic properties characterize the waves moving outward through the pond. Their **wavelength** is the distance between adjacent peaks. Their **frequency** is the number of peaks passing by any point each second. If a passing wave causes the leaf to bob up and down twice per second, then its frequency is 2 **cycles per second** (referring to the up and down "cycles" of the passing waves). Cycles per second often are called **hertz** (**Hz**), so we can also describe this frequency as 2 Hz. The third basic characteristic of the waves is the **speed** at which any peak travels across the pond. (A fourth characteristic of a wave is its amplitude, or height from trough to peak, but this will not be important to our study of light in this book.)

The wavelength, frequency, and speed of a wave are related by a simple formula, which we can understand with the help of an example. Suppose a wave has a wavelength of 1 centimeter and a frequency of 2 hertz. The wavelength tells us that each time a peak passes by, the wave peak has traveled 1 centimeter. The frequency tells us that two peaks pass by each second. Thus, the speed of the wave must be 2 centimeters per second. If you try a few more similar examples, you'll find that the general rule is

wavelength × frequency = speed

Photons and Electromagnetic Waves

In our everyday lives, waves and particles appear to be very different. After all, no one would confuse the ripples on a pond with a baseball. However, light behaves as *both* a particle and a wave. Like particles, light comes in individual "pieces," called **photons**, that can hit a wall one at a time. Like ripples on a pond, light can make (charged) particles bob up and down. We will discuss the implications of this "wave–particle duality" in Chapter S4. Here we need only discuss how we measure the wave and particle properties of light.

First, let's ignore the particle properties of light and look at its wave nature. Although waves don't carry material along with them, something must vibrate to transmit energy along a wave. For example, water waves are the up-and-down vibrations of the water surface, and sound waves are back-and-forth vibrations of the air as it responds to changing pressure. In the case of light, it is electric and magnetic *fields* that vibrate.

The concept of a **field** is a bit abstract. Fields associated with forces, such as electric and magnetic fields, describe how these forces affect a particle placed at any point in space. For example, we say that the Earth has a *gravitational field* because if you place an object above the Earth's surface, the force of gravity pulls the object to the ground. That is, any object placed in a gravitational field feels the force of gravity. In a similar way, a charged particle (such as an electron) placed in an *electric field* or a *magnetic field* feels electric or magnetic forces.

Light is an **electromagnetic wave**—a wave in which electric and magnetic fields vibrate. Like a leaf on a rippling pond, an electron will bob up and down when an electromagnetic wave passes by. If you could set up a row of electrons, they would wriggle like a snake (Figure 6.4a). The wavelength is the distance between adjacent peaks of the electric or magnetic field, and the frequency is the number of peaks that pass by any point each second (Figure 6.4b). All light travels at the same speed (in a vacuum)—about 300,000 kilometers per second, or 3×10^8 m/s—regardless of its wavelength or frequency. Therefore, light with a shorter wavelength must have a higher frequency, and vice versa.

Measuring the particle properties of light requires thinking of each photon as a distinct entity. Just as a baseball carries a specific amount of kinetic energy, each photon of light carries a specific amount of radiative energy. The shorter the wavelength of the light (or, equivalently, the higher its frequency), the higher the energy of the photons. For example, a photon with a wavelength of 100 nanometers (nm) has more energy than a photon with a 120-nm wave-

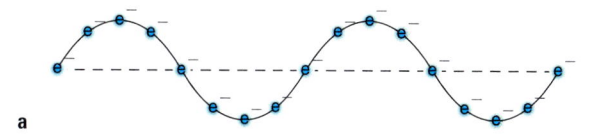

FIGURE 6.4 (a) A row of electrons would wriggle up and down as light passes by, showing that light carries a vibrating electric field. It also carries a magnetic field (not shown) that vibrates perpendicular to the direction of the electric field vibrations. (b) Characteristics of light waves. Because all light travels at the same speed, light of longer wavelength must have lower frequency.

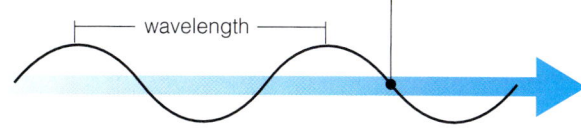

Mathematical Insight 6.1   Wavelength, Frequency, and Energy

As described in the text, the speed of any wave is the product of its wavelength and its frequency. Because all forms of light travel at the same speed (in a vacuum), $c = 3 \times 10^8$ m/s, we can write:

$$\lambda \times f = c$$

where λ (the Greek letter *lambda*) stands for wavelength and f stands for frequency. Solving this formula allows us to find the wavelength of light if we know the frequency, or vice versa:

$$\lambda = \frac{c}{f} \quad \text{or} \quad f = \frac{c}{\lambda}$$

The formula for the radiative energy carried by a photon of light is:

$$E = h \times f$$

where h is a number called *Planck's constant* ($h = 6.626 \times 10^{-34}$ joule $\times$ s). Thus, energy increases in proportion to the frequency of the photon. Because $f = c/\lambda$, we can also write this formula as:

$$E = \frac{hc}{\lambda}$$

showing that the energy decreases in proportion to the wavelength of the photon.

Example 1: The numbers on a radio dial for FM radio stations are their frequencies in megahertz (MHz), or millions of hertz. If your favorite radio station is "93.3 on your dial," it broadcasts radio waves with a frequency of 93.3 million cycles per second. What is the wavelength of these radio waves?

Solution: We know the speed of light and the frequency, so the wavelength is:

$$\lambda = \frac{c}{f} = \frac{3 \times 10^8 \frac{m}{s}}{93.3 \times 10^6 \frac{1}{s}} = 3.2 \text{ m}$$

Note that, when we work with frequency in equations, the "cycles" do not show up as a unit; that is, the units of frequency are simply 1/s, or "per second."

Example 2: The average wavelength of visible light is about 550 nanometers (1 nm = 10^{-9} m). What is the frequency of this light?

Solution: This time we know the wavelength, so the frequency of the light is:

$$f = \frac{c}{\lambda} = \frac{3 \times 10^8 \frac{m}{s}}{550 \times 10^{-9} \text{ m}} = 5.45 \times 10^{14} \frac{1}{s}$$

The frequency of visible light is about 5.5×10^{14} cycles per second, or about 550 trillion Hz.

Example 3: What is the energy of a visible light photon with wavelength 550 nm?

Solution: We know the wavelength, so the energy is:

$$E = \frac{hc}{\lambda}$$

$$= \frac{(6.626 \times 10^{-34} \text{ joule} \times \text{s}) \times (3 \times 10^8 \frac{m}{s})}{550 \times 10^{-9} \text{ m}}$$

$$= 3.6 \times 10^{-19} \text{ joule}$$

Note that this energy for a single photon is extremely small compared to, say, the energy of 100 joules used each second by a 100-watt light bulb.

length. (A nanometer (nm) is a billionth of a meter: 1 nm = 10^{-9} m. Many astronomers work with a unit called the Angstrom (Å): 1 nm = 10Å.)

6.3 The Many Forms of Light

Because light consists of electromagnetic waves, light is often called *electromagnetic radiation* and the spectrum of light is called the **electromagnetic spectrum**. Photons of light can have *any* wavelength or frequency, so in principle the complete electromagnetic spectrum extends from a wavelength of zero to infinity. For convenience, we refer to different portions of the electromagnetic spectrum by different names (Figure 6.5).

The **visible light** that we see with our eyes has wavelengths ranging from about 400 nm at the blue end of the rainbow to about 700 nm at the red end. Light with wavelengths somewhat longer than red light is called **infrared**, because it lies beyond the red end of the rainbow. Light with very long wavelengths is called **radio** (or *radio waves*)—thus, radio is a form of light, *not* a form of sound.

At the other end of the spectrum, light with wavelengths somewhat shorter than blue light is called **ultraviolet**, because it lies beyond the blue (or violet) end of the rainbow. Light with even shorter wavelengths is called **X rays**, and the shortest-wavelength light is called **gamma rays**. Note that visible light is an extremely small part of the entire electromagnetic spectrum: The reddest red that our eyes can see has only about twice the wavelength of the bluest blue, but the radio waves from your favorite radio station are a billion times longer than the X rays used in a doctor's office.

Because wavelengths decrease as we move from the radio end toward the gamma-ray end of the spectrum, the frequencies and energies must increase. Visible photons happen to have enough energy to activate the molecular receptors in our eyes. Ultraviolet photons, with a shorter wavelength than visible light, carry more energy—enough to harm our skin cells, causing sunburn or skin cancer. X-ray photons have enough energy to transmit easily through skin and muscle but not so easily through bones or teeth. That is why photographs taken with X-ray light allow doctors and dentists to see our underlying bone structures.

Interactions between light and matter depend on the types of light and matter involved. A brick wall is opaque to visible light but transmits radio waves, and glass that is transparent to visible light can be

---

Common Misconceptions: Is Radiation Dangerous?

Many people associate the word *radiation* with danger. However, the word *radiate* simply means "to spread out from a center" (note the similarity between *radiation* and *radius* [of a circle]), and *radiation* is simply energy being carried through space. If energy is being carried by particles, such as protons or neutrons, we call it *particle radiation*. If energy is being carried by light, we call it *electromagnetic radiation*. High-energy forms of radiation are dangerous because they can penetrate body tissues and cause cell damage; these forms include particle radiation from radioactive substances, such as uranium and plutonium, and electromagnetic radiation such as ultraviolet, X rays, or gamma rays. Low-energy forms of radiation such as radio waves are usually harmless. And solar radiation, the light that comes from the Sun, is necessary to life on Earth. Thus, while some forms of radiation are dangerous, others are harmless or beneficial.

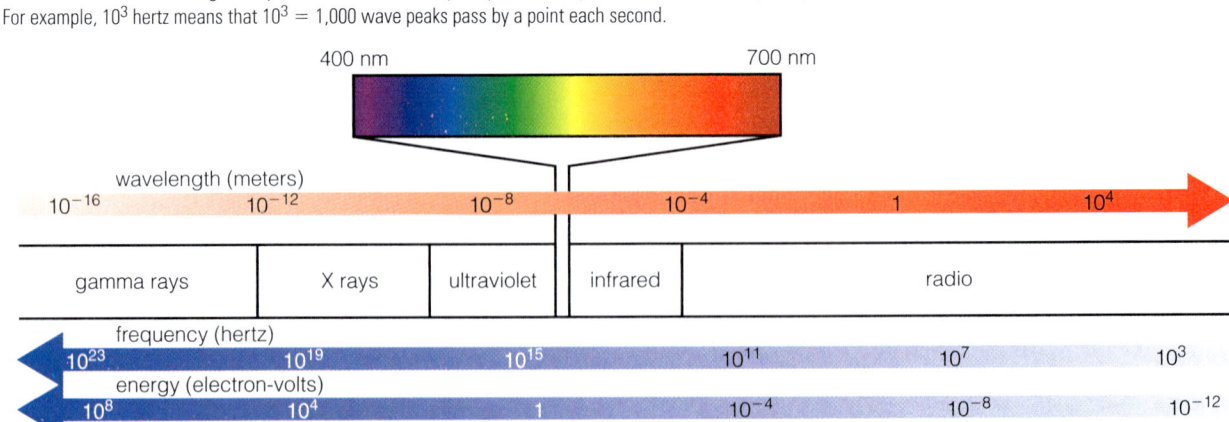

FIGURE 6.5 The electromagnetic spectrum. The unit of frequency, hertz, is equivalent to waves (or cycles) per second. For example, 10^3 hertz means that $10^3 = 1,000$ wave peaks pass by a point each second.

Movie Madness: Lois Lane's Underwear

In the 1978 movie *Superman,* the caped crusader claims to use his "X-ray vision" to determine that Lois Lane is wearing pink underwear. Sorry, but even if Superman really has "X-ray vision," his claim is impossible regardless of whether he sees X rays or whether his eyes emit them. First of all, there's no such thing as a *pink* X ray, because pink is a color in the visible part of the spectrum. Second, underwear neither emits nor reflects X rays. If it emitted X rays, then we'd all need to wear shielding to protect ourselves from its harmful effects. If it reflected X rays, then we could simply wear underwear instead of lead shields at the dentist's office. It's a good thing that Lois Lane was not aware of these problems with Superman's claim; otherwise, she might have suspected him of secretly rifling through her dresser!

Common Misconceptions: You Can't Hear Radio Waves and You Can't See X Rays

Most people associate the term *radio* with sound, but radio waves are a form of *light* with long wavelengths—too long for our eyes to see. Radio stations encode sounds (e.g., voices, music) as electrical signals, which they broadcast as radio waves. What we call "a radio" in daily life is an electronic device that receives these radio waves and decodes them to re-create the sounds played at the radio station. Television is also broadcast by encoding information (both sound and pictures) in the form of light called radio waves.

X rays are also a form of light, with wavelengths far too short for our eyes to see. In a doctor's or dentist's office, a special machine works somewhat like the flash on an ordinary camera but emits X rays instead of visible light. This machine flashes the X rays at you, and a piece of photographic film records the X rays that are transmitted through your body. Note that you never *see* the X rays; you see only an image left on film by the transmitted X rays.

opaque to ultraviolet light. In general, certain types of matter tend to interact more strongly with certain types of light, so each type of light carries different information about distant objects in the universe. Astronomers therefore seek to observe light of all wavelengths, using telescopes adapted to detecting each different form of light, from radio to gamma rays.

6.4 Light and Matter

Whenever matter and light interact, matter leaves its fingerprints. Examining the color of an object is a crude way of studying the clues left by the matter it contains. For example, a red shirt absorbs all visible photons except those in the red part of the spectrum, so we know that it must contain a dye with these special light-absorbing characteristics. If we take light and disperse it into a spectrum, we can see the spectral fingerprints in more detail.

Figure 6.6 shows a schematic spectrum of light from a celestial body such as a planet. The spectrum is a graph that shows the amount of radiation, or **intensity**, at different wavelengths. At wavelengths where a lot of light is coming from the celestial body, the intensity is high; at wavelengths where there is little light, the intensity is low. Our goal is to see how the bumps and wiggles in this graph convey a wealth of information about the celestial body in question. Let's begin by going through a short list of ways that matter interacts with light and showing the spectra that result from these interactions.

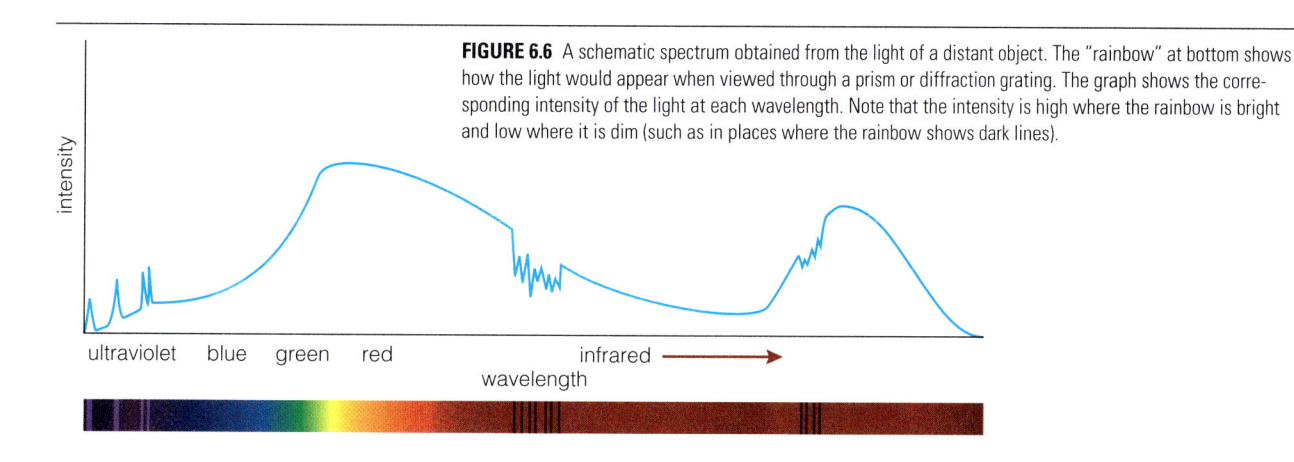

FIGURE 6.6 A schematic spectrum obtained from the light of a distant object. The "rainbow" at bottom shows how the light would appear when viewed through a prism or diffraction grating. The graph shows the corresponding intensity of the light at each wavelength. Note that the intensity is high where the rainbow is bright and low where it is dim (such as in places where the rainbow shows dark lines).

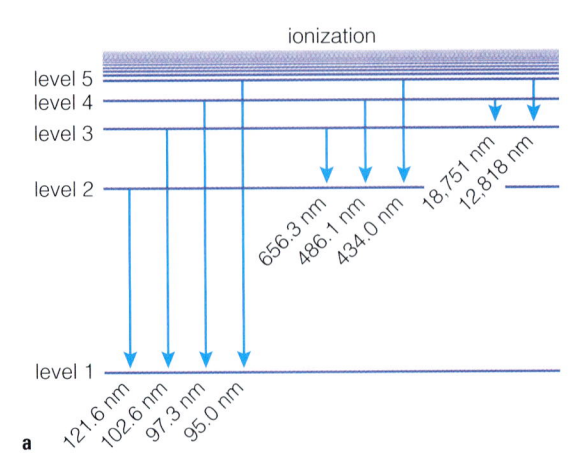

FIGURE 6.7 (a) Photons emitted by various energy level transitions in hydrogen. (b) The visible emission line spectrum from heated hydrogen gas. These lines come from transitions in which electrons fall from higher energy levels to level 2. (c) If we pass white light through a cloud of cool hydrogen gas, we get this absorption line spectrum. These lines come from transitions in which electrons jump from energy level 2 to higher levels.

Absorption and Emission by Thin Gases

You know that gases can absorb light; for example, ozone in the Earth's atmosphere absorbs ultraviolet light from space, preventing it from reaching the ground. Gases can also emit light, which is what makes neon lights and interstellar clouds glow so beautifully. But *how* do gases absorb or emit light?

Recall that the electrons in atoms can have only specific energies, somewhat like the specific heights of the rungs on a ladder [Section 4.4]. If an electron in an atom is bumped from a lower energy level to a higher one—by a collision with another atom, for example—it will eventually fall back to the lower level. The energy that the atom loses when the electron falls back down must go somewhere, and often it goes to *emitting* a photon of light. The emitted photon must have exactly the same amount of energy that the electron loses, which means that it has a specific wavelength (and frequency).

Figure 6.7a shows the allowed energy levels in hydrogen, along with the wavelengths of the photons emitted by various downward *transitions* of an electron from a higher energy level to a lower one. For example, transitions from level 2 to level 1 emit an ultraviolet photon of wavelength 121.6 nm, and transitions from level 3 to level 2 emit a red visible light photon of wavelength 656.3 nm.[1] If you heat some hydrogen gas so that collisions are continually bumping electrons to higher energy levels, you'll get an **emission line spectrum** consisting of the photons emitted as each electron falls back to lower levels (Figure 6.7b).

TIME OUT TO THINK *If nothing continues to heat the hydrogen gas, all the electrons eventually will end up in the lowest energy level (the ground state, or level 1). Use this fact to explain why we should* not *expect to see an emission line spectrum from a very cold cloud of hydrogen gas.*

---

[1]Astronomers call transitions between level 1 and other levels the *Lyman* series of transitions. The transition between level 1 and level 2 is Lyman α, between level 1 and level 3 Lyman β, and so on. Similarly, transitions between level 2 and higher levels are called *Balmer* transitions. Other sets of transitions also have names, but they are less commonly used.

FIGURE 6.8 Emission line spectra for helium, sodium, and neon. The patterns and wavelengths of lines are different for each element, giving each a unique spectral fingerprint.

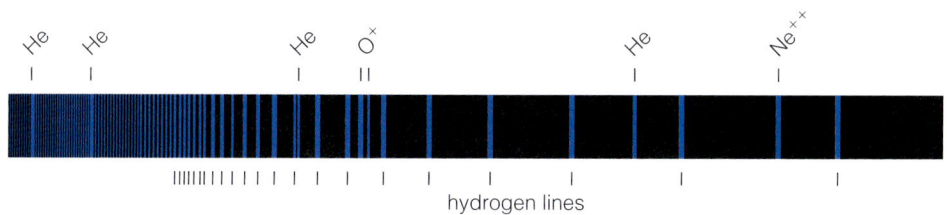

FIGURE 6.9 The emission line spectrum of the Orion Nebula. The lines are identified with the chemical elements or ions that produce them (He = helium; O = oxygen; Ne = neon).

Photons of light can also be absorbed, causing electrons to jump *up* in energy—but only if an incoming photon happens to have precisely the right amount of energy. For example, just as an electron moving downward from level 2 to level 1 in hydrogen emits a photon of wavelength 121.6 nm, absorbing a photon with this wavelength will cause an electron in level 1 to jump up to level 2.

Suppose a lamp emitting white light illuminates a cloud of hydrogen gas from behind. The cloud will absorb photons with the precise energies needed to bump electrons from a low energy level to a higher one, while all other photons pass right through the cloud. The result is an **absorption line spectrum** that looks like a rainbow with light missing at particular wavelengths (Figure 6.7c).

If you compare the bright emission lines in Figure 6.7b to the dark absorption lines in Figure 6.7c, you will see that the lines occur at the same wavelengths regardless of whether the hydrogen is absorbing or emitting light. Absorption lines simply correspond to upward jumps of the electrons between energy levels, while emission lines correspond to downward jumps.

The energy levels of electrons in each chemical element are unique [Section 4.4]. As a result, each element produces its own distinct set of spectral lines, giving it a unique "spectral fingerprint." For example, Figure 6.8 shows emission line spectra for helium, sodium, and neon. Note that the spectra of atoms such as neon are more complex than the spectra of hydrogen and helium because they have many electrons and therefore many allowed energy levels.

TIME OUT TO THINK *Three common examples of objects with emission line spectra are storefront neon signs, fluorescent light bulbs, and the yellow sodium lights used in many cities at night. When you look at objects of different colors under such lights, do you see their normal colors? Why or why not?*

The fact that each chemical element has a unique spectral fingerprint makes spectral analysis extremely useful: When you see the fingerprint of a particular element, you immediately know that the gas producing the spectrum contains this element. For example, Figure 6.9 shows the spectral fingerprints of hydrogen, helium, oxygen, and neon in an emission line spectrum from the Orion Nebula. Not only does each chemical element produce a unique spectral fingerprint, but *ions* of a particular element (atoms that are missing one or more electrons) produce fingerprints different from those of neutral atoms [Section 4.3]. For example, the spectrum of doubly ionized neon (Ne^{++}) is different from that of singly ionized neon (Ne^+), which in turn is different from that of neutral neon (Ne). This fact can help us determine the temperature of a hot gas or plasma. At higher temperatures, more highly charged ions will be present, so we can estimate the temperature by identifying the ions that are creating spectral lines.

Just as atoms and ions can absorb or emit light at particular wavelengths, so can *molecules*. Like electrons in atoms, the electrons in molecules can have only particular energies, and therefore molecules produce spectral lines when electrons change energy levels. However, because molecules are made of two

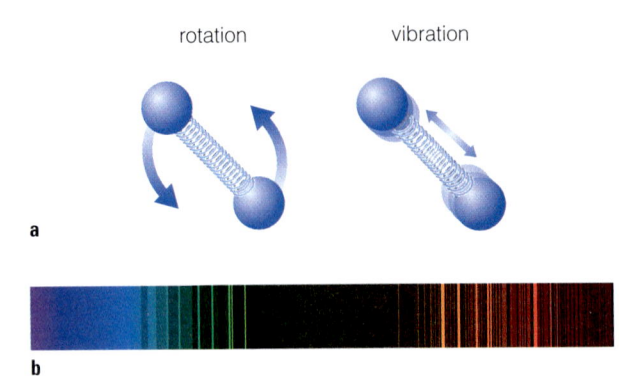

FIGURE 6.10 (a) We can think of a two-atom molecule as two balls connected by a spring. Although this model is overly simplistic, it illustrates how molecules can rotate and vibrate. (b) This spectrum of molecular hydrogen (H_2) shows that molecular spectra consist of lines bunched into broad *molecular bands*.

or more atoms bound together, they can also have energy due to vibration or rotation (Figure 6.10a). It turns out that, just as its electrons can be in only specific energy levels, a molecule can rotate or vibrate only with particular amounts of energy. Thus, a molecule can absorb or emit a photon when it changes its rate of vibration or rotation. Because molecules can change energy in three different ways, their spectra look very different from the spectra of individual atoms. Molecules produce a spectrum with many sets of tightly bunched lines, called **molecular bands** (Figure 6.10b). The energy jumps in molecules are usually smaller than those in atoms—and therefore produce lower-energy photons—so most molecular bands lie in the infrared rather than in the visible or ultraviolet. That is one reason why infrared telescopes and instruments are so important to astronomers.

Thermal Radiation: Every Body Does It

In a low-density gas, individual atoms or molecules are essentially independent of one another [Section 4.3]. That is why thin, low-density clouds of gas produce relatively simple emission or absorption spectra, with lines (or bands) in locations determined by the energy levels of their constituent atoms (or molecules). But what happens in an opaque object, such as a star, a planet, or you? (Although saying that a star is opaque may sound strange, it is true because we cannot see *through* a star.) Because photons of light cannot pass through an opaque object, they bounce randomly around among the atoms or molecules inside it, constantly exchanging energy. Recall that the "bouncing around" of atoms and molecules tends to randomize their kinetic energies, giving them an average kinetic energy that characterizes the object's *temperature*. In a similar way, the "bouncing around" of the photons inside an opaque object randomizes their radiative energies in a way that also depends only on the object's temperature. These photons can escape only when they happen to reach the surface of the object (where no more material is "in the way" to block their escape). Thus, the radiation coming from an opaque object depends only on its temperature and is called **thermal radiation** (sometimes called *blackbody* radiation).

No real object is "perfectly" opaque (absorbing and reemitting *all* radiation that strikes it), but many objects make close approximations. For example, almost all familiar objects—including the Sun, the planets, and even you—glow with light that approximates thermal radiation.

Two simple rules describe how a thermal radiation spectrum depends on the temperature of the emitting object:

Mathematical Insight 6.2 Laws of Thermal Radiation

The two rules of thermal radiation have simple mathematical formulas. Rule 1, called the *Stefan–Boltzmann law* (named after its discoverers), is expressed as:

$$\text{emitted power per square meter} = \sigma T^4$$

where σ (Greek letter *sigma*) is a constant ($\sigma = 5.7 \times 10^{-8}$ watt/($m^2 \times$ Kelvin^4)).

Rule 2, called *Wien's law*, is expressed approximately as:

$$\lambda_{max} \approx \frac{2{,}900{,}000}{T \text{ (Kelvin)}} \text{ nm}$$

where λ_{max} (lambda-max) is the wavelength (in nanometers) of maximum intensity, which is the peak of the hump in a thermal radiation spectrum.

Example: Consider a perfectly opaque object with a temperature of 15,000 K. How much power does it emit per square meter? What is its wavelength of peak intensity?

Solution: The emitted power per square meter from a 15,000-K object is:

$$\sigma T^4 = 5.7 \times 10^{-8} \frac{\text{watt}}{m^2 \times K^4} \times (15{,}000 \text{ K})^4$$

$$= 2.9 \times 10^9 \frac{\text{watt}}{m^2}$$

Its wavelength of maximum intensity is:

$$\lambda_{max} \approx \frac{2{,}900{,}000}{15{,}000 \text{ (Kelvin)}} \text{ nm} \approx 190 \text{ nm}$$

This wavelength is in the ultraviolet portion of the electromagnetic spectrum.

1. *Hotter objects emit more total radiation per unit surface area.* The radiated energy is proportional to the *fourth* power of the temperature expressed in Kelvin (*not* in Celsius or Fahrenheit). For example, a 600 K object has twice the temperature of a 300 K object and therefore radiates $2^4 = 16$ times as much total energy per unit surface area.

2. *Hotter objects emit photons with a higher average energy,* which means a shorter average wavelength.

You can see the first rule in action by playing with a light that has a dimmer switch: When you turn the switch up, the filament in the light bulb gets hotter and the light brightens; when you turn it down, the filament gets cooler and the light dims. (You can verify the changing temperature by placing your hand near the bulb.) You can see the second rule in action by observing a fireplace poker (Figure 6.11). When the poker is relatively cool, it emits only infrared radiation, which we cannot see. As it gets hot, it begins to glow red ("red hot"). If the poker continues to heat up, the average wavelength of the emitted photons gets shorter, moving toward the blue end of the visible spectrum. By the time it gets very hot, the mix of colors emitted by the poker looks white ("white hot").

Figure 6.12 shows how these two rules affect the idealized thermal radiation spectra. The spectra of hotter objects show bigger "humps" because they emit more total radiation per unit area (rule 1). Hotter objects also have the peaks of their humps at shorter wavelengths because of the higher average energy of their photons (rule 2). Note that hotter objects emit more light at *all* wavelengths but the biggest difference appears at the shortest wavelengths. An object with a temperature of 310 K, which is about human body temperature, emits mostly in the infrared and emits no visible light at all—that explains why we don't glow in the dark! A relatively cool star, with a 3,000-K surface temperature, emits mostly red light; that is why some bright stars in our sky appear reddish, such as Betelgeuse (in Orion) and Antares (in Scorpius). The Sun's 5,800-K surface emits most strongly in green light (around 500 nm), but the Sun looks yellow or white to our eyes because it also emits other colors throughout the visible spectrum. Hotter stars emit mostly in the ultraviolet, but because our eyes cannot see ultraviolet they appear blue-white in color. If an object were heated to a temperature of millions of degrees, it would radiate mostly X rays. Some astronomical objects are indeed hot enough to emit X rays, such as disks of gas encircling exotic objects like neutron stars and black holes (see Chapter 17).

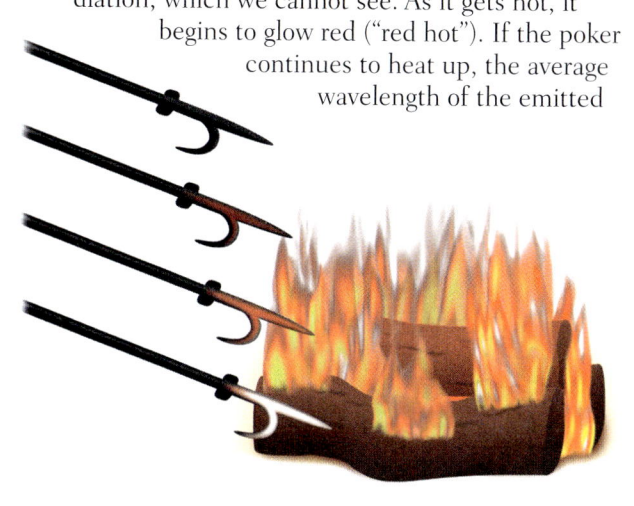

FIGURE 6.11 A fireplace poker gets brighter as it is heated, demonstrating rule 1 for thermal radiation (hotter objects emit more total radiation per unit surface area). In addition, its "color" moves from infrared to red to white as it is heated, demonstrating rule 2 (hotter objects emit photons with higher average energy).

Summary of Spectral Formation

We can now summarize the circumstances under which objects produce thermal, absorption line, or emission line spectra. (These rules are often called *Kirchhoff's laws.*)

■ Any opaque object produces thermal radiation over a broad range of wavelengths. If the object

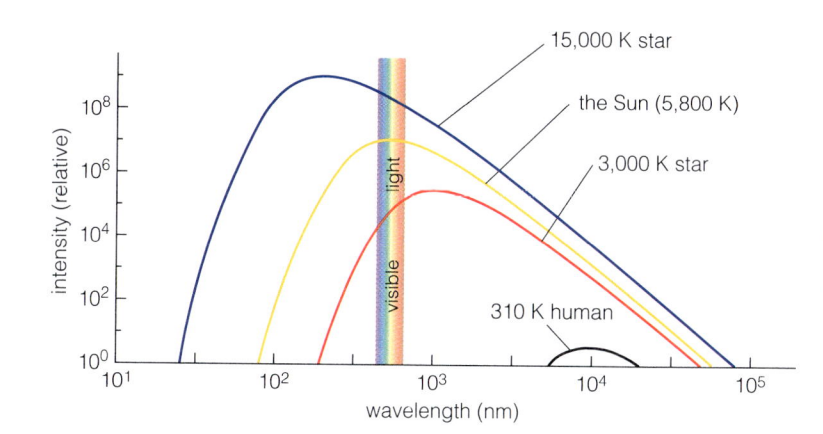

FIGURE 6.12 Graphs of idealized thermal radiation spectra. Note that, per unit surface area, hotter objects emit more radiation at every wavelength, demonstrating rule 1 for thermal radiation. The peaks of the spectra occur at shorter wavelengths (higher energies) for hotter objects, demonstrating rule 2 for thermal radiation.

is hot enough to produce visible light, as is the case with the filament of a light bulb, we see a smooth, *continuous* rainbow when we disperse the light through a prism or a diffraction grating (Figure 6.13a). On a graph of intensity versus wavelength, the rainbow becomes the characteristic hump of a thermal radiation spectrum.

- When thermal radiation passes through a thin cloud of gas, the cloud leaves fingerprints that may be either absorption lines or emission lines, depending on its temperature. If the background source of thermal radiation is hotter than the cloud, the balance between emission and absorption in the cloud's spectral lines tips toward absorption. We then see absorption lines cutting into the thermal spectrum. This is the case when the light from the hot light bulb passes through a cool gas cloud (Figure 6.13b). On the graph of intensity versus wavelength, these absorption lines create dips in the thermal radiation spectrum; the width and depth of each dip depend on how much light is absorbed by the chemical responsible for the line.

- If the background source is colder than the cloud or if there is no background source at all, the spectrum is dominated by bright emission lines produced by the cloud's atoms and molecules (Figure 6.13c). These lines create narrow peaks on a graph of intensity versus wavelength.

Reflected Light

We've now covered enough material to understand spectra emitted by objects generating their own light, such as stars and clouds of interstellar gas. But most of our daily experience involves *reflected* (or *scattered*) light. The source of the light is thermal radiation from the Sun or a lamp. After this light strikes the ground, clouds, people, or other objects, we see only the wavelengths of light that are reflected. For example, a red sweatshirt absorbs blue light and reflects red light, so its visible spectrum looks like the ther-

FIGURE 6.13 (**a**) An opaque object, such as a light bulb filament, produces a continuous spectrum of thermal radiation. (**b**) If thermal radiation passes through a thin gas that is cooler than the emitting object, dark absorption lines are superimposed on the continuous spectrum. (**c**) Viewed against a cold, dark background, the same gas produces an emission line spectrum.

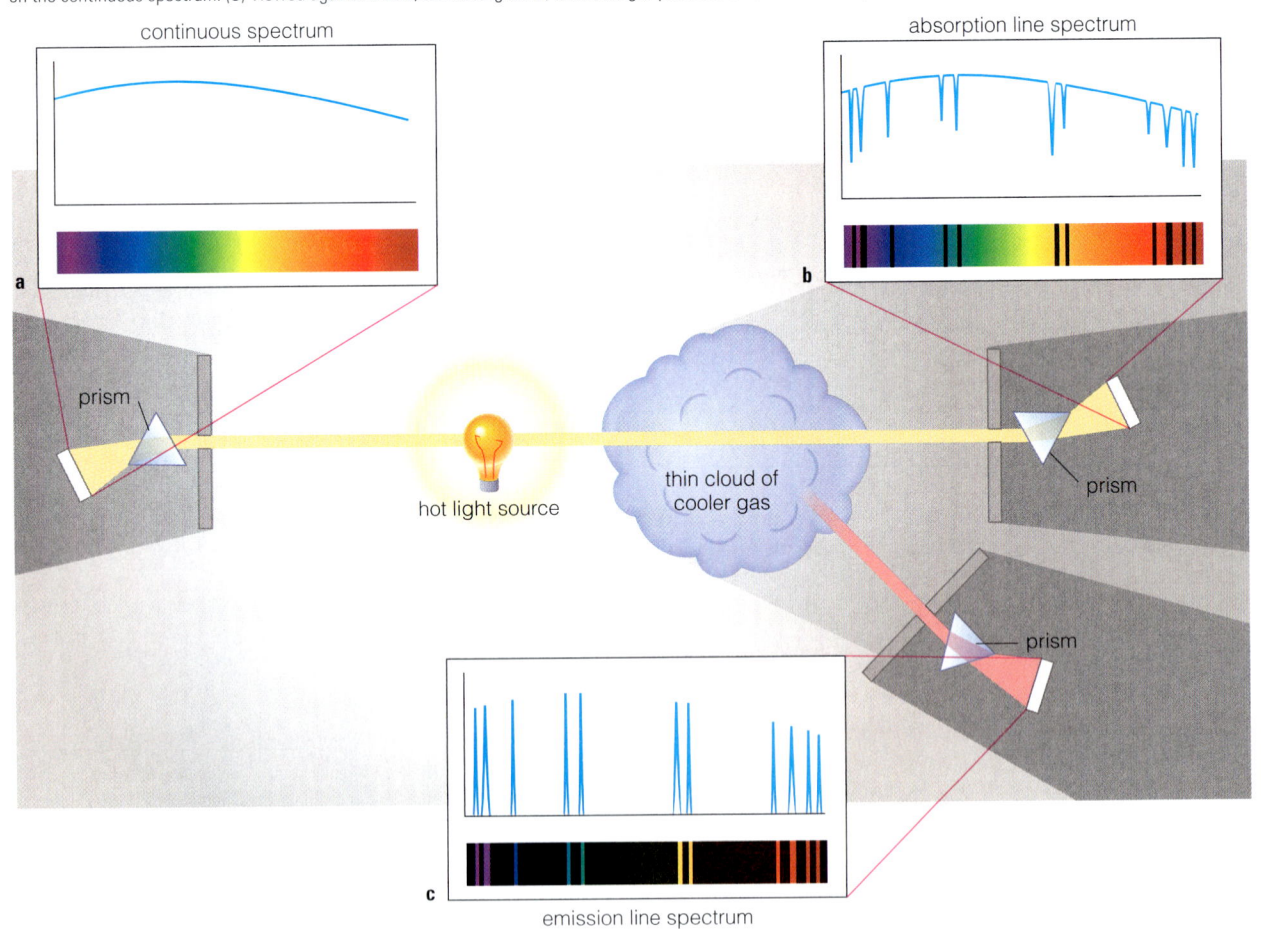

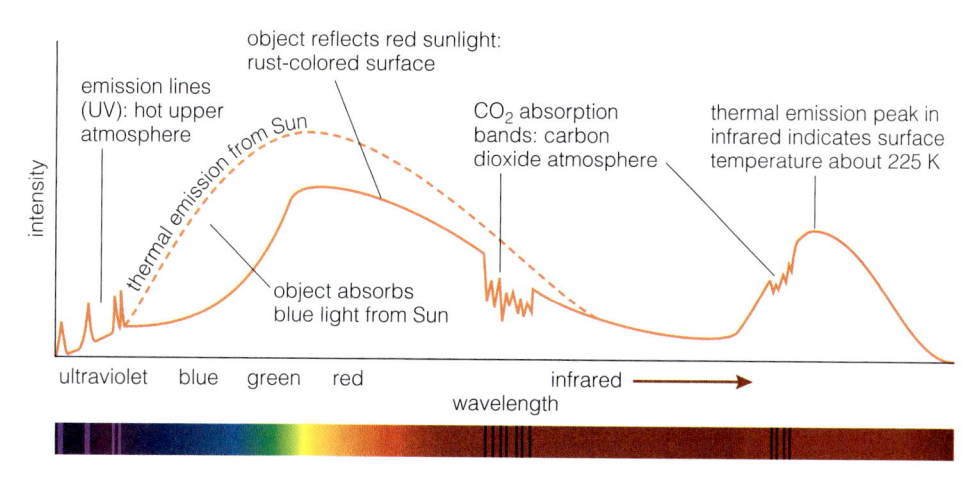

FIGURE 6.14 The spectrum of Figure 6.6, with interpretation. We can conclude that the object looks red in color because it absorbs more blue light than red light from the Sun. The absorption lines tell us that the object has a carbon dioxide atmosphere, and the emission lines tell us that its upper atmosphere is hot. The hump in the infrared (near the right of the diagram) tells us that the object has a surface temperature of about 225 K. It is a spectrum of the planet Mars.

mal radiation spectrum of its light source—the Sun—but with blue light missing.

In the same way that we distinguish lemons from limes, we can use color information in reflected light to learn about celestial objects. Different fruits, different rocks, and even different atmospheric gases reflect and absorb light at different wavelengths. Although the absorption features that show up in spectra of reflected light are not as distinct as the emission and absorption lines for thin gases, they still provide useful information. For example, the surface materials of a planet determine how much light of different colors is reflected or absorbed. The reflected light gives the planet its color, while the absorbed light heats the surface and helps determine its temperature.

Putting It All Together

Figure 6.14 again shows the complicated spectrum we began with in Figure 6.6, but this time with labels indicating the processes responsible for its various features. What can we say about this object from its spectrum? The hump of thermal emission shows that this object has a surface temperature of about 225 K, well below the freezing point of water. The absorption bands in the infrared come mainly from carbon dioxide, which tells us that the object has a carbon dioxide atmosphere. The emission lines in the ultraviolet come from hot gas in a high, thin layer of the object's atmosphere. The reflected light looks like the Sun's 5,800 K thermal radiation except that the blue light is missing, so the object must be reflecting sunlight and must look red in color. Perhaps by now you have guessed that this figure represents the spectrum of the planet Mars.

6.5 The Doppler Shift

The volume of information about temperature and composition that light contains is truly amazing, but we can learn even more once we identify the various lines in a spectrum. Among the most important pieces of information contained in light is information about motion. In particular, we can determine the radial motion (toward or away from us) of a distant object from changes in its spectrum caused by the **Doppler effect**.

You've probably noticed the Doppler effect on *sound*; it is especially easy to notice when you stand near train tracks and listen to the whistle of a train (Figure 6.15). As the train approaches, its whistle is relatively high pitched; as it recedes, the sound is relatively low pitched. Just as the train passes by, you hear the dramatic change from high to low pitch—a sort of "weeeeeeee–ooooooooooh" sound. You can visualize the Doppler effect by imagining that the train's sound waves are bunched up ahead of it, resulting in shorter wavelengths and thus the high pitch you hear as the train approaches. Behind the train, the sound waves are stretched out to longer wavelengths, resulting in the low pitch you hear as the train recedes.

The Doppler effect causes similar shifts in the wavelengths of light. If an object is moving toward us, then its entire spectrum is shifted to shorter wavelengths. Because shorter wavelengths are bluer when we are dealing with visible light, the Doppler shift of an object coming toward us is called a **blueshift**. If an object is moving away from us, its light is shifted to longer wavelengths; we call this a **redshift** because longer wavelengths are redder when we are

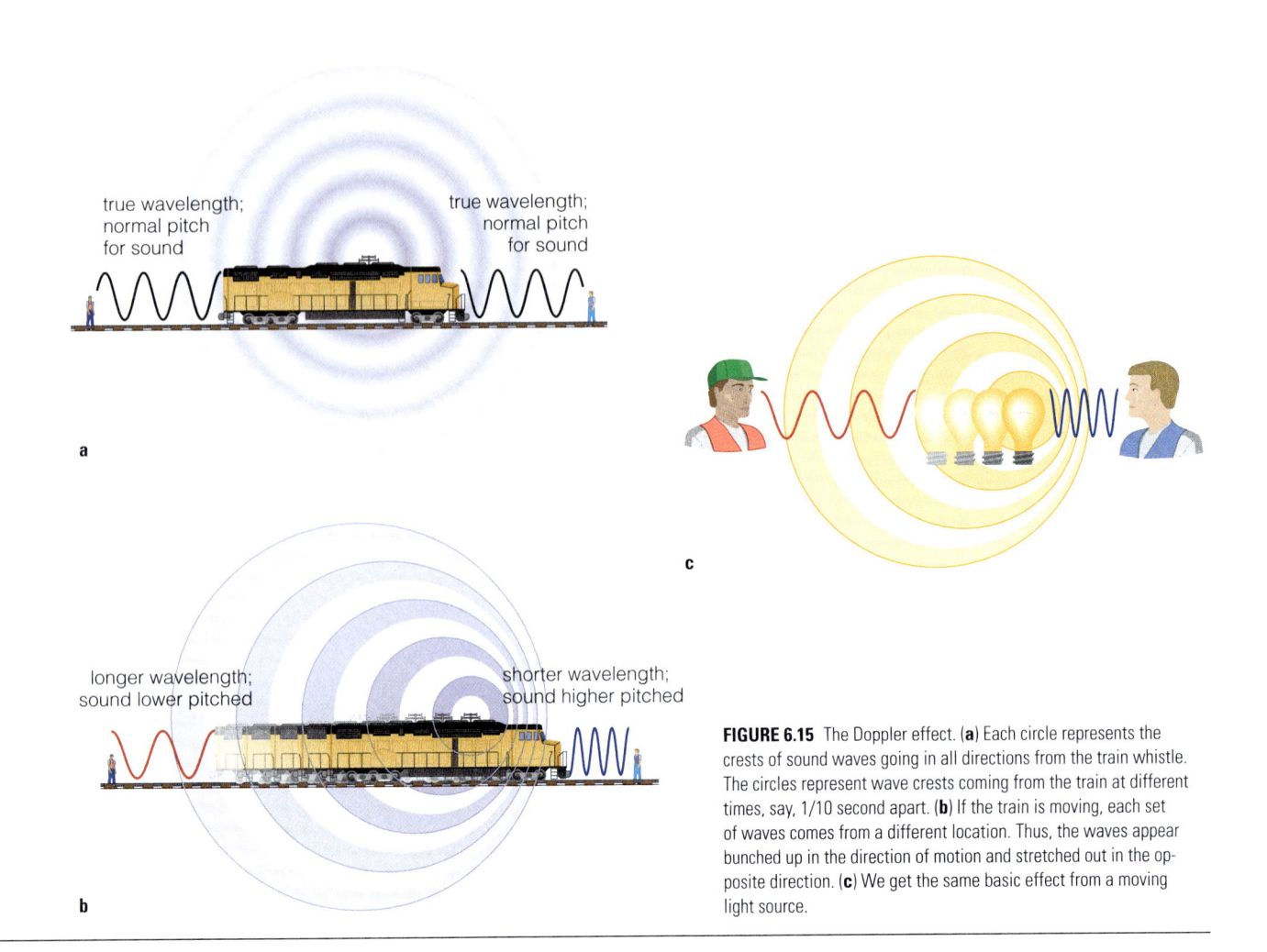

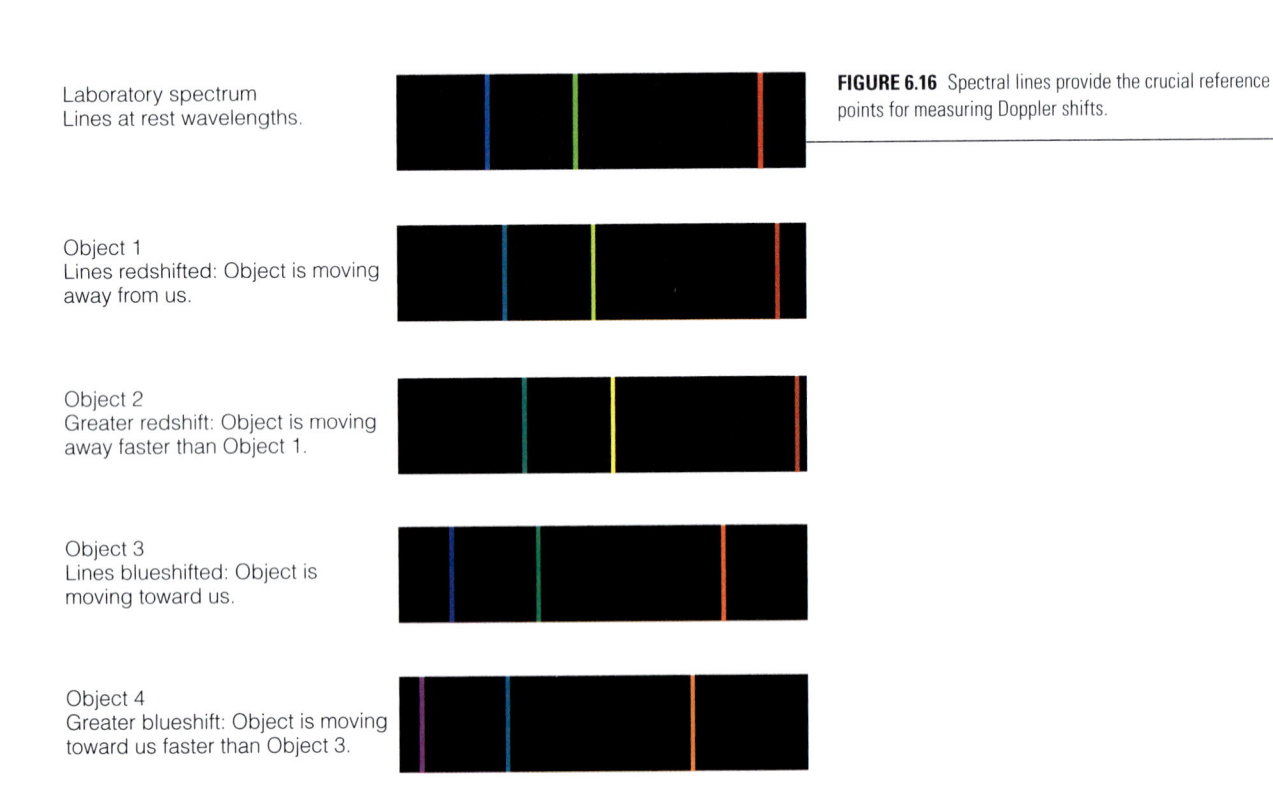

FIGURE 6.15 The Doppler effect. (**a**) Each circle represents the crests of sound waves going in all directions from the train whistle. The circles represent wave crests coming from the train at different times, say, 1/10 second apart. (**b**) If the train is moving, each set of waves comes from a different location. Thus, the waves appear bunched up in the direction of motion and stretched out in the opposite direction. (**c**) We get the same basic effect from a moving light source.

FIGURE 6.16 Spectral lines provide the crucial reference points for measuring Doppler shifts.

Laboratory spectrum
Lines at rest wavelengths.

Object 1
Lines redshifted: Object is moving away from us.

Object 2
Greater redshift: Object is moving away faster than Object 1.

Object 3
Lines blueshifted: Object is moving toward us.

Object 4
Greater blueshift: Object is moving toward us faster than Object 3.

dealing with visible light. Note that the terms *blueshift* and *redshift* are used even when we are not dealing with visible light.

Spectral lines provide the reference points we use to identify and measure Doppler shifts (Figure 6.16). For example, suppose we recognize the pattern of hydrogen lines in the spectrum of a distant object. We know the **rest wavelengths** of the hydrogen lines—that is, their wavelengths in stationary clouds of hydrogen gas—from laboratory experiments in which a tube of hydrogen gas is heated so the wavelengths of the spectral lines can be measured. If the hydrogen lines from the object appear at longer wavelengths, then we know that they are redshifted and the object is moving away from us; the larger the shift, the faster the object is moving. If the lines appear at shorter wavelengths, then we know that they are blueshifted and the object is moving toward us. Note that the Doppler shift tells us only the component of an object's velocity that is directly toward or away from us (that is, the *radial* component); it does not give us any information about the component of velocity across our line of sight.

TIME OUT TO THINK *Suppose the hydrogen emission line with a rest wavelength of 121.6 nm (the transition from level 2 to level 1) appears at a wavelength of 120.5 nm in the spectrum of a particular star. Given that these wavelengths are in the ultraviolet, is the shifted wavelength closer to or farther from blue visible light in the spectrum? Why then, do we say that this spectral line is blueshifted?*

The Doppler effect not only tells us how fast a distant object is moving toward or away from us but also can reveal information about motion *within* the object. For example, suppose we look at spectral lines of a planet or star that happens to be rotating (Figure 6.17). As the object rotates, light from the part of the object rotating toward us will be blueshifted, light from the part rotating away from us will be redshifted, and light from the center of the object won't be shifted at all. The net effect, if we look at the whole object at once, is to make each spectral line appear *wider* than it would if the object were not rotating. The faster the object is rotating, the broader in wavelength the spectral lines become. Thus, we can determine the rotation rate of distant objects by measuring the width of their spectral lines. As we will see later, this is only a small part of the information revealed through the Doppler effect on the spectra of celestial objects.

Mathematical Insight 6.3 The Doppler Shift

As long as an object's radial velocity is small compared to the speed of light (i.e., less than a few percent of c), we can use a simple formula to calculate the radial velocity (toward or away from us) of an object from its Doppler shift:

$$\frac{\text{radial velocity}}{\text{speed of light}} = \frac{\text{shifted wavelength} - \text{rest wavelength}}{\text{rest wavelength}}$$

If the result is positive, the object has a redshift and is moving away from us; a negative result means that the object has a blueshift and is moving toward us. The formula can also be written symbolically as:

$$\frac{v}{c} = \frac{\Delta\lambda}{\lambda_0} \quad \text{or} \quad v = \frac{\Delta\lambda}{\lambda_0} \times c$$

where v is the object's radial velocity, c is the speed of light, λ_0 is the rest wavelength of a particular spectral line, and $\Delta\lambda$ is its wavelength shift (positive for a redshift and negative for a blueshift).

Example: The rest wavelength of one of the visible lines of hydrogen is 656.285 nm. This line is easily identifiable in the spectrum of the bright star Vega, but it appears at a wavelength of 656.255 nm. What is the radial velocity of Vega?

Solution: The line's wavelength in Vega's spectrum is slightly shorter than its rest wavelength, so the line is blueshifted and Vega's radial motion is *toward* us. Vega's radial velocity is:

$$\frac{656.255 \text{ nm} - 656.285 \text{ nm}}{656.285 \text{ nm}} \times 300{,}000 \frac{\text{km}}{\text{s}}$$

$$= (-4.57 \times 10^{-5}) \times (3 \times 10^5) \frac{\text{km}}{\text{s}}$$

$$= -13.7 \frac{\text{km}}{\text{s}}$$

The negative answer confirms that Vega is moving *toward* us at 13.7 km/s.

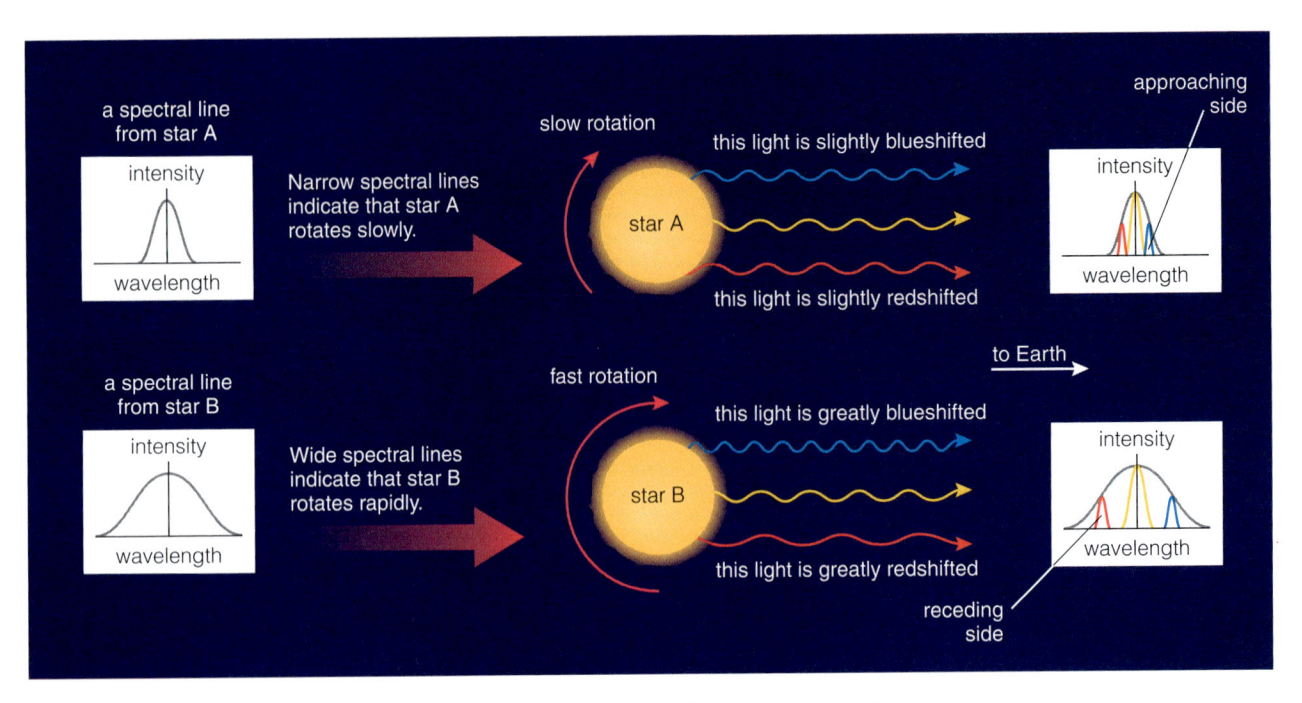

FIGURE 6.17 The Doppler effect broadens the widths of the spectral lines of rotating objects. One side of a rotating object is moving toward us, creating a blueshift, while the other side is rotating away from us, creating a redshift. The faster the rotation, the greater the spread in wavelength between the light from the two sides. When we look at a spectral line for the object as a whole, we see light from all parts of the object—that is, the parts with blueshifts as well as the parts with redshifts—at the same time. Thus, the greater Doppler shifts in a fast-rotating object make the overall spectral line wider, since the light is spread over a greater range of wavelengths.

THE BIG PICTURE

This chapter was devoted to one essential purpose: understanding how to read the messages contained in the spectra of distant objects. "Big picture" ideas that will help you keep your understanding in perspective include the following:

- There is far more to light than meets the eye. By dispersing light into a spectrum with a prism or a diffraction grating, we discover a wealth of information about the object from which the light has come. Most of what we know about the universe comes from information that we receive in the form of light.

- The visible light that our eyes can see is only a small portion of the complete electromagnetic spectrum. Different portions of the spectrum may contain different pieces of the story of a distant object, so it is important to study spectra at many wavelengths.

- The spectrum of any object is determined by interactions of light and matter. These interactions can produce *emission lines, absorption lines,* or *thermal radiation;* the spectra of most objects contain some degree of all three of these. Some spectra also include transmitted light (from a light source behind the object) and reflected light.

- By studying the spectra of a distant object, we can determine its composition, surface temperature, motion toward or away from us, rotation rate, and more.

168   PART II   KEY CONCEPTS FOR ASTRONOMY

Review Questions

1. What is the difference between *energy* and *power*, and what units do we use to measure each?

2. What do we mean when we speak of a *spectrum* produced by a prism or *diffraction grating*?

3. Describe each of the four basic ways that light can interact with matter: *emission, absorption, transmission,* and *reflection*. How is *scattering* related to reflection?

4. What does it mean for a material to be *transparent?* To be *opaque?* What is *opacity?*

5. How does a particle differ from a wave? Define each of the following terms as it applies to waves: *wavelength, frequency, cycles per second, hertz, speed.*

6. What is a *photon?* In what way is a photon like a particle? In what way is it like a wave?

7. Briefly describe the concept of a *field,* and use it to explain why we say that light is an *electromagnetic wave.*

8. Explain why light with a shorter wavelength must have a higher frequency, and vice versa.

9. Describe the *electromagnetic spectrum*. In terms of frequency, wavelength, and energy, distinguish among *radio, infrared, visible light, ultraviolet, X rays,* and *gamma rays.*

10. Explain how transitions between energy levels can cause atoms to emit or absorb photons.

11. Explain how spectral lines can be used to determine the composition and temperature of their source.

12. What is *thermal radiation?* Describe the two laws of thermal radiation. Explain how the effects of both laws can be seen in Figure 6.12.

13. Summarize the circumstances under which objects produce *thermal, emission line,* or *absorption line* spectra as shown in Figure 6.13.

14. Study Figure 6.14 carefully, and explain each feature in the spectrum.

15. Describe the *Doppler effect* for light. Explain why light from objects moving toward us shows a *blueshift,* while light from objects moving away from us shows a *redshift.* What can we say about an object with a large Doppler shift, compared to an object with a small Doppler shift?

16. Describe how the Doppler effect allows us to determine the rotation rates of distant stars.

Discussion Questions

1. *The Changing Limitations of Science*. In 1835, French philosopher Auguste Comte stated that the composition of stars could never be known by science. Although spectral lines had been seen in the Sun's spectrum by that time, not until the mid-1800s did scientists recognize that spectral lines give clear information about chemical composition (primarily through the work of Foucault and Kirchhoff). Why might our present knowledge have seemed unattainable in 1835? Discuss how new discoveries can change the apparent limitations of science. Today, other questions seem beyond the reach of science, such as the question of how life began on Earth. Do you think such questions will ever be answerable by science? Defend your opinion.

2. *Your Microwave Oven. Microwaves* is a name sometimes given to light near the long-wavelength end of the infrared portion of the spectrum. A *microwave oven* emits microwaves that happen to have just the right wavelength needed to cause energy level jumps in water molecules. Use this fact to explain how a microwave oven cooks your food. Why doesn't a microwave oven make a plastic or ceramic dish get hot?

Problems

Sensible Statements? For **problems 1–4**, decide whether the statement is sensible and explain why it is or is not.

1. If you could view a spectrum of light reflecting off a blue sweatshirt, you'd find the entire rainbow of color.

2. Because of their higher frequency, X rays must travel through space faster than radio waves.

3. If the Sun's surface became much hotter (while the Sun's size remained the same), the Sun would emit more ultraviolet light but less visible light than it currently emits.

4. If a distant galaxy has a substantial redshift (as viewed from our galaxy), then anyone living in that galaxy would see a substantial redshift in a spectrum of the Milky Way Galaxy.

5. *Spectral Summary.* Clearly explain how studying an object's spectrum can allow us to determine each of the following properties of the object.

    a. The object's surface chemical composition.

    b. The object's surface temperature.

    c. Whether the object is a thin cloud of gas or something more substantial.

    d. Whether the object has a hot upper atmosphere.

    e. The speed at which the object is moving toward or away from us.

    f. The object's rotation rate.

6. *Planetary Spectrum.* Suppose you take a spectrum of light coming from a planet that looks blue to the eye. Do you expect to see any visible light in the planet's spectrum? Is the visible light emitted by the planet, reflected by the planet, or both? Which (if any) portions of the visible spectrum do you expect to find "missing" in the planet's spectrum? Explain your answers clearly.

7. *Hotter Sun.* Suppose the surface temperature of the Sun were about 12,000 K, rather than 6,000 K.
   a. How much more thermal radiation would the Sun emit?
   b. How would the thermal radiation spectrum of the Sun be different?
   c. Do you think it would still be possible to have life on Earth? Explain.

8. *The Doppler Effect.* In hydrogen, the transition from level 2 to level 1 has a rest wavelength of 121.6 nm. Suppose you see this line at a wavelength of 120.5 nm in Star A, at 121.2 nm in Star B, at 121.9 nm in Star C, and at 122.9 nm in Star D. Which stars are coming toward us? Which are moving away? Which star is moving fastest relative to us (either toward or away)? Explain your answers without doing any calculations.

9. *The Expanding Universe.* Recall from Chapter 1 that we know the universe is expanding because (1) all galaxies outside our Local Group are moving away from us and (2) more distant galaxies are moving faster. Explain how Doppler shift measurements allow us to know these two facts.

*10. *Human Wattage.* A typical adult uses about 2,500 Calories of energy each day.
   a. Using the fact that 1 Calorie is about 4,000 joules, convert the typical adult energy usage to units of joules per day.
   b. Use your answer from part (a) to calculate a typical adult's average *power* requirement, in watts. Compare this to that of a light bulb.

*11. *Wavelength, Frequency, and Energy.*
   a. What is the frequency of a visible light photon with wavelength 550 nm?
   b. What is the wavelength of a radio photon from an "AM" radio station that broadcasts at 1,120 kilohertz? What is its energy?
   c. What is the energy (in joules) of an ultraviolet photon with wavelength 120 nm? What is its frequency?
   d. What is the wavelength of an X-ray photon with energy 10 keV (10,000 eV)? What is its frequency? (*Hint:* Recall that 1 eV = 1.60×10^{-19} joule.)

*12. *How Many Photons?* Suppose that all the energy from a 100-watt light bulb came in the form of photons with wavelength 600 nm. (This is not quite realistic; see problem 15.)
   a. Calculate the energy of a *single* photon with wavelength 600 nm.
   b. How many 600-nm photons must be emitted each second to account for all the light from this 100-watt light bulb? Based on your answer, explain why we don't notice the particle nature of light in our everyday lives.

*13. *Taking the Sun's Temperature.* The Sun radiates a total power of about 4×10^{26} watts into space. The Sun's radius is about 7×10^8 meters.
   a. Calculate the average power radiated by each square meter of the Sun's surface. (*Hint:* The formula for the surface area of a sphere is $A = 4\pi r^2$.)
   b. Using your answer from part (a) and the Stefan–Boltzmann law (see Mathematical Insight 6.2), calculate the average surface temperature of the Sun. (*Note:* The temperature calculated this way is called the Sun's *effective temperature.*)

*14. *Doppler Calculations.* Calculate the speeds of each of the stars described in problem 8. Be sure to state whether each star is moving toward or away from us.

*15. *Understanding Light Bulbs.* A standard (incandescent) light bulb uses a hot tungsten coil to produce a thermal radiation spectrum. The temperature of this coil is typically about 3,000 K.
   a. What is the wavelength of maximum intensity for a standard light bulb? Compare this to the 500-nm wavelength of maximum intensity for the Sun. Also explain why standard light bulbs must emit a substantial portion of their radiation as invisible, infrared light.
   b. Overall, do you expect the light from a standard bulb to be the same as, redder than, or bluer than light from the Sun? Why? Use your answer to explain why professional photographers use a different type of film for indoor photography than for outdoor photography.
   c. *Fluorescent* light bulbs primarily produce emission line spectra rather than thermal radiation spectra. Explain why, if the emission lines are in the visible part of the spectrum, a fluorescent bulb can emit more light than a standard bulb of the same wattage.
   d. Today, *compact fluorescent* light bulbs are designed to produce so many emission lines in the visible part of the spectrum that their light looks very similar to the light of standard bulbs. However, they are much more energy-efficient: A 15-watt compact fluorescent bulb typically emits as much visible light as a standard 75-watt bulb. Although compact fluorescent bulbs generally cost more than standard bulbs, is it possible that they could save you money? Besides initial cost and energy efficiency, what other factors must be considered?

All of this has been discovered and observed these last days thanks to the telescope that I have [built], after having been enlightened by divine grace.

GALILEO

CHAPTER 7
Telescopes and Spacecraft

We are in the midst of a great revolution in human understanding of the universe. Astonishing new discoveries about the early history of the universe, the lives of galaxies and stars, and the planets in our solar system are frequent features of the daily news.

The fuel for this astronomical revolution is the wealth of new data acquired with telescopes and spacecraft. New technologies have vastly improved the power of telescopes on the ground, and telescopes lofted into space can overcome the observational difficulties posed by the Earth's atmosphere. Whereas past astronomers could study only the messages contained in visible light, today's astronomers can read the story of the universe contained in the entire spectrum.

In addition, we've sent spacecraft to other planets for close-up study, in a few cases even sampling the surfaces or atmospheres of other worlds. In this chapter, we explore how telescopes and spacecraft are helping us to better understand our universe.

7.1 Eyes and Cameras: Everyday Light Sensors

We observe the world around us with the five basic senses: touch, taste, smell, hearing, and sight. We learn about the world by using our brains to analyze and interpret the data recorded by our senses. The science of astronomy progresses similarly. We collect data about the universe, and then we analyze and interpret the data to develop theories about how the universe works. Within our solar system, we can analyze some matter directly, by collecting samples from the Earth's surface or from meteorites that fall to Earth, or by occasionally sending spacecraft to sample the surfaces or atmospheres of other worlds. Aside from these few samples, nearly all other data about the universe come to us in the form of light.

Astronomers collect light with telescopes and record light with photographic film or electronic detectors. You are already familiar with many of the basic principles, because of your everyday experience with eyes and cameras.

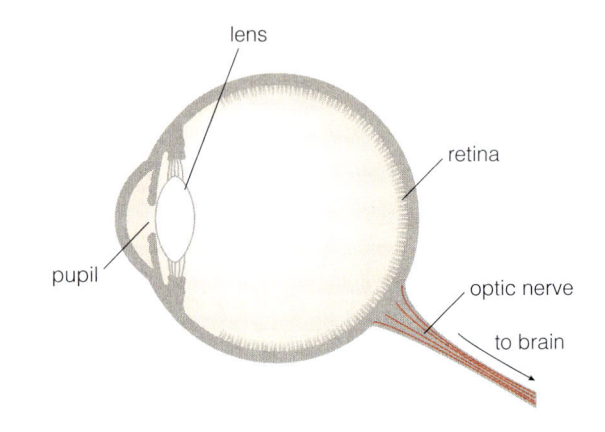

FIGURE 7.1 A simplified diagram of the human eye.

The Eye

The eye is more complex than any human-made instrument, but its basic components are a *lens*, a *pupil*, and a *retina* (Figure 7.1). The retina contains light-sensitive cells (called *cones* and *rods*) that, when triggered by light, send signals to the brain via the optic nerve.

The lens of an eye acts much like a simple glass lens. Light rays that enter the lens farther from the center are bent more, and rays that pass directly through the center are not bent at all. In this way, parallel rays of light, such as the light from a distant star, converge to a point called the **focus** (Figure 7.2a). If you have perfect vision, the focus of your lens is on your retina. That is why distant stars appear as *points* of light to our eyes or on photographs.

Light rays that come from different directions, such as from different parts of an object, converge at different points to form an **image** of the object (Figure 7.2b). The place where the image appears in focus is called the **focal plane** of the lens. Again, in a healthy eye, the focal plane is on your retina. (The retina actually is curved, rather than a flat plane, but we will ignore this detail.) Note that the image formed by a lens is upside down. It is flipped

FIGURE 7.2 (a) A glass lens bends parallel rays of light to a point called the focus of the lens. In a healthy eye, the lens focuses light on the retina. (b) Light from different parts of an object focuses at different points to make an image of the object. Note that the image formed on the retina is upside-down. The world does not look upside-down because your brain flips the image.

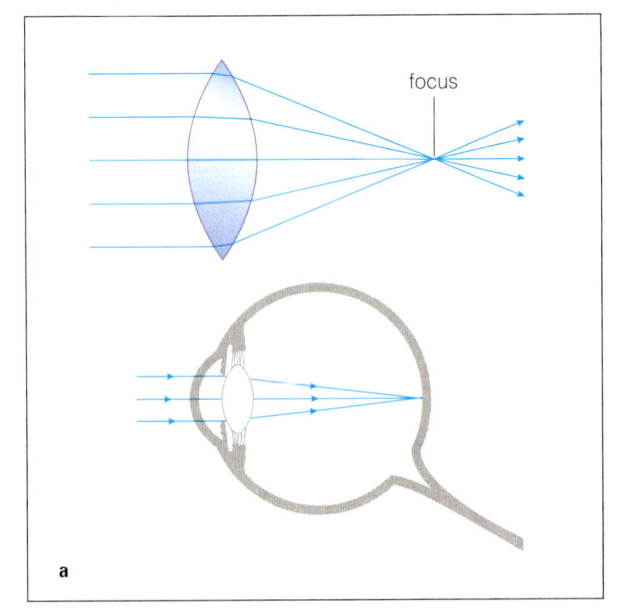

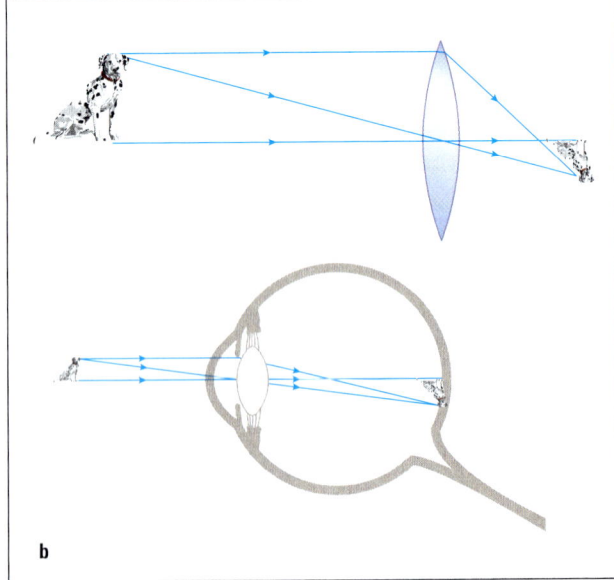

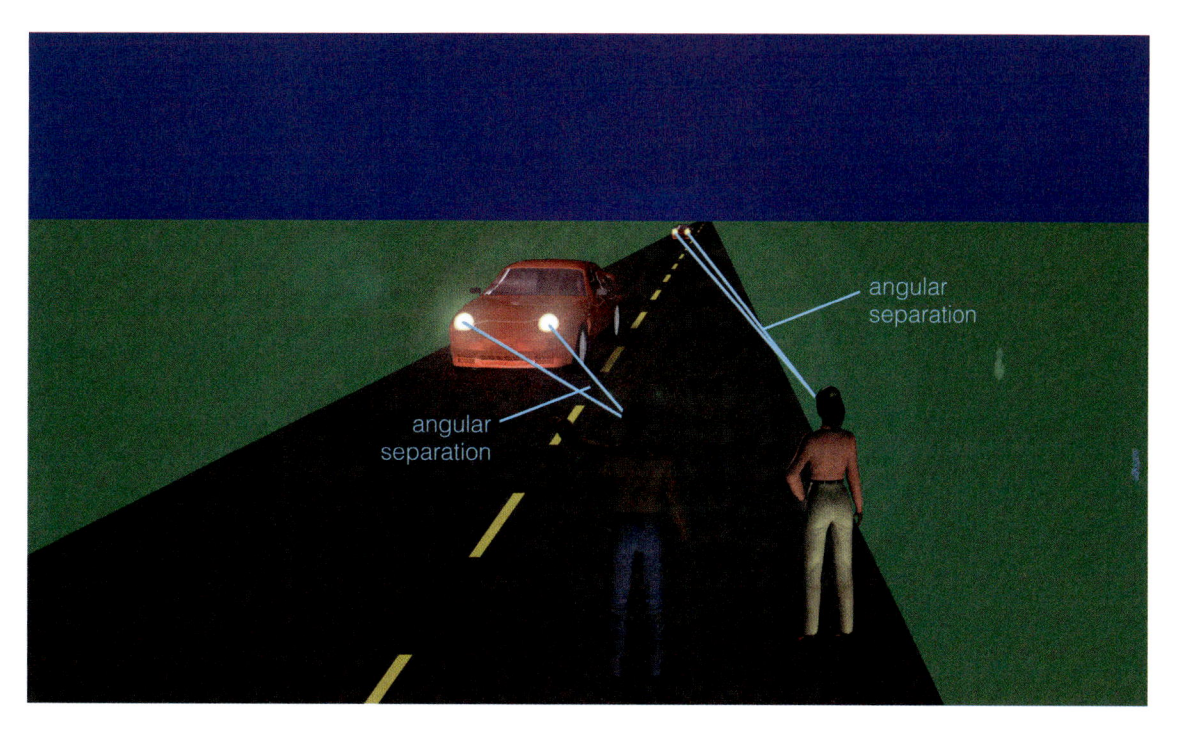

FIGURE 7.3 The angular separation of the headlights increases as the car comes closer.

right side up by your brain, where the true miracle of vision occurs.

The pupil controls the amount of light that enters the eye by adjusting the size of its opening. The pupil dilates (opens wider) in low light levels, allowing your eye to gather as many photons as possible. It constricts in bright light, so that your eye is not overloaded with light.

TIME OUT TO THINK *Cover one eye and look toward a bright light. Then look immediately into a mirror and compare the openings of your two pupils. Which one is wider, and why? Why do eye doctors dilate your pupils during eye exams?*

One way to quantify your visual acuity is to measure the smallest angle over which your eyes can tell that two dots—or two stars—are distinct. Imagine looking at a car heading toward you on a long, straight road. When the car is very far away, your eyes cannot distinguish the headlights individually, and they look like one light. But as the car comes closer, the angular separation of the two headlights gets bigger, and you eventually see two distinct headlights (Figure 7.3). The smallest angular separation at which you can tell that the headlights are distinct is the **angular resolution** of your eyes.

The human eye has an angular resolution of about 1 arcminute ($\frac{1}{60}°$), meaning that two stars will appear distinct if they lie farther than 1 arcminute apart in the sky. Two stars separated by *less* than 1 arcminute therefore appear as a single point of light. Most stars are actually members of binary or multiple-star systems, but, because they are so far away, the angular separation of the individual stars is far below the angular resolution of our eyes.

Cameras and Film

The basic operation of a camera is quite similar to that of an eye (Figure 7.4). The camera lens plays the role of the lens of the eye, and film plays the role of the retina. The camera is "in focus" when the film lies in the focal plane of the lens. Fancier cameras even have an adjustable circular opening, called the camera's *aperture,* that controls the amount of light entering the camera just as the pupil controls the amount of light entering the eye.

The chemicals in photographic film darken or change color in response to light, thereby recording an image. Recording images on film offers at least two important advantages over simply looking at them or drawing them. First, an image on film is much more reliable and detailed than a drawing. Second, whereas the eye continually and automatically sends images to the brain, we can use a camera's *shutter* to control the amount of time, called the **exposure time**, over which light collects on a single piece of film. A longer exposure time means that more photons reach the film, allowing more opportunities for the light-sensitive chemicals to change in

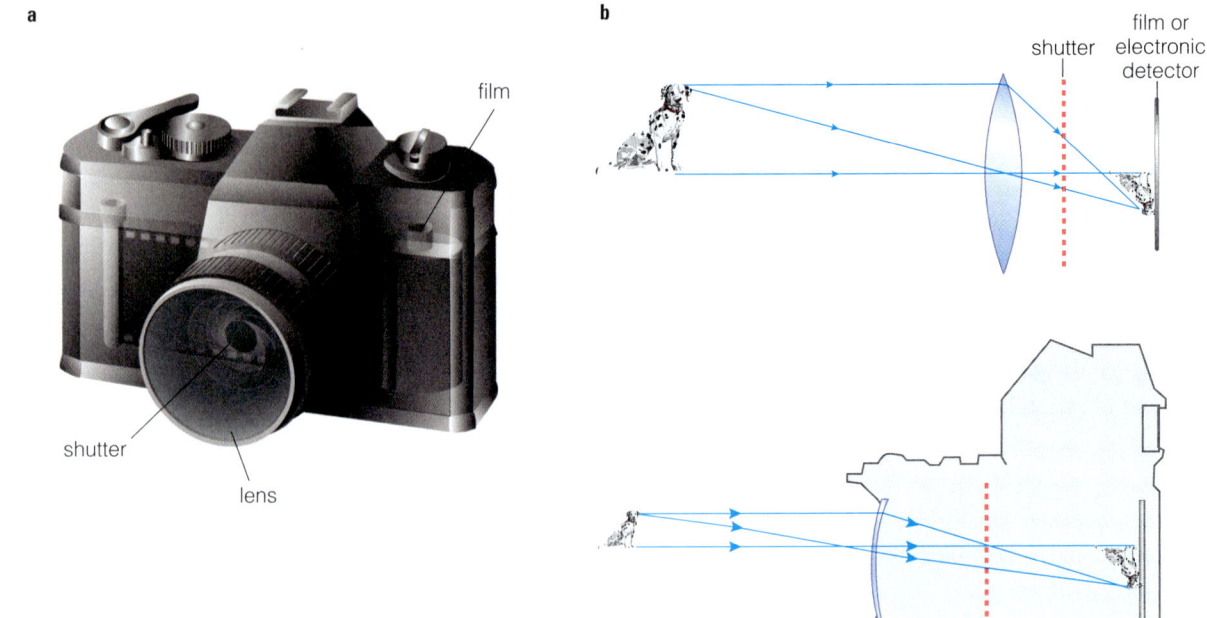

FIGURE 7.4 (a) A camera works much like an eye. When the shutter is open, light passes through the lens to form an image on the film or electronic detector. (b) Basic components of a camera.

Mathematical Insight 7.1 Angular Separation

The angular separation of two points is related to their actual separation and distance. Figure 7.5 shows two points with an actual separation s between them. The angle α is the angular separation between the two points when viewed from a distance d.

As long as d is much larger than s, we can think of s as a small portion of an imaginary circle with radius d. Recall that the circumference of a circle is $2 \times \pi \times$ radius, so the circumference of the dotted circle in the figure is $2\pi d$. Thus, the separation s represents a fraction $s/(2\pi d)$ of this circumference. We find the angle α by multiplying this fraction by the $360°$ in a full circle:

$$\alpha = \frac{s}{2\pi d} \times 360°$$

Example 1: Suppose the two headlights on a car are separated by 1.5 meters and you are looking at the car from a distance of 500 meters. What is the angular separation of the headlights? Can your eyes resolve the two headlights?

Solution: The separation of the headlights is $s = 1.5$ m, and their distance is $d = 500$ m. Thus, their angular separation is

$$\alpha = \frac{1.5 \text{ m}}{2\pi \times 500 \text{ m}} \times 360° = 0.17°$$

The angular resolution of the human eye is about $\frac{1}{60}° \approx 0.017°$, or 10 times smaller than the angular separation of the two headlights. Thus, your eyes can easily resolve the two headlights at a distance of 500 meters.

Example 2: The angular diameter of the Moon is about $0.5°$, and the Moon is about 380,000 km away. Estimate the diameter of the Moon.

Solution: We are given $\alpha = 0.5°$ and $d = 380,000$. We are looking for s—the diameter of the Moon in this case—so we first solve the angular separation equation for s:

$$\alpha = \frac{s}{2\pi d} \times 360° \quad \Rightarrow \quad s = \frac{2\pi d}{360°} \times \alpha$$

Now we substitute the given values:

$$s = \frac{2\pi \times 380,000 \text{ km}}{360°} \times 0.5° \approx 3,300 \text{ km}$$

This estimate is fairly close to the Moon's actual diameter (3,476 km), which we could find by using more precise values for the Moon's angular diameter and distance.

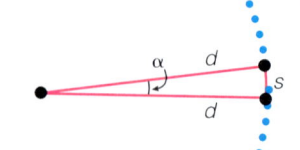

FIGURE 7.5 As long as the angle α is small, the lengths s and d are related to α by the formula $\alpha = (s/2\pi d) \times 360°$.

174 PART II KEY CONCEPTS FOR ASTRONOMY

response to light. As a result, long-exposure photographs can reveal images of objects far too faint for our eyes to see.

CCDs

Thanks to the advantages of film over the human eye, the advent of photography in the mid-1800s spurred a leap in astronomical data collection. However, photographic film is far from ideal for recording light. Most photons striking film have no effect at all: Fewer than 10% of the visible photons reaching the film cause a change in the light-sensitive chemicals.[1] Thus, your film may record nothing if too little light reaches it. At the other extreme, a long exposure time can *overexpose* (or *saturate*) your film, darkening so large a fraction of the light-sensitive chemicals that details of the image are lost.

Today, many cameras and camcorders come equipped to record images with electronic detectors called *charge-coupled devices*, or **CCDs**, that can accurately record 90% or more of the photons that strike them. A CCD is a chip of silicon carefully engineered to be extraordinarily sensitive to photons (Figure 7.6). The chip is physically divided into a grid of squares called *picture elements*, or **pixels** for short. When a photon strikes a pixel, it causes a bit of electric charge to accumulate. Each subsequent photon striking the same pixel adds to this accumulated electric charge. After an exposure is complete, a computer measures the total electric charge in each pixel, thus determining how many photons have struck each one. The overall image is stored in the computer as an array of numbers representing the results from each pixel. At this point, the image can be manipulated through techniques of *image processing* to bring out details that otherwise might be missed in analyzing an image.

7.2 Telescopes: Giant Eyes

Our naked eyes allow us to admire the beauty of the night sky, but for most astronomical purposes they are completely inadequate. Their small size limits their angular resolution and the amount of light they can collect. They are sensitive only to visible light. And they are attached to bodies that require a pleasantly warm atmosphere for survival, an atmosphere that distorts light and prevents much of the electromagnetic spectrum from reaching the ground.

---

[1] The probability that an individual photon will be detected is called the *quantum efficiency*. It is about 1% for the human eye, up to about 10% for film, and 90% or more for CCDs.

FIGURE 7.6 This photograph shows a CCD (charge-coupled device) on a circuit board. The quarter is included for size comparison.

Telescopes solve all these problems. We can design telescopes to compensate for the distorting effects of our atmosphere or launch them into space aboard spacecraft. We can build telescopes and detectors that are sensitive to light our eyes cannot see, such as radio waves or X rays. Most important, telescopes function as giant eyes, collecting far more light with far better angular resolution than our naked eyes.

Basic Telescope Design

Telescopes come in two basic designs: *refracting* and *reflecting*. A **refracting telescope** operates much like an eye, using transparent glass lenses to focus the light from distant objects (Figure 7.7a). The earliest telescopes, including those built by Galileo, were refracting telescopes. The world's largest refracting telescope, completed in 1897, has a lens that is 1 meter (40 inches) in diameter and a telescope tube that is 19.5 meters (64 feet) long (Figure 7.7b).

A **reflecting telescope** uses a precisely curved **primary mirror** to gather and focus light (Figure 7.8a). Note that the mirror focuses light at the **prime focus**, which lies *in front of* the mirror. Thus, if we place a camera at the prime focus, it blocks some of the light entering the telescope. This does not create a serious problem as long as the camera is small compared to the primary mirror. If the telescope is large enough, a "cage" in which an astronomer can work can be suspended over the prme focus (Figure 7.8b).

An alternative and more common arrangement for reflecting telescopes is to place a **secondary mirror** just below the prime focus. The secondary mirror reflects the light to a location that is more

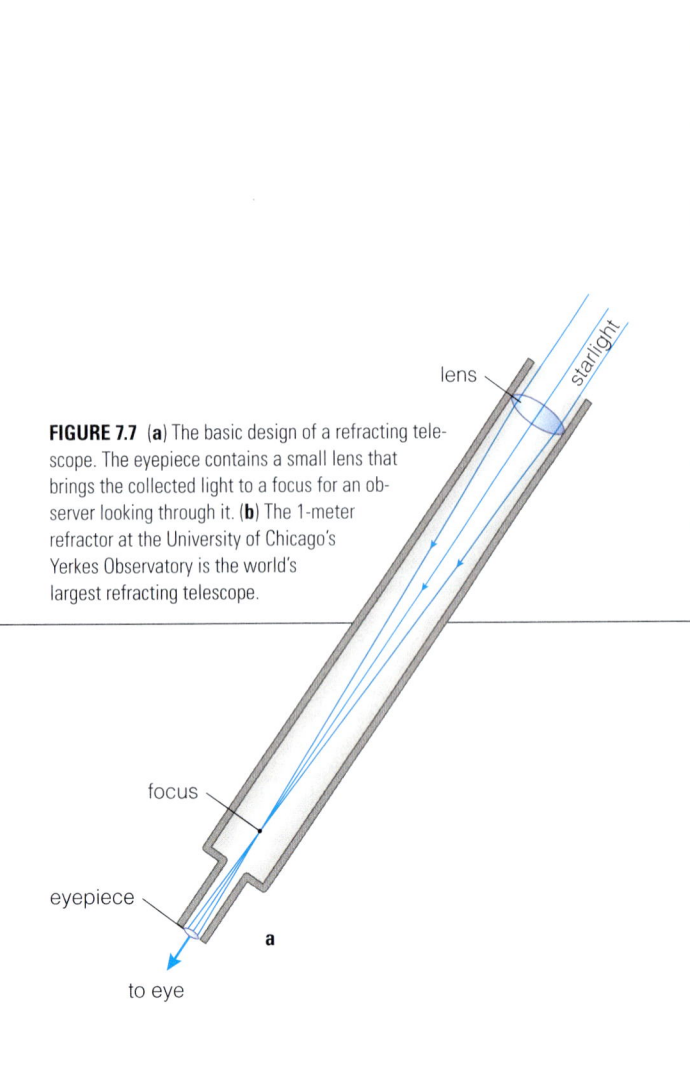

FIGURE 7.7 (**a**) The basic design of a refracting telescope. The eyepiece contains a small lens that brings the collected light to a focus for an observer looking through it. (**b**) The 1-meter refractor at the University of Chicago's Yerkes Observatory is the world's largest refracting telescope.

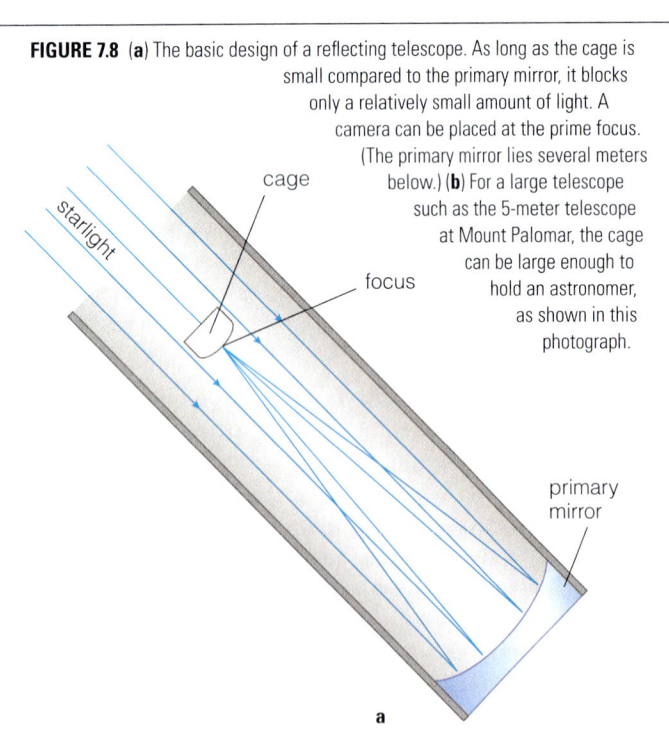

FIGURE 7.8 (**a**) The basic design of a reflecting telescope. As long as the cage is small compared to the primary mirror, it blocks only a relatively small amount of light. A camera can be placed at the prime focus. (The primary mirror lies several meters below.) (**b**) For a large telescope such as the 5-meter telescope at Mount Palomar, the cage can be large enough to hold an astronomer, as shown in this photograph.

FIGURE 7.9 Alternative designs for reflecting telescopes. Each of these designs incorporates a small, secondary mirror just below the prime focus. The secondary mirror reflects the light to a location that is more convenient for viewing or for attaching instruments.

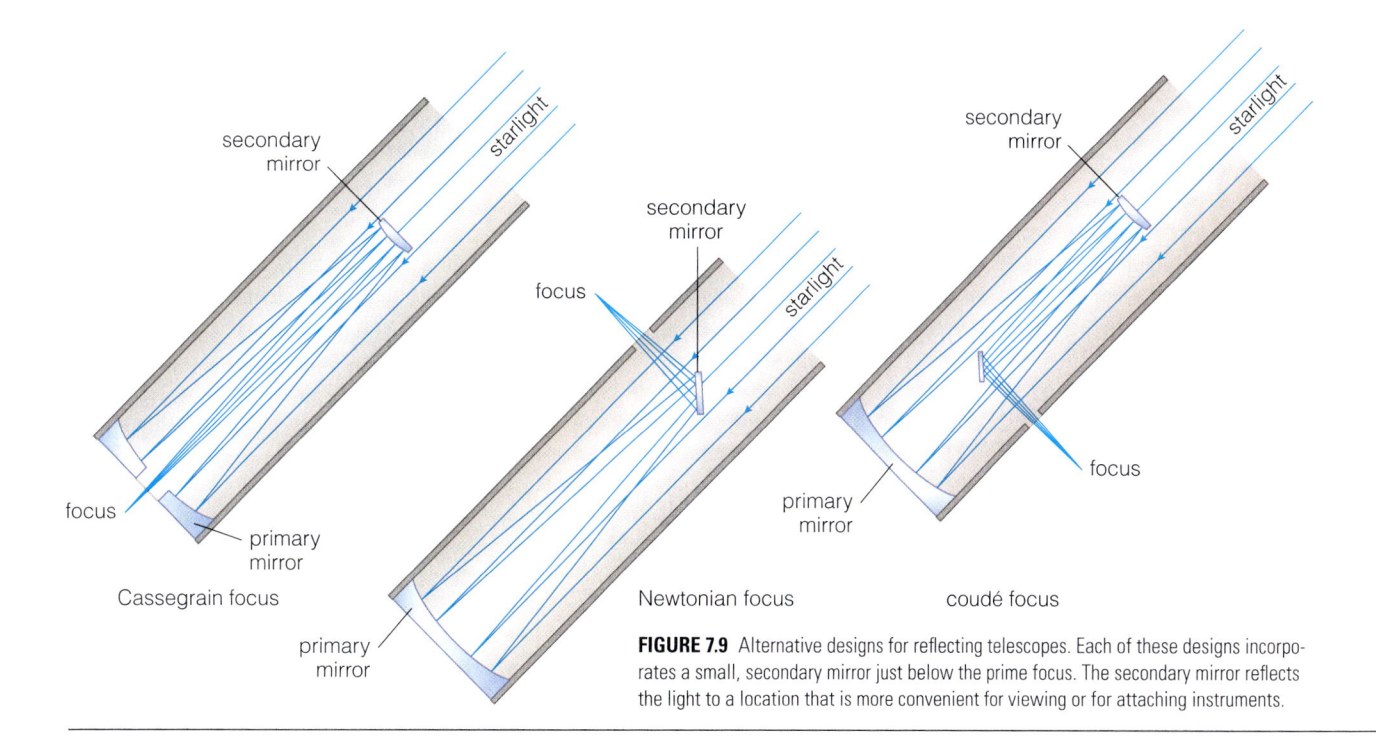

convenient for viewing or for attaching instruments (Figure 7.9): through a hole in the primary mirror (a *Cassegrain focus*), to a hole in the side (a *Newtonian focus*, particularly common in smaller telescopes), or to a third mirror that deflects light toward instruments that are not attached to the telescope (a *coudé focus* or a *Nasmyth focus*).

Today, most astronomical research involves reflecting telescopes, mainly for two practical reasons. First, because light passes *through* the lens of a refracting telescope, lenses must be made from clear, high-quality glass with precision-shaped surfaces on both sides. In contrast, only the reflecting surface of a mirror must be precisely shaped, and the quality of the underlying glass is not a factor. Second, large glass lenses are extremely heavy and can be held in place only by their edges, making it difficult to stabilize refracting telescopes, in which lenses are positioned at the top. The primary mirror of a reflecting telescope is mounted at the bottom, where its weight is a far less serious problem. (A third problematic feature of lenses, called *chromatic aberration*, occurs because a lens brings different colors of light into focus at slightly different places. This problem also plagues camera lenses, but it can be minimized by using combinations of lenses and special optical coatings.)

Fundamental Properties of Telescopes

The power of any telescope, whether refracting or reflecting, is characterized by two fundamental properties: its **light-collecting area**, which describes

> **Common Misconceptions:**
> **Magnification and Telescopes**
>
> You are probably familiar with how binoculars, magnifying glasses, and telephoto lenses for cameras make objects appear larger in size—the phenomenon we call magnification. Many people assume that astronomical telescopes are also characterized by their magnification. In fact, astronomers are much more interested in a telescope's light-collecting area and angular resolution. The magnification of most telescopes is set to achieve the maximum amount of detail given the telescope's resolution—no more, no less. After all, there's little point in magnifying an image if doing so doesn't show any more detail.

the telescope's size in terms of how much light it can collect, and its **angular resolution**, which tells us how much detail we can see in the telescope's images. Let's examine each property in a bit more detail.

Light-Collecting Area The "size" of a telescope is usually described by the diameter of its primary mirror (or lens). For example, a "5-meter telescope" is a telescope with a primary mirror measuring 5 meters in diameter. The light-collecting area is proportional to the *square* of the mirror diameter. (More specifically,

CHAPTER 7 TELESCOPES AND SPACECRAFT 177

Table 7.1 Largest Optical Telescopes

Size	Telescope Name	Sponsor	Location	Operational Date	Special Features
10 m	Keck I	Cal Tech, U. California, NASA	Mauna Kea, HI	1993	Primary mirror consists of 36 1.8-m hexagonal segments.
10 m	Keck II	Cal Tech, U. California, NASA	Mauna Kea, HI	1996	Twin of Keck I; future plans for interferometry with Keck I.
9.2 m	Hobby–Eberly	U. Texas, Penn State, Stanford, Germany	Mt. Locke, TX	1997	Consists of 91 1-m segments, for a total diameter of 11 m, but only 9.2 m can be used at a time; designed primarily for spectroscopy.
2 × 8.4 m	Large Binocular Telescope	U. Arizona, Arcetri Astrophysical Observatory (Italy)	Mt. Graham, AZ	2004*	Two 8.4-m mirrors on a common mount give light-collecting area of 11.8-m telescope.
4 × 8 m	Very Large Telescope	European Southern Observatory	Cerro Paranal, Chile	2000	Four separate 8-m telescopes can work individually or together as the equivalent of a 16-m telescope.
8.3 m	Subaru	National Observatory of Japan	Mauna Kea, HI	1999	Japan's first large telescope project.
8 m	Gemini North	U.S., U.K., Canada, Chile, Brazil, Argentina	Mauna Kea, HI	1999	Identical to Gemini South so similar observations can be made from each hemisphere.
8 m	Gemini South	U.S., U.K., Canada, Chile, Brazil, Argentina	Cerro Pachon, Chile	2000	Identical to Gemini North.
6.5 m	Magellan	Carnegie Institute, Harvard, U. Michigan, MIT	Las Campanas, Chile	2000	A second 6.5-m telescope is planned at a later date.
6.5 m	MMT	Smithsonian Institution, U. Arizona	Mt. Hopkins, AZ	2000	Replaced an older telescope in same observatory.

*Scheduled completion date.

for a round mirror surface, we can use the formula for the area of a circle: area = $\pi \times$ (radius)^2.) Thus, a 2-meter telescope has $2^2 = 4$ times the light-collecting area of a 1-meter telescope.

For more than 40 years after its opening in 1947, the 5-meter Hale telescope on Mount Palomar outside San Diego remained the world's most powerful. (A Russian telescope built in 1976 was larger, but it suffered optical quality problems that made it far less useful.) Building larger telescopes posed a technological challenge, because large mirrors tend to sag under their own weight. Recent technological innovations have made it possible to build very large, low-weight mirrors, fueling a boom in large telescope construction (Table 7.1). For example, each of the twin 10-meter Keck telescopes in Hawaii has a primary mirror consisting of 36 smaller mirrors that function together (Figure 7.10). When you realize that each Keck telescope has more than a million times the light-collecting area of the human eye, it is no wonder that modern telescopes have so dramatically enhanced our ability to observe the universe.

TIME OUT TO THINK *What is the population of your hometown? Suppose everyone in your hometown looked at the dark sky at the same time. How would the total amount of light entering everyone's eyes compare with the light collected by a 10-meter telescope? Explain.*

Angular Resolution Large telescopes can have remarkable angular resolution. For example, the Hubble Space Telescope has an angular resolution of about 0.05 arcsecond (for visible light), which would allow you to read this book from a distance of about 800 meters (a half mile).

FIGURE 7.10 (**a**) The two Keck telescopes sit atop Mauna Kea in Hawaii. (**b**) The Keck telescopes use 36 hexagonal mirrors to function as a single 10-meter primary mirror. These mirrors make the honeycomb pattern that surrounds the hole enclosing the man in this photo.

In principle, larger telescopes have better angular resolution; that is, they can distinguish finer details. However, the angular resolution may be degraded if a telescope is poorly made, and, as we'll discuss shortly, turbulence in the Earth's atmosphere can limit the angular resolution of ground-based telescopes.

The ultimate limit to a telescope's resolving power comes from the properties of light. Because light is an electromagnetic wave [Section 6.2], beams of light can interfere with one another like overlapping sets of ripples on a pond (Figure 7.11). This *interference* causes a blurring of images that limits a telescope's angular resolution even when all other conditions are perfect. That is why even a high-quality telescope in space, such as the Hubble Space Telescope, cannot have perfect angular resolution[2] (Figure 7.12).

The angular resolution that a telescope could achieve if it were limited only by the interference of light waves is called its **diffraction limit**. (*Diffraction* is a technical term for the specific effects of interference that limit telescope resolution.) It depends on both the diameter of the primary mirror and the wavelength of the light being observed. Larger telescopes have smaller diffraction limits for any

---

[2]In fact, the Hubble Space Telescope's primary mirror was made with the wrong shape, which further limited its angular resolution until corrective optics were installed in 1993. The corrective optics consist of small mirrors that essentially undo the blurring created by the primary mirror.

FIGURE 7.11 This computer-generated image shows how overlapping sets of ripples on a pond interfere with one another. The effects of the two sets of ripples add in some places, making the water rise extra high or fall extra low, and cancel in other places, making the water surface flat. Light waves also exhibit interference. (The colors in this image are for visual effect only.)

particular wavelength of light. However, achieving a particular angular resolution requires a larger telescope for longer-wavelength light. That is why, for example, a radio telescope must be far larger than an optical telescope to achieve the same angular resolution.

7.3 Uses of Telescopes

Astronomers use many different kinds of instruments and detectors to extract the information contained in the light collected by a telescope. Every astronomical observation is unique, but most observations fall into one of three basic categories:

- **Imaging** yields pictures of astronomical objects. At its most basic, an imaging instrument is simply a camera. Astronomers sometimes place *filters* in front of the camera that allow only particular colors or wavelengths of light to pass through. Most of the richly hued images in this and other astronomy books are made by combining images recorded through different filters (Figure 7.13).

- **Spectroscopy** involves spreading light into a spectrum. Instruments called *spectrographs* use diffraction gratings (or other devices) to disperse light into spectra, which are then recorded with a detector such as a CCD (Figure 7.14).

- **Timing** tracks how an object's brightness varies with time. For a slowly varying object, a timing experiment may be as simple as comparing a set

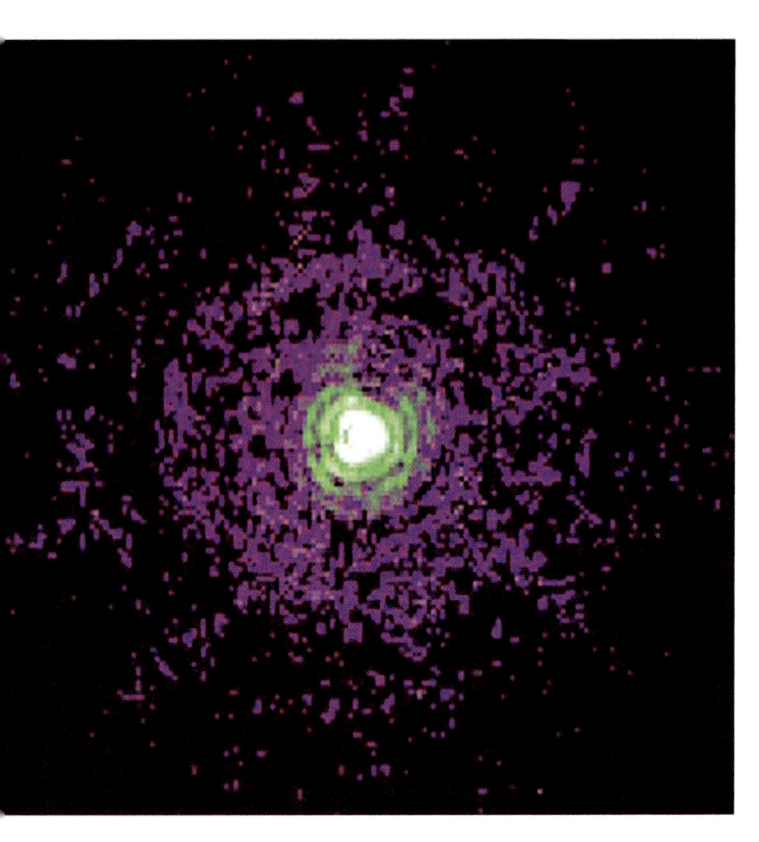

FIGURE 7.12 When examined in detail, a Hubble Space Telescope image of a star has rings resulting from the wave properties of light. With higher angular resolution, the rings would be smaller.

Mathematical Insight 7.2 The Diffraction Limit

A simple formula gives the diffraction limit of a telescope in arcseconds.

$$\text{diffraction limit} \approx 2.5 \times 10^5 \times \left(\frac{\text{wavelength of light}}{\text{diameter of telescope}}\right)$$
(arcseconds)

The wavelength of light and the diameter of the telescope must be in the same units.

Example 1: What is the diffraction limit of the 2.4-m Hubble Space Telescope for visible light with a wavelength of 500 nm?

Solution: For light with a wavelength of 500 nm (500×10^{-9} m), the diffraction limit of the Hubble Space Telescope is approximately:

$$\text{diffraction limit} = 2.5 \times 10^5 \times \left(\frac{\text{wavelength}}{\text{telescope diameter}}\right)$$
$$= 2.5 \times 10^5 \times \frac{500 \times 10^{-9} \text{ m}}{2.4 \text{ m}}$$
$$= 0.05 \text{ arcsecond}$$

Example 2: Suppose you wanted to achieve a diffraction limit of 0.001 arcsecond for visible light of wavelength 500 nm. How large a telescope would you need?

Solution: First we solve the diffraction limit equation for telescope diameter:

$$\text{diffraction limit} \approx 2.5 \times 10^5 \times \left(\frac{\text{wavelength}}{\text{telescope diameter}}\right)$$
$$\Downarrow$$
$$\text{telescope diameter} \approx 2.5 \times 10^5 \times \left(\frac{\text{wavelength}}{\text{diffraction limit}}\right)$$

Now we substitute the given values for the wavelength and diffraction limit:

$$\text{telescope diameter} = 2.5 \times 10^5 \times \frac{500 \times 10^{-9} \text{ m}}{0.001 \text{ arcsecond}}$$
$$= 125 \text{ m}$$

An optical telescope would need a mirror diameter of 125 meters—longer than a football field—to achieve an angular resolution of 0.001 arcsecond.

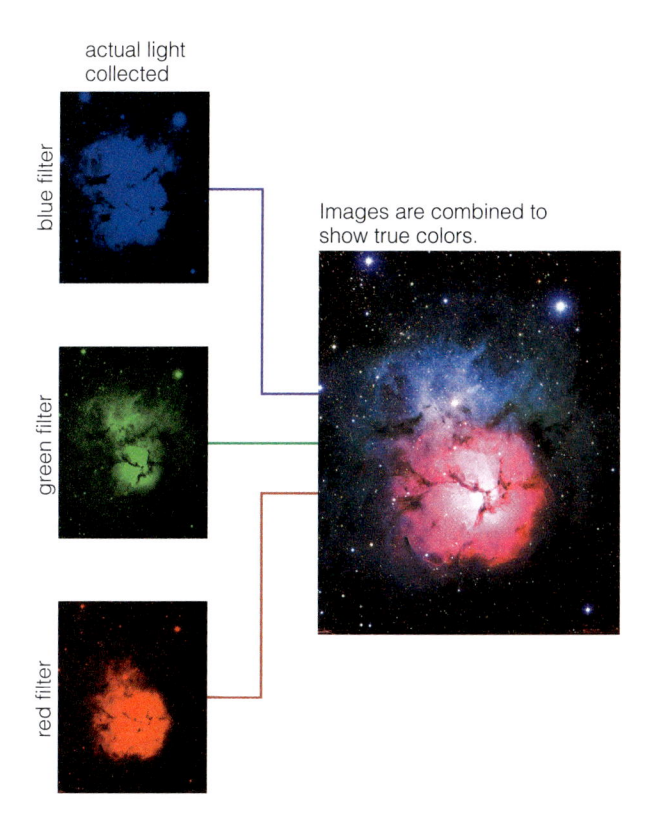

FIGURE 7.13 In astronomy, color images are usually constructed by combining several images taken through different filters.

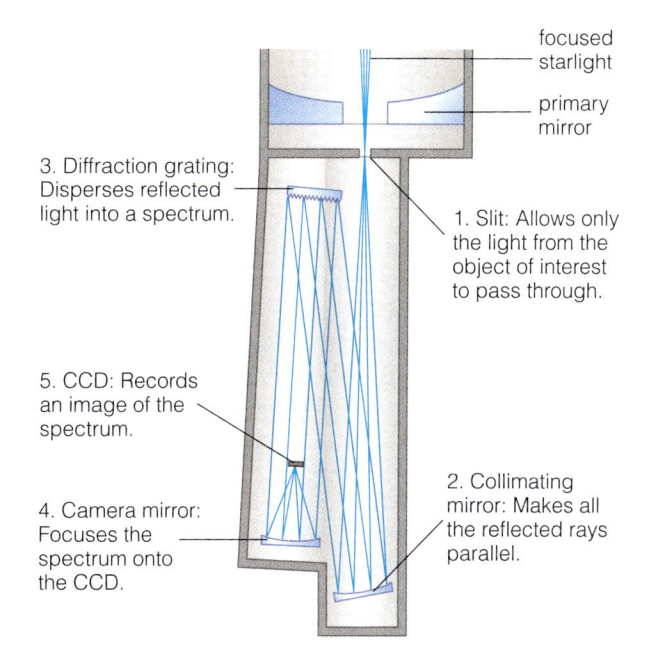

FIGURE 7.14 The basic design of a spectrograph. In this diagram, the spectrograph is attached to the bottom of a reflecting telescope, with light entering the spectrograph through a hole in the primary mirror. A narrow slit (or small hole) at the entrance to the spectrograph allows only light from the object of interest to pass through. This light bounces from the collimating mirror to the diffraction grating, which disperses the light into a spectrum. The dispersed light is then focused by the camera mirror and recorded by a CCD, giving an image of the spectrum.

of images or spectra obtained on different nights. For more rapidly varying sources, specially designed instruments essentially make rapid multiple exposures, in some cases recording the arrival time of every individual photon.

Major observatories typically have several different instruments capable of each of these tasks, and some instruments can perform all three basic tasks. (Note that some astronomers include a fourth general category called *photometry,* which is the accurate measurement of light intensity from a particular object at a particular time. We do not call this out as a separate category because today's detectors can do photometry at the same time that they are being used for imaging, spectroscopy, or timing.)

Images of Nonvisible Light

Many astronomical images show nonvisible light; for example, Figure 7.15 is an "X-ray image." But what exactly does this mean, given that X rays are invisible to the eye? The easiest way to understand the idea is to think about X rays at a doctor's office [Section 6.3]. When the doctor makes an "X ray of your arm," he or she actually uses a machine that sends X rays through your arm; those that pass through are recorded on a piece of X-ray-sensitive film. We cannot see the X rays themselves, but we can see the image left behind on

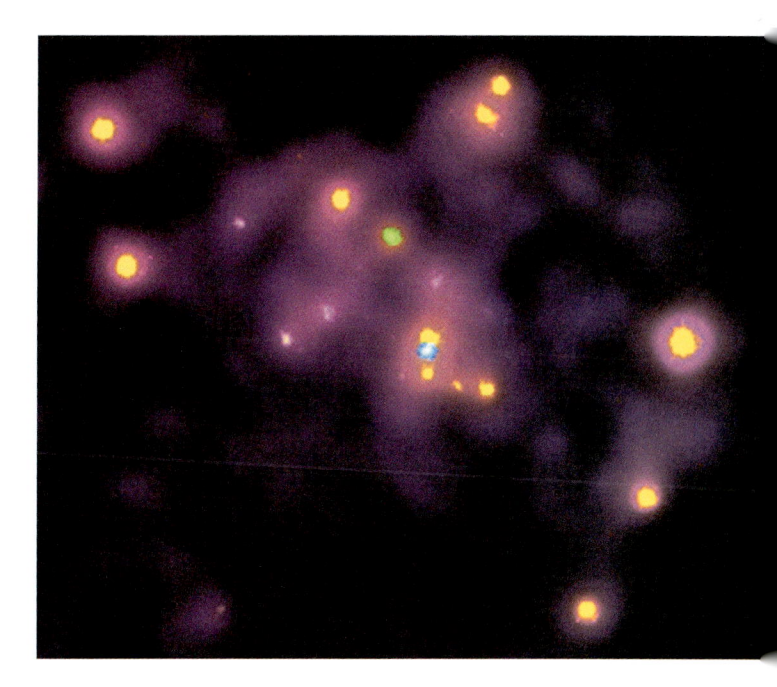

FIGURE 7.15 X rays are invisible, but we can color-code the information recorded by an X-ray detector to make an image of the object as it would appear in X rays. This image, from NASA's Chandra X-Ray Observatory, shows the region (roughly 700 light-years on a side) in the center of the Andromeda Galaxy. The blue dot in the center marks what astronomers believe to be a supermassive black hole [Section 20.5]. The rest of the image is color-coded, with yellow indicating the most intense X rays. Most of the other bright X-ray sources are probably X-ray binaries [Section 17.3], star systems in which a neutron star or black hole orbits a normal star.

Chapter 7 Telescopes and Spacecraft

the film. Astronomical images work the same way; whether a detector records visible or nonvisible light, it makes an image showing the intensity of the light.

The simplest way to display a nonvisible image is in black and white, with brighter regions of the image corresponding to brighter light coming from the object. (A doctor's X ray is actually a negative, making bones look bright because few X rays pass through them.) However, it is much easier for the human eye to perceive color differences than shades of gray. That is why nonvisible images are often color-coded. In Figure 7.15, for example, the colors (except for the blue dot) correspond to the intensity of the recorded X rays: The brightest regions are yellow, while the violet and black represent regions that emit few, if any, X rays. All images recorded in nonvisible light must necessarily use some kind of color or gray-scale coding. Because the colors in these images do not represent what our eyes actually see, they are often called **false-color** images.

TIME OUT TO THINK *False-color images are common even outside astronomy. For example, medical images from CAT scans and MRIs are usually displayed in color, even though neither type of imaging uses visible light. What do you think the colors mean in CAT scans and MRIs? How are the colors useful to doctors?*

Spectral Resolution

As we discussed in Chapter 6, a spectrum can reveal a wealth of information about an object, including its chemical composition, temperature, and rotation rate. However, just as the amount of information we can glean from an image depends on the angular resolution, the information we can glean from a spectrum depends on the **spectral resolution**: The higher the spectral resolution, the more detail we can see (Figure 7.16).

In principle, astronomers would always like the highest possible spectral resolution. However, higher spectral resolution comes at a price. A telescope collects only so much light in a given amount of time, and the spectral resolution depends on how widely this light is spread out by the spectrograph. If the spectrograph spreads out the light too much, the spectrum may become so dim that nothing can be recorded without a very long exposure time. Thus, for a particular telescope, making a spectrum of an object requires a longer exposure time than making an image of the same object, and high-resolution spectra take longer than low-resolution spectra.

7.4 Atmospheric Effects on Observations

A telescope on the ground does not look directly into space but rather looks through the Earth's sometimes murky atmosphere. In fact, the Earth's atmosphere creates several problems for astronomical observations. The most obvious problem is weather—an optical telescope is useless under cloudy skies. Another problem is that our atmosphere scatters the bright lights of cities, creating what astronomers call **light pollution**, which can obscure the view even for the best telescopes. For example, the 2.5-meter telescope at Mount Wilson, the world's largest when it was built in 1917, would still be very useful today

FIGURE 7.16 Higher spectral resolution means we can see more details in the spectrum. Both (**a**) and (**b**) show spectra of the same object for the same wavelength band, but the higher spectral resolution in (**b**) enables us to see individual spectral lines that appear merged together in (**a**). (The spectrum shows interstellar absorption lines.)

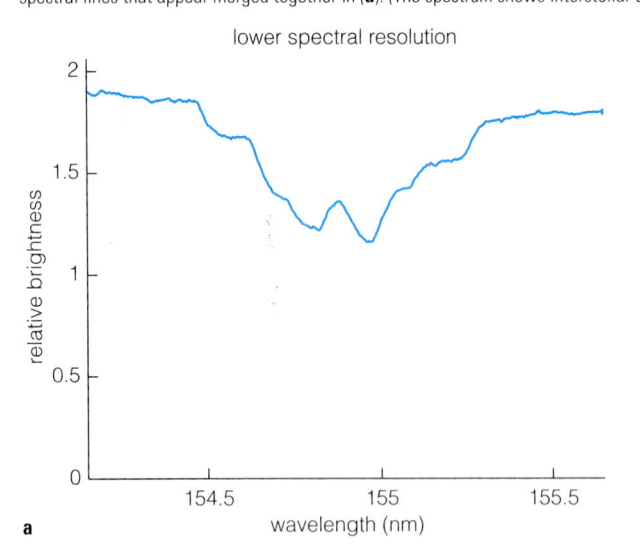

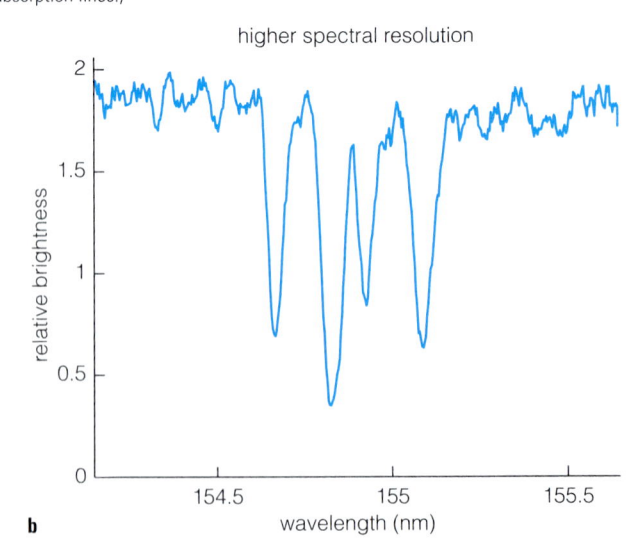

FIGURE 7.30 An artist's conception of a lunar observatory.

The Future of Astronomy in Space

It will always be cheaper and easier to do astronomy from ground-based observatories, but space is likely to play an ever-increasing role in astronomy. NASA is considering future missions to every planet, as well as to some moons, asteroids, and comets. Mars is targeted for a particularly ambitious set of missions, with at least one mission scheduled about every 2 years (when Mars is located at the position in its orbit that makes it easiest to reach from Earth). A sample return mission, bringing Mars rocks back to Earth, may occur as early as 2005. Some scientists hope to launch a human mission to Mars within the next couple of decades.

Telescopic astronomy in space also should advance spectacularly. NASA is already at work on follow-ons to the Hubble Space Telescope, such as the *Next Generation Space Telescope* (NGST), which may launch as early as 2009. For the more distant future, many astronomers dream of an observatory on the far side of the Moon (Figure 7.30). Because the Moon has no atmosphere, it offers all the advantages of telescopes in space while also offering the ease of operating on a solid surface.

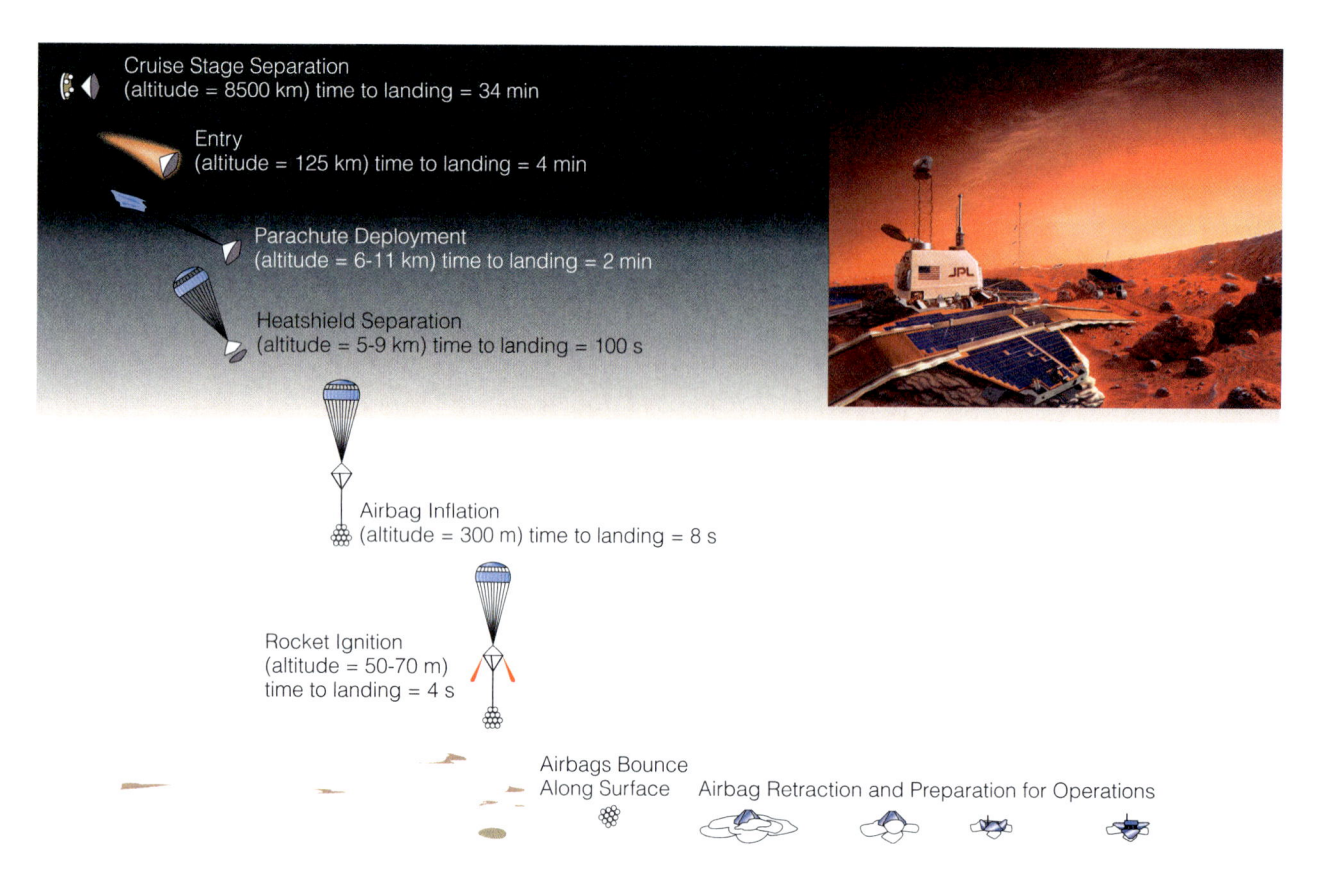

FIGURE 7.28 The *Pathfinder* lander used parachutes and air bags to land safely on Mars. The inset shows an artist's conception of the lander on Mars.

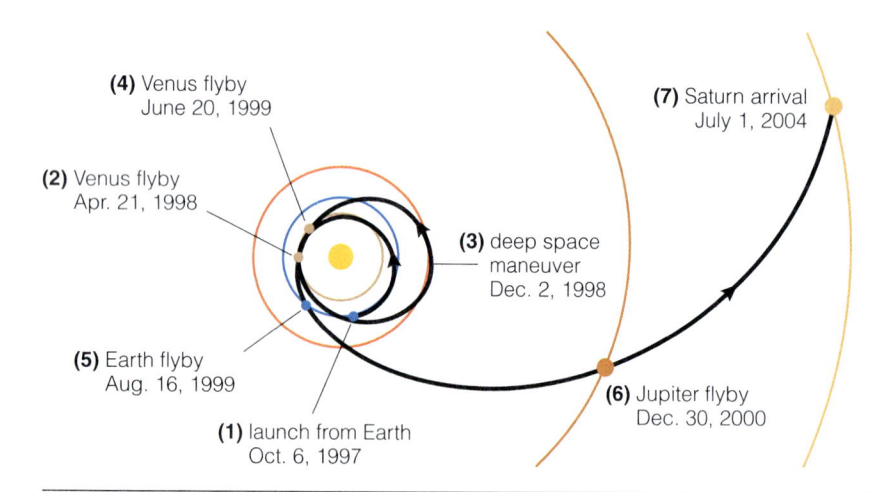

FIGURE 7.29 The trajectory of *Cassini* from Earth to Saturn.

of each elliptical orbit. Atmospheric drag slowed the spacecraft, removing its unwanted orbital energy.

Spacecraft headed to the outer solar system present another problem: The farther we wish to send a spacecraft from the Sun, the more energy it requires. To date, we've never launched a spacecraft to the outer solar system with enough energy to reach its final destination directly. Instead, we've given it only enough energy to make it to at least one planet and then have taken advantage of that planet's gravity to "slingshot" the spacecraft to higher speeds. In effect, a spacecraft gains energy by stealing a little bit of the planet's orbital energy around the Sun; because spacecraft are so small compared to planets, the effect on the planet's orbit is negligible. The Cassini mission used this slingshot technique four times: twice at Venus, then with a close flyby of Earth, and finally with a flyby of Jupiter in December, 2000. Cassini is scheduled to arrive at Saturn in 2004 (Figure 7.29).

Table 7.3 Selected Robotic Missions to Other Worlds

Name	Type	Destination	Lead Space Agency	Arrival Year	Mission Highlights
Cassini	Orbiter	Saturn	NASA	2004*	Includes lander (called Huygens) for Titan
Mars Exploration Rovers	Lander	Mars	NASA	2004*	Twin rovers to study the surface in two locations
Mars Express	Orbiter and lander	Mars	ESA**	2004*	Climate and surface studies of Mars
Nozomi	Orbiter	Mars	Japan	2004*	Japanese-led mission to study Martian atmosphere
Contour	Flyby	3 comets	NASA	2003*, 2006*, 2008*	Detailed study of three comet nuclei
Mars Odyssey 2001	Orbiter	Mars	NASA	2001	Detailed study of Martian surface features and composition
Near Earth Asteroid Rendezvous (NEAR)	Orbiter	Eros (asteroid)	NASA	2000	First spacecraft dedicated to in-depth study of an asteroid
Mars Global Surveyor	Orbiter	Mars	NASA	1997	Detailed imaging of surface from orbit
Mars Pathfinder	Lander	Mars	NASA	1997	Carried *Sojourner*, the first robotic rover on Mars
Galileo	Orbiter	Jupiter	NASA	1995	Dropped probe into Jupiter; close-up study of moons
Magellan	Orbiter	Venus	NASA	1990	Detailed radar mapping of surface of Venus
Voyager 1	Flyby	Jupiter, Saturn	NASA	1979, 1980	Unprecedented views of Jupiter and Saturn; continuing out of solar system
Voyager 2	Flyby	Jupiter, Saturn, Uranus, Neptune	NASA	1979, 1981, 1986, 1989	Only mission to Uranus and Neptune; continuing out of solar system
Vikings 1 and 2	Orbiter and lander	Mars	NASA	1976	First landers on Mars

*Scheduled arrival year.   **European Space Agency

Orbital Considerations and Cost

Because spacecraft require very high speeds to leave the Earth, spacecraft must be launched atop large rockets. The heavier the spacecraft, the larger and more expensive the rocket required. The most obvious way to reduce the weight and launch cost of a spacecraft is to make it smaller. Just as powerful computers today are smaller than ever, new technologies are reducing the size of many scientific instruments.

Once a spacecraft is on its way, it requires no more fuel unless it needs to *change* its orbit [Section 5.6]. Thus, because a spacecraft headed toward another world approaches on an unbound orbit (relative to the target world), it is destined for a flyby unless it carries rocket engines and fuel to change its orbit to a bound one (elliptical or circular). The weight of these engines and fuel tends to make orbiters heavier and more expensive to launch than flybys.

Two 1997 Mars missions used imaginative techniques to reduce the amount of fuel they needed to carry. The *Pathfinder* lander used parachutes to slow its descent through the thin Martian atmosphere. Before hitting the surface at about 50 km/hr, it deployed air bags around itself for protection (Figure 7.28). The air bags made the lander bounce several times along the surface before it settled at its landing site. The *Mars Global Surveyor*, an orbiter, saved on weight by carrying only enough fuel to change its flyby trajectory to a highly eccentric orbit around Mars. To settle into a smaller and more circular orbit, it skimmed the Martian atmosphere at the low point

interferometry becomes increasingly difficult for light with shorter wavelengths, and astronomers are only beginning to succeed at infrared and visible interferometry. Nevertheless, one reason why *two* Keck telescopes were built on Mauna Kea is so that they can be used for infrared interferometry and someday for optical interferometry. The potential value of such interferometers is enormous. In the future, infrared interferometers may be able to obtain spectra from individual planets around other stars, allowing spectroscopy that could determine the compositions of their atmospheres and help determine whether they harbor life.

7.6 Spacecraft

The space age began in earnest with the launch of the Soviet Union's *Sputnik* satellite in 1957. Within the first decade after *Sputnik,* astronomers were already using spacecraft to carry telescopes into orbit and to make the first close-up studies of the Moon, Venus, and Mars. Today, orbiting spacecraft carry telescopes that allow us to collect light from nearly every region of the electromagnetic spectrum, while other spacecraft have visited all the planets in our solar system except Pluto. Spacecraft are even used to study the Earth. In fact, we can study many facets of our planet much more easily from space than from the ground. Because spacecraft are so important to astronomy, let's briefly investigate what they do and how they work.

Spacecraft Basics

With a few exceptions, such as the Space Shuttle, the Space Station, and the Apollo missions to the Moon, nearly all spacecraft are *robotic,* meaning that no humans are aboard. Their operations are partly automatic and partly controlled by scientists who send instructions via radio signals from Earth. Broadly speaking, most robotic spacecraft fall into four main categories:

1. **Earth-orbiters**. Simply being in space is enough to overcome the observational problems presented by Earth's atmosphere, so most space-based astronomical observatories orbit the Earth. Earth orbit is also the obvious place for spacecraft that study the Earth itself.

2. **Flybys**. Close-up study of other planets requires sending spacecraft to them. Flybys follow unbound orbits past their destinations; that is, they fly past a world just once and then continue on their way. For example, *Voyager 2*'s orbit allowed it to make flybys of Jupiter, Saturn, Uranus, and Neptune (Figure 7.27).

3. **Orbiters** (of other worlds). A flyby can study a planet or moon close up—but only once. For more detailed or longer-term studies, we need a spacecraft that goes into orbit around another world.

4. **Probes and landers**. The most "up close and personal" study of other worlds comes from spacecraft that send probes into their atmospheres or landers to their surfaces.

Table 7.3 lists a few of the most significant robotic missions to other worlds.

FIGURE 7.27 The trajectory of *Voyager 2*, which made flybys of each of the four jovian planets in our solar system.

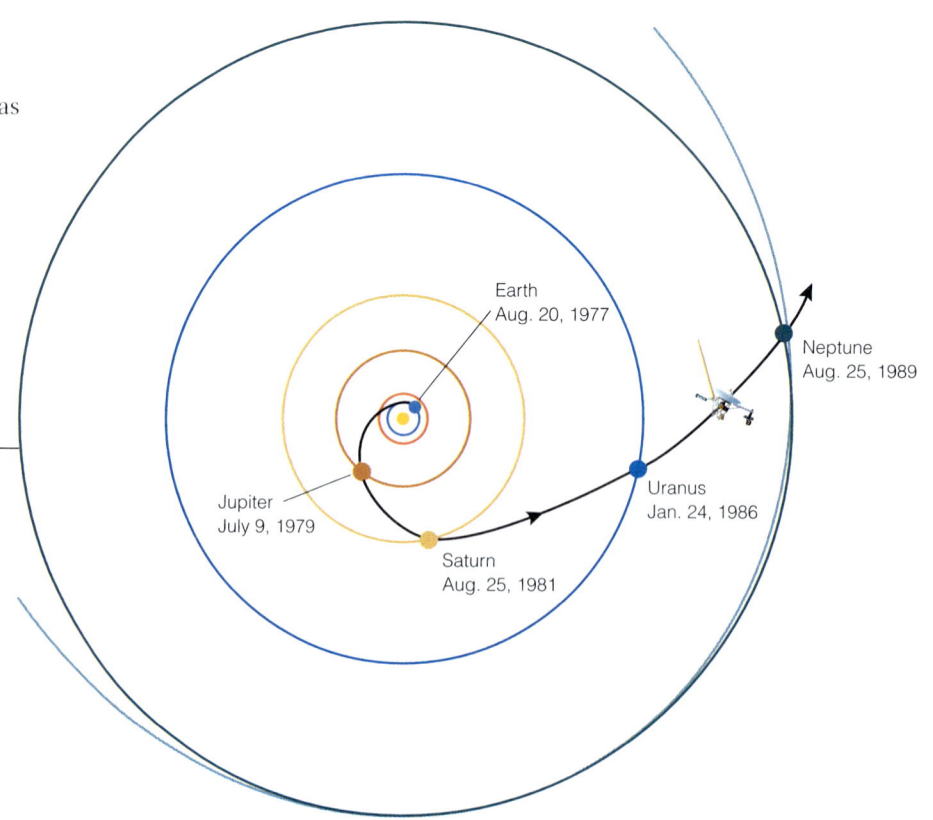

FIGURE 7.23 This photograph shows the Compton Gamma Ray Observatory being deployed from the Space Shuttle in 1991. Compton revolutionized our understanding of the gamma-ray universe before it was destroyed in 2000. Its demise came about after a gyroscope failure raised the prospect that atmospheric drag would eventually lead to an uncontrolled crash; NASA therefore used the spacecraft's on-board fuel for a controlled crash in the ocean, eliminating the possibility of major damage from the 17-ton observatory hitting the ground.

FIGURE 7.24 The Arecibo radio telescope stretches across a natural valley in Puerto Rico. At 305 meters across, it is the world's largest single radio dish.

of radio telescopes: They learned to link two or more individual telescopes to achieve the angular resolution of a much larger telescope (Figure 7.25). This technique is often called **interferometry** because it works by taking advantage of the wavelike properties of light that cause interference (see Figure 7.11). The procedure relies on precisely timing when radio waves reach each dish and supercomputers to analyze the resulting interference patterns.

One famous example of radio interferometry, the Very Large Array (VLA) near Socorro, New Mexico, consists of 27 individual radio dishes that can be moved along railroad tracks laid down in the shape of a Y (Figure 7.26). The light-gathering capability of the VLA's 27 dishes is equal to their combined area, equivalent to that of a single telescope 130 meters across. But the VLA's angular resolution, achieved by spacing the individual dishes as widely as possible, is equivalent to that of a single radio telescope with a diameter of almost 40 kilometers. Today, astronomers can achieve even higher angular resolution by linking radio telescopes around the world.

In principle, interferometry can improve angular resolution not only for radio waves but also for any other form of light. In practice,

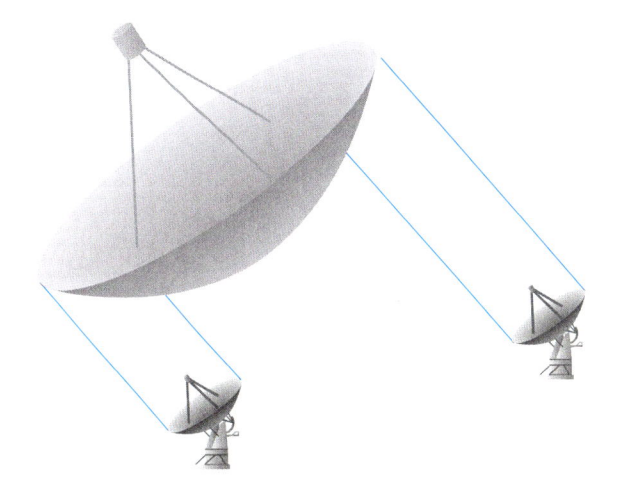

FIGURE 7.25 Interferometry gives two (or more) small radio dishes the angular resolution of a much larger dish. However, their total light-collecting area is only the sum of the light-collecting areas of the individual dishes.

FIGURE 7.26 The Very Large Array (VLA) in New Mexico consists of 27 telescopes that can be moved along train tracks. The telescopes work together through interferometry and can achieve an angular resolution equivalent to that of a single radio telescope almost 40 kilometers across.

ical doctors, it gives astronomers headaches. Trying to focus X rays is somewhat like trying to focus a stream of bullets. If the bullets are fired directly at a metal sheet, they will puncture or damage the sheet. However, if the metal sheet is angled so that the bullets barely graze its surface, it will slightly deflect the bullets. Specially designed mirrors can deflect X rays in much the same way. Such mirrors are called **grazing incidence** mirrors because X rays merely graze their surfaces as they are deflected toward the focal plane. X-ray telescopes, such as NASA's Chandra X-ray Observatory, generally consist of several nested grazing incidence mirrors (Figure 7.22).

TIME OUT TO THINK *If you look straight down at your desktop, you probably cannot see your reflection. But if you glance along the desktop surface (or another smooth surface, such as that of a book), you should see reflections of objects in front of you. Explain how these reflections represent* grazing incidence *for visible light.*

Gamma rays can penetrate even grazing incidence mirrors and therefore cannot be focused in the traditional sense. Capturing such high-energy light at all requires detectors so massive that the photons cannot simply pass through them. The largest gamma-ray observatory to date was the 17-ton Compton Gamma Ray Observatory (Figure 7.23), which was launched in 1991 and destroyed in a controlled crash to Earth in 2000.

Radio Telescopes and Interferometry

Radio telescopes use large metal dishes as "mirrors" to reflect radio waves. However, the long wavelengths of radio waves mean that very large telescopes are necessary to achieve reasonable angular resolution. The largest radio dish in the world, the Arecibo telescope, stretches 305 meters (1,000 feet) across a natural valley in Puerto Rico (Figure 7.24). Despite its large size, Arecibo's angular resolution is only about 1 arcminute at commonly observed radio wavelengths (e.g., 21 cm [Section 18.2])—a few hundred times worse than the visible-light resolution of the Hubble Space Telescope.

In the 1950s, radio astronomers developed an ingenious technique for improving the angular resolution

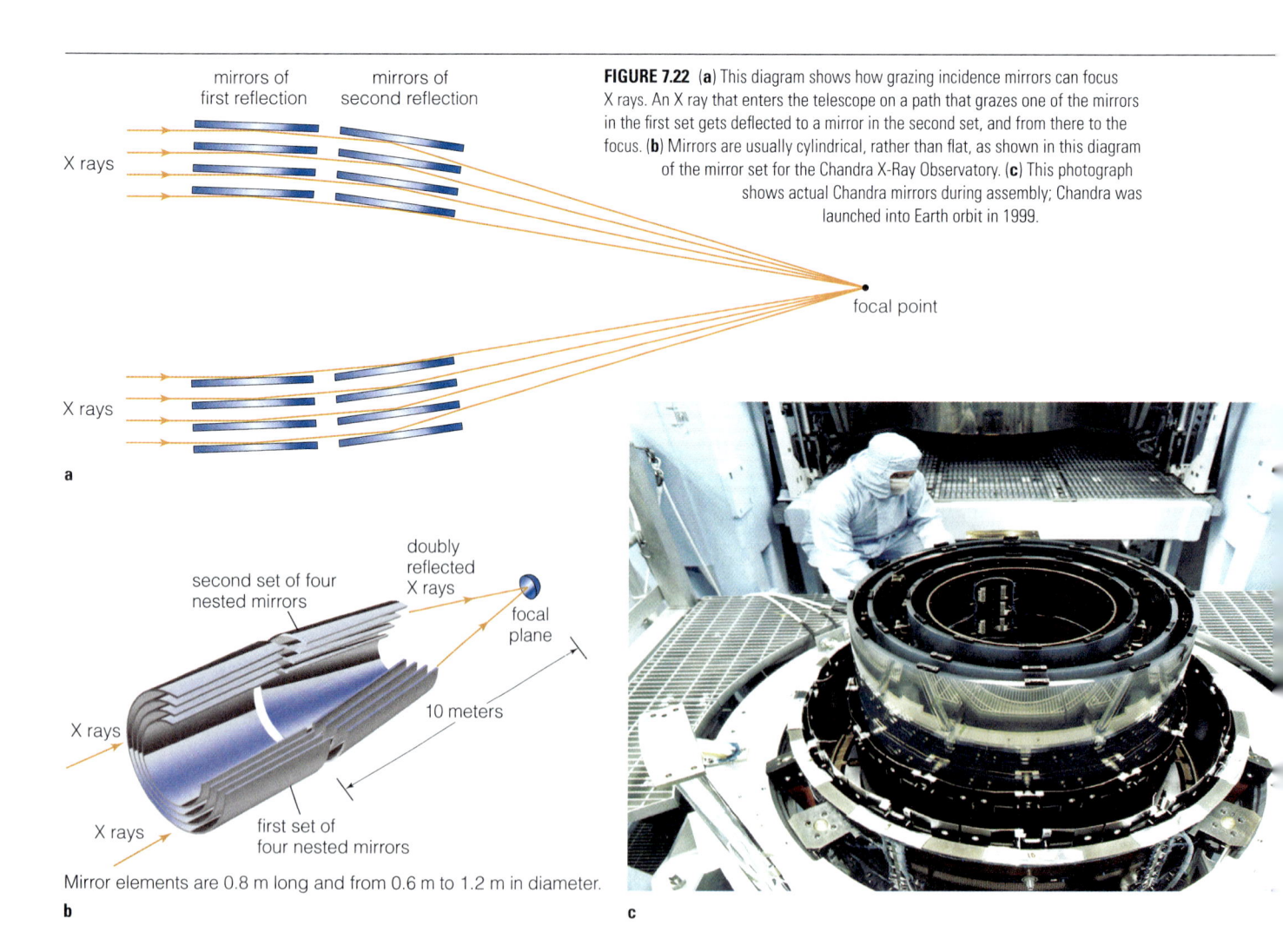

FIGURE 7.22 (**a**) This diagram shows how grazing incidence mirrors can focus X rays. An X ray that enters the telescope on a path that grazes one of the mirrors in the first set gets deflected to a mirror in the second set, and from there to the focus. (**b**) Mirrors are usually cylindrical, rather than flat, as shown in this diagram of the mirror set for the Chandra X-Ray Observatory. (**c**) This photograph shows actual Chandra mirrors during assembly; Chandra was launched into Earth orbit in 1999.

Mirror elements are 0.8 m long and from 0.6 m to 1.2 m in diameter.

> **Common Misconceptions: Closer to the Stars?**
>
> Many people mistakenly believe that space telescopes are advantageous because their locations above the Earth make them closer to the stars. You can quickly realize the error of this belief by thinking about scale. On the 1-to-10-billion scale discussed in Section 1.2, the Hubble Space Telescope is so close to the surface of the millimeter-diameter Earth that you would need a microscope to resolve its altitude. Thus, the distances to the stars are effectively the same whether a telescope is on the ground or in space. The real advantages of space telescopes all arise from their being above Earth's atmosphere and thus not subject to the many observational problems it presents.

in a few cases, very high in the atmosphere on airplanes or balloons).

Today, astronomers study light across the entire spectrum. With the exception of telescopes for radio, some infrared, and visible-light observations, most of these telescopes are in space. Indeed, while the Hubble Space Telescope is the only major visible-light observatory in space (it is also used for infrared and ultraviolet observations), many less famous observatories are in Earth orbit for the purpose of making observations in nonvisible wavelengths. Table 7.2 lists some of the most important existing and planned telescopes in space.

Infrared and Ultraviolet Telescopes

Telescopes for nonvisible wavelengths often require very different designs than optical telescopes. However, light from much of the infrared and ultraviolet portions of the spectrum behaves enough like visible light to eliminate the need for significant telescope modifications. That is, an optical telescope can record infrared or ultraviolet light as long as it is equipped with appropriate detectors and mirror coatings. Of course, the telescope must be in space to receive significant ultraviolet light.

Near the extreme-wavelength ends of the infrared or ultraviolet, telescopes need special technology. Extreme ultraviolet light (the shortest wavelengths of ultraviolet) behaves like X rays, which we'll discuss shortly. Extreme infrared light (the longest wavelengths of infrared) poses observing difficulties, because ordinary telescopes are warm enough to emit significant amounts of long-wavelength infrared light and this telescope emission would interfere with any attempt to observe these wavelengths from the cosmos. One solution to this problem is to make the telescope so cold that it emits very little infrared radiation. NASA's planned Space Infrared Telescope Facility (SIRTF), scheduled for launch in 2002 or 2003, will be cooled with liquid helium to just a few degrees above absolute zero.

X-Ray and Gamma-Ray Telescopes

X rays have sufficient energy to penetrate many materials, including living tissue and ordinary mirrors. While this property makes X rays very useful to med-

Table 7.2 Selected Major Observatories in Space

Name	Launch Year	Lead Space Agency	Special Features
Hubble Space Telescope	1990	NASA	Optical, infrared, and ultraviolet imaging and spectroscopy
Rossi X-Ray Timing Experiment (RXTE)	1995	NASA	X-ray timing and spectroscopy
Far Ultraviolet Spectroscopic Explorer (FUSE)	1999	NASA	Ultraviolet spectroscopy
Chandra X-Ray Observatory	1999	NASA	X-ray imaging and spectroscopy
X-Ray Multi-Mirror Mission (XMM)	1999	ESA[*]	European-led mission for X-ray spectroscopy
High Energy Transient Explorer (HETE-2)	2000	NASA	Study of gamma-ray bursts
Microwave Anisotropy Probe (MAP)	2001[**]	NASA	Study of the cosmic microwave background
Space Infrared Telescope Facility (SIRTF)	2002[**]	NASA	Infrared observations of the cosmos
Space Interferometry Mission (SIM)	2006[**]	NASA	First mission for optical interferometry in space
Next Generation Space Telescope (NGST)	2009[**]	NASA	Follow-on to Hubble Space Telescope
Terrestrial Planet Finder (TPF)	2011[**]	NASA	Search for Earth-like planets

[*]European Space Agency
[**]Scheduled launch year

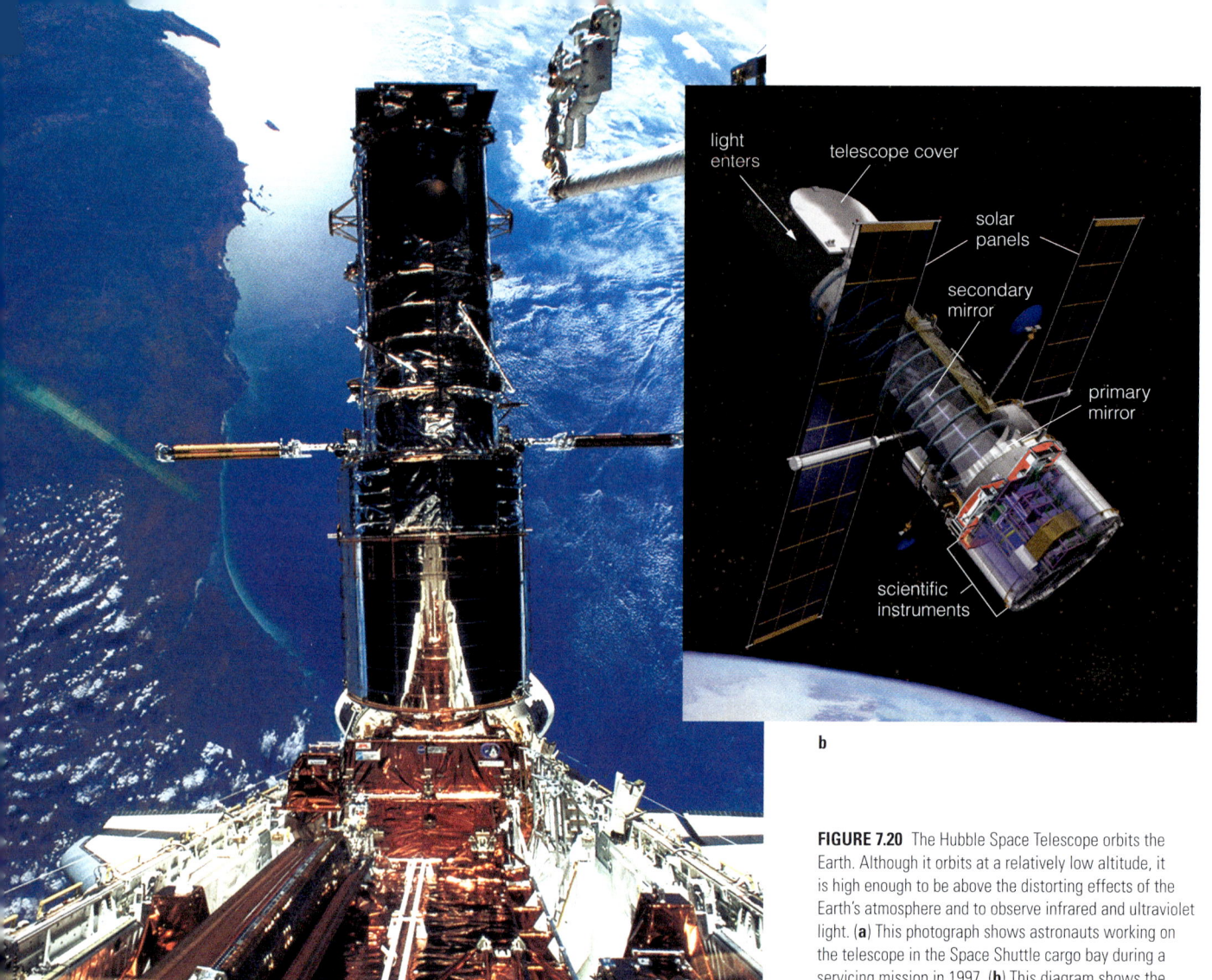

FIGURE 7.20 The Hubble Space Telescope orbits the Earth. Although it orbits at a relatively low altitude, it is high enough to be above the distorting effects of the Earth's atmosphere and to observe infrared and ultraviolet light. (**a**) This photograph shows astronauts working on the telescope in the Space Shuttle cargo bay during a servicing mission in 1997. (**b**) This diagram shows the basic components of the Hubble Space Telescope.

FIGURE 7.21 Diagram showing the approximate depths to which different wavelengths of light penetrate the Earth's atmosphere. Note that most of the electromagnetic spectrum—except for visible light, a small portion of the infrared, and radio—can be observed only from very high altitudes or from space.

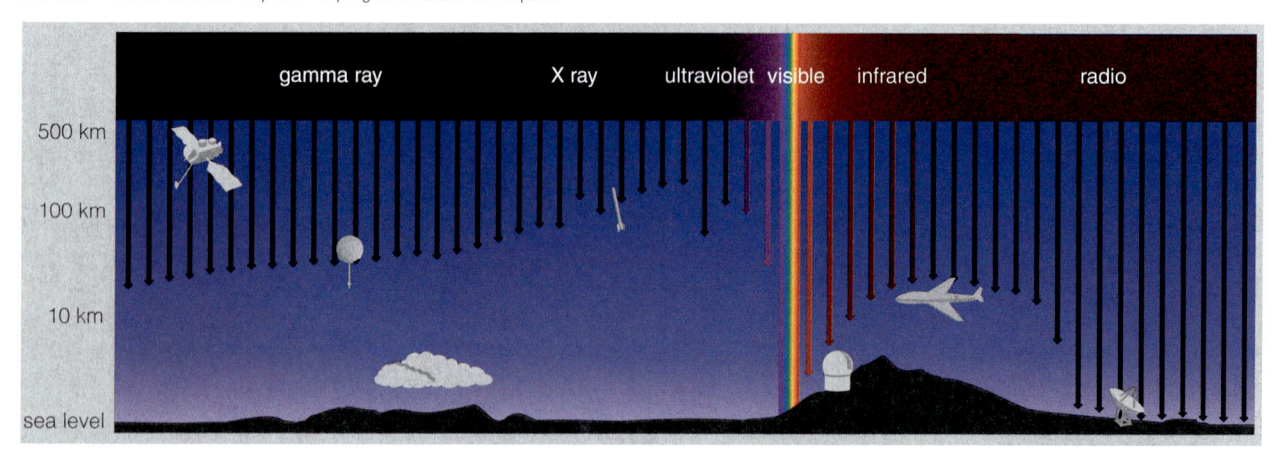

CHAPTER 7 TELESCOPES AND SPACECRAFT 185

FIGURE 7.18 This photograph shows the airplane for NASA's airborne observatory, SOFIA, during a test flight. The 2.5-meter telescope will be located in the fuselage behind the painted black rectangle, which will open so the telescope can view the heavens. SOFIA is scheduled to begin observations in late 2004.

Modern technology also provides solutions to some of the problems caused by the atmosphere. Putting a telescope on an airplane takes it above much of the atmosphere, allowing many infrared observations not possible from the ground. NASA's new airborne observatory, called SOFIA (Stratospheric Observatory for Infrared Astronomy), will have a 2.5-meter infrared telescope looking out through a large hole cut in the fuselage of a Boeing 747 airplane (Figure 7.18). Advanced technologies make it possible for the airplane to fly smoothly despite its large hole and to keep the telescope pointed accurately at observing targets during flight.

Perhaps the most amazing new technology is **adaptive optics**, which can eliminate most atmospheric distortion. Atmospheric turbulence causes a stellar image to dance around in the focal plane of a telescope. Adaptive optics essentially make the telescope's mirrors do an opposite dance, canceling out the atmospheric distortions (Figure 7.19). The mirror shape (usually the secondary mirror) must change slightly many times each second to compensate for the rapidly changing atmospheric distortions. A computer calculates the necessary changes by monitoring the distortions of a bright star near the object under study. In some cases, if there is no bright star near the object of interest, the observatory shines a laser into the sky to create an *artificial star* (a point of light in the Earth's atmosphere) that can be monitored for distortions.

Of course, the ultimate solution to atmospheric distortion is to put telescopes in space. That is the primary reason why the Hubble Space Telescope (Figure 7.20) was built, and it is why it is so successful even though its 2.4-meter primary mirror is much smaller than the mirrors of many ground-based telescopes.

7.5 Telescopes Across the Spectrum

If we studied only visible light, we'd be missing much of the picture. Planets are relatively cool and emit primarily infrared light. The hot upper layers of stars like the Sun emit ultraviolet and X-ray light. Many objects emit radio waves, including some of the most exotic objects in the cosmos—*quasars*, which may contain black holes buried in their centers [Section 20.5]. Some violent events even produce gamma rays that travel through space to the Earth. Indeed, most objects emit light over a broad range of wavelengths.

Unfortunately, the atmosphere poses a major problem that no Earth-bound technology can overcome: It prevents most forms of light from reaching the ground at all. Figure 7.21 shows the depth to which different forms of light penetrate the Earth's atmosphere. Only radio waves, visible light, parts of the infrared spectrum, and the longest wavelengths of ultraviolet light reach the ground. Observing other wavelengths requires placing telescopes in space (or,

FIGURE 7.19 (a) Atmospheric distortion makes this ground-based image of a double star look like a single star. (b) Using the same telescope, but with adaptive optics, the two stars are clearly distinguishable. The angular separation between the two stars is 0.38 arcsecond. (Both images were taken in near-infrared light with the Canada-France-Hawaii telescope and are shown in false color.)

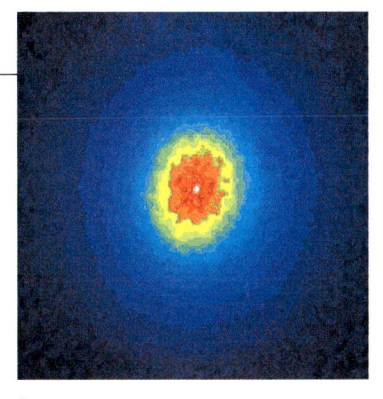

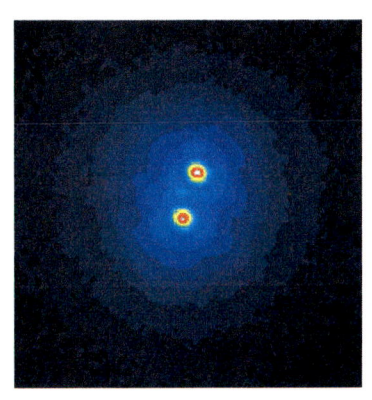

a   b

if it weren't located so close to the lights of what was once the small town of Los Angeles. Similar but less serious light pollution hinders many other telescopes, including the Mount Palomar telescopes near San Diego and the telescopes of the National Optical Astronomy Observatory on Kitt Peak near Tucson. Fortunately, many communities are working to reduce light pollution. Placing reflective covers on the tops of streetlights directs more light toward the ground, rather than toward the sky. Using "low pressure" (sodium) lights also helps. Because these lights shine most brightly in just a few wavelengths of light, special filters can be used to absorb these wavelengths while transmitting most of the light from the astronomical objects under study. It's worth noting that both reflective covers and low-pressure lights also offer significant energy savings to communities that use them.

A somewhat less obvious problem is the fact that the atmosphere distorts light. The ever-changing motion, or **turbulence**, of air in the atmosphere bends light in constantly shifting patterns. This turbulence causes the familiar twinkling of stars and blurs astronomical images.

TIME OUT TO THINK *If you look down a long, paved street on a hot day, you'll notice the images of distant cars and buildings rippling and distorting. How are these distortions similar to the twinkling of stars? Why do you think these distortions are more noticeable on hot days than on cooler days?*

Many of these atmospheric effects can be mitigated by choosing appropriate sites for observatories. The key criteria are that the sites be dark (limiting light pollution), dry (limiting rain and clouds), calm (limiting turbulence), and high (placing them above at least part of the atmosphere). Islands are often ideal, and the 4,300-meter (14,000-foot) summit of Mauna Kea on the Big Island of Hawaii is home to many of the world's best observatories (Figure 7.17).

FIGURE 7.17 Observatories on the summit of Mauna Kea in Hawaii.

Common Misconceptions: Twinkle, Twinkle Little Star

Twinkling, or apparent variation in the brightness and color of stars, is *not* intrinsic to the stars. Instead, just as light is bent by water in a swimming pool, starlight is bent by the Earth's atmosphere. Air turbulence causes twinkling because it constantly changes how the starlight is bent. Hence, stars tend to twinkle more on windy nights and at times when they are near the horizon (and therefore are viewed through a thicker layer of atmosphere). Planets also twinkle, but not nearly as much as stars: Because planets have a measurable angular size, the effects of turbulence on any one ray of light are compensated for by the effects of turbulence on others, reducing the twinkling seen by the naked eye (but making planets shimmer noticeably in telescopes). While twinkling may be beautiful, it blurs telescopic images. Avoiding the effects of twinkling is one of the primary reasons for putting telescopes in space. There, above the atmosphere, astronauts and telescopes do not see any twinkling.

THE BIG PICTURE

In this chapter, we've focused on the technological side of astronomy: the telescopes and spacecraft that we use to learn about the universe. Keep in mind the following "big picture" ideas as you continue to learn about astronomy:

- Technology drives astronomical discovery. Every time we build a bigger telescope or a more sensitive detector, open up a new wavelength region to study, or send a spacecraft to another world, we learn more about the universe than was possible before.

- Telescopes work much like giant eyes, enabling us to see the universe in great detail. New technologies for making larger telescopes, along with advances in adaptive optics and interferometry, are making ground-based telescopes more powerful than ever.

- For the ultimate in observing the universe, space is the place! Telescopes in space allow us to detect light from across the entire spectrum while also avoiding the distortion caused by Earth's atmosphere. Flybys, orbiters, and landers allow us to make detailed studies of other worlds in our solar system that are not possible from Earth.

Review Questions

1. Briefly describe how the eye collects light. How is a camera lens similar to the eye?
2. What is the *focus* of a lens? What is the *focal plane*?
3. What is *angular resolution*? Suppose you look at two stars with an angular resolution smaller than that of your eye. What will you see?
4. Briefly describe how a camera collects and records light. How do we control the *exposure time* with a camera?
5. What is a *CCD*? Briefly describe the advantages of using CCDs over using photographic film.
6. Briefly distinguish between a *refracting telescope* and a *reflecting telescope*.
7. Define and distinguish between a telescope's *light-collecting area* and its *angular resolution*. Why are these properties the two most important properties of a telescope?
8. What is the *diffraction limit* of a telescope, and how does it depend on the telescope's size and the particular wavelength of light being observed?
9. Define and differentiate between the three basic categories of astronomical observation: *imaging*, *spectroscopy*, and *timing*.
10. What is a *false-color* image? Why are false-color images useful?
11. What do we mean by *spectral resolution*? Explain why high spectral resolution is desirable but more difficult to achieve than lower spectral resolution.
12. Explain how *light pollution* and atmospheric *turbulence* affect astronomical observations made from the Earth's surface.
13. What is *adaptive optics*? What problems caused by the Earth's atmosphere can it solve?
14. Study Figure 7.21 in detail, and describe how deeply each portion of the electromagnetic spectrum penetrates the Earth's atmosphere.
15. Briefly describe the advantages of putting telescopes in space. Are there any disadvantages? Explain.
16. Why is it useful to study light from across the spectrum? Briefly describe the characteristics of telescopes used to observe different portions of the spectrum.
17. Briefly explain the purpose of using two or more telescopes together through *interferometry*.
18. Describe each of the four main categories of spacecraft and their uses.
19. Briefly explain the trade-offs between orbital considerations and cost involved in planning a spacecraft mission. How were costs reduced in the Mars Pathfinder mission, the Mars Global Surveyor mission, and the Cassini mission?

Discussion Questions

1. *Science and Technology Funding.* Technological innovation clearly drives scientific discovery in astronomy, but the reverse is also true. For example, Newton's discoveries were made in part to explain the motions of the planets, but they have had far-reaching effects on our civilization. Congress often must make decisions between funding programs with purely scientific purposes ("basic research") and programs designed to develop new technologies. If you were a member of Congress, how would you try to allocate spending between basic research and technology? Why?

2. *A Lunar Observatory.* Do the potential benefits of building an astronomical observatory on the Moon justify its costs at the present time? If it were up to you, would you recommend that Congress begin funding such an observatory? Defend your opinions.

Problems

Sensible Statements? For **problems 1–8**, decide whether the statement is sensible and explain why it is or is not.

1. The image was blurry because the photographic film was not placed at the focal plane.

2. By using a CCD, I can photograph the Andromeda Galaxy with a shorter exposure time than I would need with photographic film.

3. Thanks to adaptive optics, the telescope on Mount Wilson can now make ultraviolet images of the cosmos.

4. New technologies will soon allow astronomers to use X-ray telescopes on the Earth's surface.

5. Thanks to interferometry, a properly spaced set of 10-meter radio telescopes can achieve the angular resolution of a single, 100-kilometer radio telescope.

6. Thanks to interferometry, a properly spaced set of 10-meter radio telescopes can achieve the light-collecting area of a single, 100-kilometer radio telescope.

7. Scientists are planning a flyby of Uranus in order to do long-term monitoring of the planet's weather.

8. An observatory on the Moon's surface could have telescopes monitoring light from all regions of the electromagnetic spectrum.

9. *Angular Resolution.*
   a. Briefly describe why smaller is better when it comes to angular resolution.
   b. Suppose that two stars are separated in the sky by 0.1 arcsecond. What will you see if you look at them with a telescope that has an angular resolution of 0.01 arcsecond? What will you see if you look at them with a telescope that has an angular resolution of 0.5 arcsecond?

10. *Light-Collecting Area.*
    a. How much greater is the light-collecting area of one of the 10-meter Keck telescopes than that of the 5-meter Hale telescope?
    b. Suppose astronomers built a 100-meter telescope. How much greater would its light-collecting area be than that of the 10-meter Keck telescope?

11. *Project: Twinkling Stars.* Using a star chart, identify 5 to 10 bright stars that should be visible in the early evening. On a clear night, observe each of these stars for a few minutes. Note the date and time, and for each star record the following information: approximate altitude and direction in your sky, brightness compared to other stars, color, how much the star twinkles compared to other stars. Study your record. Can you draw any conclusions about how brightness and position in your sky affect twinkling? Explain.

*12. *Angular Separation Calculations.*
    a. Two light bulbs are separated by 0.2 meter and you look at them from a distance of 2 kilometers. What is the angular separation of the lights? Can your eyes resolve them? Explain.
    b. The diameter of a dime is about 1.8 centimeters. What is its angular diameter if you view it from across a 100-meter-long football field?

*13. *Calculating the Sun's Size.* The angular diameter of the Sun is about the same as that of the Moon (0.5°). Use this fact and the Sun's average distance of about 150 million km to estimate the diameter of the Sun. Compare your result to the Sun's actual diameter of 1.392 million km.

*14. *Viewing a Dime with the HST.* The Hubble Space Telescope (HST) has an angular resolution of about 0.05 arcsecond. How far away would you have to place a dime (diameter = 1.8 cm) for its angular diameter to be 0.05 arcsecond? (*Hint:* Start by converting an angular diameter of 0.05 arcsecond into degrees.)

*15. *Close Binary System.* Suppose that two stars in a binary star system are separated by a distance of 100 million km and are located at a distance of 100 light-years from Earth. What is the angular separation of the two stars? Give your answer in both degrees and arcseconds. Can the Hubble Space Telescope resolve the two stars?

*16. *Diffraction Limit of the Eye.* Calculate the diffraction limit of the human eye, assuming a lens size of 0.8 cm, for visible light of 500-nm wavelength. How does this compare to the diffraction limit of a 10-meter telescope?

*17. *The Size of Radio Telescopes.* What is the diffraction limit of a 100-meter radio telescope observing radio waves with a wavelength of 21 cm? Compare this to the diffraction limit of the 2.4-meter Hubble Space Telescope for visible light. Use your results to explain why radio telescopes must be much larger than optical telescopes to be useful.

*18. *Hubble's Field of View.* Large telescopes often have small fields of view. For example, the Hubble Space Telescope's (HST) "wide field" camera has a field of view that is roughly square and about 0.04° on a side.
    a. Calculate the angular area of the HST's field of view in square degrees.
    b. The angular area of the entire sky is about 41,250 square degrees. How many pictures would the HST have to take with its wide field camera to obtain a complete picture of the entire sky?
    c. Assuming that it requires an average of 1 hour to take each picture, how long would it take to acquire the number of pictures you calculated in part (b)? Use your answer to explain why astronomers would like to have more than one large telescope in space.

Web Projects

Find useful links for Web projects on the text Web site.

1. *Planetary Mission.* Choose a current mission to a planet and find as much information as you can about the mission. Write a two- to three-page summary of the mission, including discussion of its purpose, design, and cost.

2. *Major Observatories.* Take a virtual tour of one of the world's major astronomical observatories. Write a short report on why the observatory is useful to astronomy.

3. *Earth-Observing Satellites.* Research one or more satellites that are designed to study the Earth. Briefly explain the purpose of each satellite and why it is able to collect data from space that would be more difficult to collect on the ground.

PART III
Learning from Other Worlds

The evolution of the world may be compared to a display of fireworks that has just ended: some few red wisps, ashes and smoke. Standing on a cooled cinder, we see the slow fading of the suns, and we try to recall the vanished brilliance of the origin of the worlds.

G. LEMAÎTRE (1894–1966),
ASTRONOMER AND CATHOLIC PRIEST

CHAPTER 8
Formation of the Solar System

How old is the Earth, and how did it come to be? Is it unique? Our ancestors could do little more than guess, but today we can answer with reasonable confidence: The Earth and the rest of our solar system formed from a great cloud of gas and dust about 4.6 billion years ago. Whereas ancient astronomers believed the Earth to be fundamentally different from objects in the heavens, we now see the Earth as just one of many worlds.

As we examine the evidence that helped shape the modern scientific view of the formation of our solar system, we'll highlight one common technique of scientific research: sifting through seemingly unrelated facts to find the most important trends, patterns, and characteristics and then attempting to explain them. The explanation must be based on known physical laws, and the success or failure of a theory depends on how well it matches the facts. As we discover more and more planets around other stars, we hope to test our theories further and modify them as necessary.

8.1 Comparative Planetology

Galileo's telescopic observations began a new era in astronomy in which the Sun, the Moon, and the planets could be studied for the first time as *worlds*, rather than merely as lights in the sky. During the early part of this new era, astronomers studied each world independently, but in recent decades we have discovered that the similarities among worlds are often more profound than the differences.

Today, we know that the Sun, planets, moons, and other bodies in our solar system all formed at about the same time from the same cloud of interstellar gas and dust (*interstellar* means "between the stars") and in accord with the same physical laws. Thus, the differences between these worlds must be attributable to physical processes that we can study and understand by comparing the worlds to one another. This approach is called **comparative planetology**; the term *planetology* is used broadly to include moons, asteroids, and comets as well as planets. The idea is that we can learn more about the Earth, or any other world, by studying it in the context of other objects in our solar system—rather like learning about a person by studying his or her family, friends, and culture.

Before beginning the comparative study of worlds, we must be sure that the comparison will provide valid lessons. The basic premise of comparative planetology—that the similarities and differences among worlds in our solar system can be traced to common physical processes—assumes that the planets share a common origin. Thus, we begin our study of comparative planetology by seeking clues to our origins from the current state of our solar system; then we will examine the modern theory of solar system formation that explains much of what we see.

8.2 The Origin of the Solar System: Four Challenges

Learning about the origin of the solar system is easiest if we begin by looking at our solar system with a "big picture" view, rather than focusing on individual worlds. Imagine that we have the perspective of an alien spacecraft making its first scientific survey of our solar system. What would we see?

First, as we learned from our tour of a scale model of our solar system in Chapter 1, we'd find that our solar system is mostly empty. Interplanetary space is a near-vacuum containing only very low density gas and tiny, solid grains of dust. But we'd soon focus on the widely spaced planets, mapping their orbits, measuring their sizes, compositions, and densities, and taking inventory of their moons and ring systems. Figure 8.1 shows the resulting schematic maps, and Table 8.1 summarizes the planetary properties. Note that asteroids and comets are also listed in the table because these objects orbit the Sun, though with orbital properties somewhat different from those of the planets.

Careful study of the maps and the table reveals several general features of our solar system. Let's investigate by grouping our observations into four major challenges that must be met by any theory claiming to explain how our solar system formed.

Challenge 1: Patterns of Motion

The maps in Figure 8.1 show several striking patterns:

- All planets orbit the Sun in the same direction—counterclockwise as seen from high above the Earth's North Pole.

- All planetary orbits lie nearly in the same plane.

- Almost all planets travel on nearly circular orbits, and the spacing between planetary orbits increases with distance from the Sun according to a fairly regular trend. The most notable exception to this trend is an extra-wide gap between Mars and Jupiter that is populated with asteroids.

- Most planets rotate in the same direction in which they orbit—counterclockwise as seen from above the Earth's North Pole—with fairly small axis tilts (i.e., less than about 25°).

- Almost all moons orbit their planet in the same direction as the planet's rotation and near the planet's equatorial plane.

- The Sun rotates in the same direction in which the planets orbit.

These patterns show that motion in the solar system is generally quite organized. If each planet had come into existence independently, we might expect planetary motions to be much more random. (Note that the motions of asteroids are slightly more random, and those of comets even more so, as we will discuss shortly.)

The first challenge for any theory of solar system formation is to explain why motions in the solar system are generally so orderly.

Challenge 2: Categorizing Planets

Before reading any further, study the properties of the planets in Table 8.1. Can you categorize the planets into groups? How many categories do you need? What characteristics do planets within a category share?

Astronomers classify most of the planets in two distinct groups: the rocky **terrestrial planets** and the gas-rich **jovian planets**. The word *terrestrial*

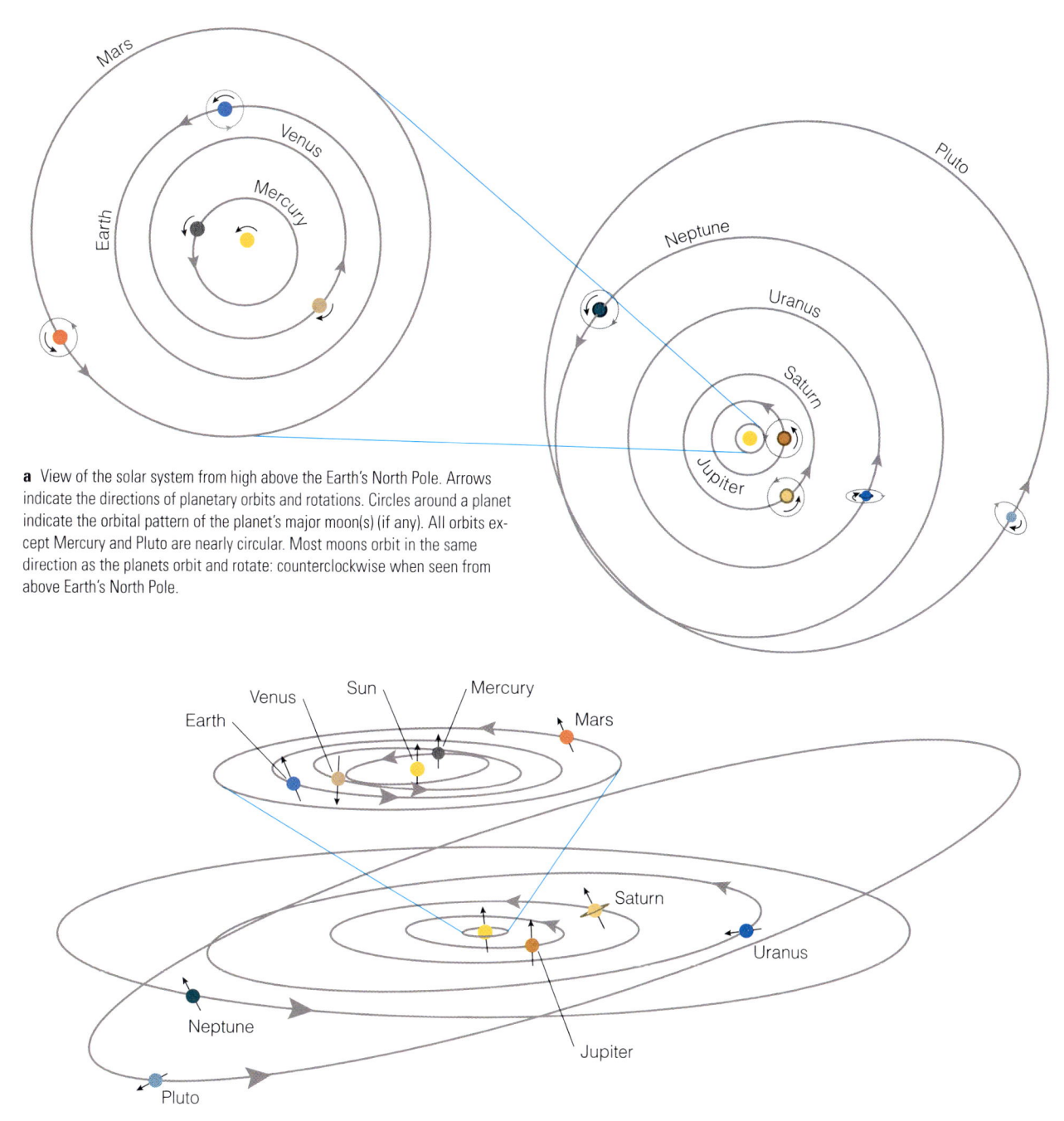

a View of the solar system from high above the Earth's North Pole. Arrows indicate the directions of planetary orbits and rotations. Circles around a planet indicate the orbital pattern of the planet's major moon(s) (if any). All orbits except Mercury and Pluto are nearly circular. Most moons orbit in the same direction as the planets orbit and rotate: counterclockwise when seen from above Earth's North Pole.

b Side view of the solar system. Arrows indicate the orientation of the rotation axes of the planets and their orbital motion. (Planetary tilts in this diagram are aligned in the same plane for easier comparison. Planets not to scale.)

FIGURE 8.1 The layout of the solar system.

means "Earth-like" (*terra* is the Latin word for Earth), and the terrestrial planets are the four planets of the inner solar system—Mercury, Venus, Earth, and Mars. These are the worlds most like our own. They are relatively small, close to the Sun, and close together. They have solid, rocky surfaces and an abundance of metals deep in their interiors. They have few moons, if any, and none have rings. Note that many scientists count our Moon as a fifth terrestrial world because it shares these general characteristics, although it's not technically a planet.

The word *jovian* means "Jupiter-like," and the four jovian planets are Jupiter, Saturn, Uranus, and Neptune. The jovian planets are much larger than the terrestrial planets, farther from the Sun, and widely separated from each other. These huge planets have little in common with Earth. They are made mostly of hydrogen, helium, and **hydrogen compounds** such as water, ammonia, and methane. They contain small amounts of rocky material only deep in their cores. (Here we mean small in a relative

CHAPTER 8 FORMATION OF THE SOLAR SYSTEM 199

Table 8.1 Planetary Facts*

Photo	Planet	Average Distance from Sun (AU)	Temperature†	Relative Size	Average Equatorial Radius (km)	Average Density (g/cm³)	Composition	Known Moons	Rings?
	Mercury	0.387	700 K	·	2,440	5.43	Rocks, metals	0	No
	Venus	0.723	740 K	•	6,051	5.24	Rocks, metals	0	No
	Earth	1.00	290 K	•	6,378	5.52	Rocks, metals	1	No
	Mars	1.52	240 K	·	3,397	3.93	Rocks, metals	2 (tiny)	No
	Most asteroids	2–3	170 K	·	≤500	1.5–3	Rocks, metals	?	No
	Jupiter	5.20	125 K	●	71,492	1.33	H, He, hydrogen compounds‡	28	Yes
	Saturn	9.53	95 K	●	60,268	0.70	H, He, hydrogen compounds‡	30	Yes
	Uranus	19.2	60 K	●	25,559	1.32	H, He, hydrogen compounds‡	21	Yes
	Neptune	30.1	60 K	●	24,764	1.64	H, He, hydrogen compounds‡	8	Yes
	Pluto	39.5	40 K	·	1,160	2.0	Ices, rock	1	No
	Most comets	10–50,000	A few K§	·	A few km?	<1?	Ices, dust	?	No

*Appendix C gives a more complete list of planetary properties.
†Surface temperatures for all objects except Jupiter, Saturn, Uranus, and Neptune, for which cloud-top temperatures are listed.
‡Includes water (H_2O), methane (CH_4), and ammonia (NH_3).
§Comets passing close to the Sun warm considerably, especially their outer layers.

Table 8.2 Comparison of Terrestrial and Jovian Planets

Terrestrial Planets	Jovian Planets
Smaller size and mass	Larger size and mass
Higher density (rocks, metals)	Lower density (light gases, hydrogen compounds)
Solid surface	No solid surface
Closer to the Sun (and closer together)	Farther from the Sun (and farther apart)
Warmer	Cooler
Few (if any) moons and no rings	Rings and many moons

sense: The cores are still probably 10 Earth masses or more.) Jovian planets do not have solid surfaces; if you plunged into the atmosphere of a jovian planet, you would sink deeper and deeper until you were crushed by overwhelming pressure. The jovian planets are sometimes said to be made mostly of gas, but in fact the intense pressures and temperatures of the jovian planet interiors transform familiar gases (and solids) to phases unlike anything we see on Earth. Each jovian planet has rings and an extensive system of moons. These solid satellites are made mostly of low-density ices and rocks. Table 8.2 contrasts the general traits of the terrestrial and jovian planets.

The second challenge for any theory of solar system formation is to explain why the inner and outer solar system planets divide so neatly into two classes.

TIME OUT TO THINK *Are the distinctions in Table 8.2 clear-cut? Examine Table 8.1 to compare, for example, the radii of the largest terrestrial planet and the smallest jovian planet.*

Note that Pluto is left out in the cold, both literally and figuratively. On one hand, it is small and solid like the terrestrial planets. On the other hand, Pluto is far from the Sun, cold, and made of low-density ices. Pluto also has an unusual orbit, more eccentric than the orbit of any other planet and substantially inclined to the plane in which the other planets orbit the Sun. For a long time, scientists considered Pluto to be a lone misfit. However, some scientists now classify Pluto with other icy bodies in the outer solar system: comets [Section 12.5].

Challenge 3: Asteroids and Comets

No formation theory is complete without an explanation for the most numerous objects in the solar system: asteroids and comets. **Asteroids** are small, rocky bodies that orbit the Sun primarily in the **asteroid belt** between the orbits of Mars and Jupiter (Figure 8.2). (Another large population of asteroids—the Trojan asteroids—shares Jupiter's orbit, falling into two clumps leading and trailing Jupiter by 60° [Section 12.2].) Asteroids orbit the Sun in the same

FIGURE 8.2 Most asteroids orbit the Sun between Mars and Jupiter. Their orbits are noticeably tilted and eccentric.

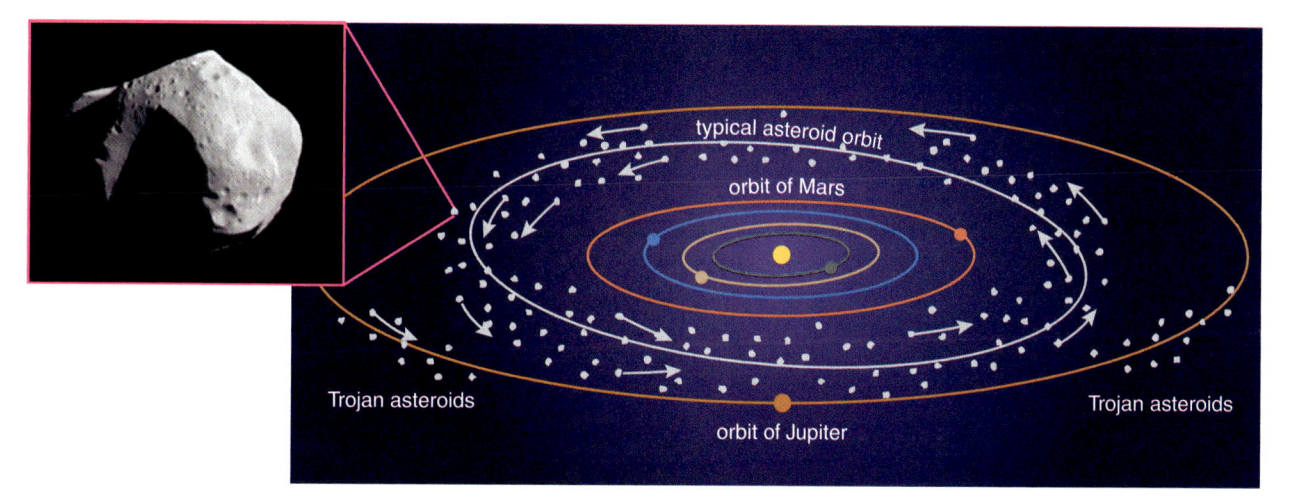

CHAPTER 8 FORMATION OF THE SOLAR SYSTEM

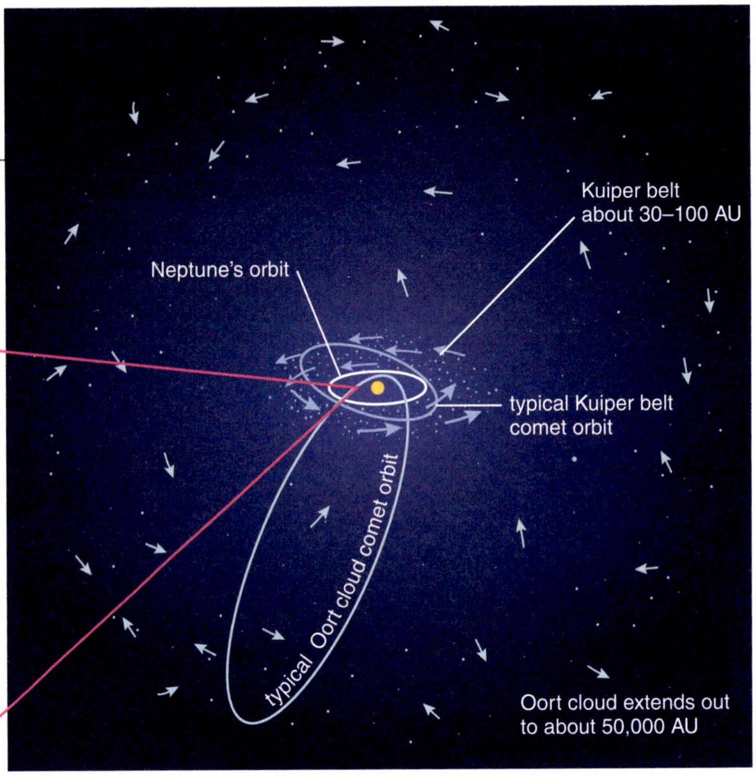

FIGURE 8.3 Small, icy comets orbit the Sun in two regions: the Kuiper belt just outside Neptune's orbit, and the Oort cloud beyond. (Diagram not to scale.)

direction as the planets. Their orbits generally lie close to the plane of planetary orbits, although they are usually a bit tilted. Some asteroids have elliptical orbits that are quite eccentric compared to the nearly circular orbits of the planets. More than 10,000 asteroids have been identified and cataloged, but these are probably only the largest among a much greater number of small asteroids. The very largest asteroids are a few hundred kilometers in radius—much less than half the Moon's radius.

Comets are small, icy bodies that spend most of their lives well beyond the orbit of Pluto; we generally recognize them only on the rare occasions when one dives into the inner solar system and grows a spectacular tail. Based on the orbits of the relatively few comets that reach the inner solar system, astronomers have determined that many billions of comets must be orbiting the Sun, primarily in two broad regions (Figure 8.3). The first region, called the **Kuiper belt** (pronounced koy-per), begins somewhere in the vicinity of the orbit of Neptune (about 30 AU from the Sun) and extends to perhaps 100 AU from the Sun. Comets in the Kuiper belt have orbits that lie fairly close to the plane of planetary orbits, and they travel around the Sun in the same direction as the planets. The second region populated by comets, called the **Oort cloud**, is a huge, spherical region centered on the Sun and extending perhaps halfway to the nearest stars. The orbits of comets in the Oort cloud are completely random, with no pattern to their inclinations, orbital directions, or eccentricities.

The third challenge for any theory of solar system formation is to explain the existence and general properties of the large numbers of asteroids and comets.

Challenge 4: Exceptions to the Rules

The first three challenges involve explaining general patterns in our solar system. But some objects don't fit these patterns. For example, Mercury and Pluto have larger orbital eccentricities and inclinations than the other planets. The rotational axes of Uranus and Pluto are substantially tilted, and Venus rotates "backward"—clockwise, rather than counterclockwise, as viewed from high above Earth's North Pole. Unlike the other terrestrial planets, Earth has a large moon. Pluto has a moon almost as big as itself. While most moons of the jovian planets orbit in the planet's equatorial plane with the same orientation as their planet's rotation, a few orbit in the opposite direction or on tilted or eccentric orbits.

Allowing for these and other exceptions to the general patterns is the fourth challenge for any theory of solar system formation.

Summarizing the Challenges

Table 8.3 summarizes the major characteristics of our solar system embodied in the four challenges. In the remainder of this chapter, we will examine the leading theory of solar system formation and put it to the test to see whether it can meet these four challenges. In this sense, we will be following the idealized scientific method by starting with the facts, putting forward a theory, and putting the theory to the test. In doing so, we have the benefit of hindsight. The historical path to our modern theory of the solar system was actually quite complex, with many competing theories being proposed and eventually discarded because they were unable to meet the challenges we have laid out.

The process of identifying the most important facts to explain is an essential first step in scientific analysis, followed by the development of a theory to explain these facts. We might tackle an environmental problem with a similar approach—for example, determining which pollutants are damaging an ecosystem and then developing a logical explanation for how these pollutants get there. A correct explanation of the facts is the first step in finding a solution.

8.3 The Nebular Theory of Solar System Formation

Over the past three centuries, several models have been put forth in attempts to meet the four challenges for a solar system formation theory. In the past few decades, a tremendous amount of evidence has accumulated in support of one model. This model, called the **nebular theory**, holds that our solar system formed from a giant, swirling **interstellar cloud** of gas and dust; such a cloud is also called a **nebula** (the Latin word for "cloud"). (We use the term *nebular theory* to encompass the entire formation process described in this chapter; some texts use this term more narrowly.) More generally, we now have evidence that all star systems (e.g., a star and its planets or a binary star system) form similarly from interstellar clouds (Figure 8.4).

Our galaxy, like the entire universe, originally contained only hydrogen and helium (and trace amounts of lithium) [Section 1.1]. All the heavier elements are produced in stars. The galactic recycling process (see Figure 8.5) gradually enriches the galaxy with heavier elements so that later generations of stars are born with a greater proportion of heavier elements than earlier generations.

Table 8.3 Four Major Characteristics of the Solar System

Large bodies in the solar system have orderly motions. All planets and most satellites have nearly circular orbits going in the same direction in nearly the same plane. The Sun and most of the planets rotate in this same direction as well.

Planets fall into two main categories: small, rocky terrestrial planets near the Sun and large, hydrogen-rich jovian planets farther out. The jovian planets have many moons and rings made of rock and ice.

Swarms of asteroids and comets populate the solar system. Asteroids are concentrated in the asteroid belt, and comets populate the regions known as the Kuiper belt and the Oort cloud.

Several notable exceptions to these general trends stand out, such as planets with unusual axis tilts or surprisingly large moons, and moons with unusual orbits.

FIGURE 8.4 This photograph shows the central region of the Orion Nebula, an interstellar cloud in which star systems—possibly including planets—are forming. The photo is actually a composite of more than a dozen separate images taken with the Hubble Space Telescope. (See Figure 18.14 for a complete view of the Orion Nebula.)

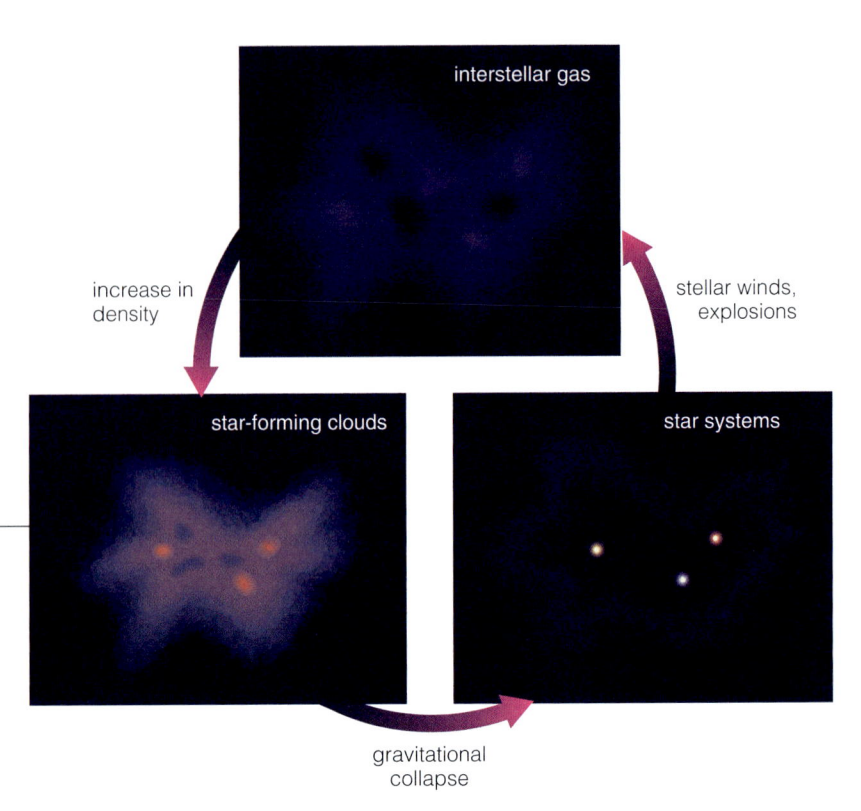

FIGURE 8.5 The galactic recycling process: Interstellar gas contracts under gravity to star-forming clouds. These clouds fragment into smaller clouds, like the solar nebula, that give birth to individual star systems. Stellar winds and explosions return gas from stars to interstellar space; some of this returned gas consists of heavy elements produced in the stars.

By the time our solar system formed, 4.6 billion years ago, about 2% of the original hydrogen and helium in the galaxy had been converted into heavier elements. Although 2% sounds like a small amount, it was more than enough to form the rocky terrestrial planets—and us. Nevertheless, hydrogen and helium remain the most abundant elements by far and make up the bulk of the Sun and the jovian planets.

TIME OUT TO THINK *Could a solar system like ours have formed with the first generation of stars after the Big Bang? Explain.*

Collapse of the Solar Nebula

An individual star system forms from just a small part of a giant interstellar cloud. We refer to the collapsed piece of cloud that formed our own solar system as the **solar nebula**. The solar nebula collapsed under its own gravity; the collapse may have been triggered by a cataclysmic event such as the impact of a shock wave from a nearby exploding star. Before the collapse, the low-density gas may have spanned a few light-years in diameter. As it collapsed to a diameter of about 200 AU, roughly twice the present-day diameter of Pluto's orbit, three important processes gave form to our solar system (Figure 8.6).

First, the temperature of the solar nebula increased as it collapsed. Such heating represents energy conservation in action [Section 4.2]. As the cloud shrank, its gravitational potential energy was converted to the kinetic energy of individual gas particles falling inward. These particles crashed into one another, converting the kinetic energy of their inward fall into the random motions of thermal energy. Some of this energy was radiated away as thermal radiation. The solar nebula became hottest near its center, where much of the mass collected to form the **protosun** (the prefix *proto* comes from a Greek word meaning "earliest form of"). The protosun eventually became so hot that nuclear fusion ignited in its core—at which point our Sun became a full-fledged star.

Second, like an ice skater pulling in her arms as she spins, the solar nebula rotated faster and faster as it shrank in radius. This increase in rotation rate represents the conservation of angular momentum in action [Section 5.2]. The rotation helped ensure that not all of the material in the solar nebula collapsed onto the protosun: The greater the angular momentum of a rotating cloud, the more spread out it will be.

Third, the solar nebula flattened into a disk—the **protoplanetary disk** from which the planets eventually formed. This flattening is a natural consequence of collisions between particles, which explains why flat disks are so common in the universe (e.g., the disks of spiral galaxies like the Milky Way, ring systems around planets, and *accretion disks* around neutron stars or black holes [Section 17.3]). A cloud may start with any size or shape, and different clumps of gas within the cloud may be moving in random directions at random speeds. When the cloud collapses,

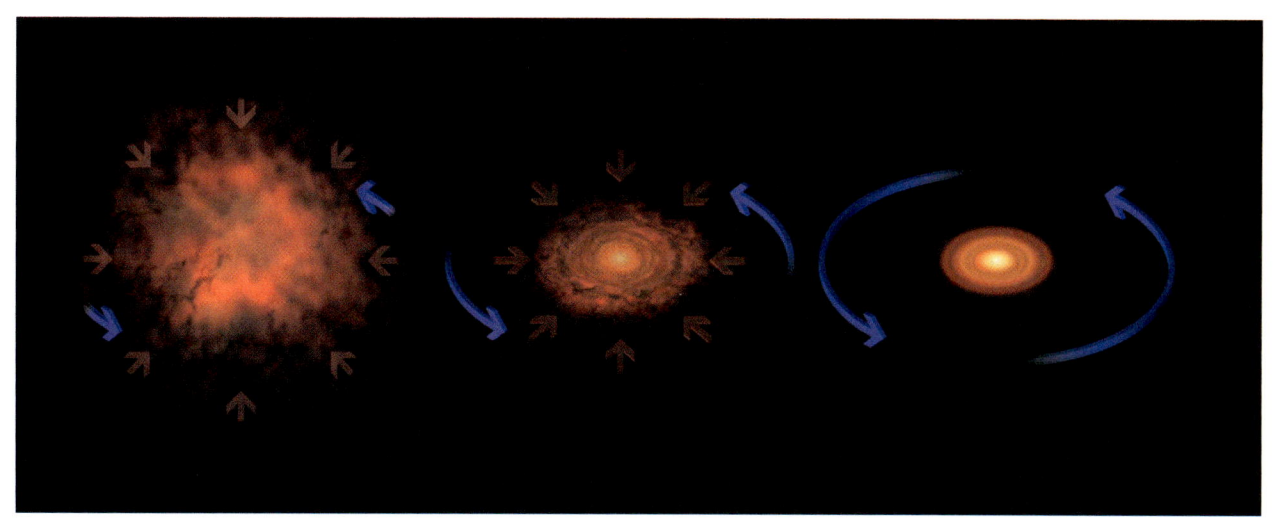

a The original cloud is large and diffuse, and its rotation is almost imperceptibly slow.

b The cloud heats up and spins faster as it contracts.

c The result is a spinning, flattened disk, with mass concentrated near the center.

FIGURE 8.6 This sequence of paintings shows the collapse of an interstellar cloud. In our solar nebula, the hot, dense central bulge became the Sun, and the planets formed in the disk.

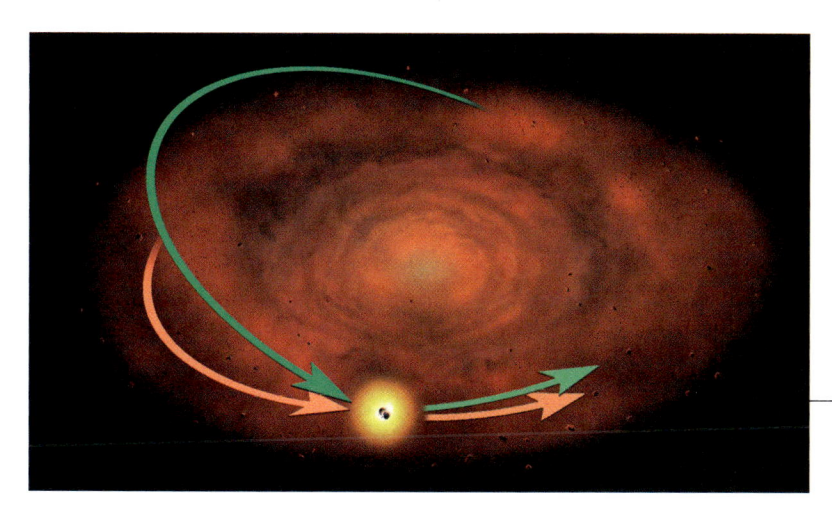

FIGURE 8.7 As shown in this painting, collisions between particles in the solar nebula average out their random motions and flatten the cloud into a disk. The green arrow represents the path of a particle that originally had a tilted orbit. After the collision, its orbit lies closer to the plane of the other particles. If the particles had started on an eccentric orbit, collisions would have made its orbit more circular. (Particle sizes are highly exaggerated.)

these different clumps collide and merge, giving the new clumps the average of their differing velocities (Figure 8.7). The result is that the random motions of the clumps in the original cloud become more orderly as the cloud collapses, changing the cloud's original lumpy shape into a rotating, flattened disk. Similarly, collisions between clumps of material in highly elliptical orbits reduce their ellipticities, making their orbits more circular. You can see a similar effect if you shake some pepper into a bowl of water and quickly stir it in a random way. Because the water molecules are always colliding with one another, the motion of the pepper grains will settle down into a slow rotation representing the average of the original, random velocities.

TIME OUT TO THINK *Try the experiment with the pepper and bowl of water. Note that, no matter how you stir the water, it almost always settles down into a slow rotation in one direction or the other. How is this process similar to what took place in the solar nebula? How is it different?*

These three processes—heating, spinning, and flattening—explain the tidy layout of our solar system. The flattening of the protoplanetary disk explains why all planets orbit in nearly the same plane. The spinning explains why all planets orbit in the same direction and also plays a role in making most planets rotate in this same direction. The fact that collisions in the protoplanetary disk tend to make highly elliptical orbits more circular explains why most planets in our solar system have nearly circular orbits today.

Evidence Concerning Nebular Collapse

The theory that our solar system formed from interstellar gas may sound reasonable, but we cannot accept it without hard evidence. Fortunately, we have strong observational evidence that the same processes are occurring elsewhere in our galaxy. A collapsing nebula emits thermal radiation [Section 6.4], primarily in the infrared. We've detected such infrared radiation from many other nebulae where star systems appear to be forming. We've even seen structures around other stars that look similar to protoplanetary disks (Figure 8.8).

Other support for the nebular theory comes from sophisticated computer models that simulate formation processes. A simulation begins with a set of data representing the conditions that we observe in interstellar clouds. Then, with the aid of a computer, we apply the laws of physics to these data to simulate how the conditions in a real cloud would change with time. These computer simulations successfully reproduce most of the general characteristics of motion in our solar system, suggesting that the nebular theory is on the right track. To date, however, the simulations do not explain the observed patterns of spacing between planets; much work lies ahead before we will fully understand the early history of our solar system. All in all, though, the nebular theory meets the challenge of explaining most of the orderly motions of the planets and satellites. Now let's turn to our second challenge: explaining the strikingly different characteristics of the terrestrial and jovian planets.

8.4 Building the Planets

The churning and mixing of the gas in the solar nebula ensured that its composition was about the same throughout: roughly 98% hydrogen and helium, and 2% heavier elements such as carbon, nitrogen, oxygen, silicon, and iron. How did the planets and other bodies in our solar system end up with such a wide variety of compositions when they came from such uniform material? To answer this question, we must investigate how the material in the protoplanetary disk came together to form the planets.

Condensation: Sowing the Seeds of Planets

In the center of the collapsing solar nebula, gravity drew much of the material together into the protosun. In the rest of the spinning protoplanetary disk, however, the gaseous material was so spread out that, at first, gravity was not strong enough to pull it to-

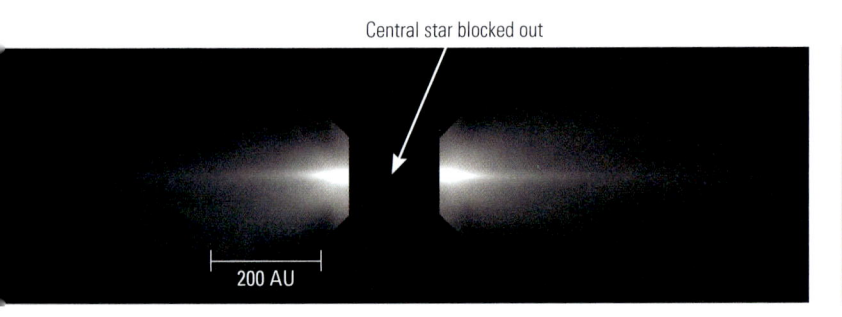

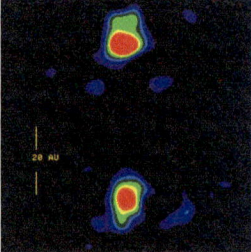

a (Left) A dust disk around the star Beta Pictoris taken by the Hubble Space Telescope at optical wavelengths. (Right) Twin dust disks around binary stars, taken by the VLA at radio wavelengths.

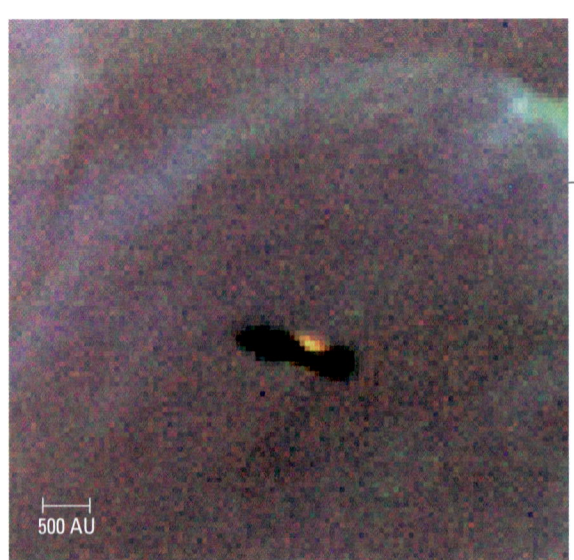

b Protoplanetary disks around stars in the constellation Auriga (left) and in the Orion Nebula (right). Hubble Space Telescope images.

FIGURE 8.8 Evidence for disks around other stars.

Materials in the Solar Nebula				
	Metals	**Rocks**	**Hydrogen Compounds**	**Light Gases**
Examples	iron, nickel, aluminum	silicates	water (H_2O) methane (CH_4) ammonia (NH_3)	hydrogen, helium
Typical Condensation Temperature	1,000–1,600 K	500–1,300 K	<150 K	(do not condense in nebula)
Relative Abundance (by mass)	(0.2%)	(0.4%)	(1.4%)	(98%)

gether to form planets. The formation of planets therefore required "seeds"—solid chunks of matter that gravity could eventually draw together. Understanding these seeds is the key to explaining the differing compositions of the planets.

The vast majority of the material in the solar nebula was gaseous: High temperatures kept virtually all the ingredients of the solar nebula vaporized near the protosun. Farther out the nebula was still primarily gaseous because hydrogen and helium are gases at nearly all temperatures. But the 2% of material consisting of heavier elements could form solid seeds where temperatures were low enough. The formation of solid or liquid particles from a cloud of gas is called **condensation**. Pressures in the solar nebula were so low that liquid droplets rarely formed, but solid particles could condense in the same way that snowflakes condense from water vapor in our atmosphere. We refer to such solid particles as **condensates**. The different kinds of planets and satellites formed out of the different kinds of condensates present at different locations in the solar nebula.

The ingredients of the solar nebula fell into four categories based on their condensation temperatures (Figure 8.9):

- **Metals** include iron, nickel, aluminum, and other materials that are familiar on Earth but less common on the surface than ordinary rock. Most metals condense into solid form at temperatures between 1,000 K and 1,600 K. Metals made up less than 0.2% of the solar nebula's mass.

- **Rocks** are materials common on the surface of the Earth, primarily silicon-based minerals (silicates). Rocks are solid at temperatures and pressures found on Earth but typically melt or vaporize at temperatures of 500–1,300 K depending on their type. Rocky materials made up about 0.4% of the nebula by mass.

- **Hydrogen compounds** are molecules such as methane (CH_4), ammonia (NH_3), and water (H_2O) that solidify into **ices** below about 150 K. These compounds were significantly more abundant than rocks and metals, making up 1.4% of the nebula's mass.

- **Light gases** (hydrogen and helium) never condense under solar nebula conditions. These gases made up the remaining 98% of the nebula's mass.

The order of condensation temperatures is easy to remember: It's the same as the order of densities.

Temperature differences between the hot inner regions and the cool outer regions of the solar nebula determined what kinds of condensates were available to form planets (Figure 8.10). Very near the protosun, where the nebula temperature was above 1,600 K, there were no condensates—everything remained gaseous. Slightly farther out, where the temperature was slightly lower, metal flakes appeared. Near the distance of Mercury's orbit, flakes of rock joined the mix. Moving outward past the orbits of Venus and Earth, more varieties of rock minerals condensed. Near the location of the future asteroid belt, temperatures were low enough to allow minerals containing small amounts of water to condense as well. Dark, carbon-rich materials also condensed here.

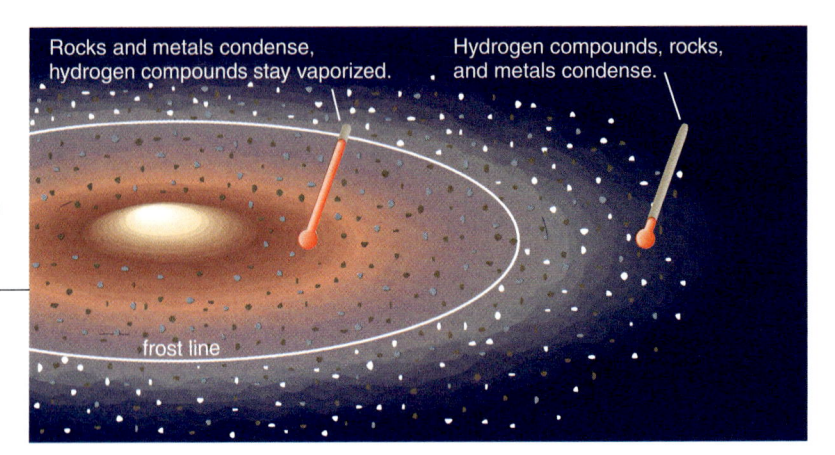

FIGURE 8.10 Temperature differences in the solar nebula led to different kinds of condensed materials, sowing the seeds for two different kinds of planets.

Only beyond the **frost line**, which lay between the present-day orbits of Mars and Jupiter, were temperatures low enough (150 K ≈ −123°C) for hydrogen compounds to condense into ices. (Water ice did not form at the familiar 0°C because the pressures in the nebula were 10,000 times lower than on Earth.) Thus, the outer solar system contained condensates of all kinds: rocks, metals, and ices. However, ice flakes were nearly three times more abundant than flakes of rock and metal because of the greater abundance of hydrogen compounds in the solar nebula.

Accretion: Assembling the Planetesimals

The first solid flakes that condensed from the solar nebula were microscopic in size. They orbited the protosun in the same orderly, circular paths as the gas from which they condensed. Individual flakes therefore moved with nearly the same speed as their neighboring flakes, allowing them to collide very gently. At this point, the flakes were far too small to attract one another by gravity, but they were able to stick together through electrostatic forces—the same "static electricity" that makes hair stick to a comb. Thus, the flakes grew slowly into larger particles. As the particles grew in mass, gravity began to aid the process of their sticking together, accelerating their growth. This process of growing by colliding and sticking is called **accretion**. The growing objects formed by accretion are called **planetesimals**, which essentially means "pieces of planets." Small planetesimals probably came in a variety of shapes, such as we see in many small asteroids today. Larger planetesimals (i.e., those several hundred kilometers across) became spherical because the force of gravity was strong enough to overcome the strength of rock and pull everything toward the center (Figure 8.11).

The growth of planetesimals was rapid at first: As planetesimals grew larger, they had both more surface area with which to collide and more gravity to attract other planetesimals. Some probably grew to

Common Misconceptions:
Solar Gravity and the Density of Planets

You might think that the dense rocky and metallic materials were simply pulled to the inner part of the solar nebula by the Sun's gravity or that light gases simply escaped from the inner nebula because gravity couldn't hold them. But this is not the case; all the ingredients were orbiting the Sun together under the influence of the Sun's gravity. The orbit of a particle or a planet does *not* depend on its size or density, so the Sun's gravity is *not* the cause of the different kinds of planets. Rather, the different temperatures in the solar nebula are the cause.

FIGURE 8.11 Gravity is not strong enough to alter the irregular shapes of small objects. A larger object with the same shape will eventually be compressed to a sphere by the greater strength of gravity.

Weak gravity is unable to deform small objects.

Stronger gravity makes larger objects spherical.

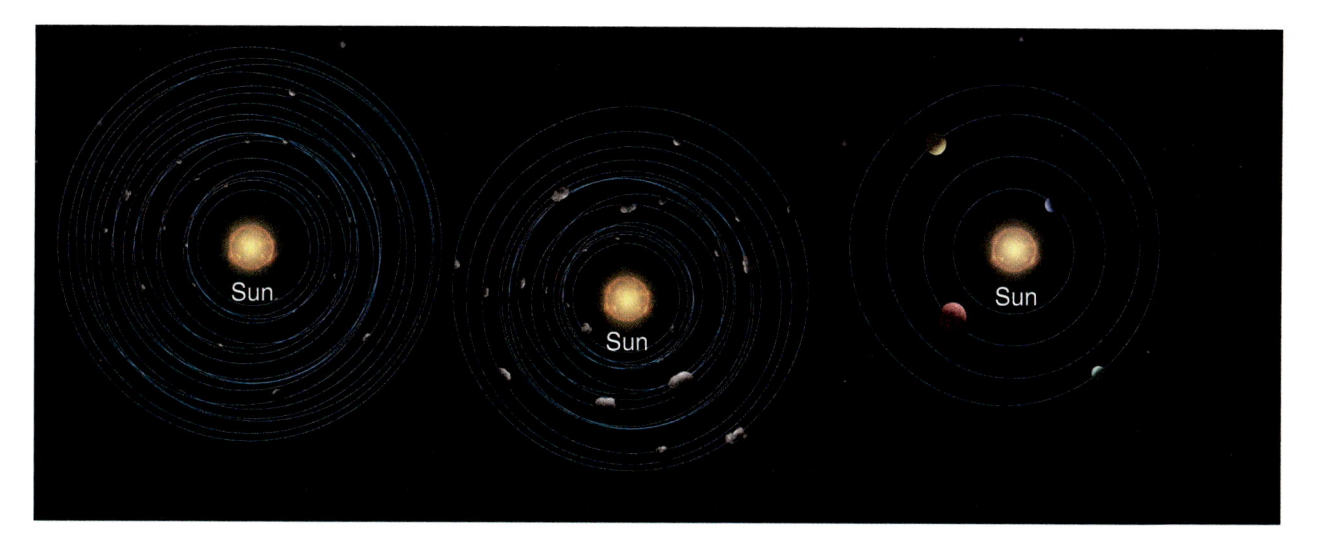

FIGURE 8.12 These diagrams show how planetesimals gradually accrete into the terrestrial planets. Early in the accretion process, there are many Moon-size planetesimals on crisscrossing orbits (left). As time passes, a few planetesimals grow larger by accreting smaller ones, while others shatter in collisions (center). Ultimately, only the largest planetesimals avoid shattering and grow into full-fledged planets (right). Diagram not to scale.

hundreds of kilometers in size in only a few million years. (This might sound like a long time, but it is only about 1/1,000 the age of the solar system.) However, once the planetesimals reached these relatively large sizes, further growth became more difficult. Gravitational encounters between planetesimals tended to alter their orbits, particularly those of the smaller planetesimals [Section 5.6]. With different orbits crossing each other, collisions between planetesimals tended to occur at higher velocities and hence were more destructive. Collisions started to produce fragmentation more often than accretion. Only the largest planetesimals avoided such shattering and continued to grow into full-fledged planets (Figure 8.12).

The sizes and compositions of the planetesimals depended on the temperature of the surrounding solar nebula. In the inner solar system, where only rocky and metallic flakes condensed, planetesimals were made of rock and metal. That is why the terrestrial planets ended up being composed of rocks and metals. Moreover, because rocky and metallic elements made up only about 0.6% of the material in the solar nebula, the planetesimals in the inner solar system could not grow very large, which explains why the terrestrial planets are relatively small.

Beyond the frost line between the orbits of Mars and Jupiter, where temperatures were cold enough for ices to condense, planetesimals were built from ice flakes in addition to flakes of rock and metal. Because ice flakes were much more abundant, these planetesimals were made mostly of ices and could grow to much larger sizes than could planetesimals in the inner solar system. The largest icy planetesimals of the outer solar system became the cores of the jovian planets.

The densities of various objects in the solar system support this theory. The high density of planets

FIGURE 8.13 Shiny flakes of metal are clearly visible in this meteorite, mixed in among the rocky material. Such metallic flakes are just what we would expect to find if condensation really occurred in the solar nebula as described by the nebular theory.

near the Sun (4–5 g/cm^3) shows that they are made of rock and metal. In the outer solar system, solid objects such as moons and comets have much lower densities (1–3 g/cm^3), revealing a high proportion of ices and a lower proportion of rocks and metals.

Further evidence of accretion comes from study of *meteorites*—pieces of our solar system that have fallen to Earth. Many meteorites contain metallic grains embedded in a variety of rocky minerals (Figure 8.13). Meteorites thought to come from greater

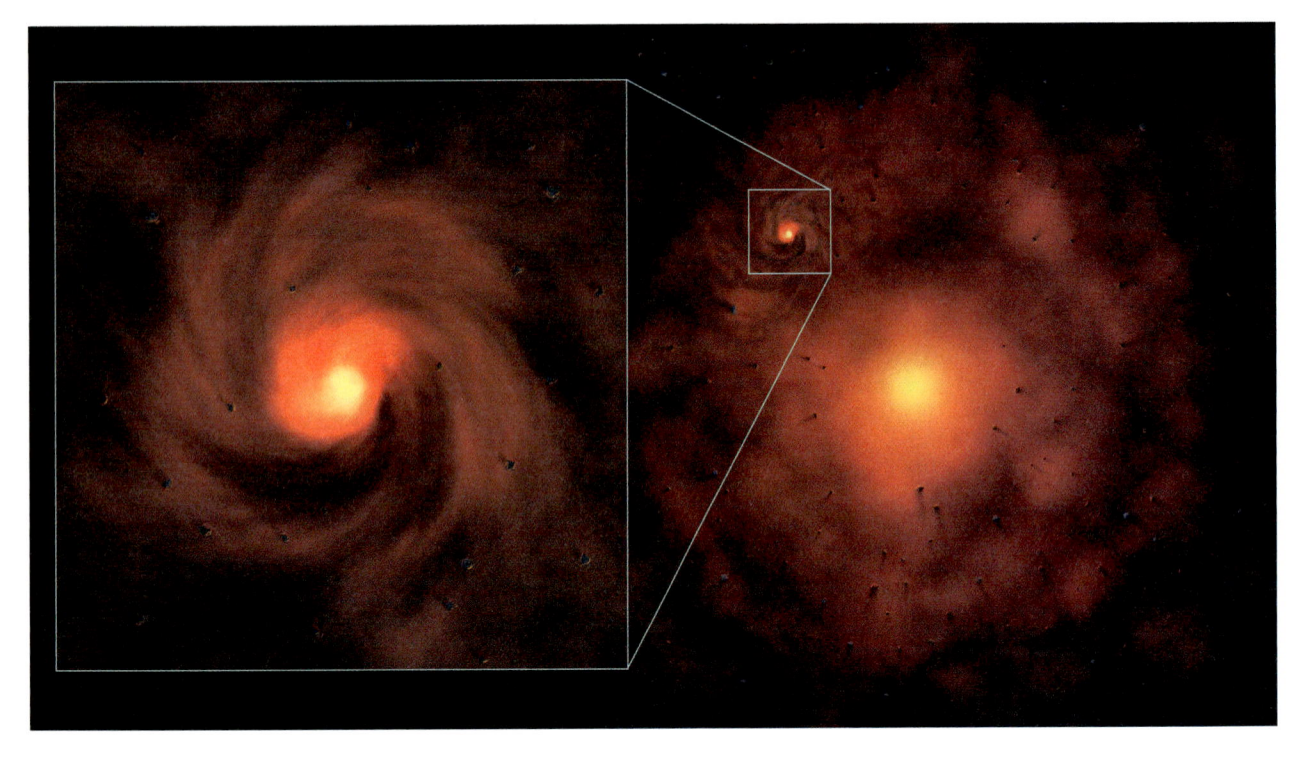

FIGURE 8.14 Large icy planetesimals in the cold outer regions of the solar nebula captured significant amounts of hydrogen and helium gas. This process of nebular capture created jovian nebulae, resembling the solar nebula in miniature, in which the jovian planets and satellites formed. This painting shows a jovian nebula (enlarged at left) located within the entire solar nebula.

distances contain abundant carbon-rich materials, and some contain water [Section 12.3].

Nebular Capture: Making the Jovian Planets

Accretion proceeded rapidly in the outer solar system, because the presence of ices meant much more material in solid form. Some of the icy planetesimals of the outer solar system quickly grew to sizes many times larger than the Earth. At these large sizes, their gravity was strong enough to capture the far more abundant hydrogen and helium gas from the surrounding nebula. As they accumulated substantial amounts of gas, the gravity of these growing planets grew larger still—allowing them to capture even more gas. The process by which icy planetesimals act as seeds for capturing far larger amounts of hydrogen and helium gas, called **nebular capture**, led directly to the formation of the jovian worlds (Figure 8.14). It explains their huge sizes and the large abundance of hydrogen and helium reflected in their low average densities.

Nebular capture also explains the formation of the diverse satellite systems of the jovian planets. As the early jovian planets captured large amounts of gas from the solar nebula, the same processes that formed the protoplanetary disk—heating, spinning, and flattening—formed similar but smaller disks of material around these planets. Condensation (of metals, rocks, and a lot of ice) and accretion took place within these **jovian nebulae**, essentially creating a miniature solar system around each jovian planet. The spinning disks of the jovian nebulae explain why most of the jovian planet satellites orbit in nearly circular paths lying close to the equatorial plane of their parent planet and also why they orbit in the same direction in which their planet rotates. The composition of the jovian nebulae explains why the jovian planets possess systems of large, icy satellites. Their densities of 1–3 g/cm^3 reflect their mixtures of icy and rocky condensates. Temperature differences within the jovian nebulae may have led to density differences between satellites analogous to the density differences between planets.

Our theory of condensation, accretion, and nebular capture meets the second challenge for our solar system formation theory: It explains the general differences between terrestrial and jovian planets. (Our theory can also explain the presence of planetary rings, but we postpone this interesting discussion until Chapter 11.) However, our theory does not yet explain why the spacing between the planets increases with distance from the Sun, although we can come up with some reasonable hypotheses. For example, the rapid accretion in the outer solar system may have allowed a few protoplanets to gobble up all their neighbors, leaving the jovian planets widely spaced.

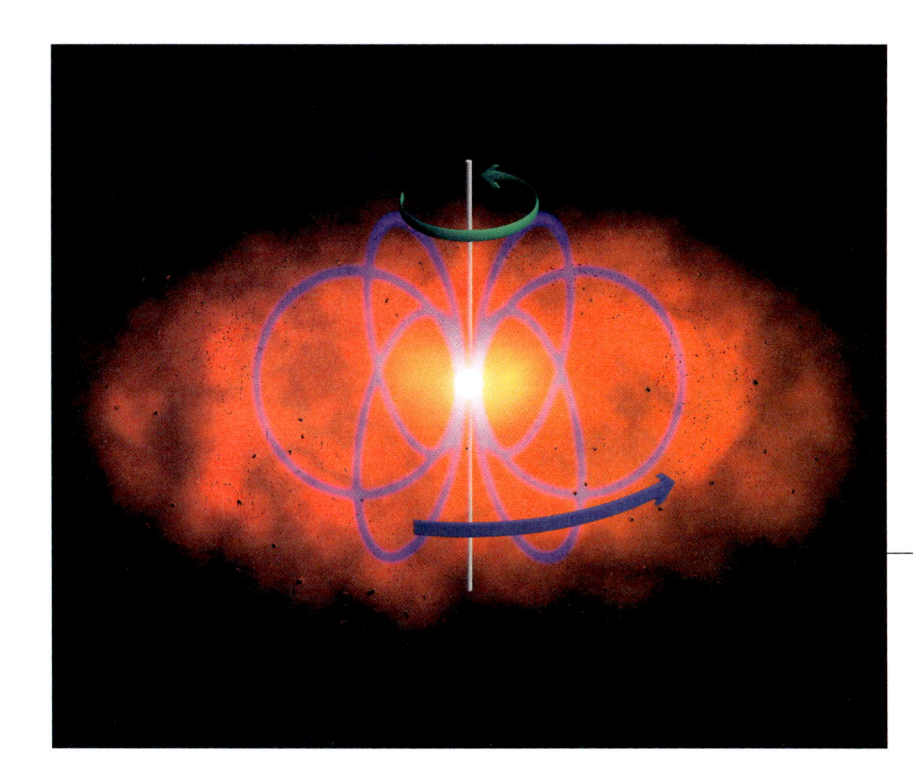

FIGURE 8.15 The magnetic braking process: Charged particles in the solar nebula tend to move with the Sun's magnetic field (represented by the purple loops). As the magnetic field rotates with the Sun, these charged particles are dragged through the disk. Friction between the charged particles and the rest of the disk slows the Sun's rotation (and also slightly speeds up the disk). The Sun's relatively slow rotation probably resulted from this process. (Particle sizes are highly exaggerated.)

The Solar Wind: Clearing Away the Nebula

What happened to the remaining gas of the solar nebula? Apparently, this gas was blown into interstellar space by the **solar wind**, a flow of energetic particles ejected by the Sun in all directions. Although the solar wind is fairly weak today, we have evidence that it was much stronger when the Sun was young—strong enough to have swept huge quantities of gas out of our solar system. The clearing of the gas interrupted the cooling process in the nebula. Had the cooling continued longer, ices might have condensed in the inner solar system. Instead, the solar wind swept the still-vaporized hydrogen compounds away from the inner solar system, along with the remaining hydrogen and helium throughout the solar system. When the gas cleared, the compositions of objects in the early solar system were essentially set.

We've seen that the nebular theory accounts quite well for the motions of the planets and moons and the compositional trends in the solar system. However, the nebular theory also makes one prediction that at first seems to contradict our observations: Conservation of angular momentum in the collapsing solar nebula means that the young Sun should have been rotating very fast, but the Sun actually rotates quite slowly, with each full rotation taking about a month.

Fortunately, this apparent contradiction between theory and observation has a simple resolution. Angular momentum cannot simply disappear, but it is possible to transfer angular momentum from one object to another—and then get rid of the second object. A spinning skater can slow her spin by grabbing her partner and then pushing him away. Beginning in the 1950s, scientists realized that the young Sun's rapid rotation would have generated a magnetic field far stronger than that of the Sun today. It made the Sun highly *active* (e.g., more and larger sunspots on its surface), a circumstance under which the Sun emits larger amounts of ultraviolet and X-ray light [Section 14.5]. This high-energy radiation ionized gas in the solar nebula, creating many charged particles.

As we will discuss in more detail in Chapter 14, charged particles and magnetic fields tend to stick together. As the sun rotated, its magnetic field dragged the charged particles along and added to their angular momentum. Because the particles were gaining angular momentum, the Sun was losing it. The strong solar wind then blew these particles into interstellar space, leaving the Sun with greatly diminished angular momentum and hence the much slower rotation that we see today. The process by which the Sun's magnetic field helped slow its rotation is called **magnetic braking**, because it effectively "applied the brakes" to the Sun's rapid rotation (Figure 8.15).

Although we cannot look into the past to see if magnetic braking really did slow the Sun's rotation, we can look for evidence of the magnetic braking process in other star systems. When we look at young stars that have recently formed in interstellar clouds, we find that nearly all of them rotate rapidly and have strong magnetic fields and strong stellar winds [Section 16.2]. In contrast, older stars almost invariably rotate slowly, just like our Sun. This fact suggests that nearly all stars have their original rapid rotations slowed by magnetic braking, just as we should expect from the nebular theory.

8.5 Leftover Planetesimals

We are now ready to turn to our third and fourth challenges: explaining the existence of asteroids and comets and explaining the various exceptions to general trends. Both challenges involve the planetesimals that remained between the forming planets in the early solar system.

Origin of Asteroids and Comets

The strong wind from the young Sun cleared excess gas from the solar nebula, but many planetesimals remained scattered between the newly formed planets. These "leftovers" became comets and asteroids. Like the planetesimals that formed the planets, they formed from condensation and accretion in the solar nebula. Thus, their compositions followed the pattern determined by condensation: planetesimals of rock and metal in the inner solar system, and icy planetesimals in the outer solar system.

Leftover planetesimals originally must have had nearly circular orbits in the same plane as the orbits of the planets. But gravitational encounters with the newly formed planets soon made their orbits more random [Section 5.6]. When small planetesimals passed near a large planet, the planet was hardly affected, but the planetesimals were flung off at high speed in random directions. The strong gravity of the jovian planets, in particular, tugged and nudged the orbits of leftover planetesimals, even at great distances. The result was that the remaining planetesimals ended up with more highly elliptical orbits than the planets, and sometimes with large tilts relative to the plane of planetary orbits. Most of these planetesimals eventually either crashed into one of the planets or were flung out of the solar system, but many others still survive today.

Asteroids are the rocky, leftover planetesimals of the inner solar system. Some asteroids are scattered throughout the inner solar system, but most are concentrated in the "extra-wide" gap between Mars and Jupiter that contains the *asteroid belt*. This region probably once contained enough rocky planetesimals to form another terrestrial planet. However, the gravity of Jupiter (the largest and closest jovian planet) tended to nudge the orbits of these planetesimals, sending most of them on collision courses with the planets or with one another. The present-day asteroid belt contains the remaining planetesimals from this "frustrated planet formation." Although thousands of asteroids remain in the asteroid belt, their *combined* mass is less than 1/1,000 of Earth's mass. Jupiter's gravity continues to nudge these asteroids, changing their orbits and sometimes leading to violent, shattering collisions. Debris from these collisions often crashes to Earth in the form of meteorites.

Comets are the icy, leftover planetesimals of the outer solar system. The icy planetesimals that cruised the space between Jupiter and Neptune couldn't grow to more than a few kilometers in size before suffering either a collision or a gravitational encounter with one of the jovian planets. Many were flung into distant, randomly oriented orbits, becoming the comets of the spherical *Oort cloud* (see Figure 8.3). Beyond the orbit of Neptune, the icy planetesimals were much less likely to be destroyed by collisions or cast off by gravitational encounters. Instead, they remained in orbits going in the same direction as planetary orbits and concentrated near the plane of planetary orbits (but with somewhat more randomness than the orbits of the planets). They were also able to continue their accretion, and many may have grown to hundreds or even thousands of kilometers in diameter. These are the comets of the *Kuiper belt* (see Figure 8.3); Pluto is probably the largest member of this class. (Important discoveries often lead to confusion in nomenclature. The Kuiper belt is sometimes called the Kuiper disk, and its residents are sometimes called comets, Kuiper belt objects [KBOs], Kuiperoids, iceteroids, or Plutinos. To astronomers, they are technically just distant asteroids, even though they are icy instead of rocky.)

Evidence that asteroids and comets really are leftover planetesimals comes from analysis of meteorites, spacecraft visits to comets and asteroids, and computer simulations of solar system formation. Note also that our theory predicts the existence of comets in both the Oort cloud and the Kuiper belt—a prediction first made in the 1950s. Not until the early 1990s did astronomers verify the existence of objects orbiting within the Kuiper belt. The nebular theory has thus met our third challenge: explaining the existence of asteroids and comets. Moreover, it has suggested an explanation for the seemingly anomalous planet Pluto. Now we turn to our fourth and final challenge: explaining the exceptions to the general trends in our solar system.

The Early Bombardment: A Rain of Rock and Ice

The collision of a leftover planetesimal with a planet is called an **impact**, and the responsible planetesimal is called the **impactor**. On planets with solid surfaces, impacts leave the scars we call **impact craters**. Impacts were extremely common in the young solar system; in fact, impacts were part of the accretion process in the late stages of planetary formation. The vast majority of impacts occurred in the first few hundred million years of our solar system's history.

The heavy bombardment of planetary surfaces in the early solar system would have resembled a rain of rock and ice from space (Figure 8.16). The surface of the Earth was once scarred like the Moon's surface, but most impact craters on Earth were erased long ago by erosion and other geological processes. Only the outlines of a few large craters are recognizable on Earth. But vast numbers of craters remain on worlds that experience less erosion, such as the Moon and Mercury. Indeed, one way of estimating the age of a planetary surface (the time since the surface last changed in a substantial way) is to count the number of craters: If there are many craters, the surface must still look much as it did when the early bombardment ended, about 4 billion years ago.

The early rain of rock and ice did more than just scar planetary surfaces. It also brought the materials from which atmospheres, oceans, and polar caps eventually formed. Remember that the terrestrial planets were built from planetesimals of metal and rock. But Earth's oceans, the polar caps of Earth and Mars, and the atmospheres of Venus, Earth, and Mars are all derived from hydrogen compounds that remained gaseous in the inner solar nebula. These materials must have arrived on the terrestrial planets after their initial formation, most likely brought by impacts of planetesimals formed farther out in the solar system. We don't yet know whether the impactors came from the asteroid belt, where rocky planetesimals contained small amounts of water and other hydrogen compounds that had condensed as ices, or whether they were comets containing huge amounts of ice. Either way, the water we drink and the air we breathe probably once were part of planetesimals floating beyond the orbit of Mars.

TIME OUT TO THINK *What was Jupiter's role in bringing water (and other materials that could not condense in the inner solar nebula) to Earth? How might Earth be different if Jupiter had never formed?*

FIGURE 8.16 Around 4 billion years ago, Earth, its Moon, and the other planets were heavily bombarded by leftover planetesimals. This painting shows the young Earth and Moon glowing with the heat of accretion, and an impact in progress on the Earth.

Captured Moons

We can easily explain the orbits of most jovian planet satellites by their formation in a jovian nebula that swirled around the forming planet. But our fourth challenge reminds us that some moons have unusual orbits—orbits in the "wrong" direction (opposite the rotation of their planet) or with large inclinations to the planet's equator. These unusual moons are probably leftover planetesimals that were *captured* into orbit around a planet.

It's not easy for a planet to capture a moon. An object cannot switch from an unbound orbit (e.g., an asteroid whizzing by Jupiter) to a bound orbit (e.g., a moon orbiting Jupiter) unless it somehow loses orbital energy [Section 5.6]. Captures probably occurred when the capturing planet had a very extended atmosphere or, in the case of the jovian planets, its own miniature solar nebula. Passing planetesimals could be slowed by friction with the gas, just as artificial satellites are slowed by drag with the Earth's atmosphere. If friction reduced a planetesimal's orbital energy enough, it could have become an orbiting moon. Because of the random nature of the capture process, the captured moons would not necessarily orbit in the same direction as their planet or in its equatorial plane.

FIGURE 8.17 The two moons of Mars, shown here in photos taken by the Viking spacecraft, are probably captured asteroids. Phobos is only about 13 km across and Deimos is only about 8 km across—making each of these two moons small enough to fit within the boundaries of a typical large city.

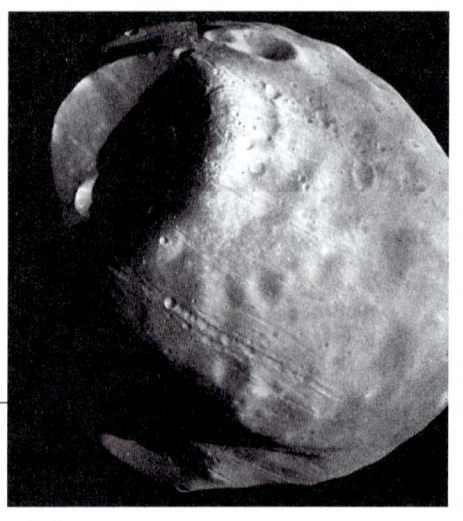

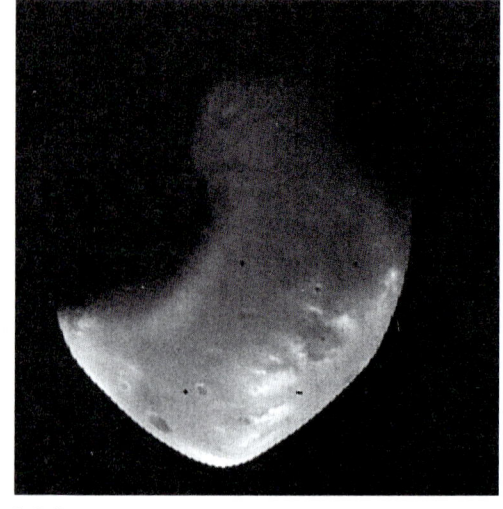

a Phobos

b Deimos

The two small moons of Mars—Phobos and Deimos—probably were asteroids captured by this process (Figure 8.17). They resemble asteroids seen in the asteroid belt and are much darker and lower in density than Mars. Jupiter probably also captured several of its moons. Two groups of four satellites circle Jupiter in unusual orbits: All are noticeably eccentric and tilted, and one of the two groups orbits Jupiter in the "wrong" direction. Astronomers speculate that two asteroids may have been captured by Jupiter and that each broke into several pieces in the process. Other jovian planets may also have captured moons, including one that is particularly large—Triton, the largest moon of Neptune. Triton is considerably larger than Pluto and orbits Neptune in a direction opposite to Neptune's rotation. It may be a captured object from the Kuiper belt [Section 11.5]. The one unusual moon that cannot be explained by this process is our own. Our Moon is much too large to have been captured by Earth.

Giant Impacts and the Formation of Our Moon

The largest planetesimals remaining as the planets formed may have been huge; some may have been the size of Mars. When one of these planet-size planetesimals collided with a planet, the spectacle would have been awesome. Such a **giant impact** could have significantly altered the planet's fate and may explain some of the remaining mysteries in our fourth challenge.

What if an object the size of Mars had collided with Earth? Computer simulations address this interesting question. The Earth would have shattered from the impact, and material from the outer layers would have ended up in orbit around the Earth (Figure 8.18). There this material could have reaccreted to form a large satellite. Depending on exactly where

FIGURE 8.18 Artist's conception of the impact of a Mars-size object with Earth, as may have occurred soon after Earth's formation. The ejected material comes mostly from the outer rocky layers and accretes to form the Moon, which is poor in metal.

and how fast the giant impactor struck the Earth, the blow might also have tilted the Earth's axis, changed its rotation rate, or completely torn it apart. Today, such a giant impact is the leading hypothesis for explaining the origin of our Moon. The Moon's composition is similar to that of the Earth's outer layers, exactly what we would predict for the kind of collision shown in Figure 8.18. The Moon is also depleted in easily vaporized ingredients, as would be expected from the tremendous heating that would have occurred during the collision. Moreover, we can rule out the idea of the Moon forming simultaneously with Earth. In that case, Earth and the Moon would have formed from the same material and therefore should have the same density, but the Moon's density is considerably lower than the Earth's.

In fact, many of the unusual properties of specific planets—properties that defy the general trends expected by the nebular theory—may be the results of giant impacts. Mercury may have lost much of its outer, rocky layer in a giant impact, leaving it with a huge metallic core. A giant impact might even have contributed to the slow, backward rotation of Venus, which may have had a "normal" rotation until the arrival of the giant impactor. Giant impacts probably also were responsible for the axis tilts of many planets (including Earth) and for tipping Uranus on its side. Pluto's moon Charon may have formed in a giant-impact process similar to the one that formed our Moon.

Unfortunately, we can do little to test whether a particular giant impact really occurred billions of years ago. The difficulty in proving the giant-impact hypothesis makes the idea controversial, and most planetary scientists didn't take such ideas seriously when they were first proposed. But no other idea so effectively explains the formation of our Moon and other "oddities" that we've discussed. Moreover, giant impacts certainly should have occurred, given the number of large, leftover planetesimals predicted by the nebular theory. Random giant impacts are the most promising explanation for the many exceptional circumstances noted in the fourth challenge.

Summary: Meeting the Challenges

The nebular theory explains the great majority of important facts contained in our four challenges. But you should not be left with the impression that competing theories were never put forth or that solar system formation is now a "solved problem." Theories have evolved hand-in-hand with the discovery of the nature of our solar system, and what we have presented here is the culmination of that endeavor to date. Planetary scientists are still struggling with more quantitative aspects, seeking reasons for the exact sizes, locations, and compositions of the planets. Perhaps as we study other planetary systems and examine their properties, we will be able to improve our understanding of solar system formation.

Assuming that the nebular theory is correct, it is interesting to ask whether our solar nebula was "destined" to form the solar system we see today. The first stages of planet formation were orderly and inevitable according to the nebular theory. Nebular collapse, condensation, and the first stages of accretion were relatively gradual processes that probably would happen all over again if we turned back the clock. But the final stages of accretion, and giant impacts in particular, are inherently random in nature and probably would not happen again in just the same way. A larger or smaller planet might form at Earth's location and might suffer from a larger giant impact or from no impact at all. We don't yet know whether these differences would fundamentally alter the solar system or simply change a few "minor" details—such as the possibility of life on Earth.

8.6 The Age of the Solar System

The nebular theory accounts for the major physical properties of our solar system, supporting the idea that all the planets formed at about the same time from the same cloud of gas. But *when* did it all happen, and how do we know? The answer is that the solar system began forming about 4.6 billion years ago, a fact we learned by determining the age of the oldest rocks in the solar system.

The concept of a rock's age is tricky: The rock's atoms were forged in stars and are therefore much older than the Earth. Atoms are not stamped with any date of manufacture, and old atoms are truly indistinguishable from young ones. By the age of a rock, we mean the time since those atoms became locked together, that is, *the time since the rock last solidified.*

Radioactive Dating

Most rocks contain minute amounts of **radioactive elements**—atoms whose nuclei have a tendency to break apart. (The term *radioactive* has nothing to do with radio waves, though both terms come from the concept of particles or waves radiating away from a source.) A radioactive element starts with a certain number of protons and neutrons in its nucleus. When that nucleus breaks apart, or **decays**, it ejects some subatomic particles and leaves behind a **decay product**—a different element or isotope with a different number of protons and/or neutrons [Section 4.3]. Over time, the amount of the original radioactive material (the *parent*) in the rock decreases, and the amount of the decay product (the *daughter*) increases.

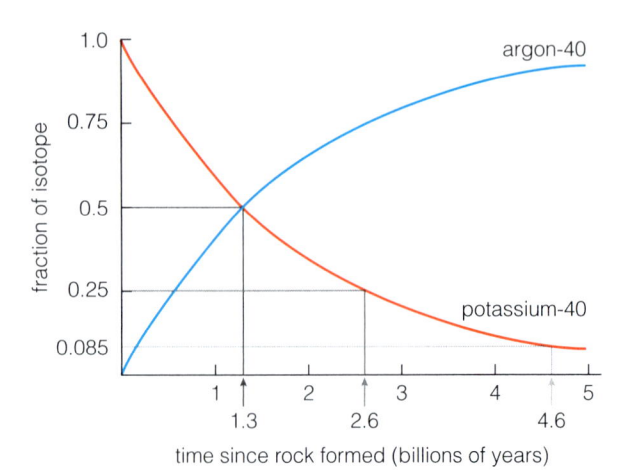

FIGURE 8.19 Radioactive decay of potassium-40 to argon-40. The red line shows the decreasing amount of potassium-40, and the blue line shows the increasing amount of argon-40. The half-life for this reaction is 1.3 billion years. At 2.6 billion years, only a quarter of the original potassium-40 remains. After 4.6 billion years, only 0.085 of the original amount remains.

Most nuclei are stable, but certain combinations of neutrons and protons in a nucleus are unstable and prone to break apart sooner or later. One kind of radioactive decay is called *nuclear fission*. The ejected particles are very energetic; this is the energy that drives nuclear power plants and that generates much of the Earth's internal heat.

The rate at which a particular radioactive substance decays is characterized by its **half-life**, the time it takes for half of the parent nuclei to decay. Each isotope of each element has its own half-life, ranging from a fraction of a second to billions of years. For example, potassium-40 (i.e., potassium with atomic weight 40) has a half-life of about 1.3 billion years. If a rock begins with a certain amount of potassium-40, half this amount will remain after 1.3 billion years, one-fourth the original amount will remain after 2.6 billion years, one-eighth the original amount will remain after 3.9 billion years, and so on (Figure 8.19). As the potassium-40 decays, it leaves behind its decay product, argon-40. Thus, by comparing the amounts of potassium-40 and argon-40 in the rock, we can determine its age. This process of determining the age of an object by studying the results of radioactive decay is called **radioactive dating**.

Mathematical Insight 8.1 Radioactive Decay

The amount of a radioactive substance decays by a factor of 2 with each half-life, so radioactivity is an example of *exponential decay*. We can express this decay process with a simple formula relating the current amount of a radioactive substance in a rock to the original amount:

$$\frac{\text{current amount}}{\text{original amount}} = \left(\frac{1}{2}\right)^{t/T_{\text{half}}}$$

where t is the time since the rock formed and T_{half} is the half-life of the radioactive material. This equation is graphed in Figure 8.19 for the decay of potassium-40 into argon-40. Note that the steady decline in the amount of potassium-40 with time is matched by a steady rise in the amount of argon-40. If you examine the graph over any 1.3-billion-year period (the half-life of potassium-40), you'll see that the amount of potassium declines by half.

Example: The famous Allende meteorite lit up the skies of Mexico as it shattered during its fall to Earth on February 8, 1969. Scientists collected pieces of the meteorite for study. To determine the age of the meteorite, they heated and chemically analyzed a small bit of meteorite. They found both potassium-40 and argon-40 present in a ratio of approximately 0.85 unit of potassium-40 atoms to 9.15 units of gaseous argon-40 atoms. (The units are unimportant, since only the relative amounts of the parent and daughter materials matter.) How old is the Allende meteorite?

Solution: Since no argon gas should have been present in the meteorite when it formed, the 9.15 units of argon must originally have been potassium. Thus, the sample started with $0.85 + 9.15 = 10$ units of potassium-40, of which 0.85 unit remains. Thus, the *current amount* of potassium-40 is 0.85 unit, and the *original amount* is 10 units. With these values, the radioactive decay equation becomes:

$$\left(\frac{1}{2}\right)^{t/T_{\text{half}}} = \frac{\text{current amount}}{\text{original amount}} = \frac{0.85}{10} = 0.085$$

You can get a fairly accurate estimate of the age t by reading the graph in Figure 8.19 or by testing some guesses in the equation above using your calculator. The most accurate method is to solve the equation for the age t by using logarithms:

$$t = T_{\text{half}} \times \frac{\log_{10}\left(\frac{\text{current amount}}{\text{original amount}}\right)}{\log_{10}\left(\frac{1}{2}\right)}$$

$$= 1.3 \text{ billion yr} \times \frac{\log_{10}(0.085)}{\log_{10}\left(\frac{1}{2}\right)} = 4.6 \text{ billion yr}$$

Thus, from the ratio of potassium-40 to argon-40, we conclude that the Allende meteorite solidified from the solar nebula about 4.6 billion years ago.

Radioactive dating is simple in principle, but in practice it requires careful laboratory work and a good understanding of "rock chemistry." For example, suppose you find a rock that contains equal numbers of atoms of potassium-40 and argon-40. If you assume that all the argon came from potassium decay, then the rock must be 1.3 billion years old (because half the original potassium atoms must have decayed to yield equal numbers of potassium and argon atoms). But this age is correct only if the assumption is valid; that is, the rock must not have contained any argon-40 when it formed. In this case, knowing a bit of chemistry helps. Potassium-40 is a natural ingredient of many minerals in rocks, but argon-40 is a gas that never combines with other elements and did not condense in the solar nebula. Therefore, if you find argon-40 gas trapped inside minerals, you can be sure that it came from the radioactive decay of potassium-40. Another assumption is that none of the argon-40 gas has escaped from the rock. Rock chemistry tells us that as long as the rock has not been significantly heated, even atoms of gases will remain trapped within it.

As another example, consider rocks from the ancient lunar highlands [Section 9.5]. These rocks contain minerals with a very small amount of uranium-238, which decays (in several steps) to lead-206 with a half-life of about 4.5 billion years. Lead and uranium have very different chemical behaviors, and some minerals start with virtually no lead. Laboratory analysis of such minerals in lunar rocks shows that they now contain almost equal proportions of uranium-238 and lead-206. We conclude that half the original uranium-238 has decayed, turning into the same number of lead-206 atoms. The lunar rock must be about one half-life old, or almost 4.5 billion years. (More precisely, the oldest lunar rocks are about 4.4 billion years old.)

TIME OUT TO THINK *If future scientists examine the lunar rocks 4.5 billion years from now, what proportions of uranium-238 and lead-206 will they find?*

Radioactive dating is possible with many other materials having a variety of different half-lives. In some cases it can be very difficult, particularly if some of the daughter material may have been present when the rock first formed. In that case, sophisticated chemistry may be needed to determine how much material is original and how much is the product of decay. Fortunately, rocks often contain several different radioactive materials, so scientists can date a rock by analyzing several different parent–daughter ratios. If the ages found from analyzing the different materials agree, we can be very confident that we know the true age of the rock.

Earth Rocks, Moon Rocks, and Meteorites

How on Earth can we measure the age of the solar system? In fact, nothing on Earth's surface remains from the formation era. Geological activity, such as plate tectonics and volcanoes, has ensured that virtually all the rocks present on the early Earth have since melted and resolidified. Because dating tells us only how long it has been since a rock last solidified, we cannot determine the age of our planet from its rocks, let alone the age of the solar system. Nevertheless, Earth rocks tell us that the solar system is old: The oldest rocks on the Earth date back about 4 billion years, although most of the Earth is covered in rocks only hundreds of millions of years old.

TIME OUT TO THINK *Suppose you do radioactive dating on a chunk of lava recently spewed out of Kilauea, an active volcano on the island of Hawaii. How old would it be? Explain.*

Determining the age of the solar system requires finding rocks from beyond the Earth—rocks that might not have melted or vaporized since the birth of the solar system. But how do we get rocks from space? Astronauts brought back numerous rocks from the Moon, and the oldest of these date to about 4.4 billion years ago. But this still is not the age of the solar system, because the Moon's surface melted and resolidified as a result of the early bombardment. Our other main source of rocks from space is meteorites that have fallen to Earth.

Many meteorites appear to be unchanged since their formation and are therefore samples of the condensation and accretion processes in the early solar system. Careful analysis of radioactive elements in meteorites shows that the oldest ones formed about 4.6 billion years ago, marking the beginning of accretion from the solar nebula. (Museums and specialty catalogs sell meteorite samples. For a few dollars, you can buy a small, solidified piece of the solar nebula. It will be the oldest object you'll ever touch.) Calculations show that accretion probably lasted only about 0.1 billion years (100 million years), so Earth and the other planets formed about 4.5 billion years ago.

The age of 4.6 billion years means our solar system is less than half as old as the universe. Ours is a middle-aged solar system in an old universe.

A Trigger for the Collapse?

The very existence of radioactive elements in meteorites is profoundly important, quite apart from the age information those elements provide. Radioactive elements are made only deep inside stars or in violent stellar explosions (supernovae). Thus, the presence of these elements in meteorites and on Earth underscores the fact that our solar system is made from the remnants of past generations of stars.

The discovery of daughter elements of short-lived radioactive elements has further implications. The rare isotope xenon-129 is found in some meteorites. Xenon is gaseous even at extremely low temperatures, so it could not have condensed and become trapped in planetesimals forming in the solar nebula. Thus, any xenon present in meteorites must be a product of radioactive decay. In fact, xenon-129 is a decay product of iodine-129. But iodine-129 has a half-life of just 17 million years, so this iodine must have traveled from the star in which it was produced to our solar nebula fairly quickly. (If the supernova had occurred a billion years earlier, the iodine would have decayed entirely into xenon, which would not have condensed or accreted.) This suggests that a nearby star exploded only a few million years (or less) before our solar system formed. The shock wave emanating from the exploding star may even have triggered the collapse of the solar nebula by giving gravity a little extra push. Once gravity got started, the rest of the collapse was inevitable.

8.7 Other Planetary Systems

How common are planetary systems? What are the odds of another Earth-like planet? Are habitable planets as common in the real universe as they are in science fiction? We cannot yet answer these questions definitively, but we've made great progress in our understanding of planetary systems.

As we've discussed, observations have confirmed that protoplanetary disks are common, as predicted by our theory of solar system formation. Even binary and multiple star systems, which many astronomers once thought unlikely to harbor planets, now seem to be reasonable candidates for planets. Observations show protoplanetary disks around at least some binary star systems (see Figure 8.8a), and recent calculations show that planets can safely orbit in such systems either by being close to one star or by being in a large orbit around both stars.

More importantly, rapid advances in observational technology now allow us to search for actual planets, not just for protoplanetary disks. At the beginning of the 1990s, we had no conclusive proof that planets existed around any star besides our Sun. By 2001, dozens of planetlike objects had been detected around other stars. We cannot always determine precise masses for these objects, and some may turn out to be more like small stars than the planets of our solar system. In addition, our current technology is not yet capable of detecting planets as small as Earth, so we still have no observational evidence concerning the existence of Earth-like planets. Nevertheless, we can now be quite sure that our solar system is not the only planetary system, and we know of many more planets outside our solar system than within it.

Detecting Extrasolar Planets

You might think that the easiest way to discover **extrasolar planets**, or planets around other stars, would be simply to photograph them through powerful telescopes. Unfortunately, current observational technology cannot produce such images. The primary problem arises from the fact that any light from an orbiting planet would be overwhelmed by light from the star it orbits. For example, a Sun-like star would be a *billion times* brighter than the reflected light from an Earth-like planet. Because even the best telescopes blur the light from stars at least a little, finding the small blip of planetary light amid the glare of scattered starlight would be very difficult. Astronomers are working on technologies that may overcome this problem, but for now we rely primarily on techniques that observe the star itself to find indirect evidence of planets.

To date, the most commonly used strategy involves searching for a planet by watching for the small gravitational tug it exerts on its star. In most cases, the easiest way to find this tug is to identify small Doppler shifts in the star's spectrum [Section 5.5]. An orbiting planet causes its star to alternately move slightly toward and away from us, which makes the star's spectral lines alternately shift toward the blue and toward the red (Figure 8.20). Remarkably, current techniques can measure a star's velocity to better than 3 m/s—jogging speed—which is good enough to find gravitational tugs caused by planets the size of Jupiter, and in some cases even smaller. For example, the data in Figure 8.20b reveal a planet with roughly 60% of the mass of Jupiter orbiting the star called 51 Pegasi. This planet lies so close to its star that its "year" lasts only four of our days and its temperature is probably over 1000 K.

By 2001, the Doppler-shift method had been used to identify more than 50 planets around other stars (Figure 8.21), and discoveries are coming so rapidly that the number may be far greater by the time you read this. Highlights of recent discoveries include a bona fide "planetary system" of three planets orbiting the star Upsilon Andromedae, a handful of planets as small as Saturn around various stars (see Figure 8.21), and a dust disk plus a planet orbiting the star Epsilon Eridani.

Interestingly, astronomers have recently discovered some planet-sized objects that appear to be floating freely between stars. These objects seem to

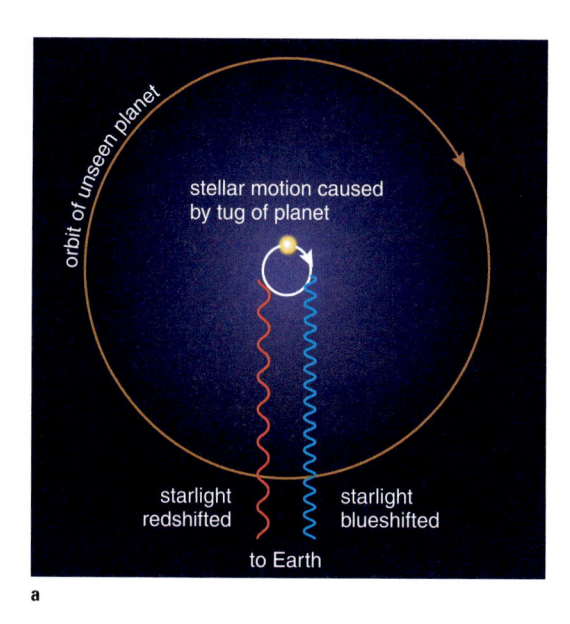

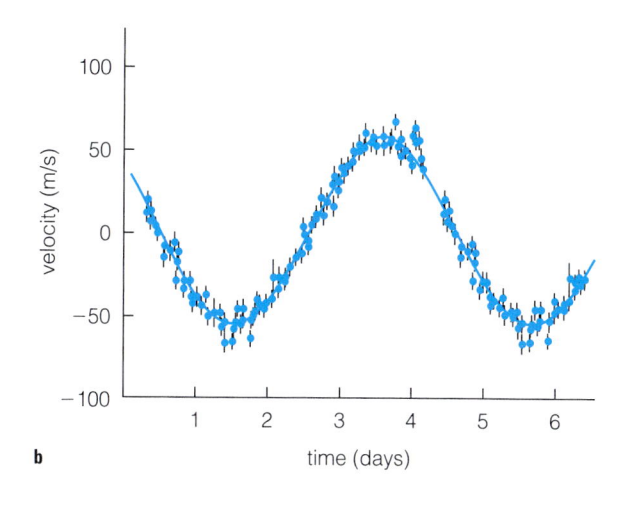

FIGURE 8.20 (a) Doppler shifts allow us to detect the slight motion of a star caused by an orbiting planet. (b) A periodic Doppler shift in the spectrum of 51 Pegasi shows the presence of a large planet with an orbital period of about 4 days. Dots are actual data points; bars through dots represent measurement uncertainty.

be Jupiter-like in mass; because they do not orbit stars, astronomers are still debating whether or not they should be called planets. The first such objects were found in 2000 in the Orion Nebula. We see these objects not in reflected sunlight (since they have no suns), but in the afterglow of the heat of their formation. These objects probably did not form in planetary systems as we have discussed throughout this chapter. Instead, they probably formed in much the same way that stars form in interstellar clouds but ended up much smaller in size.

The Nature of Extrasolar Planets

Although the Doppler-shift technique allows us to discover orbiting objects, by itself it cannot tell us the nature of the orbiting bodies. The first problem is that the amount of a Doppler shift depends both on the mass of the orbiting object and on the tilt of its orbit: Orbits seen edge-on (from our vantage point on Earth) produce the largest Doppler shifts, while face-on orbits produce no Doppler shift at all. Therefore, an observed Doppler shift could be produced either by a low-mass planet in an edge-on orbit or by a more massive planet (or even a *brown dwarf* [Section 16.2] or small star) in a more tilted orbit. The masses listed in Figure 8.21 are therefore underestimates in some cases, but probably by no more than a factor of 2. The second problem is that, even with precise mass estimates, we still would not know the size of the planet or whether the planet is more terrestrial or jovian in nature.

For the planet orbiting HD209548, the uncertainties in mass and size vanished in fall 1999 when meticulous observations by two teams of astronomers revealed a 1.7% drop in the brightness of the star (Figure 8.22). The drop occurred at exactly the time that a newly discovered planet was expected to pass on the side of the orbit nearest Earth. The planet must have passed directly on the line of sight between the Earth and the star, temporarily blocking some of the star's light from reaching the Earth. At that moment, we learned that the orbit was truly edge-on and therefore that the mass estimate was correct. We also learned that the planet's disk covers about 1.7% of the star's disk which allowed us to estimate the planet's radius and confirm that it is a jovian planet.

Lessons for Solar System Formation

The discovery of extrasolar planets presents us with opportunities to test our theory of solar system formation. Can our existing theory explain other planetary systems, or will we have to go back to the drawing board? So far, the discoveries have presented at least one significant challenge.

The new challenge arises from the fact that most of the recently discovered planets are quite different from those of our solar system. Many have highly elliptical orbits rather than the nearly circular orbits of planets in our solar system, and most of the planets are more massive than Jupiter. Most intriguing is the fact that many of these large planets lie quite close to their stars—a very different situation from

Planets Around Sun-like Stars

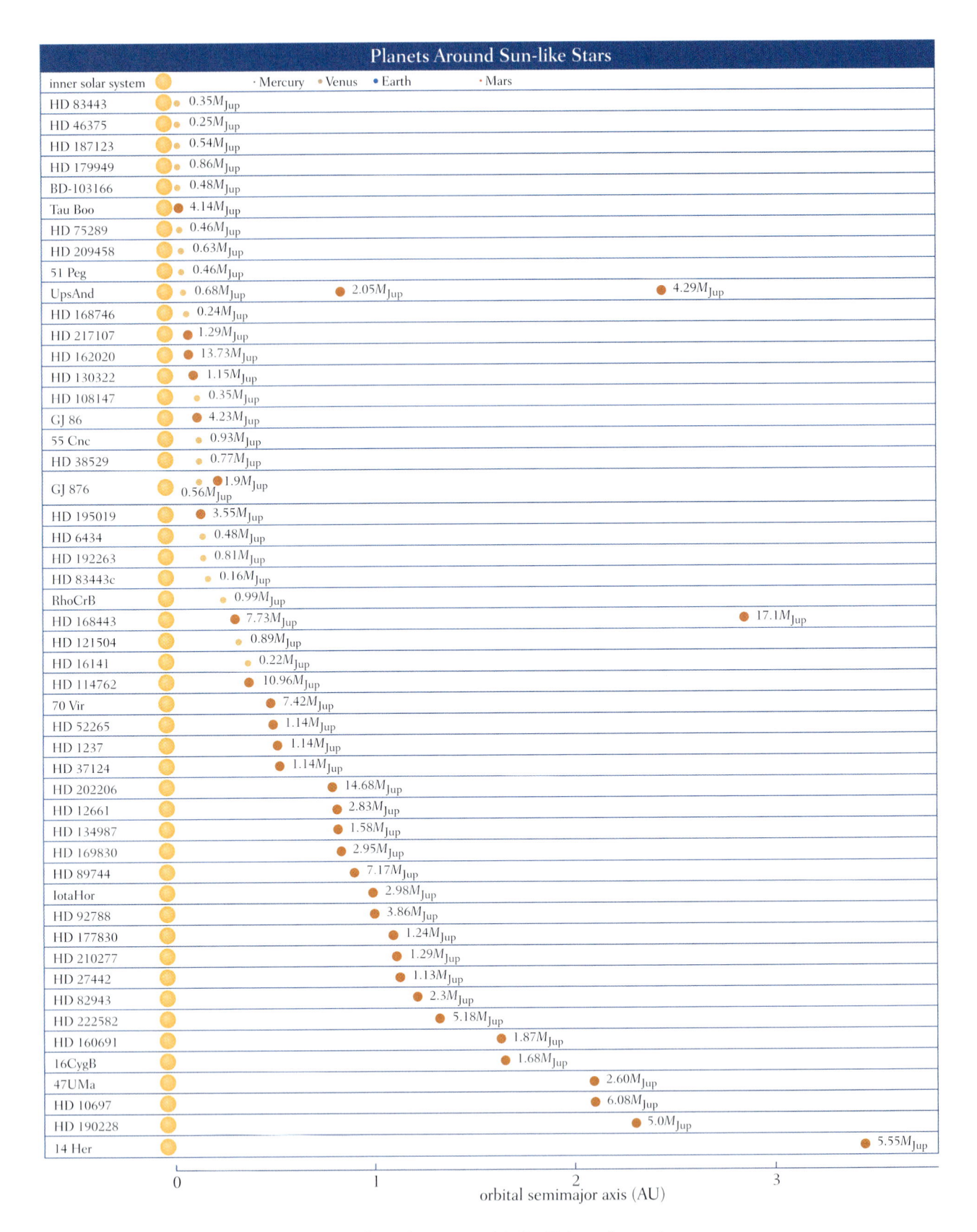

Star	Planets (mass, approx. orbital distance in AU)
inner solar system	· Mercury · Venus · Earth · Mars
HD 83443	0.35 M_{Jup}
HD 46375	0.25 M_{Jup}
HD 187123	0.54 M_{Jup}
HD 179949	0.86 M_{Jup}
BD-103166	0.48 M_{Jup}
Tau Boo	4.14 M_{Jup}
HD 75289	0.46 M_{Jup}
HD 209458	0.63 M_{Jup}
51 Peg	0.46 M_{Jup}
UpsAnd	0.68 M_{Jup}, 2.05 M_{Jup}, 4.29 M_{Jup}
HD 168746	0.24 M_{Jup}
HD 217107	1.29 M_{Jup}
HD 162020	13.73 M_{Jup}
HD 130322	1.15 M_{Jup}
HD 108147	0.35 M_{Jup}
GJ 86	4.23 M_{Jup}
55 Cnc	0.93 M_{Jup}
HD 38529	0.77 M_{Jup}
GJ 876	0.56 M_{Jup}, 1.9 M_{Jup}
HD 195019	3.55 M_{Jup}
HD 6434	0.48 M_{Jup}
HD 192263	0.81 M_{Jup}
HD 83443c	0.16 M_{Jup}
RhoCrB	0.99 M_{Jup}
HD 168443	7.73 M_{Jup}, 17.1 M_{Jup}
HD 121504	0.89 M_{Jup}
HD 16141	0.22 M_{Jup}
HD 114762	10.96 M_{Jup}
70 Vir	7.42 M_{Jup}
HD 52265	1.14 M_{Jup}
HD 1237	1.14 M_{Jup}
HD 37124	1.14 M_{Jup}
HD 202206	14.68 M_{Jup}
HD 12661	2.83 M_{Jup}
HD 134987	1.58 M_{Jup}
HD 169830	2.95 M_{Jup}
HD 89744	7.17 M_{Jup}
IotaHor	2.98 M_{Jup}
HD 92788	3.86 M_{Jup}
HD 177830	1.24 M_{Jup}
HD 210277	1.29 M_{Jup}
HD 27442	1.13 M_{Jup}
HD 82943	2.3 M_{Jup}
HD 222582	5.18 M_{Jup}
HD 160691	1.87 M_{Jup}
16CygB	1.68 M_{Jup}
47UMa	2.60 M_{Jup}
HD 10697	6.08 M_{Jup}
HD 190228	5.0 M_{Jup}
14 Her	5.55 M_{Jup}

orbital semimajor axis (AU): 0, 1, 2, 3

FIGURE 8.21 This diagram shows the orbital distances and approximate masses of the first 55 planets discovered around other stars. Most of the planets found so far are closer to their stars and more massive than the planets in our solar system. (Planet sizes are not to scale.)

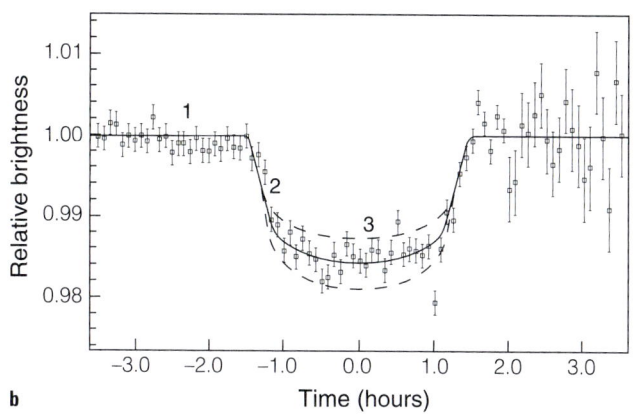

FIGURE 8.22 Careful measurements of the brightness of the star called HD209548 revealed that an orbiting planet passes directly in front of it as seen from Earth, which means that the planet's orbit must be edge-on as seen from Earth. (**a**) Artist's conception of the planet as it passes directly in front of its star as seen from Earth. (**b**) These data show the 1.7% drop in the star's brightness that proved the planet is passing in front of the star as seen from Earth.

that in our solar system, where the large planets are found in the outer solar system. Moreover, as we discussed in Section 8.4, our theory has a good explanation for why large planets should form only in the outer solar system: That is the only place where the solar nebula was cool enough for ices to condense and grow into the large seeds needed for jovian planets. Thus, the existence of dozens of systems with large, close-in planets seems to contradict our theory.

Fortunately, there are at least two reasonable ways out of this dilemma. First, recent theoretical work has shown that the gas in a forming solar system can exert drag on young jovian planets, tending to make them migrate closer to their stars. Thus, if a star's wind does not sweep the remaining nebular gas into space soon enough, the jovian planets may end up close to their stars—which could explain the large, close-in planets we've detected. In some cases, the planets may even end up being consumed by their parent stars! If this idea proves correct, the layout of our solar system suggests that the solar wind blew out the remaining gas of the solar nebula relatively early compared to other star systems.

Second, the extrasolar planets discovered to date may turn out to be relatively rare exceptions to general rules. Astronomers have succeeded in detecting planets around only a few percent of the stars they've studied so far. It's possible that planets also circle most of the other stars but have eluded detection because they are more like the planets in our own solar system. The Doppler technique makes it much easier to find large planets with close-in orbits, because such planets exert greater gravitational tugs on their stars than smaller or more distant planets. Thus, our early discoveries of extrasolar planets may be like glancing at animals in the rain forest: The jungle appears full of brightly colored parrots and frogs, but far more animals fail to catch our eye. In the same way, we may have discovered the rare exceptions but so far missed the more common cases that are like our own solar system.

As we discover more planets, we should gain a better understanding of how well our theory of solar system formation holds up. The discoveries should come rapidly, thanks to new technologies and planned space missions. In particular, NASA has placed a high priority on building ambitious orbital telescopes for the detection and study of extrasolar planets. Within a couple of decades, NASA hopes to launch high-resolution interferometers [Section 7.5] that will enable us to obtain images and spectra of planets in other solar systems. Then, at last, we will know whether solar systems like ours—and planets like Earth—are rare or common.

THE BIG PICTURE

We've seen that the nebular theory accounts for the major characteristics of our solar system. As you continue your study of the solar system, keep in mind the following "big picture" ideas.

- Close examination of the solar system unveils a wealth of patterns, trends, and groupings, leading us to the conclusion that all the planets formed from the same cloud of gas at about the same time.

- Chance events may have played a large role in determining how individual planets turned out. No one knows how different the solar system might be if it started over.

- Planet-forming processes are apparently universal. The discovery of protoplanetary disks and full-fledged planets around other stars has brought planetary science to the brink of an exciting new era.

CHAPTER 8 FORMATION OF THE SOLAR SYSTEM 221

Review Questions

1. What is *comparative planetology*? What is its basic premise? What are its primary goals?

2. Briefly summarize the observed patterns of motion in our solar system.

3. Summarize the differences between *terrestrial planets* and *jovian planets*. Why is the Moon grouped with the terrestrial planets? Where does Pluto fit in?

4. What are *asteroids*? Where are they found? What are *comets*, and how do those of the *Oort cloud* and the *Kuiper belt* differ in terms of their orbits?

5. Summarize the four challenges that any theory of the solar system must explain.

6. What is the *nebular theory*? How does it get its name? What do we mean by the *solar nebula*?

7. Describe the processes that led the solar nebula to collapse into a spinning disk. Also describe the evidence supporting the idea that our solar system had a *protosun* and a *protoplanetary disk* early in its history.

8. Distinguish among *metals*, *rocks*, *hydrogen compounds*, and *light gases*. What were their relative abundances in the solar nebula?

9. Explain how temperature differences in the solar nebula led to the condensation of different materials at different distances from the protosun. What do we mean by the *frost line* in the solar nebula?

10. What is *accretion*? What are *planetesimals*? Explain the role of electrostatic forces in getting accretion started and the role of gravity in accelerating accretion as planetesimals grow larger.

11. Briefly explain why accretion was able to produce much larger planetesimals in the outer solar system than in the inner solar system.

12. Describe the process of *nebular capture* that allowed the jovian planets to grow to large sizes. How does this process explain the satellite systems of jovian planets?

13. What is the *solar wind*? How did its strength in the past differ from its strength today? What role did it play in ending the growth of the planets?

14. Why does the Sun's slow rotation rate seem, at least at first, to contradict the nebular theory? Explain how the process of *magnetic braking* accounts for this apparent contradiction.

15. Why do we think that asteroids and comets are leftover planetesimals? Briefly describe how gravitational encounters affected their orbits.

16. What was the early bombardment? How can we use *impact craters* to estimate the age of a planetary surface?

17. What clues suggest that a moon was captured instead of forming with its planet? How do we think such captures occurred? Give a few examples of moons thought to have been captured.

18. Describe the process by which a giant impact may have led to the formation of our Moon. What other oddities of our solar system might be explained by giant impacts?

19. Summarize how the nebular theory meets the four challenges laid out in this chapter. Describe a few of the remaining unanswered questions.

20. Briefly describe the process of *radioactive dating* and how we use it to establish the age of our solar system. Why can't we determine the age of our solar system by dating Earth rocks? What kinds of rocks can we use?

21. What evidence suggests that a nearby stellar explosion may have triggered the collapse of the solar nebula?

22. Describe how we can detect *extrasolar planets* today, and summarize the current evidence concerning planets in other solar systems.

23. What have we learned about solar system formation from discoveries of extrasolar planets?

Discussion Questions

1. *Theory and Observation.* Discuss the interplay between theory and observation that has led to our modern theory of the formation of the solar system. What role does technology play in allowing us to test this theory?

2. *Random Events in Solar System History.* According to our theory of solar system formation, numerous random events, such as giant impacts, had important consequences for the way our solar system turned out. Can you think of other random events that might have caused the planets to form very differently? If there had been a different set of random events, what important properties of our solar system would have turned out the same and what ones might be different? Discuss.

3. *Lucky to Be Here?* In considering the overall process of solar system formation, do you think it was very likely for a planet like Earth to have formed? Could random events in the early history of the solar system have prevented us from even being here today? Defend your opinion. Do your opinions on these questions have any implications for your belief in the possibility of Earth-like planets around other stars?

Problems

Surprising Discoveries? For **problems 1–8**, suppose we found a solar system with the property described. (These are *not* real discoveries.) In light of our theory of solar system formation, decide whether the discovery should be considered reasonable or surprising. Explain.

1. A solar system has five terrestrial planets in its inner solar system and three jovian planets in its outer solar system.

2. A solar system has four large jovian planets in its inner solar system and seven small planets made of rock and metal in its outer solar system.

3. A solar system has ten planets that all orbit the star in approximately the same plane. However, five planets orbit in one direction (e.g., counterclockwise), while the other five orbit in the opposite direction (e.g., clockwise).

4. A solar system has 12 planets that all orbit the star in the same direction and in nearly the same plane. The 15 largest moons in this solar system orbit their planets in nearly the same direction and plane as well. However, several smaller moons have highly inclined orbits around their planets.

5. A solar system has six terrestrial planets and four jovian planets. Each of the six terrestrial planets has at least five moons, while the jovian planets have no moons at all.

6. A solar system has four Earth-size terrestrial planets. Each of the four planets has a single moon nearly identical in size to Earth's Moon.

7. A solar system has many rocky asteroids and many icy comets. However, most of the comets orbit the star in a belt much like the asteroid belt of our solar system, while the asteroids inhabit regions much like the Kuiper belt and Oort cloud of our solar system.

8. A solar system has several planets similar in composition to our jovian planets but similar in mass to our terrestrial planets.

9. *Two Classes of Planets.* Explain in terms a friend or roommate would understand why the jovian planets are lower in density than the terrestrial planets even though they all formed from the same cloud.

10. *A Cold Solar Nebula.* Suppose the entire solar nebula had cooled to 50 K before the solar wind cleared it away. How would the composition and sizes of the terrestrial planets be different from what we see today? Explain your answer in a few sentences.

11. *No Nebular Capture.* Suppose the solar wind had cleared away the solar nebula before the process of nebular capture was completed in the outer solar system. How would the jovian planets be different? Would they still have satellites? Explain your answer in a few sentences.

12. *Angular Momentum.* Suppose our solar nebula had begun with much more angular momentum than it did. Do you think planets could still have formed? Why or why not? What if the solar nebula had started with zero angular momentum? Explain your answers in one or two paragraphs.

13. *Jupiter's Action.* Suppose that, for some reason, the planet Jupiter had never formed. How do you think the distribution of asteroids and comets in our solar system would be different? How would these differences have affected Earth? Explain your answer in a few sentences.

14. *Satellite Systems as Mini–Solar Systems.* The nebula surrounding the forming jovian planets was hotter near its center, so satellites forming from the nebula could be built from different types of condensates, just as the planets were. Would satellite densities increase or decrease with increasing distance from the planet? According to Appendix C, which jovian planet satellite system best matches this prediction? Explain your answers in a few sentences.

*15. *Dating Lunar Rocks.* Suppose you are analyzing Moon rocks that contain small amounts of uranium-238, which decays into lead with a half-life of 4.5 billion years.

   a. In one rock from the "lunar highlands," you determine that 55% of the original uranium-238 remains; the other 45% has decayed into lead. How old is the rock?

   b. In a rock from the "lunar maria," you find that 63% of its original uranium-238 remains; the other 37% has decayed into lead. Is this rock older or younger than the highlands rock? By how much? (The significance of these ages will become clear in the next chapter.)

*16. *Radioactive Dating with Carbon-14.* The half-life of carbon-14 is about 5,700 years.

   a. You find a piece of cloth painted with organic dyes. By analyzing the dye in the cloth, you find that only 77% of the carbon-14 originally in the dye remains. When was the cloth painted?

   b. A well-preserved piece of wood found at an archaeological site has 6.2% of the carbon-14 that it must have had when it was alive. Estimate when the wood was cut.

   c. Is carbon-14 useful for establishing the age of the Earth? Why or why not?

*17. *Unusual Meteorites.* Some unusual meteorites thought to be chips from Mars [Section 12.3] contain small amounts of the radioactive element thorium-232 and its decay product lead-208. The half-life for this decay process is 14 billion years. A detailed analysis shows that 94% of the original thorium remains. How old are these meteorites? Compare your answer to the age of the solar system and comment briefly.

*18. *51 Pegasi.* The star 51 Pegasi has about the same mass as our Sun, and the planet discovered around it has an orbital period of 4.23 days. The mass of the planet is estimated to be 0.6 times the mass of Jupiter.

   a. Use Kepler's third law to find the planet's average distance (semimajor axis) from its star. (*Hint:* Because the mass of 51 Pegasi is about the same as the mass of our Sun, you can use Kepler's third law in its original form, $p^2 = a^3$ [see Chapter 5]; be sure to convert the period into years.)

   b. Briefly explain why, according to our theory of solar system formation, it is surprising to find a planet the size of the 51 Pegasi planet orbiting at this distance.

   c. Hypothesize as to how the 51 Pegasi planet might have come to exist. Explain your hypothesis in a few sentences.

*19. *Transiting Planets.* The star HD209548 is about the same size as our Sun, and the planet discovered around it blocks 1.7% of the star's area when it passes in front of the star. (Appendix E contains useful data for this problem.)

   a. How large a planet is required to block the observed fraction of the star's area? Give your answer in kilometers. (*Hint:* Remember that the brightness drop tells us that the planet blocked 1.7% of the star's visible *area.*)

   b. The mass of the planet is estimated to be 0.6 times the mass of Jupiter. What is the density of the planet?

   c. Compare the density of this planet to that of planets in our solar system, and comment on whether this density makes the planet terrestrial or jovian in nature.

Web Projects

Find useful links for Web projects on the text Web site.

1. *New Planets.* Find up-to-date information about discoveries of planets in other solar systems. Create a personal "Web journal," complete with pictures from the Web, describing at least three recent discoveries of new planets. For each case, write one or two paragraphs in your journal describing the method used in the discovery, comparing the discovered planet to the planets of our own solar system, and summarizing whether the discovery poses any new challenges to our theory of solar system formation.

2. *Missions to Search for Planets.* Learn about one proposed space mission to search for planets around other stars, such as Kepler, SIM, or Terrestrial Planet Finder. Write a short report about the plans for the mission and its current status.

Nothing is rich but the inexhaustible wealth of nature. She shows us only surfaces, but she is a million fathoms deep.

RALPH WALDO EMERSON

CHAPTER 9
Planetary Geology
Earth and the Other Terrestrial Worlds

Think back to a time when you went for a walk or a drive through the open countryside. Did you see a valley? If so, the creek or river that carved it probably still flows at the bottom. Were the valley walls composed of lava, telling of an earlier time when molten rock gushed from volcanoes? Or were they made of neatly layered sedimentary rocks that formed when ancient seas covered the area? Were the rock layers tilted? That's a sign that movements of the Earth's crust have pushed the rocks around.

Since the beginning of the space age, we've been able to make similar observations and ask similar questions about geological features on other worlds. As a result, we can now make detailed geological comparisons among the different planets. We've learned that, even though all the terrestrial worlds are similar in composition and formed at about the same time, their geological histories have differed because of a few basic properties of each world. Thus, by comparing the geologies of the terrestrial worlds, we can learn much more about the Earth.

FIGURE 9.1 Global views of the terrestrial planets to scale and representative surface close-ups, each a few hundred kilometers across. The global view of Venus shows its surface without its atmosphere, based on radar data from the Magellan spacecraft; all other images are photos (or composite photos) taken from spacecraft.

9.1 Comparative Planetary Geology

The five terrestrial worlds—Mercury, Venus, Earth, the Moon, and Mars—share a common ancestry in their birth from the solar nebula, but their present-day surfaces show vast differences (Figure 9.1). Mercury and the Moon are battered worlds densely covered by craters except in areas that appear to be volcanic plains. Venus has volcanic plains that appear to have been twisted and torn by internal stresses, leaving bizarre bulges and odd volcanoes that dot the surface. Mars, despite its middling size, has the solar system's largest volcanoes and is the only planet other than Earth where running water played a major role in shaping the surface. Earth has surface features similar to all those on other terrestrial worlds and more—including a unique layer of living organisms that covers almost the entire surface of the planet. Our purpose in this chapter is to understand how the profound differences among the terrestrial surfaces came to be.

The study of surface features and the processes that create them is called **geology**. The root *geo* means "Earth," and *geology* originally referred only to the study of the Earth. Today, however, we speak of *planetary geology*, the extension of geology to include all the solid bodies in the solar system, whether rocky or icy.

Geology is relatively easy to study on Earth, where we can examine the surface in great detail. People have scrambled over much of the Earth's surface, identifying rock types and mapping geological features. We've pierced the Earth's surface with 10-kilometer-deep drill shafts and learned about the deeper interior by studying *seismic waves* generated by earthquakes. Thanks to these efforts, today we have a fairly good understanding of the history of our planet's surface and the processes that shaped it.

Studying the geology of other planets is much more challenging. The Moon is the only alien world from which we've collected rocks, some gathered by the Apollo astronauts and others by Russian robotic landers in the 1970s. We can also study the dozen or so meteorites from Mars that have landed on Earth. Aside from these few rocks, we have little more than images taken from spacecraft from which to decode the history of the other planets over the past 4.6 billion years. It's like studying people's facial expressions to understand what they're feeling inside—and what their childhood was like! Fortunately, this decoding is considerably easier for planets than for people.

Spacecraft have visited and photographed all the terrestrial worlds. Patterns of sunlight and shadow on photographs taken from orbit reveal features such as cliffs, craters, and mountains. Besides ordinary, visible-light photographs, we also have spacecraft images

Earth

Earth's Moon

Mars

taken with infrared and ultraviolet cameras, spectroscopic data, and in some cases three-dimensional data compiled with the aid of radar. Finally, we have close-up photographs of selected locations on all the terrestrial worlds but Mercury, taken by spacecraft that have landed on their surfaces (Figure 9.2). The result is that we now understand the geology of the terrestrial worlds well enough to make detailed and meaningful comparisons among them.

Comparative planetary geology (one part of comparative planetology) hinges on the principle that a planet's surface features can be traced back to its fundamental properties. For example, we saw in Chapter 8 that a planet's composition depends primarily on how far from the Sun it formed. In the rest of this chapter, we will look for relationships between surface features and the fundamental properties and processes of the planets. But because surface geology depends largely on a planet's interior, we must first look inside the terrestrial worlds.

9.2 Inside the Terrestrial Worlds

Planets are all approximately spherical and quite smooth relative to their size. For example, compared to the Earth's radius of 6,378 km, the tallest mountains could be represented by grains of sand on a typical globe.

TIME OUT TO THINK *Find a globe of the Earth that shows mountains in raised relief (i.e., you can feel the mountains as bumps on the globe). Are the heights of the mountains correctly scaled relative to the Earth's size? Explain.*

It's easy to understand why the gaseous jovian planets ended up nearly spherical: Gravity always acts to pull material together, and a sphere is the most compact shape possible. It's a bit subtler when we deal with the rocky terrestrial worlds, because rock can resist the pull of gravity. Indeed, many small moons and asteroids are "potato-shaped" precisely because their weak gravity is unable to overcome the rigidity of their rocky material. Before we can understand why the terrestrial worlds ended up smooth and spherical and why they all have similar internal structures, we must first investigate the behavior of rock.

Solid as a Rock?

We often think of rocks as the very definition of strength, but rocks are not always as solid as they may seem. Most rocks are a hodgepodge of different minerals, each characterized by its chemical compo-

CHAPTER 9 PLANETARY GEOLOGY: EARTH AND THE OTHER TERRESTRIAL WORLDS 227

FIGURE 9.2 Surface views of the terrestrial worlds. No spacecraft have landed on Mercury, so an artist's conception is shown; all other images are photos. The Venus photo shows only a small patch of surface at the base of the Russian Venera lander, visible in the lower right.

a Mercury

b Venus

c Earth

d Moon

e Mars

228   PART III LEARNING FROM OTHER WORLDS

sition. For example, *granite* is a common rock composed of crystals of quartz (SiO_2), feldspar ($NaAlSi_3O_8$ is one of several varieties), and other minerals. Rocks are solid because of electrical bonds between the mineral molecules. But when these bonds are subjected to sustained stress over millions or billions of years, they can break and re-form, gradually allowing rocky material to deform and flow. In fact, the long-term behavior of rock is much like that of Silly Putty™, which stretches when you pull it slowly but breaks if you pull it sharply (Figure 9.3).

The strength of a rock depends on its composition, its temperature, and the surrounding pressure. Rocks of different composition are held together by molecular bonds of different strength. Even two very similar rocks may differ in strength if one contains traces of water: The water can act as a lubricant to reduce the rock's strength. Higher temperatures also make rocks weaker: Just as Silly Putty becomes more pliable if you heat it, warm rocks are weaker and more deformable than cooler rocks of the same type. Pressure can increase rock strength: The very high pressures found deep in planetary interiors can compress rocks so much that they stay solid even at temperatures high enough to melt them under ordinary conditions. (Most substances are denser as solids than as liquids, so the compression caused by high pressure tends to make things solid. Water is a rare exception, becoming *less* dense when it freezes—which is why ice floats in water.)

At high enough temperatures (usually over 1,000 K), some or all of the minerals in a rock may melt, making it *molten* (or partially molten, if only some minerals have melted). Just as different types of solid rock are more deformable than others, different types of rock behave differently when molten. Some molten rocks are runny like water, while others flow slowly like honey or molasses. We describe the "thickness" of a liquid with the technical term **viscosity**: Water has a low viscosity because it flows easily, while honey and molasses have much higher viscosities because they flow slowly. As we'll soon see, molten rock viscosities are very important to understanding volcanoes.

Terrestrial-World Layering

The rocky terrestrial worlds became spherical because of rock's ability to flow. Even when rock is solid, the strength of gravity in a moderately large world (over about 500 km in diameter) can overcome the strength of solid rock and make the world spherical in less than a billion years. If the world is molten, as was the case with the terrestrial worlds in their early histories, it can become spherical much more quickly. (For similar reasons, Earth's ocean surface is smoother and more spherical than its rocky surface.)

FIGURE 9.3 Silly Putty stretches when pulled slowly but breaks cleanly when pulled rapidly. Rock behaves just the same, but on a longer time scale.

Gravity also gave the terrestrial worlds layered interiors. You know that in a mixture of oil and water, the less dense oil rises to the top while the denser water sinks to the bottom. This process in which gravity separates materials by density is called **differentiation** (because it results in layers made of *different* materials). The terrestrial worlds underwent differentiation early in their histories, when they were molten throughout their interiors. As a result, each terrestrial world now has three interior layers of differing composition (Figure 9.4). Dense metals such as iron and nickel sank through molten rocky material to form a **core**. Rocky material composed of *silicates*—minerals that contain silicon and oxygen (and other elements)—came to rest above the core, forming a thick **mantle**. A low-density "scum" of rocks composed of the lightest silicates rose to form a **crust**.

The terms *core, mantle,* and *crust* are defined by the composition within each layer. But characterizing the layers by *rock strength* rather than composition

rocky crust
(lower density)

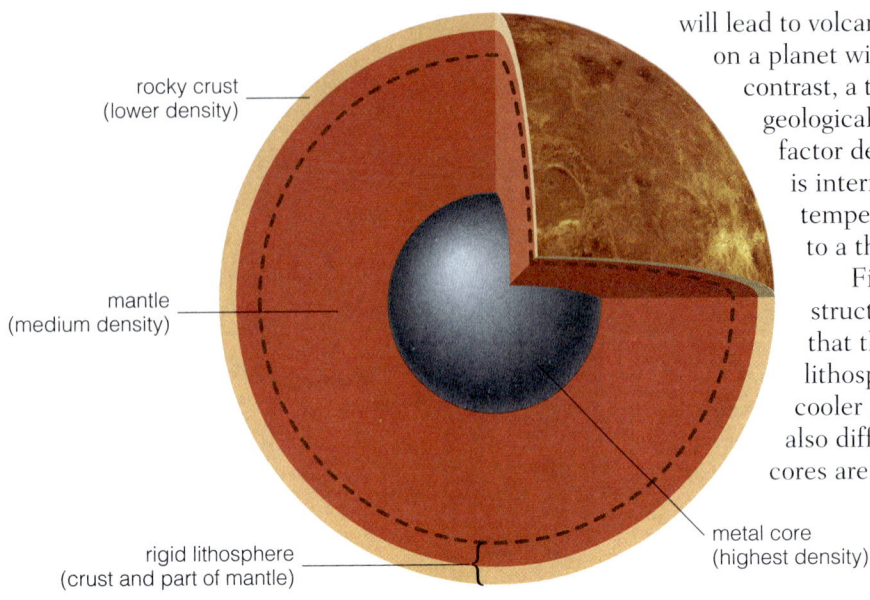

mantle
(medium density)

rigid lithosphere
(crust and part of mantle)

metal core
(highest density)

FIGURE 9.4 Interior structure of a generic terrestrial world. The thicknesses of the crust and lithosphere are enlarged for clarity.

turns out to be more useful for understanding geological activity. From this point of view, we speak of an outer layer of relatively rigid rock called the **lithosphere** (*lithos* means "stone" in Greek). The lithosphere generally encompasses the crust and the uppermost portion of the mantle. Beneath the lithosphere, the higher temperatures allow rock to deform and flow much more easily. Thus, the lithosphere is essentially a layer of rigid rock that "floats" on the softer rock below. (The underlying layer of softer rock is called the *asthenosphere* (literally, the "weak-o-sphere"). Although rock in the asthenosphere deforms more easily, it is not actually molten.) On the Earth, the lithosphere is broken into *plates* that move as the underlying rock flows gradually, creating the phenomenon called *continental drift* [Section 13.2].

The thickness of the lithosphere plays an important role in surface geology. Interior heat and motion will lead to volcanoes and other geological activity on a planet with a relatively thin lithosphere. In contrast, a thick, strong lithosphere inhibits geological activity. The most important factor determining lithospheric thickness is internal temperature: Higher internal temperature makes rocks softer, leading to a thinner rigid lithosphere.

Figure 9.5 compares the interior structures of the terrestrial worlds. Note that the smaller worlds have thicker lithospheres, indicating that they have cooler interiors. The cores and mantles also differ in their relative sizes: Because cores are made from metals and mantles are made from rock, their relative sizes depend on the proportions of metal and rock in a planet. These proportions were determined primarily by condensation in the solar nebula, but they may also have been affected by giant impacts [Section 8.5]. Planetary cores may be partially or completely molten, depending on the particular combination of internal temperature and pressure.

You may be wondering how we know what the interiors of the terrestrial worlds look like. After all, even on Earth our deepest drill shafts barely prick the lithosphere. Fortunately, several techniques allow us to study planetary interiors without actually having to drill into them. First, we can measure a planet's size from telescopic or spacecraft images, and its mass by applying Newton's version of Kepler's third law to the orbital properties of a natural or artificial satellite [Section 5.4]. Together, these measurements tell us the average density of the planet, which gives clues about its composition. Second, the strength of gravity on a planet may differ slightly from place to place depending on the underlying interior structure, so we can learn about that structure through gravity measurements made from orbiting satellites. Third, a planet's magnetic field is generated in its interior, so measurements of the magnetic-field strength also

FIGURE 9.5 Interior structures of the terrestrial worlds, in order of size. The crust and lithosphere are enlarged on Venus and Earth for greater visibility.

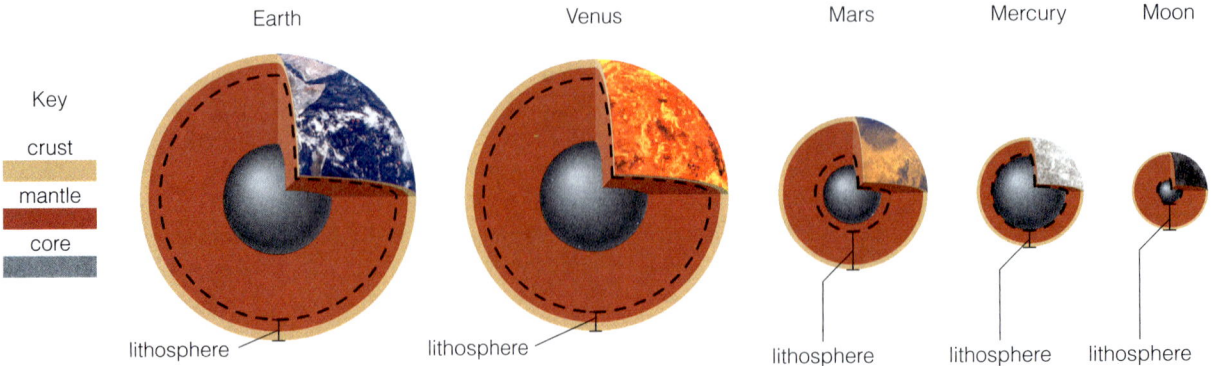

Earth   Venus   Mars   Mercury   Moon

Key
crust
mantle
core

lithosphere   lithosphere   lithosphere   lithosphere   lithosphere

230   PART III   LEARNING FROM OTHER WORLDS

THINKING ABOUT . . .

Seismic Waves

Earthquakes (or "planet-quakes") generate vibrations, or **seismic waves**, that propagate through a planet's interior. The word *seismic* comes from the Greek word for "shake," and seismic waves shake the surface when they arrive—even if they've traveled all the way through the interior from the location where a quake originated.

On Earth, *seismographs* located in many places around the world record the shaking of the surface as seismic waves arrive after an earthquake. By comparing the recordings from many different places, we can reconstruct the paths that the seismic waves took through the Earth's interior. Seismic waves generally travel through the Earth at speeds of thousands of kilometers per hour, but the precise speeds and directions of the waves depend on the composition, density, pressure, temperature, and phase (solid or liquid) of the material they pass through. Thus, we can deduce all of these internal conditions through careful study of seismic waves.

Seismic waves come in two basic types, and the difference between the two types helps us study Earth's interior. We can demonstrate the two wave types by imagining that the bonds holding rocks together are like Slinkys (Figure 9.6a). Pushing and pulling on one end of a Slinky generates a wave in which the Slinky is bunched up in some places and stretched out in others. Waves like this in rock are called P waves. (The *P* stands for *primary*, because these waves travel fastest and therefore are the first to arrive, but is more easily remembered as *pressure* or *pushing*.) P waves are essentially a type of sound wave and therefore can travel through liquid or gas as well as through solid rock. In the second type of wave, shaking one end of the Slinky side-to-side or up and down sends a wave down its length. Such waves are called S waves in rock, and they travel through rock a bit more slowly than P waves. (The *S* stands for *secondary* but is easier to remember as *shear* or *shake*.) Unlike P waves, S waves cannot travel through liquids (or gases): Molecules in liquid are not connected by strong bonds like those that hold Slinkys together, so they cannot exert sideways forces on their neighbors. Thus, the fact that S waves do not reach the side of the world opposite an earthquake tells us that the Earth's interior must have a liquid layer in the outer region of the core (Figure 9.6b).

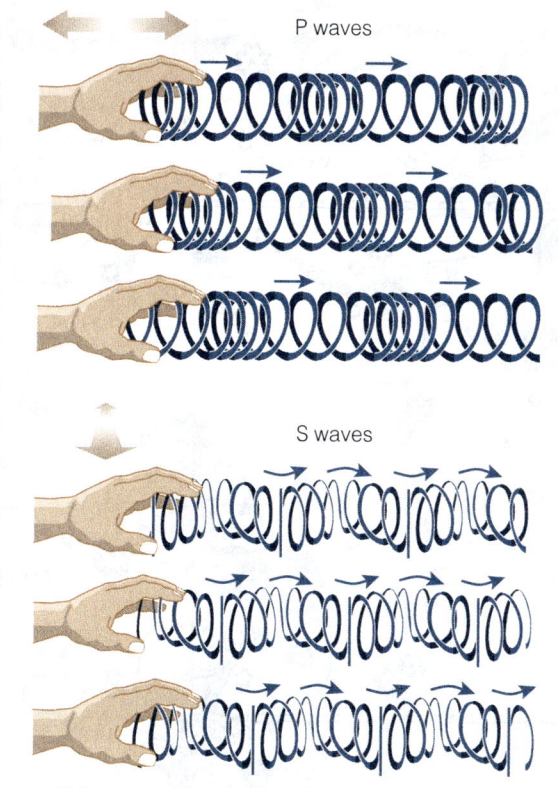

a Slinky examples demonstrating P and S waves.

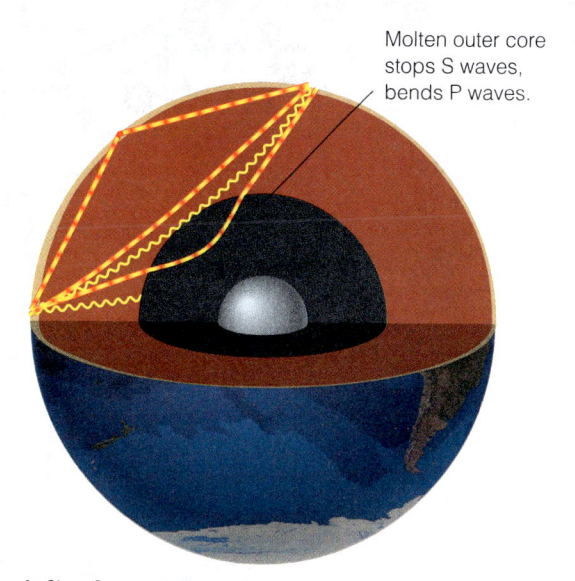

b Since S waves do not reach the side of the Earth opposite the earthquake, we infer that part of Earth's core is molten.

FIGURE 9.6 Seismic waves provide a probe of a planet's interior.

---

provide information about interior structure. Fourth, lava brought up from deep inside a planet can tell us something about the interior composition. Finally, planetary vibrations (*seismic waves* from "planet-quakes") tell us about interior structure in much the same way that shaking a present offers a few hints on what's inside. These diverse types of measurements are combined with computer models based on known laws of physics to calculate a planet's interior structure. Our measurements are much more extensive for Earth than for any other world, so we know the most

CHAPTER 9 PLANETARY GEOLOGY: EARTH AND THE OTHER TERRESTRIAL WORLDS 231 about Earth's interior. But scientists have applied some of the same techniques in more limited ways to other worlds, giving us confidence that the general pictures shown in Figure 9.5 are correct.

9.3 How Interiors Work

Geological activity depends on a planet's interior structure, and interior structure depends on internal temperature. In this section, we investigate how heat (thermal energy) is deposited in the planetary interior and how it leaks outward over time.

How Interiors Get Hot

It's often said that the Sun is the ultimate source of all our energy. On the surface, this statement is true: Sunlight bathes the Earth's surface with a power of roughly 1,000 watts per square meter (in the daytime), while the heat leaking outward from the Earth's interior contributes only about 0.05 watt per square meter. Thus, the Earth's surface receives more than 10,000 times as much energy from the Sun as from its own interior. In general, the energy a world receives from sunlight depends on its distance from the Sun. This energy, along with properties of the planet's atmosphere, determines the surface temperature [Section 10.2].

In contrast, the Sun contributes very little to a planet's *internal* temperature. In fact, the Sun's only role is in helping to determine the starting point for the interior temperature structure: A terrestrial world is coolest at its surface and gets warmer as you go into its interior. The three most important sources of internal heat are *accretion, differentiation,* and *radioactivity* (Figure 9.7). (A fourth process, called *tidal heating,* is not important for the terrestrial worlds but is for some of the jovian moons, particularly Io [Section 11.5].)

Accretion was the earliest major source of internal heat for the terrestrial worlds. The many violent impacts that occurred during the latter stages of accretion deposited so much energy that planet-wide melting occurred. Imagine the entire Earth engulfed in molten rock! This melting enabled the process of differentiation, in which the densest materials (metals) settled to the planet's core, while lighter rocks rose to the surface.

Both accretion and differentiation yield heat by converting *gravitational potential energy* into thermal energy [Section 4.2]. An object that hits an accreting world starts with a lot of gravitational potential energy when it is far away. This gravitational potential energy is converted to kinetic energy as gravity makes the object accelerate toward the planet's surface. Upon impact, the kinetic energy is converted to sound

accretion
- Gravitational potential energy is converted into kinetic energy.
- Kinetic energy is converted into thermal energy.

differentiation
- Dense materials fall to the core, converting gravitational potential energy into thermal energy.
- Light materials rise to the surface.

radioactivity
- Nuclear energy is converted into thermal energy.

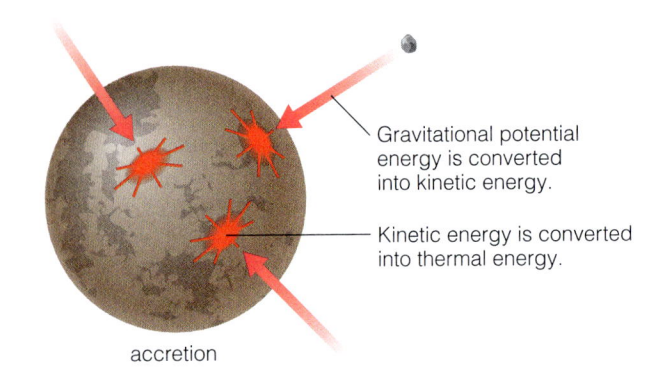

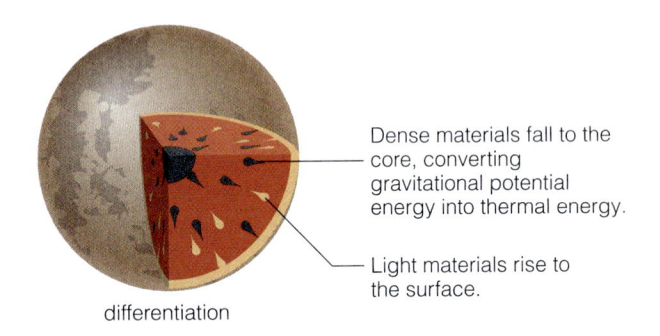

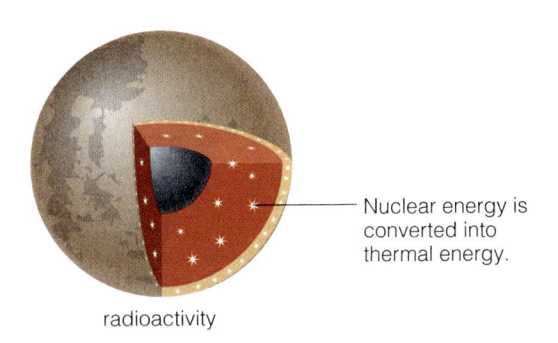

FIGURE 9.7 The three main internal energy sources in terrestrial planets. Only radioactivity is a major heat source today.

and heat, thus adding to the thermal energy of the planet. Differentiation converts gravitational potential energy to thermal energy just like a brick falling to the bottom of a swimming pool. The brick loses gravitational potential energy as it falls, and this energy ultimately appears as thermal energy in the water.

The third internal heat source is the decay of radioactive elements such as uranium, potassium, thorium, and others. When radioactive nuclei decay, subatomic particles fly off at high speeds, colliding with neighboring atoms and heating them. In essence, this transfers some of the mass-energy of the radioactive element ($E = mc^2$ [Section 4.2]) to the thermal energy of the planetary interior. Most radioactive elements are locked in minerals that make up the

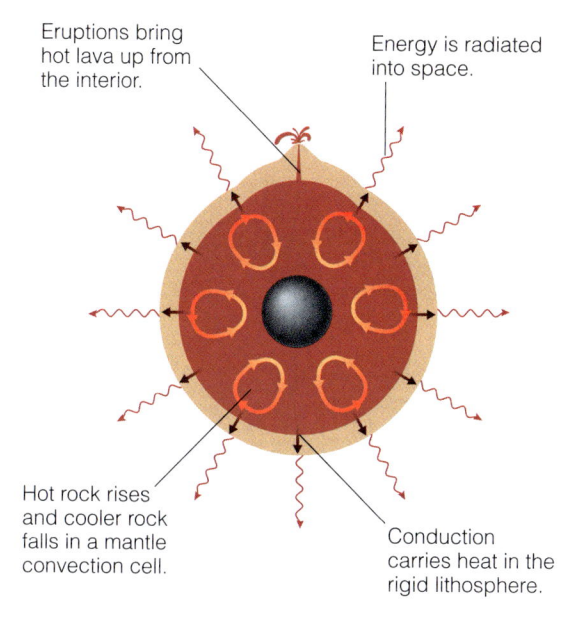

FIGURE 9.8 Heat escapes a planet's interior through conduction, convection, and eruption.

- Eruptions bring hot lava up from the interior.
- Energy is radiated into space.
- Hot rock rises and cooler rock falls in a mantle convection cell.
- Conduction carries heat in the rigid lithosphere.

Common Misconceptions: Pressure and Temperature

You might think that Earth's interior is hot just because the pressures are high. After all, if we compress a gas from low pressure to high pressure, it heats up. But the same is not necessarily true of rock. High pressure hardly compresses rock, so the compression causes little increase in temperature. Thus, while high pressures and temperatures sometimes go together in planets, they don't have to. In fact, after all the radioactive elements decay (billions of years from now), Earth's deep interior will become quite cool even though the pressure will be the same as it is today. The temperatures inside Earth and the other planets can remain high only if there is an internal source of heat, such as accretion, differentiation, or radioactivity.

crust and mantle, but some end up in the core as well. Note that accretion and differentiation deposited heat in planetary interiors billions of years ago, but radioactivity continues to heat the terrestrial planets to this very day.

How Interiors Cool Off

Some of a planet's internal heat, or thermal energy, is always escaping. Heat flows outward from the hot interior toward the cooler surface through three main processes: *conduction, convection,* and *eruption* (Figure 9.8). **Conduction** is the process that makes heat flow from your hand to a glass of ice water. It occurs because the microscopic jiggling of molecules is more energetic (faster) at higher temperatures. Some of the thermal energy of this motion is transferred from a warm rock to its cooler neighbors. Conduction is the main process by which heat flows upward through the lithosphere.

Some planetary interiors also lose heat through **convection**, in which hot material expands and rises, while cooler material contracts and falls. Convection can occur any time a substance is strongly heated from underneath; you can see convection whenever you heat a pot of soup on the stove, and you're probably also familiar with it in the context of weather, in which warm air rises, while cool air falls in our atmosphere. A planet's hot interior also convects, but much more slowly. The rising hot material transfers heat from deep in the planetary interior toward the surface. When the material fully cools, it begins to descend back down until the heating from below heats it enough to make it rise upward again. Convection is an ongoing process, and each small region of rising and falling material is called a **convection cell**.

The third major process by which heat can escape a planet's interior is volcanic **eruption**. An eruption directly transfers heat outward by depositing hot lava on the surface.

Regardless of whether heat reaches the surface through the slow leakage of convection and conduction or through the sudden burst of an eruption, this heat eventually radiates away into space. Recall that all objects emit *thermal radiation* characteristic of their temperatures [Section 6.4]. Because of their relatively low surface temperatures, planets radiate almost entirely in the infrared portion of the spectrum.

The interiors of the terrestrial planets are slowly cooling as their heat escapes. By now, 4.6 billion years after their formation, much of the original heat from accretion and differentiation has leaked away. Most of the heat today comes from radioactive decay, and even the rate of radioactive decay declines as a planet ages (because a particular nucleus can decay only once). Meanwhile, interior cooling gradually thickens the lithosphere and leaves any molten rock lying deeper inside.

Ultimately, a planet's size is the single most important factor in determining how long its interior remains hot: Larger planets stay hot longer, just as larger baked potatoes stay hot longer than small ones. You can see why size is the critical factor by picturing a large planet as a smaller planet wrapped in extra layers of rock. The extra rock acts as insulation, so heat from the center takes longer to reach the surface on the larger planet than it would on the smaller one. Because of their relatively large sizes, Earth and Venus have scarcely cooled off since their births and therefore have thin lithospheres and substantial geological activity. The smallest terrestrial worlds—Mercury and the Moon—have mostly cooled off and therefore have very thick lithospheres that essentially make them geologically "dead." Mars, intermediate in size, has cooled significantly, resulting in an intermediate interior temperature.

TIME OUT TO THINK *Give an example from everyday life of a small object cooling off faster than a large one and of a smaller object warming up more quickly than a large one. How do these examples relate to the issue of geological activity on the terrestrial worlds?*

Planetary Cores and Magnetic Fields

Interior structure also determines whether a planet has a **magnetic field** through which it can influence charged particles or magnetic materials. You are probably familiar with the general pattern of a bar magnet's magnetic field, which we can see by placing it among small iron filings (Figure 9.9a). A planet's magnetic field is generated by a process more similar to that of an *electromagnet,* in which the magnetic field arises as a battery forces charged particles (electrons) to move along a coiled wire (Figure 9.9b). Planets do not have batteries, but they may have charged particles in motion in their metallic cores (Figure 9.9c). Deep within an electrically conducting core, molten metals rise and fall in convection cells. At the same time, the molten material spins with the planet's rotation. If this combination of convection and rotation is strong enough, it moves electrons in the same way they move in an electromagnet. The

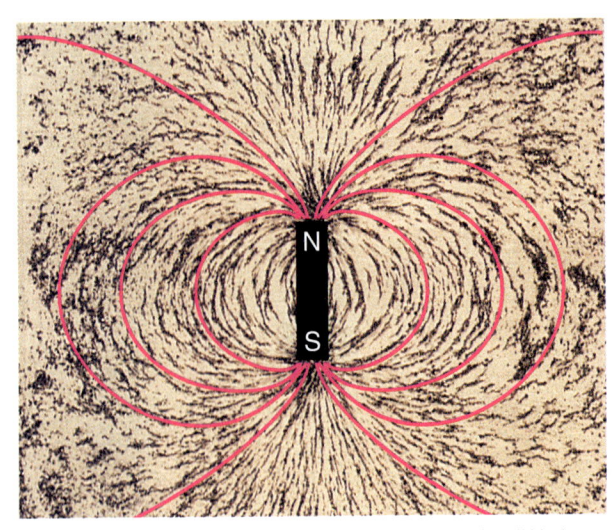

a This photo shows how a bar magnet influences iron filings (small black specks) around it. The *magnetic field lines* (red) represent this influence graphically.

FIGURE 9.9 Sources of magnetic fields.

Mathematical Insight 9.1   The Surface Area–to–Volume Ratio

We can see why large planets cool more slowly than smaller ones by thinking about their relative surface areas and volumes. Consider the Earth and the Moon. Both started out hot inside but continually radiate heat away from their surfaces to space. As heat escapes from the surface, more heat flows upward from the interior to replace it. This process will continue until the interior is no hotter than the surface. Because all the heat escapes from the surface, the key factor in a world's heat-loss rate is its *surface area*. The surface area of a sphere is given by the formula $4\pi \times (\text{radius})^2$, so the ratio of Earth's surface area to that of the Moon is:

$$\frac{\text{Earth surface area}}{\text{Moon surface area}} = \frac{4\pi \times (r_{\text{Earth}})^2}{4\pi \times (r_{\text{Moon}})^2}$$

$$= \left(\frac{r_{\text{Earth}}}{r_{\text{Moon}}}\right)^2 = \left(\frac{6{,}378 \text{ km}}{1{,}738 \text{ km}}\right)^2 \approx 13$$

Thus, at a given temperature, Earth has about 13 times more area from which to radiate its heat away.

However, the total amount of heat deposited inside a world by radioactivity roughly depends on its mass or, if we assume that we're dealing with worlds of similar density, its volume. The formula for the volume of a sphere is $\frac{4}{3}\pi \times (\text{radius})^3$, so the volume ratio for Earth and the Moon is:

$$\frac{\text{Earth volume}}{\text{Moon volume}} = \frac{\frac{4}{3}\pi \times (r_{\text{Earth}})^3}{\frac{4}{3}\pi \times (r_{\text{Moon}})^3}$$

$$= \left(\frac{r_{\text{Earth}}}{r_{\text{Moon}}}\right)^3 = \left(\frac{6{,}378 \text{ km}}{1{,}738 \text{ km}}\right)^3 \approx 50$$

Thus, Earth's interior contains some 50 times more heat than the Moon's interior. This larger amount of internal heat more than offsets the larger radiating area. That is why the Earth is still hot inside while the Moon has mostly cooled off.

The general idea that larger objects cool more slowly than smaller ones is described by the mathematical idea of the *surface area–to–volume ratio*. As the Earth–Moon example shows, larger objects have relatively smaller surface areas compared to their volumes. Thus, they lose internal heat more slowly and also take longer to be heated from the outside. This principle explains many everyday phenomena. For example, crushing a cube of ice into smaller pieces increases the total amount of surface area, while the total volume of ice remains the same. Thus, the crushed ice will cool a drink more quickly than would the original ice cube.

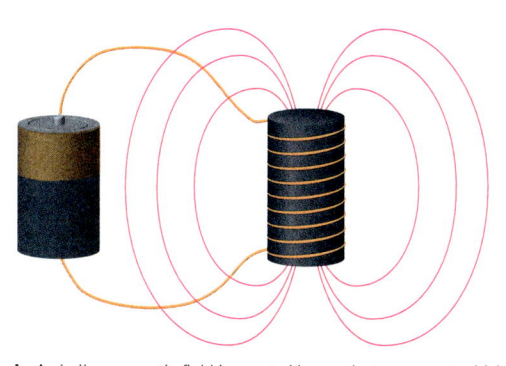

b A similar magnetic field is created by an electromagnet, which is essentially just a coiled wire attached to a battery. The field is created by the battery-forced motion of charged particles (electrons) along the wire.

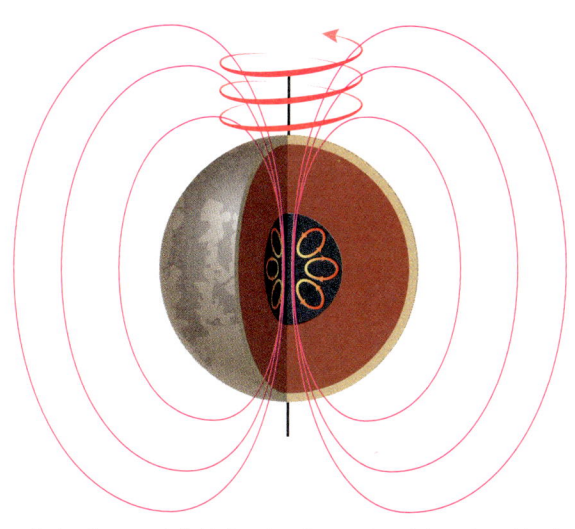

c A planet's magnetic field also arises from motion of charged particles. For a terrestrial planet, the charged particles are in a molten metallic core, and their motion arises from the planet's rotation and interior convection.

---

result is a planetary magnetic field. Similar combinations of convection and rotation of conducting materials (such as metallic hydrogen or ionized plasma) give rise to magnetic fields in the jovian planets and in stars.

This simple analysis explains why Earth has the strongest magnetic field of the terrestrial worlds: It is the only one that has both a partially molten metallic core (which can convect) and reasonably rapid rotation. The Moon may lack a metallic core; even if it has one, it has certainly solidified and ceased convecting. In either case, it generates no magnetic field. Mars also has virtually no magnetic field, probably because of similar core solidification. In contrast, Venus probably has a molten metal layer in its core, but either its convection or its 243-day rotation period is too slow to generate a magnetic field. Mercury remains an enigma: It possesses a measurable magnetic field despite its small size and slow, 59-day rotation. The solution to this enigma may lie in Mercury's high density, which tells us that it has a huge metal core—a larger core than that of Mars, even though it is a smaller planet overall. Perhaps Mercury's core is still partly molten and convecting.

Although convection and rotation of electrically conducting material basically explain why planetary magnetic fields exist, many details remain mysterious. For example, we don't know why the Earth's magnetic axis intersects the surface in northern Canada, 11° away from the rotation axis at the North Pole. Nor do we know why Earth's magnetic field varies in strength over time. Even more mysterious is the fact, discovered from geological evidence, that Earth's entire magnetic field completely reverses its orientation (magnetic north becomes magnetic south, and vice versa) every half-million years or so.

Magnetic fields can be very useful. On Earth, they help us navigate. Their presence or absence and their strength on any planet provide important information about the planetary interior. But aside from this utility, magnetic fields have virtually no effect on the structure of a planet itself. They can, however, form a protective *magnetosphere* that surrounds a planet, with potentially profound effects on a planet's atmosphere and any inhabitants [Section 10.3].

9.4 Shaping Planetary Surfaces

When we look around the Earth, we find an apparently endless variety of geological surface features. The diversity increases when we survey the other planets. But on closer examination, geologists have found that almost all the features observed result from just four major **geological processes** that affect planetary surfaces:

- **Impact cratering**: the excavation of bowl-shaped depressions (*impact craters*) by asteroids or comets striking a planet's surface.

- **Volcanism**: the eruption of molten rock, or *lava*, from a planet's interior onto its surface.

- **Tectonics**: the disruption of a planet's surface by internal stresses.

- **Erosion**: the wearing down or building up of geological features by wind, water, ice, and other phenomena of planetary weather.

Planetary geology once consisted largely of cataloging the number and kinds of geological features found on the planets. The field has advanced remarkably in recent decades, however, and it is now possible to describe how planets work in general and what geological features we expect to find on different planets. This progress has been made possible by a detailed examination of the geological processes and the determination of what factors affect and control them.

Understanding Geological Relationships

We can understand the fundamental ideas of comparative geology by exploring just a few key cause-and-effect relationships. Some of the important relationships are fairly easy to see. For example, a planet can have active volcanoes only if it has a sufficiently hot interior; and water, a major agent of erosion, remains liquid only if the planet has a temperature in the proper range and an atmosphere with sufficient pressure. Other relationships are subtler, such as that between the presence of an atmosphere and the planetary surface temperature. We can use flowcharts (or "concept maps") to help us keep track of the most important geological relationships.

Figure 9.10 shows one such planetary flowchart. The first row lists four **formation properties** with which each planet was endowed at birth: *size* (mass and radius), *distance from the Sun, chemical composition,* and *rotation rate*. These properties generally remain unchanged throughout a planet's history, unless the planet suffers a giant impact. The second row lists four **geological controlling factors** that drive a planet's geological activity: *surface gravity, internal temperature, surface temperature,* and the characteristics of the planet's *atmosphere*. Before we move on, let's briefly discuss how to interpret blocks and arrows on the flowchart:

- Each block represents a particular property, controlling factor, or process.

- Each arrow represents a cause-and-effect relationship between one block and another. (There may be other connections besides those shown; the figure includes only those that have substantial effects on geology.)

Arrows from the first row to the second row show how a planet's formation properties help determine the geological controlling factors. The single arrow pointing from *mass and radius* to *surface gravity* shows that the strength of gravity at the surface is determined solely by size, in accord with Newton's law of gravity. Two arrows point to *internal temperature*, indicating that the formation properties of *size* and

FIGURE 9.10 Major cause-and-effect relationships between a planet's formation properties and its geological controlling factors. Each block represents a particular formation property, controlling factor, or geological process. Each arrow represents a cause-and-effect relationship between one block and another. This diagram forms the foundation for several to follow.

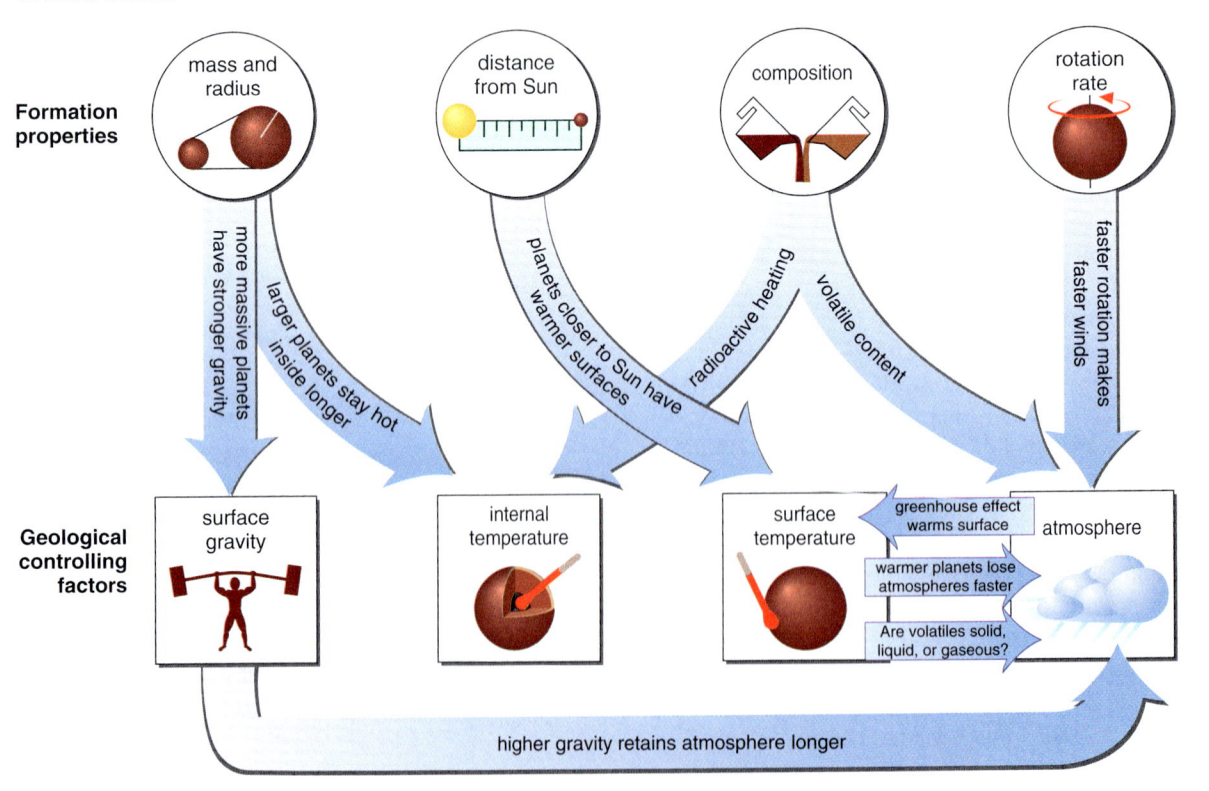

composition both affect it: Size is the most important factor in determining how rapidly a planet loses its internal heat, and composition—particularly the amount of radioactive elements—determines how much new heat is released in the interior. Other arrows from the first to the second row in Figure 9.10 show that a planet's *surface temperature* depends in part on its *distance from the Sun* and that its *atmosphere* is affected by both its *composition* and its *rotation rate*.

Arrows between the geological controlling factors show how they affect one another. Both *surface gravity* and *surface temperature* have arrows pointing to *atmosphere*, because a planet's ability to hold atmospheric gases depends on both how fast the gas molecules move (which depends on temperature) and the strength of the planet's gravity [Section 10.5]. A second arrow from *surface temperature* to *atmosphere* shows that the temperature also influences what substances exist as gases (versus solid or liquid). The *atmosphere*, in turn, points back to *surface temperature* because a thick atmosphere can make a planet's surface warmer than it would be otherwise (through the *greenhouse effect* [Section 10.2]).

TIME OUT TO THINK *To make sure you understand the planetary flowchart, answer the following questions: (1) What planetary formation properties determine whether a planet's interior is hot, and why? (2) How does surface gravity affect a planet's atmosphere, and why?*

The planetary flowcharts in this chapter contain all the information needed for a conceptual model of how planetary geology works—a model that allows us to answer questions about why different planets have different geological features. Figure 9.10 represents the beginning of this model. In the sections that follow, we will add each of the four geological processes (impact cratering, volcanism, tectonics, and erosion) to the set of relationships we have just covered, ending up with a fairly complete model for planetary geology. You do not need to memorize the model in this graphical form; instead, use it to help you understand planetary geology. You should also think about how you could adapt this technique of developing a conceptual model to any other problem that interests you.

Impact Cratering

Impact cratering occurs when a leftover planetesimal (such as a comet or an asteroid) crashes into the surface of a terrestrial world. Impacts can have devastating effects on planetary surfaces, which we can see both from the impact craters left behind and from laboratory experiments that reproduce the impact process (Figure 9.11). Impactors typically hit planets at speeds between 30,000 and 250,000 km/hr (10–70 km/s) and pack enough energy to vaporize solid rock and excavate a *crater* (the Greek word for "cup"). Craters are generally circular because the impact blasts out material in all directions, no matter

FIGURE 9.11 Artist's conception of the impact process. The last frame shows how, in a larger crater, the center can rebound just as water does after you drop a pebble into it.

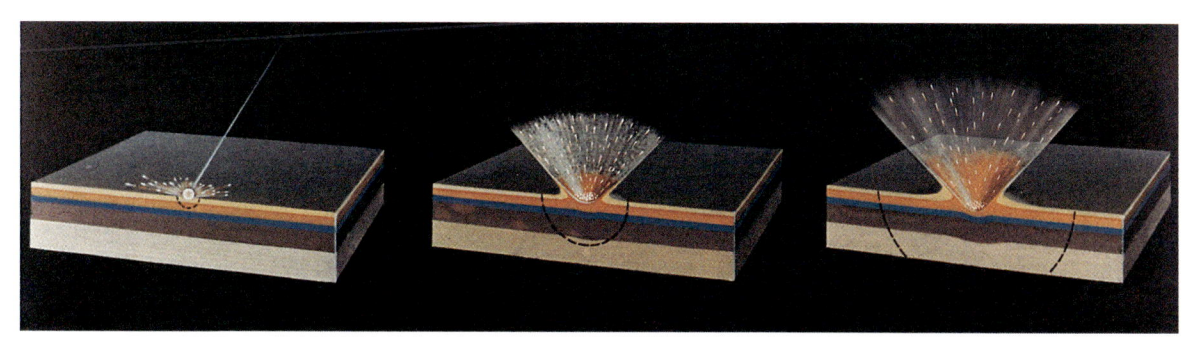

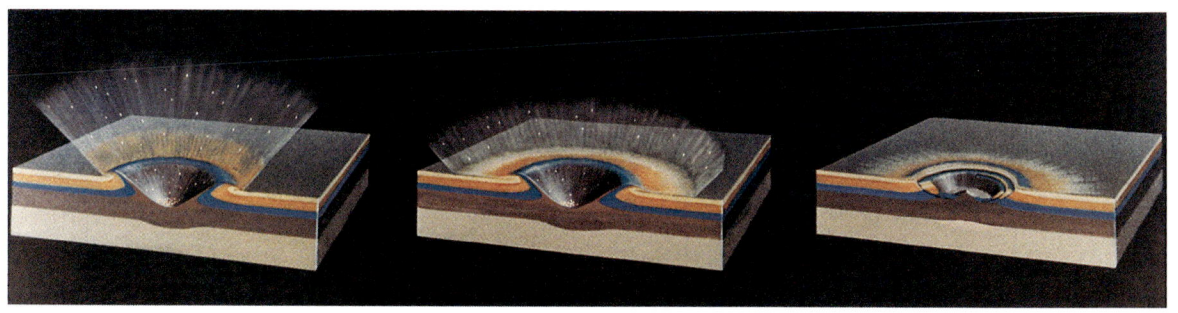

Lunar maria are huge impact basins that were flooded by lava. Only a few small craters appear on the maria.

Lunar highlands are ancient and heavily cratered.

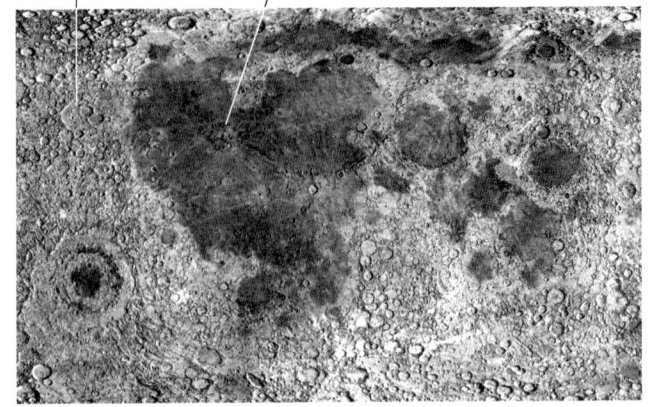

a The Moon's surface, as seen in this Apollo photograph, shows both heavily and lightly cratered areas.

b This map of the entire lunar surface shows the differences between the lightly cratered maria (dark areas) and the heavily cratered highlands. At the lower left is a multi-ring basin on the far side of the moon known as Mare Orientale. The mare in the center and right portions of the map make the "man-in-the-moon" face we see from Earth.

FIGURE 9.12 Geology of the Moon.

---

which direction the impactor came from. A typical crater is about 10 times wider than the impactor that created it, with a depth about 10–20% of the crater width. Thus, for example, a 1-kilometer-wide impactor creates a crater about 10 kilometers wide and 1–2 kilometers deep. Debris from the blast, called **ejecta**, shoots high into the atmosphere and then rains down over a large area. If the impact is large enough, some of the atmosphere may be blasted away into space, and some of the rocky ejecta may completely escape from the planet.

Craters come in all sizes, but small craters far outnumber large ones because there are far more small objects orbiting the Sun than large ones. Very large impacts can form **impact basins**. The **lunar maria**, easily visible with binoculars, are impact basins up to 1,100 kilometers across (Figure 9.12a). (*Maria* is the Latin word for "seas"; they got their name because their smooth appearance reminded early observers of oceans.) These large impacts violently fractured the Moon's lithosphere, making cracks through which molten lava later escaped to flood the impact basins; when the lava cooled and solidified, it left a smooth surface in the region of the lava flood. The impacts that made the largest basins were so violent they sent out ripples that left tremendous *multi-ring basins* shaped like bull's-eyes (Figure 9.12b).

The present-day abundance of craters on a planet's surface tells us a great deal about its geological history. Even though impacts still occur today, the vast majority of craters formed during the "rain of rock and ice" that ended around 3.8 billion years ago [Section 8.5]. At that time, the terrestrial worlds were saturated (completely covered) with impact craters. A surface region that is still saturated with craters, such as the *lunar highlands*, must have remained essentially undisturbed for the last 3.8 billion years (Figure 9.12a). In contrast, the original craters must have been somehow "erased" in regions that now have few craters, such as the lunar maria. The flood of lava that formed the lunar maria covered any craters that had formed inside the impact basin, and the few craters that exist today within the maria must have formed from impacts occurring after the lava flows solidified. These craters tell us that the impact rate since the end of heavy bombardment has been quite small: Radioactive dating of moon rocks shows that the maria are 3–3.5 billion years old, but they have only 3% as many craters as the lunar highlands.

TIME OUT TO THINK *Earth must also have been saturated with impact craters early in its history, but we see relatively few impact craters on Earth today. What processes erase impact craters on Earth?*

Craters with unusual shapes provide additional clues about surface conditions, as we can see by comparing craters on Mars. Craters in rocky surfaces usually have a simple bowl shape (Figure 9.13a).

a Many craters are bowl-shaped.

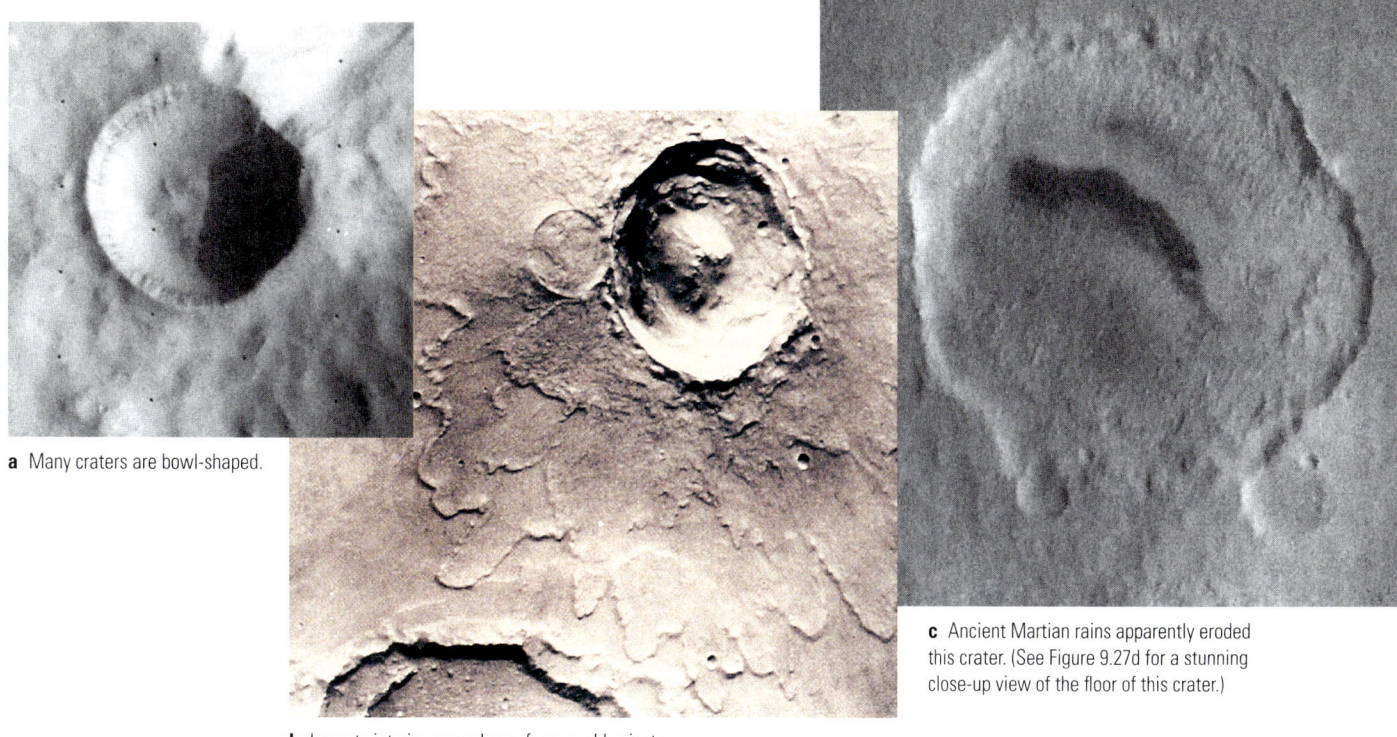

b Impacts into icy ground may form muddy ejecta.

c Ancient Martian rains apparently eroded this crater. (See Figure 9.27d for a stunning close-up view of the floor of this crater.)

FIGURE 9.13 Crater shapes on Mars tell us about Martian geology. These photos were taken from orbit by the Viking spacecraft.

However, some Martian craters look as if they were formed in mud (Figure 9.13b), suggesting that underground water or ice vaporized upon impact and lubricated the flow of ejecta away from the crater. Other craters lack a sharp rim and bowl-shaped floor, suggesting that geological processes such as erosion have altered their shape over time (Figure 9.13c). Planetary geologists must be cautious in interpreting unusually shaped craters, because impacts are not the only process that can form craters. Fortunately, craters created by other processes, such as volcanism, tend to have distinctly different shapes than impact craters (see Figure 9.17b).

The most common impactors are sand-size particles called *micrometeorites* when they impact a surface. Such tiny particles burn up as meteors in the atmospheres of Venus, Earth, and Mars. But on worlds that lack significant atmospheres, such as Mercury and the Moon, the countless impacts of micrometeorites gradually pulverize the surface rock to create a layer of powdery "soil." On the Moon, the Apollo astronauts and their rovers left marks in this powdery surface (Figure 9.14). Because of the lack of wind and rain, the footprints and tire tracks will last millions of years, but they will eventually be erased by micrometeorite impacts. In fact, over millions and billions of years, these tiny impacts smooth out rough crater rims much as erosion processes do more rapidly on Earth.

FIGURE 9.14 Reminders of the Apollo missions to the Moon, including this footprint, will last for millions of years.

FIGURE 9.15 Major connections between impact cratering and other geological processes, controlling factors, and formation properties. This diagram includes all the blocks and arrows from Figure 9.10, but those that are not directly relevant to impact cratering are shaded lighter to clarify this illustration. The destruction of volcanic, tectonic, and erosional features by impacts is not shown.

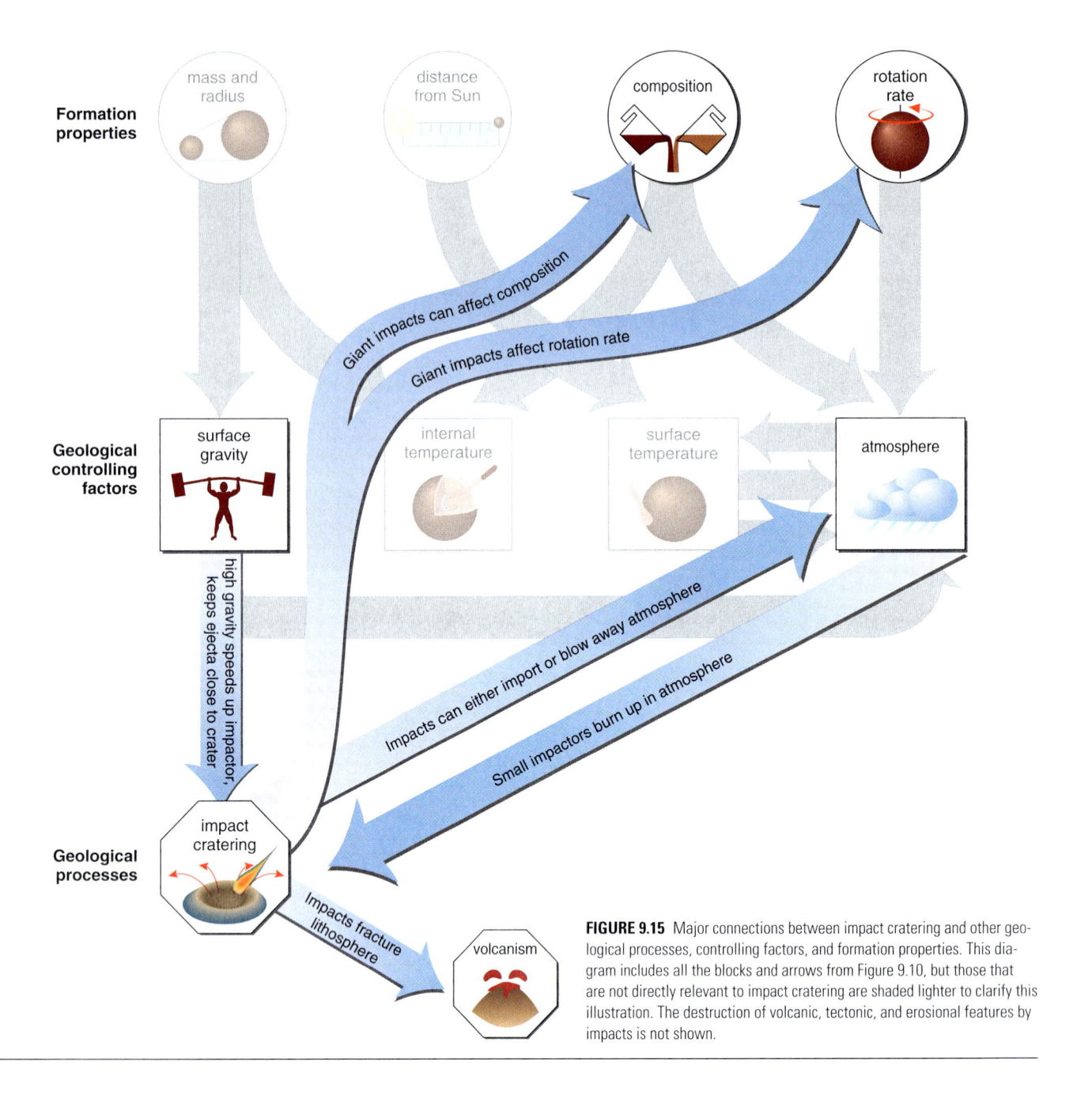

Figure 9.15 summarizes the important cause-and-effect relationships linking impact cratering with other geological processes, controlling factors, and formation properties. Note that the planet's formation properties do not greatly affect impact cratering, because the impactors do not come from the planet itself. Craters formed on all the terrestrial worlds, especially early in their histories. But giant impacts can affect formation properties by altering a planet's composition or rotation rate. In addition, impacts affect features made by other geological processes. Apart from simply obliterating volcanoes, cliffs, or riverbeds, impacts can fracture the lithosphere, creating a path for volcanic lava to reach the surface (as occurred in the lunar maria). The relationships between impacts and the atmosphere are particularly interesting: Not only can an atmosphere burn up small impactors, but the impactors themselves can either bring in atmospheric ingredients or blast away some of the atmosphere. Before you read on, take a bit of time to analyze the remaining arrows and make sure all the connections make sense to you.

Volcanism

We find evidence for volcanoes and lava flows on all the terrestrial planets, as well as on a few of the

a Hot magma erupts to the surface.

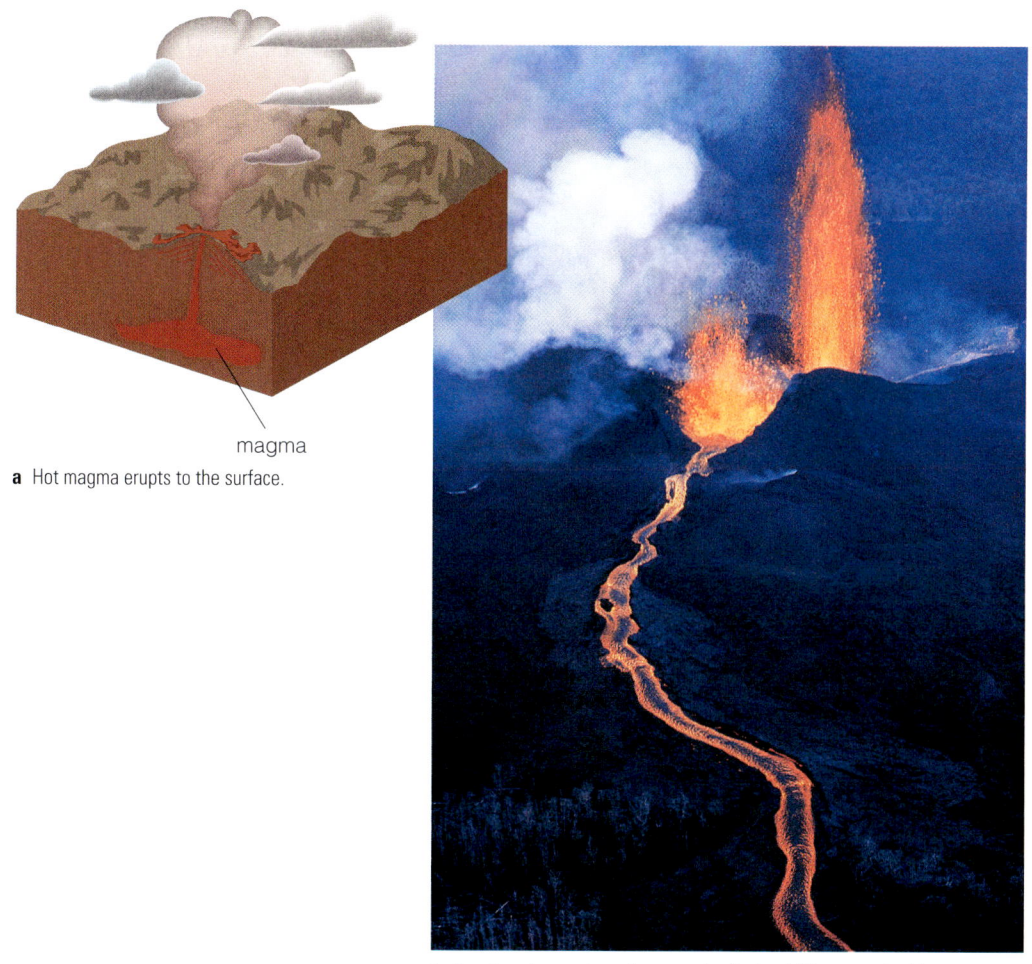

b Eruption of an active volcano on the flanks of Kilauea on the Big Island in Hawaii.

FIGURE 9.16 Volcanism.

---

moons of the outer solar system. (The moons may have "lavas" of water or other normally icy materials.) Volcanic surface features differ from one world to another, but some general principles apply.

Volcanoes erupt when underground molten rock, or **magma**, finds a path through the lithosphere to the surface (Figure 9.16). Magma rises for two main reasons: First, molten rock is generally less dense than solid rock, so it has a natural tendency to rise. Second, a *magma chamber* may be squeezed by tectonic forces, driving the magma upward under pressure. Any trapped gases expand as magma rises, sometimes leading to dramatic eruptions.

The structure of a volcanic flow depends on the viscosity of the lava that erupts onto the surface. The rock type known as **basalt** makes relatively low-viscosity lava when molten because it is made of relatively short molecular chains that don't tangle with one another. Temperature affects lava viscosity as well: The hotter the lava, the more easily the chains can jiggle and slide along one another and the lower the viscosity. The amount of water or gases trapped in the lava also affects viscosity. Such materials can act as a lubricant, decreasing the viscosity of subsurface magma. But when lavas erupt, materials such as water may form gas bubbles that *increase* the viscosity. (You can understand why bubbles increase viscosity by thinking about how much higher a "mountain" you can build with bubble-bath foam than you can with soapy water.)

The runniest basalt lavas flow far and flatten out before solidifying, creating vast *volcanic plains* such as the lunar maria (Figure 9.17a). Somewhat more viscous basalt lavas solidify before they can completely spread out, resulting in **shield volcanoes** (so-named because they are shield-shaped). Shield volcanoes can be very tall, but they are not very steep; most have slopes of only 5°–10° (Figure 9.17b). The mountains of the Hawaiian Islands are shield vola Low-viscosity lava makes flat lava plains.

Lava plains (maria) on the Moon

b Medium-viscosity lava makes shallow-sloped shield volcanoes.

Olympus Mons (Mars)

c High-viscosity lava makes steep-sloped stratovolcanoes.

Mount St. Helens

FIGURE 9.17 Volcanoes produce different types of features depending primarily on the viscosity of the lava erupted.

canoes; measured from the ocean floor to their summits, the Hawaiian mountains are the tallest (and widest) on Earth. Tall, steep **stratovolcanoes** such as Mount St. Helens are made from much more viscous lavas that can't flow very far before solidifying (Figure 9.17c).

Figure 9.18 summarizes the geological relationships that involve volcanism. The main point to keep in mind is that internal temperature exerts the greatest control over volcanism: A hotter interior means that lava lies closer to the surface and can erupt more easily. The formation property of composition is also very important, because it determines the melting temperatures and viscosities of lavas that erupt and therefore what types of volcanoes form. Volcanism is connected to other geological processes in several ways beyond the simple fact that lava flows can cover up or fill in previous geological features. Tectonics and volcanism often go together: Tectonic stresses can force lava to the surface or cut off eruptions. Another connection reveals far-reaching effects: Volcanism affects erosion indirectly, because volcanoes are the primary source of gases in the current terrestrial planet atmospheres [Section 10.5]. Before continuing, be sure you understand all the relationships shown in Figure 9.18.

Tectonics

We are now ready to discuss tectonics, the third major geological process in our model. The root of the word *tectonics* comes from Greek legend, in which Tecton was a carpenter. In geology, *tectonics* refers to the processes that do "carpentry" on planetary surfaces—that is, to the internal forces and stresses that act on the lithosphere to create surface features.

Several types of internal stress can drive tectonic activity. On planets with internal convection, the strongest type of stress often comes from the circulation of the convection cells themselves. The tops of convection cells can drag against the lithosphere, sometimes forcing sections of the lithosphere together or apart. On Earth, these forces have broken the lithosphere into *plates* that move over, under, and around each other in what we call *plate tectonics*; we'll discuss plate tectonics further in Chapter 13. Even if the crust does not break into plates, stresses from underlying convection cells may create vast mountain ranges, cliffs, and valleys. A second type of stress associated with convection comes from individual rising *plumes* of hot mantle material that push

FIGURE 9.18 Major connections between volcanism and other geological processes, controlling factors, and formation properties. This diagram includes all the blocks and arrows from Figure 9.10, but those that are not directly relevant to volcanism are shaded lighter to clarify this illustration. Volcanoes also cover up impact, tectonic, and erosional features, but these connections are not shown.

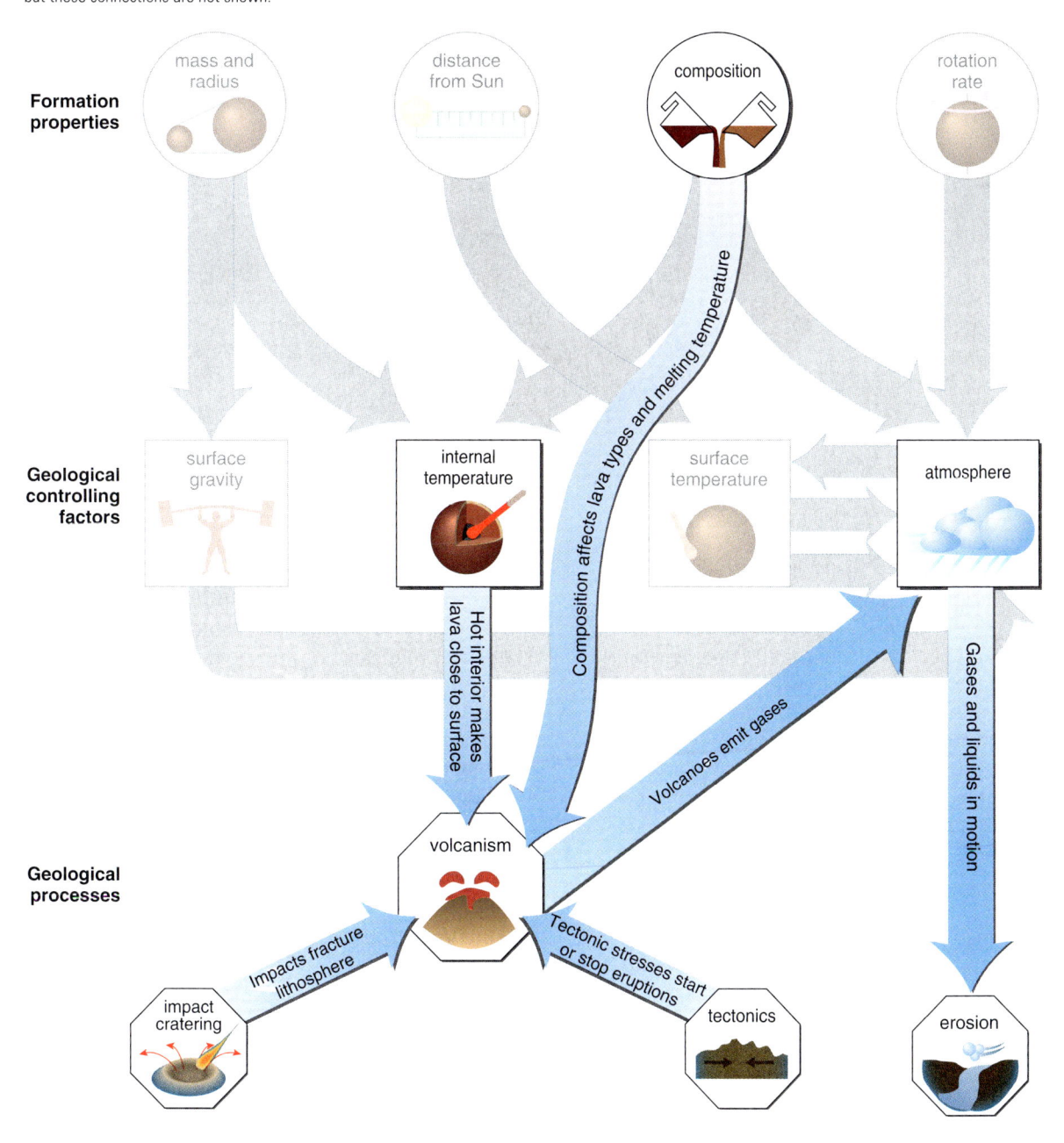

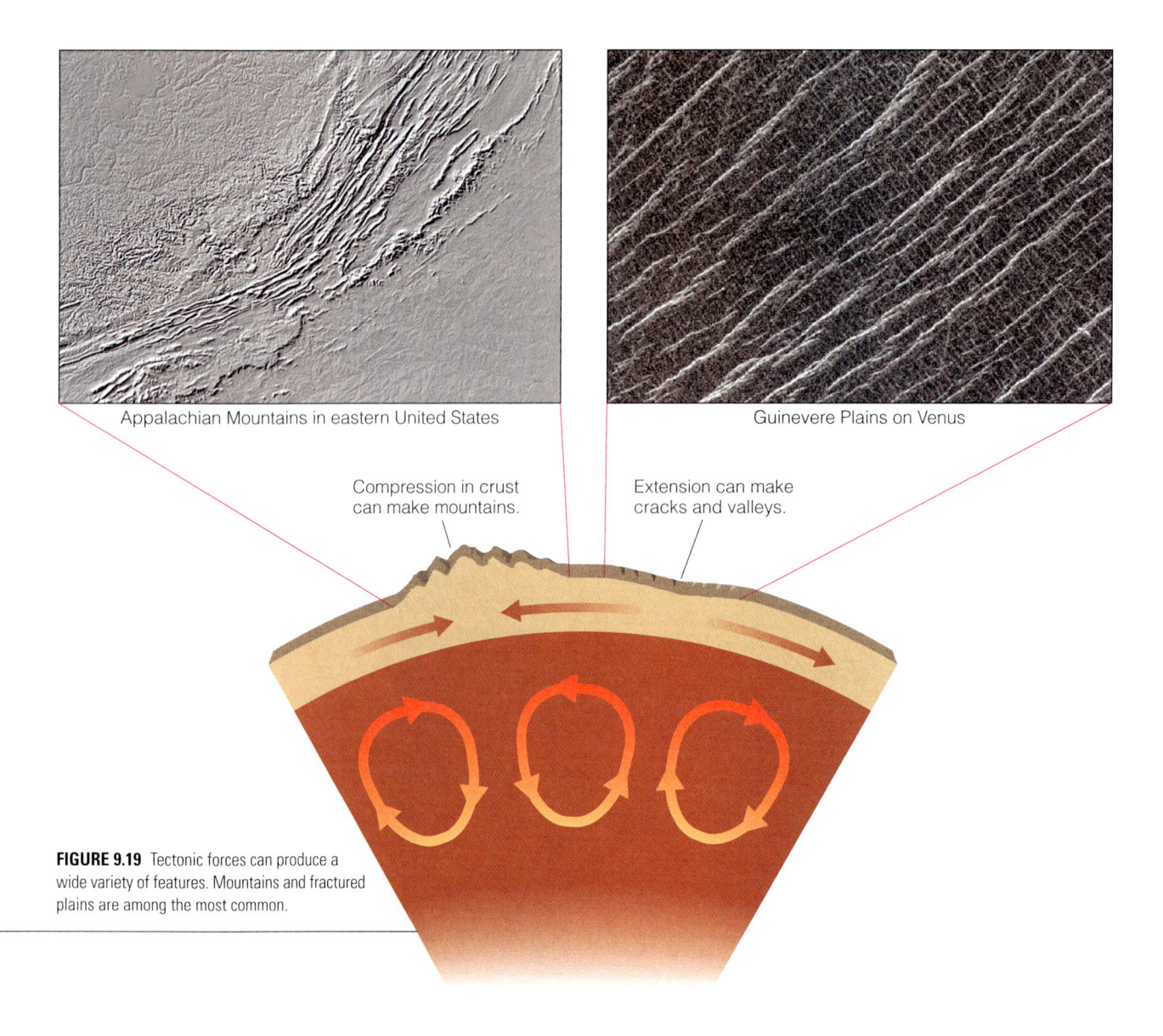

FIGURE 9.19 Tectonic forces can produce a wide variety of features. Mountains and fractured plains are among the most common.

Appalachian Mountains in eastern United States

Guinevere Plains on Venus

Compression in crust can make mountains.

Extension can make cracks and valleys.

up on the lithosphere. (The Hawaiian Islands result from such a plume [Section 13.2].) Internal stress can also arise from temperature changes in the planetary interior. For example, the crust may be forced to expand and stretch if the planetary interior heats up from radioactive decay. Conversely, the mantle and lithosphere must respond when a planetary core cools and contracts, leading to planet-wide compression forces. Tectonic stresses can even occur on more local scales; for example, the weight of a newly formed volcano can bend or crack the lithosphere beneath it.

Tectonic features take an incredible variety of forms (Figure 9.19). Mountains may rise where the crust is compressed. Such crustal compression helped create the Appalachian Mountains of the eastern United States. Huge valleys and cliffs may result where the crust is pulled apart; examples include the Guinevere Plains on Venus and New Mexico's Rio Grande Valley. (The river named Rio Grande came *after* the valley formed from tectonic processes.) Tectonic forces may bend or break rocks, and tectonic activity on Earth is always accompanied by earthquakes. Other worlds undoubtedly experience "planetquakes," and planetary geologists are eager to plant seismometers on them to probe their interiors.

Figure 9.20 summarizes the connections between tectonics and other geological processes and properties. Like volcanism, tectonics is most likely on larger worlds that remain hot inside. Tectonics may also have been important in the past on smaller worlds that underwent major shrinking or expansion.

Erosion

The last of the four major geological processes to be added to our model is erosion, which encompasses a

244 PART III LEARNING FROM OTHER WORLDS

Formation properties: mass and radius; distance from Sun; composition; rotation rate

Geological controlling factors: surface gravity; internal temperature; surface temperature; atmosphere

Geological processes: tectonics; volcanism

Internal temperature changes disrupt crust

Warm interior causes convection

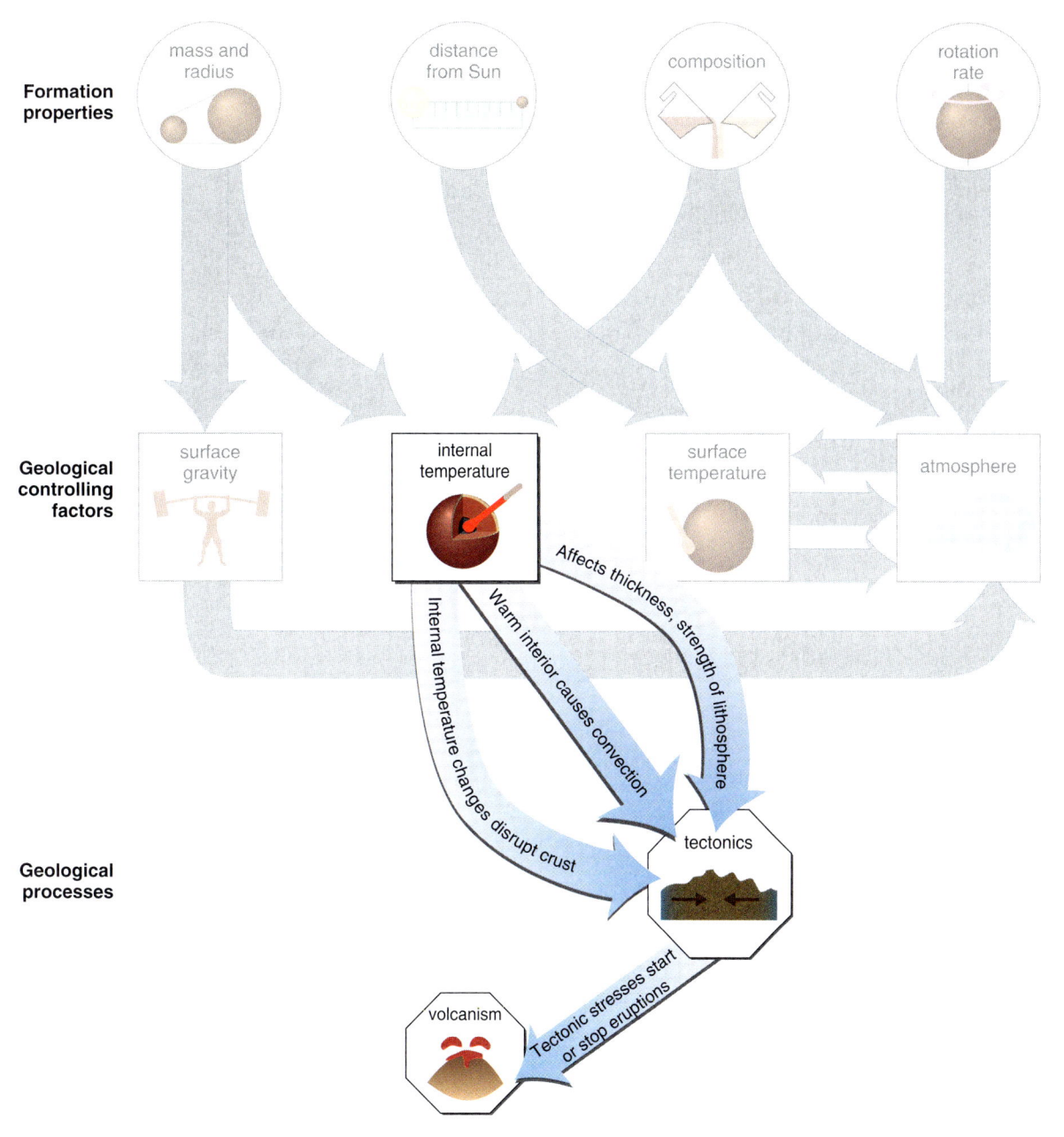

Affects thickness, strength of lithosphere

Tectonic stresses start or stop eruptions

FIGURE 9.20 Connections between tectonics and other geological processes, controlling factors, and formation properties. This diagram includes all the blocks and arrows from Figure 9.10, but those that are not directly relevant to tectonics are shaded lighter to clarify this illustration. Tectonics can also destroy impact, volcanic, and erosional features, but these connections are not shown.

---

variety of processes connected by a single theme: the breakdown and transport of rocks by **volatiles**. The term *volatile* means "evaporates easily" and refers to substances—such as water, carbon dioxide, and methane—that are usually found as gases, liquids, or surface ices on the terrestrial worlds. Wind, rain, rivers, flash floods, and glaciers are just a few examples of processes that contribute to erosion on Earth. Erosion not only breaks down existing geological features (wearing down mountains and forming gullies, riverbeds, and deep valleys), but also builds new ones (such as sand dunes, river deltas, and lakebed deposits). If enough material is deposited over time, the layers can compact to form **sedimentary rocks**.

Virtually no erosion takes place on worlds without a significant atmosphere, such as Mercury and the Moon. But planets with atmospheres can have significant erosional activity. Larger planets are better able to generate atmospheres by releasing gases trapped in their interiors; their stronger gravity also tends to prevent their atmospheres from escaping to space

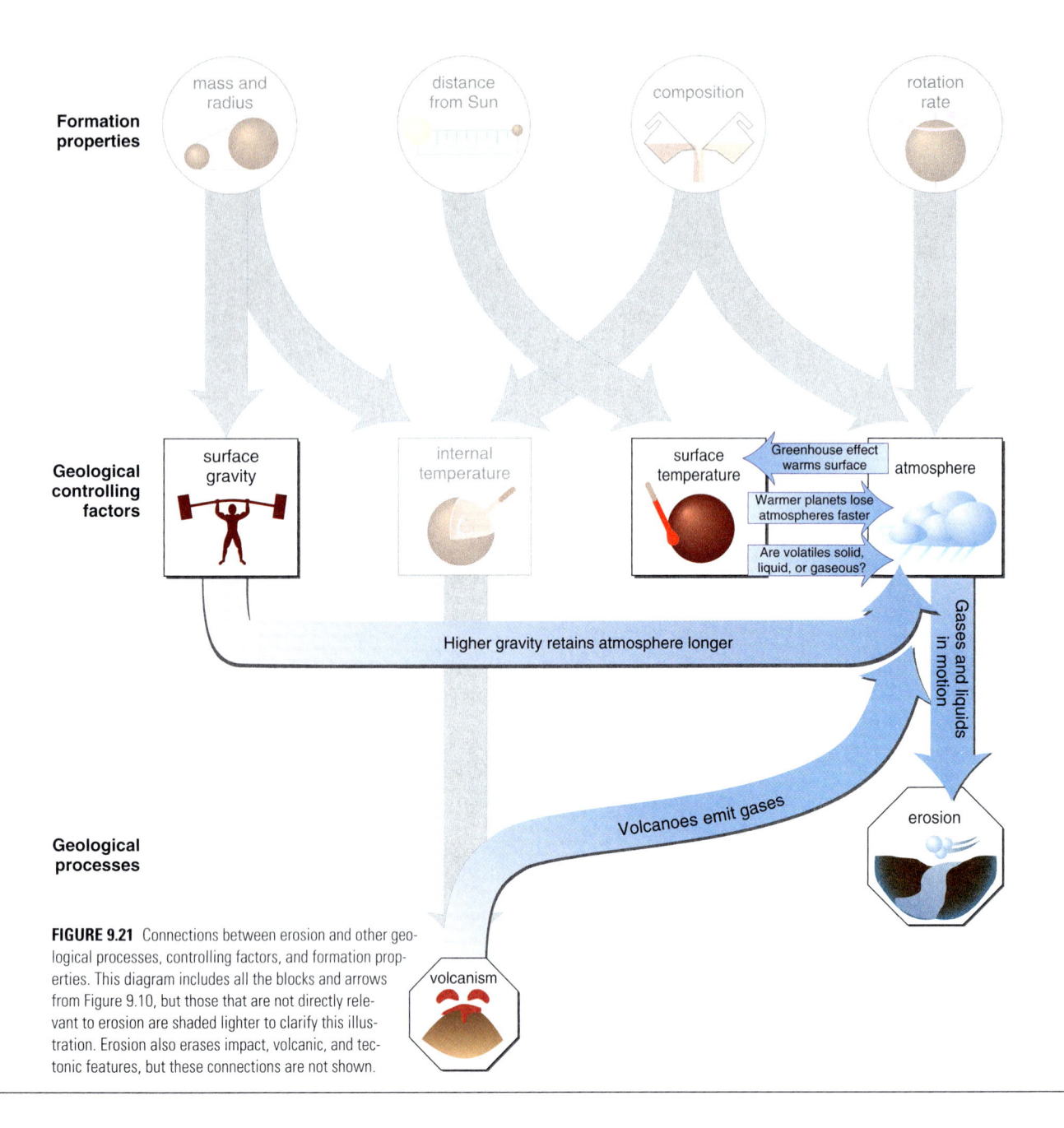

FIGURE 9.21 Connections between erosion and other geological processes, controlling factors, and formation properties. This diagram includes all the blocks and arrows from Figure 9.10, but those that are not directly relevant to erosion are shaded lighter to clarify this illustration. Erosion also erases impact, volcanic, and tectonic features, but these connections are not shown.

[Section 10.5]. In general, a thick atmosphere is more capable of driving erosion than a thin atmosphere, but a planet's rotation rate is also important. Slowly rotating planets like Venus have correspondingly slow winds and therefore weak or nonexistent wind erosion. Surface temperature also matters: Erosion isn't very effective if most of the volatiles stay frozen on the surface, but it can be very powerful when volatiles are able to evaporate and then recondense as rain or snow. Figure 9.21 summarizes the connections between erosion and other geological processes and properties.

9.5 A Geological Tour of the Terrestrial Worlds

Now that we've examined the processes and properties that shape the geology of the terrestrial worlds, we're ready to return to the issue of why the terrestrial worlds ended up so geologically different from one another. We'll take a brief "tour" of the terrestrial worlds. We could organize the tour by distance from the Sun, by density, or even by alphabetical order. But we'll choose the planetary property that has the strongest effect on geology: size, which controls a planet's internal heat. We'll start with our Moon, the smallest terrestrial world.

a Astronaut explores a small crater.

b Site of an ancient lava river (also seen in Figure 9.17a).

c Lava flows filled impact basins to create maria like this one (Mare Imbrium). Wrinkles on the maria are evidence of minor tectonic forces.

FIGURE 9.22 Impact cratering and volcanism are the most important geological processes on the Moon.

The Moon (1,731-km radius, 1.0 AU from Sun)

On a clear night, you can see much of the Moon's global geological history with your naked eye. The Moon is unique in this respect; other planets are too far away for us to see any surface details, and Earth is so close that we see only the local geological history. Twelve Apollo astronauts visited the lunar surface between 1969 and 1972, taking photographs, making measurements, and collecting rocks. Thanks to these visits and more recent observations of the Moon from other spacecraft, we know more about the Moon's geology than that of any other object, with the possible exception of the Earth.

The Moon's composition differs somewhat from the Earth's. According to the leading theory, our Moon formed early in the history of our solar system when a giant impact blasted away some of Earth's rocky outer layers [Section 8.5]. The impact did not dredge up metallic core material, which explains why the Moon has a lower metal content than Earth and a smaller metal core. In addition, lunar rocks collected by astronauts contain a much smaller proportion of volatiles (such as water) than Earth rocks, presumably because heat from the giant impact evaporated the volatiles and allowed them to escape into space before the Moon formed from the debris.

As the Moon accreted following the giant impact, the heat of accretion melted its outer layers. The lowest-density molten rock rose upward through the process of differentiation, forming what is sometimes called a *magma ocean*. This magma ocean cooled and solidified during the heavy bombardment that took place early in the history of the solar system, leaving the Moon's surface crowded with craters. We still see this ancient, heavily cratered landscape in the lunar highlands (Figure 9.22a). Lunar samples confirm that the highlands are composed of low-density rocks. As the bombardment tailed off, around 3.8 billion years ago, a few larger impactors struck the surface and formed impact basins.

Even as the heat of accretion leaked away, radioactive decay kept the Moon's interior molten long enough for volcanism to reshape parts of its surface. Between about 3 and 4 billion years ago, molten rock welled up through cracks in the deepest impact basins, forming the lunar maria. Because the lava that filled the maria rose up from the Moon's mantle, the maria contain dense, iron-rich rock that is darker in color than the rock of the lunar highlands. The contrasts between the light-colored rock of the highlands and the dark rock of the maria make the "man-in-the-Moon" pattern that some people imagine when they look at the full moon.

The lunar lavas must have been among the least viscous (runniest) in the solar system, perhaps because the lack of volatiles meant a lack of bubbles (which increase viscosity) in the erupting lava. The lava plains of the maria cover a large fraction of the lunar surface. Only a few small shield volcanoes exist on the Moon, and no steeper-sided stratovolcanoes have been found. Low-viscosity lavas also carved out long, winding channels. These channels must once have been rivers of molten rock that helped fill the lunar maria (Figure 9.22b).

Almost all the geological features of the Moon were formed by impacts and volcanism, although a few small-scale tectonic stresses wrinkled the surface as the lava of the maria cooled and contracted (Figure 9.22c). The Moon has virtually no erosion because of the lack of atmosphere, but the continual rain of micrometeorites has rounded crater rims. The heyday of lunar volcanism is long gone—3 billion years gone. Over time, the Moon's small size allowed its interior to cool, thickening the lithosphere. Today, the Moon's lithosphere probably extends to a depth of 1,000 km, making it far too thick to allow further volcanic or tectonic activity. The Moon has probably been in this geologically "dead" state for 3 billion years.

Despite its geological inactivity, the Moon is of prime interest to those hoping to build colonies in space. The *Lunar Prospector* arrived at the Moon in early 1998 and began mapping the surface in search of promising locations for a permanent human base. Although no concrete plans are yet in place, several nations are exploring the possibility of building a human outpost on the Moon within the next couple of decades.

Mercury (2,439-km radius, 0.39 AU from Sun)

Mercury (Figure 9.23) is the least studied of the terrestrial worlds. Its proximity to the Sun makes it difficult to study through telescopes, and it has been visited by only one spacecraft: *Mariner 10,* which collected data during three rapid flybys of Mercury in 1974–1975. *Mariner 10* obtained images of only one hemisphere of Mercury (Figure 9.23b), but the influence of Mercury's gravity on *Mariner 10*'s orbit allowed planetary scientists to determine Mercury's mass and average density. Based on its relatively high density, Mercury must be about 61% iron by mass, with a core that extends to perhaps 75% of its radius. Mercury's high metal content is due to its formation close to the Sun, possibly enhanced by a giant impact that blasted away its outer, rocky layers [Section 8.6].

Mercury has craters almost everywhere, indicating an ancient surface (Figure 9.23c). A huge impact basin, called the *Caloris Basin* (Figure 9.23a), covers a large portion of one hemisphere. Like the large

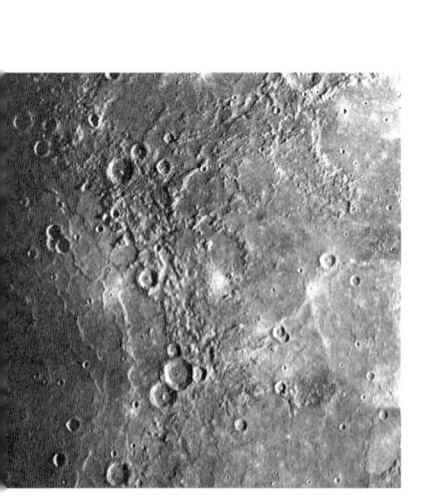

FIGURE 9.23 Mercury, like the Moon, is a heavily cratered object with evidence of volcanism.

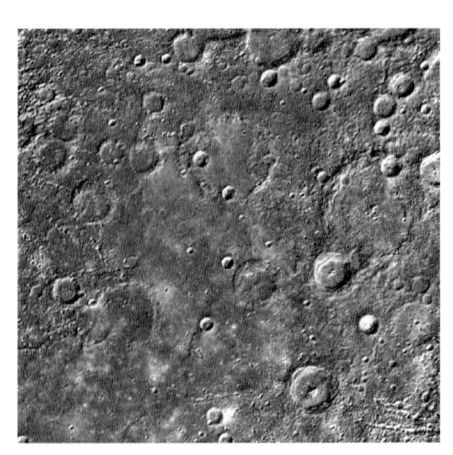

a Edge of Caloris Basin

b Mercury

c Close up view showing small lava plains that have covered up craters

lunar basins, the Caloris Basin has few craters within it, indicating that it must have been formed toward the end of the solar system's early period of heavy bombardment.

Despite the superficial similarity between their cratered surfaces, Mercury and the Moon also exhibit significant differences. Mercury has noticeably fewer craters than the lunar highlands (compare Figure 9.12a to Figure 9.23b). Smooth patches between many of the craters tell us that volcanic lava once flowed, covering up small craters. Mercury has no vast maria but has smaller lava plains almost everywhere, suggesting that Mercury had at least as much volcanism as the Moon.

TIME OUT TO THINK *Geologists have discovered that the ejecta didn't travel quite as far away from craters on Mercury as it did on the Moon. Why not? (Hint: What planetary formation properties affect impact craters?)*

Tectonic processes played a more significant role on Mercury than on the Moon, as evidenced by tremendous cliffs. The cliff shown in Figure 9.24a is several hundred kilometers long and up to 3 kilometers high in places; the central crater apparently crumpled during the cliff's formation. This cliff, and many others on Mercury, probably formed when tectonic forces compressed the crust (Figure 9.24b).

FIGURE 9.24 (**a**) The tall cliff (indicated by the arrow) on Mercury is several hundred kilometers long. (**b**) Cliffs formed from tectonic stresses (indicated by arrows) caused by Mercury's contraction.

However, nowhere on Mercury do we find evidence of extension (stretching) to match this crustal compression. Can it be that the whole planet simply shrank? Apparently so. Early in its history, Mercury gained much more internal heat from accretion and differentiation than did the Moon, owing to its larger proportion of iron and stronger gravity. Then, as the core cooled, it contracted by perhaps as much as 20 kilometers in radius. This contraction crumpled and compressed the crust, forming the many cliffs. The contraction probably also closed off volcanic vents, ending Mercury's period of volcanism.

Mercury lost its internal heat relatively quickly because of its small size. Its days of volcanism and tectonic shrinking probably ended within its first billion years. Mercury, like the Moon, has been geologically dead for most of its existence.

Mars (3,396-km radius, 1.52 AU from Sun)

Early telescopic observations of Mars revealed several uncanny resemblances to Earth. A Martian day is just over 24 hours, and the Martian rotation axis is tilted about the same amount as Earth's. Mars also has polar caps, which we now know to be composed primarily of frozen carbon dioxide, with smaller amounts of water ice. Telescopic observations also showed seasonal variations in surface coloration over the course of the Martian year (about 1.9 Earth years). All these discoveries led to the perception that Mars and Earth were at least cousins, if not twins. By the early 1900s, many astronomers—as well as the public—envisioned Mars as nearly Earth-like, possessing water, vegetation that changed with the seasons, and possibly intelligent life.

In 1877, Italian astronomer Giovanni Schiaparelli reported seeing linear features across the surface of Mars through his telescope. He named these features *canali*, the Italian word for "channels." (It's not clear whether he really thought the channels contained water or was merely continuing the tradition, started with the Moon, of naming dark features after bodies of water.) This report inspired the American astronomer Percival Lowell to build an observatory in Flagstaff, Arizona, for the study of Mars. In 1895, Lowell claimed the canali were *canals* being used by intelligent Martians to route water around their planet. He suggested that Mars was falling victim to unfavorable climate changes and that water was a precious resource that had to be managed carefully. Lowell's studies drove rampant speculation about the nature of the Martians, fueling science fiction fantasies. The public mania drowned out the skepticism of astronomers who saw

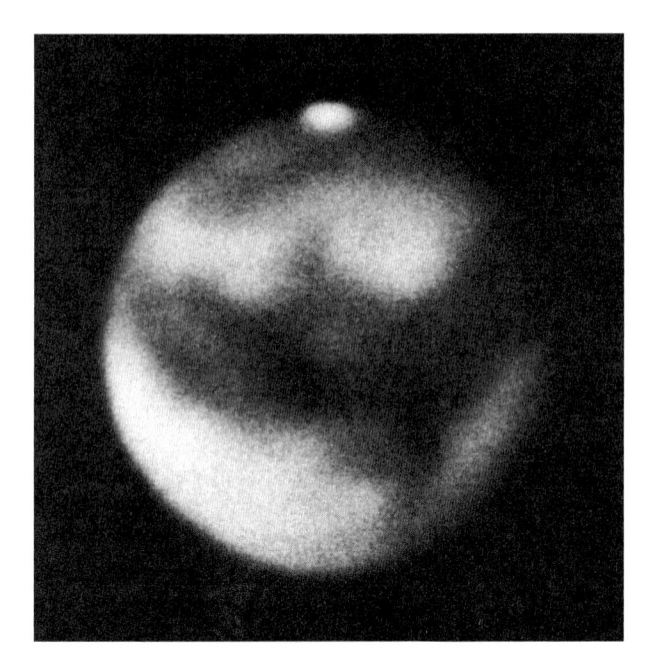

FIGURE 9.25 Can you see how the markings on Mars in the telescopic photo on the left might have resembled the geometrical features in the drawing by Percival Lowell on the right? Try blurring your eyes.

no canals through their telescopes or in their photographs (Figure 9.25).

The debate about Martian canals and cities was finally put to rest in 1965 with photos of the barren, cratered surface taken during the *Mariner 4* flyby. Schiaparelli's *canali* turned out to be nothing more than dark dust deposits redistributed by seasonal winds. Several other spacecraft studied Mars in the 1960s and 1970s, culminating with the impressive Viking missions in 1976. Each of the two Viking spacecraft deployed an orbiter that mapped the Martian surface from above and landers that returned surface images and regular weather reports for almost 5 years. More than 20 years passed before the next successful missions to Mars, the Mars Pathfinder mission and Mars Global Surveyor mission in 1997 [Section 7.6].

Martian Geology The Viking and Mars Global Surveyor images show a much wider variety of geological features than seen on the Moon or Mercury. Some areas of the southern hemisphere are heavily cratered (Figure 9.26a), while the northern hemisphere is covered by huge volcanoes surrounded by extensive volcanic plains. Volcanism was expected on Mars—40% larger than Mercury, it should have retained a hot interior much longer. But no one knows why volcanism affected the northern hemisphere so much more than the southern hemisphere, or why the northern hemisphere on average is lower in altitude than the southern hemisphere.

Mars has a long, deep system of valleys called *Valles Marineris* running along its equator (Figure 9.26b). Named for the *Mariner 9* spacecraft that first imaged it, Valles Marineris is as long as the United States is wide and almost four times deeper than the Grand Canyon. No one knows exactly how it formed; parts of the canyon are completely enclosed by high cliffs on all sides, so neither flowing lava nor water could have been responsible. Nevertheless, the linear "stretch marks" around the valley are evidence of tectonic stresses on a very large scale.

More cracks are visible on the nearby *Tharsis Bulge,* a continent-size region that rises well above the surrounding Martian surface (Figure 9.26c). Tharsis was probably created by a long-lived plume of rising mantle material that bulged the surface upward while stretching and cracking the crust. The plume was also responsible for releasing vast amounts of basaltic lava that built up several gigantic shield volcanoes, including *Olympus Mons,* the largest shield volcano in the solar system (Figure 9.26d). Olympus Mons has roughly the same shallow slope as the volcanic island of Hawaii, but it is three times larger in every dimension. Its base is some 600 kilometers across, making it large enough to cover an area the size of Arizona; it stands 26 kilometers high—some three times higher than Mount Everest.

How old are the volcanoes? Are they still active? In the absence of red-hot lava (none has been seen so far), the abundance of impact craters provides the best evidence. The Martian volcanoes are almost devoid of craters, but not quite. Planetary geologists

a The southern hemisphere of Mars is heavily cratered. The image spans several hundred kilometers.

b Valles Marineris is a huge valley created in part by tectonic stresses.

c A mantle plume below the Tharsis Bulge region was probably responsible for the volcanoes near the bottom of the image and the network of valleys near the top.

d Olympus Mons—the largest shield volcano in the solar system—covers an area the size of Arizona and rises higher than Mt. Everest.

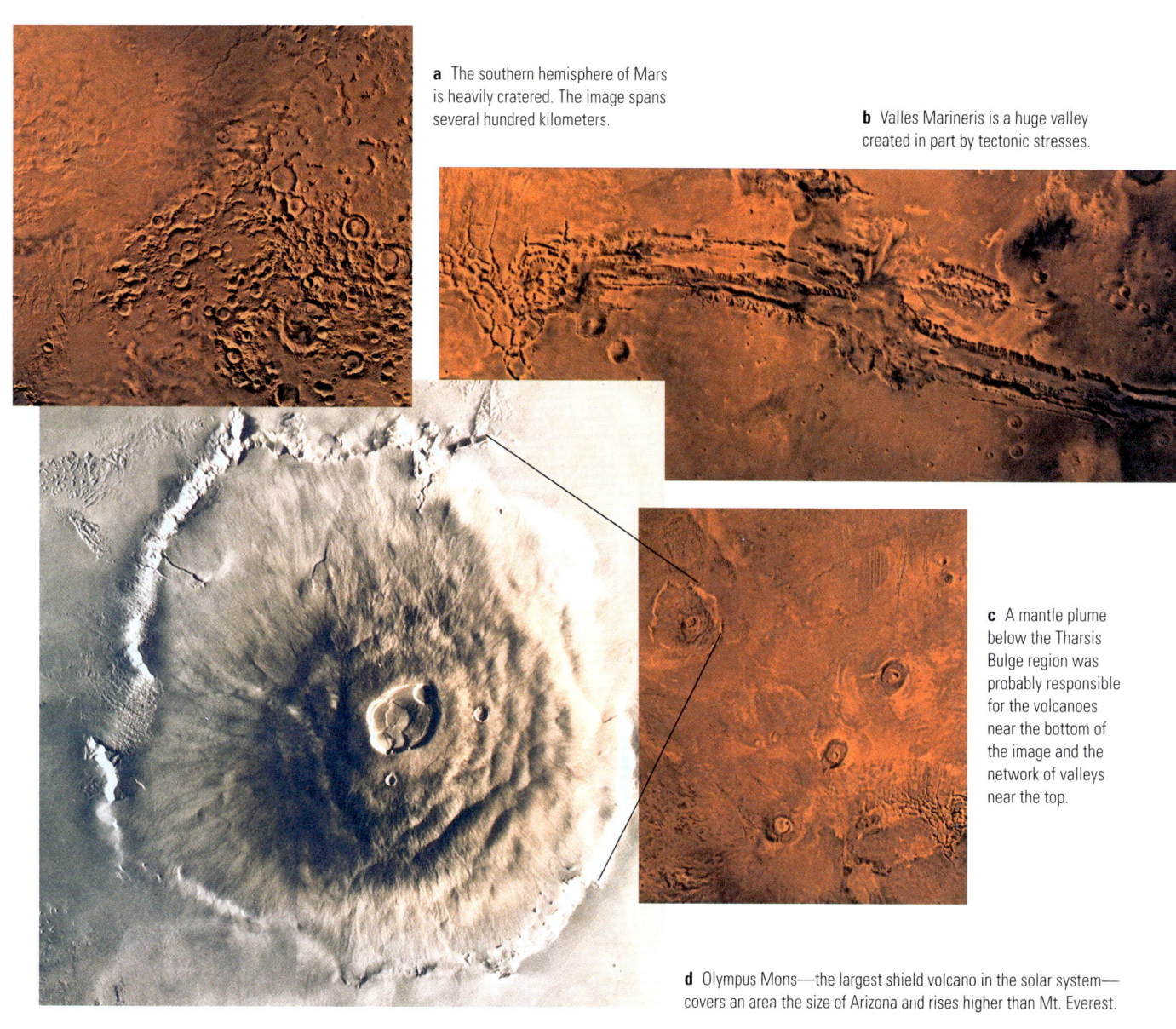

FIGURE 9.26 These photos of the Martian surface, taken from orbiting spacecraft, show that impact cratering has been an important process on Mars, but volcanism and tectonics have been even more important in shaping the current Martian surface.

---

therefore estimate that the volcanoes went dormant about a billion years ago. Geologically speaking, a billion years is not that long, and it remains possible that the Martian volcanoes will someday come back to life. However, the Martian interior is presumably cooling and its lithosphere thickening. At best, it's only a matter of a few billion more years before Mars becomes as geologically dead as the Moon and Mercury.

Martian Water The Martian robotic explorers have also revealed geological features unlike any seen on the Moon or Mercury—features caused by erosion. Figure 9.27a no doubt reminds you of a dry riverbed on Earth seen from above. The eroded craters visible on some of the oldest terrain suggest that rain fell on the Martian surface billions of years ago (see Figure 9.13c). In some places, erosion has even acted below the surface, where underground water and ice have broken up or dissolved the rocks. When the water and ice disappeared, a wide variety of odd pits and troughs were left behind. This process may have contributed to the formation of parts of Valles Marineris.

Underground ice may be the key to one of the biggest floods in the history of the solar system. Figure 9.27b shows a landscape hundreds of kilometers across that was apparently scoured by rushing water. Geologists theorize that heat from a volcano melted great quantities of underground ice, unleashing a catastrophic flood that created winding channels and large sandbars along its way.

The *Mars Pathfinder* spacecraft went to this interesting region to see the flood damage close up.

FIGURE 9.27 Unlike Mercury and the Moon, Mars shows unmistakable evidence of widespread erosion. The latest results point to the action of liquid water.

a This Viking photo (from orbit) shows ancient riverbeds that were probably created billions of years ago.

b This photo shows winding channels and sandbars that were probably created by catastrophic floods.

c View from the floodplain (see **b** above) from the *Mars Pathfinder*; the *Pathfinder* landing site is now known as Carl Sagan Memorial Station. The *Sojourner* rover is visible near the large rock.

d Close-up view of the floor of the eroded crater shown in Figure 9.13c. Billions of years ago, the crater was apparently a pond, and layer upon layer of sediment was deposited on the bottom. The water is long gone, and winds have since sculpted the layers into the astonishing patterns captured in this photograph from the *Mars Global Surveyor*.

e Topographic map of Mars with low-lying regions in blue and higher elevations in red, brown, and white. The white patches are tall volcanoes; Valles Marineris cuts through the terrain to their right. Some planetary geologists believe that an ocean once filled the smooth, low-lying northern regions shown in dark blue.

f This photograph from *Mars Global Surveyor* shows gullies on a crater wall; scientists suspect they were formed by water seeping out from the ground during episodic flash floods. The gullies are geologically young, but no one yet knows whether similar gullies may still be forming today.

The lander released the *Sojourner* rover, named after Sojourner Truth, an African-American heroine of the Civil War era who traveled the nation advocating equal rights for women and blacks. The six-wheeled rover, no larger than a microwave oven, carried cameras and instruments to measure the chemical composition of nearby rocks. Together the lander and rover confirmed the flood hypothesis: Gentle, dry, winding channels were visible in stereo images; rocks of many different types had been jumbled together in the flood; and the departing waters had left rocks stacked against each other in just the same manner that floods do here on Earth (Figure 9.27c).

In the decades since we first learned of past water on Mars, scientists have debated when water last flowed. Recent images from the *Mars Global Surveyor* have only deepened the mystery. Figure 9.27d shows a close-up view of the floor of a crater (the same crater shown in Figure 9.13c). The sculpted patterns in this picture were probably created as erosion exposed layer upon layer of sedimentary rock, much like the sedimentary layers visible in Earth's Grand Canyon. This suggests that ancient rains once filled the crater like a pond, with the sedimentary layers built up by material that settled to the bottom. Figure 9.27e, made from the spacecraft's precise topographic measurements, shows evidence of even more widespread water, suggesting that much of Mars's northern hemisphere was once covered by an ocean. Perhaps the most surprising finding is the presence of small gullies on the walls of many craters and valleys (Figure 9.27f). The gullies probably formed when underground water broke out in episodic flash floods, carrying boulders and soil into a broad fan at the bottom of the slope. Moreover, the gullies must be geologically young, because they have not been covered over by ongoing dust storms, nor have small craters formed on their surfaces. Unfortunately, "young" can be quite ambiguous in a geological sense: No one knows if these gullies formed "just" millions of years ago or if some might still be forming today.

Even if water still flows on occasion, it seems clear that Mars was much wetter in the past than it is today. Ironically, Percival Lowell's supposition that Mars was drying up has turned out to be basically correct, even though he did not have real evidence to support it. We'll discuss reasons for Mars's dramatic climate change in Chapter 10.

The Mars Global Surveyor and Pathfinder missions are paving the way for more adventurous missions. *Mars Global Surveyor* is searching for optimal landing sites for future rovers—one will search for evidence of life past and present, and another will gather rock samples for return to Earth. For the latest news on Mars missions, check the book Web site and Web sites listed in Appendix H.

Venus (6,051-km radius, 0.72 AU from Sun)

Venus is a big step up in size from Mars—it is almost as big as Earth. Thus, we would expect it to have had an early geological history much like Earth's, with impact cratering, volcanism, and tectonics.

Venus's thick cloud cover prevents us from seeing through to its surface, but we can study its geological features with radar. *Radar mapping* involves bouncing radio waves off the surface from a spacecraft and using the reflections to create three-dimensional images of the surface. From 1990 to 1993, the *Magellan* spacecraft used radar to map the surface of Venus, discerning features as small as 100 meters across. During its 4 years of observations, *Magellan* returned more data than all previous planetary missions combined. Scientists have named almost all the geological features for female goddesses and famous women.

Venusian Impacts and Volcanism Most of Venus is covered by relatively smooth, rolling plains with few mountain ranges. The surface has some impact craters (Figure 9.28a), but very few compared to the Moon or even Mars. Venus lacks very small craters, because small impactors burn up in its dense atmosphere. But large craters are also rare, indicating that they must be erased by other geological processes. Volcanism is clearly at work on Venus, as evidenced by an abundance of volcanoes and lava flows. Some are shield volcanoes, indicating familiar basaltic eruptions (Figure 9.28b). Some volcanoes have steeper sides, probably indicating eruptions of a higher-viscosity lava (Figure 9.28c).

Venusian Tectonics The most remarkable features on Venus are tectonic in origin. Its crust is quite contorted; in some regions, the surface appears to be fractured in a regular pattern (Figure 9.28d; see also Figure 9.19). Many of the tectonic features are associated with volcanic features, suggesting a strong linkage between volcanism and tectonics on Venus. One striking example of this linkage is a type of roughly circular feature called a *corona* (Latin for "crown") that probably resulted from a hot, rising plume in the mantle (Figure 9.28e). The plume pushed up on the crust, forming rings of tectonic stretch marks on the surface. The plume also forced lava to the surface, dotting the area with volcanoes.

Does Venus, like Earth, have plate tectonics that moves pieces of its lithosphere around? Convincing evidence for plate tectonics might include deep trenches and linear mountain ranges like those we see on Earth, but *Magellan* found no such evidence on Venus. The apparent lack of plate tectonics suggests that Venus's lithosphere is quite different from Earth's. Some geologists theorize that Venus's lithosphere is thicker and stronger, preventing its surface

a Impact craters like these are rare on Venus.

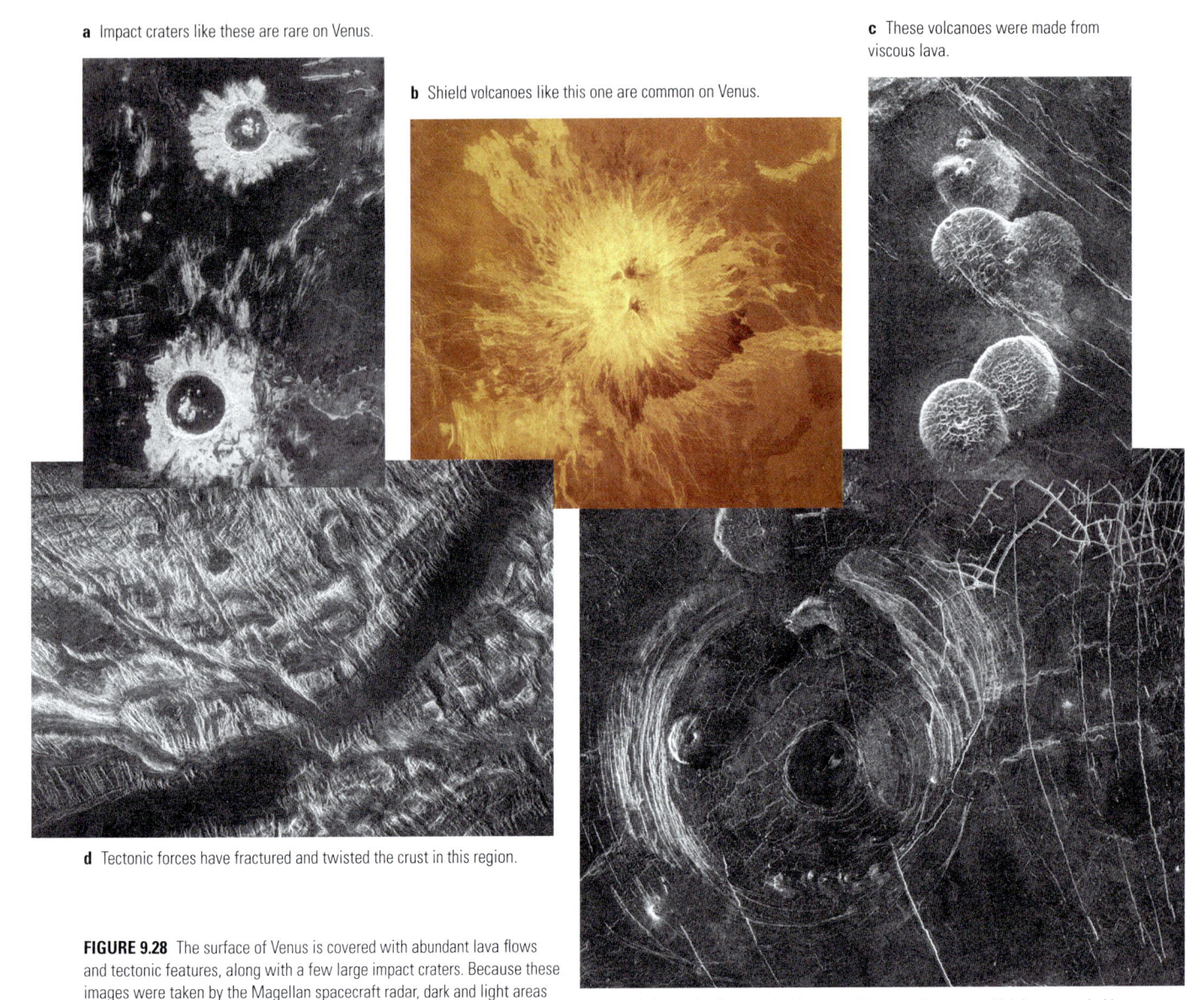

b Shield volcanoes like this one are common on Venus.

c These volcanoes were made from viscous lava.

d Tectonic forces have fractured and twisted the crust in this region.

e A mantle plume probably created the round corona, which is surrounded by tectonic stress marks.

FIGURE 9.28 The surface of Venus is covered with abundant lava flows and tectonic features, along with a few large impact craters. Because these images were taken by the Magellan spacecraft radar, dark and light areas correspond to how well radio waves are reflected, not visible light. Nonetheless, geological features stand out well. All images are several hundred kilometers across.

---

from fracturing into plates. Another possibility is that plate tectonics happens only occasionally on Venus. The uniform but low abundance of craters suggests that most of the surface formed around a billion years ago. Plate tectonics is one candidate for the process that wiped out older surface features and created a fresh surface at that time. For the last billion years, only volcanism, a bit of impact cratering, and small-scale tectonics have occurred.

Venusian Erosion? One might expect Venus's thick atmosphere to drive strong erosion, but the view both from orbit and from the surface suggests otherwise. The Soviet Union landed two probes on the surface of Venus in 1975. Before being destroyed by the 700 K heat, they returned images of a bleak, volcanic landscape with little evidence of erosion (see Figure 9.2). Venus apparently lacks the rains and winds that drive erosion, probably because of its high surface temperature and slow rotation rate.

The lack of strong erosion on Venus leaves the tectonic contortions of its surface exposed to view, even though some probably approach a billion years in age. Earth's terrain might look equally stark and rugged from space if not for the softening influences of wind, rain, and life. Like Earth, Venus probably has ongoing tectonic and volcanic activity, although we have no direct proof of it.

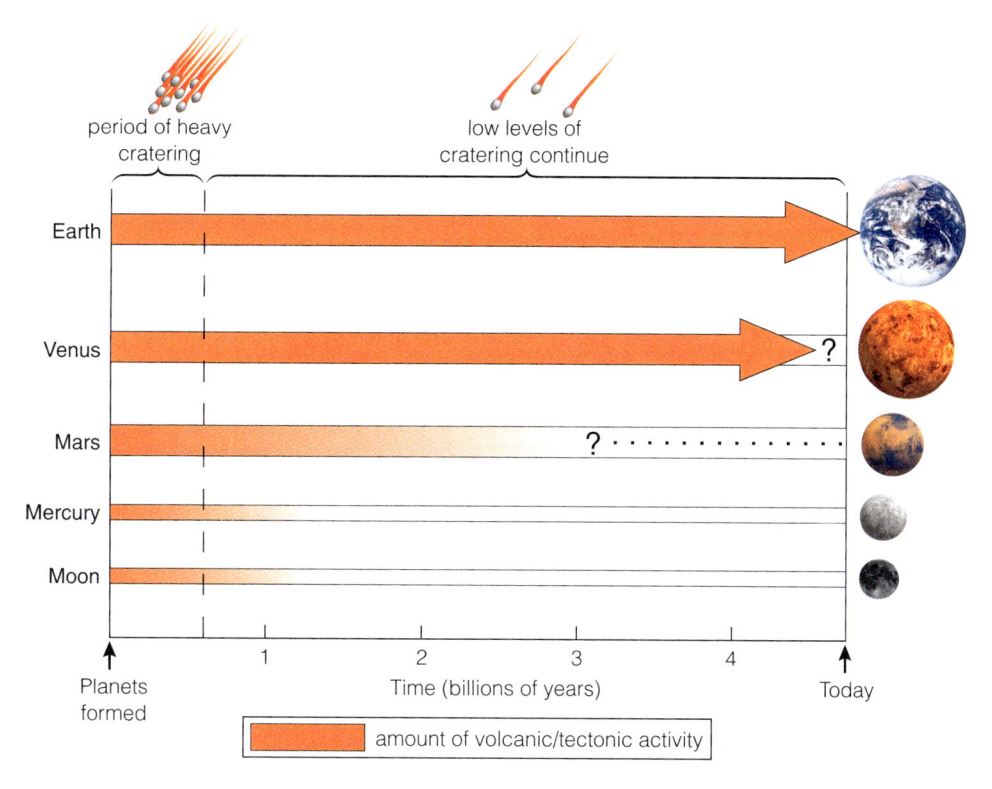

FIGURE 9.29 Volcanic/tectonic histories of the ancient worlds. This diagram summarizes the relationship between planetary size and volcanic/tectonic activity for the terrestrial worlds. The length of the red arrow shows the duration of activity: Earth is the only terrestrial world that we know to be active today, but we expect that Venus is still active as well. Most ancient craters have been erased on both. Mars may still have low levels of volcanism, and some ancient craters still remain. The narrower width of the arrows for Mercury and the Moon indicates that their volcanic and tectonic activity was never strong enough to wipe out all their heavily cratered surfaces. The diagram also shows the period of heavy cratering in the early history of the solar system. Erosion is not shown; it is negligible everywhere except Earth (where it is significant) and Mars (where it was once significant and endures to the present day at low levels).

Earth (6,378-km radius, 1.0 AU from Sun)

We've toured our neighboring terrestrial worlds and found a clear relationship between size and volcanic and tectonic activity. Thus, it should come as no surprise that Earth, the largest of the terrestrial worlds, has a high level of ongoing volcanism and tectonics. We've already discussed some of the volcanic and tectonic features on Earth, and we'll explore others in Chapter 13, where we'll also see how plate tectonics has affected Earth's climate and hence its biology.

Figure 9.29 summarizes the trends among the terrestrial worlds for volcanism and tectonics. It shows one of the key rules in planetary geology: Larger planets stay volcanically and tectonically active much longer than smaller planets. The situation for erosion is somewhat more complex. Earth has far more erosion than any other terrestrial world, but the most similar erosion is found on Mars, whose surface is sculpted by flowing water despite its small size. To understand erosion, we must look more closely at planetary atmospheres—the topic of our next chapter.

TIME OUT TO THINK *Suppose another star system has a rocky terrestrial planet that is much larger than Earth. What kind of geology would you expect it to have: a surface saturated with craters, or widespread volcanic and tectonic activity? Explain.*

THE BIG PICTURE

In this chapter, we have seen how the images returned from our robotic explorers have helped decipher the histories of the terrestrial worlds and have yielded clues to what their interiors must be like. As you continue your study of the solar system, keep the following "big picture" ideas in mind:

- Much of planetary geology can be distilled down to a few basic geological processes that depend on a handful of planetary properties.

- Every terrestrial world was once as heavily cratered as the Moon is today, but craters have been erased on the other worlds to varying degrees—mainly depending on each planet's size.

- Volcanism and tectonics depend primarily on a world's size, but erosion depends on characteristics of planetary atmospheres.

- Our model of planetary geology explains much of what we see on terrestrial worlds in our solar system and therefore might help us learn what happens on planets in other solar systems.

Review Questions

1. Briefly explain how the terrestrial planets ended up spherical in shape. Would you also expect small asteroids to be spherical? Why or why not?

2. What is *viscosity*? Give examples of substances with low viscosity and substances with high viscosity.

3. How does *differentiation* occur, and under what circumstances? Explain how differentiation led to the *core–mantle–crust* structure of the terrestrial worlds.

4. What is a *lithosphere*? What determines the strength and thickness of a planet's lithosphere?

5. Describe and distinguish among the three major processes by which planetary interiors get hot: *accretion, differentiation,* and *radioactivity.*

6. Describe and distinguish among the three major processes by which planetary interiors transfer heat to planetary surfaces: *conduction, convection,* and *eruption.* How does the heat escape from the planet's surface into space?

7. What is the primary factor in determining how long a planetary interior remains hot? Why?

8. What is a *magnetic field*? Contrast the magnetic fields of terrestrial planets.

9. Briefly describe the four major geological processes: *impact cratering, volcanism, tectonics,* and *erosion.*

10. Describe and distinguish between what we call *formation properties* and *geological controlling factors* in the context of planetary geology. Summarize in your own words the relationships shown in Figure 9.10.

11. Briefly describe how impacts affect planetary surfaces, and explain how we can estimate the geological age of a planetary surface from its number of impact craters.

12. What created the "soil" on the lunar surface? Explain why the footprints of the astronauts on the Moon will last a long time, but not forever.

13. Explain how differences in lava viscosity lead to *volcanic plains, shield volcanoes,* or *stratovolcanoes.*

14. Describe several types of internal stress that can drive tectonic activity. Under what conditions does a lithosphere break into *plates* and show *plate tectonics?*

15. What are *volatiles*, and how are they important to erosion? Briefly explain why Mercury and the Moon have virtually no erosion.

16. Briefly summarize the geological history of the Moon. How and when did the maria form? Why is the Moon geologically "dead" today?

17. Briefly summarize the geological history of Mercury. What is the Caloris Basin? How did Mercury get its many huge cliffs?

18. Briefly discuss the superficial similarities between Mars and Earth that led many people in the early 1900s to believe that Mars must harbor life. What were the *canali* that some astronomers reported on Mars? Do they really exist?

19. Briefly describe major geological features on Mars and their possible origins.

20. Describe the evidence for past water flow on Mars, including catastrophic floods and rainfall. Is it possible that water still flows on Mars today? Explain.

21. Briefly describe the evidence for volcanic and tectonic activity on Venus. Does Venus have plate tectonics? Does it show signs of significant erosion?

22. Briefly explain how Earth's geology fits the general patterns we should expect based on study of other terrestrial worlds.

Discussion Questions

1. *Making Models.* In this chapter, we developed a relatively simple conceptual model to describe the complex interactions that govern planetary geology. Use a similar approach to make a model for a complex problem of interest to you. The problem need not be astronomical; for example, you might consider global warming, changes in the stock market, or ways that class size could be reduced at your college or university. Your model need not use flowcharts, but it should include (1) some phenomenon you wish to understand, (2) the processes that affect it, and (3) the various factors those processes depend on.

2. *Geological Connections.* Consider the four geological processes: impact cratering, volcanism, tectonics, and erosion. Which two do you think are most closely connected to each other? Explain your answer, giving several ways in which the processes are connected.

3. *Testing the Model.* Our theories of planetary geology seem to explain the terrestrial worlds fairly well, but do they work elsewhere in the solar system?
   a. According to the model based on the terrestrial worlds, which of the four geological processes would you predict to be most important on Jupiter's moon Io (1,815-km radius) and Uranus's moon Miranda (243-km radius)?
   b. Based on the photos of these moons (Figure 11.17 and Figure 11.25), are your predictions correct?
   c. In a strictly logical sense, what would you conclude about our model of planetary geology based on your answer to part (b)? Explain.

Problems

Surprising Discoveries? For **problems 1–4,** suppose a spacecraft were to make the observations described. (These are *not* real discoveries.) In light of your understanding of planetary geology, decide whether the discovery should be considered reasonable or surprising. Explain your answer, tracing your logic back to the planet's formation properties. Explain.

1. The next mission to Mercury photographs part of the surface never seen before and detects vast fields of sand dunes.

2. A radar mapper in orbit around Venus detects a new volcano larger than any other on Venus but not seen by the Magellan spacecraft.

3. A Venus radar mapper discovers extensive regions of layered sedimentary rocks similar to those found on Earth and Mars.

4. Clear-cutting in the Amazon rain forest exposes vast regions of ancient terrain that is as heavily cratered as the lunar highlands.

5. *Earth vs. Moon.* Why is the Moon heavily cratered, but not Earth? Explain in a paragraph or two, relating your answer to their planetary formation properties.

6. *Erosion.* Consider erosion on Mercury, Venus, the Moon, and Mars. Which of these four worlds shows the greatest erosion? Why? Write a paragraph for each world explaining why the other three worlds do not have significant erosion and relating erosional activity to the four planetary formation properties.

7. *Miniature Mars.* Suppose Mars had turned out to be significantly smaller than its current size—say, the size of our Moon. How would this have affected the number of geological features due to each of the four major geological processes (impact cratering, volcanism, tectonics, and erosion)? Do you think Mars would still be a good candidate for harboring extraterrestrial life? Why or why not? Summarize your answers in two or three paragraphs.

8. *Mystery Planet.* It's the year 2098, and you are designing a robotic mission to a newly discovered planet around a star that is nearly identical to our Sun. The planet is as large in radius as Venus, rotates with the same daily period as Mars, and lies 1.2 AU from its star. Your spacecraft will orbit but not land.

    a. Some of your colleagues believe that the planet has no metallic core. How could you support or refute their hypothesis? (*Hint:* Remember that metal is dense and electrically conducting.)

    b. Other colleagues believe that the planet formed with few if any radioactive elements. How could images of the surface support or refute their hypothesis?

    c. Others suspect it has no atmosphere, but the instruments designed to study the planet's atmosphere fail due to a software error. How could you use the spacecraft's photos of geological features to determine whether a significant atmosphere is (or was) present on this planet?

9. *Dating Planetary Surfaces.* We have discussed two basic techniques for determining the age of a planetary surface: studying the abundance of impact craters, and radioactive dating of surface rocks [Section 8.6].

    a. Which technique seems more reliable? Why?

    b. Which technique is easier to use for planets besides Earth? Why?

*10. *Impact Energies.* A relatively small impact crater 10 km in radius could be made by a comet 1 km in radius traveling at 30 km/s.

    a. Assume the comet has a density of 1 g/cm^3 (1,000 kg/m^3), and calculate its total kinetic energy. *Hints:* First find the volume of the comet in m^3, then find its mass by multiplying its volume by its density, and finally calculate its kinetic energy. You'll need the following formulas; if you use mass in kg and velocity in m/s, the answer for kinetic energy will have units of joules:

    volume of sphere = $\frac{4}{3}\pi \times$ (radius)^3

    kinetic energy = $\frac{1}{2} \times$ mass $\times$ (velocity)^2

    b. Convert your answer from part (a) to megatons of TNT, the unit used for nuclear bombs. Comment on the degree of devastation that the impact of such a comet could cause if it struck a populated region on Earth. (*Hint:* One megaton of TNT releases 4.2×10^{15} joules of energy.)

11. *Project: Geological Properties of Silly Putty.*

    a. Roll room-temperature Silly Putty into a ball and measure its diameter. Place the ball on a table and gently place one end of a heavy book on it. After 5 seconds, measure the height of the squashed ball.

    b. Warm the Silly Putty in hot water and repeat the experiment in part (a). Then chill the Silly Putty in ice water and repeat. How did the different temperatures affect the rate of "squashing"?

    c. Did heating and cooling have a small or large effect on the rate of "squashing"? How does this experiment relate to planetary geology?

12. *Project: Planetary Cooling in a Freezer.* To simulate the cooling of planetary bodies of different sizes, use a freezer and two small *plastic* containers of similar shape but different size. Fill them with cold water and place them in the freezer. Checking every hour or so, record the time and your estimate of the thickness of the "lithosphere" (the frozen layer) in the two tubs.

(continued)

a. On a graph, plot the lithospheric thickness versus time for each tub. In which tub does the lithosphere thicken faster?

b. How long does it take each tub to freeze completely? What is the ratio of the two freezing times?

c. Describe in a few sentences the relevance of your experiment to planetary geology.

13. *Amateur Astronomy: Observing the Moon.* Any amateur telescope is adequate to identify geological features on the Moon. The light highlands and dark maria should be evident, and the shading of topography will be seen in the region near the terminator (the line between night and day). Try to observe near first- or third-quarter phase. Sketch the Moon at low magnification, and then zoom in on a region of interest, noting its location on your sketch. Sketch your field of view, label its features, and identify the geological process that created them. Look for craters, volcanic plains, and tectonic features. Estimate the horizontal size of the features by comparing them to the size of the whole Moon (radius = 1,738 km).

Web Projects

Find useful links for Web projects on the text Web site.

1. *Current Mars Exploration.* Find the latest results from current Mars missions and briefly explain how the latest results have changed or improved our understanding of Mars.

2. *Future Planetary Missions.* Learn about at least one upcoming mission designed to study a terrestrial world. Write a short summary of the mission and what it is designed to accomplish.

For the first time in my life I saw the horizon as a curved line. It was accentuated by a thin seam of dark blue light—our atmosphere. Obviously this was not the ocean of air I had been told it was so many times in my life. I was terrified by its fragile appearance.

Ulf Merbold, astronaut (Germany)

CHAPTER 10
Planetary Atmospheres
Earth and the Other Terrestrial Worlds

Life as we know it would be impossible on Earth without our atmosphere. In addition to supplying the oxygen we breathe, it shields us from harmful ultraviolet and X-ray radiation coming from the Sun, protects us from continual bombardment by micrometeorites, generates rain-giving clouds, and traps just enough heat to keep Earth habitable. We couldn't have designed a better atmosphere if we'd tried.

Despite forming under similar conditions, our neighboring planets ended up with atmospheres utterly inhospitable to us. No other planet in our solar system has air that we could breathe, or temperature and pressure conditions in which we could survive without a spacesuit. Mars, only slightly farther from the Sun than Earth, has an atmosphere that leaves its surface cold, dry, and barren. Venus, only slightly closer to the Sun, has an atmosphere that makes its surface conditions resemble a classical view of hell.

How did Earth end up with such fortunate conditions? To fully appreciate what happened on Earth, we must understand what happened on all the terrestrial worlds.

FIGURE 10.1 Views of the terrestrial worlds and their atmospheres from orbit and from the surface. The surface views for Mercury and Venus are artist's conceptions; the others are photos. The global views are photos taken from spacecraft—although the Venus image shows the clouds as they appear in ultraviolet light rather than visible light.

10.1 Planetary Atmospheres

All the planets of our solar system have atmospheres to varying degrees, as do several of the solar system's larger moons. The jovian planets are essentially atmosphere throughout—they are largely made of gaseous material. In contrast, the atmospheres of solid bodies—terrestrial worlds, jovian moons with atmospheres, and Pluto—are relatively thin layers of gas that contain only a minuscule fraction of a world's mass. Although we will concentrate on understanding the atmospheres of the terrestrial worlds, similar principles apply to the atmospheres of all the planets and satellites.

The terrestrial atmospheres are even more varied than the terrestrial geologies. The Moon and Mercury have very little atmosphere at all—even in broad daylight you would seem to be facing the blackness of space when standing on their surfaces. What little atmosphere they possess probably comes from materials vaporized as the surface is bombarded by micrometeorites, the solar wind, or energetic solar photons. These materials consist mostly of individual atoms of potassium, sodium, and oxygen, among others. If you could condense the entire atmosphere of the Moon into solid form, you would have so little material that you could store it in a basement.

At the other extreme, Venus is completely enshrouded by a thick atmosphere. If you stood on the surface of Venus, you'd feel a crushing pressure 90 times greater than that on Earth and a searing temperature higher than that in the hottest ovens used for cooking. Moving through the air on Venus would feel something like a cross between swimming and flying, because its density at the surface is nearly 10% of the density of liquid water. You couldn't breathe the air, because it consists almost entirely of carbon dioxide and does not contain any of the molecular oxygen (O_2) on which we depend. The surface views are perpetually overcast as weak sunlight filters though the thick clouds above. Violent storms on Venus's surface are virtually nonexistent, and corrosive chemicals that rain downward from high in the atmosphere evaporate before ever hitting the ground.

The atmosphere of Mars is also made mostly of carbon dioxide, but much less of it than on Venus. The result is very thin air with a pressure so low that your body tissues would bulge painfully if you stood on the surface without wearing a full space suit. In fact, the pressure and temperature are both so low that liquid water would rapidly disappear, some evaporating and some freezing into ice. Thus, the fact that we see dried-up riverbeds on Mars tells us that its atmosphere must have been very different in the past [Section 9.5]. Mars has seasons like Earth because of its similar axis tilt, and it is occasionally engulfed in planet-wide dust storms.

Earth's atmosphere of nitrogen and oxygen is the only one under which human life is possible, at least without the creation of an artificial environment.

What makes Earth's atmosphere so special? The best way to answer this question is to compare planetary atmospheres. Each atmosphere is unique, but, as we saw with geology in Chapter 9, similar properties and processes determine the characteristics of all atmospheres. Table 10.1 summarizes the properties of the terrestrial atmospheres, and Figure 10.1 shows representative views from above the worlds and from their surfaces. We'll spend the rest of this chapter learning how and why the terrestrial atmospheres came to differ so profoundly. Before we begin our comparative study, however, we need to discuss a few basic properties of atmospheres. Let's begin with the most basic question: What is an atmosphere?

Table 10.1 Atmospheres of the Terrestrial Worlds

World	Composition	Surface Pressure*	Winds, Weather Patterns	Clouds, Hazes
Mercury	helium, sodium, oxygen	10^{-14} bar	None: too little atmosphere	None
Venus	96% CO_2 3.5% N_2	90 bars	Slow winds, no violent storms, acid rain	Sulfuric acid clouds
Earth	77% N_2 21% O_2 1% argon H_2O (variable)	1 bar	Winds, hurricanes	H_2O clouds, pollution
Moon	helium, sodium, argon	10^{-14} bar	None: too little atmosphere	None
Mars	95% CO_2 2.7% N_2 1.6% argon	0.007 bar	Winds, dust storms	H_2O and CO_2 clouds, dust

*1 bar ≈ the pressure at sea level on Earth.

CHAPTER 10 PLANETARY ATMOSPHERES: EARTH AND THE OTHER TERRESTRIAL WORLDS   261

What Is an Atmosphere?

An *atmosphere* is a layer of gases that surrounds a world. The air that makes up any atmosphere is a mixture of many different gases that may consist either of individual atoms or of molecules. At the relatively low temperatures found throughout most planetary atmospheres, most atoms in gases combine to form molecules. For example, the air that we breathe consists of *molecular* nitrogen (N_2) and oxygen (O_2), as opposed to individual atoms (N or O). The wide variety of gases in planetary atmospheres falls into the category called *volatiles* [Section 9.4], materials that are usually gaseous or liquid at temperatures of a few hundred Kelvin. Common volatiles include nitrogen (N_2), oxygen (O_2), and water (H_2O) on Earth; carbon dioxide (CO_2) on Venus and Mars; and hydrogen (H_2), helium (He), methane (CH_4), and ammonia (NH_3) on the jovian planets.

Even at moderate temperatures, individual air molecules whiz around at high speeds and frequently collide with one another. On Earth, the nitrogen and oxygen molecules fly around at average speeds of about 500 meters per second—fast enough to cross your bedroom a hundred times in a second. Given that a single breath of air contains some 10^{22} molecules, you can imagine how frequently molecules collide. In fact, each molecule in the air around you will suffer a million collisions in the time it takes to read this paragraph.

All these collisions in a gas create the **gas pressure**. An easy way to understand pressure is to think about a balloon. The air molecules inside a balloon are constantly colliding with the balloon walls, exerting a pressure that tends to make the balloon expand. Because the particles in a gas move every which way, the pressure acts equally in all directions. At the same time, air molecules outside the balloon collide with the walls from the other side, tending to make the balloon contract. A sealed balloon remains stable when the inward pressure and the outward pressure are balanced. (We are neglecting any tension in the balloon material.) Now imagine that you blow more air into the balloon. The extra molecules inside mean more collisions with the balloon wall, momentarily making the pressure inside greater than the pressure outside. The balloon therefore expands until the inward and outward pressures are again in balance. If you heat the balloon, the gas molecules begin moving faster and therefore collide harder with the walls. Thus, heating a balloon also makes the inside pressure momentarily greater than the outside pressure, causing the balloon to expand. Conversely, cooling a balloon makes it contract, because the outside pressure momentarily exceeds the inside pressure.

We can understand *atmospheric pressure* using similar principles. Gas in an atmosphere is held down by gravity. The atmosphere over any area has some weight that presses downward, tending to compress the atmosphere beneath it. At the same time, gas pressure pushes in all directions, including upward, which tends to make the atmosphere expand. Planetary atmospheres exist in a perpetual balance between the downward weight of their gases and the upward push of their gas pressure. (We call this balance *gravitational equilibrium* [or *hydrostatic equilibrium*]; it is also very important in stars [Section 14.1].) The higher you go in an atmosphere, the less the weight of the gas above you. Thus, the pressure must also become less as you go upward, which explains why the pressure decreases as you climb a mountain or ascend in an airplane. You can visualize what happens by imagining the atmosphere as a very big stack of pillows. The pillows at the bottom are very compressed because of the weight of all the pillows above. As you go upward, the pillows are less and less compressed because less weight lies on top of them. The standard unit of pressure is the **bar** (as in *barometer*) and is roughly equal to Earth's atmospheric pressure at sea level.[1]

TIME OUT TO THINK *There is pressure under water just as there is pressure under air. Use this fact to explain why the pressure increases dramatically with depth in the oceans. How does this pressure affect deep-sea divers and submarines?*

Where Does an Atmosphere End?

The higher you go in an atmosphere, the lower the pressure becomes. Atmospheres therefore do not have clear upper boundaries but instead gradually fade away with increasing altitude. Nevertheless, there are a number of ways to characterize the thickness of an atmosphere surrounding a planet. One simple way is to ask when you would seem to have left the atmosphere and entered "space." Atmospheres scatter sunlight, which is what prevents us from seeing stars in the daytime. Thus, you might say that you had reached "space" when you had risen high enough that you could see stars in the daytime. The terrestrial atmospheres are remarkably thin when viewed in this way. For example, stars become visible in the daytime at an altitude of a few tens of kilometers above the Earth—less than 1% of the Earth's radius (Figure 10.2a). But even the Space Shuttle, orbiting hundreds of kilometers above the Earth's surface, runs into the upper vestiges of Earth's atmosphere (Figure 10.2b). So the question of where an atmosphere ends is more one of semantics than of science.

---

[1] 1 bar ≈ 14.7 pounds/inch^2 (≈ 100,000 newtons/m^2). That is, if you gathered up all the air directly above 1 square inch of the Earth's surface, it would weigh 14.7 pounds. You don't feel this weight on your shoulders because the pressure is also pushing up under your arms, inward against your sides, and so on in all directions. Also, the fluids in your body push outward with the same pressure.

altitude. The uppermost portion of the atmosphere is called the **exosphere**. The exosphere is the region where the atmosphere "fades away" into space; it is hot because the gases absorb X rays from the Sun. The remaining X rays are absorbed in the next layer down, called the **thermosphere**. These X rays ionize atoms in the part of the thermosphere that we call the **ionosphere**. Ultraviolet photons are absorbed by certain gases in the **stratosphere**.[2] Visible photons penetrate the entire atmosphere, and some are absorbed by the ground. The ground radiates away this energy as infrared light, which is trapped in the lowest atmospheric layer, called the **troposphere**. In the rest of this section, we will investigate in more detail the processes that make these basic atmospheric layers, from the ground up.

TIME OUT TO THINK *As we will discuss shortly, some planetary atmospheres lack ultraviolet-absorbing gases and therefore lack stratospheres. How would the structure of such a planetary atmosphere differ from that of the generic atmosphere shown in Figure 10.6?*

Visible Light: Warming the Surface and Coloring the Sky

The Sun emits most of its energy in the form of visible light, and atmospheres are generally transparent to visible light. Thus, it is primarily visible light from the Sun that reaches and warms a planetary surface.

Although most visible light passes directly through the atmosphere, some is scattered by molecules in the atmosphere. This scattering of sunlight prevents us from seeing stars in the daytime. If our atmosphere did not scatter light, you'd be able to block out the light of the Sun simply by holding your hand in front of it, and then you'd have no trouble seeing other stars. Scattering of light also prevents shadows from being pitch black. On the Moon, where there is no air to scatter light, you can see stars in the daytime, and shadowed regions are extremely dark.

The scattering of light by the atmosphere also explains why the sky is blue. It turns out that molecules scatter blue light (higher frequencies) much more effectively than red light (lower frequencies). Thus, although the Sun illuminates the atmosphere with all colors of light, effectively only the blue light gets scattered. When the Sun is overhead, this scattered blue light reaches your eyes from all directions, explaining why the sky appears blue (Figure 10.7). At sunset (or sunrise), the sunlight must pass through

---

[2] Technically, the stratosphere is only the region where the temperature rises with altitude above the troposphere. The region where the temperature falls again is called the *mesosphere*. We will not make this distinction in this book.

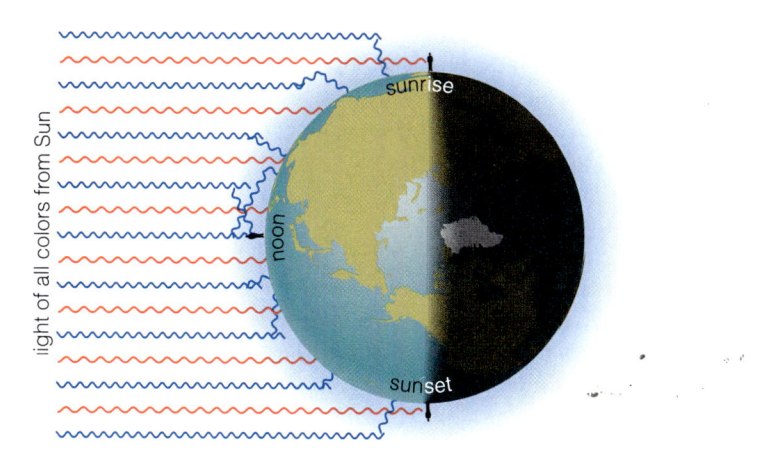

FIGURE 10.7 Atmospheric gases scatter blue light more than they scatter red light. During most of the day, you therefore see blue photons coming from most directions in the sky, making the sky look blue. But only the red photons reach your eyes at sunrise or sunset, when the light must travel a longer path through the atmosphere to reach you.

a greater amount of atmosphere on its way to you. Thus, most of the blue light is scattered away from you altogether (making someone else's sky blue), leaving only red light to color your sunset.

TIME OUT TO THINK *Suppose atmospheric molecules didn't scatter light at all. In that case, what color would the sky be? Explain.*

Infrared Light, the Greenhouse Effect, and the Troposphere

The relatively small amount of sunlight emitted farther in the infrared is mostly absorbed in our generic atmosphere before it reaches the ground. But the planet itself emits significant amounts of infrared radiation, because the planet must radiate exactly as much energy as it absorbs from the Sun to maintain a constant temperature. Herein lies the key to how atmospheres make planetary surfaces warmer than they would be otherwise.

Figure 10.8 shows how the **greenhouse effect** works.[3] Carbon dioxide, water vapor, and other *greenhouse gases* in the atmosphere absorb some of the infrared radiation emitted upward from the planet's surface. These gases therefore warm up and emit infrared thermal radiation themselves—but in all directions. Some of this radiation is directed back

---

[3] The name *greenhouse effect* comes from botanical greenhouses, usually built mostly of glass, which keep the air inside them warmer than it would be otherwise. But it turns out that greenhouses trap heat simply by not letting hot air rise, a very different way of trapping heat than the infrared absorption that leads to the "greenhouse effect" in planetary atmospheres.

down toward the surface, making the surface warmer than it would be from absorbing visible sunlight alone. The more greenhouse gases present, the greater the degree of greenhouse warming.

TIME OUT TO THINK *Clouds on Earth are made of water, and H₂O is a very effective greenhouse gas—even when the water vapor condenses into droplets in clouds. Use this fact to explain why clear nights tend to be colder than cloudy nights. (Hint: Sunlight warms the surface only in the daytime, but the surface radiates its heat away both day and night.)*

Air is denser at lower altitudes, so the greenhouse effect primarily affects temperatures in the troposphere. Above the troposphere, there is so little air that greenhouse gases cannot effectively trap infrared radiation. The fact that the greenhouse effect makes air warmer near the ground drives *convection* [Section 9.3] in the troposphere. The warm air near the ground expands and rises, while cooler air near the top of the troposphere contracts and falls. The descending cooler air gets heated, and the cycle repeats. The troposphere gets its name from this convection: *Tropos* is Greek for "turning." The process of convection carries heat upward and continually churns the air in the troposphere, causing the ever-changing conditions that we call the weather.

Local weather patterns called *inversions* occasionally make the air colder near the surface than higher up in the troposphere—the opposite of the usual condition, in which the troposphere is warmer at the bottom. Over cities on Earth, inversions often lead to smog problems by inhibiting the convection that normally carries pollution away to higher altitudes.

Ultraviolet Light and the Stratosphere

The primary source of heat in the troposphere—the absorption of infrared radiation from the ground—is not important in higher layers of the atmosphere. Once infrared radiation from the ground reaches the top of the troposphere, it travels essentially unhindered to space. Thus, we need to consider only the effects of sunlight in investigating atmospheric structure above the troposphere. As we move upward through the stratosphere, the primary source of atmospheric heating is absorption of ultraviolet light from the Sun. This heating tends to be stronger at higher altitudes because the ultraviolet light gets absorbed before it reaches lower altitudes.

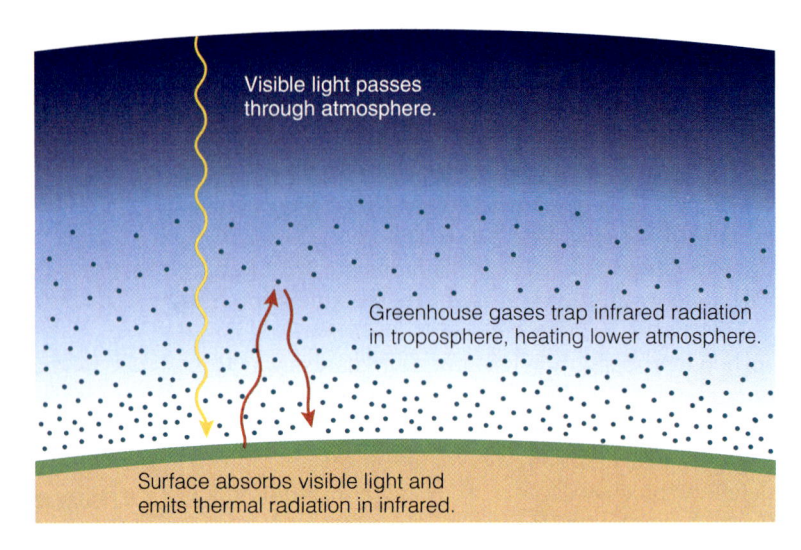

FIGURE 10.8 The greenhouse effect: The troposphere becomes warmer than it would be if it had no greenhouse gases. Note that the surface emits thermal radiation at longer wavelengths than the visible light that heated it. (See Section 7.2.)

As a result, the stratosphere gets warmer with increasing altitude—the opposite of the situation in the troposphere. Convection therefore cannot occur in the stratosphere: Heat cannot rise if the air is even hotter higher up. (Convection lessens even in the upper troposphere, where temperature changes slowly with altitude. Thus, airplane rides become smoother at high altitudes because less convection means less turbulence.) The stratosphere gets its name because the lack of convection makes its air relatively stagnant and *stratified* (layered), rather like a sitting jar of oil and water. The lack of convection also means that the stratosphere essentially has no weather and no rain. That is why pollutants that reach the stratosphere—such as ozone-destroying chlorofluorocarbons (CFCs) [Section 13.5]—remain there for decades.

Note that a planet can have a stratosphere *only* if its atmosphere contains molecules that are particularly good at absorbing ultraviolet photons. Ozone plays this role on Earth. In fact, Earth is the only terrestrial planet with such an ultraviolet-absorbing layer and hence the only terrestrial planet with a stratosphere—at least in our solar system. (We'll discuss why Earth is unique in this way in Chapter 13.) The jovian planets also have stratospheres, as we'll discuss in the next chapter.

X Rays and the Thermosphere and Ionosphere

Virtually all gases are good X-ray absorbers, because X rays have sufficient energy to ionize almost any atom and more than enough to dissociate almost any molecule. Solar X rays are absorbed by the first gases they encounter as they enter the atmosphere. This

Common Misconceptions: Higher Altitudes Are Always Colder

Many people think that the low temperatures in the mountains are just the result of lower pressures, but Figure 10.9 shows that it's not that simple. The higher temperatures near sea level on Earth are a result of the greenhouse effect trapping more heat at lower altitudes. If Earth had no greenhouse gases, mountaintops wouldn't be so cold—or, more accurately, sea level wouldn't be so warm.

X-ray absorption strongly heats the upper atmosphere in the daytime, usually to much higher temperatures than those on the surface. That is why this upper region of the atmosphere is called the *thermo*sphere. Despite its high temperatures, the gas in the thermosphere would not feel hot to your skin because its density and pressure are so low [Section 4.2].

The thermosphere contains a small but important fraction of its gas as charged ions and free electrons—the result of ionization. The layer within the thermosphere that contains most of these ions is called the *ionosphere*. The ionosphere is very important for radio communications on Earth. Most radio broadcasts are completely reflected back to Earth's surface by the ionosphere, almost as though the Earth were wrapped in aluminum foil. Without this reflection by the ionosphere, radio communication would work only between locations in sight of each other.

The Exosphere

The exosphere is the region that marks the fuzzy "boundary" between the atmosphere and space. (The prefix *exo* means "outermost" or "outside.") The gas density in the exosphere is so low that collisions between atoms or molecules are very rare. The high temperatures in the exosphere give the individual atoms or molecules high speeds, sometimes allowing them to escape from the planet. The Space Station and many satellites actually orbit the Earth within its exosphere, and the small amount of gas in the exosphere can exert a bit of atmospheric drag. That is why satellites in low-Earth orbit eventually spiral deeper into the atmosphere and burn up [Section 5.4].

Comparative Structure of the Terrestrial Atmospheres

Now that we understand the structure of our generic atmosphere, let's see how it corresponds to actual terrestrial atmospheres. The atmospheres of the Moon and Mercury have so little gas that they essentially possess only the top layer, the exosphere. Venus, Earth, and Mars have denser atmospheres and are more comparable (Figure 10.9). The similarities and differences among these three planetary atmospheres make sense in light of our study of the generic atmosphere. All three have a warm troposphere at the bottom, created by the greenhouse effect, and a warm thermosphere at the top, where solar X rays are absorbed. The greatest difference is the extra "bump" of Earth's stratosphere, where ultraviolet light is absorbed. Without the warming in our unique layer of stratospheric ozone, Earth's middle altitudes would be almost as cold as those on Mars.

Although the structures of the tropospheres are similar on Venus, Earth, and Mars, the tropospheric temperatures on the three planets are dramatically different. On Venus, the thick CO_2 atmosphere causes a greenhouse effect that raises the surface temperature about 500 K above what it would be without greenhouse gases. The thin CO_2 atmosphere of Mars creates only a weak greenhouse effect, making its surface temperature only 5 K warmer than its "no greenhouse" temperature. Earth has a moderate greenhouse effect that increases its temperature about 40 K, making it comfortable for life. The greenhouse effect on Earth is almost entirely due to gases that make up less than 2% of our atmosphere: mainly carbon dioxide, water vapor, and methane. The major constituents of Earth's atmosphere, N_2 and O_2, have no effect on infrared light and do not contribute to the greenhouse effect. (Molecules with only two atoms—and especially molecules with two of the

FIGURE 10.9 Temperature profiles for Venus, Earth, and Mars. Note that Venus and Earth are considerably warmer than they would be without the greenhouse effect. (The thermospheres and exospheres, not shown, are qualitatively alike.)

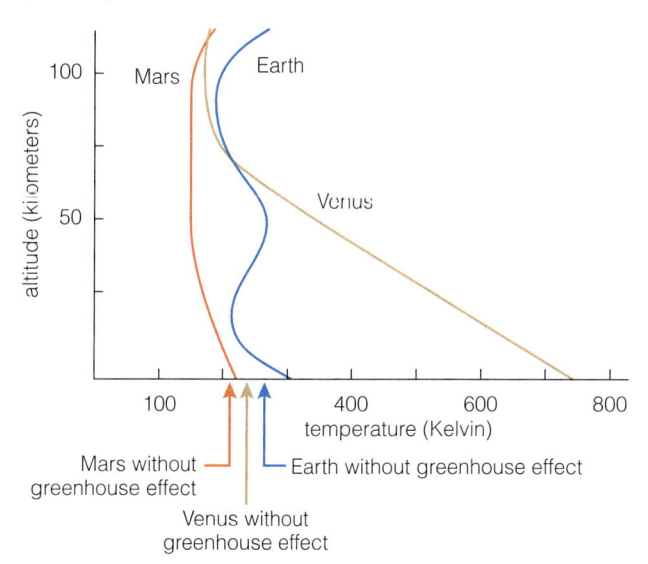

same kind of atom—are poor infrared absorbers because they have very few ways to vibrate and rotate.) If they did contribute to the greenhouse effect, Earth might be too warm for life.

10.3 Magnetospheres and the Solar Wind

One more type of energy that we haven't yet considered comes from the Sun: the low-density breeze of charged particles that we call the *solar wind* [Section 8.4]. Although the solar wind does not significantly affect atmospheric structure, it can have other important influences. Among the terrestrial planets, only Earth has a strong enough magnetic field to divert the charged particles of the solar wind. Thus, solar wind particles can impact the exospheres of Venus and Mars—and the surfaces of Mercury and the Moon. In contrast, Earth's strong magnetic field creates a **magnetosphere** that acts like a protective bubble surrounding the planet (Figure 10.10). (Jovian planets also have magnetospheres [Section 11.4].) Charged particles cannot easily pass through a magnetic field, so a magnetosphere diverts most of the solar wind around a planet.

Some solar wind particles manage to infiltrate Earth's magnetosphere at its most vulnerable points, near the magnetic poles. Once inside, they have a difficult time escaping. Charged particles are restricted by magnetic forces to following magnetic field lines (Figure 10.10a). In many magnetospheres, the ions and electrons accumulate to make **charged particle belts** encircling the planet. (The charged particle belts around the Earth are called the *Van Allen belts* after their discoverer.) The high energies of the particles in charged particle belts can be very hazardous to spacecraft and astronauts passing through them.

Particles trapped in magnetospheres also cause the beautiful spectacle of light called the **aurora** (Figure 10.10b). If a trapped charged particle has enough energy, it can follow the magnetic field all the way down to the upper parts of a planet's atmosphere. The charged particles collide with atmospheric atoms and molecules, causing them to radiate and produce auroras (Figure 10.10c,d). Because the charged particles follow the magnetic field, auroras are most common near the magnetic poles. On Earth, auroras therefore are best viewed from Canada, Alaska, and Russia in the Northern Hemisphere (the *aurora borealis*, or northern lights) and from Australia, Chile, and Argentina in the Southern Hemisphere (the *aurora australis*, or southern lights).

Common Misconceptions: The Greenhouse Effect Is Bad

The greenhouse effect is often in the news, usually in discussions about environmental problems. But the greenhouse effect itself is not a bad thing. In fact, we could not exist without it. Remember that the "no greenhouse" temperature of the Earth is well below freezing. Thus, the greenhouse effect is the only reason why our planet is not frozen over. Why, then, is the greenhouse effect discussed as an environmental problem? [Section 13.5] It is because human activity is adding more greenhouse gases to the atmosphere—which might change the Earth's climate. After all, while the greenhouse effect makes the Earth livable, it is also responsible for the searing 740 K temperature of Venus—proving that it's possible to have too much of a good thing.

10.4 Weather and Climate

So far, we've mostly talked about a planet's atmosphere as if it were the same at all locations and at all times. The atmospheric structure profiles show only *average* temperatures at different altitudes in the atmosphere. But we know from experience on Earth that surface and atmospheric conditions constantly change. In any particular location, some days may be hotter or cooler than others, some may be clearer or cloudier, some may be calmer or stormier. This ever-varying combination of winds, clouds, temperature, and pressure is what we call **weather**.

Local weather can vary dramatically from one day to the next, or even from hour to hour. The weather can also vary greatly between places just a short distance apart. For example, temperatures on a warm day may be several degrees hotter just a few kilometers inland from a coastal city. The complexity of weather makes it difficult to predict, although modern-day *meteorologists*[4] (people who study weather) often can predict weather quite accurately a few days in advance. On longer time scales, local weather is fundamentally unpredictable, at least on Earth.

Climate is the long-term average of weather and is generally stabler than weather. Deserts remain deserts and rain forests remain rain forests over periods of hundreds or thousands of years, while the day-to-day and even year-to-year weather may vary dramatically. Weather and climate can be hard to distinguish on a human time scale: Do a few hot sum-

---

[4]The word *meteor* comes from a Greek word meaning "a thing in the air," which explains why the very different concepts of *meteorology* and *meteors* have the same word origin.

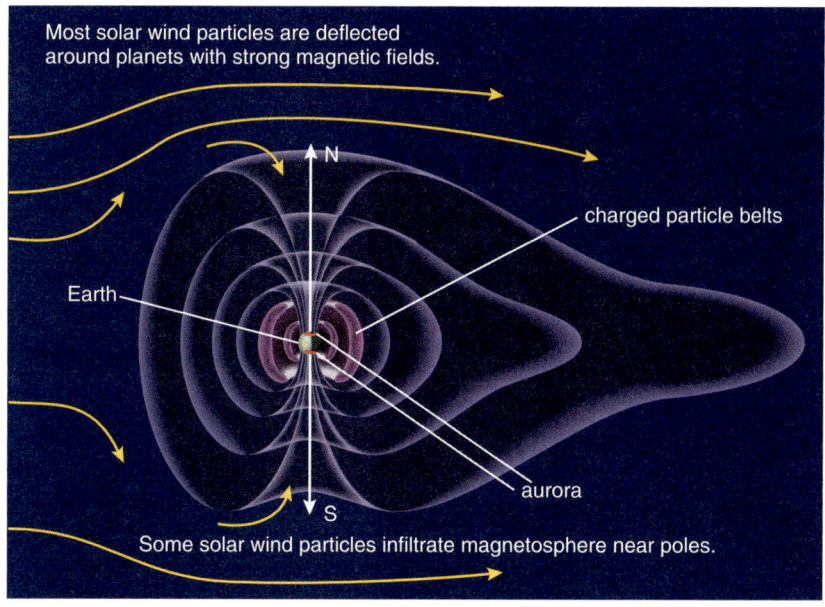

a Earth's magnetosphere and its interaction with the solar wind.

b Aurora along the coast of Norway.

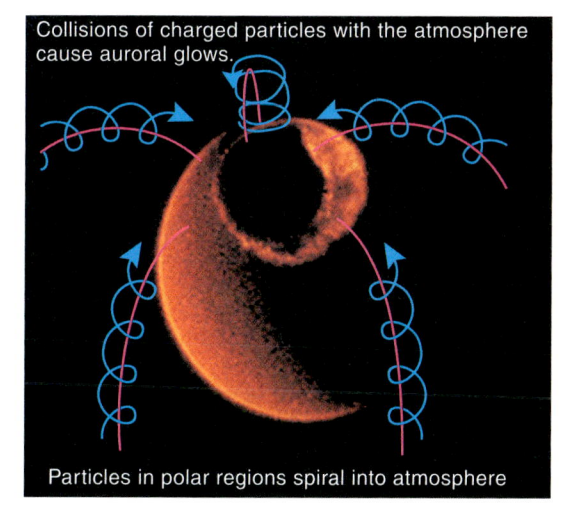

c Origin of the aurora.

FIGURE 10.10 A planet's magnetosphere acts like a protective bubble that shields the surface from charged particles coming from the solar wind. Among the terrestrial planets, only the Earth has a strong enough magnetic field to create a magnetosphere. The Earth's magnetosphere allows charged particles to strike the atmosphere only near the poles, thereby creating the phenomena of the aurora borealis and aurora australis.

d Earth's aurora as seen from the Space Shuttle.

---

mers in a row imply a change in climate, or are they merely coincidence? We'll return to this question when we study the issue of global warming on Earth in Chapter 13. For now, let's concentrate on understanding what drives planetary weather.

Seasonal Weather Patterns

The most obvious weather pattern on a planet is the change in the seasons. In general, a planet has seasons only if it has a significant axis tilt, so that over the course of its orbit around the Sun, first one hemisphere and then the other receives more direct sunlight [Section 1.3]. Otherwise, both hemispheres receive the same amount of sunlight year-round. For example, Venus has no seasons because its axis is nearly perpendicular to the ecliptic. Mars has seasonal changes like Earth because its 25° axis tilt is comparable to Earth's $23\frac{1}{2}°$ tilt. Thus, when one Martian hemisphere is in summer, the other is in winter, and vice versa. However, Martian seasons last almost

CHAPTER 10 PLANETARY ATMOSPHERES: EARTH AND THE OTHER TERRESTRIAL WORLDS 271

THINKING ABOUT . . .

Weather and Chaos

Scientists today have a very good understanding of the physical laws and mathematical equations that govern the behavior and motion of atoms in the air, oceans, and land. Why, then, do we have so much trouble predicting the weather? For a long time, most scientists assumed that the difficulty of weather prediction would go away once we had enough weather stations to collect data from around the world and sufficiently powerful computers to deal with all the data. However, we now know that weather is fundamentally unpredictable on time scales longer than a few weeks. To understand why, we must look at the nature of scientific prediction.

Suppose you want to predict the location of a car on a road 1 minute from now. You need two basic pieces of information: where the car is now, and how it is moving. If the car is now passing Smith Road and heading north at 1 mile per minute, it will be 1 mile north of Smith Road in 1 minute.

Now, suppose you want to predict the weather. Again, you need two basic types of information: (1) the current weather and (2) how weather changes from one moment to the next.

You could attempt to predict the weather by creating a "model world." For example, you could overlay a globe of the Earth with graph paper and then specify the current temperature, pressure, cloud cover, and wind within each square. These are your starting points, or initial conditions. Next, you could input all the initial conditions into a computer, along with a set of equations (physical laws) that describe the processes that can change weather from one moment to the next.

Suppose the initial conditions represent the weather around the Earth at this very moment and you run your computer model to predict the weather for the next month in New York City. The model might tell you that tomorrow will be warm and sunny, with cooling during the next week and a major storm passing through a month from now. Now suppose you run the model again but make one minor change in the initial conditions—say, a small change in the wind speed somewhere over Brazil. For tomorrow's weather, this slightly different initial condition will not change the weather prediction for New York City. For next week's weather, the new model may yield a slightly different prediction. But for next month's weather, the two predictions may not agree at all!

The disagreement between the two predictions arises because the laws governing weather can cause very tiny changes in initial conditions to be greatly magnified over time. This extreme sensitivity to initial conditions is sometimes called the *butterfly effect*: If initial conditions change by as much as the flap of a butterfly's wings, the resulting prediction may be very different.

The butterfly effect is a hallmark of *chaotic systems*. Simple systems are described by linear equations in which, for example, increasing a cause produces a proportional increase in an effect. In contrast, chaotic systems are described by nonlinear equations, which allow for subtler and more intricate interactions. For example, the economy is nonlinear because a rise in interest rates does not automatically produce a corresponding change in consumer spending. Weather is nonlinear because a change in the wind speed in one location does not automatically produce a corresponding change in another location. Many (but not all) nonlinear systems exhibit chaotic behavior.

Despite their name, chaotic systems are not completely random. In fact, many chaotic systems have a kind of underlying order that explains the general features of their behavior even while details at any particular moment remain unpredictable. In a sense, many chaotic systems are "predictably unpredictable." Our understanding of chaotic systems is increasing at a tremendous rate, but much remains to be learned about them.

---

twice as long as Earth seasons because a Martian year is almost twice as long as an Earth year.

A planet's changing distance from the Sun can affect the severity of seasons. Earth's nearly circular orbit makes this effect unimportant. But Mars has a more elliptical orbit that puts it significantly closer to the Sun during southern hemisphere summer, making it warmer than summer in the northern hemisphere, and farther from the Sun during southern hemisphere winter, making it colder than winter in the northern hemisphere (Figure 10.11). Mars therefore has more extreme seasons in its southern hemisphere than in its northern hemisphere.

Global Wind Patterns

Winds are flows of air caused by pressure differences between different regions of a planet. On a local scale, winds vary as much as the weather, blowing in different directions and with different strengths at different times. But certain wind patterns are fixed on a global scale, creating what we call **global wind patterns** (or *global circulation*).

The first major factor affecting global wind patterns is atmospheric heating. In general, the equatorial regions of a planet receive more heat from the Sun than do polar regions. The excess heat makes the atmosphere expand above the equator, and air spills northward and southward at high altitudes (Figure 10.12). By itself, this process creates two huge **circulation cells** resembling convection cells but much larger. (They are often called *Hadley cells* after the person who first suggested their existence in 1735.) The circulation cells serve to transport heat both from lower to higher altitudes and from the equator to the poles. If this circulation is very efficient and if the atmosphere is dense enough to carry a lot of thermal energy, the resulting winds can equalize planetary temperatures from equator to pole. That is the case on Venus, where temperatures near the poles are essentially the same as those at the equator.

TIME OUT TO THINK *Suppose a planet rotated synchronously with its orbit around the Sun, so that it kept the same face toward the Sun at all times. What kind of circulation pattern would you expect on this planet? Explain. (Hint: Which regions of the planet receive the most sunlight, and which regions receive the least sunlight?)*

The second major factor affecting global wind patterns is the planet's rotation. It's easiest to understand this effect by analogy. The movement of air over a planet is like the motion of a ball rolled on a merry-go-round (Figure 10.13). Imagine that you are sitting near the edge of a merry-go-round. Your speed around the axis will be higher than the speeds of the inner parts of the merry-go-round. If you roll a ball toward the center, the ball begins with your relatively high speed around the axis. As it rolls inward, this high speed around the axis makes it move ahead of the slower-moving inner regions. If your merry-go-round rotates counterclockwise, the ball therefore deviates to the right instead of heading straight inward. The deviation caused by the merry-go-round's rotation is called the **Coriolis effect**. It also works in reverse: If you roll the ball outward from the center, the ball lags behind the faster-moving outer regions, again deviating to the right. If your merry-go-round instead rotates clockwise, the deviations are to the left.

The Coriolis effect occurs on rotating planets because equatorial regions travel faster around the rotation axis than do polar regions (see Figure 1.10). On Earth, air moving away from the equator deviates ahead of the Earth's rotation to the east, and air moving toward the equator deviates behind the Earth's rotation to the west (Figure 10.14a). Note that, either way, the deviation is to the *right* in the Northern Hemisphere and to the *left* in the Southern Hemisphere. One result is that air flowing toward low-pressure zones ends up circulating around them counterclockwise in the Northern Hemisphere and clockwise in the Southern Hemisphere (Figure 10.14b). Low-pressure regions ("L" on weather maps) are often associated with storms, including hurricanes. Because the circulating winds prevent air from flowing in to equalize the pressure, low-

Seasons on Mars

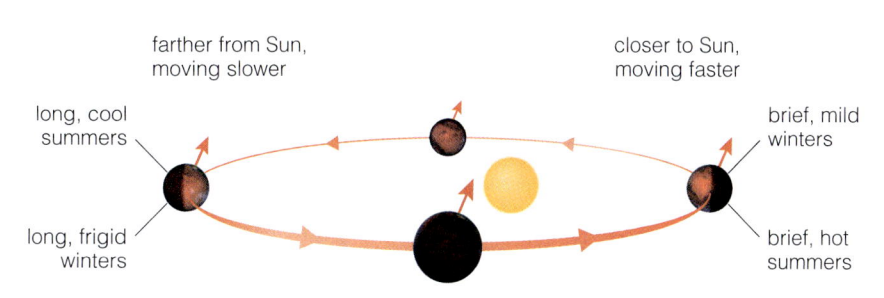

FIGURE 10.11 The ellipticity of Mars's orbit makes the severity of seasons different in the northern and southern hemispheres.

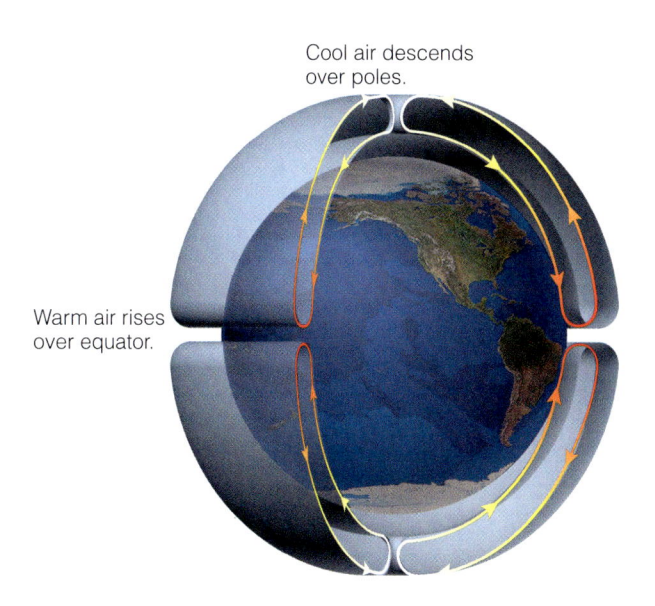

FIGURE 10.12 Circulation cells. Heat rises above the equator, setting up a flow of warm air toward the poles at high altitudes and a flow of cool air toward the equator near the surface. The planet's rotation is neglected for the moment. (Figure 10.15 shows how rotation affects circulation cells on Earth.)

FIGURE 10.13 The Coriolis effect on a merry-go-round rotating counterclockwise: A ball rolled inward starts with a speed around the center faster than that of regions nearer the center; this "extra" speed makes the ball deviate to the right. A ball rolled outward starts with a speed around the center slower than that of regions farther out; this "lag" also makes the ball deviate to the right. If the merry-go-round were rotating clockwise, both balls would veer to the left.

CHAPTER 10 PLANETARY ATMOSPHERES: EARTH AND THE OTHER TERRESTRIAL WORLDS 273

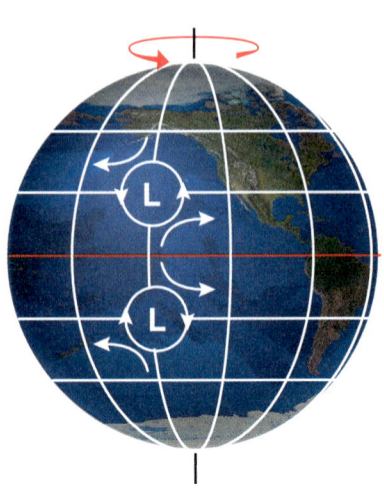

a The Coriolis effect causes air moving across the surface toward low-pressure regions to be diverted into a circular flow around the regions. The direction of rotation is opposite in the two hemispheres.

b You can see the direction of flow in a satellite photo of storms in the Pacific.

FIGURE 10.14 The Coriolis effect drives wind patterns on rotating planets.

Common Misconceptions: Do Toilets Flush Backward in the Southern Hemisphere?

A common myth holds that water circulates "backward" in the Southern Hemisphere; that is, toilets flush the opposite way, water spirals down sink drains the opposite way, and so on. If you visit an equatorial town, you may even find charlatans who, for a small fee, will demonstrate how to change the direction of water spiraling down a drain simply by stepping across the equator. This myth sounds similar to the Coriolis effect—which really does make hurricanes spiral in opposite directions in the two hemispheres—but it is completely untrue. The Coriolis effect is noticeable only when material moves significantly closer to or farther from the Earth's rotation axis—which means scales of hundreds of kilometers. Its effect is completely negligible on small scales, such as the scale of toilets or drains. Apart from the shape of the basin, the only thing that affects the direction in which the water in a toilet or sink swirls is the initial angular momentum, which you can affect by swirling the water in one direction or the other. Conservation of angular momentum [Section 5.1] then dictates that the water will swirl faster and faster as it gets closer to the drain. You can find water in toilets and sinks swirling in either direction, and even tornadoes twisting in either direction, in both hemispheres.

pressure regions tend to remain stable for many days, during which time prevailing winds can carry them thousands of kilometers across the planet. High-pressure regions ("H" on weather maps) can also last for days and travel great distances, but they are usually storm-free.

The Coriolis effect also diverts the global wind patterns. On Earth, this diversion causes each of the two huge circulation cells (shown in Figure 10.12) to split into three smaller cells (Figure 10.15a). Note that surface winds in the cells near the equator and near the poles flow toward the equator and hence are diverted westward by the Coriolis effect. In contrast, surface winds in the mid-latitude cells flow toward the poles and are diverted eastward by the Coriolis effect (Figure 10.15b). You've probably noticed that surface winds generally blow from west to east at mid-latitudes, which is why storms generally first hit the west coast of North America and then progress eastward.

TIME OUT TO THINK *Study Figure 10.15 a bit more. If you want to take maximum advantage of the prevailing winds, what course should you plot for a sailing trip from Europe to North America and back?*

The strength of the Coriolis effect depends on a planet's size and rotation rate: It is weakest on planets that are small and rotating slowly. Venus has a

FIGURE 10.15 The Coriolis effect breaks up the circulation pattern shown in Figure 10.12.

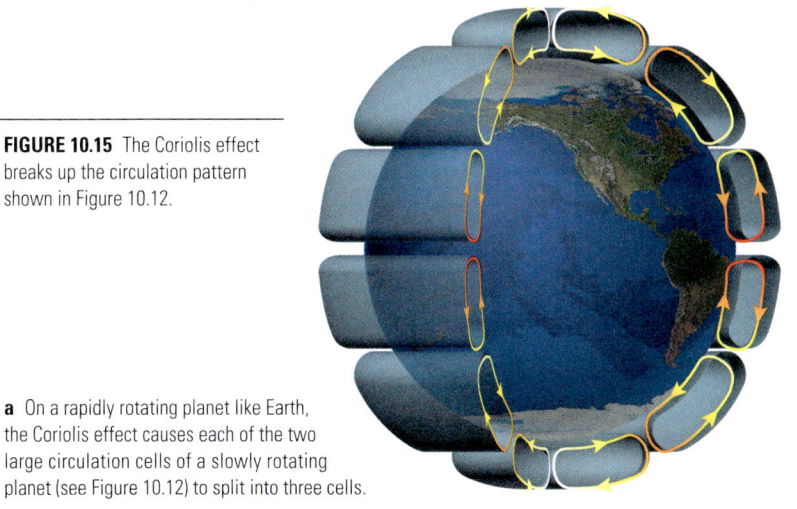

a On a rapidly rotating planet like Earth, the Coriolis effect causes each of the two large circulation cells of a slowly rotating planet (see Figure 10.12) to split into three cells.

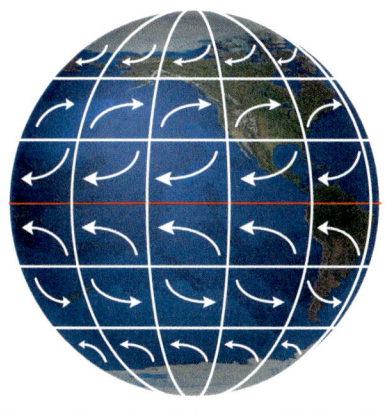

b The resulting surface wind patterns. Note that, at mid-latitudes such as over North America, the surface winds blow from the west. These motions combine with those of Figure 10.14a to give the overall wind patterns.

very weak Coriolis effect because of its slow rotation and hence has very slow surface winds; it lacks hurricane-like storms. The top wind speeds measured by the Soviet Union's Venera landers were no more than about 6 kilometers per hour. Apart from faster winds near the top of the atmosphere (of unknown cause), Venus is a rather dull place in terms of weather. Its temperature is nearly the same everywhere, and the surface winds are calm.

Mars has a rotation period very close to that of Earth, but its smaller size means a weaker Coriolis effect. Its circulation cells therefore do not split like those on Earth. Moreover, its thin atmosphere means that the circulation cells transport very little heat, so the poles remain much colder than the equator. Polar temperatures in the winter are so low (about 145 K, or −130°C) that carbon dioxide condenses into "dry ice" at the polar caps (Figure 10.16a). Because carbon dioxide is the major constituent of the Martian atmosphere, this condensation changes the atmospheric pressure by as much as 20% from season to season. The condensation occurring at the winter pole draws CO_2 out of the atmosphere, which is replenished by winds blowing from the opposite polar cap, where the dry ice *sublimes* into CO_2 gas [Section 4.3]. The combination of these pole-to-pole winds, circulation cells, and the Coriolis effect gives Mars very dynamic weather.

The direction of the pole-to-pole winds on Mars changes with the alternating seasons. Sometimes these winds initiate huge dust storms (Figure 10.16b), particularly when the more extreme summer approaches in the southern hemisphere. At times, the surface is so shrouded by airborne dust that surface markings are hidden from view and large dunes form and shift with strong winds. As the dust settles out onto the surface, it can change the color or albedo over vast areas—thereby creating the seasonal changes in appearance that fooled some astronomers in the late 1800s and early 1900s into thinking they saw

FIGURE 10.16 Polar ice caps and dust storms on Mars.

a This photo, taken by the *Viking Orbiter*, shows the ice cap at a Martian pole during the Martian winter. (Image is 400 kilometers across.)

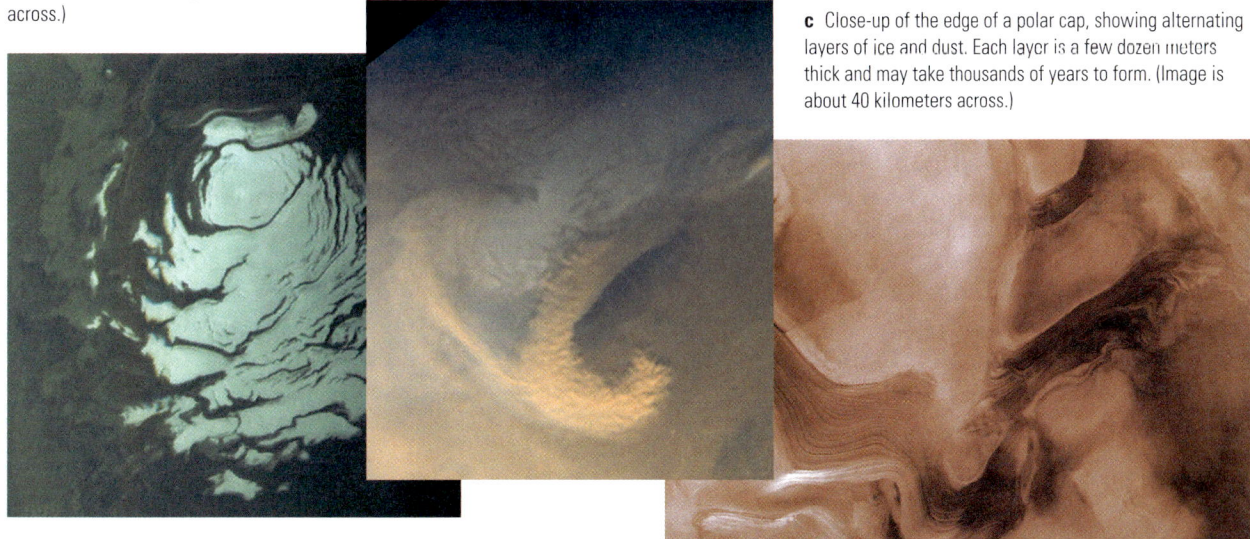

b Mars Global Surveyor photo of a dust storm (curving clouds) brewing over Mars's northern polar cap. (Image is 1000 kilometers across.)

c Close-up of the edge of a polar cap, showing alternating layers of ice and dust. Each layer is a few dozen meters thick and may take thousands of years to form. (Image is about 40 kilometers across.)

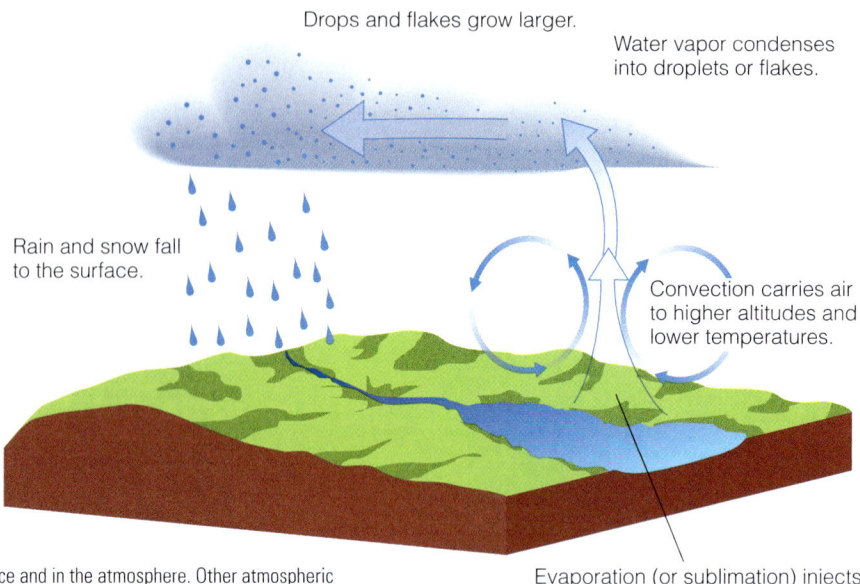

FIGURE 10.17 The cycle of water on Earth's surface and in the atmosphere. Other atmospheric ingredients on Earth and other planets have important cycles but may not be as apparent to the eye.

changes in vegetation [Section 9.5]. Dust also settles onto the polar caps, making alternating layers of darker dust and brighter frost (Figure 10.16c). The dust storms leave Mars with a perpetually dusty sky, giving it a pale pink color.

Clouds and Precipitation

We may think of clouds as imperfections on a sunny day, but they have profound effects on a planet. Clouds fundamentally change the appearance of a planet, reflecting sunlight back to space and thus reducing the amount of sunlight that warms the surface. Clouds are also important to planetary geology because they are a prerequisite for rain, a major cause of erosion. Despite these significant effects, clouds on most planets form from very minor ingredients of the atmosphere (see Table 10.1).

Clouds form when one of the gases in the air condenses into liquid or solid form, a process usually resulting from convection (Figure 10.17). On Earth, evaporation of surface water (or sublimation of ice and snow) releases water vapor into the atmosphere. Convection carries the water vapor to high, cold regions of the troposphere, where it condenses into droplets or flakes of ice. Thus, clouds on Earth consist of countless tiny water droplets or ice flakes, a fact you can feel if you walk through a cloud on a mountaintop. The droplets or flakes grow within the clouds until they become so large that the upward convection currents can't hold them aloft. They then begin to fall toward the surface as rain, snow, or hail.

Strong convection means more clouds and precipitation. Thunderstorms often form on late summer afternoons when the sunlight-warmed surface

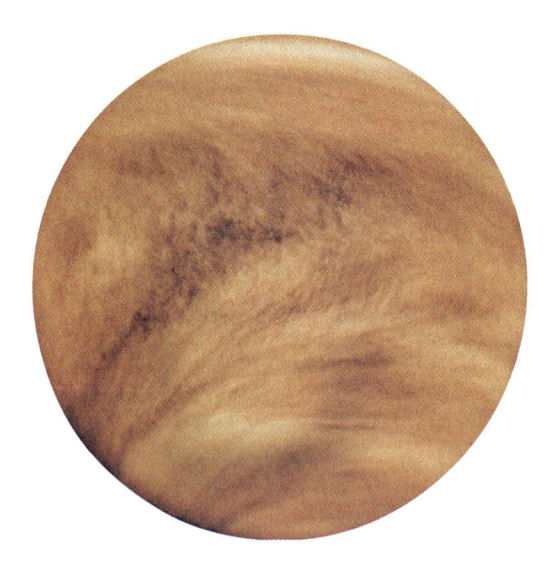

FIGURE 10.18 Clouds on Venus are all that can be seen in this ultraviolet image taken by the *Pioneer Venus Orbiter*.

drives stronger convection. Earth's equatorial regions also experience more convection and rain than other regions. When you combine this fact with the global wind patterns (see Figure 10.15b), you can understand why jungles lie at the equator and deserts lie at latitudes 20°–30° north and south of the equator. The equatorial rain depletes the moisture in the air as the winds carry it north or south, leaving little moisture to rain on the deserts.

Venus is covered with clouds because its high surface temperature drives strong convection and circulation (Figure 10.18). In contrast to Earth's water clouds, the clouds of Venus are made of sulfuric acid (H_2SO_4) dissolved in water droplets. These

276 PART III LEARNING FROM OTHER WORLDS

droplets form high in Venus's troposphere, where temperatures are 400 K cooler than on the surface. The droplets grow in the clouds until they form a sulfuric acid rain that falls toward the surface. But the droplets evaporate completely long before they reach the ground. Thus, it never rains on the surface of Venus, and the lower 30 kilometers of the atmosphere is very clear. The sulfuric acid is formed by chemical reactions involving sulfur dioxide (SO_2) and water, both of which come from volcanic eruptions. Because Venus lacks an ultraviolet-absorbing stratosphere, water molecules in its atmosphere are broken apart by ultraviolet light from the Sun. The atmosphere also loses sulfur dioxide through chemical reactions with surface rocks. The presence of sulfuric acid clouds therefore suggests that volcanoes on Venus must still be active, erupting at least occasionally to replenish sulfur dioxide and water.

Clouds can also form when winds blowing over mountains carry air parcels to higher altitudes. Droplets form in an air parcel when it is pushed to high altitude over a mountain. These droplets then evaporate as the air parcel descends to the valley or plains below. The result is that droplets constantly form in the air over the mountain, creating a cloud that remains stationary even though strong winds are blowing. This process forms clouds over terrestrial mountain ranges like the Rockies, and also over tall Martian mountains like Olympus Mons (Figure 10.19). The clouds over the Martian mountains consist of water-ice crystals, and water-ice fog sometimes fills Martian canyons. But the total amount of water vapor in the Martian atmosphere is so small that it would make a frost layer less than 1/100 millimeter thick on the planet. Carbon dioxide clouds can also form on Mars at much higher altitudes, but they are very rare.

Interestingly, Mars possesses far more water than we ever see in its clouds. The polar caps contain large amounts of water ice in addition to frozen carbon dioxide. In fact, all the carbon dioxide at the northern polar cap sublimes during the northern summer, exposing a frozen-water polar cap. So much carbon dioxide is deposited on the southern polar cap during the long, deep winter that it does not all sublime during the relatively short summer. We therefore do not know how much water is hidden below. At high latitudes, water ice probably also permeates the Martian soil 1–2 meters below the surface, where not even the heat of a summer day can penetrate. This *ground ice* probably has had important geological effects. The Martian craters that look as if they were made in mud (see Figure 9.13b) probably formed when impacts melted the ground ice, and catastrophic flooding in the Martian past may have occurred when sudden volcanic heating melted the ground ice into liquid water.

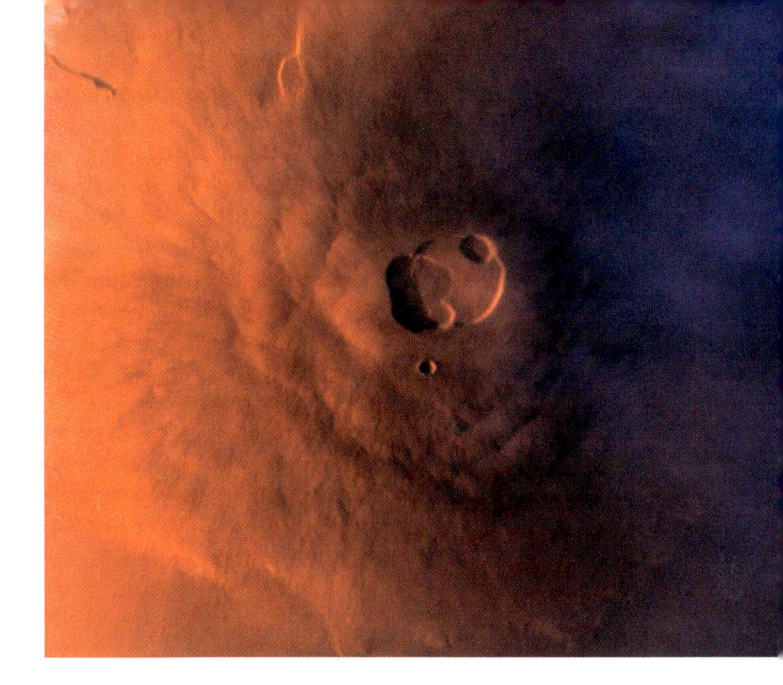

FIGURE 10.19 High-altitude clouds (the blue streaks in the upper right corner) over Olympus Mons, photographed by the *Mars Global Surveyor*.

Long-Term Climate Change

We know that long-term climate change can occur. For example, the Earth has been in and out of ice ages throughout its history. What can change the climate of an entire planet?

At least four factors can cause climate change. First, the Sun slowly grows in brightness as it ages [Section 14.3]. The Sun was about 30% fainter when the solar system first formed than it is today. Planets therefore may have been cooler early in the history of the solar system, or perhaps greenhouse gases were more abundant.

Second, changes in the tilt of a planet's axis can cause climate change. The axis tilt may slowly change over thousands or millions of years because of small gravitational tugs from moons, other planets, or the Sun. Although *average* planetary temperature may not be affected by the axis tilt, temperatures in different regions may change a great deal. A smaller axis tilt means year-round reduced sunlight in polar regions. Temperatures therefore drop in the polar regions, and perhaps even in mid-latitudes, causing *ice ages*. Earth and Mars probably both experience ice ages for this reason, and several of the icy satellites of the outer solar system undergo analogous cycles.

A third influence on climate change is changes in a planet's albedo. If a planet reflects more sunlight, it absorbs less—which can lead to planet-wide cooling. Ice ages may lead to further planetary cooling because the extra ice covering can increase the planet's albedo. Microscopic dust particles, or *aerosols*, released by volcanic eruptions can have the same effect. If the dust particles are blasted all the way into the stratosphere, they will remain suspended there for years, reflecting sunlight back to space. Human activity may be changing Earth's albedo. Smog particles can act like volcanic dust, reflecting sunlight before it reaches the ground. Deforestation can also increase Earth's albedo by removing sunlight-absorbing plants. But some human activity decreases albedo, such as paving a road with blacktop. Clearly, the overall human effect on Earth's albedo is not easy to gauge.

The fourth major cause of climate change is varying atmospheric conditions, particularly the abundance of greenhouse gases. If the abundance of greenhouse gases increases, the planet generally will warm. If the planet warms enough, increased evaporation and sublimation may add substantial amounts of gases to the planet's atmosphere, leading to an increase in atmospheric pressure. Conversely, if the abundance of greenhouse gases decreases, the planet generally will cool, and atmospheric pressure may decrease as gases freeze. We'll apply all four of these concepts to the planets in the next section to understand their evolving climates.

10.5 Atmospheric Origins and Evolution

Today's headlines are full of stories about the human effect on Earth's atmosphere, such as ozone depletion, global warming, and air pollution. But the atmospheres of Earth and the other terrestrial planets have seen much larger changes in their billions of years of history. What causes such large changes? To answer this question, we must examine both how a planet gets its atmospheric gases and how it can lose them.

How Atmospheres Are Created

The jovian planets obtained their atmospheres by capturing gas from the solar nebula, but the terrestrial worlds were too small to capture significant amounts of gas before the solar wind cleared away the nebula [Section 8.4]. Any gas that the terrestrial worlds did capture from the solar nebula consisted primarily of hydrogen and helium (just like the solar nebula as a whole). But these light gases escape easily from the terrestrial worlds and by now are long gone. The atmospheres of the terrestrial worlds therefore must have formed after the worlds themselves. Terrestrial atmospheres can get their gas from three different processes: *outgassing, evaporation/sublimation,* and *bombardment* (Figure 10.20).

The most important process is **outgassing**, the release of gases from the planetary interior by volcanic eruptions. Recall that interior gases were originally brought to the terrestrial planets by comets and volatile-rich asteroids during the early bombardment that ended about 4 billion years ago [Section 8.5]. The high pressures inside the planets trapped these gases in rocks in much the same way a pressurized bottle traps the bubbles of carbonated beverages. When the rocks melt and erupt onto the surface as lava, the release of pressure expels the gases. The gases may be released with the lava itself or from nearby vents (Figure 10.21). The most common gases expelled are water (H_2O), carbon dioxide (CO_2), nitrogen (N_2), and sulfur-bearing gases (H_2S or SO_2). (Aside from water, none of the gases existed in the solar nebula. Instead, they were created by chemical reactions that occurred *inside* the planets.) Because outgassing depends on volcanism, it is most important on larger planets that retain their internal heat longer. Outgassing supplied most of the gas that became the atmospheres of Venus, Earth, and Mars.

FIGURE 10.20 Three processes by which atmospheres gain gas.

outgassing · evaporation/sublimation · bombardment

Once outgassing creates an atmosphere, some of the gases may condense onto the surface. Earth's first volcanoes may have belched out water vapor over a dry, barren planet. As more and more water vapor filled the atmosphere, some of it began to condense and fall as rain—eventually filling our oceans. Similarly, Martian volcanoes outgassed large volumes of carbon dioxide, some of which is now frozen in the Martian polar caps. Surface liquids and ices continually exchange gas with the atmosphere. Sometimes the amount of gas released by *evaporation* (from liquids) or *sublimation* (from ices) exceeds the amount returning by condensation (from the atmosphere back into liquid or ice). Evaporation/sublimation is considered the second most important way in which gases are supplied to an atmosphere. Even though the gases first entered the atmosphere through outgassing, sublimation and evaporation can resupply gases that the atmosphere had lost.

Sublimation is most important on relatively cold bodies where ices form. It adds gas to the Martian atmosphere with the changing Martian seasons, as the warmer southern summer releases vast amounts of carbon dioxide from the south polar cap. It is even more important on some of the icy bodies of the outer solar system and for comets that happen to enter the inner solar system [Section 12.4].

Planets and satellites where outgassing and sublimation are negligible may still have thin atmospheres created by a third process: *bombardment* by micrometeorites, microscopic particles in the solar wind, or energetic solar photons. Bombardment can make only a thin atmosphere (a denser atmosphere would protect the surface from the bombarding particles), but it is the primary source of the atmospheres of the

FIGURE 10.21 Volcanoes National Park, Hawaii. Volcanoes give off H_2O, CO_2, N_2, and sulfur-bearing gases. Some gas is given off from the lava itself, and some leaks out from nearby gas vents.

Moon and Mercury. The atmosphere may be composed both of vaporized surface materials and of any gas released by the material doing the bombardment.

How Atmospheric Gases Are Lost

Just as several processes can add gas to a planetary atmosphere, an atmosphere can lose gas in several ways: *thermal escape, bombardment, atmospheric cratering, condensation,* and *chemical reactions* (Figure 10.22). The terrestrial planets lost any original

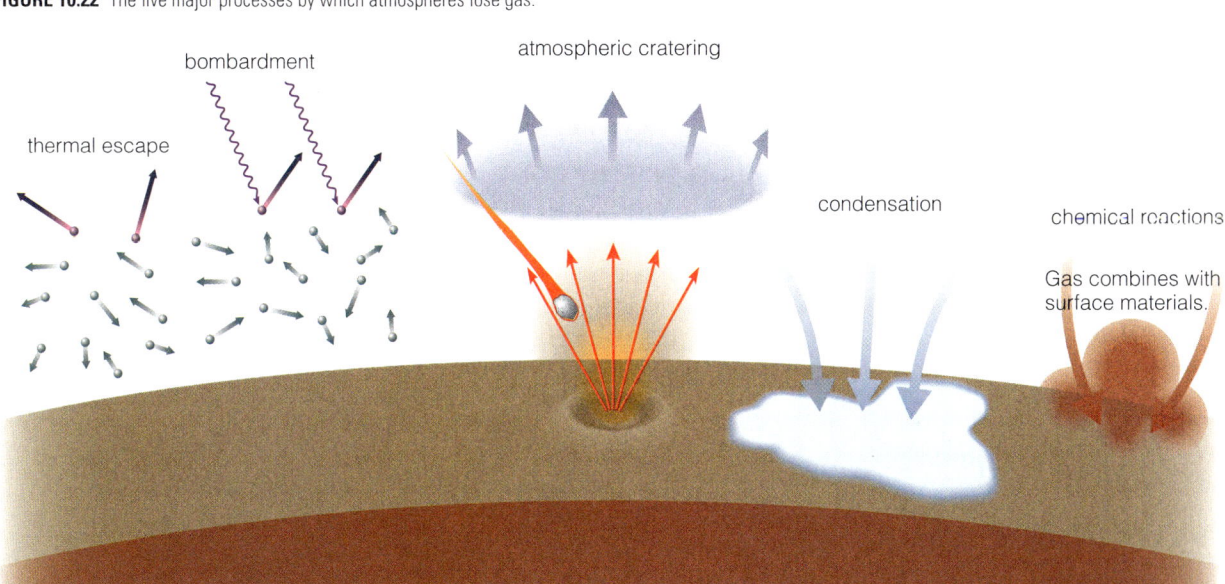

FIGURE 10.22 The five major processes by which atmospheres lose gas.

hydrogen and helium they had captured from the solar nebula through the process of **thermal escape**, in which an atom or molecule in the exosphere moves fast enough to escape the pull of gravity. Three factors determine whether an atmospheric gas can be lost to space by thermal escape:

1. The planet's *escape velocity*, or the speed at which a particle must travel to escape the pull of gravity—assuming it is traveling in the right direction and doesn't collide with anything else. The more massive the planet, the higher its escape velocity [Section 5.6].

2. The temperature. Individual particles in a gas move with a wide range of random speeds [Section 4.2], but the higher the temperature, the faster their *average* speed.

3. The mass of the gas particles. At a particular temperature, lighter particles (such as H or He) move around at faster average speeds than heavier ones (such as O_2 or CO_2) and therefore are more likely to achieve escape velocity.[5]

---

[5] At a particular temperature, all gas particles have the same *average* kinetic energy. Because the kinetic energy is described by the formula $\frac{1}{2}mv^2$, low-mass particles must have higher average speeds (v) than high-mass particles of the same kinetic energy.

Mercury cannot hold much of an atmosphere because its small size and high temperature mean that nearly all gas particles eventually achieve escape velocity. Gases escape about as easily from the Moon because it is smaller than Mercury, even though it is also somewhat cooler. Light gases such as hydrogen and helium escape any planet much more easily than heavier gases. That is why Venus, Earth, and Mars retained heavier gases but lost light gases—including any hydrogen released when molecules such as H_2O were broken apart by solar ultraviolet radiation.

TIME OUT TO THINK *Suppose Mercury had formed at a much greater distance from the Sun, say, somewhere beyond the orbit of Mars. How would its lower surface temperature affect its ability to hold an atmosphere? Considering all you know about planetary formation and how atmospheric gases are created and lost, do you think Mercury would have a thicker atmosphere in this case? Explain.*

A second process that can cause atmospheric loss is *bombardment* by microscopic particles and photons—the same process that creates the atmospheres of the Moon and Mercury. Just as bombardment can eject atoms from a surface into a thin atmosphere, it can also eject atmospheric particles from a planet's exosphere altogether. Bombardment by solar wind

---

Mathematical Insight 10.2 Thermal Escape from an Atmosphere

Whether an atmosphere loses gas by thermal escape depends on how many of the gas particles (atoms or molecules) are moving faster than the escape velocity. In general, particles of a particular type in a gas move at a wide range of speeds that depend on the temperature. For example, Figure 10.23 shows the range of speeds of sodium atoms at the temperature of the Moon's extremely thin atmosphere. The peak in the figure represents the most common speed of the sodium atoms. This speed, called the *peak thermal velocity* of the sodium atoms, is given by the following formula:

$$v_{thermal} = \sqrt{\frac{2kT}{m}}$$

where m is the mass of a single atom, T is the temperature in Kelvin, and $k = 1.38 \times 10^{-23}$ joule/Kelvin is *Boltzmann's constant*. This formula shows that higher temperatures mean higher thermal velocities for the atoms or molecules in a gas. It also shows that, for a given temperature, particles of lighter gases (smaller m) move at faster speeds than particles of heavier gases.

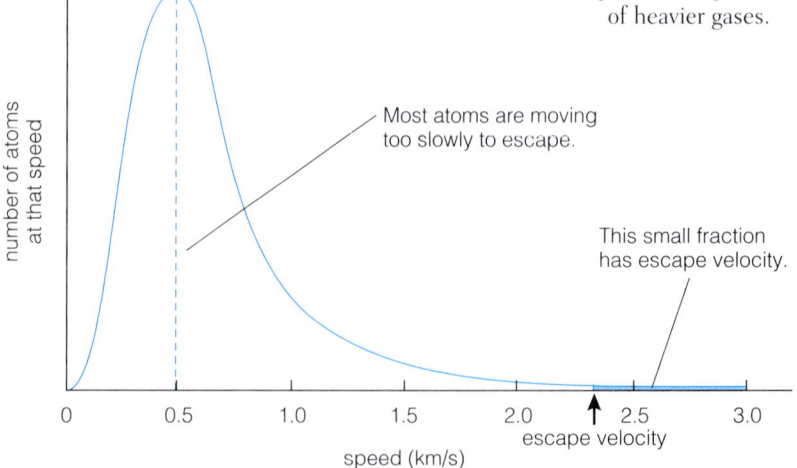

FIGURE 10.23 In a gas at a given temperature, different atoms are always moving at different speeds. This plot shows the range of speeds of sodium atoms in the lunar atmosphere.

particles is particularly important because it can sweep ionized atoms from high in the atmosphere into space. Bombardment by energetic solar photons can also be important, because ultraviolet photons can break apart molecules and send fragments flying off at speeds faster than escape velocity. Note that bombardment can eject atoms that are too heavy for thermal escape. In general, planets with protective magnetic fields—such as Earth—are less susceptible to this type of atmospheric loss.

Larger impacts can also cause atmospheric escape. The same impacts that create craters on the surface may blast away a significant amount of the atmosphere as well, a process sometimes called *atmospheric cratering*. For the most part, atmospheric cratering happened only early in the solar system's history. We expect it to have been most important on smaller worlds, where gravity's hold on the atmosphere is weaker. Mars probably lost a significant amount of atmospheric gas through atmospheric cratering.

Thermal escape, bombardment, and atmospheric cratering all cause a planet to lose gas permanently into space. Two other processes allow atmospheric gas to be absorbed back into the solid planet. Gases may simply *condense* into liquid or solid form if a planet cools down, especially in the cold polar regions. On Mars, carbon dioxide condenses into the polar caps during winter. Condensation probably also explains the presence of water ice in craters near the Moon's poles, discovered by the *Lunar Prospector* orbiter in 1998. The water probably came from comet impacts that each briefly created a thin lunar atmosphere. The water molecules from these impacts bounced around the surface at random until they condensed into ice in the polar craters. The bottoms of these craters lie in nearly perpetual shadow, keeping them so cold that the water remains perpetually frozen. Recent radar observations of Mercury suggest that it, too, may have water ice in its polar craters, probably for the same reason.

Finally, gases may become locked up in the surface through *chemical reactions* with rocks or liquids. On Earth, for example, carbon dioxide dissolves in the oceans, where it ultimately undergoes reactions that create *carbonate* rocks, such as limestone, on the ocean floor [Section 13.3]. Volcanoes have outgassed huge amounts of carbon dioxide into Earth's atmosphere over its history, but nearly all this gas has been converted into limestone by chemical reactions.

Putting It All Together

We've seen that the processes by which atmospheric gases are created and lost are closely tied to planetary formation properties and other planetary processes. We now have the perspective to link what we have learned about planetary atmospheres with our

---

If the peak thermal velocity of a particular type of gas is higher than the escape velocity, then most of the gas particles will be traveling fast enough to escape the atmosphere, and the planet will lose this atmospheric gas to space fairly quickly. However, even if the peak thermal velocity is lower than the escape velocity, the wide range of particle speeds in a gas means that some small fraction of the gas particles may be traveling faster than the escape velocity. These fast-moving particles will escape, as long as they're moving upward and don't hit another particle on their way out. Collisions among the remaining gas particles will continually ensure that a few always attain escape velocity, and thus the atmosphere can slowly leak away into space. In general, most of the gas particles of a particular type will escape over the age of the solar system if the peak thermal velocity is 20% or more of the escape speed.

Example: The temperature on the dayside of the Moon is $T = 400$ K. Calculate the peak thermal velocities of both hydrogen atoms and sodium atoms. Then explain why the Moon (escape velocity ≈ 2.4 km/s) cannot retain a hydrogen atmosphere for very long but retains a thin exosphere of sodium. The mass of a hydrogen atom is 1.67×10^{-27} kg; the mass of a sodium atom is 23 times greater, or 3.84×10^{-26} kg.

Solution: From the formula above, the peak thermal velocity of the hydrogen atoms on the dayside of the Moon is:

$$v_{\text{thermal}} = \sqrt{\frac{2 \times (1.38 \times 10^{-23}\frac{\text{joule}}{\text{K}}) \times (400 \text{ K})}{1.67 \times 10^{-27} \text{ kg}}}$$

$$\approx 2{,}600 \text{ m/s} = 2.6 \text{ km/s}$$

Similarly, the thermal velocity of the sodium atoms is:

$$v_{\text{thermal}} = \sqrt{\frac{2 \times (1.38 \times 10^{-23}\frac{\text{joule}}{\text{K}}) \times (400 \text{ K})}{3.84 \times 10^{-26} \text{ kg}}}$$

$$\approx 540 \text{ m/s} = 0.54 \text{ km/s}$$

Note that the peak thermal velocity of the hydrogen atoms is slightly greater than the Moon's escape velocity of about 2.4 km/s. Thus, the Moon cannot retain a hydrogen atmosphere. However, the peak thermal velocity of sodium is only about one-fifth, or about 20%, of the escape velocity. Thus, the Moon cannot hold a sodium atmosphere indefinitely, but it can hold some of the sodium atoms that are continually entering its atmosphere through bombardment.

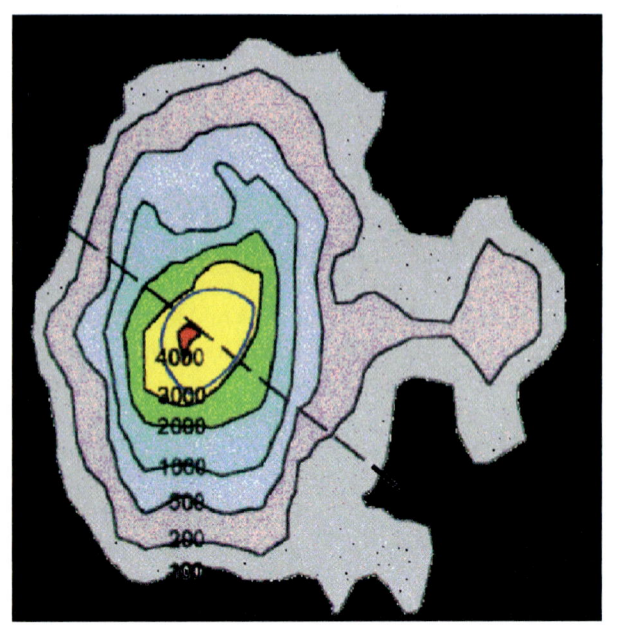

a Mercury's atmosphere. Color-coded contours represent the gas density from lowest (black) to highest (red). Mercury's partially illuminated disk is indicated as the gibbous-shaped blue contour near the center (just within the yellow region); the Sun lies to the left along the direction of the dashed line.

b The Moon's atmosphere. This composite image represents the gas density from lowest (green) to highest (red). The Moon itself was blocked to make this image, resulting in the central black area of no data. The Moon's size is shown schematically by the white circle.

FIGURE 10.24 The atmospheres of Mercury and the Moon—which are essentially exospheres only—viewed through instruments sensitive to emission lines from sodium atoms. Although these are extremely low density atmospheres, they extend quite high.

understanding of planetary geology from the previous chapter. In particular, we are in a better position to appreciate how planetary size affects the amount of outgassing and therefore the abundance of atmospheric gases. A closer look at atmospheric change will allow us to understand the different erosional histories of Earth, Mars, and Venus, which we left unresolved in the last chapter. Before you read further, you should refresh your memory of the "planetary flowcharts" in Chapter 9 and make your own analysis of atmospheric change on each of the terrestrial worlds.

10.6 History of the Terrestrial Atmospheres

We are now ready to return to the issue with which we began the chapter: how and why the terrestrial atmospheres came to differ so profoundly. Let's take a "historical tour" of the terrestrial atmospheres, in order of increasing planetary size.

The Moon and Mercury

The extremely thin atmospheres of the Moon and Mercury are the easiest to explain (Figure 10.24). Volcanic outgassing ceased long ago on these small worlds, and any early atmosphere was lost. Both Mercury and the Moon may have some ice in craters near their poles, but it remains perpetually frozen. Thus, the only source of atmospheric gas is bombardment. The relatively few particles blasted into the atmosphere by bombardment collide more often with the surface than with one another. They bounce around the surface like tiny rubber balls, arcing hundreds of kilometers into the sky before crashing back down. Sometimes these particles travel fast enough to escape the pull of gravity (thermal escape). Other times they are stripped away by bombardment by solar wind particles. Thus, gas does not accumulate in these atmospheres, but a small amount is always present because bombardment occurs continuously. Both atmospheres are relatively hot and extend high above the surface.

Mars

Mars is only 40% larger in radius than Mercury, but its surface tells of a much more fascinating and complex atmospheric history. Pictures show clear evidence that water once flowed on the Martian surface. The abundant geological signs of water erosion on Mars present a major enigma to our understanding of the planet. The evidence includes not only obvious features such as dried-up streambeds (see Figure 9.27a), but also subtler features such as eroded crater rims and floors. Further evidence comes from the study of meteorites believed to have come to Earth from Mars [Section 12.3]. These *Martian meteorites* contain types of minerals (clays and carbonates) that can form only in the presence of liquid water. Some planetary geologists even believe they have found evidence in Mars images for ancient oceans and glaciers on Mars, although these conclusions are controversial (Figure 9.27).

The fact that liquid water once flowed on Mars means that its past must have been very different from its current "freeze-dried" state. The surface temperature must have been much warmer, and therefore Mars must have had a much stronger greenhouse effect and much higher atmospheric pressure. Computer simulations show that these requirements can be met by a carbon dioxide atmosphere about 400 times denser (i.e., having greater pressure) than the current Martian atmosphere—or about twice as thick as Earth's atmosphere.

The idea that Mars had a much denser atmosphere in the past is reasonable. Mars has plenty of ancient volcanoes, and calculations show that these volcanoes could have supplied all of the carbon dioxide needed to warm the planet. Even CO_2 ice clouds could have helped hold heat in the Martian troposphere. Moreover, if Martian volcanoes outgas carbon dioxide and water in the same proportions as do volcanoes on Earth, Mars would have had enough water to fill oceans tens or even hundreds of meters deep.

The bigger question is not whether Mars once had a denser atmosphere but where those atmospheric gases are today. Mars must somehow have lost most of its carbon dioxide gas. This loss reduced the strength of the greenhouse effect until the planet essentially froze over. The fate of the lost carbon dioxide is not completely clear. Some is locked up in the polar caps, and some in carbonate minerals like those identified in the Martian meteorites. Some of the gas in the early Martian atmosphere may have been blasted away by large impacts (atmospheric cratering). But, according to magnetic field measurements from the *Mars Global Surveyor* [Section 9.3], the final blow to Mars's atmosphere was probably bombardment by solar wind particles. Early in Mars's history, the warm convecting interior produced both volcanic outgassing and a magnetospheric bubble to protect the atmosphere. As the small planet cooled, the magnetic field weakened, and eventually bombardment by the solar wind stripped away the atmosphere. Had Mars kept its magnetic field, it might have retained much of its atmosphere and kept its moderate climate.

The vast amounts of water once present on Mars are also gone. Some is tied up in the polar caps and underground ice, but most was probably lost forever. Mars lacks an ultraviolet-absorbing stratosphere, and water molecules are easily broken apart by ultraviolet photons. Once water molecules were broken apart, the hydrogen atoms would have been rapidly lost to space through thermal escape. Some of the remaining oxygen was probably lost by solar wind bombardment, and the rest was probably drawn out of the atmosphere through chemical reactions with surface rock. This oxygen literally rusted the Martian rocks, giving the "red planet" its distinctive tint.

The history of the Martian atmosphere holds important lessons for us on Earth. Mars was apparently once a world with pleasant temperatures and streams, rain, glaciers, lakes, and possibly oceans. It had all the necessities for "life as we know it." But the once hospitable planet turned into a frozen and barren desert at least 3 billion years ago, and it is unlikely that Mars will ever be warm enough for its frozen water to flow again (without human intervention). If life once existed on Mars, it is either extinct or hidden away in a few choice locations, such as hot springs around not-quite-dormant volcanoes. As we think about the possibility of future climate change on Earth, Mars presents us with an ominous example of how much things can change.

Venus

Venus presents a stark contrast to Mars. Its thick carbon dioxide atmosphere creates an extreme greenhouse effect that bakes the surface to unbearable temperatures. Venus's larger size allowed it to retain more interior heat than Mars, leading to greater volcanic activity [Section 9.5]. The associated outgassing produced the vast quantities of carbon dioxide in Venus's atmosphere, as well as the much smaller amounts of nitrogen and sulfur dioxide. But the outgassing presents a quandary when we consider water. We expect similar gases to come out of volcanoes on Earth and Venus, since both planets have similar overall compositions. Water vapor is the most common gas from volcanic outgassing on Earth, but Venus is incredibly dry, with only minuscule amounts of water vapor in its atmosphere and not a drop on the surface. Did Venus once have huge amounts of water? If so, what happened to it?

The leading theory suggests that Venus did indeed outgas plenty of water. But Venus's proximity to the Sun kept surface temperatures too high for oceans to form as they did on Earth. Outgassed water vapor accumulated in the atmosphere along with carbon dioxide and other gases. Because both water and carbon dioxide are strong greenhouse gases, Venus rapidly grew warmer still. Before the water disappeared, the surface temperature may have been double its current high value. Solar ultraviolet photons broke water molecules apart, and the high temperature allowed rapid escape of hydrogen (just as on Mars). The remaining oxygen was lost to a combination of chemical reactions with surface rocks and bombardment by the solar wind. Thus, the water that once was on Venus is gone forever. In fact, Venus may have lost so much water that its crust and mantle also dried out.

It's difficult to prove that Venus lost an ocean's worth of water billions of years ago, but some evidence comes from the gases that didn't escape. Recall that most hydrogen nuclei contain just a single proton. But a tiny fraction of all hydrogen atoms in the universe contain a neutron in addition to the proton. This isotope of hydrogen, called *deuterium*, can be present in water molecules in place of one of the hydrogen atoms (making molecules of "heavy water." The deuterium atom can be released into the atmosphere when an ultraviolet photon breaks apart such a water molecule. Both hydrogen and deuterium can escape from the atmosphere, but less deuterium escapes due to its higher mass. If Venus once had huge amounts of water molecules that broke apart, the more rapid escape of ordinary hydrogen should have left its atmosphere enriched in deuterium. And, indeed, measurements show that deuterium is at least a hundred times more abundant (relative to ordinary hydrogen) on Venus than anywhere else in the solar system.[6]

The next time you see Venus shining brightly as the morning or evening "star," consider the radically different path it has taken from that of Earth—and thank your lucky star. If Earth had formed a bit closer to the Sun or if the Sun had been a shade hotter, our planet might have suffered the same greenhouse-baked fate.

Earth

In this chapter, we have used Earth as the "typical" planet for understanding the structure of a terrestrial world's atmosphere and weather. Astronomers began their studies of Mars and Venus expecting these planets to be variants of Earth. Instead we've discovered that they are fundamentally different from Earth, particularly in the areas of atmospheric composition and climate. Rather than being a "role model" for understanding other planets, Earth looks more like the odd one out. The lesson of planetary exploration is that understanding Earth requires understanding the solar system as a whole: its formation, geological processes, atmospheres, and even the jovian planets, asteroids, and comets. We are more than halfway through this list already, and after two more chapters we will come full circle, ready to explore and explain Earth's unique circumstances in Chapter 13.

---

[6]Some scientists remain skeptical that Venus lost huge amounts of water, instead suggesting that it was "born dry"—perhaps because it never received as much water as Earth or Mars. In that case, Venus's extraordinary abundance of deuterium relative to ordinary hydrogen requires an alternative explanation.

THE BIG PICTURE

With what we have learned in this chapter and the previous chapter, we now have a complete "big picture" view of how the terrestrial planets started out so similar yet ended up so different. As you continue, keep the following important ideas in mind:

- Planetary atmospheres are collections of gas particles in search of balance. The vertical spread of an atmosphere is a balance between gravity and random thermal motions of molecules. Temperatures at different altitudes are in balance with the amount of light absorbed at each altitude.

- Weather and climate change are the planets' vain attempts to equalize temperatures and pressures over their surfaces. It's a losing battle, because solar heating can never be uniform. But it's the very nature of gases to move toward equilibrium, even if they never attain it.

- The many levels of balance rely on an active, not static, atmosphere. From the unseen vibrations and rotations of speeding molecules to planet-wide winds and weather, atmospheres are in motion.

- The long-term evolution of planetary atmospheres reflects an imbalance between atmospheric gains and losses, which vary over billions of years. Complete atmospheric transformation over the age of the solar system appears to be the rule for large terrestrial planets, not the exception.

Review Questions

1. What is an atmosphere? List a few of the most common atmospheric gases. Where does an atmosphere end?

2. Explain the microscopic origin of *gas pressure*. What is 1 *bar* of pressure?

3. Briefly describe the three factors that would determine planetary temperatures in the absence of *greenhouse gases*. How do the "no greenhouse" temperatures of the terrestrial planets compare to their actual temperatures? Why?

4. What types of gases absorb infrared light? Ultraviolet light? X rays? Explain how these interactions between light and matter lead to the generic atmospheric structure profile shown in Figure 10.5.

5. Briefly explain why the sky is blue in the daytime and why sunrises and sunsets are red.

6. Give a few examples of greenhouse gases. Briefly describe how the *greenhouse effect* makes a planetary surface warmer than it would be otherwise.

7. What is convection? Briefly explain why it occurs in the *troposphere* but not in the *stratosphere*.

8. Briefly explain why a planet can have a stratosphere only if it has ultraviolet-absorbing molecules in its atmosphere. What molecule plays this role on Earth?

9. Explain how X-ray absorption in the *thermosphere* creates an *ionosphere* and how the ionosphere affects radio communications.

10. Briefly describe the actual atmospheric structure of each of the five terrestrial worlds; contrast these structures with the generic structure in Figure 10.5.

11. What is a *magnetosphere*? What are *charged particle belts*? Briefly describe how the solar wind affects magnetospheres and how auroras are produced.

12. Briefly distinguish between *weather* and *climate*.

13. Explain why Earth and Mars have seasons but Venus does not. How do seasons on Mars differ from seasons on Earth? Why does atmospheric pressure on Mars change during the seasons, while atmospheric pressure on Earth remains steady year-round?

14. What do we mean by *global wind patterns*? Briefly explain why terrestrial planets tend to have huge *circulation cells* carrying warm air toward the poles and cool air toward the equator.

15. What is the origin of the *Coriolis effect?* How does it affect global wind patterns?

16. Briefly describe how clouds form and how they affect planetary weather. What causes rain?

17. Briefly describe the four factors that can lead to long-term climate change and the effects of each factor.

18. Briefly describe the three processes that can add gas to a planetary atmosphere and the five processes by which a planetary atmosphere can lose gas.

19. Why do Mercury and the Moon have only thin exospheres? Why may there be ice in some polar craters on these worlds?

20. Briefly describe the atmospheric history of Mars.

21. Why should we expect Venus and Earth to have had very similar early atmospheres? How did the two planetary atmospheres become so different?

Discussion Questions

1. *Charting Atmospheric Evolution.* How would you develop "planetary flowcharts" for the atmospheric source and loss processes analogous to those developed for geological processes in Chapter 9? Develop two charts, one for a source process and one for a loss process. Be sure your flowchart includes all the relevant connections described in this chapter.

2. *Mars: Past and Future.* Summarize the evolution of the Martian atmosphere since planetary formation, emphasizing the important source and loss processes. What is your prediction for the future of the Martian atmosphere?

Problems

Plausible Claims? For **problems 1–7**, suppose you heard someone make the given claim. Decide whether the claim seems plausible. Explain.

1. If the Earth's atmosphere did not contain molecular nitrogen, X rays from the Sun would reach the surface.

2. When Mars had a thicker atmosphere in the past, it probably also had a stratosphere.

3. If the Earth rotated faster, storms would probably be more common and more severe.

4. Mars would not have seasons if it had a circular rather than an elliptical orbit around the Sun.

5. If the solar wind were much stronger, Mercury might develop a carbon dioxide atmosphere.

6. The Earth's oceans probably formed at a time when no greenhouse effect operated on Earth.

7. Mars once may have been warmer, but because it is farther from the Sun than is Earth, it's not possible that it could ever have been warmer than the Earth.

8. *Cool Venus.* Table 10.2 shows that Venus's temperature in the absence of the greenhouse effect is lower than Earth's, even though it is closer to the Sun.

    a. Explain this unexpected result in a sentence or two.

    b. Now suppose that Venus had neither clouds nor greenhouse gases. What do you think would happen to the temperature of Venus? Why?

    c. How are clouds and volcanoes linked on Venus? What change in volcanism might result in the disappearance of clouds? Explain.

9. *Atmospheric Structure.* Study Figure 10.5, which shows the atmospheric structure for a generic terrestrial planet. For each of the following cases, make a sketch similar to Figure 10.5 showing how the atmospheric structure would be different and explain the differences in words.

    a. Suppose the planet had no greenhouse gases.

    b. Suppose the Sun emitted no solar ultraviolet light.

    c. Suppose the Sun had a higher output of X rays.

10. *Inversions.* Consider a local inversion, in which the air is colder near the surface than higher up in the troposphere.

    a. Explain why convection is suppressed in an inversion.

    b. Cities that experience frequent inversions, such as Los Angeles and Denver, often have more serious problems with air pollution than other cities of similar size. Explain how an inversion can trap pollutants, keeping them close to the city in which they are generated.

11. *Coastal Winds.* During the daytime, heat from the Sun tends to make the air temperature warmer over land near the coast than over the water offshore. But at night, when land cools off faster than the sea, the temperatures tend to be cooler over land than over the sea. Use these facts to predict the directions in which winds generally blow during the day and at night in coastal regions; for example, do the winds blow out to sea or in toward the land? (*Hint:* How are these conditions similar to those that cause planetary circulation cells?) Explain your reasoning in a few sentences. Diagrams may help.

12. *A Swiftly Rotating Venus.* Suppose Venus had rotated as rapidly as Earth throughout its history. Briefly explain how and why you would expect it to be different in terms of each of the following: geological processes, atmospheric circulation, magnetic field, and atmospheric evolution. Write a few sentences about each.

13. *Sources and Losses.* Choose one atmospheric source process and one atmospheric loss process. Describe for each the ways in which the process is related to the four planetary formation properties discussed in Chapter 9. (For example, how is the creation of atmospheric gas by bombardment related to a planet's size, distance from the Sun, composition, and rotation rate?) (*Hint:* A process may not depend on all four formation properties.)

14. *Project: Atmospheric Science in the Kitchen.*
    a. Find an empty plastic bottle, such as a water bottle with a screwtop that makes a good seal. Warm the air inside by filling the bottle partway with hot water, and then shaking and emptying the bottle. Seal the bottle and place it in the refrigerator or freezer. What happens after 15 minutes or so? Explain why this happens, imagining that you could see the individual air molecules.
    b. You may have noticed that loose ice cubes in the freezer gradually shrink away to nothing or that frost sometimes builds up in old freezers. What are the technical terms used in this chapter for these phenomena? On what terrestrial planet do these same processes play a major role in controlling the atmosphere?

*15. *Habitable Planet Around 51 Peg?* A recently discovered planet orbits the star 51 Pegasi at a distance of only 0.051 AU. The star is approximately as bright as our Sun. The planet has a mass 0.6 times that of Jupiter, but no one knows if it is more similar to our jovian or to our terrestrial planets (if either).
    a. Suppose it is a terrestrial-type planet with an albedo of 0.15 (i.e., without clouds). Calculate its "no greenhouse" temperature, assuming it rotates fast enough to have the same temperatures on its dayside and its nightside. How does this temperature compare to that of Earth?
    b. Repeat part (a), but this time assume that the planet is covered in very reflective clouds, giving it an albedo of 0.8.
    c. Based on your answers to parts (a) and (b), do you think it is likely that the conditions on this planet are conducive to life? Explain your answer in one or two paragraphs.

*16. *Escape from Venus.*
    a. Calculate the escape velocity from Venus's exosphere (about 200 km above Venus's surface). (*Hint:* See Mathematical Insight 5.3.)
    b. Calculate and compare the thermal speeds of hydrogen and deuterium atoms at the exospheric temperature of 350 K. The mass of a hydrogen atom is 1.67×10^{-27} kg, and the mass of a deuterium atom is about twice the mass of a hydrogen atom.
    c. Comment in a few sentences on the relevance of the calculations in parts (a) and (b) to atmospheric evolution on Venus.

Web Projects

Find useful links for Web projects on the text Web site.

1. *Spacecraft Study of Atmospheres.* Learn about a current or planned mission to study the atmosphere of one of the terrestrial worlds. Write a one- to two-page essay describing the mission and what it may teach us about terrestrial planet atmospheres.

2. *Martian Weather.* Find the latest weather report for Mars from spacecraft and other satellites. What season is it in the northern hemisphere? When was the most recent dust storm? What surface temperature was most recently reported from Mars's surface, and where was the lander located? Summarize your findings by writing a 1-minute script for a television news update on Martian weather.

3. *Terraforming Mars.* Some people have proposed that we might someday *terraform* Mars, making it more Earth-like so that we might live there more easily. Research some proposals for terraforming Mars. Do you think the proposals could work? Do you think it would be a good idea to terraform Mars? Write a one- to two-page essay summarizing your findings and defending your opinions.

Do there exist many worlds, or is there but a single world? This is one of the most noble and exalted questions in the study of Nature.

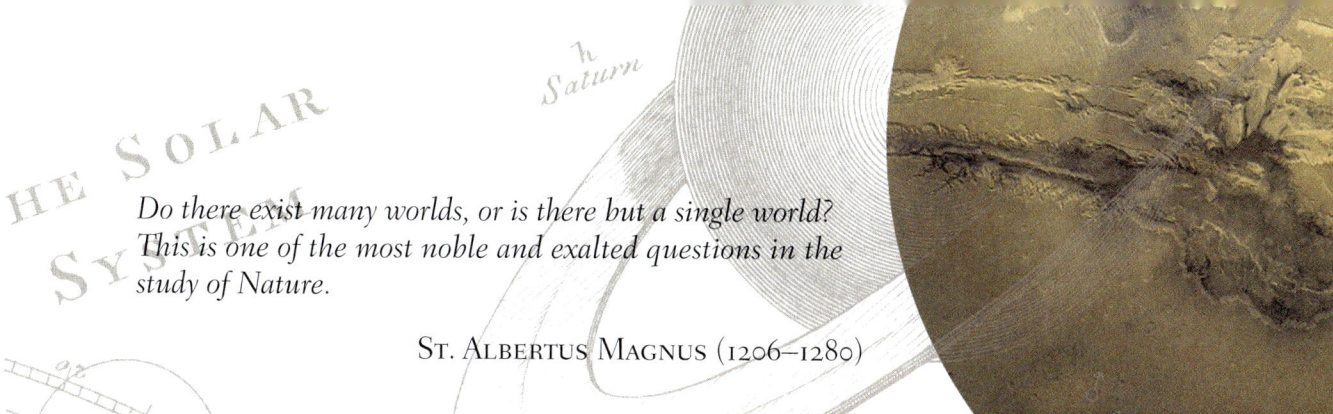

St. Albertus Magnus (1206–1280)

CHAPTER 11
Jovian Planet Systems

In Roman mythology, the namesakes of the jovian planets are rulers among gods: Jupiter is the king of the gods, Saturn is Jupiter's father, Uranus is the lord of the sky, and Neptune rules the sea. But even the most imaginative of our ancestors did not foresee the true majesty of the jovian planets. The smallest, Neptune, is still large enough to contain the volume of more than 50 Earths. The largest, Jupiter, has a volume some 1,400 times that of Earth. These worlds are totally unlike the terrestrial planets. They are essentially giant balls of gas, with no solid surface on which to stand.

Why should we care about a set of worlds so different from our own? Apart from satisfying natural curiosity about the diversity of the solar system, the jovian planet systems serve as a testing ground for our general theories of comparative planetology. Are our theories good enough to predict the nature of these planets and their elaborate satellite systems, or are different processes at work? We'll find out in this chapter as we investigate the nature of the jovian planets, along with their intriguing moons and beautifully complex rings.

FIGURE 11.1 Jupiter, Saturn, Uranus, and Neptune, shown to scale with Earth for comparison.

11.1 The Jovian Worlds: A Different Kind of Planet

The jovian planets have long held an important place in many human cultures. Jupiter is the third-brightest object that appears regularly in our night skies, after the Moon and Venus, and ancient astronomers carefully charted its 13-year trek through the constellations of the zodiac. Saturn is fainter and slower-moving, but people of many cultures also mapped its 27-year circuit of the sky. Ancient skywatchers did not know of the planets Uranus and Neptune, but astronomers charted their orbits soon after their discoveries in 1781 and 1846, respectively.

Scientific study of the jovian worlds dates back to Galileo in the early 1600s; however, scientists did not fully appreciate the differences between terrestrial and jovian planets until they established the absolute scale of the solar system (that is, in units such as kilometers instead of AU). Once distances were established, scientists could calculate the truly immense sizes of the jovian planets from their *angular* sizes (as measured through telescopes) [Section 7.1]. Knowing the distance scale also allowed calculating jovian planet masses by applying Newton's version of Kepler's third law [Section S.4] to observed orbital characteristics of jovian moons. Together, the measurements of size and mass revealed the low densities of the jovian planets, proving that these worlds are very different from the Earth.[1]

The slow pace of discovery about the jovian planets gave way to revolutionary advances with the era of spacecraft exploration. The first spacecraft to the outer planets, *Pioneer 10* and *Pioneer 11*, flew past Jupiter and Saturn in the early 1970s. The Voyager missions followed less than a decade later [Section 7.6]. *Voyager 2* continued on a "grand tour" of the jovian planets, flying past Uranus in 1986 and Neptune in 1989. The results were spectacular. Figure 11.1 shows a montage of the jovian planets compiled by the Voyager spacecraft. Exploration of the jovian planet systems continued with the Galileo spacecraft, which went into orbit around Jupiter in 1995. Still ahead is the Cassini mission, scheduled for arrival at Saturn in 2004.

Jupiter, Saturn, Uranus, and Neptune are so different from the terrestrial planets that we must create an entirely new mental image of the term *planet*. Table 11.1 summarizes the bulk properties of these

---

[1] Isaac Newton proved that Jupiter was *relatively* less dense than the Earth before anyone could measure the *absolute* density of either.

Table 11.1. Comparison of Bulk Properties of the Jovian Planets

Planet	Average Distance from Sun (AU)	Mass (Earth masses)	Radius (Earth radii)	Average Density (g/cm³)	Bulk Composition
Jupiter	5.20	317	11.2	1.33	Mostly H, He
Saturn	9.53	90	9.4	0.70	Mostly H, He
Uranus	19.2	14	4.11	1.32	Hydrogen compounds and rocks, H and He
Neptune	30.1	17	3.92	1.64	Hydrogen compounds and rocks, H and He

giant worlds. The most obvious difference between the jovian and terrestrial planets is size. Earth's mass in comparison to that of Jupiter is like the mass of a squirrel in comparison to that of a full-grown person. By volume, Earth in comparison to Jupiter is like a pea in a cup of soup. (Recall that volume is proportional to the cube of the radius. Thus, because Jupiter's radius is 11.2 times Earth's, its volume is $11.2^3 = 1,405$ times Earth's.)

The overall composition of the jovian planets—particularly Jupiter and Saturn—is more similar to that of the Sun than to any of the terrestrial worlds. Their primary chemical constituents are the light gases hydrogen and helium, although their atmospheres also contain many hydrogen compounds and their cores consist of a mixture of rocks, metals, and hydrogen compounds. Although these cores are small in proportion to the jovian planet volumes, each is still more massive than any terrestrial planet. Moreover, the pressures and temperatures near the cores are so extreme that the "surfaces" of the cores do not resemble terrestrial surfaces at all. In a sense, the general structures of the jovian planets are opposite those of the terrestrial planets: Whereas the terrestrial planets have thin atmospheres around rocky bodies, the jovian planets have proportionally small, rocky cores surrounded by massive layers of gas (Figure 11.2). A spacecraft descending into Jupiter, for example, would have to travel *tens of thousands* of kilometers before reaching the core.

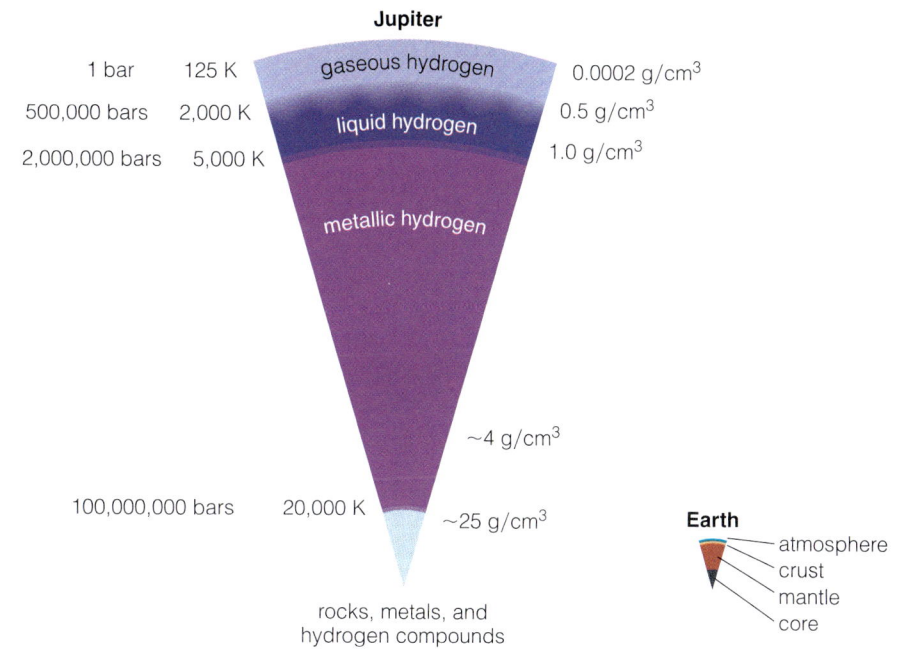

FIGURE 11.2 Jupiter's interior structure, labeled with the pressure, temperature, and density at various depths. Earth's interior structure is shown to scale for comparison. Note that Jupiter's core is only slightly larger than Earth but is about 10 times more massive.

CHAPTER 11 JOVIAN PLANET SYSTEMS 289

The jovian planets may be Sun-like in composition, but their masses are far too low to provide the interior temperatures and densities needed for nuclear fusion. Calculations show that nuclear fusion is possible only with a mass at least 80 times that of Jupiter. (Some people have called Jupiter a "failed star" for this reason, but we prefer to think of it as a very successful planet.) The jovian planets have undoubtedly lost internal heat during the more than 4 billion years since their formation and thus must have been much warmer in the distant past. This heat may have "puffed up" their atmospheres, making them larger and brighter in the past. If so, Jupiter would have been even more prominent in Earth's sky billions of years ago than it is today.

The jovian planets rotate much more rapidly than any of the terrestrial worlds, but precisely defining their rotation rates can be difficult because they are not solid. We can measure a terrestrial rotation period simply by watching the apparent movement of a mountain or crater as a planet rotates, but on jovian planets we can observe only the movements of clouds. Cloud movements can be deceptive, because their apparent speeds may be affected by winds as well as by planetary rotation. Nevertheless, observations of clouds at different latitudes suggest that the jovian planets rotate faster near their equators than near their poles. (The Sun also exhibits this type of *differential rotation* [Section 14.5].) We can measure the rotation rates of the jovian interiors by tracking emissions from charged particles trapped in their *magnetospheres* [Section 10.3]. This technique allows us to observe the rotation period of a magnetosphere, which should be the same as the rotation period deep in the interior, where the magnetic fields are generated. These measurements show that the jovian "days" range from only about 10 hours on Jupiter and Saturn to 16–17 hours on Uranus and Neptune.

TIME OUT TO THINK *How would you measure the rotation period of Earth if you made telescopic observations from Mars? How might Earth's clouds mislead you if you focused on them?*

Even the shapes of the jovian planets are somewhat different from the shapes of the terrestrial planets. Gravity makes the jovian planets approximately spherical, but their rapid rotation rates make them slightly squashed (Figure 11.3). Material near the equator, where speeds around the rotation axis are highest, is flung outward in the same way that you feel yourself flung outward when you ride a merry-go-round. The result is that jovian planets bulge noticeably around the equator. The size of the equatorial bulge depends on the balance between the strength of gravity pulling the material inward and the rate of rotation pushing the material outward. The balance tips most strongly toward flattening on Saturn, with its rapid 10-hour rotation period and its relatively weak surface gravity. Saturn is about 10% wider at its equator than at its poles. In addition to altering a planet's appearance, the equatorial bulge itself exerts an extra gravitational pull that helps keep satellites and rings aligned with the equator.

We can trace almost all the major characteristics of the jovian planets back to their formation [Section 8.4]. The jovian planets formed far from the Sun where icy grains condensed in the solar nebula, yielding far more solid matter to accrete into planetesimals. As the massive ice/rock cores of the future jovian planets accreted, their strong gravity captured hydrogen and helium gas from the surrounding neb-

FIGURE 11.3 The jovian planets are not quite spherical.

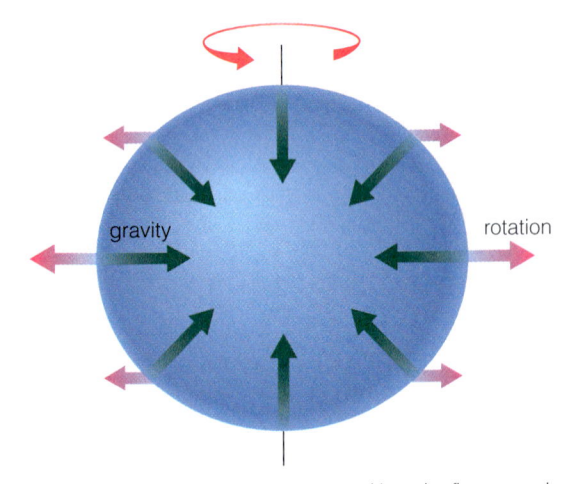

a Gravity alone makes a planet spherical, but rapid rotation flattens out the spherical shape by flinging material near the equator outward.

b Saturn is clearly not spherical. Compare its actual shape to the dashed circle.

ula. Some of this gas formed flattened "miniature solar nebulae" (or *jovian nebulae*) around the jovian planets. Just as solid grains condensed and combined into planets in the full solar nebula, solid grains in the nebulae surrounding the jovian planets condensed to form satellite systems. The deviations from the general patterns, such as the large 98° axis tilt of Uranus,[2] probably arose from giant impacts as the planets were forming. In the rest of this chapter, we will explore the features of the jovian planets and their satellites in greater depth, putting our theories of planetary development to the test.

11.2 Jovian Planet Interiors

As was the case with the terrestrial planets, developing a true understanding of the jovian planets requires looking deep inside them. Probing jovian planet interiors is even more difficult than probing terrestrial interiors, but several techniques allow us to learn what lies below the clouds. We've already discussed how Earth-based observations yielded the sizes and average densities of the jovian planets and how spacecraft measurements of magnetic and gravitational fields provide additional clues to their interior structure. Detailed observations of planetary shapes also provide information about interior structure; for example, computer models using shape data tell us that Saturn's core makes up a larger fraction of its mass than Jupiter's core.

Spectroscopy from Earth and from spacecraft reveals the chemical compositions of the jovian upper atmospheres. Their deeper atmospheres are generally more difficult to probe, but two recent events provided rare opportunities. In July 1994, fragments of comet Shoemaker–Levy 9 slammed into Jupiter, blasting material from deeper in the atmosphere out into the open, where astronomers could study it through telescopes [Section 12.6]. Then, in December 1995, NASA's Galileo spacecraft dropped a scientific probe into Jupiter's atmosphere. The Galileo probe survived to a depth of about 300 kilometers—a significant distance, but still much less than 1% of Jupiter's 70,000-kilometer radius—providing our first direct data from within a jovian world.

Beneath the depths at which we have compiled direct data, we learn about the jovian interiors by combining laboratory studies and theoretical models. These studies tell us how the ingredients of the jovian planets (especially hydrogen and helium) act under the tremendous temperatures and pressures expected deep below the cloud tops. Nowadays, elaborate computer models successfully match the observed sizes, densities, atmospheric compositions, and even shapes of the jovian planets. We therefore believe that we have a fairly clear understanding of their interiors. We'll begin our discussion of the jovian interiors by using Jupiter as a prototype; then we'll apply the principles of comparative planetology to understand the other jovian planet interiors.

Inside Jupiter

Imagine plunging head-on into Jupiter in a futuristic space suit that allows you to survive the incredible interior conditions. Near the cloud tops, you'll find the temperature to be a brisk 125 K ($-148°C$), the density to be a low 0.0002 g/cm^3, and the atmospheric pressure to be about 1 bar [Section 10.1]—the same as the pressure at sea level on Earth. The deeper you go, the higher the temperature, density, and pressure become (see Figure 11.2). By a depth of 7,000 km—about 10% of the planet's radius—you'll find that the temperature has increased to a scorching 2,000 K, the density has reached 0.5 g/cm^3 (half that of water), and the pressure is about 500,000 bars. Under these conditions, hydrogen acts more like a liquid than a gas.

At a depth of 14,000 km, you'll find a density about the same as that of water, a temperature near 5,000 K, and a pressure of 2 million bars. This extreme pressure forces hydrogen into an even more compact form: *metallic hydrogen,* in which all the atoms share electrons, just as happens in everyday metals.[3] The metallic and hence electrically conducting nature of this interior hydrogen is important in generating Jupiter's strong magnetic field.

Continuing your descent, you'll reach Jupiter's core at a depth of 60,000 km, about 10,000 km from the center. The core is a mix of hydrogen compounds, rocks, and metals, but these materials bear little resemblance to familiar solids or liquids because of the extreme 20,000-K temperature and 100-million-bar pressure. The core materials probably remain mixed together, with a density of about 25 g/cm^3, rather than separating into layers of different composition as in the terrestrial planets. The total mass of Jupiter's core is about 10 times that of Earth, so the core alone would make an impressive planet.

---

[2]The 98° tilt is measured by assuming that Uranus was "supposed to" rotate counterclockwise as seen from above Earth's North Pole—just like most other planets. We could equivalently say that Uranus has an 82° tilt (180° − 98° = 82°) but rotates backward.

[3]This phase change from nonmetal to metal does not occur in any everyday substance under normal conditions on Earth, and for a long time it was a theoretical prediction without experimental verification. Metallic hydrogen was finally produced under extreme pressure in laboratories in the early 1990s.

Our study of the terrestrial planets taught us that internal heat sources can have a strong effect on the behavior of planets, so it is natural to ask if Jupiter and the other jovian planets are also affected by internal heat sources. Jupiter has a tremendous amount of internal heat—so much that it emits almost twice as much energy as it receives from the Sun. (For comparison, Earth's internal heat adds only 0.005% as much energy to the surface as does sunlight.) This internal heating is much too large to be explained by radioactive decay alone, so we must search for other explanations. The most probable explanation is that Jupiter is still slowly contracting, as if it has not quite finished forming from a jovian nebula. This *gravitational contraction* [Section 14.1] is so gradual that we cannot measure it directly, but it converts gravitational potential energy to thermal energy, keeping Jupiter hot inside.

Comparing Jovian Planet Interiors

Now that we've discussed the interior of Jupiter, what can we say about the other jovian planets? Look back at the bulk properties of the jovian planets in Table 11.1. Although there are no obvious correlations between size, density, and composition, we can explain the properties by examining the behavior of gaseous materials in more detail.

Building a planet of hydrogen and helium is a bit like making one out of very fluffy pillows (Figure 11.4a). Imagine assembling a planet pillow by pillow. As each new pillow is added, those on the bottom are compressed more by those above. As the lower layers are forced closer together, their mutual gravitational attraction increases, compressing them even further. At first the stack grows substantially with each additional pillow, but eventually the growth slows until adding pillows hardly increases the height of the stack. This analogy explains why Jupiter is only slightly larger than Saturn in radius even though it is more than three times more massive. The extra mass of Jupiter compresses its interior to a much greater extent, making Jupiter's average density almost twice that of Saturn. In fact, Saturn's average density of 0.7 g/cm^3 is less than that of water. More precise calculations show that Jupiter's radius is almost the maximum possible for a jovian planet: If more gas were added to Jupiter, its weight would actually compress the interior enough to make the planet *smaller* instead of larger (Figure 11.4b). In fact, the smallest stars are significantly smaller in radius than Jupiter, even though they are 80 times more massive.

The pillow analogy suggests that a hydrogen/helium planet less massive than Saturn should have an even lower density. Because Uranus and Neptune have significantly higher densities than Saturn, we conclude that they cannot have the same hydrogen/helium composition. Instead, they must have a much larger fraction of higher-density material, such as hydrogen compounds and rocks, and a smaller fraction of "plain" hydrogen and helium.

FIGURE 11.4 The growth of jovian planets.

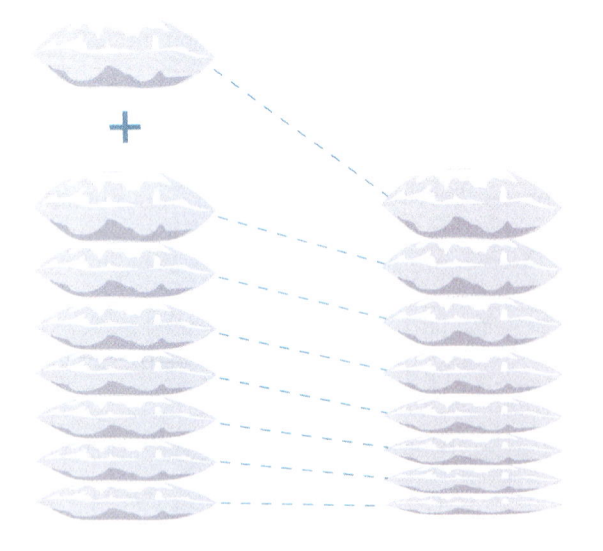

a Adding pillows to a stack may increase its height at first, but eventually it just compresses all the pillows in the stack. In a similar way, adding mass to a jovian planet would eventually just compress the planet to higher density without increasing the planet's radius.

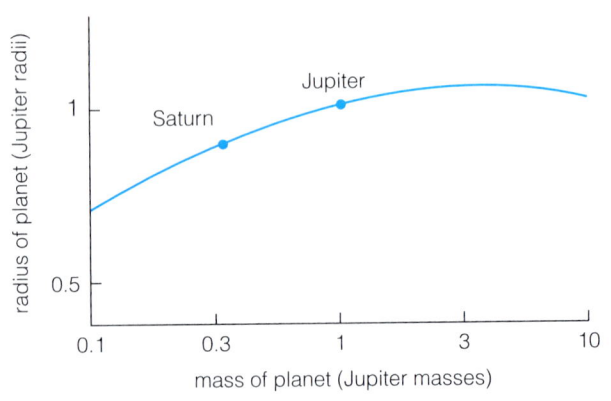

b This graph shows how the radius of a hydrogen/helium planet depends on the planet's mass. Jupiter's radius is only slightly larger than Saturn's, although it is three times more massive. For a planet much more massive than Jupiter, gravitational compression would actually make it smaller in size.

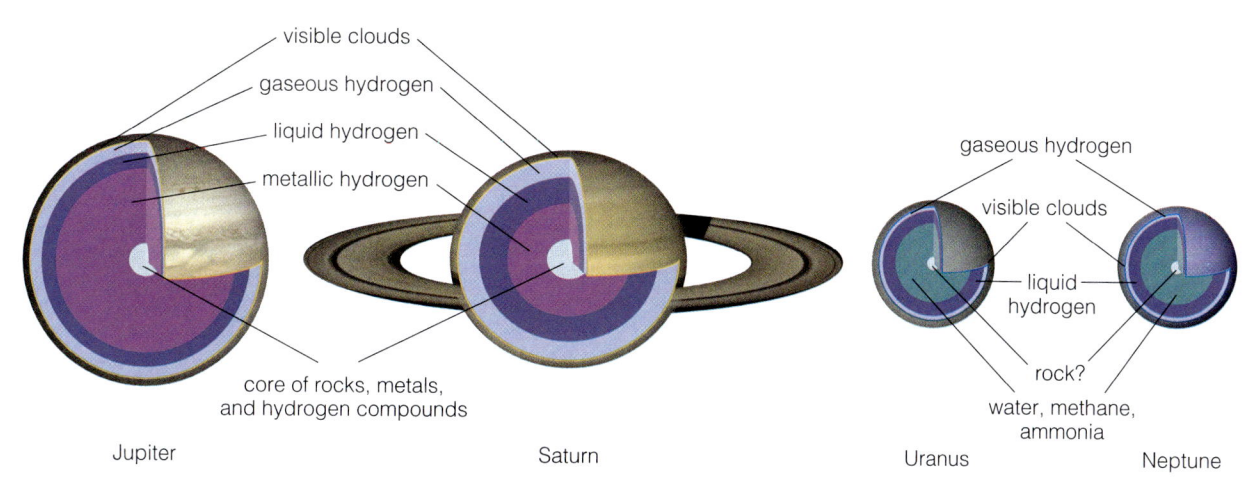

FIGURE 11.5 These diagrams compare the interior structures of the jovian planets (shown approximately to scale). All four planets have cores equal to about 10 Earth masses of rock, metal, and hydrogen compounds, and they differ primarily in the hydrogen/helium layers that surround the cores.

Computer models of the jovian interiors yield the somewhat surprising result that the cores of all four jovian planets are quite similar—about 10 Earth masses of rock, metal, and hydrogen compounds. This suggests that all four jovian planets began with about the same size "seed" from accretion and that their differences stem from their capturing different amounts of additional gas from the solar nebula. Jupiter captured more than 300 Earth masses of gas from the solar nebula, Saturn captured about one-fourth as much gas as Jupiter, and Uranus and Neptune captured only a few Earth masses of solar-nebula gas. This general pattern makes sense, because icy planetesimals took longer to accrete in the outer solar system, where planetesimals were more spread out. Thus, the more distant jovian planets didn't have as much time as Jupiter to capture solar-nebula gas before the nebula was cleared by the solar wind [Section 8.4].

The similar cores also mean that the interior structures of the jovian planets differ mainly in the hydrogen/helium layers that surround the cores (Figure 11.5). Saturn is large enough for interior densities, temperatures, and pressures to be high enough to force gaseous hydrogen into liquid and then metallic form, just as in Jupiter. But these conditions occur relatively deeper in Saturn than in Jupiter, because of its smaller size. Pressures within Uranus and Neptune are not high enough to form liquid or metallic hydrogen. However, their cores of rock, metal, and hydrogen compounds may be liquid, making for very odd "oceans" buried deep inside them.

Like Jupiter, Saturn emits nearly twice as much energy as it receives from the Sun. Saturn's mass is too small for it to be generating all its excess heat by contracting like Jupiter. Instead, its lower temperatures probably allow helium to condense and "rain down" from higher regions in the interior. The gradual rain of these relatively dense helium droplets resembles the process of *differentiation* that once occurred in terrestrial planet interiors [Section 9.2]. Voyager observations confirmed that Saturn's atmosphere is somewhat depleted of helium (compared to Jupiter's atmosphere), just as we would expect if helium has been raining down into Saturn's interior for billions of years.

Neither Uranus nor Neptune has conditions that allow helium rain to form, and most of their original heat from accretion should have escaped long ago. This explains why Uranus emits virtually no excess internal energy. Neptune is much more mysterious: Like Jupiter and Saturn, it emits nearly twice as much energy as it receives from the Sun. The only reasonable explanation for Neptune's internal heat source is that the planet is somehow still contracting, rather like Jupiter, thereby converting gravitational potential energy into thermal energy. But no one knows why a planet of Neptune's size would still be contracting more than 4 billion years after its formation.

11.3 Jovian Planet Atmospheres

The jovian planets lack the geology of the terrestrial planets because they do not have solid surfaces. But whatever they lack in geology they more than make up for with their atmospheres. Fortunately, their atmospheric processes are quite similar to those we've already discussed for the terrestrial atmospheres, despite their much greater extent and very different compositions. Let's again begin with Jupiter.

Jupiter's Atmosphere

Jupiter's atmosphere is almost entirely hydrogen and helium (about 75% hydrogen and 24% helium by mass), but it also contains trace amounts of many

a Artist's conception of the Galileo probe entering Jupiter's atmosphere. This view, looking outward from beneath the clouds, shows the suitcase-size probe falling with its parachute extended above it.

FIGURE 11.6 Jupiter's atmosphere.

other hydrogen compounds. Because oxygen, carbon, and nitrogen are the most common elements besides hydrogen and helium, the most common hydrogen compounds are methane (CH_4), ammonia (NH_3), and water (H_2O). Spectroscopy reveals the presence of more complex compounds, including acetylene (C_2H_2), ethane (C_2H_6), propane (C_3H_8), and larger molecules that can act like haze particles. Although hydrogen compounds make up only a minuscule fraction of Jupiter's atmosphere, they are responsible for virtually all aspects of its appearance. Some of these compounds condense to form the clouds so prominent in telescope and spacecraft images, and others are responsible for Jupiter's great variety of colors. Without these compounds, Jupiter would be a uniform, colorless ball of gas.

TIME OUT TO THINK *Jupiter possesses substantial amounts of gases like methane, propane, and acetylene, highly flammable fuels used here on Earth. There is plenty of lightning on Jupiter to provide sparks, so why aren't we concerned about Jupiter exploding? (Hint: What's missing from Jupiter's atmosphere that's necessary for ordinary fire?)*

Jupiter is the only jovian planet that we've directly sampled. On December 7, 1995, following a 6-year trip from Earth, the Galileo spacecraft released a scientific probe (the size of a large suitcase) into Jupiter's atmosphere (Figure 11.6a). The probe collected temperature, pressure, composition, and radiation measurements for about an hour as it descended, until it was destroyed by the ever-increasing pressures and temperatures. The probe sent its data via radio signals back to the Galileo spacecraft in orbit around Jupiter, which relayed the data back to Earth.

The Galileo probe helped confirm that the thermal structure of Jupiter's atmosphere is very similar to that of terrestrial atmospheres (Figure 11.6b). Let's follow its fiery descent. The probe plunged into Jupiter at a speed of over 200,000 km/hr, with a robust heat shield facing forward to protect the scientific instruments. Above the cloud tops, the probe

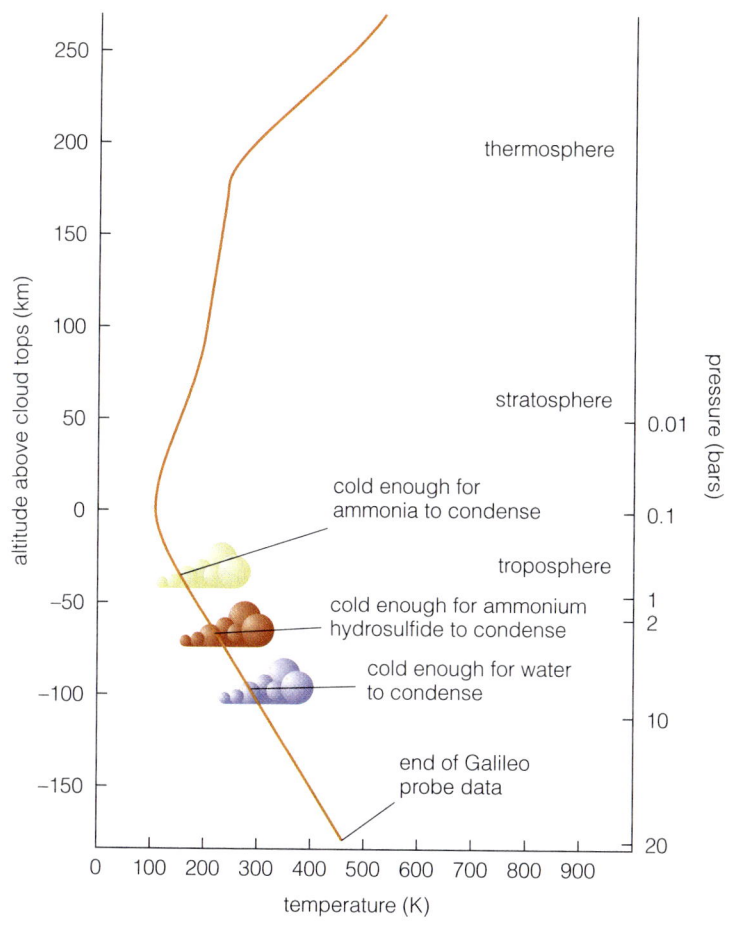

b This graph shows the temperature structure of Jupiter's atmosphere (compare to Figure 10.5). Jupiter has at least three distinct cloud layers because different atmospheric gases condense at different temperatures and hence at different altitudes.

found very low density gas that is heated to perhaps 1,000 K by solar X rays and by energetic particles from Jupiter's magnetosphere. This thin, hot gas makes up Jupiter's *thermosphere*. Next, the probe encountered Jupiter's *stratosphere*, where solar ultraviolet photons are absorbed by a few minor ingredients in the atmosphere. This absorption gives the stratosphere a peak temperature of about 170 K. Chemical reactions driven by the solar ultraviolet photons also create a smoglike haze that masks the color and sharpness of the clouds below. Below the stratosphere lies Jupiter's *troposphere*, where the increased density greatly slowed the probe's descent; the probe then slowed further by jettisoning its heat shield and releasing a parachute. The temperature at the top of the troposphere is close to 125 K; it rises with depth because greenhouse gases trap both solar heat and Jupiter's own internal heat. The similarities between the atmospheric structures of Earth and Jupiter may seem surprising for two planets that are so different in other ways, but these similarities con-firm that atmospheres are governed primarily by interactions between sunlight and gases.

The Galileo probe found tremendous winds and turbulence in the troposphere. As on the terrestrial planets, Jupiter's weather occurs mostly in its troposphere, where higher temperatures at lower altitudes drive vigorous convection. This convection is responsible for the thick clouds that enshroud Jupiter. As a parcel of gas rises upward through the troposphere, it encounters gradually lower temperatures. At relatively low altitudes, the surrounding temperature is low enough for water vapor to condense into liquid droplets or flakes of ice, forming water clouds similar to those on Earth. At higher altitudes, temperatures become low enough for another minor atmospheric ingredient, ammonium hydrosulfide (NH_4SH), to condense and form clouds. Still higher, the temperatures fall to the point at which ammonia (NH_3) condenses to form clouds of ammonia crystals—the prominent white clouds visible on Jupiter. Thus, Jupiter has several layers of clouds of different compositions (see Figure 11.6b). The relentless motion of Jupiter's atmosphere continuously regenerates the extensive clouds.

Jupiter also has planet-wide *circulation cells* similar to those on Earth, where solar heat causes equatorial air to expand and spill northward and southward toward the poles [Section 10.4]. However, whereas Earth's rotation splits its circulation cells into just three separate cells in each hemisphere, Jupiter's much more rapid rotation creates a strong Coriolis effect that causes its circulation cells to split into many huge bands encircling the entire planet at fixed latitudes (Figure 11.7a,b). The bands of rising air are called **zones**, and they appear white because of the ammonia clouds that form as the air rises to high, cool altitudes. Within the zones, ammonia "snowflakes" rain downward against the rising convective motions. The adjacent **belts** of falling air are depleted in the cloud-forming ingredients and do not contain any white ammonia clouds. They appear dark because we can see down to the red or tan ammonium-hydrosulfide clouds that form at lower altitudes. The distinction between Jupiter's belts and zones is analogous to the distinction on Earth between the cloudy, rainy equatorial zone (a region of generally rising air) and the clear desert skies found roughly 20°–30° north

a Belts and zones correspond to clouds of different composition at different altitudes. The Coriolis effect diverts motions in the circulation cells into strong easterly and westerly winds.

belt zone belt

The Coriolis effect diverts winds.

Rising air forms white ammonia cloud.

Snow depletes air of ammonia.

No cloud or snow in descending air; clouds below are visible.

ammonium hydrosulfide cloud

Belts are warm, red, low-altitude clouds.

Zones are cool, white, high-altitude clouds.

b The color difference between belts and zones is evident in this Hubble Space Telescope image.

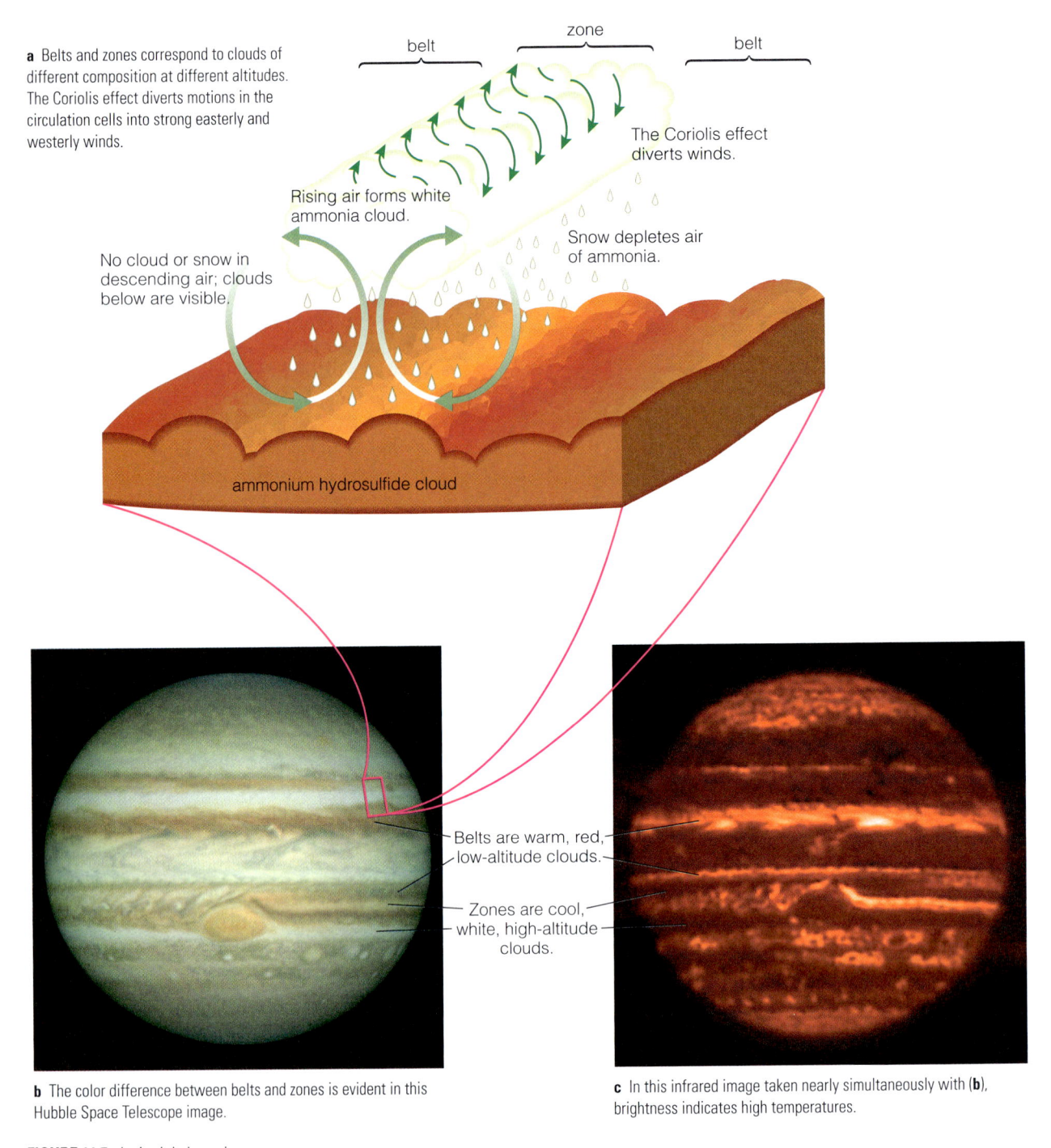

c In this infrared image taken nearly simultaneously with (**b**), brightness indicates high temperatures.

FIGURE 11.7 Jupiter's belts and zones.

---

and south of the equator (regions of descending air). Infrared images confirm the temperature differences between Jupiter's warm belts and cool zones (Figure 11.7b,c).

Jupiter's global wind patterns are shaped by these alternating bands of rising air in zones and falling air in belts. The rising air in the zones indicates higher pressures than in the adjacent belts, so winds must flow from the high-pressure zones toward the low-pressure belts. Figure 11.7a shows the northward winds flowing from a zone to a belt in the northern hemisphere. The northward winds are quickly diverted into fast eastward-flowing winds by the very strong Coriolis effect. Similarly, the southward-flowing winds from zones to belts in the northern hemisphere are diverted to the west. (Note that both these diversions of winds are to the right in the northern hemisphere; in the southern hemisphere, winds are diverted to the left [Section 10.4].) The winds are generally strongest

296 PART III LEARNING FROM OTHER WORLDS

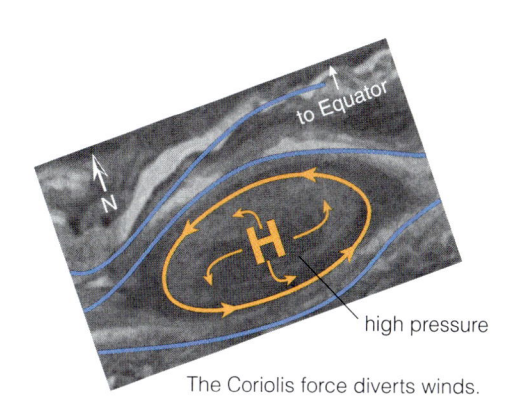

FIGURE 11.8 This photograph shows Jupiter's Great Red Spot, a huge high-pressure storm that is large enough to swallow two or three Earths. The smaller photo (right) of the Great Red Spot is overlaid with a weather map of the region.

at the equator and at the boundaries between belts and zones. Peak wind speeds exceed 400 km/hr.

Jupiter's **Great Red Spot** is perhaps the most dramatic weather pattern in the solar system (Figure 11.8). We know that it has been prominent in Jupiter's southern hemisphere for at least three centuries, because it has been seen ever since telescopes became powerful enough to detect it. The Great Red Spot is huge—more than twice as wide as the Earth. Could it be a tremendous low-pressure storm like a hurricane on Earth? Not quite. Hurricane winds circulate around low-pressure regions and therefore circulate *clockwise* in the southern hemisphere [Section 10.4]. The winds in the Great Red Spot circulate counterclockwise, indicating that it is a *high*-pressure storm, possibly kept spinning by the influence of nearby belts and zones.

Other, smaller storms are always brewing in Jupiter's atmosphere—small only in comparison to the Great Red Spot. Brown ovals are low-pressure storms with their cloud tops deeper in Jupiter's atmosphere, and white ovals are high-pressure storms topped with ammonia clouds. No one knows what drives Jupiter's storms, why Jupiter has only one Great Red Spot, or why the Great Red Spot lasts so much longer than storms on Earth. Storms on Earth lose their strength when they pass over land, so perhaps Jupiter's biggest storms last for centuries because there is no solid surface below to sap their energy.

Jupiter's vibrant colors remain perplexing despite decades of intense study. Observations tell us that the high clouds of ammonia in the zones are usually white and that the lower clouds of ammonium hydrosulfide in the belts are brown or red. But pure ammonium-hydrosulfide crystals are as white as ammonia crystals, so the colors of the belt clouds must come from ingredients besides the ammonium-hydrosulfide crystals themselves. Sulfur compounds are a plausible suspect, because they form a variety of reds and tans (also seen on the surface of Jupiter's moon Io). Phosphorus compounds (including phosphine, PH_3) are another possibility. Whatever their origin, the red and tan compounds are found primarily at lower (and therefore warmer) altitudes. The chemical reactions that form them evidently require the extra energy of Jupiter's internal heat, or possibly lightning deep below the clouds. The bright red color of the very high altitude clouds in the Great Red Spot remains an even greater mystery. Perhaps chemical reactions caused by solar ultraviolet radiation produce the compounds responsible for these colors.

As far as we know, Jupiter's climate is steady and unchanging. Jupiter has no appreciable axis tilt and therefore has no seasons. In fact, Jupiter's polar temperatures are quite similar to its equatorial temperatures. Solar heating alone might leave the poles relatively cool, but Jupiter's internal heat source keeps the planet uniformly warm.

Comparing Jovian Planet Atmospheres

The most striking difference among the jovian atmospheres is their colors. Starting with Jupiter's distinct red colors, the jovian planets turn to Saturn's more subdued yellows, to Uranus's faint blue-green tinge, and finally to Neptune's pronounced blue hue (see Figure 11.1). What's responsible for this regular progression of colors? The primary constituents in all the jovian atmospheres (hydrogen and helium) are colorless, so the planetary colors must come from trace gases or chemical reactions that create colored compounds. In fact, we can understand the color differences by comparing the atmospheric structures and compositions of the four planets. All four are quite similar, except for the effects of the lower temperatures and lower gravities on the more distant planets (Figure 11.9).

Saturn's reds and tans almost certainly come from the same compounds that produce these colors on Jupiter—whatever those compounds may be. As on Jupiter, these compounds are probably created by chemical reactions beneath the cloud layers and carried upward by convection. However, because of Saturn's lower temperatures, its cloud layers lie deeper in its atmosphere than Jupiter's. As a result, a thicker layer of tan "smog" overlies the clouds, washing out Saturn's colors. Saturn's lower gravity also means that its atmosphere is less compressed, so cloud layers are more spread out vertically. These thicker cloud layers prevent us from seeing down to the lower, more richly hued cloud levels that we see on Jupiter.

The blue colors of Uranus and Neptune have a completely different explanation: methane gas, which is at least 20 times more abundant (by percentage) on these planets than on Jupiter or Saturn. The highest-altitude clouds of Uranus and Neptune are made from flakes of methane ice; Jupiter and Saturn lack such clouds because their warmer temperatures prevent methane ice from condensing. Methane gas above these clouds absorbs red light but transmits blue light (Figure 11.10). The clouds then reflect (scatter) the blue light upward, where it is again transmitted through the gas above. Thus, we see this reflected blue light when we look at the atmospheres of Uranus and Neptune. Uranus and Neptune may also have red and tan cloud layers like those of Jupiter and Saturn, but if so they are hidden deep below the methane gas and clouds.

The fainter blue of Uranus compared to Neptune indicates that less sunlight penetrates the atmosphere to the level of the clouds. Instead, the light is probably scattered by abundant smoglike haze. The extra haze on Uranus may arise because its extreme axis tilt keeps one hemisphere in bright sunlight for decades at a time during its 84-year orbit, allowing more ultraviolet-driven chemical reactions. This continuous sunlight in one hemisphere may also explain why Uranus has a surprisingly hot thermosphere that extends thousands of kilometers above its cloud tops.

Voyager, Galileo, Cassini, and the Hubble Space Telescope cameras have monitored the meteorology of the four jovian planets with excellent resolution (Figure 11.11). Jupiter's weather phenomena (clouds, belts, zones, and other storms) are by far the strong-

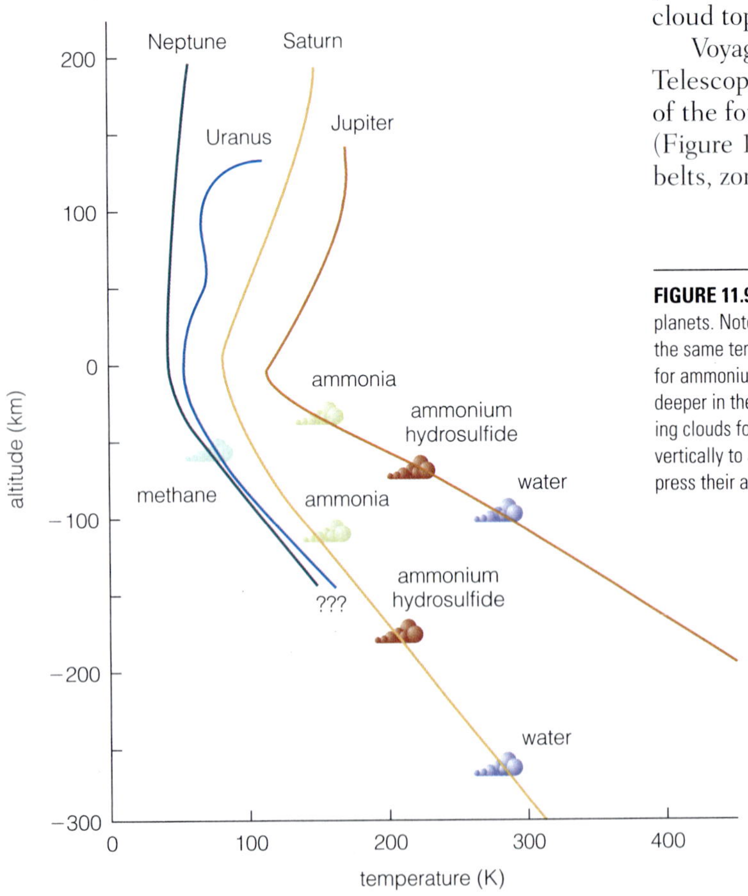

FIGURE 11.9 Temperature variation with altitude for each of the jovian planets. Note that clouds of a particular composition always form at about the same temperature: less than 100 K for methane, 150 K for ammonia, 200 K for ammonium hydrosulfide, and 270 K for water. These temperatures occur deeper in the atmospheres of planets farther from the Sun, so the corresponding clouds form deeper as well. Note also that the cloud layers are separated vertically to a greater extent on smaller planets with weaker gravity to compress their atmospheres. ("???" indicates uncertainty beneath this level.)

est and most active. Saturn also possesses belts and zones (in more subdued colors), along with some small storms and an occasional large storm, but it lacks any weather feature as prominent as Jupiter's Great Red Spot. (For unknown reasons, Saturn's winds are even stronger than Jupiter's.) Neptune's atmosphere is also banded and has a high-pressure storm, called the Great Dark Spot, that acts much like Jupiter's Great Red Spot.

Seasonal changes play a relatively small role on most jovian planets. We might expect seasons on Saturn and Neptune because they have axis tilts similar to that of Earth, and some seasonal weather changes have been observed. But their internal heat sources maintain similar temperatures all over, just as Jupiter's internal heat keeps its polar and equatorial temperatures about the same. The greatest surprise in the weather patterns on the jovian planets is the weather on Uranus, the planet tipped on its side. When the Voyager spacecraft flew past in 1986, it was midsummer in Uranus's northern hemisphere, with the summer pole pointed almost directly at the Sun. Photographs revealed virtually no clouds, belts,

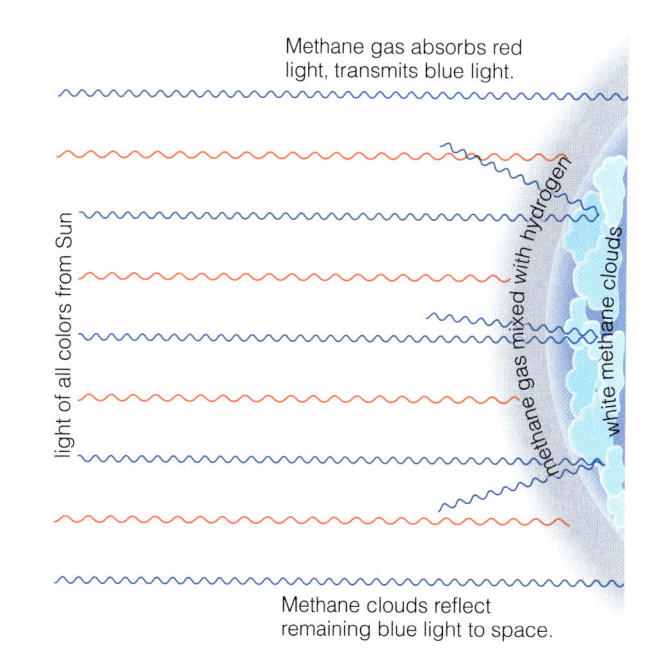

FIGURE 11.10 Neptune and Uranus are blue because methane gas absorbs red light but transmits blue light. Clouds of methane-ice flakes reflect the transmitted blue light back to space.

FIGURE 11.11 Close-ups of cloud patterns on the four jovian planets. Scientists combine series of images like these to make weather movies similar to those for Earth shown on the evening news. The reddish spots in the Uranus image are high clouds, and the Great Dark Spot is visible in the Neptune image to the left of center. Images for Jupiter, Saturn, and Neptune at visible wavelengths, Uranus at infrared wavelengths.

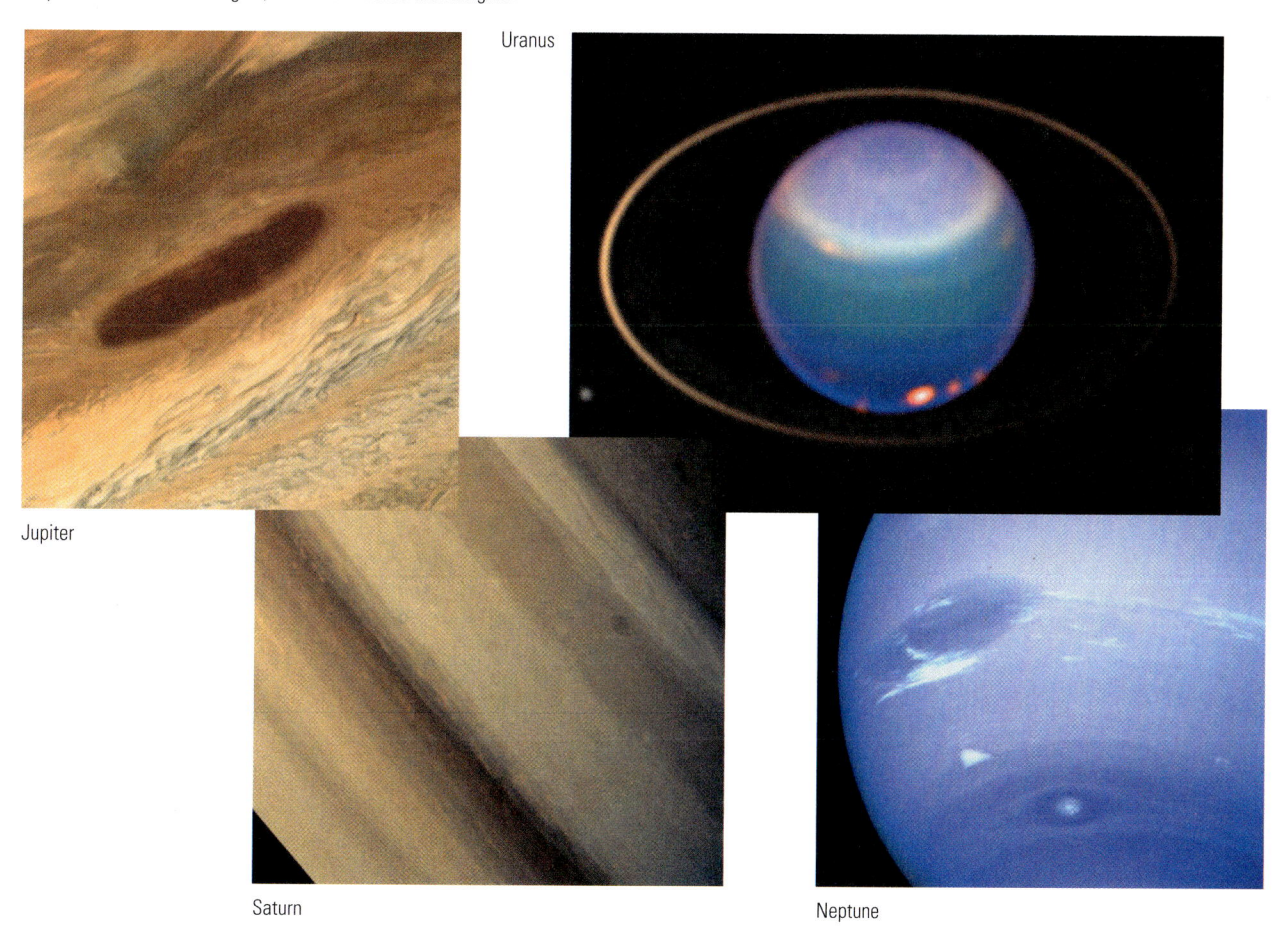

Jupiter

Uranus

Saturn

Neptune

CHAPTER 11 JOVIAN PLANET SYSTEMS 299

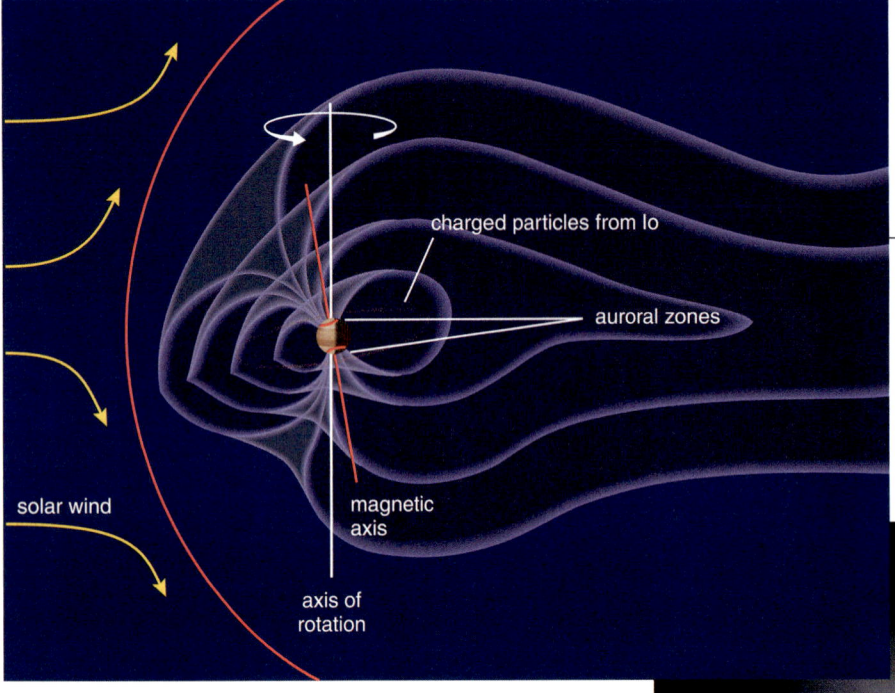

FIGURE 11.12 Auroras on Jupiter.

a Like Earth's magnetosphere (but much larger), Jupiter's magnetosphere has charged particle belts and auroras. In Jupiter's case, the charged particles come from the volcanically active moon Io.

or zones, which scientists attributed to the lack of a significant internal heat source on Uranus. But recently the Hubble Space Telescope has detected storms raging on the planet (bright spots in Figure 11.11). The seasons are changing, and the winter hemisphere is seeing sunlight for the first time in decades.

11.4 Jovian Planet Magnetospheres

Each jovian planet is surrounded by a bubblelike magnetosphere consisting of the planet's magnetic field and the particles trapped within it. The jovian planets themselves dwarf the terrestrial planets, and their magnetospheres are even more impressive. The Voyager spacecraft carried sophisticated instruments to measure magnetic fields and observe the charged particle belts encircling the planets. We learned not only about the magnetospheres themselves, but also about their interactions with the planets beneath them and the satellite systems within them.

Jupiter's Magnetosphere

Jupiter's magnetic field is awesome—about 20,000 times stronger than Earth's. This strong magnetic field deflects the solar wind some 3 million km (about 40 Jupiter radii) before it even reaches Jupiter (Figure 11.12a). If our eyes could see this part of Jupiter's

b Images of Jupiter's aurora in the ultraviolet are overlaid on a visible-wavelength image of Jupiter for reference (Hubble Space Telescope photo).

magnetosphere, it would be larger than the full moon in our sky. The solar wind sweeps around Jupiter's magnetosphere, stretching it out on the far side of Jupiter all the way to Saturn's orbit. Just as on Earth, the magnetosphere traps charged particles and makes them spiral along the magnetic field lines. The more energetic particles cause *auroras* as they follow the magnetic field into Jupiter's upper atmosphere, colliding with atoms and molecules and causing them

300 PART III LEARNING FROM OTHER WORLDS

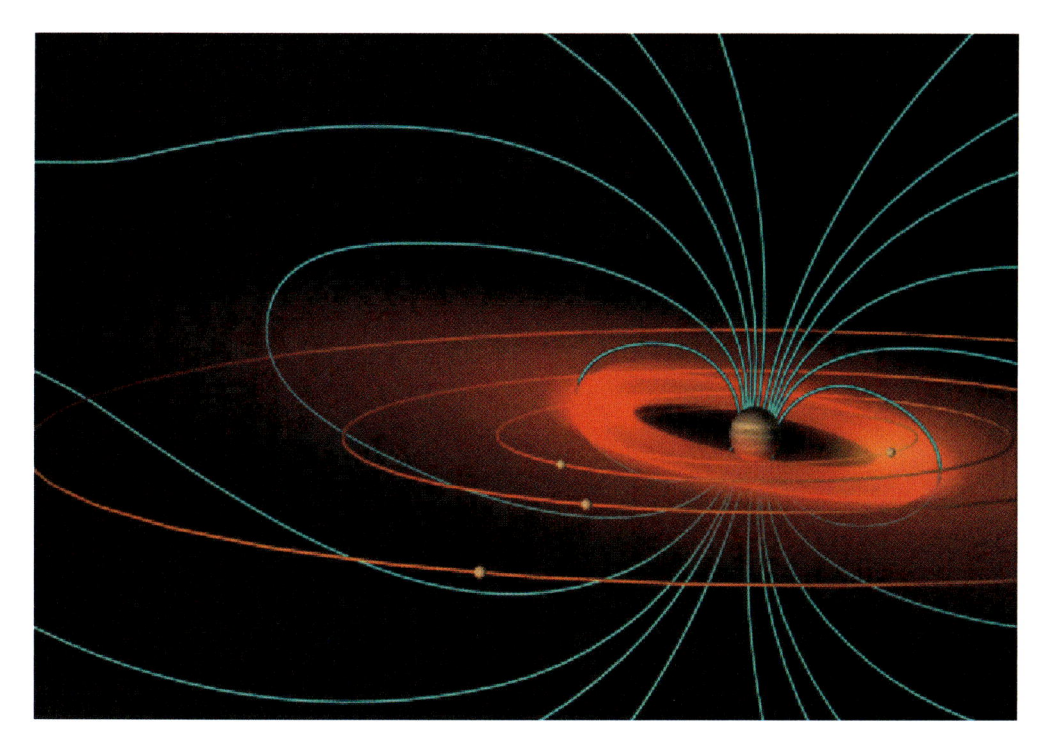

FIGURE 11.13 Artist's conception of Jupiter's magnetosphere, with Io embedded deep within. An orange glow around Io represents its escaping atmosphere, and the red donut represents the Io torus—the charged particle belt formed by Io. The blue lines show Jupiter's strong magnetic field.

to radiate (Figure 11.12b). Jupiter's magnetosphere contains a great deal of plasma, some captured from the solar wind but most originating from Jupiter's volcanically active moon, Io.

Not only does Io help feed plasma into the magnetosphere, but the magnetosphere in turn has important effects on Io and other satellites of Jupiter. The charged particles bombard the surfaces of Jupiter's icy moons, with each particle blasting away a few atoms or molecules. This process alters the surface materials and can even generate thin bombardment atmospheres—just as bombardment creates the thin atmospheres of Mercury and the Moon [Section 10.5]. On Io, the bombardment leads to the continuous escape of the atmospheric gases released by volcanic outgassing. As a result, Io loses atmospheric gases faster than any other object in the solar system. The escaping gases (sulfur, oxygen, and a hint of sodium) are ionized and feed a donut-shaped charged particle belt, called the *Io torus,* that approximately traces Io's orbit (Figure 11.13).

Comparing Jovian Planet Magnetospheres

The strength of each jovian planet's magnetic field depends primarily on the size of the electrically conducting layer buried in its interior. Jupiter, with its huge metallic-hydrogen layer, has by far the strongest magnetic field. Saturn has a thinner layer of metallic hydrogen and hence a correspondingly weaker magnetic field. Uranus and Neptune have no metallic hydrogen at all. Their relatively weak magnetic fields must be generated in their core "oceans" of hydrogen compounds, rock, and metal. The actual size of a planet's magnetosphere depends on the pressure of the solar wind pushing on a planet's magnetic field, as well as on the magnetic-field strength itself. The pressure of the solar wind is weaker at greater distances from the Sun. Thus, the magnetospheric "bubbles" surrounding the outer planets, particularly Uranus and Neptune, are larger than they would be if these planets were closer to the Sun. This effect is not sufficient to offset the weak magnetic fields of Uranus and Neptune, however, so these planets have small magnetospheres (Figure 11.14).

While the trend of decreasing magnetic-field strengths makes sense, the magnetic fields still pose some unsolved problems. The magnetic fields of Jupiter and Saturn are fairly closely aligned with their rotation axes (10° tilt and 0° tilt, respectively), just as we might expect given that the magnetic fields are generated in their rotating interiors. But Voyager observations showed that the magnetic field of Uranus is tipped by a whopping 60° relative to its rotation axis (see Figure 11.14); its center is also significantly offset from the planet's center. Scientists briefly speculated that whatever giant impact caused the large

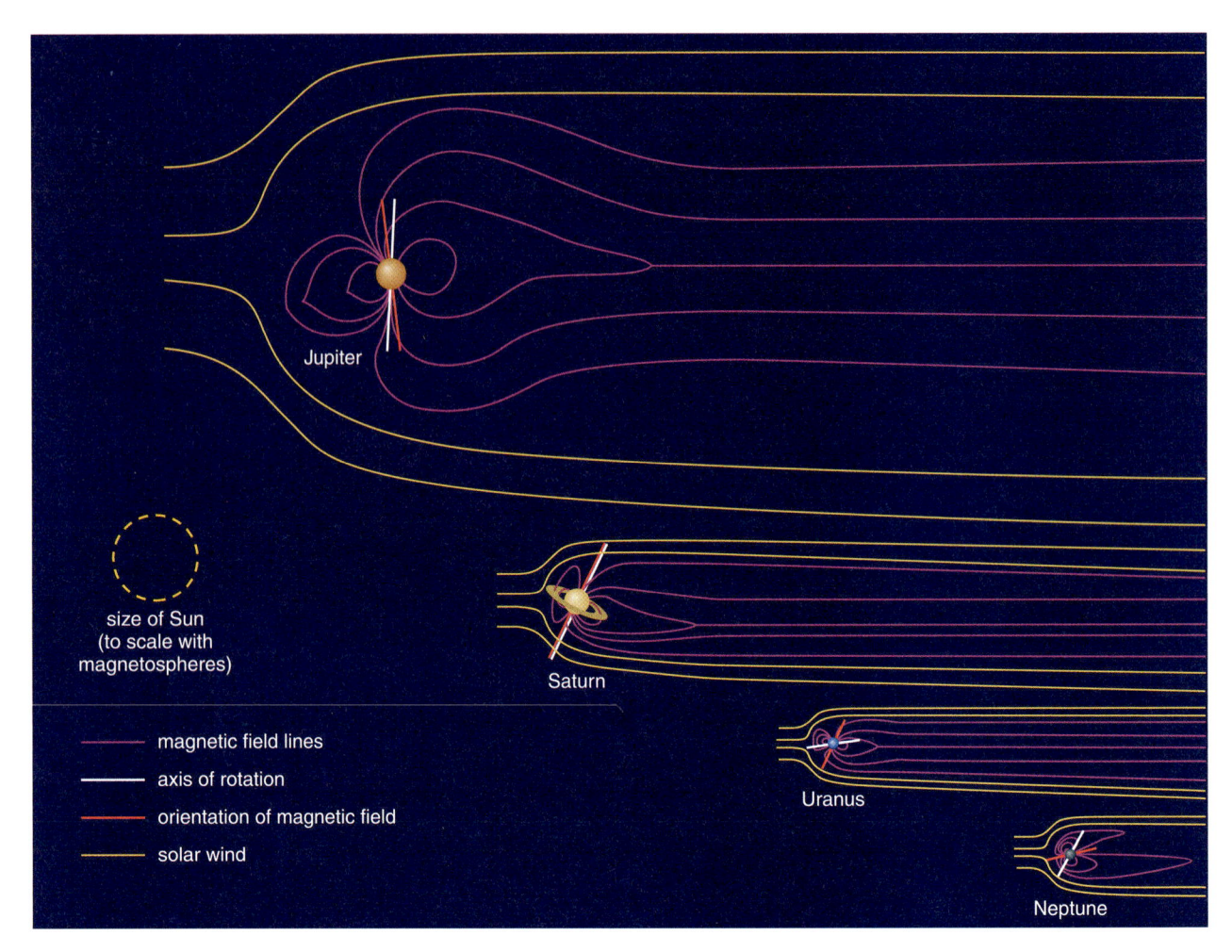

FIGURE 11.14 Comparison of jovian planet magnetospheres. The size of Jupiter's magnetosphere is particularly impressive in light of the greater pressure from the solar wind nearer the Sun. (Planets enlarged for clarity.)

tilt of Uranus's rotation axis also distorted the magnetic field. But this idea was discarded when Voyager discovered a similarly large magnetic field tilt (46°) for Neptune. Scientists still cannot explain the magnetic-field tilts of Uranus and Neptune.

No magnetosphere is as full of charged particles as Jupiter's, primarily because no other jovian planet has a satellite like Io. All magnetospheres trap particles from the solar wind, but Jupiter again captures the most. These trapped particles give Jupiter the brightest auroras; the more distant jovian planets have progressively weaker auroras.

11.5 A Wealth of Worlds: Satellites of Ice and Rock

Nearly 90 known moons (*satellites*) orbit the jovian planets. Figure 11.15 shows the larger ones, and Appendix C lists fundamental data for the known moons. It's worth taking a few minutes to study the figure and the appendix, looking for patterns, before you read on.

Broadly speaking, the jovian moons can be divided into three groups: small moons less than about 300 km across, medium-size moons ranging from about 300 to 1,500 km in diameter, and large moons more than 1,500 km in diameter. Most of the known moons fall into the small category, and additional small moons may yet be discovered. The small moons generally look more like potatoes than like spheres (Figure 11.16), because their gravities are too weak to force their rigid material into spheres [Section 8.4]. Many of these moons (particularly those farther from their planets) also have unusual orbits, with significant eccentricities, large orbital tilts, or orbits that go backward relative to the planet's rotation. Such orbits are signs that the moons are captured asteroids rather than moons that formed from the "miniature solar nebulae" that surrounded each jovian planet. In a few cases, several small moons share similar orbits. For example, four moons orbit Jupiter backward at distances between 21 and 24 million kilometers and

302 PART III LEARNING FROM OTHER WORLDS

FIGURE 11.15 The larger moons of the jovian planets, with sizes (but not distances) shown to scale. Mercury, the Moon, and Pluto are included for comparison.

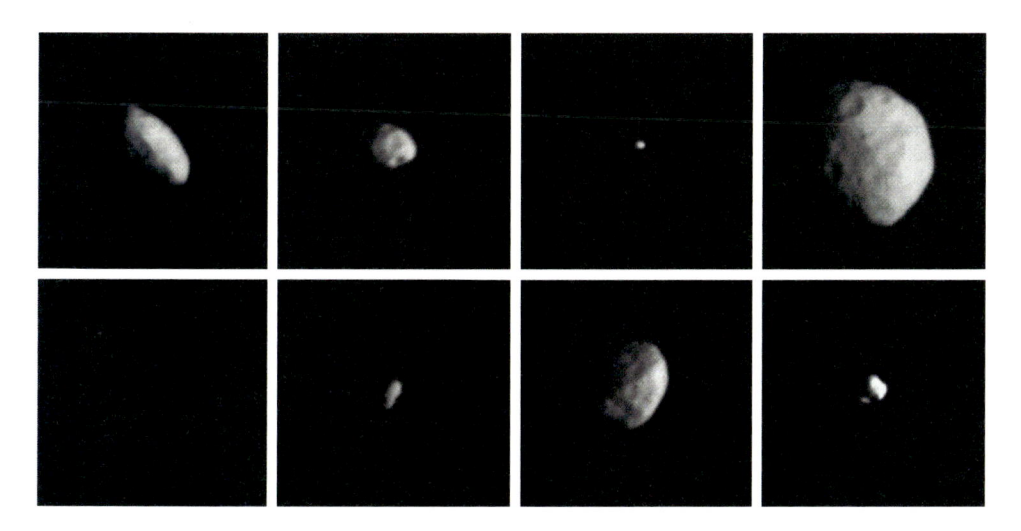

FIGURE 11.16 A montage of the small moons of Saturn, shown to scale. The small moons of the other jovian planets also are probably not spherical. These photographs were taken by the Voyager spacecraft.

CHAPTER 11 JOVIAN PLANET SYSTEMS 303

probably once were part of a larger captured asteroid that fragmented into several pieces.

Nearly all the medium-size and large moons follow the orbital patterns we expect from formation in "miniature solar nebulae": They have approximately circular orbits that lie close to the equatorial plane of their parent planet, and they orbit in the same direction in which their planet rotates. These moons are planetlike in almost all ways. They are approximately spherical, each has a solid surface with its own unique geology, and some possess atmospheres, hot interiors, and even magnetic fields. A few are planetlike in size as well. The two largest moons—Jupiter's moon Ganymede and Saturn's moon Titan—are larger than the planet Mercury. Four others are larger than Pluto: Jupiter's moons Io, Europa, and Callisto, and Neptune's moon Triton.

Nearly all jovian moons share an uncanny trait: They always keep the same face turned toward their planet, just as our Moon always shows the same face to Earth (see Figure 2.20). This *synchronous rotation* arose from the strong tidal forces [Section 5.5] exerted by the jovian planets, which caused each moon to end up with equal rotational and orbital periods regardless of how fast the moon rotated when it formed.

Ice Geology

Prior to the Voyager missions, most scientists expected the jovian moons to be cold and geologically dead. After all, only two of the moons are even as large as Mercury, and Mercury has been geologically dead for a long time. Instead, Voyager provided many surprises, revealing worlds with spectacular past and present geological activity. We've since learned even more—about Jupiter's moons in particular—thanks primarily to data from the Galileo spacecraft.

How can moons have so much more geological activity than similar-size (or larger) planets? In a few cases, the answer involves unforeseen heat sources that we'll discuss shortly. But another important factor is the icy composition of most jovian moons. These moons accreted from solid particles that condensed in the "miniature solar nebulae" around the jovian planets. Because they were so far from the Sun, temperatures in these nebulae were quite cool and were rich with ice crystals of various hydrogen compounds (mostly water, but also small amounts of ammonia and methane). As a result, the jovian planet satellites contain much higher proportions of ice relative to rock than do the terrestrial planets. We even find distinctions among the jovian satellite systems: Jupiter's satellites have the highest proportion of rock, while the satellites of more distant planets contain higher proportions of the more volatile methane and ammonia ices.

We can understand why icy moons have more geological activity than similar-size rocky moons by investigating how ices compare to rocks in three geologically important properties: strength (or rigidity), radioactive content, and melting point. In terms of strength, the differences might be smaller than you would expect. Ice at a temperature of 100 K is almost as rigid as rock at a few hundred Kelvin. Nevertheless, the difference is great enough to cause ice mountains and ice cliffs on the jovian moons to sag and fade away more quickly than similar features made of rock on terrestrial planets. Most radioactive elements are found in rocks, not ices, so icy moons have relatively little internal heat generated by radioactive decay and hence lower internal temperatures. But ices also have much lower melting points than rocks, so the heat of accretion in many medium-size and large moons was enough to cause differentiation of their interiors. Thus, most of these moons have internal structures similar to those of the terrestrial planets but with an additional, icy layer on the outside. Despite the icy composition, no satellite surface is "clean as the driven snow"; the ices are "dirtied" by rocky and carbon-rich compounds. Each satellite has its own proportions of rock and ice. Some are cleaner, some are dirtier, and some are clean on one side and dirty on the other.

All in all, the *ice geology* of the jovian moons bears many similarities to the rock geology of the terrestrial planets [Section 9.4]. *Impact cratering*, occurring mostly during the time of the early bombardment more than 4 billion years ago, probably left all the satellites with battered surfaces. *Volcanism* is certainly present on some jovian moons, and *tectonics* seems likely as well. Both of these processes are driven by internal heat and hence take place more easily on icy moons than on rocky planets, because ice becomes deformable and melts at lower temperatures than rock. We know very little about *erosion* in the outer solar system, although we expect it to be relatively rare. Most satellites possess either no atmosphere or a very thin one, so in most cases wind or rain erosion is unlikely. Nevertheless, many of the jovian moons have interesting features that go beyond simple generalities. There are far too many moons to study all of them in depth here, but as we undertake our brief tour of the jovian moon systems, we will stop to investigate a few of the more intriguing moons in detail.

The Galilean Satellites of Jupiter: Io, Europa, Ganymede, and Callisto

The first stops on our tour are the Galilean satellites of Jupiter—the four moons that Galileo saw through his telescope and that disproved the idea that all celestial bodies circle the Earth [Section 5.3]. These four moons are large enough that they would count as planets if they orbited the Sun. They bear the names of four mythological lovers of the Roman god Jupiter: Io, Europa, Ganymede, and Callisto. Their densities decrease with distance from Jupiter, just as the densities of the planets decrease with distance from the Sun, suggesting that similar condensation and accretion patterns occurred within the "miniature solar nebula" around Jupiter and in the solar nebula. Io, the innermost Galilean satellite, formed from rocky and metallic condensates with little or no icy condensates. Europa is mostly rocky, but it formed with enough ice to give it an icy outer shell. Ganymede and Callisto, the outermost Galilean moons, formed from a mix of icy and rocky condensates.

Io Just a few months before the Voyager spacecraft reached Jupiter, a group of scientists made the astonishing prediction that we would find active volcanism on Io. This prediction went against the common belief that only much larger bodies could have substantial geological activity. But the prediction proved correct.[4] Direct proof came from the discovery of towering volcanic plumes reaching hundreds of kilometers above the surface and spreading fallout over almost a million square kilometers of Io's surface (Figure 11.17a). The indirect proof was just as mind-boggling: the complete lack of impact craters on the surface—not a single one, to the resolution of Voyager's cameras. Io's eruptions have buried *all* its impact craters.

Despite Io's unusual colors and frost-covered surface, planetary scientists found volcanic eruptions remarkably similar to those of basalt volcanoes on Earth (Figure 11.17b). When the flowing lava comes in contact with frost, the resulting "steam" jets off at high speed to make the plume. The process is similar to lava flowing into the ocean on Earth, but Io's relative lack of atmosphere and low gravity let the plumes reach great altitude (Figure 11.17c). Not all of Io's volcanoes erupt so violently. Lava flows out of some vents, and sulfur gas expelled from the volcano coats the surface in red patches (Figure 11.17d). Some of the higher mountains were probably built by basaltic lava flows similar to those of terrestrial volcanoes (Figure 11.17e). Many of Io's volcanic vents glow red-hot (Figure 11.17f). Tectonic processes are probably also active on Io, but evidence for them is covered by the frequent lava flows and plume deposits everywhere on Io's surface. Galileo and Voyager images showed that fallout from volcanic plumes can cover an area the size of Arizona in a matter of months (Figure 11.17g).

Io's volcanoes also produce a very thin sulfur-dioxide atmosphere about a billion times less dense than our own atmosphere. It is too thin to cause erosion, and Io's weak gravity allows a steady escape of the atmospheric gases. These volcanic gases supply the plasma in the Io torus (see Figure 11.13) and in Jupiter's magnetosphere.

Why is Io so geologically active? After all, it is about the size of our own geologically dead Moon, and it is not as icy in composition as most other jovian satellites. Io must have an additional internal heat source besides the usual combination of accretion, differentiation, and radioactivity that heats the interiors of the terrestrial worlds [Section 9.3]. This fourth internal heat source is **tidal heating**, and its theory was the basis for the pre-Voyager prediction of volcanic activity on Io.

How does tidal heating work? Tidal forces lock Io in synchronous rotation so that it keeps the same face toward Jupiter. But Io's orbit is slightly elliptical, so its orbital speed and distance from Jupiter vary. As a result, the strength and even the direction of the tidal force change very slightly over its orbit: Io is continuously flexed by Jupiter (Figure 11.18a). The flexing of Io's rocky crust and interior releases heat, just as flexing warms Silly Putty. In fact, Io's tidal heating releases more than 200 times more heat (per gram of mass) than the radioactive heat driving Earth's geology, which explains why Io is the most volcanically active body in the solar system.

But why is Io's orbit slightly elliptical, when almost all other large satellites' orbits are virtually circular? Io's neighbors are responsible: Gravitational nudges between Io, Europa, and Ganymede maintain **orbital resonances** between them (Figure 11.18b). During the time in which Ganymede completes one orbit of Jupiter, Europa completes exactly two orbits and Io completes exactly four orbits. The satellites therefore line up periodically, and the gravitational tugs these satellites exert on one another add up over time. The satellites are always tugged in exactly the same direction, and the satellite orbits become very slightly elliptical as a result. (If the satellites didn't line up periodically, the tugs would be exerted in random directions at random times—as ineffective as pushing a child on a swing at random times.) The resulting tidal forces are strongest on Io because it is closest to Jupiter, but they are also important on Europa and Ganymede.

---

[4]The heroes of this extraordinary discovery are theoreticians Pat Cassen, Stan Peale, and Ray Reynolds, and Linda Morabito, the Voyager navigation engineer who first spotted the plumes in calibration images.

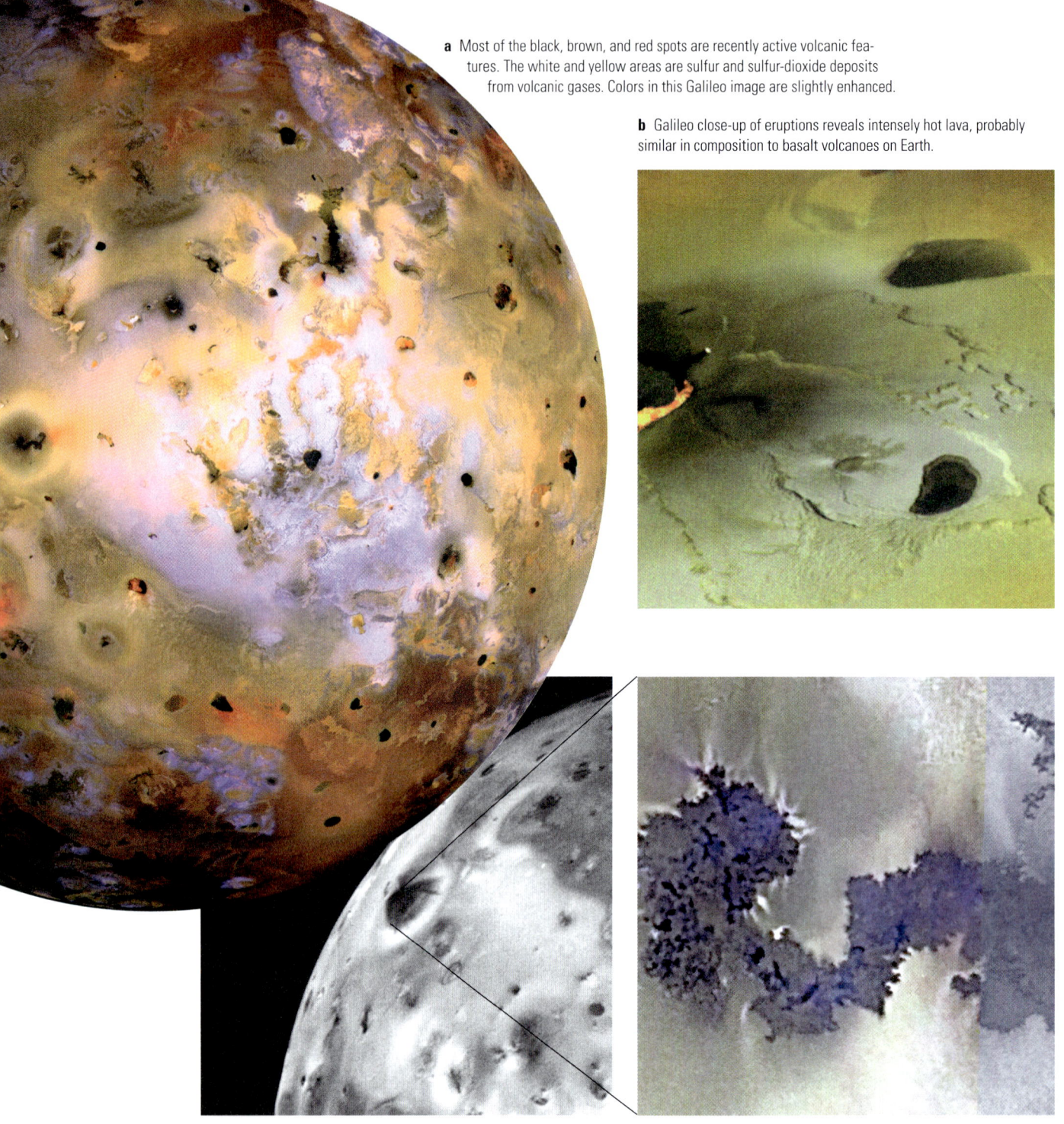

a Most of the black, brown, and red spots are recently active volcanic features. The white and yellow areas are sulfur and sulfur-dioxide deposits from volcanic gases. Colors in this Galileo image are slightly enhanced.

b Galileo close-up of eruptions reveals intensely hot lava, probably similar in composition to basalt volcanoes on Earth.

c When basaltic lava flows over sulfur-dioxide ice, the explosive sublimation creates huge plumes. This plume rises 80 km high.

FIGURE 11.17 Io is the most volcanically active body in the solar system.

Europa Europa's bizarre, fractured crust is proof enough that tidal heating has taken hold there (Figure 11.19a). The icy surface is nearly devoid of impact craters and may be only a few million years old. Jumbled icebergs (Figure 11.19b) suggest that an ocean of liquid water lies just a few kilometers below the surface, extending down to perhaps 100 km deep. The icy shell overlying the ocean is repeatedly flexed by tidal forces, resulting in a global network of cracks. Close-up images of the cracks taken by the Galileo orbiter show that most have a remarkable double-ridged pattern, perhaps the result of debris piling up around a crack that is repeatedly opened and closed by tidal flexing (Figure 11.19c,d). On Earth, changing winds form similar double ridges in

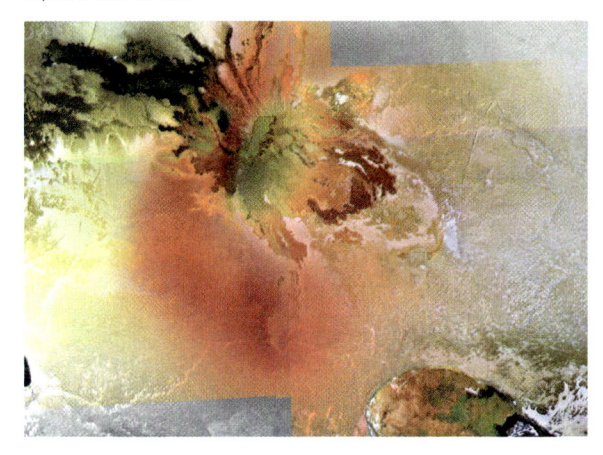

d The reddish color surrounding this volcano comes from sulfur gas expelled from the lava.

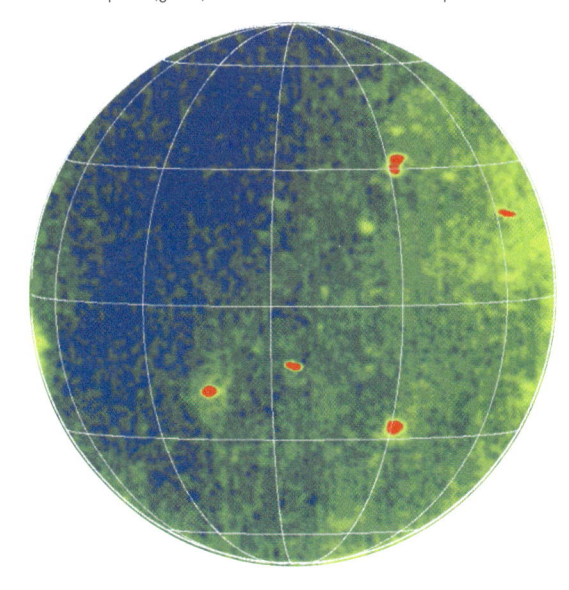

f This false-color photo shows the glow of Io's volcanic vents (red) and atmosphere (green) when Io is in the darkness of Jupiter's shadow.

g This enhanced color photo shows fallout (dark patch) from a volcanic plume on Io. The fallout region covers an area the size of Arizona. (The orange ring is the fallout from another volcano.)

e This photo shows a shield volcano on Io that may be made of basaltic lava.

sea ice in the Arctic Ocean, though on much smaller scales. (We do not expect winds on Europa, although it does have a very thin atmosphere created by magnetospheric bombardment.)

In some ways, Europa resembles a rocky terrestrial planet wrapped in an icy crust. (Some scientists suggest that the icy shell has not always rotated at exactly the same rate at the rocky center, leading to even greater cracking of the crust.) The rocky object that makes up most of Europa is undoubtedly geologically active below the layer of ocean and ice. Perhaps basaltic lavas erupt on Europa's seafloors, sometimes violently enough to jumble up the icy crust above (Figure 11.20). And, just possibly, tidal heating makes Europa's oceans as hospitable for life as our own oceans [Section 13.6].

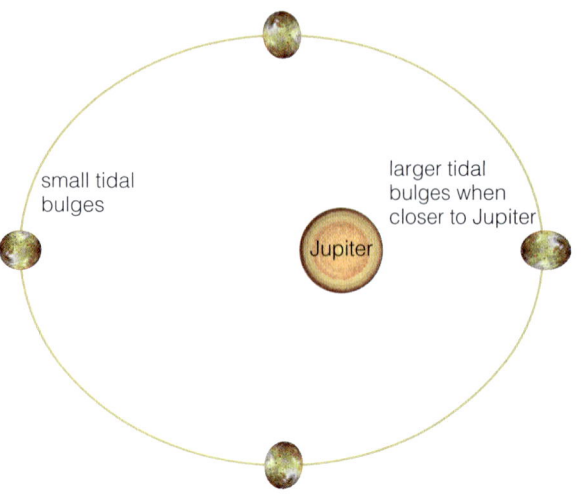

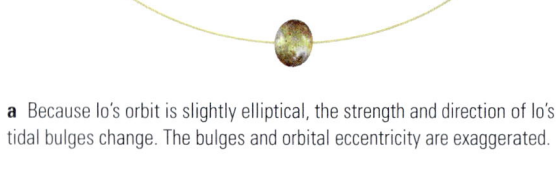

a Because Io's orbit is slightly elliptical, the strength and direction of Io's tidal bulges change. The bulges and orbital eccentricity are exaggerated.

FIGURE 11.18 Orbital resonances cause tidal heating on Io, Europa, and Ganymede.

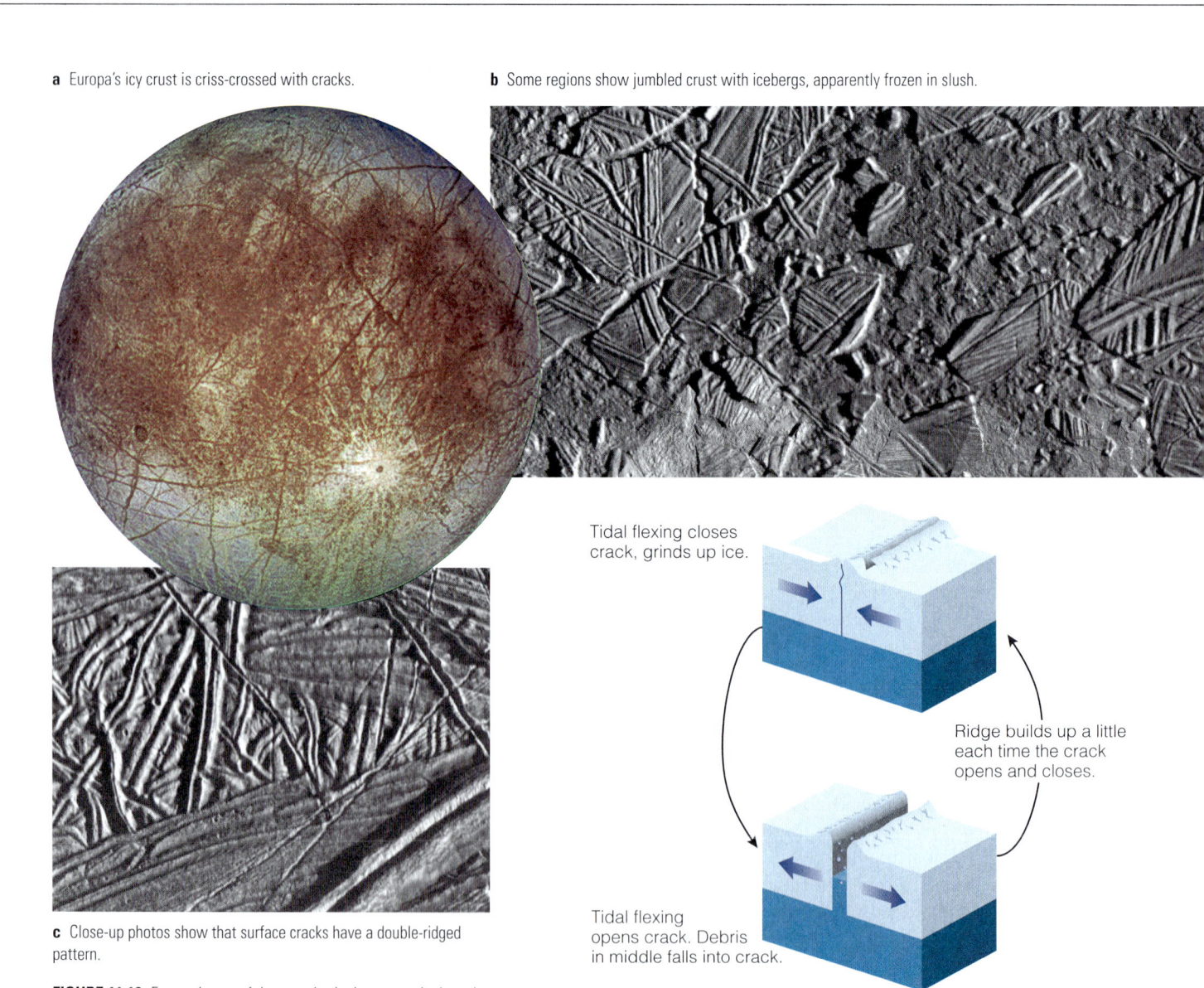

b About every seven Earth days (one Ganymede orbit, two Europa orbits, and four Io orbits), the three moons line up as shown. The small gravitational tugs repeat and make all three orbits slightly elliptical.

a Europa's icy crust is criss-crossed with cracks.

b Some regions show jumbled crust with icebergs, apparently frozen in slush.

c Close-up photos show that surface cracks have a double-ridged pattern.

FIGURE 11.19 Europa is one of the most intriguing moons in the solar system.

Tidal flexing closes crack, grinds up ice.

Ridge builds up a little each time the crack opens and closes.

Tidal flexing opens crack. Debris in middle falls into crack.

d A possible mechanism for making the double-ridged surface cracks.

FIGURE 11.20 Tidal heating may give Europa a subsurface ocean beneath its icy crust. This artist's conception imagines a region where the crust has been disrupted by an undersea volcano.

a Ganymede's numerous craters (bright spots) show that its surface is older than Europa's.

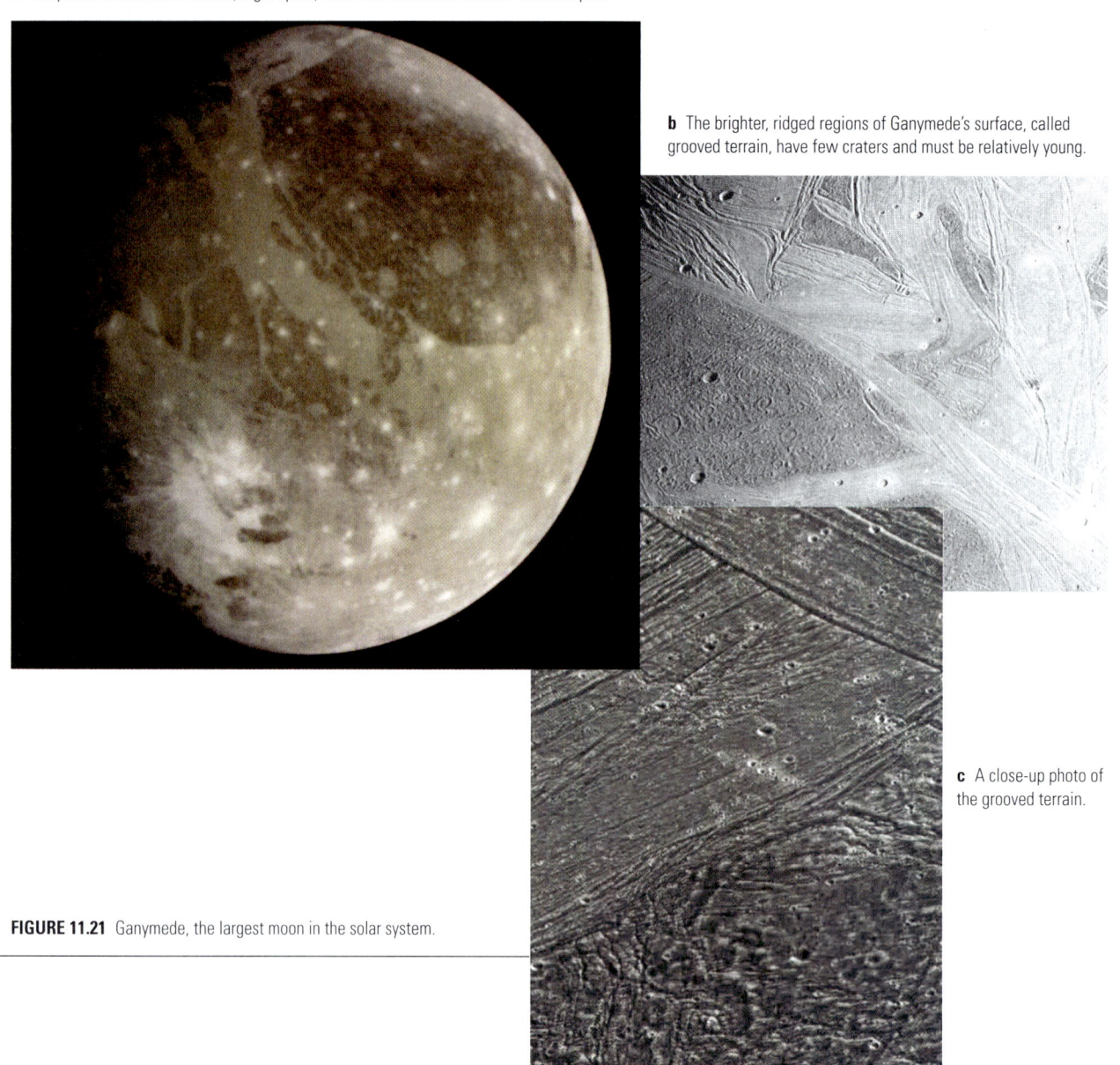

b The brighter, ridged regions of Ganymede's surface, called grooved terrain, have few craters and must be relatively young.

c A close-up photo of the grooved terrain.

FIGURE 11.21 Ganymede, the largest moon in the solar system.

Ganymede Photographs of Ganymede tell the story of an interesting geological history. Parts of the surface have many impact craters, indicating that these regions are billions of years old (Figure 11.21a). But other areas show unusual "grooved terrain" unlike anything we've seen in terrestrial geology (Figure 11.21b,c). The grooves may form either from tectonic stresses or from water erupting along a crack in the surface; because ice expands when it solidifies (unlike rock), the grooves may form as the surface refreezes. A freezing surface patch on Ganymede therefore pushes outward, perhaps creating the grooves and ridges seen on its surface. Ganymede also shows some evidence of tectonics similar to plate tectonics on Earth: Some chunks of the surface appear to have shifted relative to one another. Erosion is unimportant on Ganymede, because it has just a hint of an atmosphere generated by bombardment. The precise nature of Ganymede's internal heating remains mysterious, although it certainly is some combination of tidal heating and radioactive decay. Surprisingly, Ganymede has its own magnetic field—perhaps indicating a molten, convecting core. It may even harbor a liquid-water layer below the ice, but evidence for such an ocean is still quite tentative.

Callisto The outermost Galilean satellite, Callisto, looks most like what scientists first expected for outer solar system satellites: a heavily cratered iceball (Figure 11.22a). The bright patches on its surface are the result of cratering: Large impactors dug up cleaner ice from deep down and spread it over the surface. The odd rings in the upper left of Callisto's image

a Heavy cratering indicates an ancient surface.

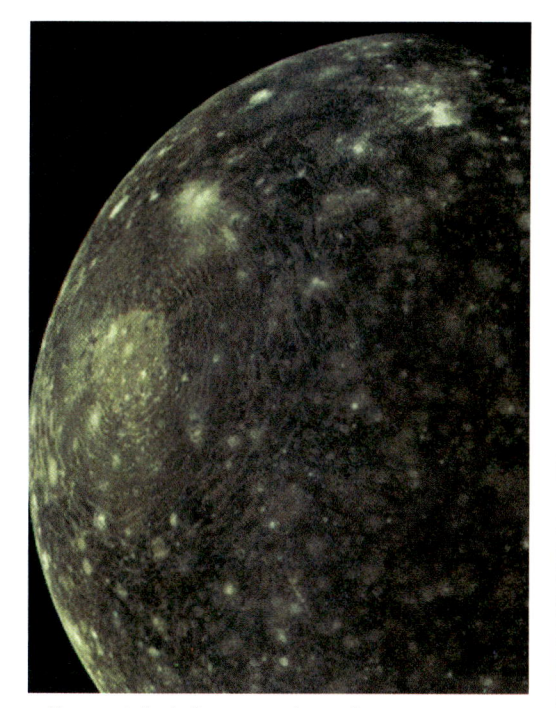

b Close-up photos show a dark powder overlying the low areas of the surface.

FIGURE 11.22 Callisto shows no evidence of volcanic or tectonic activity.

also came from impacts: The largest impactors shattered the icy sphere in a way that produced concentric rings around the impact site.

Despite its relatively large size (the third-largest moon in the solar system), Callisto lacks volcanic and tectonic features. Internal heat sources must be very small. In fact, gravity measurements by the Galileo spacecraft show that Callisto never underwent differentiation: Dense rock and lighter ice are still thoroughly mixed throughout. Callisto has no tidal heating because it shares no orbital resonances with other satellites. Nonetheless, Callisto appears to have a magnetic field, which could be generated in a salty ocean below the ice. The surface also holds some surprises. Close-up images show the surface to be covered by a dark, powdery substance concentrated in low-lying areas, leaving ridges and crests bright white (Figure 11.22b). No one knows the nature of this material or how it got there.

Titan and the Medium-Size Moons of Saturn

Leaving Jupiter's moons behind, our satellite tour takes us next to Saturn. Here we find an amazing moon shrouded in a thick atmosphere: Titan. It is Saturn's only large moon and is the second-largest moon in the solar system after Ganymede.

Titan's hazy and cloudy atmosphere hides its surface (Figure 11.23a). The atmosphere is about 90% nitrogen, making it the only world besides Earth where nitrogen is the dominant atmospheric constituent. However, on Earth the rest of the atmosphere is mostly oxygen, while the rest of Titan's atmosphere consists of argon, methane, and other hydrogen compounds, including ethane (C_2H_6). Thus, we could not breathe the air on Titan and live. The methane and ethane actually give Titan an appreciable greenhouse effect, but the surface temperature is still a frigid 93 K ($-180°C$). The surface pressure on Titan is only somewhat higher than that on Earth—about 1.5 bars at the surface, which would be fairly comfortable if not for the lack of oxygen and the cold temperatures.

How did Titan end up with such an unusual atmosphere? Titan is composed mostly of ices, including methane and ammonia ice. Some of this ice sublimed long ago to form an atmosphere on Titan. Over billions of years, solar ultraviolet light broke apart the ammonia molecules (NH_3) into hydrogen and nitrogen. The light hydrogen escaped from the atmosphere (through thermal escape [Section 10.5]), but the heavier nitrogen molecules remained and accumulated.

We encounter a particularly intriguing idea when we examine what happened to the methane gas that sublimed into Titan's atmosphere long ago. Methane

a Titan is enshrouded by a hazy, cloudy atmosphere.

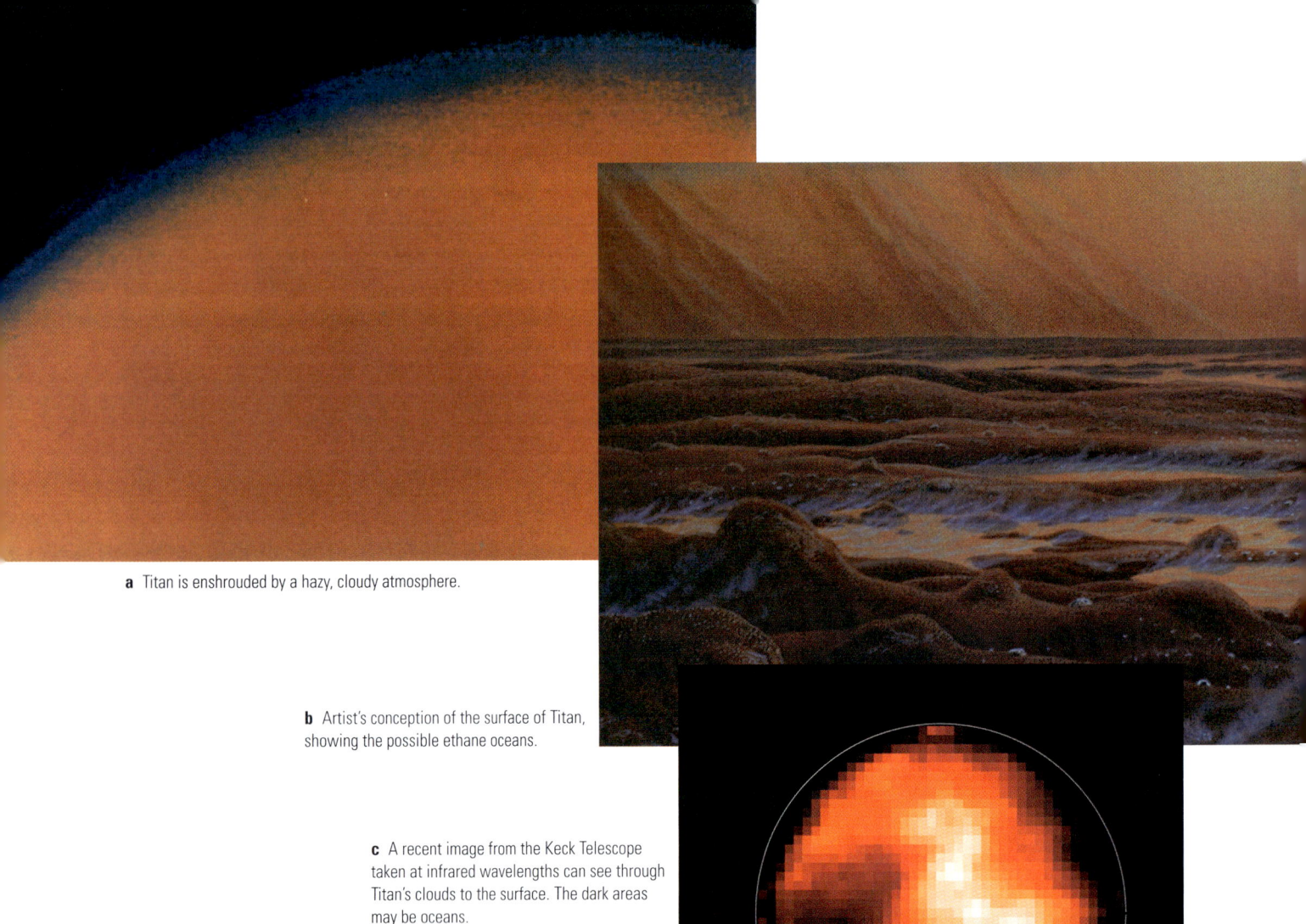

b Artist's conception of the surface of Titan, showing the possible ethane oceans.

c A recent image from the Keck Telescope taken at infrared wavelengths can see through Titan's clouds to the surface. The dark areas may be oceans.

FIGURE 11.23 Saturn's moon Titan.

is no longer present in large quantities, so some process must have removed it from the atmosphere. The most likely process is chemical reactions, triggered by solar ultraviolet light, that transform methane into ethane. If this idea is correct, atmospheric ethane may form clouds and rain on Titan, perhaps creating oceans of liquid ethane on the surface (Figure 11.23b). According to some calculations, Titan may have produced enough ethane over its history to create ethane oceans a kilometer deep. Recent infrared pictures (Figure 11.23c) have probed through the clouds to the surface, but we cannot yet determine whether such oceans really exist.

Because we cannot see its surface clearly, we do not know whether Titan is geologically active. However, given its relatively large size and icy composition, it seems likely that Titan has a rich geological history. Our understanding of Titan should improve dramatically when the Cassini spacecraft reaches Saturn in 2004 [Section 7.6]. In addition to cameras and spectrographs, Cassini carries a radar mapper similar to the Magellan radar that mapped the surface of Venus [Section 9.5]. Thus, we will see the surface of Titan at high resolution and at last will learn about its geology—and determine whether ethane oceans (or even lakes) exist. *Cassini* will also drop a probe (named *Huygens*, for a seventeenth-century pioneer in astronomy and optics) into Titan's atmosphere that will take pictures and measure atmospheric properties on the way down. Just in case, the probe is designed to float in liquid ethane.

312 PART III LEARNING FROM OTHER WORLDS

Enceladus   Tethys   Dione

Rhea   Iapetus   Mimas

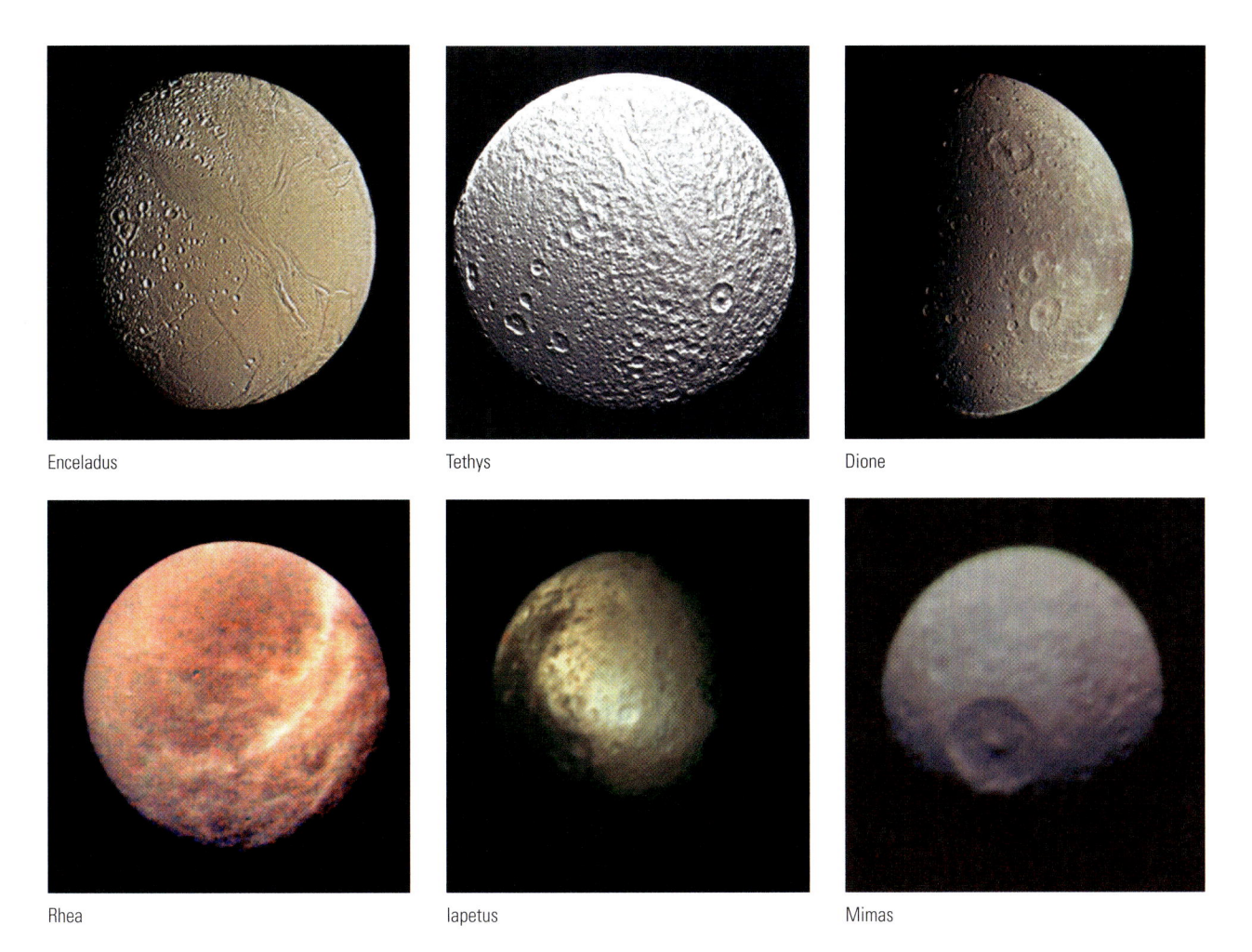

FIGURE 11.24 Saturn's six medium-size moons, which range in diameter from 390 km (Mimas) to 1,530 km (Rhea). All but Mimas show some evidence of volcanism and/or tectonics.

TIME OUT TO THINK *Based on your understanding of the four geological processes on icy satellites, what kind of surface features do you think might be present on Titan?*

Given that Titan and Ganymede are similar in both size and composition, it's natural to wonder why Titan has so much atmosphere and Ganymede has so little. One possibility is that cooler temperatures in the solar nebula at Saturn's greater distance from the Sun allowed more methane and ammonia ice to condense. These ices are more volatile than the water ice that makes up Jupiter's satellites and hence are more likely to sublime and form an atmosphere. Alternatively, atmospheric cratering [Section 10.5] may have stripped away Ganymede's atmosphere but not Titan's. Atmospheric cratering was probably more important on Ganymede than on Titan because Jupiter's stronger gravity accelerates impactors to higher speeds. No one yet knows which of these two processes—condensation in the solar nebula or atmospheric cratering—bears more responsibility for the differences between Ganymede and Titan.

Saturn's medium-size moons (Rhea, Iapetus, Dione, Tethys, Enceladus, and Mimas) show evidence of substantial geological activity (Figure 11.24). While all show some cratered surfaces, Enceladus has grooved terrain similar to Ganymede's, and the others show evidence of flows of "ice lava." But these satellites are considerably smaller than Ganymede: Enceladus is barely 500 kilometers across—small enough to fit inside the borders of Colorado. How can such small moons support so much geological activity? The answer probably lies with "ice geology," in which icy combinations of water, ammonia, and methane can melt, deform, and flow at remarkably low temperatures. Thus, even small objects with relatively little internal heat can sustain fascinating geological activity. Iapetus is particularly bizarre, with some regions distinctly bright and others distinctly dark. The dark regions appear to be covered by a thin veneer of dark material, but we still don't know what it is or how it got there.

Mimas (radius = 195 km), the smallest of the "medium-size" moons of Saturn, is essentially a heav-

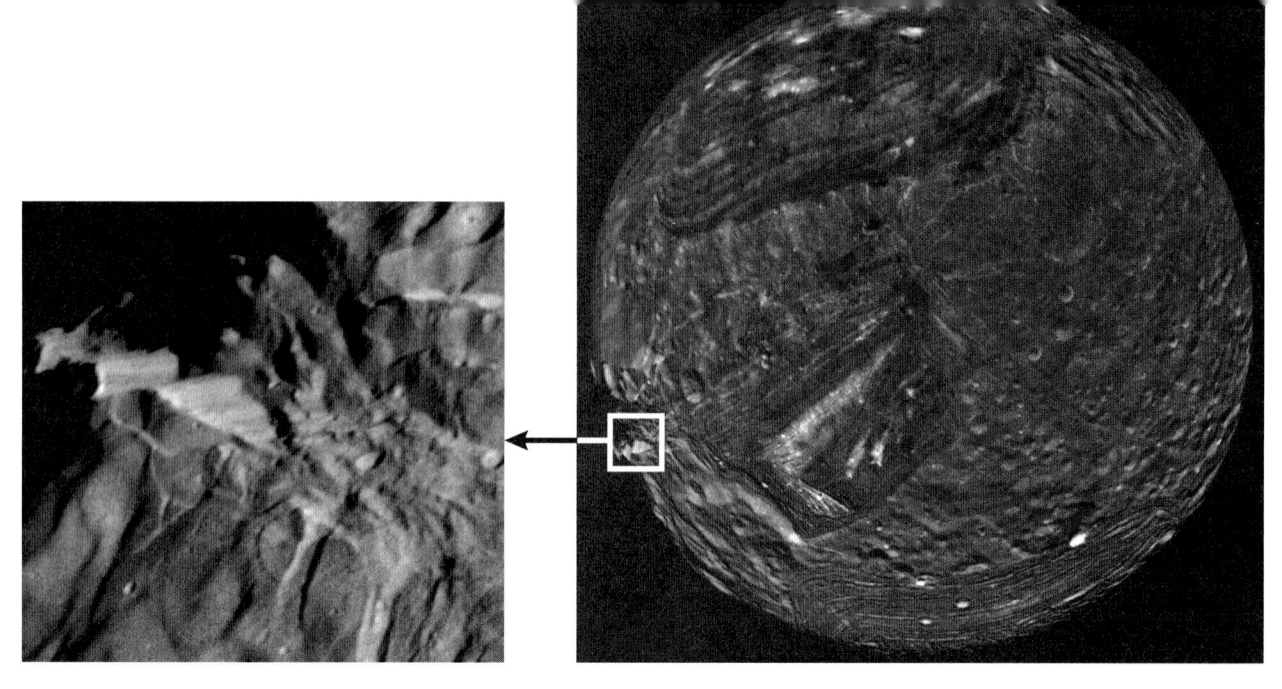

FIGURE 11.25 The surface of Miranda shows astonishing tectonic activity despite its small size. The cliff walls seen in the inset are higher than those of the Grand Canyon on Earth.

ily cratered iceball. One huge crater is sometimes called "Darth Crater" because of Mimas's resemblance to the Death Star in the *Star Wars* movies. (The official name of the crater is Herschel.) The impact that created this crater probably came close to breaking Mimas apart.

The Medium-Size Moons of Uranus

Uranus is orbited by 5 medium-size moons (see Figure 11.15) and 10 smaller moons discovered by Voyager. Astronomers have recently discovered 6 more moons, including several that orbit the planet backward.

The medium-size moons pose several puzzles. For example, Ariel and Umbriel are virtual twins in size, yet Ariel shows evidence of volcanism and tectonics, while the heavily cratered surface of Umbriel suggests a lack of geological activity. Titania and Oberon also are twins in size, but Titania appears to have had much more geological activity than Oberon. No one knows why these two pairs of similar-size moons should vary so greatly in geological activity.

Miranda, the smallest of Uranus's medium-size moons (radius = 235 km), is the most surprising (Figure 11.25). Prior to the Voyager flyby, scientists expected Miranda to be a cratered iceball like Saturn's similar-size moon Mimas (see Figure 11.24). Instead, Voyager images of Miranda show tremendous tectonic features and relatively few craters. Why should Miranda be so much more geologically active than Mimas? Our best guess is an episode of tidal heating from a temporary orbital resonance with another satellite of Uranus billions of years ago. Evidently, Mimas never underwent a similar period of tidal heating.

Triton, the Backward Moon of Neptune

Last stop on our satellite tour is Neptune, so distant that only two of its moons (Triton and Nereid) were known to exist before the Voyager visit. Triton is the only large moon in the Neptune system, and only one other moon (Proteus) even makes it into our "medium-size" category.

Triton may appear to be a typical satellite, but it is not. Its orbit is retrograde (it travels in a direction opposite to Neptune's rotation) and highly inclined to Neptune's equator. These are telltale signs of a captured satellite. But Triton is not small and potato-shaped like most captured asteroids. Instead it is large, spherical, and icy. In fact, it is larger than the planet Pluto. As such, Triton presents many challenges to our understanding of satellite capture. Nonetheless, it is almost certain that Triton once orbited the Sun instead of orbiting Neptune.

A quick study of the Triton photos in Figure 11.26 reveals a surface unlike any other in the solar system, and one that poses many unanswered questions. Are the flat surfaces like the lunar maria? Are the wrinkly ridges (nicknamed "cantaloupe terrain") tectonic in nature? Are the bright polar caps similar to those on Mars? Why is Triton's surface composition (a mixture of ices of nitrogen, methane, carbon dioxide, and carbon monoxide) unlike that of any other satellite? Clearly, Triton is more than a cratered iceball. The undisturbed craters suggest that major geological activity has subsided, but Triton appears to have undergone some sort of icy volcanism in the past. Such geological activity is astonishing in light of Triton's extremely low (40 K) surface temperature,

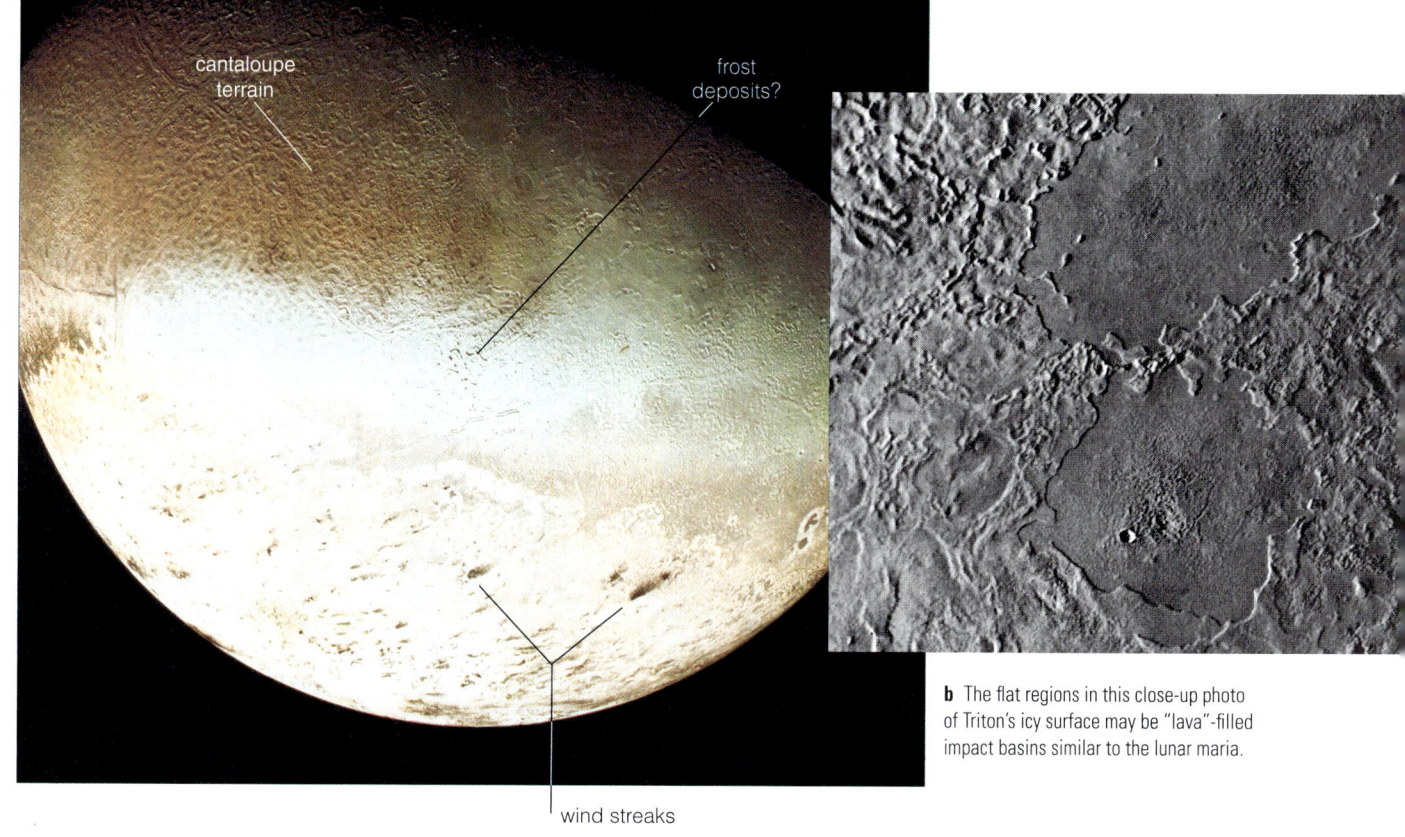

a Triton's southern hemisphere as seen by *Voyager 2*.

b The flat regions in this close-up photo of Triton's icy surface may be "lava"-filled impact basins similar to the lunar maria.

FIGURE 11.26 Neptune's moon Triton.

---

but it probably involved low-melting-point mixtures of water, methane, and ammonia ices. The internal heating source for Triton's geological activity is not known, but it may have involved tidal heating. Triton probably had a very elliptical orbit and a more rapid rotation when it was first captured by Neptune, but tidal forces would have circularized its orbit, slowed it to synchronous rotation, and possibly heated its interior enough to cause geological activity.

To cap off Triton's odd geology, the sublimation of surface ices creates a thin atmosphere. Thin as it is, the atmosphere creates wind streaks on the surface, and unknown processes pump unusual plumes of gas and particles into the atmosphere. The combination of Triton's large orbital inclination and the substantial tilt of Neptune's rotation axis leads to extreme seasonal swings, and polar caps probably grow, shrink, and migrate from pole to pole. All of this takes place on a satellite that, because of its high albedo, is even colder than Pluto. In fact, Triton is the coldest of all the planets and satellites in the solar system.

If Triton really is a captured satellite, where did it come from? Could other objects like it still be orbiting the Sun? Intriguingly, we know of at least one object that appears to be quite similar to Triton: the planet Pluto. But that's a story to be discussed in the next chapter.

Satellite Summary: The Active Outer Solar System

Our brief tour has taken us to most of the medium-size and large moons in the solar system. The major lesson of the tour is that icy-satellite geology is harder to predict than terrestrial-planet geology. Size isn't everything when it comes to geological activity. Ices form lavas more easily than rocks, so icy worlds can have more geological activity than a rocky world of the same size. Satellites made of ices that include methane and ammonia may have even greater potential geological activity. In addition, we saw how tidal heating can supply added heat even when other internal heat sources are no longer important, as is the case on Io and Europa and as occurs to lesser extents on Ganymede, Triton, and possibly Miranda.

Satellite atmospheres also revealed surprises. We found examples of all three atmospheric sources (outgassing, sublimation, and bombardment [Section 10.5]). Io's atmosphere comes from volcanic outgassing, while Europa's and Ganymede's atmospheres are attributed to bombardment. Titan's thick atmosphere probably came from sublimation, followed by extensive chemistry aided by solar ultraviolet light. Sublimation is also responsible for Triton's thin atmosphere, even out at the frigid edges of our solar system. The jovian planet satellites are far more interesting and instructive than anyone could have imagined before the Voyager and Galileo expeditions.

a Earth-based telescopic view of Saturn.

b Voyager image of Saturn's rings against the disk.

c Artist's conception of particles in a ring system. All the particles are moving slowly relative to one another and occasionally collide.

FIGURE 11.27 Looking closer at Saturn's rings.

11.6 Jovian Planet Rings

We have completed our comparative study of the jovian planets themselves and of their major moons, but we have one more topic left to cover: their amazing rings. Saturn's rings have dazzled and puzzled astronomers since Galileo first saw them through his small telescope and suggested that they resembled "ears" on Saturn. For a long time, Saturn's rings were thought to be unique in the solar system, but we now know that all four jovian planets have rings. As we did for the planets themselves, we'll look at one ring system in detail and then examine the others for important similarities and differences. Saturn's rings are the clear choice as the standard for comparison.

Saturn's Rings

You can see Saturn's rings through a backyard telescope, but learning about their nature requires higher resolution. Even through large telescopes on Earth, the rings appear to be continuous, concentric sheets of material separated by gaps (Figure 11.27a). Spacecraft images reveal these "sheets" to be made of many more individual rings, each broken up into even smaller concentric *ringlets,* and different regions of a ring to range from transparent to opaque (Figure 11.27b). But even these appearances are somewhat deceiving. If we could wander into the rings, we'd see that they are made of countless individual particles orbiting Saturn together, each obeying Kepler's laws and occasionally colliding with nearby particles (Figure 11.27c). The particles range in size from large boulders to dust grains—far too small to be photographed, even when spacecraft like *Voyager* or *Cassini* pass nearby.

Spectroscopy reveals that Saturn's ring particles are made of relatively reflective (high-albedo) water ice. The rings look bright where there are enough particles to intercept most of the Sun's light and scatter it back toward us, and they appear more transparent where there are fewer particles. In regions where light cannot easily pass through, neither can another ring particle. In the densest parts of the rings, each particle collides with another every few hours.

TIME OUT TO THINK *Which ring particles travel faster, those at the inner edge of Saturn's rings or those at the outer edge? (Hint: Think about Kepler's third law.) Can you think of a way to confirm your answer with telescopic observations?*

Saturn's rings lie in a thin plane directly above the planet's bulging equator, so they share Saturn's 27° axis tilt. Over the course of Saturn's year, the rings present a changing appearance to the Earth and Sun, appearing wide open at Saturn's solstices and edge-on at its equinoxes. The rings are perhaps the thin-

FIGURE 11.28 The dark patches in Saturn's rings are called spokes.

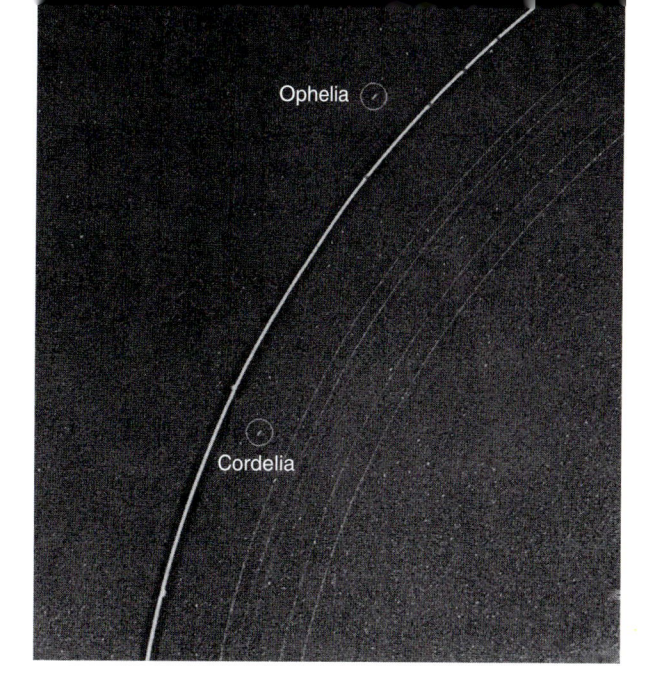

FIGURE 11.29 Voyager's best photograph of shepherd moons came during the *Voyager 2* Uranus flyby. Two satellites shepherd one of Uranus's rings. The satellite images are smeared out by their motion during the exposure.

nest known astronomical structure: They are over 270,000 kilometers across but only a few tens of *meters* thick. Collisions help keep the rings thin. You can see why by imagining adding a new ring particle with an orbit tilted relative to the other particles. (The existence of many particles on tilted orbits would make the ring thicker.) The new particle would collide with other particles each time its orbit intersected the ring plane, and its orbital tilt would be reduced with every collision. It wouldn't be long before these collisions would force the particle to conform to the orbital pattern of the other particles. Similarly, a new ring particle with a highly elliptical orbit would soon end up with a circular orbit.

Where do the rings come from? An important clue is that the rings lie close to the planet in a region where tidal forces are very influential. Within two to three planetary radii (of any planet), the tidal forces tugging an object apart become comparable to the gravitational forces holding it together. This region is called the **Roche tidal zone**. Only relatively small objects held together by nongravitational forces (such as the electrostatic forces that hold solid rock—or spacecraft or human beings—together) can survive within the Roche tidal zone. Thus, there are two basic scenarios for the origin of the rings. One possibility is that a wandering moon strayed too close to Saturn and was torn apart. A more likely scenario is that tidal forces prevented the material in Saturn's rings from accreting into a single large moon, instead forming many smaller moons.

Among the many mysteries of Saturn's rings are unusual dusty patches, called *spokes,* that can appear and change dramatically in a matter of hours (Figure 11.28). They are probably particles of microscopic dust that have been levitated out of the ring plane by forces associated with Saturn's magnetic field.

Rings and Gaps

By the time the Voyager spacecraft approached Saturn, astronomers knew of at least six distinct rings around Saturn. Voyager scientists had hoped to find reasons for the six rings, but instead they found that the total number of individual rings and gaps may be as high as 100,000. Theorists are still struggling to explain all the rings and gaps, but some general ideas are now clear.

Rings and gaps are caused by particles bunching up at some orbital distances and being forced out at others. This bunching happens when gravity nudges the orbits of ring particles in some particular way. One source of nudging is tiny moons—called *gap moons*—located within the rings themselves. The gravity of a gap moon tugs gently on nearby particles, effectively nudging them farther away to other orbits.[5] This clears a gap in the rings around the moon's orbit. Voyager spacecraft photographed a few gap moons, and there may be thousands more that are too small to see in the Voyager images. In some cases, Voyager images show two gap moons forcing particles trapped between them into line; the best such example comes from Uranus (Figure 11.29), but we find similar examples in Saturn's ring system. The gap moons are said to act as *shepherd moons* in such cases, because they shepherd the ring particles.

---

[5]It may seem counterintuitive that a gap moon's gravitational attraction should nudge a ring particle *farther away,* but theoretical models show that this happens as a result of the combined gravitational interactions between the gap moon, the ring particles, and the planet.

a The largest gap in Saturn's rings, called the Cassini division, is caused by an orbital resonance with the moon Mimas.

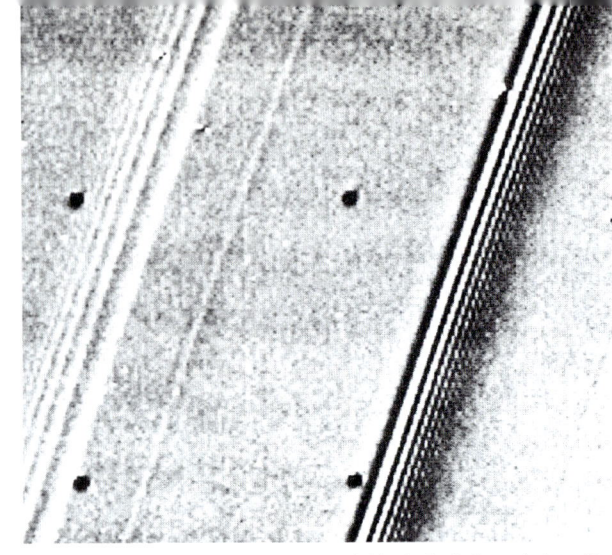

b Another Mimas resonance creates remarkable ripples in Saturn's rings. The dark spots in the image are calibration marks for the camera.

FIGURE 11.30 Gaps in Saturn's rings.

In some cases, ring particles orbiting Saturn may also be nudged by the gravity from Saturn's larger and more distant moons. For example, a ring particle orbiting about 120,000 km from Saturn's center will circle the planet in exactly half the time it takes the moon Mimas to orbit. Every time Mimas returns to a certain location, the ring particle will also return to its original location and will experience the same gravitational nudge from Mimas. The nudges reinforce one another and eventually clear a gap in the rings (Figure 11.30a). This "Mimas 2:1 resonance" is essentially the same type of *orbital resonance* that affects the orbits of Io, Europa, and Ganymede. It is responsible for a large gap in Saturn's rings that is easily visible from Earth (called the *Cassini division*). Many similar resonances are responsible for other gaps, and resonances can even create beautiful ripples within the rings (Figure 11.30b).

Although we now know three causes of rings and gaps—gap moons, shepherd moons, and orbital resonances—we have not yet identified specific causes for the vast majority of the features within Saturn's rings. Perhaps the Cassini spacecraft will identify more gap moons and shepherd moons, or perhaps it will reveal even more puzzling structures.

Comparing Planetary Rings

The ring systems of Jupiter, Uranus, and Neptune are so much fainter than Saturn's ring system that it took almost four centuries longer to discover them. Ring particles in these three systems are far less numerous, generally smaller, and much darker. Saturn's rings were the only ones known until 1977, when the rings of Uranus were discovered during observations of a *stellar occultation*—a star passing behind Uranus as seen from the Earth. During the occultation, the star "blinked" on and off nine times before it disappeared behind Uranus, and nine more times as it emerged. Scientists concluded that these nine "blinks" were caused by nine thin rings encircling Uranus. Similar observations of stars passing behind Neptune gave more confounding results: Rings appeared to be present some of the time, but not at other times. Could Neptune's rings be incomplete or transient?

The Voyager spacecraft provided some answers. Voyager cameras first discovered thin rings around Jupiter in 1979. Then, after providing incredible images of Saturn's rings, *Voyager 2* confirmed the existence of rings around Uranus as it flew past in 1986. In 1989, *Voyager 2* passed by Neptune and found that it does, in fact, have partial rings—at least when seen from Earth. The space between the ring segments is filled with dust not detectable from Earth.

How do Voyager images distinguish between boulders and dust? The secret is to look at how the ring particles scatter light (Figure 11.31a). Tiny particles like dust are not much bigger than the wavelengths of visible light, so they scatter sunlight only by very small angles from its original direction. Thus, rings that scatter light mostly forward must be made of tiny dust-size particles. Dust on your windshield scatters light in the same way, which is why it's so hard to see the road when you're driving into the Sun. Larger particles—marble- or boulder-size—mostly reflect light back toward the Sun. Thus, the Voyager spacecraft saw mostly the larger particles on their way toward each planet and saw mostly dust-size particles when they looked back toward the Sun after passing each planet (Figure 11.31b). In fact,

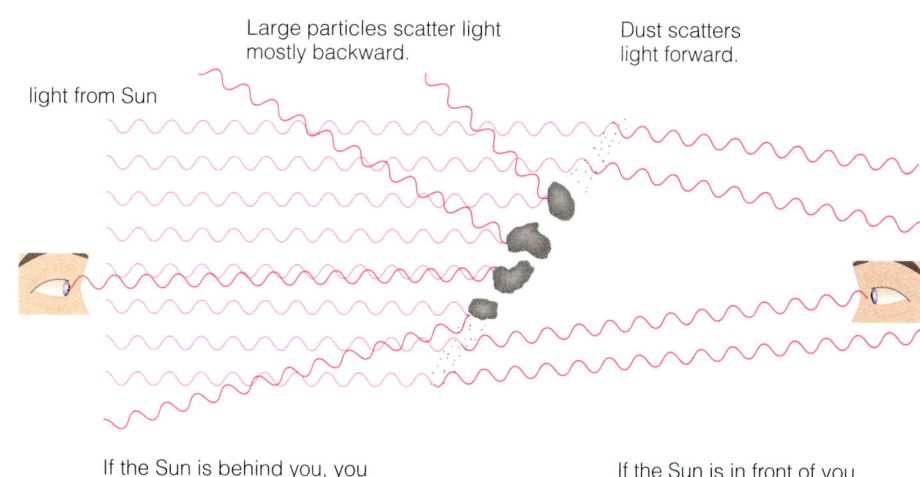

FIGURE 11.31 Boulders and dust in planetary rings.

light from Sun — Large particles scatter light mostly backward. — Dust scatters light forward.

a You can tell whether ring particles are large or small even if you can't see an individual particle.

If the Sun is behind you, you see mostly large particles.

If the Sun is in front of you, you see mostly dust.

Voyager 2 detected vast dust sheets between the widely separated rings of Uranus and Neptune.

TIME OUT TO THINK *Turn on a flashlight or projector in a dark room. Set it down and walk a few steps away from it. Explain why you see so much airborne dust when you look back toward the projector but not when you look forward in the direction of the beam. How is this phenomenon similar to the scattering of light by ring particles? (Note: You can see this effect particularly well if you look back toward the projector in a movie theater.)*

A family portrait of the jovian ring systems shows many similarities (Figure 11.32). All rings lie in their planet's equatorial plane within the Roche tidal zone. Particle orbits are fairly circular, with small orbital tilts relative to the equator. Gaps and ringlets are probably due to gap moons, shepherd moons, and orbital resonances. The differences also offer important insights: The larger size, higher reflectivity, and much greater number of particles in Saturn's rings compel us to wonder if different processes are at work there. The newly discovered rings brought challenges as well. The reasons for the slight tilt and eccentricity of Uranus's thin rings and the arcs within Neptune's rings are not yet understood.

Origin of the Rings

Where do rings come from? How old are they? Did they form with their planet out of the nebula, or are they a more recent phenomenon? These questions long puzzled astronomers when they had only one case (Saturn) to study, but the discovery that all jovian planets have rings has made the answers clearer.

Part of the answer lies in determining how long rings will last. Unlike planets and large satellites, ring particles may not last very long. Ring particles orbiting Saturn bump into one another every few hours, and even with low-velocity collisions the particles are chipped away over time. Basketball-size ring particles

b The rings of Uranus seen before the spacecraft arrived (top; Sun behind the spacecraft) and after it passed by (bottom; looking back toward the Sun). Only the larger particles that scatter light back toward the Sun are visible in the top image. Abundant dust particles become visible looking back toward the Sun (bottom).

around Saturn are ground away to dust in a few million years. Other processes dismantle rings around the other jovian planets. The thermosphere of Uranus extends into its Roche tidal zone, so atmospheric drag slowly causes ring particles to spiral into the planet. Dust orbiting Jupiter is slightly slowed by the flood of solar photons passing by, and the particles we now see in Jupiter's ring will fall into the planet in only a few million years. In summary, particles in the four ring systems cannot last as long as the 4.6-billion-year age of our solar system.

If ring particles are rapidly disappearing, some source of new particles must keep the rings supplied. The most likely source is collisions. As ring particles are ground away by collisions, small moons (perhaps the gap moons) occasionally collide and create more ring particles. Meteorites may also strike these moons, with the impacts releasing even more ring particles. We now believe that the "miniature solar nebulae" that surrounded the jovian planets formed many

CHAPTER 11 JOVIAN PLANET SYSTEMS 319

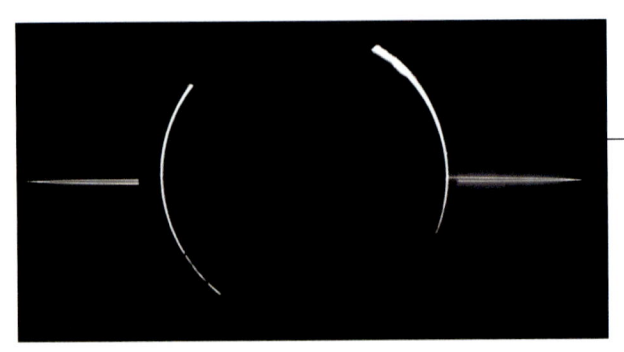

Jupiter

Saturn

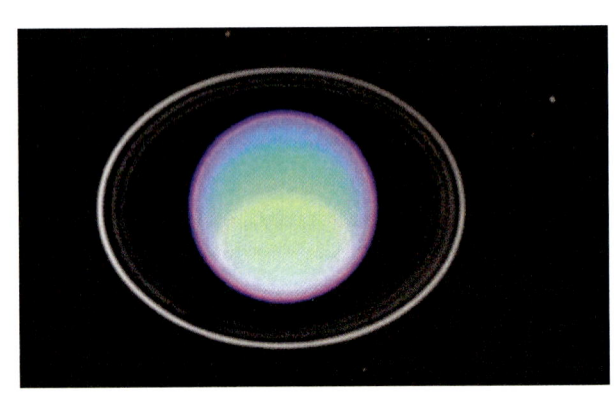

Uranus

Neptune

FIGURE 11.32 Four ring systems. The planets are not shown to scale. Uranus's rings were photographed by the Hubble Space Telescope, the others by Voyager. The Neptune frame is made of two images taken on either side of the bright planet.

small satellites and that the gradual dismantling of these satellites creates the dramatic ring systems we see today. The appearance of any particular ring system probably changes dramatically over millions and billions of years. The spectacle of Saturn's brilliant rings may be a special treat of our epoch, one that could not have been seen a billion years ago and that may not last long on the time scale of our solar system.

THE BIG PICTURE

In this chapter, we saw that the jovian planets really are a "different kind of planet"—and, indeed, a different kind of planetary system. The jovian planets dwarf the terrestrial planets, and some of their moons are as large as terrestrial worlds. As you continue your study of the solar system, keep the following "big picture" ideas in mind:

- The jovian planets may lack solid surfaces on which geology can work, but they are interesting and dynamic worlds with rapid winds, huge storms, strong magnetic fields, and interiors in which common materials behave in unfamiliar ways.

- Despite their relatively small sizes and frigid temperatures, many jovian satellites are geologically active by virtue of their icy compositions. Ironically, it was cold temperatures in the solar nebula that led to icy compositions and hence geological activity.

- Ring systems owe their existence to small satellites formed from the "miniature solar nebulae" that produced the jovian planets billions of years ago. The rings we see today are composed of particles liberated from those satellites surprisingly recently.

- Understanding the jovian planets forced us to modify many of our earlier ideas about the solar system, in particular by adding the concepts of ice geology, tidal heating, and orbital resonances. Each new set of circumstances offers further opportunities to learn how the universe works.

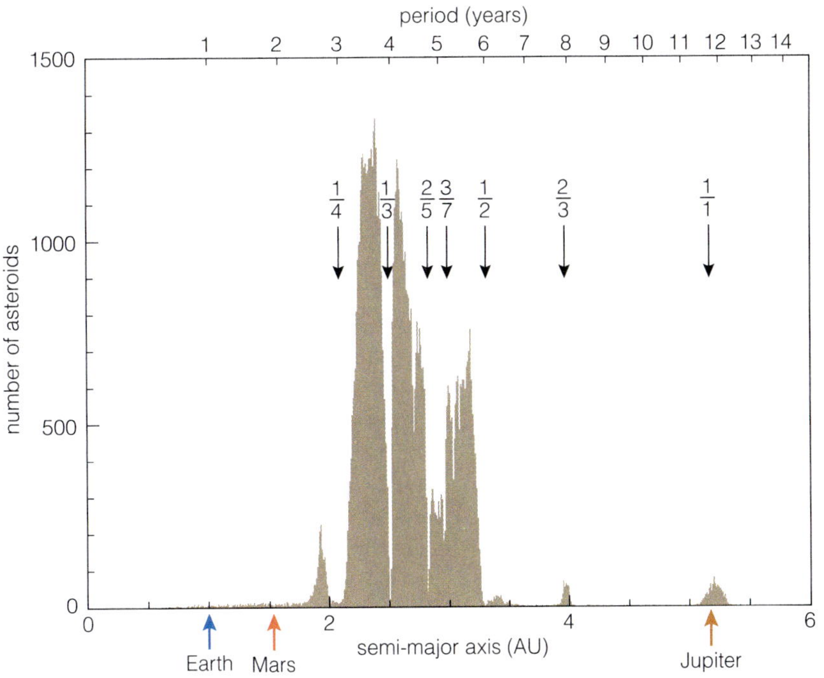

FIGURE 12.2 This graph shows the numbers of asteroids with different average distances (semimajor axes) from the Sun. Note the gaps—distances where we see few if any asteroids. The ratio labeling each gap shows how the orbital period of objects at this distance compares to the 12-year orbital period of Jupiter. For example, the label $\frac{1}{3}$ means that the orbital period of objects with this average distance is $\frac{1}{3}$ of Jupiter's orbital period, or about 4 years. Most orbital resonances result in gaps, though some (such as $\frac{1}{1}$ and $\frac{2}{7}$) happen to gather up asteroids.

TIME OUT TO THINK *How is the process that prevented asteroids from accreting into a planet similar to that which prevented Saturn's ring particles from accreting into a single moon? How is it different?*

The overlapping elliptical orbits of the remaining asteroids still lead to a major collision somewhere in the asteroid belt every 100,000 years or so. (By comparison, collisions occur somewhere within Saturn's rings many times per second, which has circularized the orbits and flattened the rings.) Thus, larger asteroids continue to be broken into smaller ones, with each collision also creating numerous dust-size particles. The asteroid belt has been grinding itself down in this way for over 4 billion years and will continue to do so for as long as the solar system exists.

Most of the asteroids found outside the asteroid belt—including the *Earth-approaching asteroids* that pass near Earth's orbit—are probably "impacts waiting to happen." But asteroids can safely congregate in two stable zones outside the main belt. These zones are found along Jupiter's orbit 60° ahead of and behind Jupiter (see Figure 12.1). The asteroids found in these two zones are called the *Trojan asteroids,* and the largest are named for the mythological Greek heroes of the Trojan War. The Trojan asteroids are stable because of a different type of orbital resonance with Jupiter. In this case, any asteroid that wanders away from one of these zones is nudged back *into* the zone by Jupiter's gravity. The existence of such stable orbital resonances was first predicted by the French mathematician Joseph Lagrange more than 200 years ago—135 years before the discovery of the first asteroid in such an orbit.[1] It is possible that the population of Trojan asteroids is as large as that of main-belt asteroids, but the greater distance to the Trojan asteroids makes them more difficult to study or even count from Earth.

Learning About Asteroids

The general locations and orbits of asteroids tell us a lot about their origin, but we would like to know many other properties, such as masses, sizes, densities, shapes, and compositions. The first step in learning about an asteroid is determining its precise orbit. Asteroids are recognizable in telescopic images because they move noticeably relative to the stars in just a short time (Figure 12.3). Once we detect an asteroid, we can calculate its orbit by applying Kepler's laws [Section 5.3]—even if we've observed only a small part of one complete orbit.

Most asteroids appear as little more than points of light even to our largest Earth-based telescopes, so the best way to study an asteroid in depth is to send a spacecraft out to meet it. Spacecraft have visited

---

[1]The locations of these stable zones are called *Lagrange points*. Similar Lagrange points lie 60° ahead of and behind the Moon in its orbit around the Earth. These locations have been suggested for space colonies because of their stability. Three additional Lagrange points lie along the line connecting any two massive objects (such as the Earth and the Moon). They are not quite stable but also are good places to "park" spacecraft because staying in place doesn't take much fuel.

of 1 kilometer or more. Despite their large numbers, asteroids don't add up to much: If they could all be put together, they'd make an object less than 1,000 kilometers in radius—far smaller than any terrestrial planet.

Painstaking observations have revealed the precise orbits of thousands of asteroids. The main **asteroid belt** lies between 2.2 and 3.3 AU from the Sun, though some interesting asteroids lie outside the belt (Figure 12.1). Asteroid orbits are distinctly elliptical and can be inclined relative to the ecliptic by as much as 20°–30°, but all orbit the Sun in the same direction as the planets. Science fiction movies often show the asteroid belt as a crowded and hazardous place, but the average distance between asteroids is millions of kilometers. Spacecraft must be carefully guided to fly close enough to study an asteroid, and accidental collisions with spacecraft are extremely unlikely.

Origin and Evolution of the Asteroid Belt

Why are asteroids concentrated in the asteroid belt, and why didn't a full-fledged planet form in this region? The answers probably lie with *orbital resonances*. Recall that an orbital resonance occurs whenever one object's orbital period is a simple ratio of another object's period, such as $\frac{1}{2}$, $\frac{1}{4}$, or $\frac{5}{3}$. In such cases, the two objects periodically line up with each other, and the extra gravitational attractions at these times can affect the objects' orbits. Consider, for example, the orbital resonances between Saturn's moons and particles in its rings: The moons are much larger than the ring particles and therefore force the ring particles out of resonant orbits, clearing *gaps* in the rings [Section 11.6]. In the asteroid belt, similar orbital resonances occur between asteroids and Jupiter, the most massive planet by far. For example, any asteroid in an orbit that takes 6 years to circle the Sun—half of Jupiter's 12-year orbital period—would receive the same gravitational nudge from Jupiter every 12 years and thus would soon be pushed out of this orbit. The same is true for asteroids with orbital periods of 4 years ($\frac{1}{3}$ of Jupiter's period) and 3 years ($\frac{1}{4}$ of Jupiter's period). (Music offers an elegant analogy: When a vocalist sings into an open piano, the strings of notes "in resonance" with the voice are "nudged" and begin to vibrate—not just the note being sung, but also notes with a half or a quarter of the note's frequency.)

Today, we see the results of these orbital resonances when we graph the number of asteroids with various orbital periods. Just as we would expect, there are virtually no asteroids with periods exactly $\frac{1}{2}$, $\frac{1}{3}$, or $\frac{1}{4}$ of Jupiter's. Thus, there are no asteroids with average distances from the Sun (semimajor axes) that correspond to these orbital periods. Figure 12.2 shows several other gaps in the asteroid belt (often called *Kirkwood gaps*, after their discoverer), corresponding to orbital resonances that make other simple ratios with Jupiter's orbital period. (By a *gap* in the asteroid belt we mean a gap in *average* distance; because asteroid orbits are elliptical, some asteroids pass through the gaps on parts of their orbits.)

The gravitational tugs from Jupiter that created the gaps in the asteroid belt probably also explain why no planet ever formed in this region. Early in the history of the solar nebula, the asteroid belt probably contained enough rocky planetesimals to form a planet as large as Earth or Mars. Just as in other parts of the solar nebula, collisions among these planetesimals frequently disrupted their orbits. However, the asteroid belt had an additional source of disruption: gravitational resonances with the forming planet Jupiter. The result was that planetesimals could not collect together to form a planet. Over the past 4.5 billion years, the ongoing orbital disruptions have gradually kicked pieces of this "unformed planet" out of the asteroid belt altogether. Once booted from the main asteroid belt, these objects eventually either crash into a larger object (Sun, planet, or moon) or are flung out of the solar system. (In extremely rare cases, they may become "captured" moons.) Thus, the asteroid belt slowly loses mass, explaining why the total mass of all its asteroids is now less than that of any terrestrial planet.

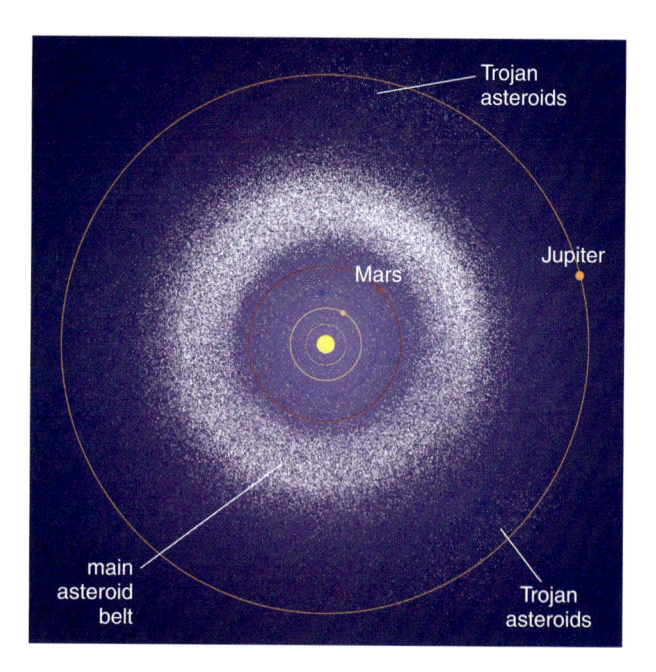

FIGURE 12.1 Positions of 8,777 asteroids for midnight, 1 January 2002. The asteroids themselves are much smaller than the dots on this scale. The Trojan asteroids are found 60° ahead of and behind Jupiter in its orbit.

12.1 Remnants from Birth

The objects we've studied so far—terrestrial planets, jovian planets, and large moons—have changed dramatically since their formation. Most planets have been transformed in almost every way. Their interiors have differentiated, their surfaces have been reshaped by geological processes, and their atmospheres have changed as some gases escape while others are added. But many small bodies remain virtually unchanged since their formation some 4.5 billion years ago. Thus, comets, asteroids, and meteorites carry the history of our solar system encoded in their compositions, locations, and numbers.

Using these small bodies to understand planetary formation is a bit like picking through the trash in a carpenter's shop to see how furniture is made. By studying the scraps, sawdust, paint chips, and glue, we can develop a rudimentary understanding of how the carpenter builds furniture. Similarly, the "scraps" left over from the formation of our solar system allow us to understand how the planets and larger moons came to exist. Indeed, much of our modern theory of solar system formation was developed from studies of asteroids, comets, and meteorites.

As we discussed in Chapter 8, the leftovers from the process of accretion in the early solar system fall into three fairly distinct groups, distinguished by their physical and orbital properties: *asteroids, Kuiper belt comets,* and *Oort cloud comets.* Asteroids are rocky or metallic in composition, and most orbit the Sun between the orbits of Mars and Jupiter. In contrast, comets are mostly icy in composition and spend most of their time well outside the orbits of the planets. This compositional difference is directly related to where the objects formed: Asteroids formed inside the *frost line* of the solar nebula, and comets formed outside [Section 8.4]. Kuiper belt comets have orbits like those of the planets, lying close to the ecliptic plane and orbiting the Sun counterclockwise as viewed from above Earth's North Pole. Their distance from the Sun ranges between about 30 AU (about the distance of Neptune) and 100 AU (more than twice the average distance of Pluto). Oort cloud comets have orbits that are randomly inclined to the ecliptic plane, and they generally lie at much greater distances from the Sun than the Kuiper belt comets—in some cases, perhaps halfway to the nearest stars.

In this chapter, we will investigate what we have learned from these remnants from the birth of our solar system. We'll also discuss the cosmic collisions that sometimes occur between leftover planetesimals and the existing planets. Before we continue, however, it's worth noting that the terminology used to describe these objects has a long history that can sometimes be confusing. The word *comet* comes from the Greek word for "hair," and comets get their name from the long, hairlike tails they display on the rare occasions when they come close enough to the Sun to be visible in our sky. *Asteroid* means "starlike," but asteroids are starlike only in their appearance through a telescope, not in their fundamental properties. The term *meteor* refers to a bit of interplanetary dust burning up in our atmosphere; literally, it means "a thing in the air." Meteors are also sometimes called *shooting stars* or *falling stars* because some people once believed that they really were stars falling from the sky. Larger chunks of rock that survive the plunge through the atmosphere and hit the ground are called *fallen stars* or *meteorites,* which means "associated with meteors." To keep potential confusion to a minimum, we will use only the following terms and definitions in this book:

- *Asteroid*: a rocky leftover planetesimal orbiting the Sun.

- *Comet*: an icy leftover planetesimal orbiting the Sun—regardless of its size or whether or not it has a tail.

- *Meteor*: a flash of light in the sky caused by a particle entering the atmosphere, whether the particle comes from an asteroid or a comet.

- *Meteorite*: any piece of rock that fell to the ground from space, whether from an asteroid, a comet, or even another planet.

12.2 Asteroids

Asteroids are virtually undetectable to the naked eye and went unnoticed for almost two centuries after the invention of the telescope. The first asteroids were discovered about 200 years ago when astronomers were searching the "extra wide" gap between the orbits of Mars and Jupiter in hopes of discovering a previously unknown planet. It took 50 years to discover the first 10 asteroids (all large and bright); today, modern telescopes with advanced detectors discover that many on a typical night. Once an asteroid's orbit is calculated and verified, the asteroid is assigned a number (in order of verification), and the discoverer chooses a name (subject to the approval of the International Astronomical Union). The earliest discovered asteroids bear names of mythological figures; more recent discoveries carry names of scientists, cartoon heroes, pets, and rock stars.

The radius of the largest asteroid, Ceres, is about 500 kilometers—about half that of Pluto. About a dozen others are large enough that we would call them medium-size moons if they orbited a planet. Most asteroids are much smaller. There are probably more than 100,000 small asteroids with diameters

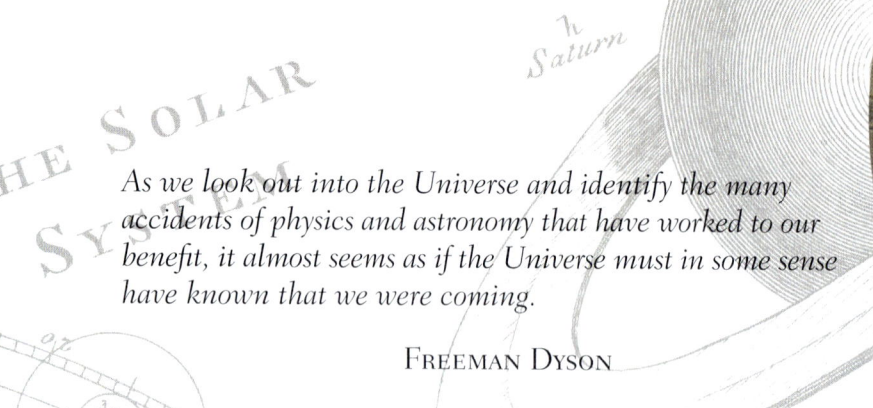

As we look out into the Universe and identify the many accidents of physics and astronomy that have worked to our benefit, it almost seems as if the Universe must in some sense have known that we were coming.

FREEMAN DYSON

CHAPTER 12
Remnants of Rock and Ice
Asteroids, Comets, and Pluto

Asteroids and comets might seem insignificant in comparison to the planets and satellites we've been discussing. But there is strength in numbers, and the trillions of small bodies orbiting our Sun are far more important than their small size might suggest.

Few topics in astronomy have the potential to touch our lives more directly than asteroids and comets. The appearance of comets has more than once altered the course of human history, when our ancestors acted upon superstitions related to comet sightings. More profoundly, asteroids or comets falling to the Earth have scarred our planet with impact craters and altered the course of biological evolution.

Asteroids and comets are also important for a very different reason: They are remnants from the birth of our solar system and therefore provide an important testing ground for our theories of how our solar system came to be. In this chapter, we will explore asteroids, comets, and those bodies that fall to Earth as meteorites. We will also examine the smallest planet, Pluto, and see that, while it may be a misfit among planets, it may be right at home among the smaller objects of our solar system.

4. An extrasolar planet is discovered that is made primarily of hydrogen and helium; it has approximately the same mass as Jupiter but is 10 times larger than Jupiter in radius.

5. A new small moon is discovered to be orbiting Jupiter. It is smaller than any other of Jupiter's moons and orbits slightly farther from Jupiter than any of the other moons, but it has several large, active volcanoes.

6. A new moon is discovered to be orbiting Neptune. The moon orbits in Neptune's equatorial plane and in the same direction in which Neptune rotates, but it is made almost entirely of metals such as iron and nickel.

7. An icy, medium-size moon is discovered to be orbiting a jovian planet in a star system that is only a few hundred million years old. The moon shows evidence of active tectonics.

8. A jovian planet is discovered in a star system that is much older than our solar system. The planet has no moons at all, but it has a system of rings as spectacular as the rings of Saturn.

9. *The Importance of Rotation.* Suppose the material that formed Jupiter came together without any rotation, so that no "jovian nebula" formed and the planet today wasn't spinning. How else would the jovian system be different? Think of as many effects as you can, and explain each in a sentence.

10. *The Great Red Spot.* Based on the infrared and visible-wavelength images in Figure 11.7, is Jupiter's Great Red Spot warmer or cooler than nearby clouds? Does this mean it is higher or lower in altitude than the nearby clouds? (*Hint:* Use what you know about the belts and zones as you study these images.)

11. *Comparing Jovian Planets.* You can do comparative planetology armed only with telescopes and an understanding of gravity.
    a. The satellite Amalthea orbits Jupiter at just about the same distance in kilometers at which Mimas orbits Saturn. Yet Mimas takes almost twice as long to orbit. What can you deduce from this difference, qualitatively?
    b. Jupiter and Saturn are not very different in radius. When you combine this information with your answer to part (a), what can you conclude?

12. *Minor Ingredients Matter.* Suppose the jovian planet atmospheres were composed 100% of hydrogen and helium rather than 98% of hydrogen and helium. How would the atmospheres be different in terms of color and weather? Explain.

*13. *Disappearing Satellite.* Io loses about a ton (~1,000 kg) of sulfur dioxide per second to Jupiter's magnetosphere.
    a. At this rate, what fraction of its mass would Io lose in 5 billion years?
    b. Suppose sulfur dioxide currently makes up 1% of Io's mass. When will Io run out of this gas at the current loss rate?

*14. *Ring Particle Collisions.* Each ring particle in the densest part of Saturn's rings collides with another about every 5 hours. If a ring particle survived for the age of the solar system, how many collisions would it undergo? Do you consider it likely that the ring particles we see in Saturn's rings today have been there since the formation of the solar system? Explain the ramifications of your answer for understanding the origin of the jovian planet rings.

15. *Project: Jupiter's Moons.* Using binoculars or a small telescope, view the moons of Jupiter. Make a sketch of what you see, or take a photograph. Repeat your observations several times (nightly, if possible) over a period of a couple of weeks. Can you determine which moon is which? Can you measure their orbital periods? Can you determine their approximate distances from Jupiter? Explain.

16. *Project: Saturn and Its Rings.* Using binoculars or a small telescope, view the rings of Saturn. Make a sketch of what you see, or take a photograph. What season is it in Saturn's northern hemisphere? How far above Saturn's atmosphere do the rings extend? Can you identify any gaps in the rings? Describe any other features you notice.

Web Projects

Find useful links for Web projects on the text Web site.

1. *Galileo and Cassini Missions.* Find the latest news on the Galileo and Cassini missions.
    a. Print out (or describe) an image from the Galileo mission of interest to you, and interpret it using the techniques developed in this chapter.
    b. In a few sentences, describe the main scientific goals of the Cassini mission. What is the current health of the spacecraft, and how is the mission progressing?

2. *Oceans of Europa.* The possibility of subsurface oceans on Europa holds great scientific interest and may even mean that life could exist on Europa. Find some of the latest research concerning Europa and learn about proposed missions to Europa. Write a one- to two-page summary of what you learn.

Review Questions

1. Briefly describe the past and present space exploration of the jovian worlds. Summarize what we've learned about the jovian planets in terms of composition, rotation rate, and shape.

2. Describe the interior structure of Jupiter. Contrast Jupiter's interior structure with the structures of the other jovian planets. Why is Jupiter denser than Saturn? Why is Neptune denser than Saturn?

3. How do we think Jupiter generates its internal heat? How do other jovian planets generate internal heat?

4. Describe the composition and temperature structure of Jupiter's atmosphere and contrast it with Earth's. Compare the temperature structures of the other jovian planets to that of Jupiter.

5. Why does Jupiter have several distinct layers of clouds?

6. How are the circulation cells on Jupiter similar to and different from those on Earth? What are *belts* and *zones*?

7. What is the *Great Red Spot*? Is it the same type of storm as a hurricane on Earth? Explain.

8. Describe the possible origins of Jupiter's vibrant colors. Contrast these with the origins of the colors of the other jovian planets.

9. Does Jupiter have seasons? Why or why not? Do other jovian planets have seasons? Explain.

10. Why does Jupiter have auroras? Contrast Jupiter's magnetosphere with those of the Earth and the other jovian planets.

11. What is the Io torus? Why doesn't Saturn, Uranus, or Neptune have a similar "torus"?

12. Briefly describe some of the general characteristics of the jovian moons.

13. How is the ice geology of the jovian moons similar to the rock geology of the terrestrial planets? How is it different?

14. Describe the volcanoes of Io and the mechanism of *tidal heating* that generates Io's internal heat.

15. What are *orbital resonances*? Explain how the resonance between Io, Europa, and Ganymede makes their orbits slightly elliptical.

16. Describe the evidence suggesting that Europa may have a subsurface ocean of liquid water.

17. Describe the general surface characteristics of Ganymede and Callisto. What do these characteristics tell us about internal heating on these worlds?

18. Describe the atmosphere of Titan. What evidence suggests that Titan may have oceans of liquid ethane?

19. Describe a few general features of Saturn's medium-size moons. Why is it surprising to find evidence of geological activity on some of these moons, and how do we explain it?

20. Describe a few general features of Uranus's medium-size moons. Why is the geological activity on Ariel and Titania puzzling? How do we explain the odd appearance of Miranda?

21. How do we know that Triton is not a typical large jovian satellite? Describe Triton's general characteristics and give possible ideas of where Triton came from.

22. Describe the appearance and structure of Saturn's rings. Why are the rings so thin? Where do the rings lie in relation to Saturn's *Roche tidal zone?*

23. Briefly describe how gap moons and orbital resonances can lead to gaps within ring systems.

24. Contrast the rings of the other jovian planets with Saturn's rings.

25. Describe two possible scenarios for the origin of planetary rings. What makes us think that ring systems must be continually replenished?

Discussion Questions

1. *A Miniature Solar System?* In what ways is the Jupiter system similar to a miniature solar system? In what ways is it different?

2. *Jovian Mission.* We can study terrestrial planets up close by landing on them, but jovian planets have no surfaces to land on. Suppose you are in charge of planning a long-term mission to "float" in the atmosphere of a jovian planet. Describe the technology you will use and how you will ensure survival for any people assigned to this mission.

3. *Extrasolar Planets.* Many of the newly discovered planets orbiting other stars are more massive than Jupiter and much closer to their stars. Assuming that they are in fact jovian (as opposed to terrestrial or something else), how would you expect these new planets to differ from the jovian planets of our solar system? How would any magnetospheres, satellites, or rings be different? Explain.

Problems

Surprising Discoveries? For **problems 1–8**, suppose someone claimed to make the discovery described. (These are *not* real discoveries.) Decide whether the discovery should be considered reasonable or surprising. More than one right answer may be possible, so explain your answer clearly.

1. Saturn's core is pockmarked with impact craters and dotted with volcanoes erupting basaltic lava.

2. Neptune's deep blue color is not due to methane, as previously thought, but instead is due to its surface being covered with an ocean of liquid water.

3. A jovian planet in another star system has a moon as big as Mars.

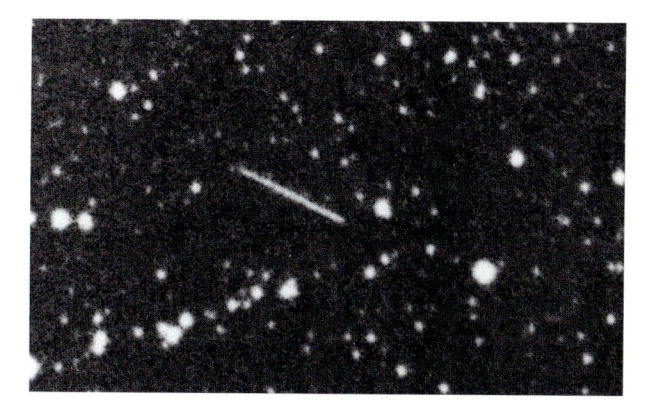

FIGURE 12.3 Because asteroids orbit the Sun, they move relative to the stars just as planets do. In this photograph, stars show up as white dots, and the motion of an asteroid makes it show up as a short streak.

only a handful of asteroids close-up in their native environment of the asteroid belt and a few other presumed asteroids "in captivity" in orbit around other planets (e.g., Phobos and Deimos orbiting Mars, and the backward outer satellites of Jupiter). The Galileo spacecraft on its way to Jupiter photographed the asteroid Gaspra (Figure 12.4a) and an asteroid pair known as Ida and Dactyl (Figure 12.4b). The Near-Earth Asteroid Rendezvous mission (NEAR) photographed the asteroid Mathilde (Figure 12.4c) on its way to the asteroid Eros. NEAR orbited Eros (Figure 12.4d) for a full year, studying its entire surface at close range, before setting down on its surface. These spacecraft images reveal battered worlds that are probably fragments of larger asteroids shattered by collisions.

a Gaspra (16 km across). Photographed by the Galileo spacecraft on its way to Jupiter.

FIGURE 12.4 Close-up views of asteroids studied by spacecraft.

b Ida (53 km) and its tiny moon, Dactyl. Photographed by the Galileo spacecraft.

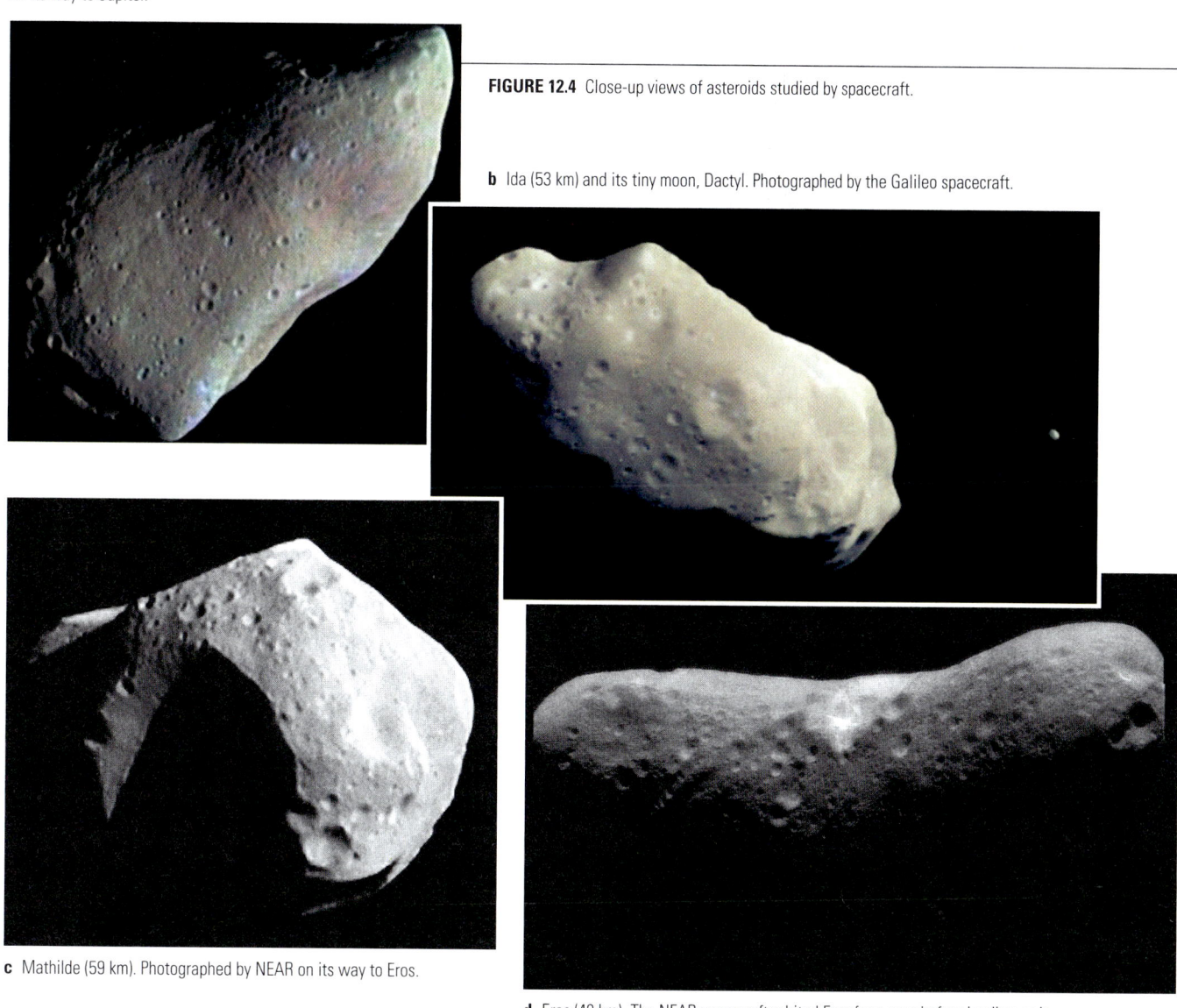

c Mathilde (59 km). Photographed by NEAR on its way to Eros.

d Eros (40 km). The NEAR spacecraft orbited Eros for a year before landing on it.

CHAPTER 12 REMNANTS OF ROCK AND ICE: ASTEROIDS, COMETS, AND PLUTO

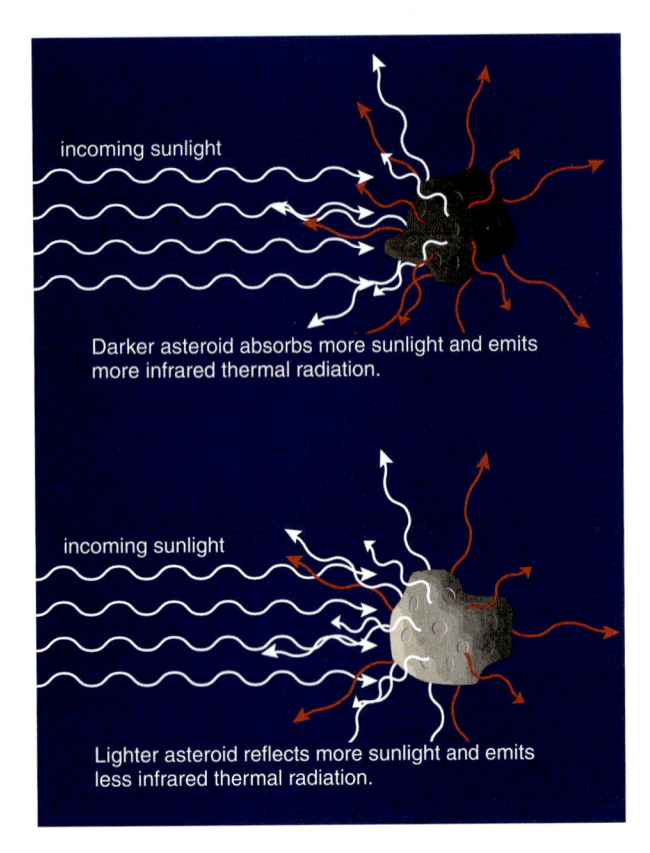

FIGURE 12.5 By comparing measurements of reflected sunlight and thermal emission in the infrared, we can learn about asteroid albedos. Once we know an asteroid's albedo, we can determine its size.

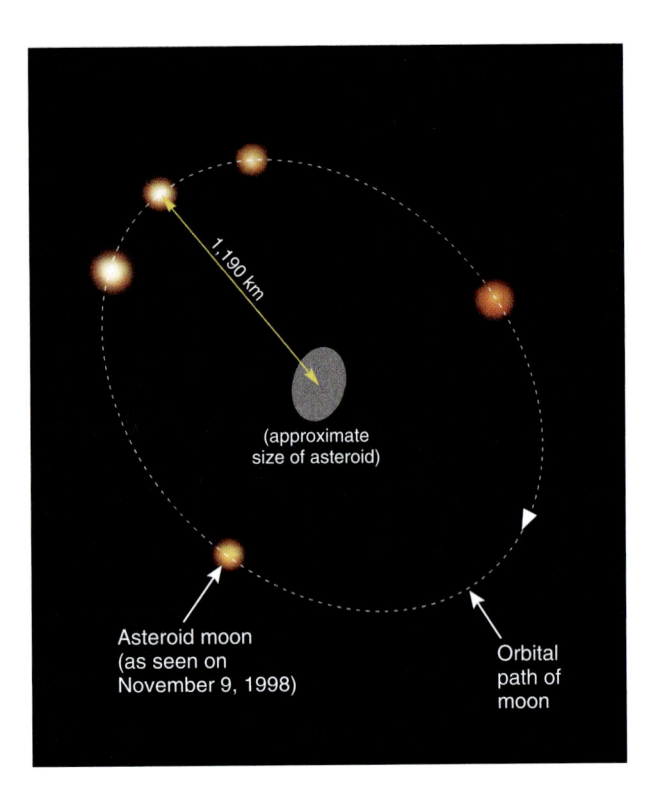

FIGURE 12.6 Asteroid masses can be measured only if we have observed the asteroid's gravitational effect on a passing spacecraft or, in very rare cases, if the asteroid has its own tiny moon. One case of an asteroid with a moon is Ida, with its moon Dactyl (see Figure 12.4b). Another is shown here, in an image made by a telescope using adaptive optics [Section 7.4]. The image shows a small moon observed in five positions in its orbit around asteroid Eugenia. The asteroid itself was blocked out during the observations and is shown as the white oval in the image.

We can directly measure the size of an asteroid only when we've seen it close up or in the rare cases when an asteroid is large enough to be resolved in a telescope. (We calculate physical size from the asteroid's angular size and distance [Section 7.1].) In many other cases, we can indirectly determine asteroid sizes by analyzing their light. The total amount of sunlight that an asteroid reflects depends on its size and albedo [Section 10.2]. For example, if two asteroids have the same albedo, the one that reflects more light must be larger. We can determine an asteroid's albedo by taking advantage of the fact that asteroids not only reflect visible light from the Sun, but also emit their own infrared thermal radiation. Because the temperature of an asteroid depends primarily on how much sunlight it absorbs, comparing its infrared thermal emission to its visible light reflection tells us its albedo. For example, an asteroid that is very reflective (high albedo) will not absorb much solar energy, so it will be cold and will give off very little infrared thermal emission (Figure 12.5). By using this technique, we have compiled good size estimates for more than a thousand asteroids.

TIME OUT TO THINK *Imagine that you discover an asteroid that is fairly bright through your telescope. What additional observations would you need to make to determine whether it is bright because it is very large, very close, or simply light-colored?*

Determining an asteroid's density can offer valuable insights into its origin and makeup but is possible only if we can measure the asteroid's mass as well as its size. Mass measurements have been made only in a few cases, because they require observing an asteroid's gravitational effect on another object, either a passing spacecraft or a tiny "moon" orbiting the asteroid (Figure 12.6). But these few cases have already revealed that different asteroids can have very different densities. For example, observations from the NEAR flyby of Mathilde yielded a density of 1.5 g/cm^3, which is so low that the asteroid cannot be a solid rocky object. Indeed, the impact that formed the huge crater on Mathilde's surface (see Figure 12.4c) would have disintegrated a solid ob-

a Gaspra's irregular shape leads to large brightness variations as it rotates. The diagrams at the top show its appearance from Earth at different times during its rotation, and the graph shows the corresponding brightness at those times. Note that Gaspra appears brighter when we are looking at a larger face because it reflects more sunlight.

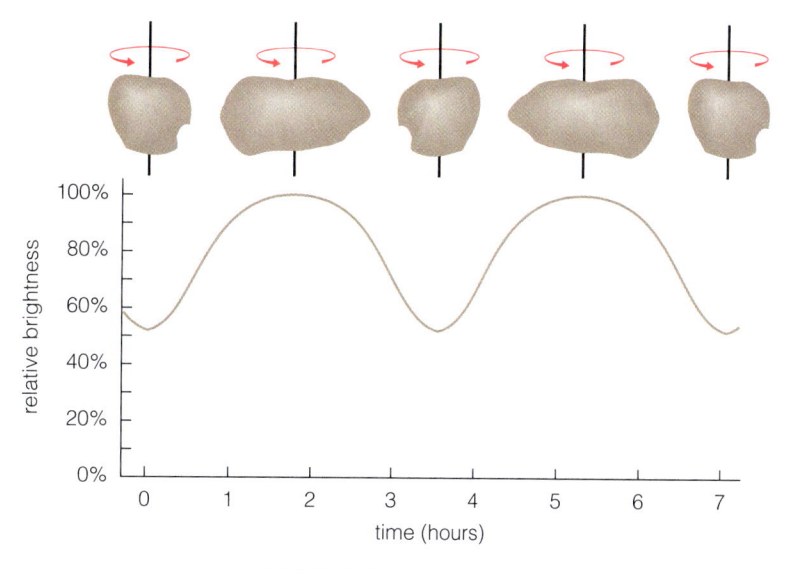

FIGURE 12.7 In some cases, we can determine asteroid shapes from measurements made from Earth.

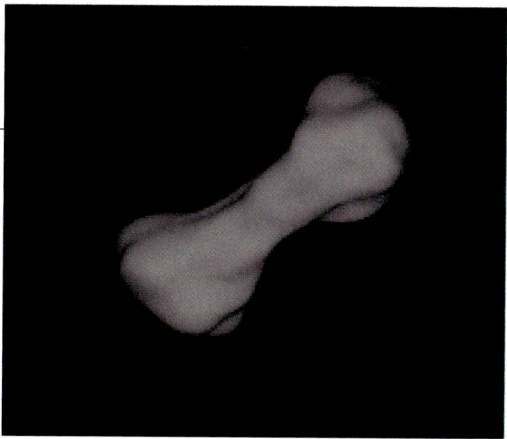

b The asteroid Kleopatra is shaped like a bone. The image shows the shape based on calculations from the results of radar observations.

ject. Mathilde must be a loosely bound "rubble pile" that could absorb the shock of impact without completely disintegrating. In contrast, NEAR found a relatively high density of 2.4 g/cm^3 for Eros, which is close to the value expected for solid rock. Eros also has a subtle pattern of surface ridges and troughs (see Figure 12.4d), which hint at a set of parallel faults and fractures in its interior.

Determining asteroid shapes is even more challenging. One technique involves looking for brightness variations as an asteroid rotates. A uniform spherical asteroid will not change its brightness as it rotates, but a potato-shaped asteroid will reflect more light when it presents its larger side toward the Sun and our telescope (Figure 12.7a). Astronomers can also determine asteroid shapes by examining the shape of the shadows cast in starlight when an asteroid happens to pass right in front of a star. By combining the exact times of the star's disappearance and reappearance as clocked with telescopes in several different locations around the world, they can calculate the shape and size of the shadow and therefore the outline of the asteroid itself. Shape measurements show that only the largest asteroid (Ceres, 940-km diameter) is approximately spherical. The next two largest asteroids are somewhat oblong (Pallas, 540 km across, and Vesta, 510 km across). Smaller asteroids like Gaspra and Ida have still odder shapes. The strength of gravity on these smaller asteroids is evidently less than the strength of the rock.

In a few cases, astronomers have used radar to obtain more detailed shape information about asteroids that have passed close to the Earth. A radio dish sends out radio "pings" that bounce off the asteroid and reflect back to Earth.[2] By analyzing the reflected signals returned as the asteroid rotates, we can create a three-dimensional model of the asteroid. Such measurements revealed the odd shape of the asteroid Kleopatra (Figure 12.7b); in fact, the radar signals were reflected so strongly that scientists believe Kleopatra to be made almost entirely of metals. Radar images indicate that other asteroids are actually two objects held in contact with each other by a weak gravitational attraction. Apparently, this attraction is strong enough to hold the objects together, but not strong enough to force them to merge into a single body.

---

[2] The intense radar beacons beamed at these asteroids are probably the strongest human signals sent from Earth. Perhaps a distant alien civilization will someday pick up these signals and wonder what we were doing.

Finally, we can determine the compositions of asteroids through spectroscopy [Section 7.3]. Thousands of asteroids have been analyzed in this way, and they fall into three main categories: About 75% of asteroids are very dark and show absorption bands from carbon-rich materials. Another 15% are more reflective and show absorption bands characteristic of rocky materials. The remaining 10% include asteroids with spectral characteristics of metals such as iron. Overall, the wide range of asteroid properties reveals their varied histories and underscores planetary scientists' interest in visiting more asteroids.

12.3 Meteorites

Spectroscopy is a pretty good method of determining the composition of an asteroid, but wouldn't it be nice to study an asteroid sample in a laboratory? No spacecraft has yet returned an asteroid sample, but we have pieces of asteroids nonetheless—the rocks called *meteorites* that fall from the sky.

The reality of rocks falling from the sky wasn't always accepted. Stories of such events arose occasionally in human history, sometimes even influencing history. For example, the ancient Greek scientist Anaxagoras (500–428 B.C.) came up with his idea that planets and stars are flaming rocks in the heavens after hearing of "fallen stars." Later scientists were skeptical of such stories. Thomas Jefferson (who worked in science as well as politics), upon hearing of a meteorite fall in Connecticut, reportedly said, "It is easier to believe that Yankee professors would lie than that stones would fall from heaven."

Today we know that rocks really do sometimes fall from the heavens. Some are large enough to do significant damage, blasting huge craters upon impact. Despite the forces of erosion, more than 100 impact craters can still be seen on Earth (Figure 12.8).

The precise origin of meteorites was long a mystery, but today we know that most come from the asteroid belt. In recent decades, a few meteorite falls have been captured on film, allowing scientists to calculate the orbits that led them to crash to Earth.

FIGURE 12.8 Large craters on Earth show that major impacts have played an important role in even relatively recent geological history.

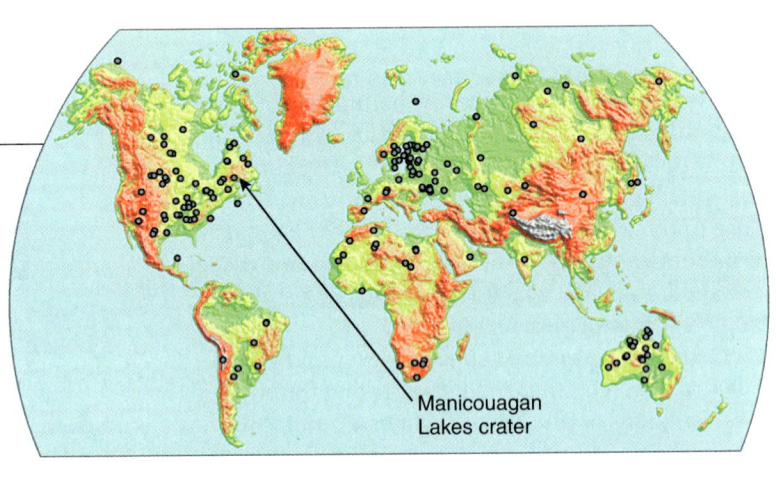

a This map shows the locations of known impact craters.

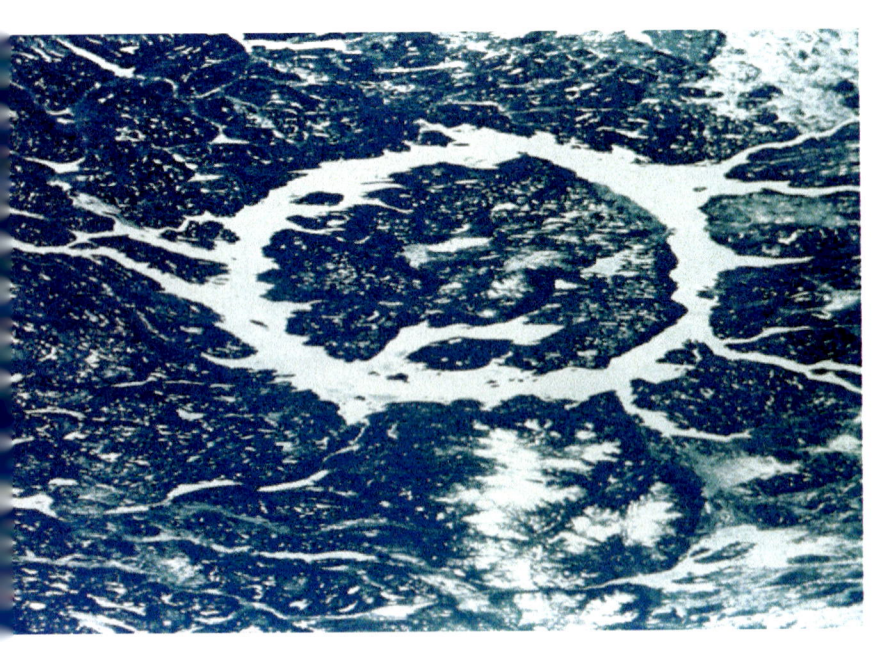

b The Manicouagan Lakes crater in Canada formed about 200 million years ago and has been heavily eroded by glaciers since then.

The results clearly show that the meteorites originated in the asteroid belt.

Rocks from space continue to rain down on Earth, and observers find a few meteorites every year by following the trajectories of very bright meteors called *fireballs*. Meteorites are often blasted apart in their fiery descent through our atmosphere, scattering fragments over an area several kilometers across. A direct hit on the head by a meteorite would be fatal, but there are no reliable accounts of human deaths from meteorites. However, meteorites have injured at least one human (by a ricochet, not a direct hit), damaged houses and cars, and killed animals (Figure 12.9). Most meteorites fall into the oceans, which cover three-quarters of Earth's surface area.

Many more meteorites—ones that fell to Earth at some time in the past—are found by accident or by organized meteorite searches. Antarctica offers the greatest treasure trove of meteorites, not because more fall there but because the icy surface makes meteorites easier to identify and the movement of glaciers carries rocky debris into a few small areas. The majority of meteorites in museums and laboratories come from Antarctica.

Comparing Meteorites

Most meteorites are difficult to distinguish from terrestrial rocks without detailed scientific analysis, but a few clues can help. Meteorites are usually covered with a dark, pitted crust resulting from their fiery passage through the atmosphere. Some can be readily distinguished from terrestrial rocks by their high metal content. (Most types of meteorites contain enough metal to attract a magnet hanging on a string. If you suspect that you have found a meteorite, most science museums will analyze a small chip free of charge.) The ultimate judge of extraterrestrial origin is laboratory analysis of the rock's composition. Terrestrial rocks may differ from one another in composition, but all contain isotopes of particular elements in roughly the same ratios. Meteorites can be identified by their very different isotope ratios. The presence of certain rare elements, such as iridium, also indicates extraterrestrial origin; nearly all of Earth's iridium sank to the core long ago and hence is absent from surface rocks.

The tens of thousands of known meteorites fall into two basic categories. The vast majority of meteorites appear to be composed of a random mix of flakes from the solar nebula, and radioactive dating shows that they formed at the same time as the solar system itself—about 4.6 billion years ago [Section 8.6]. These **primitive meteorites** are our best source of information about conditions in the solar nebula

FIGURE 12.9 Two perspectives on the risks of being hit by a meteorite.

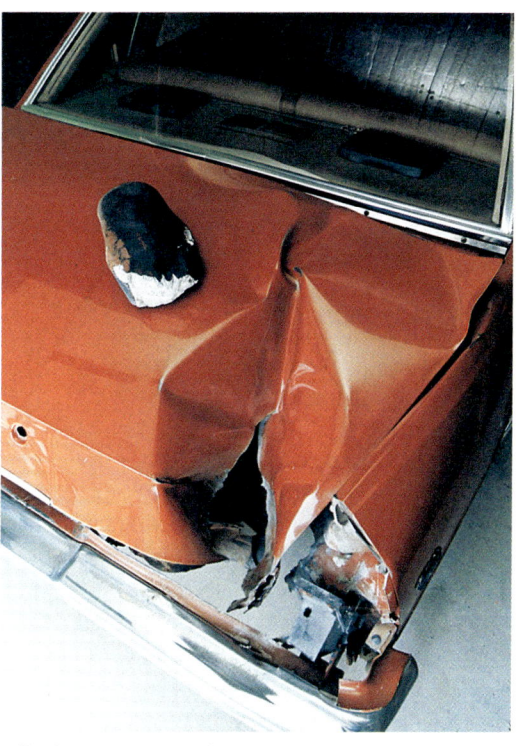

a The impact crater in this Chevrolet was created by a 12-kg meteorite. Analysis of videotapes of the meteor's fiery passage over several East Coast states confirmed its origin in the asteroid belt.

b No humans have been killed by meteorite falls, but . . . .

Stony primitive meteorite: metal flakes intermixed with rocky material.

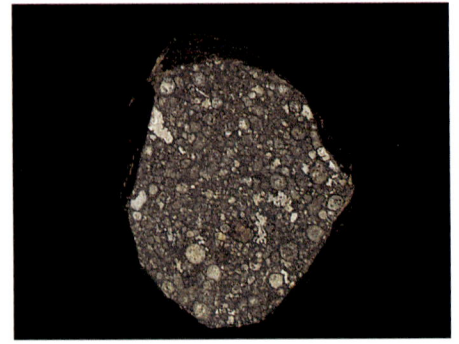

Carbon-rich primitive meteorite: similar to meteorite at left, but with carbon compounds added in.

Differentiated iron meteorite: similar to planetary core.

Differentiated stony meteorite: similar to volcanic rocks.

FIGURE 12.10 (**a**) Two examples of primitive meteorites. (**b**) Two examples of processed meteorites.

(Figure 12.10a). Most primitive meteorites are composed of rocky minerals with an important difference from Earth rocks: A small but noticeable fraction of pure metallic flakes is mixed in. (Iron in Earth rocks is chemically bound in minerals.) Some primitive meteorites also contain substantial amounts of carbon compounds, and some even contain a small amount of water.

A much smaller group of meteorites appears to have undergone substantial change since the formation of the solar system, and radioactive dating shows some of these meteorites to be younger than the primitive meteorites. We call this group **processed meteorites**, because they apparently were once part of a larger object that "processed" the original material of the solar nebula into another form (Figure 12.10b).[3] Some processed meteorites resemble the Earth's core in composition: an iron/nickel mixture with trace amounts of other metals. These meteorites have a relatively high density of about 7 g/cm^3 and played an important role in human history: As the most accessible form of metallic iron available on the Earth's surface, they helped humans make the transition from the bronze age to the iron age. Other processed meteorites have much lower densities and are made of rocks more similar to the Earth's crust or mantle. A few even have a composition remarkably close to that of the basalts erupted from terrestrial volcanoes [Section 9.4].

The Origin of Meteorites

We've seen that primitive meteorites may be either rocky or carbon-rich. Why are there two varieties? The answer lies in where they formed in the solar nebula. Carbon compounds condensed only at the relatively low temperatures that were found in the solar nebula beyond about 3 AU from the Sun. Thus, carbon-rich meteorites must come from the outer portion of the asteroid belt (beyond about 3 AU from the Sun). Laboratory spectra of these carbon-rich primitive meteorites confirm that they are a good match to the compositions of asteroids in the outer part of the asteroid belt. The remainder of the primi-

---

[3]The technical term for primitive meteorites is *chondrites*, because they contain seed-shaped "chondrules" that may be solidified droplets splashed out in the accretion process. Processed meteorites underwent major changes that destroyed the chondrules, so this category is called *achondrites*.

tive meteorites lack carbon compounds, so they presumably formed in the warmer, inner part of the asteroid belt. Curiously, most of the meteorites collected on Earth are of the rocky variety, even though most asteroids are of the carbon-rich variety. Evidently, orbital resonances are more effective at pitching asteroids our way from the inner part of the asteroid belt. All the primitive meteorites apparently survived untouched since the beginning of the solar system—at least until they came crashing down to Earth.

The processed meteorites tell a more complex story. Their compositions appear similar to the cores, mantles, or crusts of the terrestrial worlds. Thus, they must be fragments of worlds that underwent *differentiation* [Section 9.2]. That is, they are fragments of worlds that must have been heated to high enough temperatures to melt inside, allowing metals to sink to the center and rocks to rise to the surface. The processed meteorites with basaltic compositions must come from lava flows that occurred on some of the larger asteroids during a time when they had active volcanism. Perhaps as many as a dozen large asteroids were geologically active shortly after the formation of the solar system, but radioactive dating of processed meteorites suggests that these worlds soon became inactive.

Basaltic meteorites may simply have been chipped off the surface of a large asteroid by relatively small collisions. (By comparing meteorite and asteroid spectra, scientists have identified the asteroid Vesta as the likely source of some of these meteorites.) In contrast, the processed meteorites with corelike or mantlelike compositions must be fragments of worlds that completely shattered in collisions. These shattered worlds not only lost a chance to grow into planets, but they sent many fragments onto collision courses with other asteroids and other planets, including Earth. Thus, processed meteorites essentially offer us an opportunity to study a "dissected planet."

If fragments such as the basaltic meteorites can be chipped off the surface of an asteroid, is it possible that they might also be chipped off a larger object such as a moon or a planet? In fact, a few processed meteorites have been found that don't appear to match the compositions of asteroids but instead appear to match the composition either of the Moon or of Mars. We now believe that some of these meteorites in fact were chipped off the Moon in impacts and that others were chipped off Mars. The analysis of these *lunar meteorites* and *Martian meteorites* is providing new insights into the conditions on the Moon and Mars. In at least one case, a Martian meteorite may be offering us clues about whether life once existed on Mars [Section 13.6].

12.4 Comets

Humans have watched comets in awe for millennia. The occasional presence of a bright comet in the night sky was hard to miss before the advent of big cities, inexpensive illumination, and television. In some cultures, these rare intrusions into the unchanging heavens foretold bad or good luck, and in most cultures there was little attempt to interpret the event in astronomical terms. (Recall that comets were generally thought to be atmospheric phenomena until proved otherwise by Tycho Brahe [Section 5.3].) Nowadays, these leftover icy planetesimals can teach us about formation processes in the outer solar system, just as asteroids teach us about the inner solar system.

After Kepler developed his laws of planetary motion, other astronomers applied them to comets and found extremely eccentric orbits. Some comets visit the inner solar system only once before being ejected into interstellar space, others return on orbits spanning thousands of years, and a few come by more frequently. The most famous is Halley's comet, named for the English scientist Edmund Halley (1656–1742). Halley did not "discover" his comet but rather gained fame for recognizing that a comet seen in 1682 was the same one seen on a number of previous occasions. He used Newton's law of gravitation to calculate the comet's 76-year orbit and, in a book published in 1705, predicted that the comet would return in 1758. The comet was given his name when it reappeared as predicted, 16 years after his death. Halley's comet has returned on schedule ever since, including twice in the twentieth century—spectacularly in 1910 when it passed near the Earth, and unimpressively in 1986 when it passed by at a much greater distance. It will next visit in 2061.

Today, new comets carry a form of good luck. Thousands of amateur astronomers eagerly scan the skies with their telescopes and binoculars in hopes of discovering a new comet. The first discoverers (up to three) who report their findings to the International Astronomical Union have the comet named after them. Several comets are discovered every year, and most have never been seen before by human eyes. In 1996 and 1997, we were treated to back-to-back brilliant comets: comet Hyakutake and comet Hale–Bopp (Figure 12.11). The surprise champion of comet discovery is the SOHO spacecraft, an orbiting solar observatory. SOHO has detected more than 200 "Sun grazing" comets, most on their last pass by the Sun (Figure 12.12).

The Flashy Lives of Comets

Comets are icy planetesimals from the outer solar system. Their composition has been described as "dirty snowballs": ices mixed with rocky dust. Comets

FIGURE 12.11 Brilliant comets can appear at almost any time, as demonstrated by the back-to-back appearances of (**a**) comet Hyakutake in 1996 and (**b**) comet Hale–Bopp in 1997, photographed at Mono Lake in California.

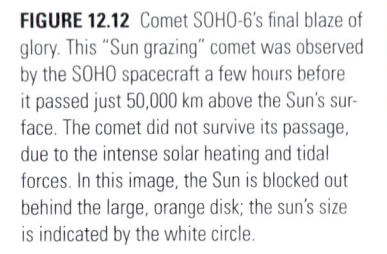

FIGURE 12.12 Comet SOHO-6's final blaze of glory. This "Sun grazing" comet was observed by the SOHO spacecraft a few hours before it passed just 50,000 km above the Sun's surface. The comet did not survive its passage, due to the intense solar heating and tidal forces. In this image, the Sun is blocked out behind the large, orange disk; the sun's size is indicated by the white circle.

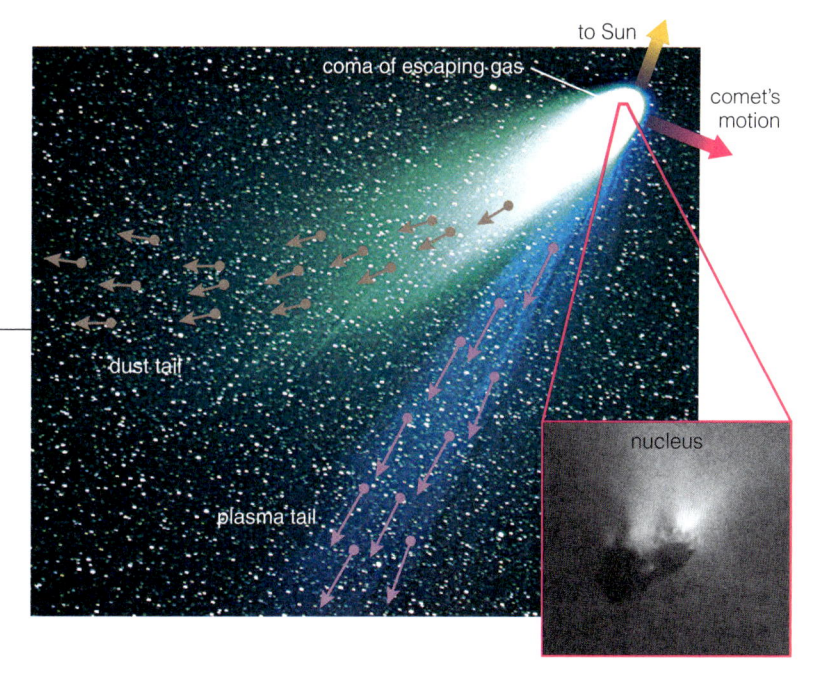

FIGURE 12.13 Anatomy of a comet. The inset photo is the nucleus of Halley's comet photographed by the Giotto spacecraft; the coma and tails shown are those of comet Hale–Bopp.

are completely frozen when they are far from the Sun, and most are just a few kilometers across. But a comet takes on an entirely different appearance when it comes closer to the Sun (Figure 12.13). The dirty snowball is the comet's **nucleus**. The sublimation of ices in the nucleus creates a rapidly escaping dusty atmosphere called the **coma**, along with a **tail** that points away from the Sun regardless of which way the comet is moving through its orbit. In most cases, we actually see two tails: a **plasma tail** made of ionized gas, and a **dust tail** made of small solid particles. Although much of comet science focuses on the dramatic tail and coma, we will begin our closer look at comets with the humbler nucleus.

Comet nuclei are rarely observable with telescopes, because they are quite small and are shrouded by the dusty coma. The European Space Agency's Giotto spacecraft provided our best view of a nucleus when it passed within 600 kilometers of comet Halley's nucleus in 1986. (The spacecraft was named for the Italian painter Giotto, who apparently was inspired by a passage of Halley's comet to include a comet in a 1304 religious painting.) The flyby revealed a dark, lumpy, potato-shaped nucleus about 16 kilometers long and 8 kilometers in width and depth. Despite its icy composition, comet Halley's nucleus is darker than charcoal, reflecting only 3% of the light that falls on it. (It doesn't take much rocky soil or carbon-rich material to darken a comet.) The ice is not packed very solidly; some estimates of the density of Halley's nucleus are considerably less than 1 g/cm^3—suggesting that the nucleus is part ice and part empty space.

How can such a small nucleus put on such a spectacular show? Comets spend most of their lives in the frigid outer limits of our solar system. As a comet accelerates toward the Sun, its surface temperature increases and ices begin to sublime into gaseous form. By the time the comet comes within about 5 AU of the Sun, sublimation begins to form a noticeable atmosphere that easily escapes the comet's weak gravity. The escaping atmosphere drags away dust particles that were mixed with the ice and begins to create the dusty coma around the nucleus. Sublimation rates increase as the comet approaches the Sun, and gases jet away from patches on the nucleus at speeds of hundreds of meters per second (Figure 12.14). The jets can make faint pinwheel patterns within the coma, due to the slow rotation of comet nuclei. The rapid sublimation carries away so much dust that a spacecraft flyby is dangerous. The Giotto spacecraft, which approached Halley's nucleus at a relative speed of nearly 250,000 km/hr (70 km/sec), was sent reeling by dust impacts just minutes before closest approach, breaking radio contact with Earth.

As gas and dust move away from the comet, they are influenced by the Sun in different ways. The gases are ionized by ultraviolet photons from the Sun and then carried straight outward away from the Sun with the solar wind at speeds of hundreds of kilometers per second. These ionized gases form the *plasma tail*, which may extend hundreds of millions of kilometers away from the Sun. Dust-size particles experience a much weaker push caused by the pressure of sunlight itself (called *radiation pressure* [Section 16.4]), so the *dust tail* is swept away from the Sun in a slightly different direction (see Figure 12.13). The comet

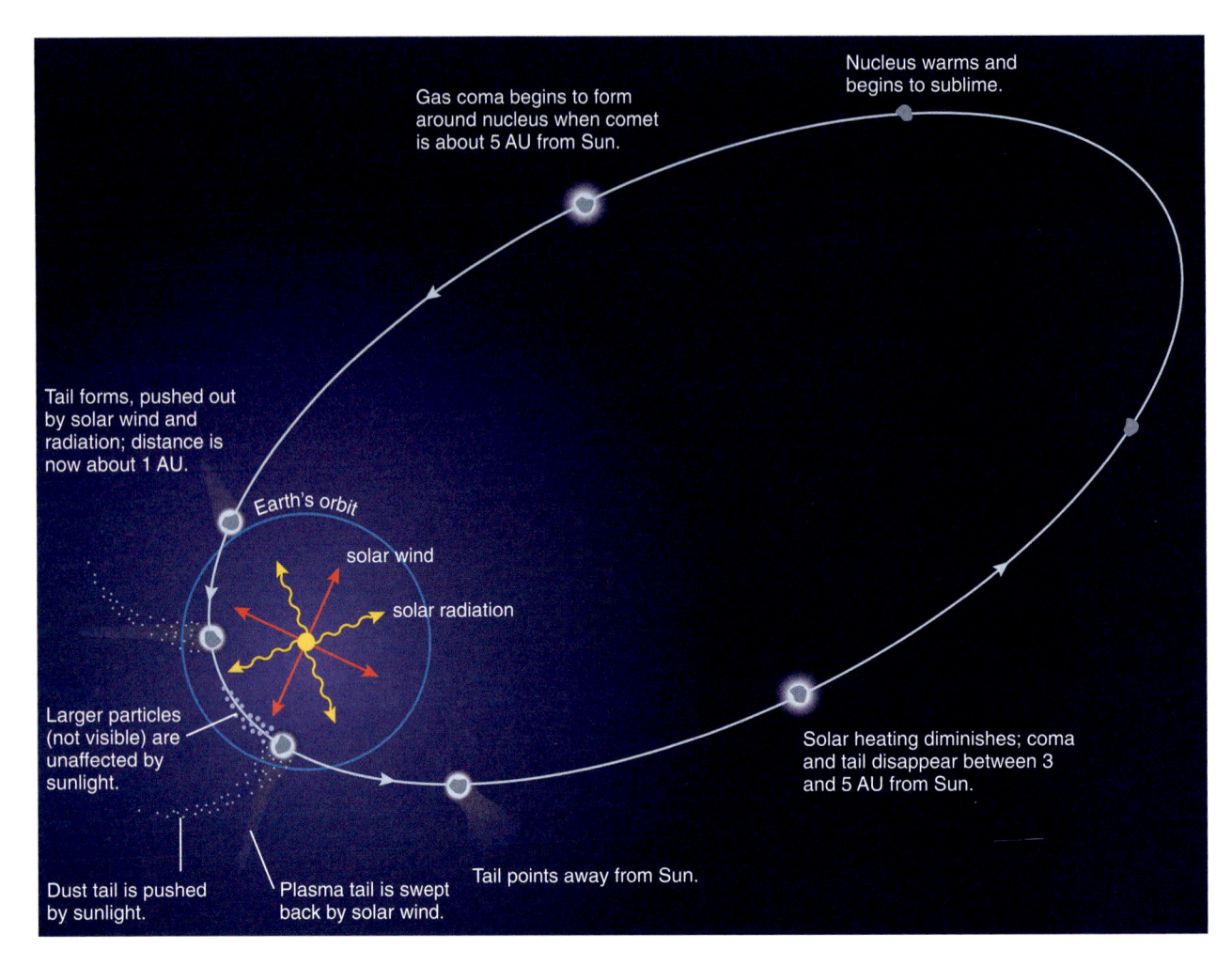

FIGURE 12.14 Comets exist as bare nuclei over most of their orbits and grow a coma and tails only when they approach the Sun. (Diagram not to scale.)

wheels around the Sun, always keeping its tails of dust and plasma pointed roughly away from the Sun. Comets also eject some larger, pebble-size particles that are not pushed away from the Sun. These form another, invisible "tail" extending along the comet's orbit; as we'll discuss in Section 12.6, these particles are responsible for meteor showers.

After the comet loops around the Sun and begins to head back outward, sublimation declines, the coma dissipates, and some of the dust settles back to the surface. Nothing happens until the comet again comes sunward—in a century, a millennium, a million years, or perhaps never.

Active comets cannot last forever. In a close pass by the Sun, a comet may shed a layer 1 meter thick. Dust that is too heavy to escape accumulates on the surface. This thick dusty deposit helps make comets dark and may eventually block the escape of interior gas that makes comets so phenomenal. A comet probably loses about 0.1% of its volatiles on every pass around the Sun. No one is certain what happens after the volatiles stop escaping. Either the remaining dust disguises the dead comet as an asteroid, or it comes "unglued" without the binding effects of ices and simply disintegrates along its orbit.

Spectra of the coma and tail show emission lines and bands from water molecules and other hydrogen compounds that condensed in the outer regions of the solar nebula. They also show emission from carbon dioxide and carbon monoxide that condensed only in the coldest regions of the solar nebula, as well as a wide variety of more complex molecules. These emissions do not give a complete picture of a comet's ingredients, since sunlight and chemical reactions have altered the composition. Organic chemicals (containing carbon, hydrogen, oxygen, and nitrogen) may also be present, suggesting that comets carry the necessary building blocks for life. Comets may even contain some grains of interstellar dust that formed *before* our solar system. NASA's Stardust mission will gather some of these materials from comet Wild 2 and will return them to Earth in 2006. A complete understanding of the composition of comets still eludes us, but understanding comets' composition may have critical implications for understanding their formation in the outer solar nebula and their importance to the planets.

The Origin of Comets

Comets are a rarity in the inner solar system, with only a handful coming within the orbit of Jupiter at any one time. However, because comets cannot survive very many passes by the Sun, comets must come from enormous reservoirs at much greater distances: the Kuiper belt and the Oort cloud (Figure 12.15). The existence of these reservoirs was first inferred by analysis of the orbits of comets that pass close to the Sun. Most comet orbits fit no pattern—they do not orbit the Sun in the same direction as the planets, and their elliptical orbits can be pointed in any direction. Comets Hyakutake and Hale–Bopp both fall into this class. It is believed that such comets are visitors from the Oort cloud. Other comets have a pattern to their orbits: They travel around the Sun in the same plane and direction as the planets, though on very elliptical orbits that take them beyond the orbit of Pluto. They also return much more frequently—typically every few decades or centuries—and are thought to be visitors from the Kuiper belt. Comets from either the Oort cloud or the Kuiper belt may occasionally pass near enough to a planet to have their orbits significantly altered, sometimes becoming trapped in the inner regions of the solar system. Halley's comet must have passed near a jovian planet in the distant past, causing it to end up on its current 76-year orbit of the Sun [Section 5.6]. Gravitational encounters with planets can also eject comets from the solar system altogether or send them crashing into the planets or the Sun.

Based on the number of comets that fall into the inner solar system, the Oort cloud must contain about a trillion (10^{12}) comets. However, Oort cloud comets normally orbit the Sun at such great distances—up to about 50,000 AU—that we have so far seen only the rare visitors to the inner solar system. The comets of the Oort cloud probably formed in the vicinity of the jovian planets and were flung into their large, random orbits by gravitational encounters with these planets. Oort cloud comets are so far from the Sun that the gravity of neighboring stars can alter their orbits, preventing some comets from ever returning near the planets and sending others plummeting toward the Sun.

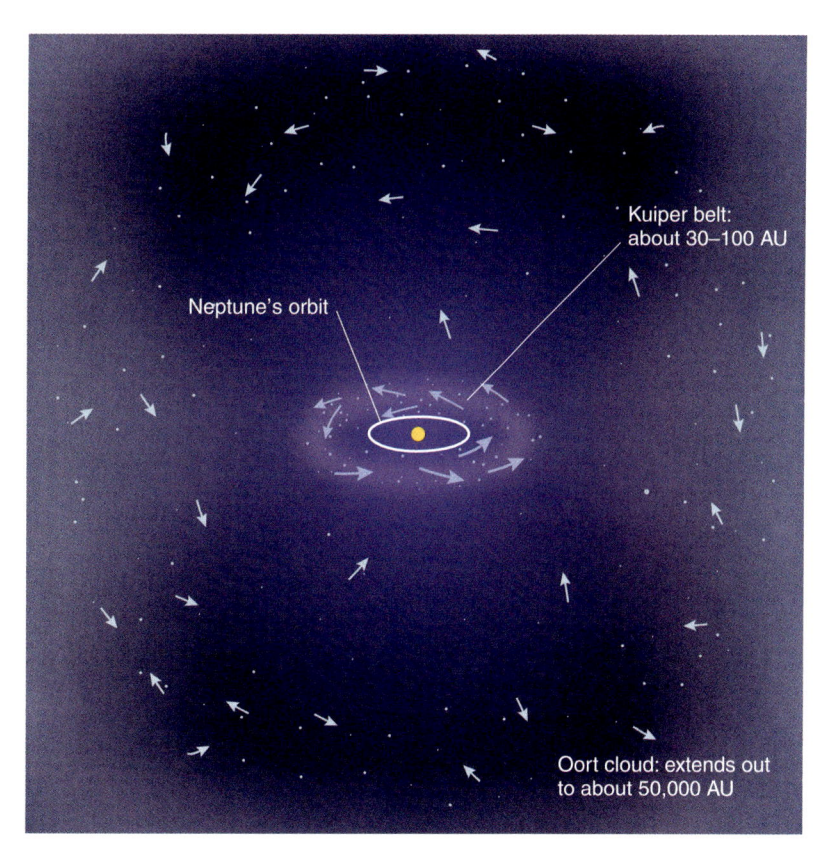

FIGURE 12.15 The Kuiper belt and the Oort cloud. Arrows indicate representative orbital motions of objects in the two regions. (Figure not to scale.)

In contrast, the comets of the Kuiper belt probably still lie in the same general region in which they formed, which explains why their orbits match the pattern of the solar nebula as a whole. In the 1990s, astronomers began using the largest telescopes to search for Kuiper belt objects in their native environment rather than on their excursions into the inner solar system. The first discovery, given the name 1992QB1, orbits beyond Pluto with an orbital period of 296 years. It is a comet approximately 280 kilometers across. Subsequent searches have identified dozens of other objects that match the predicted orbital properties of Kuiper belt comets: fairly circular orbits in the same direction as the planets, with small orbital tilts. Nearly 400 Kuiper belt objects have been discovered so far, suggesting that at least 100,000 comets more than 100 km across must populate the Kuiper belt, along with a billion objects around 10 km across (the typical size of comets that enter the inner solar system). Thus, the total mass of the Kuiper belt is far greater than that of the asteroid belt. Spectroscopy has provided some preliminary information on the composition of Kuiper belt objects. Although they are undoubtedly rich in ices, the

Kuiper belt comets are covered with dark, carbon-rich compounds.

Planetary scientists would dearly love to compare samples of the ancient materials frozen inside the two types of comet. But comet samples falling to Earth burn up in the atmosphere and do not become meteorites, so we are limited to distant spectroscopic analysis of the gases sublimed off the nucleus. We cannot yet tell if Oort cloud comets show signs of having formed in closer, warmer, and denser regions of the solar nebula than their Kuiper belt counterparts. Spacecraft missions to these frigid regions of the solar system are prohibitive, so we must be content to wait for comets to come to us and watch as the Sun dissects them meter by meter.

How big can Kuiper belt objects be? The largest Kuiper belt comet discovered so far, designated 2000WR106, may be nearly 1,000 km across. If a comet can be 1,000 km across, could one be larger still? It's time to consider the odd planet Pluto.

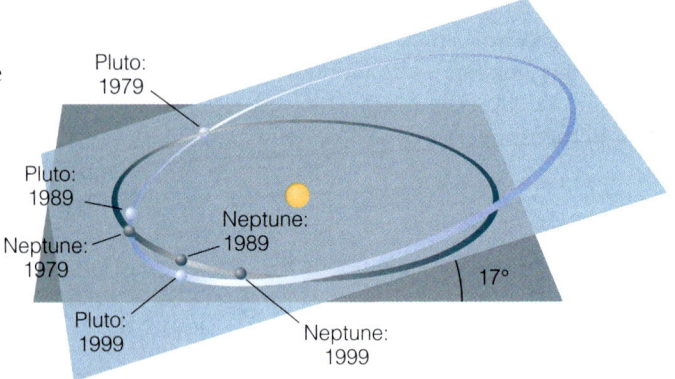

FIGURE 12.16 Pluto's orbit is significantly more elliptical and more tilted relative to the ecliptic than that of any other planet. It comes closer to the Sun than Neptune for 20 years in each 248-year orbit, as was the case between 1979 and 1999. But there's no danger of a collision between Pluto and Neptune, because they share an orbital resonance in which Neptune completes three orbits for every two orbits by Pluto.

12.5 Pluto: Lone Dog or Part of a Pack?

Pluto was discovered in 1930 by American astronomer Clyde Tombaugh, culminating a search that began not long after the discovery of Neptune in 1846. Recall that Neptune's existence and location were predicted before its actual discovery by mathematical analysis of irregularities in the orbit of Uranus (see p. 20). Further analysis of Uranus's orbit suggested the existence of a "ninth planet," and Pluto was found only 6° from the predicted position of this planet. Initial estimates suggested that Pluto was much larger than Earth, but successively more accurate measurements derived ever-smaller sizes. We now know that Pluto has a radius of only 1,195 kilometers and a mass of just 0.0025 Earth mass—making it far too small to affect the orbit of Uranus. In fact, the discovery of Pluto near the predicted position of the "ninth planet" was coincidental; Uranus's supposed orbital irregularities were apparently just errors in measurement. Pluto might have escaped detection for decades without those errors.

Pluto has long seemed to be a misfit among the planets, fitting into neither the terrestrial nor the jovian category. Its 248-year orbit is unusually elliptical and significantly tilted relative to the ecliptic (Figure 12.16). Near perihelion, it actually comes closer to the Sun than Neptune—such was the case between 1979 and 1999. At aphelion, it is 50 AU from the Sun, putting it far beyond the realm of the jovian planets. Despite the fact that Pluto sometimes comes closer to the Sun than Neptune, there is no danger of the two planets colliding: For every two Pluto orbits, Neptune circles the Sun three times. Because of this stable *orbital resonance,* when Pluto is near Neptune's orbit, Neptune is always a safe distance away. The two will probably continue their dance of avoidance until the end of the solar system.

Pluto and Its Moon

Pluto's great distance and small size make it difficult to study, but the task became much easier with the discovery of Pluto's moon Charon in 1978. The original discovery was made when astronomers noticed a "bump" on Pluto's blurry image that moved from side to side over a 6.4-day period (Figure 12.17a). Today we have much clearer pictures of Pluto and Charon from the Hubble Space Telescope (Figure 12.17b). The discovery of the moon enabled astronomers to accurately determine the mass of Pluto by applying Newton's version of Kepler's third law.[4] They also learned that Pluto's rotation axis is tipped 118° relative to its orbit, making it the third planet (in addition to Venus and Uranus) that rotates "backward."

---

[4]If you look in an astronomy text published before 1978, you'll often see Pluto's mass listed with some estimate followed by "?" because of the uncertainty. After the discovery of Charon in 1978, this uncertainty disappeared, providing a clear example of the power of Newton's version of Kepler's third law.

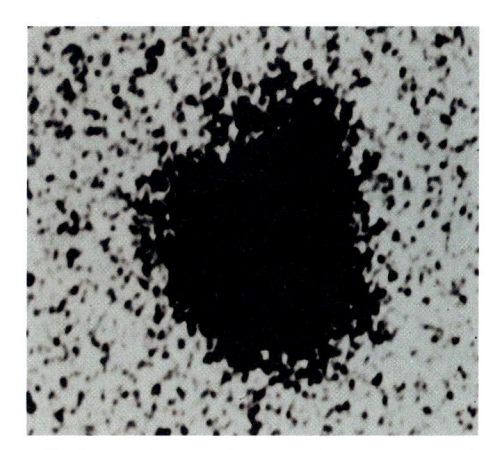

a The bump on the upper right of this fuzzy, black image of Pluto might not seem significant, but astronomers noticed that the bump moved from one side of Pluto to the other and back in six days.

b The bump (in **a**) is actually Pluto's moon Charon, clearly resolved in this Hubble Space Telescope image. The object on the left is Pluto, with Charon on the right.

FIGURE 12.17 An improving view of Pluto and Charon.

Pluto's moon was discovered just in time, because astronomers soon learned that Charon's orbit was about to go edge-on as seen from the Earth, something that happens only every 124 years. From 1985 to 1990, astronomers monitored the combined brightness of Pluto and Charon as they alternately eclipsed each other every few days. Detailed analysis of brightness variations allowed calculation of accurate sizes, masses, and densities for both Pluto and Charon, as well as the compilation of rough maps of their surface markings. Charon's diameter is more than half Pluto's, and its mass is about one-eighth the mass of Pluto. Furthermore, Charon orbits only 20,000 kilometers from Pluto. (For comparison, our Moon has a mass 1/80th that of Earth and orbits 400,000 kilometers away.) Some astronomers argue that Pluto and Charon qualify as a "double planet," but others argue that neither is large enough to qualify as a planet at all.

Curiously, Pluto and Charon have slightly different densities and presumably slightly different compositions. Pluto's density of about 2 g/cm^3 is a bit high for the expected ice/rock mix in the outer solar system, while Charon's density of 1.6 g/cm^3 is a bit low. The leading explanation is that Charon was created by a giant impact similar to that thought to have formed our Moon. A large icy object impacting Pluto may have blasted away its low-density outer layers, which then formed a ring around Pluto and eventually reaccreted into the low-density moon Charon.

Pluto currently has a thin atmosphere of nitrogen and other gases formed by sublimation of surface ices. However, the atmosphere is steadily thinning because Pluto's elliptical orbit is now carrying it farther from the Sun. (Pluto will not reach aphelion until 2113.) As Pluto recedes, its atmospheric gases are refreezing onto its surface. Maps of surface markings generated from observation of eclipses and Hubble Space Telescope images show that the surface has bright and dark patches, but we don't yet have enough information to understand why (Figure 12.18).

Despite the cold, the view from Pluto would be stunning. Charon would dominate the sky, appearing almost 10 times larger in angular size than our Moon. Pluto and Charon's mutual tidal pulls long ago made them rotate synchronously with each other (see Figure 5.22). Thus, Charon is visible from only one side of Pluto and would hang motionless in the sky, always showing the same face. Moreover, the synchronous rotation means that Pluto's "day" is the same length as Charon's "month" (orbital period) of 6.4 Earth days. Charon would neither rise nor set in Pluto's skies but instead would cycle through its phases in place. The Sun would appear more than a thousand times fainter than it appears here on Earth, and it would be no larger in angular size than Jupiter is in our skies.

Pluto: Planet or Kuiper Belt Object?

Pluto is an object in search of a category. It is a misfit among the planets, but it seems somewhat less out of place when considered in the context of the Kuiper belt. After all, it orbits in the same vicinity as Kuiper belt comets, it has a cometlike composition of icy and rocky materials, and it has a cometlike atmosphere that grows as it comes nearer the Sun in its orbit and fades as it recedes. Could Pluto be simply the largest known member of the Kuiper belt?

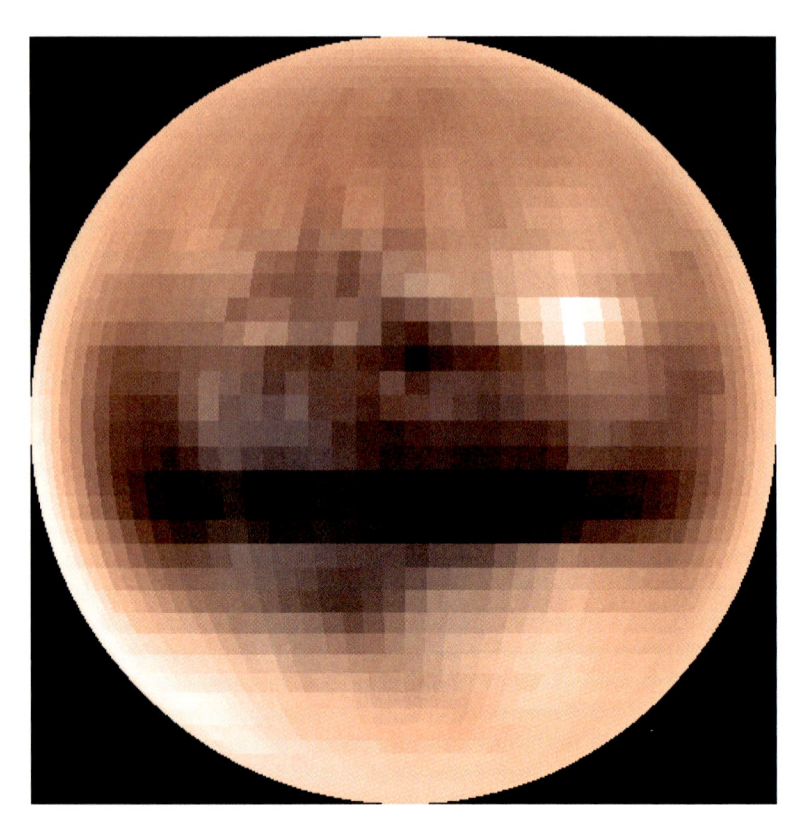

FIGURE 12.18 The surface of Pluto in approximate true color, as derived from brightness measurements made during mutual eclipses between Pluto and Charon. The brown color probably comes from methane ice, but the origin of the dark equatorial band is unknown.

A closer look at the orbits of Kuiper belt objects uncovers an uncanny resemblance to the orbit of Pluto. Like Pluto, many Kuiper belt objects lie in stable orbital resonances with Neptune, and more than a dozen Kuiper belt objects have the same period and semimajor axis as Pluto itself (and are nicknamed "Plutinos"). Aside from Pluto's large size compared to the known comets, the most obvious difference between Pluto and Kuiper belt comets is their surface brightness. Comet nuclei generally have surfaces that are quite dark from carbon compounds, while Pluto's high reflectivity indicates an icy surface. However, this difference is easily explained. Whenever either Pluto or a smaller Kuiper belt comet approaches the Sun, the most volatile ices sublimate into an atmosphere. These atmospheric gases escape completely from the smaller Kuiper belt objects, but Pluto's gravity holds them until they refreeze onto the surface. Over time, the smaller Kuiper belt objects have become depleted in the most volatile ices, leaving behind the dark, carbon-rich compounds on their surfaces. Pluto has retained its volatile ices, giving it a bright, icy surface.

All in all, Pluto stands apart from Kuiper belt objects only in its size. But Pluto may not have been so unusual in the early solar system. Recall that Neptune's moon Triton is quite similar to (and larger than) Pluto, and its orbit indicates that it must be a captured object. Thus, Triton was once a "planet" orbiting the Sun. Giant impacts of other large icy objects may have created Charon and might even have given Uranus its large tilt. Many Pluto-size objects probably roamed the Kuiper belt in its early days. Those that did not become trapped in stable resonances may have fallen inward to be accreted by the jovian planets or been thrown outward and ejected from the solar system. Pluto may be the largest surviving member of the Kuiper belt. However, it remains possible that we will someday discover other objects with sizes similar to that of Pluto, though probably farther away or considerably darker. As of 2001, only a small percentage of the sky had been thoroughly searched for large Kuiper belt objects.

TIME OUT TO THINK *What is your definition of a planet? Be as rigorous as possible. Exactly how many planets are there, according to your criteria? Does your list include Pluto, Charon, Ceres or other asteroids, or Triton (a captured cousin of Pluto)? How big would a new Kuiper belt object have to be to meet your definition? Try to convince a classmate of your definition. Is your definition useful?*

Planet or not, Pluto remains the largest known object in the solar system never visited by spacecraft. Close-up observations of its surface, atmosphere, and unusual moon are sure to teach us as much about icy bodies as past missions did about rocky bodies. NASA scientists and engineers are struggling with the difficult task of designing an affordable Pluto mission. Pluto presents real challenges: At typical spacecraft speeds it may take a decade to reach Pluto, but the faster a probe goes the less time it will have to study Pluto during its flyby. (A Pluto orbiter is simply too expensive.) Fortunately, recent advances in miniaturization are paying off. The current Pluto mission design fits on a coffee table but has instruments about as good as those on the Cassini spacecraft to Saturn, which is as large as a typical small classroom. (The Pluto probe design is also about 5–10 times cheaper than that of *Cassini*.) NASA's current goal is to send the mission to Pluto before its atmosphere completely refreezes onto the surface, but budget constraints may keep the mission from becoming a reality. (Check the text Web site for updates on the status of a Pluto mission.)

a This photo shows comet Shoemaker–Levy 9 after it was broken apart by tidal forces after passing close to Jupiter.

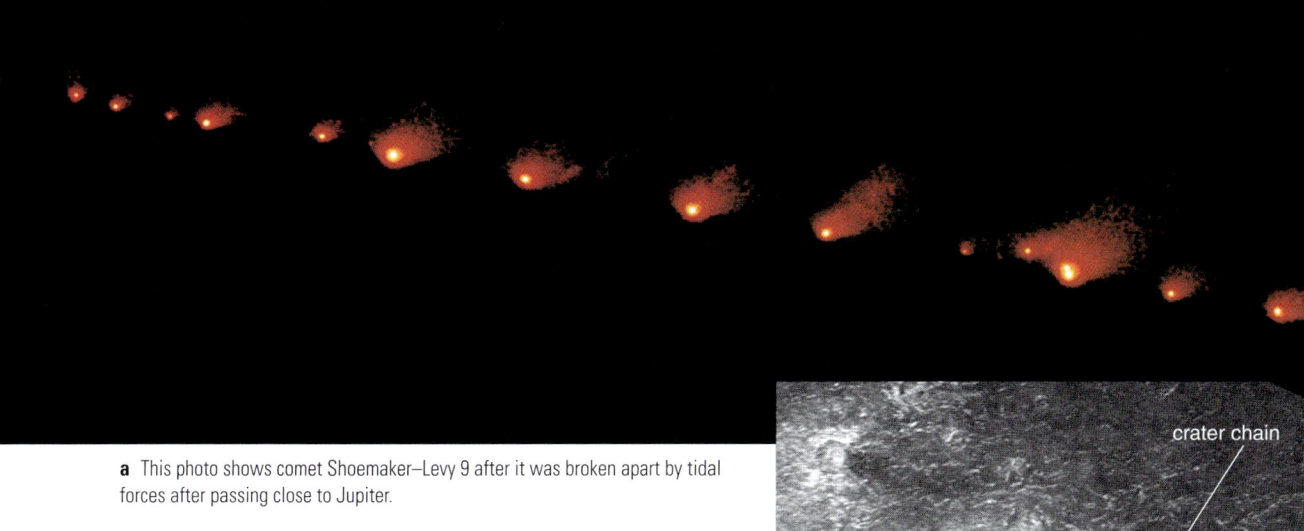

FIGURE 12.19 Fragmented comets.

b This chain of craters on Callisto probably formed long ago from impacts by fragments of a comet that, like comet Shoemaker–Levy 9, was broken apart by tidal forces while passing close to Jupiter. These impacts probably occurred just hours after the comet broke apart.

12.6 Cosmic Collisions: Small Bodies Versus the Planets

The hordes of small bodies orbiting the solar system are slowly shrinking in number through collisions with the planets and ejection from the solar system. Many more must have roamed the solar system in the days of the early bombardment, when most impact craters were formed [Section 8.5]. But there are still plenty left, and cosmic collisions still occur on occasion, with important ramifications for Earth as well as for other planets.

Shoemaker–Levy 9 Impacts on Jupiter

We usually think about impacts in the context of solid bodies, because such impacts leave long-lasting scars in the form of impact craters. But impacts must be even more common on the jovian planets than on the terrestrial planets because of their larger size and stronger gravity, although no impact craters can form on gaseous worlds. It's estimated that a major impact occurs on Jupiter about once every 1,000 years. We were privileged to witness one in 1994.

The husband-and-wife team of Gene and Carolyn Shoemaker, along with their colleague David Levy, are among the premier "comet hunters" of modern times. During a routine search of the sky in March 1993, this team came across an unusual object that Carolyn Shoemaker described as a "squashed comet." The trio's ninth joint discovery, the comet was called *Shoemaker–Levy 9*, or *SL9* for short. On closer examination, the "squashed comet" turned out to be a string of comet nuclei all in a row (Figure 12.19a). The reason for SL9's odd appearance became clear when astronomers calculated its orbit backward in time. The calculations showed that it had passed very close to Jupiter only a few months earlier (in July 1992). Because it had passed well within Jupiter's Roche tidal zone [Section 11.6], it apparently was ripped apart by tidal forces. Prior to that, SL9 was probably a single comet nucleus orbiting Jupiter, unnoticed by human observers. Similar breakups of comets near Jupiter have occurred before, as evidenced by a chain of craters on Callisto that must have formed when a string of comet nuclei crashed into its icy surface (Figure 12.19b).

The orbital calculations also showed that, thanks to gravitational nudges from the Sun, SL9 was on a collision course with Jupiter. Astronomers had more than a year to plan for observations of the impacts due in July 1994. The impacts of the 22 identified nuclei would take almost a week, and virtually every telescope on Earth and in space would be ready and watching. The only uncertainty was whether or not anything would happen. Even the best Hubble Space Telescope images were unable to determine whether the comet nuclei were massive kilometer-size chunks or merely insubstantial puffs of dust.

The answer came within minutes of the first impact: Infrared cameras recorded an intense fireball

a The Galileo spacecraft, on its way to Jupiter at the time, got a direct view of the impacts on Jupiter's night side.

b In under 20 minutes, the Hubble Space Telescope observed an impact plume rise thousands of kilometers above Jupiter's clouds and then collapse back down.

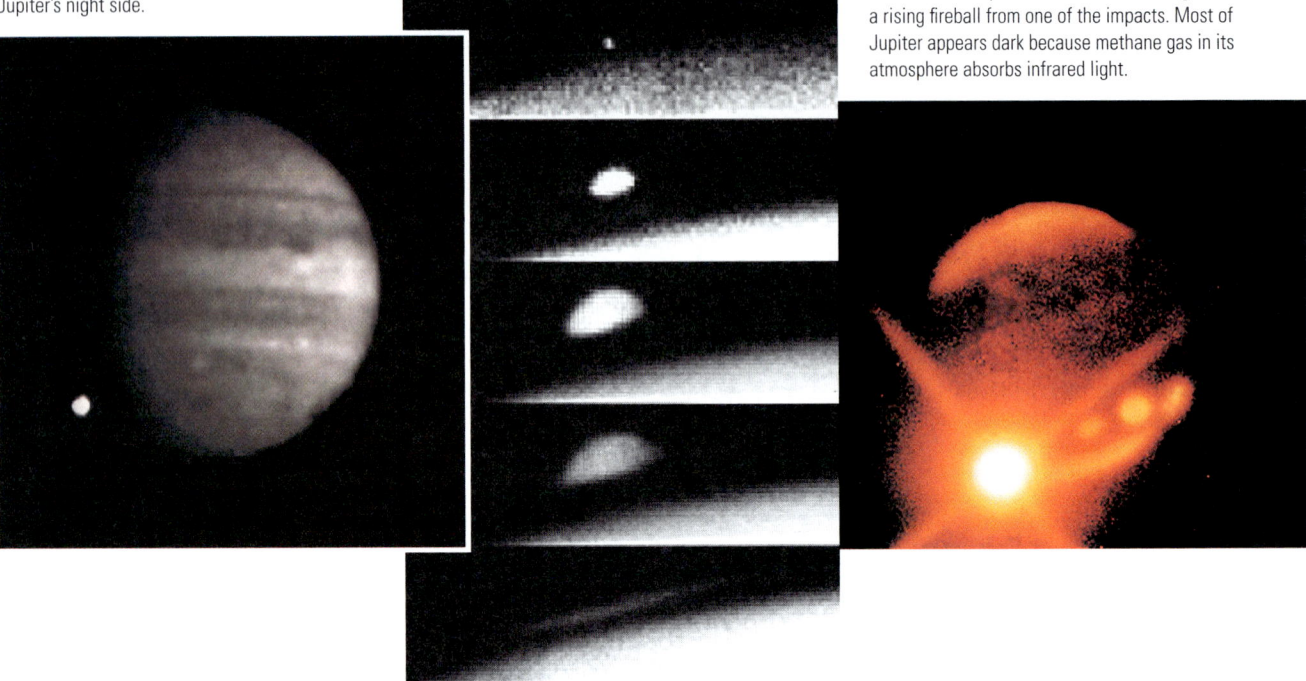

c This infrared photo shows the brilliant glow of a rising fireball from one of the impacts. Most of Jupiter appears dark because methane gas in its atmosphere absorbs infrared light.

FIGURE 12.20 The SL9 impacts allowed astronomers their most direct view ever of cosmic collisions.

---

of hot gas rising thousands of kilometers above the impact site, which lay just barely on Jupiter's far side (Figure 12.20). As Jupiter's rapid rotation carried the scene into view, the collapsing plume smashed down on Jupiter's atmosphere, leading to another episode of heating and more infrared glow. The Hubble Space Telescope caught impact plumes in action as they rose into sunlight and collapsed back down, leaving dust clouds high in Jupiter's stratosphere. The impact scars lingered for months, but Jupiter has recovered—for now.

Material splashed out by the impacts provided us with a unique opportunity to study material from well beneath the layers of Jupiter that are ordinarily visible. More important, we learned much about the impact process itself. Many scientists correctly predicted the behavior of a rising plume of gas from an impact but did not consider the effects of the plume reimpacting on Jupiter. The reimpact of the plumes turned out to have even more significant global effects than the impact itself, heating an area 10,000 kilometers across and leaving dark, dusty clouds that eventually encircled the planet.

The SL9 impacts on Jupiter also provided two important sociological lessons. First, "Comet Crash Week" proved to be one of the best examples of international collaboration in history. With the aid of the Internet, scientists quickly and effectively shared data from observatories around the world. Second, extensive media coverage helped the event capture the public imagination, providing awareness that impacts are not relegated only to ancient geological history. Each individual fragment of SL9 crashing into Jupiter carried the equivalent force of a million hydrogen bombs. If such violent impacts can happen on other planets in our lifetime, could they also happen on Earth?

Comet Tails and Meteor Showers

Far smaller impacts happen on Earth all the time, lighting up the sky as *meteors*. If you watch the sky on a clear night, you'll typically see a few meteors each hour. Most meteors are created by single pieces of comet dust, each no larger than a pea, that enter our atmosphere at speeds of up to 250,000 km/hr (70 km/s). (A small fraction of meteors may be dust from asteroids rather than comets, and the brightest meteors come from larger particles.) The particles and the surrounding atmosphere are heated so much that we see a brief but brilliant flash. Meteor particles are completely vaporized at altitudes of 50–100 kilometers, so we never get complete cometary

d This infrared photo shows high clouds created by several impacts. These clouds reflect infrared light from the Sun because they lie high above the methane gas that absorbs infrared light.

e In this visible-light photo from the Hubble Space Telescope, the high, dusty clouds left by the impacts appear as dark "scars" on Jupiter.

f This Hubble Space Telescope photo shows how the impact scars became smeared due to atmospheric winds. The scars disappeared completely within a few months.

g Artist's conception of the impacts viewed from the surface of Io.

"meteorites" to study. An estimated 25 million meteors occur worldwide every day, adding hundreds of tons of comet dust to the Earth daily.

Comet orbits tend to be filled with small particles ejected as the comets pass near the Sun. When Earth passes through a comet orbit, the particles striking our atmosphere produce a *meteor shower,* during which we may see dozens of meteors per hour. Earth's orbit intersects many comet orbits over the course of the year, and each produces a meteor shower (Table 12.1). The most famous meteor shower, the *Perseids,* occurs every August when the Earth passes through the orbit of comet Swift–Tuttle. Although meteors all strike the Earth on parallel tracks, meteor showers appear to radiate from a particular direction in the sky. You may have experienced a similar illusion while driving into a blizzard at night: The snowflakes seems to diverge from a single direction that depends on the combination of your speed and the wind speed (Figure 12.21a,b). (This illusion occurs even before the air flow around the car deflects the motion of the snowflakes.) The Perseids get their name because they appear to radiate from the constellation Persens. (Figure 12.21c shows the Leonids from Leo.) The Earth runs into more meteors on the side facing in the direction of Earth's orbital motion (just as more snow hits the front windshield while the car is moving), so the predawn sky is the best place to watch for meteors.

Table 12.1 Major Annual Meteor Showers

Shower Name	Approximate Date
Quadrantids	January 3
Lyrids	April 22
Eta Aquarids	May 5
Delta Aquarids	July 28
Perseids	August 12
Orionids	October 22
Taurids	November 3
Leonids	November 17
Geminids	December 14
Ursids	December 23

TIME OUT TO THINK *The associated "comet" for the Geminid meteor shower is an object called Phaethon, which is classified as an asteroid because we've never seen a coma or tail associated with it. How is it possible that Phaethon looks like an asteroid today but once shed the particles that create the Geminid meteor shower?*

FIGURE 12.21 Snowflakes and meteor showers.

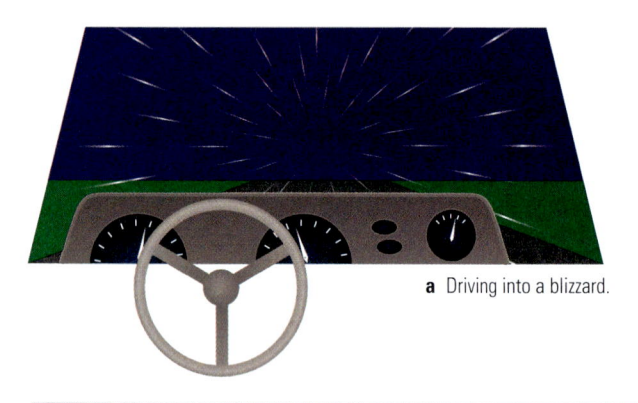

a Driving into a blizzard.

b Geometry of a meteor shower: The Earth is moving through a swarm of comet debris.

c Meteors photographed during the intense Leonid meteor shower in 1966. Star images trailed across the photo from lower left to upper right during the time exposure. Meteors appear to streak away from the top center of the frame. Note one meteor at top center made a dot instead of a streak; it's headed straight for the camera.

FIGURE 12.22 Meteor Crater in Arizona was created about 50,000 years ago by an asteroid impact. The crater is more than a kilometer across and almost 200 meters deep, but the asteroid that made it was only about 50 meters across.

Impacts and Mass Extinctions

Meteorites and impact craters bear witness to the fact that much larger impacts occasionally occur on Earth. Meteor Crater in Arizona (Figure 12.22) formed about 50,000 years ago when a metallic impactor roughly 50 meters across crashed to Earth with the explosive power of a 20-megaton hydrogen bomb. Although the crater is only a bit more than 1 kilometer across, an area covering hundreds of square kilometers was probably battered by the blast and ejecta. Far bigger impacts have occurred, sometimes with catastrophic consequences for life on Earth.

Collecting geological samples in Italy in 1978, the father–son team of Luis and Walter Alvarez discovered a thin layer of dark sediments that had apparently been deposited 65 million years ago—about the same time that the dinosaurs and many other organisms suddenly became extinct. Subsequent studies found similar sediments deposited at the same time at many sites around the world (Figure 12.23). Careful analysis showed this worldwide sediment layer to be rich in the element iridium, which is rare on Earth's surface. But iridium is common in primitive meteorites, which led the Alvarezes to a stunning conclusion: The extinction of the dinosaurs was caused by the impact of an asteroid or comet. This conclusion was not immediately accepted and still generates some controversy, but it now seems clear that a major impact coincided with the death of the dinosaurs. Moreover, while the dinosaurs were the most famous victims of this **mass extinction**, it seems that up to 99% of all living things were killed and that 75% of all *species* living on Earth were wiped out at that time.

How could an impact lead to mass extinction? The amount of iridium deposited worldwide suggests that the impactor must have been about 10 kilometers across. After a decade-long search, scientists identified what appears to be the impact crater from the event. Located off the coast of Mexico's Yucatán peninsula (Figure 12.24), it is 200 kilometers across, which is close to what we expect for a 10-kilometer impactor, and dates to 65 million years ago. Further evidence that the Yucatán crater is the right one comes from the distribution of small glassy spheres that formed when the molten impact ejecta solidified as it rained back to Earth. More of these glassy spheres are found in regions near the crater, and careful study of their distribution suggests that the impactor crashed to Earth at a slight angle. The impact almost immediately sent a shower of debris raining across much of North America and generated huge waves that may have sloshed more than 1,000 kilometers inland (Figure 12.25). Many North American species may have been wiped out shortly after impact. For the

FIGURE 12.23 The arrow points to a layer of sediment laid down by the impact of 65 million years ago. At the time, the rock layers above the arrow did not exist. Dust from the impact and soot from global wildfires settled down through the atmosphere onto the seafloor that once occupied this location in Colorado.

FIGURE 12.24 This false-color image, made with the aid of precise measurements of the strength of gravity, shows an impact crater with its center near the northwest coast of Yucatán. This crater, known as the Chicxulub crater, is thought to have been formed by the impact that killed the dinosaurs and most other species present 65 million years ago. The red box on the inset map shows the region covered by the image; the white lines on the image correspond to the coastlines and borders of Mexican states.

rest of the world, death may have come more slowly. Heat from the impact and returning ejecta probably ignited wildfires in forests around the world. Evidence of such wildfires is found in the large amount of soot that is also present in the iridium-rich sediments from 65 million years ago. The impact sent huge quantities of dust high into the stratosphere, where it remained for several years, blocking out sunlight, cooling the surface, and affecting atmospheric chemistry. Plants died for lack of sunlight, and effects propagated through the food chain.

Perhaps the most astonishing fact is not that 75% of all species died, but that 25% survived. Among the survivors were a few small, rodentlike mammals. These mammals may have survived because they lived in underground burrows and managed to store enough food to outlast the long spell of cold, dark days. Small mammals had first arisen at about the same time as the dinosaurs, more than 100 million years earlier. But the sudden disappearance of the dominant dinosaurs made these mammals dominant. With an evolutionary path swept wide open, they rapidly evolved into a large assortment of much larger mammals—including us.

The event 65 million years ago was but one of at least a dozen mass extinctions in Earth's distant past. An even larger extinction event 250 million years ago has also been linked to an impact and the ensuing catastrophic effects. Could a similar event occur in the future, wiping out all that we have accomplished?

The Asteroid Threat: Real Danger or Media Hype?

The first step in analyzing the threat of future impacts is to examine the past evidence more thoroughly. Only one significant impact has clearly occurred in modern times. In 1908, an unusual explosion occurred in a sparsely inhabited region of Siberia. (The impact is known as the *Tunguska event*.) Entire forests were flattened and set on fire, and air blasts knocked over people, tents, and furniture up to 200 kilometers away (Figure 12.26). Seismic disturbances were recorded at distances of up to 1,000 kilometers, and atmospheric pressure fluctuations were detected almost 4,000 kilometers away. Scientific studies were delayed for almost 20 years (due to the remoteness of the impact area and the politics of the time), but investigators eventually found meteoritic dust at the site—but no impact crater. These wide-reaching effects were apparently caused by a weak, stony meteorite only 30 meters across that exploded in the air, before reaching the surface. Objects of this size probably strike our planet every century or so, usually over the ocean. But the death toll could be quite high if such an object struck a densely populated area.

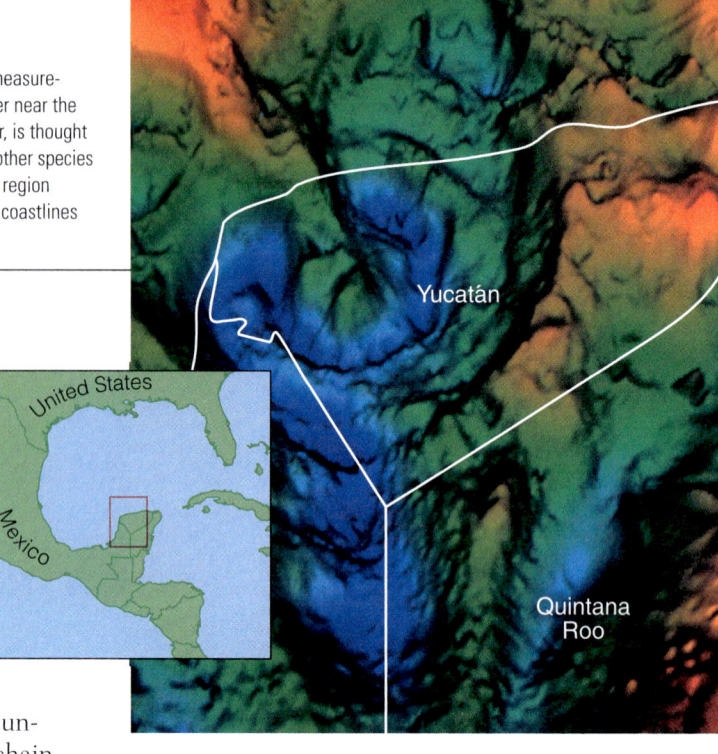

Another way to gauge the threat of impacts is to look at asteroids that might strike Earth. The largest-known *Earth-approaching asteroid,* Eros, is about 40 kilometers long (see Figure 12.4d). Hundreds of smaller Earth-approaching asteroids are known, down to sizes of tens of meters across. Orbital calculations show that none of these known objects will impact Earth in the foreseeable future. However, the vast majority of Earth-approaching asteroids probably have not yet been detected. Astronomers estimate that some 1,000 undiscovered members of this class may be larger than 1 kilometer across, and another 100,000 may be larger than 100 meters across. Many search programs are under way, detecting hundreds of objects every year. NASA has set a goal of detecting 90% of all large Earth-approaching asteroids by 2010. Comets also pose a threat, particularly because their rapid plunge from the outer solar system gives little warning.

By combining observations of asteroids with information from past impact craters, we can estimate the frequency of impacts of various sizes (Figure 12.27). Larger impacts are obviously more devastating but thankfully are rare. Impactors a few meters across probably break up high in the Earth's atmosphere

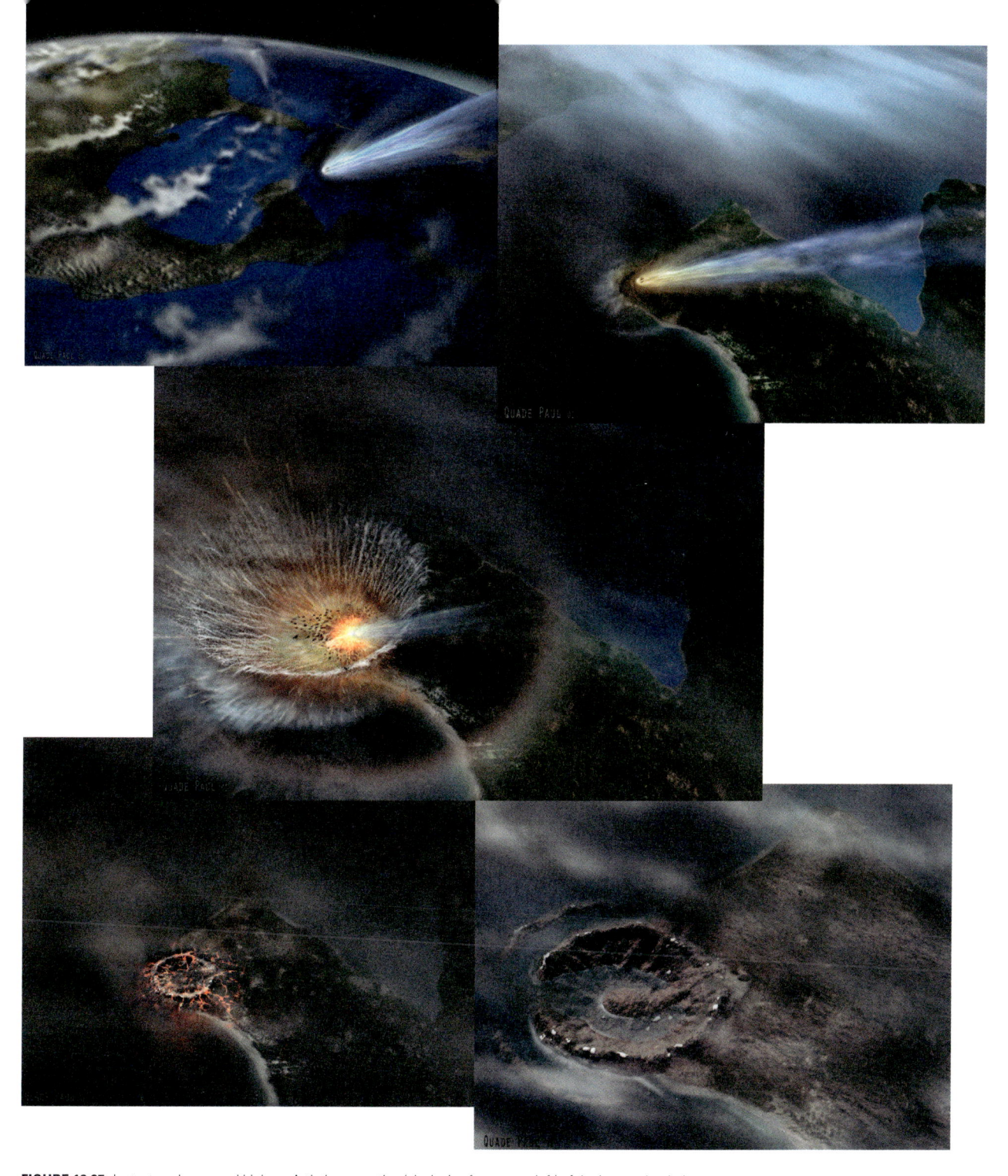

FIGURE 12.25 Instant geology—and biology: Artist's conception (clockwise from upper left) of the impact that led to a mass extinction 65 million years ago.

every few days, with each liberating the energy of an atomic bomb. However, the effects are dissipated before they reach the surface, so we do not notice them. (Military technology designed to detect nuclear bomb tests has recorded many such events.) Hundred-meter impactors that forge craters similar to Meteor Crater probably strike only about every 10,000 years or so. Impacts large enough to cause mass extinctions occur tens of millions of years apart.

Thus, while it seems a virtual certainty that the Earth will be battered by many more large impacts, the chance of a major impact happening in our lifetime is quite small. Nevertheless, the chance is not zero, and until we know the orbits of *all* large Earth-approaching asteroids we cannot predict when the next impact might occur. Some people advocate a thorough search to find all potentially dangerous as-

CHAPTER 12 REMNANTS OF ROCK AND ICE: ASTEROIDS, COMETS, AND PLUTO 347

FIGURE 12.26 Damage from the 1908 impact over Tunguska, Siberia.

teroids, a relatively cheap proposition. But others wonder whether the information would be useful: If you knew an asteroid would hit next year, could you do anything about it? Schemes to save Earth by using nuclear weapons to demolish or divert an asteroid abound, but no one knows whether current technology is up to the task. For the time being, we can only hope that our number does not come up soon.

While some people see danger in Earth-approaching asteroids, others see opportunity, because Earth-approaching asteroids bring valuable resources tantalizingly close to Earth. Iron-rich asteroids are particularly enticing, since they probably contain many precious metals that mostly sank to the core on Earth. In the not-too-distant future, it may prove technically feasible and financially profitable to mine metals from asteroids and bring these resources to Earth. It may also be possible to gather fuel and water from asteroids for use in missions to the outer solar system.

THE BIG PICTURE

In this chapter, we focused on the solar system's smallest objects and found that they can have big consequences. Asteroids, comets, and meteorites teach us much about the evolution of the solar system. And, in the case of the impact 65 million years ago, one such object may have made our existence possible. Keep in mind the following "big picture" ideas as you continue your studies:

- The smallest bodies in the solar system—asteroids and comets—are our best evidence of how the solar system formed.

- The small bodies are subjected to the gravitational whims of the planets, particularly the jovian planets. The subtleties of resonances play a major role in sculpting the outer solar system.

- The interplay of large and small bodies brings us meteorites to teach us our origins, comets and meteor showers to light up the sky, and impacts that can alter or obliterate life as we know it.

- Pluto is called the ninth planet, but it bears much more similarity to the thousands of Kuiper belt comets than to the other eight planets.

FIGURE 12.27 This graph shows how the frequency of impacts—as well as the magnitude of their effects—depends on the size of the impactor; note that smaller impacts are much more frequent than larger ones.

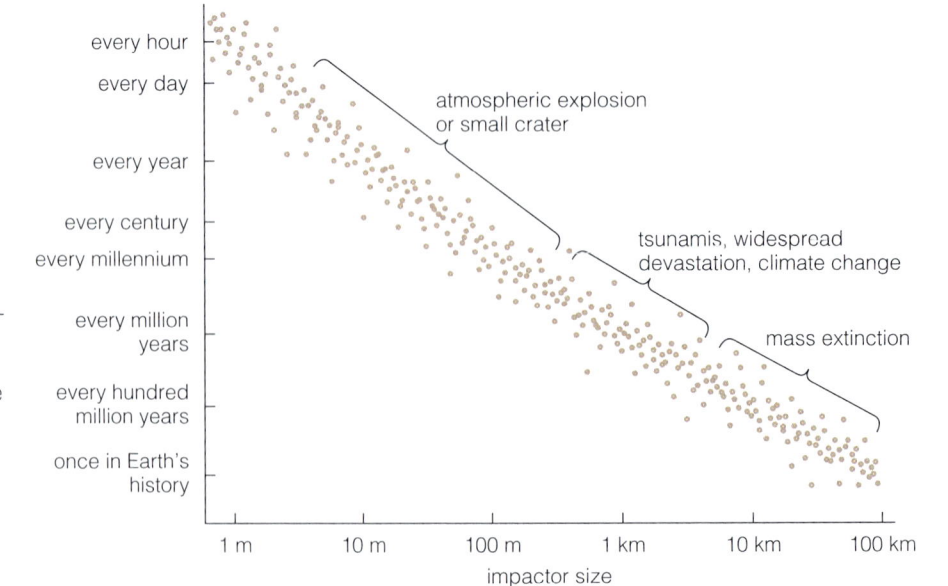

Review Questions

1. Briefly explain why comets, asteroids, and meteorites are so useful in terms of helping us understand the formation and development of our solar system.

2. What is the *asteroid belt,* and where is it located? How does the total mass of all the asteroids compare to the mass of a terrestrial world?

3. Briefly describe how orbital resonances with Jupiter have affected the asteroid belt.

4. How do we detect asteroids with telescopic images? How do we determine the orbit of an asteroid?

5. Briefly describe how we can estimate an asteroid's size, mass, density, and shape.

6. Describe the compositions of the three main groups of asteroids, as distinguished by spectroscopy.

7. How can we distinguish a meteorite from a terrestrial rock? Why do we find so many meteorites in Antarctica?

8. Distinguish between *primitive meteorites* and *processed meteorites* in terms of composition. How do the origins of these two groups of meteorites differ?

9. Why do primitive meteorites come in carbon-rich and rocky varieties?

10. How is the study of a processed meteorite an opportunity to look at a piece of a "dissected planet"?

11. How did Halley's comet get its name? How are comets named today?

12. Describe the anatomy of a comet. How does gas escape from the interior of the *nucleus* into space? Why do the *tails* point away from the Sun?

13. Why can't an active comet last forever? What happens to comets after many passes near the Sun?

14. Describe the Kuiper belt and the Oort cloud in terms of location and comet orbits. Explain how we infer the existence of these two comet reservoirs. How do we think comets came to exist in these two distinct reservoirs?

15. Describe Pluto and Charon. How were they discovered? Why won't Pluto collide with Neptune even though their orbits cross? How did Charon probably form?

16. Briefly summarize the evidence suggesting that Pluto is a Kuiper belt comet.

17. Describe the impact of comet Shoemaker–Levy 9 on Jupiter. How was the impact studied? What did we learn from this impact?

18. Explain how meteor showers are linked to comets. Why do meteor showers seem to originate from a particular location in the sky?

19. Describe the evidence suggesting that the *mass extinction* that killed off the dinosaurs was caused by the impact of an asteroid or comet. How did the impact lead to the mass extinction?

20. How often should we expect impacts of various sizes on Earth? How could we alleviate the risk of major impacts?

Discussion Questions

1. *Impact Risk.* How can we compare the risk of death from an impact to the risk of death from other hazards? One proposed way to evaluate the risk is to multiply the probability of an impact by the number of people it would kill. For example, the probability of a major impact that could kill half the world's population, or some 3 billion people, is estimated to be about 0.0000001 (10^{-7}) in any given year. If we multiply this low probability by the 3 billion people that would be killed, we find the "average" risk from such an impact to be $10^{-7} \times (3 \times 10^9) = 300$ people killed per year—about the same as the average number of people killed in airplane crashes. Do you think it is valid to say that the risk of being killed by a major impact is about the same as the risk of being killed in an airplane crash? Should we pay to limit this risk, as we pay to limit airline crashes? Defend your opinions.

2. *Observational Bias.* Meteorites and long-tailed comets represent "free samples" of more distant objects.
   a. Carbon-rich asteroids make up about 5% of meteorite falls on the Earth. How does this compare with the observed fraction of carbon-rich asteroids in the asteroid belt? The odds of a particular comet we observe originating from the Oort cloud versus the Kuiper belt are about 50–50. How does this compare with the relative number of objects in these two regions?
   b. Your answers to part (a) should show that meteorite samples and comet observations suffer from *bias*—that is, what we learn from their study may not accurately reflect the complete populations of the objects they represent. How might a similar bias play a role in more down-to-earth situations? For example, do you and your classmates form a representative sample of your city, nation, or planet? How might this affect the accuracy of a report on world events you might write together or the fairness of a decision you might vote on? In what ways can the effects of bias be removed?

3. *Rise of the Mammals.* Suppose the impact 65 million years ago had not occurred. How do you think our planet would be different? For example, do you think that mammals would still have eventually come to dominate the Earth? Would we be here? Defend your opinions.

Problems

Surprising Discoveries? For **problems 1–8**, suppose we made the discoveries described. (These are *not* real discoveries.) Decide whether the discovery should be considered reasonable or surprising. Explain.

1. A small asteroid that orbits within the asteroid belt has an active volcano.

2. Scientists discover a meteorite that, based on radioactive dating, is 7.9 billion years old.

3. An object that resembles a comet in size and composition is discovered to be orbiting in the inner solar system.

4. Studies of a large object in the Kuiper belt reveal that it is made almost entirely of rocky (as opposed to icy) material.

5. Astronomers discover a previously unknown comet that will produce a spectacular display in Earth's skies about 2 years from now.

6. A mission to Pluto finds that it has lakes of liquid water on its surface.

7. Geologists discover a crater from a 5-km object that impacted the Earth more than 100 million years ago.

8. Archaeologists learn that the fall of ancient Rome was caused in large part by an asteroid impact in southern Africa.

9. *Orbital Resonances.* How would our solar system be different if orbital resonances had never been important (for example, if Jupiter and asteroids continued to orbit the Sun but did not affect each other through resonances)? Describe at least five ways in which our solar system would be different. Which differences do you consider superficial, and which differences more profound? Consider the effects of orbital resonances discussed both in this chapter and in Chapter 11.

10. *Life Story of an Iron Atom.* Imagine that you are an iron atom in an iron meteorite recently fallen to Earth. Tell the story of how you got here, beginning from the time when you were in a gaseous state in the solar nebula 4.6 billion years ago. Include as much detail as possible. Your story should be scientifically accurate but also creative and interesting.

11. *Asteroid Discovery.* You have discovered two new asteroids and have named them Barkley and Jordan. Both lie at the same distance from the Earth and have the same brightness when you look at them through your telescope. But Barkley is twice as bright as Jordan at infrared wavelengths. What can you deduce about the two asteroids' relative reflectivities and sizes? Which would make a better target for a mission to mine metal? Which would make a better target for a mission to obtain a sample of a carbon-rich planetesimal? Explain your answers.

*12. *The "Near Miss" of Toutatis.* The 5-km asteroid Toutatis passed a mere 3 million km from the Earth in 1992.

   a. In one or two paragraphs, describe what would have happened to the Earth if Toutatis had hit it.

   b. Suppose Toutatis was destined to pass *somewhere* within 3 million km of Earth. Calculate the probability that this "somewhere" would have meant that it slammed into Earth. Based on your result, do you think it is fair to say that the 1992 passage was a "near miss"? Explain. (*Hint:* You can calculate the probability by considering an imaginary dartboard of radius 3 million km on which the bull's-eye has the Earth's radius of 6,378 km.)

*13. *Comet Temperatures.* Compare the strength of sunlight at 50,000 AU in the Oort cloud with its strength at 3 AU and 1 AU. Compare the "no greenhouse" temperatures for comets (see Mathematical Insight 10.1) at the three positions. At which location will the temperature be high enough for water ice to sublime (about 150 K)? Assume that the comets rotate rapidly and have an albedo of 3%.

14. *Project: Tracking a Meteor Shower.* Armed with an expendable star chart and a flashlight covered in red plastic, set yourself up comfortably to watch a meteor shower listed in Table 12.1. Each time you see a meteor, record its path on your star chart. Record at least a dozen, and try to determine the *radiant* of the shower—the point in the sky from which the meteors appear to radiate. Also record the direction of the Earth's motion, which you can find by determining which sign of the zodiac rises at midnight.

15. *Project: Dirty Snowballs.* If there is snow where you live or study, make a filthy snowball. (The ice chunks that form behind tires work well.) How much dirt does it take to darken snow? Find out by allowing your dirty snowball to melt and measuring the approximate proportions of water and dirt afterward.

Web Projects

Find useful links for Web projects on the text Web site.

1. *The NEAR Mission.* Learn about the NEAR mission to the asteroid Eros. How did the mission reach Eros? How did it end? What did it accomplish? Write a one- to two-page summary of your findings.

2. *Stardust.* Learn about NASA's Stardust mission (launched in 1999) to bring back cometary material to Earth. What is the current status of the spacecraft? What do we hope to learn after it returns? Write a report about the mission status and its science.

3. *Asteroid and Comet Missions.* Learn about another proposed space mission to study asteroids or comets. Write a short report about the mission plans, goals, and prospects for success.

4. *Pluto Express.* NASA has gone back and forth for years about a mission to Pluto, the only planet not yet visited by spacecraft. Find the latest on plans for a mission to Pluto. Does the mission seem likely to take place anytime soon? Write a short report about the current status of a Pluto mission. Discuss the scientific goals of such a mission and the political realities that may dictate whether it occurs.

Looking outward to the blackness of space, sprinkled with the glory of a universe of lights, I saw majesty—but no welcome. Below was a welcoming planet. There, contained in the thin, moving, incredibly fragile shell of the biosphere is everything that is dear to you, all the human drama and comedy. That's where life is; that's where all the good stuff is.

LOREN ACTON, U.S. ASTRONAUT

CHAPTER 13
Planet Earth
And Its Lessons on Life in the Universe

Perhaps you've heard the Earth described as the "third rock from the Sun." At first glance, the perspective of comparative planetology may seem to support this dispassionate view of the Earth. However, a deeper comparison between Earth and its neighbors reveals a far more remarkable planet.

Earth is unique in the solar system in many ways. It is the only planet with surface oceans of liquid water and substantial amounts of oxygen in its atmosphere, and its geological activity is more diverse than that of any other terrestrial world. Most important, Earth is the only world known to harbor life, which has played a considerable role in shaping the Earth's surface and atmosphere.

So far, we have found that our theories of solar system formation and planetary evolution explain most of the general features of our solar system. In this chapter, we will apply and extend what we've learned in order to form a coherent picture of the combined geological, atmospheric, and biological evolution of our planet. We'll also discuss a few important lessons that the solar system teaches us about our future on Earth and that the Earth teaches us about the possibility of life elsewhere in the universe.

FIGURE 13.1 Dawn over the Atlantic Ocean, photographed from the Space Shuttle.

13.1 How Is Earth Different?

Take a look at Figure 13.1. The abundant white clouds overlying the expanse of oceans show a planet unlike any other in our solar system. It is, of course, our own planet Earth.

In Chapters 9 and 10, we discussed many features of the Earth's geology and atmosphere. We found many similarities between the Earth and other terrestrial worlds, and also some important differences. For example, the Earth's surface is shaped by the same four geological processes (impact cratering, volcanism, tectonics, and erosion) that shape other worlds, but Earth is the only planet on which the lithosphere is clearly broken into plates that move around in what we call plate tectonics. The Earth's atmosphere shows even more substantial differences from the atmospheres of its neighbors: It is the only planet with significant atmospheric oxygen and the only terrestrial world with an ultraviolet-absorbing stratosphere.

But the greatest differences between Earth and other worlds lie in two features totally unique to Earth. First, the surface of the Earth is covered by huge amounts of water. Oceans cover nearly three-fourths of the Earth's surface, with an average depth of about 3 kilometers (1.8 miles). Water is also significant on land, where it flows through streams and rivers, fills lakes and underground water tables, and sometimes lies frozen in glaciers. Frozen water covers nearly the entire continent of Antarctica in the form of the southern ice cap. The northern ice cap sits atop the Arctic Ocean and covers the large island of Greenland. Water plays such an important role on Earth that some scientists treat it as a distinct planetary layer, called the **hydrosphere**, between the lithosphere and the atmosphere.

The second totally unique feature of Earth is its diversity of life. We find life nearly everywhere on Earth's surface, throughout the oceans, and even underground. The layer of life on Earth is sometimes called the **biosphere**. As we will see shortly, the biosphere helps shape many of the Earth's physical characteristics. For example, the biosphere explains the presence of oxygen in Earth's atmosphere. Without life, Earth's atmosphere would be very different.

Comparative planetology compels us to explore these unique properties of Earth closely. If we find that the formation and evolution of our Earth violated the rules we've derived for other planets—if Earth is "irregular"—life might be unique to Earth. If Earth does follow the rules—if Earth is "regular" or even "standard"—then life may be common in the universe.

13.2 Our Unique Geology

It's not surprising that Earth shows much more geological activity than Mercury or the Moon, since those small worlds have cooled since their formation and are now geologically "dead." But why have Earth and Venus ended up so different, despite very similar sizes? And, although Earth is significantly larger than Mars, both planets probably once had similar surface conditions that allowed liquid water to flow. Why did Earth remain hospitable, while Mars is now dry and barren? These mysteries warrant a closer look.

Figure 13.2 shows shaded relief maps comparing the surfaces of Earth, Venus, and Mars. While the three planets have many superficial similarities, they also have important differences. We find fewer impact craters on Venus than on Mars, and fewer still on Earth. Most of Earth's volcanic mountains are

Venus

Earth

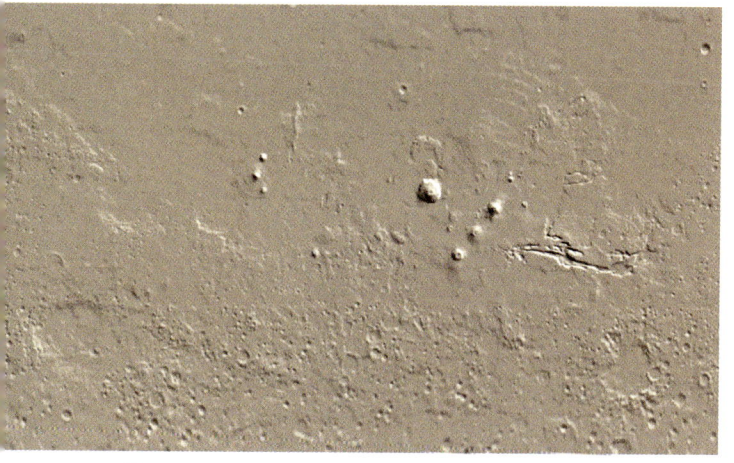

Mars

FIGURE 13.2 These shaded relief maps compare the surfaces of Venus, Earth, and Mars; the map for Earth shows the seafloor as it would appear if there were no oceans. Despite superficial similarities, there are important differences. For example, Mars has the most craters, and Earth has the fewest. Also note the unique distinction between the two kinds of surface (seafloor and continent) on Earth.

seafloor crust (5–10 km thick, dense, young)

upper mantle

continental crust (20–70 km thick, lower-density, old)

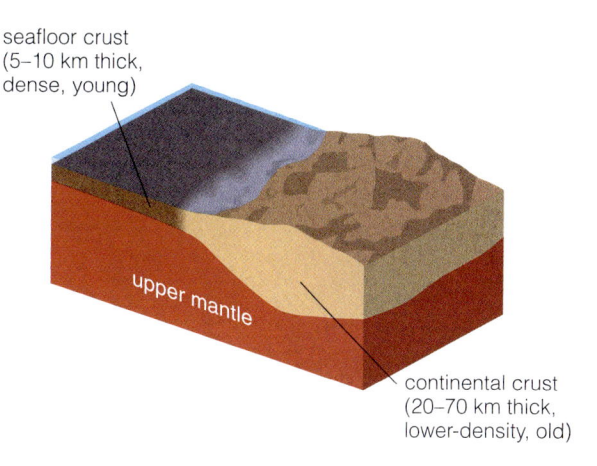

FIGURE 13.3 Earth has two distinct kinds of crust.

steep-sided and therefore must have been made from a much more viscous lava than that found on the other planets. And Earth shows far more evidence of tectonic activity—surface reshaping driven by internal stresses. To our current knowledge, Earth is the only terrestrial planet on which the lithosphere is split into distinct *plates*. Earth also has a crust that comes in two very different varieties: a relatively thick, low-density crust underlying the *continents*, and a thinner, denser crust underlying the *seafloors* (Figure 13.3). Seafloor crust is made of basalt and is typically only 5–10 km thick, whereas continental crust is made of lower-density rock (such as granite) and is 20–70 km thick. To understand these and other unique features of Earth's geology, we must look in more detail at how the four geological processes affect our planet.

Before we discuss the geological processes themselves, remember that they depend in many ways on a planet's interior structure. We know much more about Earth's interior structure than we do about that of any other planet, because geologists have created a three-dimensional "picture" of the interior by analyzing the propagation of seismic waves from earthquakes (Figure 13.4). The Earth's thin crust and the uppermost portion of the mantle make up a relatively cool and rigid *lithosphere* about 100 km thick [Section 9.2]. Below the lithosphere, the mantle is warm and partially molten in places; it supplies the magma for volcanic eruptions. Deeper in the mantle, higher pressures force the rock into the solid phase despite even higher temperatures. But even in solid form the mantle slowly flows, and convection continually carries heat up from below. The Earth's metallic core underlies the lower mantle. The outer region of the core is molten, but high pressures in the inner core force the metal into solid form. Thus, the Earth has all the

CHAPTER 13 PLANET EARTH AND ITS LESSONS ON LIFE IN THE UNIVERSE 353

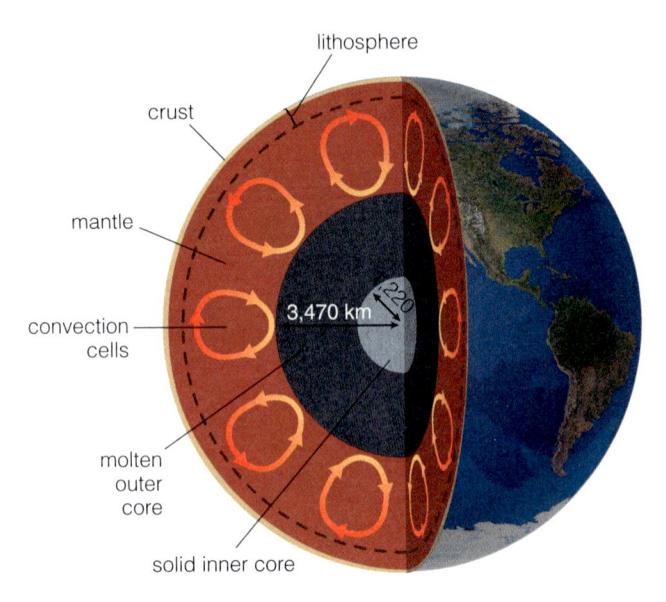

FIGURE 13.4 Internal structure of the Earth, based on the study of seismic waves (Figure 9.6).

ingredients needed for exciting geology: a hot interior to supply energy for volcanism, convection to help generate tectonic stresses, and a relatively thin lithosphere that does not inhibit geological activity.

The Four Geological Processes on Earth

The first of the four geological processes, *impact cratering*, should have occurred at roughly similar rates on all the terrestrial worlds. All were subject to the intense early bombardment during the solar system's first few hundred million years. Four billion years ago, the Earth's surface may have been as saturated with craters as the densely cratered lunar highlands are today [Section 9.5]. We find no 4-billion-year-old craters on Earth now, so these ancient impacts must have been erased by other geological processes. The impact rate fell substantially after the early bombardment, but occasional impacts still occur—sometimes with devastating consequences [Section 12.6]. The effects of erosion make impact craters surprisingly difficult to identify on Earth. Nevertheless, remnants of more than 100 impact craters have been found (see Figure 12.8). Large impacts should have occurred more or less uniformly over the Earth's surface, including both continents and seafloors, because large impactors are not stopped by the oceans. The relative absence of craters on the seafloor suggests that seafloor crust is much younger than is continental crust; it must have formed later than most of the impacts.

The rampant *erosion* on Earth arises primarily from processes involving water, but atmospheric winds also contribute. Wind tears away at geological features created by other processes, blows dust and sand across the globe, and builds features such as sand dunes. Note that wind erosion is more effective on Earth than on either Venus or Mars: Venus has very little surface wind because of its slow rotation, and wind on Mars does little damage because of the low atmospheric pressure. On Earth, water contributes to erosion through a remarkable variety of processes (Figure 13.5). Perhaps the most obvious are rain and rivers breaking down mountains, carving canyons, and transporting sand and silt across continents to deposit them in the sea. But water erosion occurs on microscopic levels too: Water seeps into cracks and crevices in rocks and breaks them down from the inside, especially when this water freezes and expands. Further examples of water-erosion processes include the slow movement of glaciers, the flow of underground rivers, and the pounding of the ocean surf.

Volcanism and *tectonics* are also very important processes on Earth. The many active volcanoes prove that volcanism still reshapes the surface, and many small-scale cliffs and valleys offer evidence of tectonics similar to that found on other terrestrial worlds. But plate tectonics may be unique to Earth. To gain a deeper understanding of our planet's closely linked volcanism and tectonics, we must study plate tectonics in more detail.

Plate Tectonics

If you cut up a map and rearrange the continents, you'll find that they fit together in surprising ways. For example, the east coast of South America fits quite nicely into the west coast of Africa (Figure 13.6). You might chalk this up to coincidence, but we also find similar types of distinct rocks and rare fossils in eastern South America and western Africa—suggesting that these two regions were once near each other. Based on such evidence, in 1912 German meteorologist and geologist Alfred Wegener proposed the idea of *continental drift*: that the continents gradually drift across the surface of the Earth, over time scales of tens of millions of years.

For decades after Wegener made his proposal, most geologists rejected the idea that continents could move, and many ridiculed the idea openly. However, the idea of continental drift gained favor in the 1950s as supporting evidence began to accumulate. Mapping of the seafloors revealed surprising structures (Figure 13.7): high *mid-ocean ridges* extending a total length of about 60,000 km along the ocean floors, and *trenches* in which the ocean depth can reach more than 8 kilometers. Geologists soon recognized that these features represented boundaries between **plates**—pieces of the lithosphere that apparently float upon the denser mantle below. Convection cells in the upper mantle move the plates around the surface, forcing them together, apart, or sideways at

a The Colorado River continues to carve the Grand Canyon after millions of years.

b Glaciers created Yosemite Valley during the ice ages.

d Erosional debris also creates geological features, as seen in this river delta.

c Surf pounds away at the California coastline.

FIGURE 13.5 A few examples of water erosion.

FIGURE 13.6 The "fit" of South America into Africa.

355

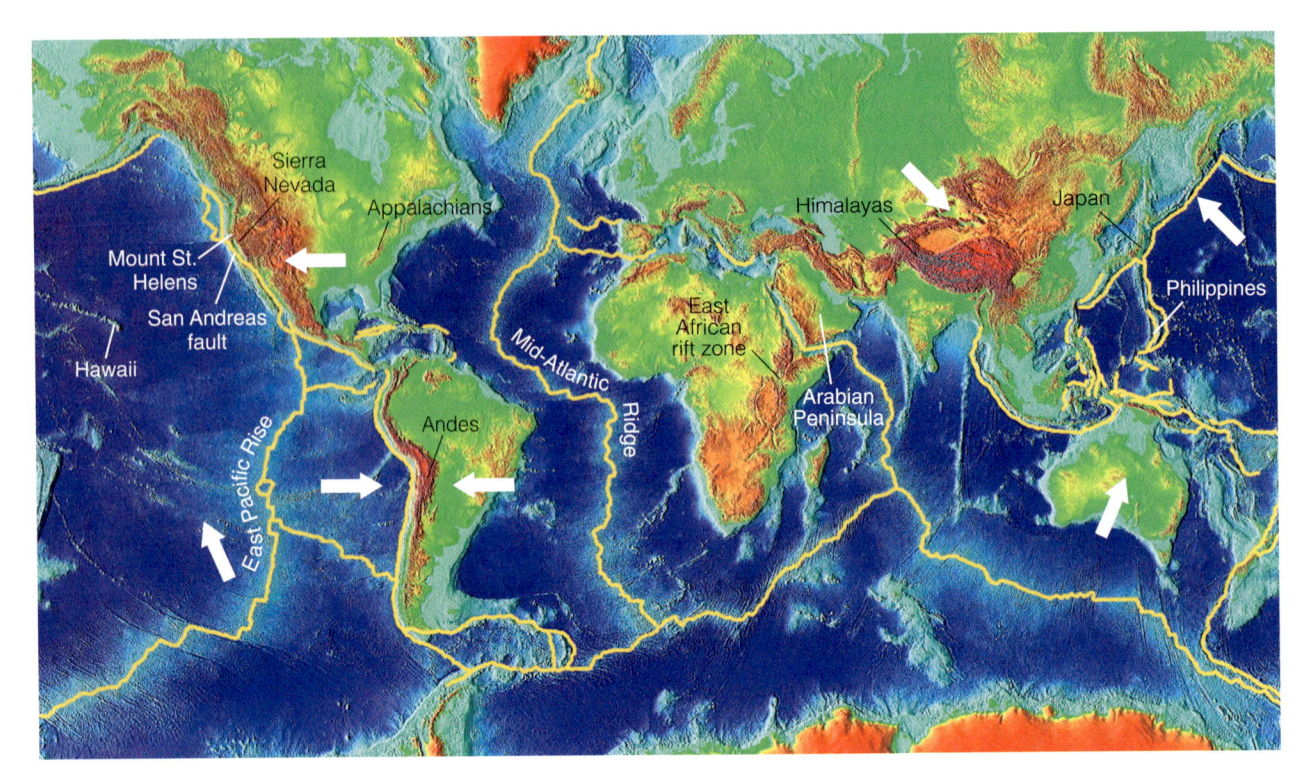

FIGURE 13.7 The discovery of mid-ocean ridges and deep trenches provided important evidence for continental drift. The relief map uses color to show elevation, progressing from blue (lowest) to red (highest). Solid lines show plate boundaries, and arrows represent directions in which the plates are moving. Important geological features discussed later in the text are identified.

their boundaries—with profound geological consequences. This discovery offered a model to explain how continents could drift about, and Wegener's idea finally gained acceptance under the new name **plate tectonics**. (Recall that tectonics is geological activity driven by internal stresses, so *plate tectonics* refers to the motion of plates driven by internal stresses.) Today we know that the Earth's lithosphere is broken into more than a dozen plates (Figure 13.7). Most major earthquakes and volcanic eruptions occur along plate boundaries.

TIME OUT TO THINK *Study the plate boundaries in Figure 13.7. Based on this diagram, explain why the West Coast states of California, Oregon, and Washington are more prone to earthquakes and volcanoes than other parts of the United States.*

Over millions of years, plate tectonics acts like a giant conveyor belt for the Earth's crust, carrying rock up from the mantle, transporting it across the seafloor, and then returning it down into the mantle (Figure 13.8). The mid-ocean ridges mark **spreading centers** between plates—places where hot mantle material rises upward and then spreads sideways, pushing the plates apart. New seafloor crust forms as the partially molten mantle material emerges through a long string of underwater volcanoes, forming crust made of basalt [Section 9.4]. The newly formed basaltic crust cools and contracts as it spreads sideways from the central ridge, giving mid-ocean ridges their characteristic shape (see Figure 13.7). The new crust spreads throughout the area that opens as plates separate, forming only a relatively thin layer of crust over the mantle (see Figure 13.3). Worldwide along the mid-ocean ridges, new crust covers an area of about 2 square kilometers every year—enough to replace the entire seafloor within a geologically short time of about 200 million years. Meanwhile, as new crust spreads over the surface, older crust must be returned to the mantle. This return occurs at the locations of the deep ocean trenches, which mark places where one plate slides under another in a process called **subduction**. As we would expect from this model of seafloor crust formation, samples of seafloor crust are composed of basaltic material, and radioactive dating [Section 8.6] shows their age to be generally less than 200 million years.

The conveyor-like process of plate tectonics is undoubtedly driven by convection, although the precise nature of the convection remains uncertain. The crust may be simply dragged along by convection in the upper mantle. Alternatively, the tops of the convection cells may actually be the crust itself, with warm rock rising up under the ocean ridges and cooler,

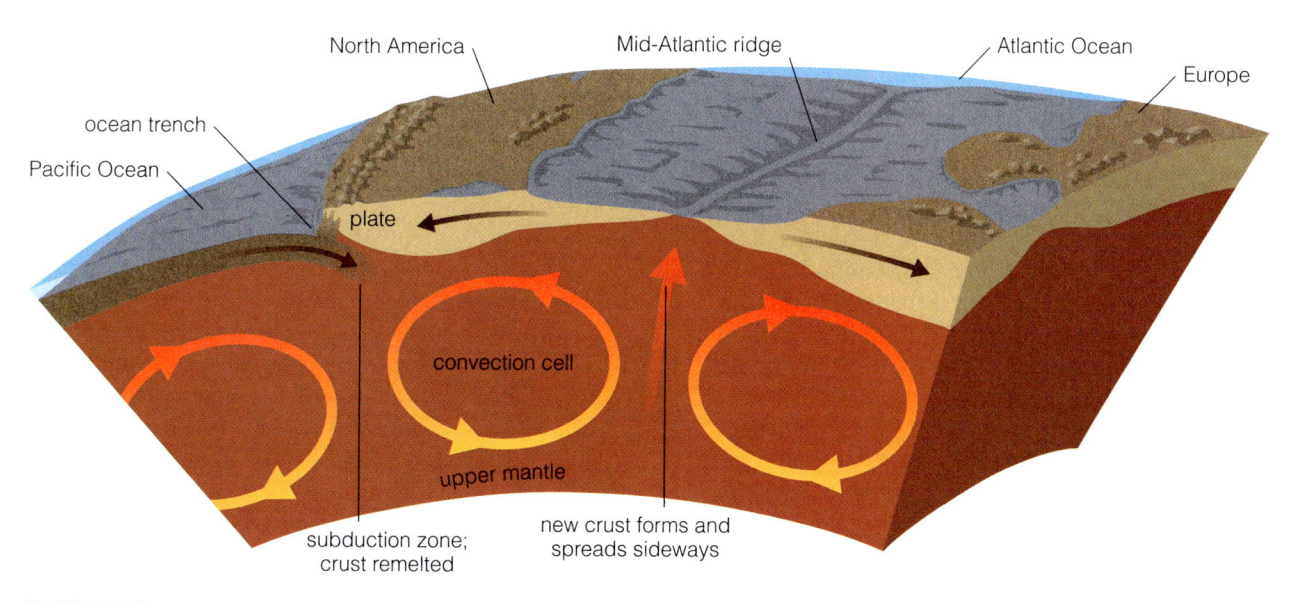

FIGURE 13.8 Plate tectonics acts like a giant conveyor belt, driven by mantle convection, that carries rock up from the mantle at mid-ocean ridges, transports it across the seafloor, and returns it down into the mantle at ocean trenches.

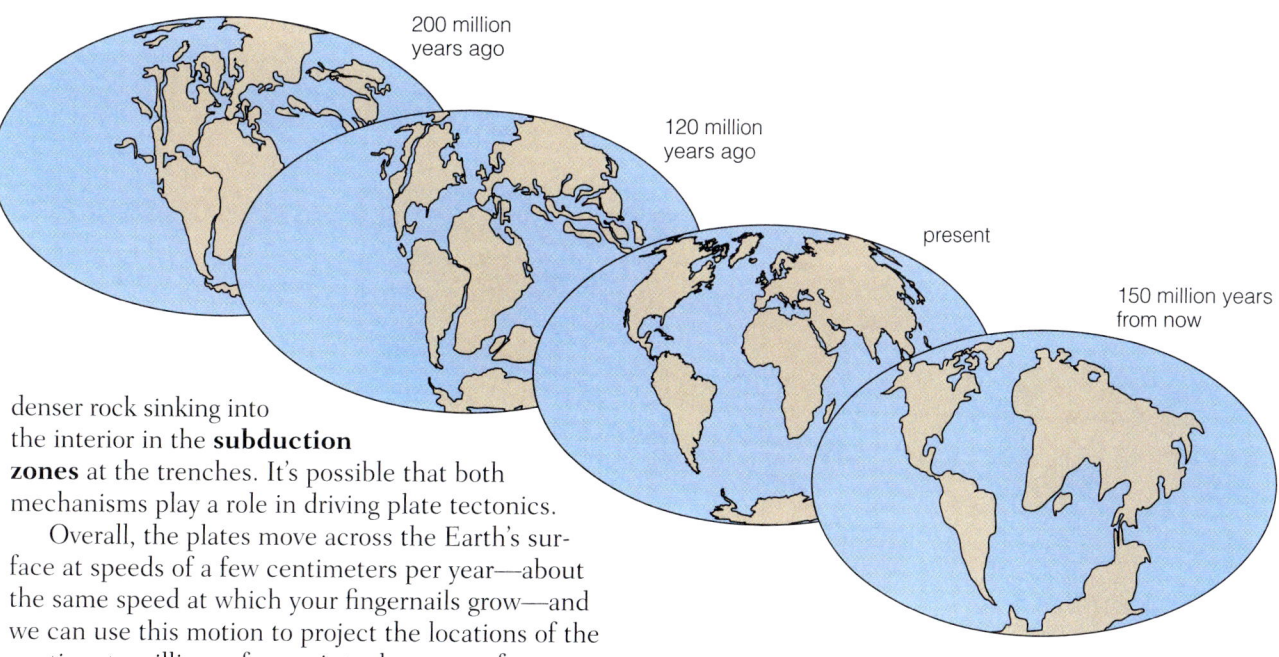

denser rock sinking into the interior in the **subduction zones** at the trenches. It's possible that both mechanisms play a role in driving plate tectonics.

Overall, the plates move across the Earth's surface at speeds of a few centimeters per year—about the same speed at which your fingernails grow—and we can use this motion to project the locations of the continents millions of years into the past or future. For example, at a speed of 2 centimeters per year, a plate will travel 2,000 kilometers in 100 million years. Over longer time scales, the plates may have altered their speeds and directions, but we can still project past locations of continents through careful comparisons of rock and fossil samples found in different places. For example, if 500-million-year-old rocks and fossils are very similar in two places, the two places were probably close together at that time, even if they are far apart now. Figure 13.9 shows several past arrangements of the continents, along with one future arrangement. Over the past billion years, the continents have slammed together, pulled apart, spun around, and changed places on the globe. Central

The Mediterranean Sea will become mountains, Australia will merge with Antarctica, and California will slide northward to Alaska.

FIGURE 13.9 Past, present, and predicted future arrangements of the Earth's continents.

Africa once lay at Earth's South Pole, and Antarctica was once nearer the equator. At various times, such as was the case about 200 million years ago, the continents were all merged into a single giant continent, often called *Pangaea* (which means "all lands"). The current arrangement of the continents is no more permanent than any other.

CHAPTER 13 PLANET EARTH AND ITS LESSONS ON LIFE IN THE UNIVERSE 357

Subduction and the Origin of Continents Where plates collide, subduction of seafloor crust creates continental crust. The seafloor crust remelts as subduction carries it back into the mantle, and the resulting magma separates by density. The lowest-density magma—which consists of rocks such as rhyolite, andesite, and granite—pushes upward to form continental crust. Thus, continental crust is made of lower-density material than seafloor crust. Plate tectonics has gradually built up thick continental crust over about 45% of the Earth's surface, and the fraction is still increasing. (Some of this continental crust is underwater—continental shelves—so land areas represent *less* than 45% of the Earth's surface.) Billions of years ago, Earth was even more of an ocean planet than it is today.

Despite being tens of kilometers thicker than seafloor crust, on average the continental crust rises only a few kilometers higher in altitude. The sheer weight of the continents presses down on the mantle, so most of their extra thickness goes deeper into the Earth. Thus, the continents poke up only a few kilometers higher than the seafloors (see Figure 13.3).

The present-day continents have been built up over billions of years; radioactive dating shows some rocks in continental crust to be as much as 4 billion years old.

The rising magma along plate boundaries creates tremendous volcanic activity. Where one seafloor plate subducts under another, the resulting eruptions form a long string of volcanic islands. Alaska's Aleutian Islands exemplify this process at an early stage. As the process continues, islands can grow and merge; this occurred with both Japan and the Philippines, each of which once contained many small islands that merged into the fewer islands we see today. As these islands continue to grow and merge, they may eventually create a new continent. In other cases, a plate carrying islands may run up against an existing continental plate. The dense seafloor crust surrounding the islands subducts below the continents, but the islands resist subduction because they are made of low-density rock. The islands are essentially scraped off the seafloor and stuck onto the continent. Alaska, British Columbia, Washington, Oregon, and most of California began their existence as numerous Pacific islands that later attached to the growing North American continent (Figure 13.10). Seafloor continues to subduct below these states and provinces.

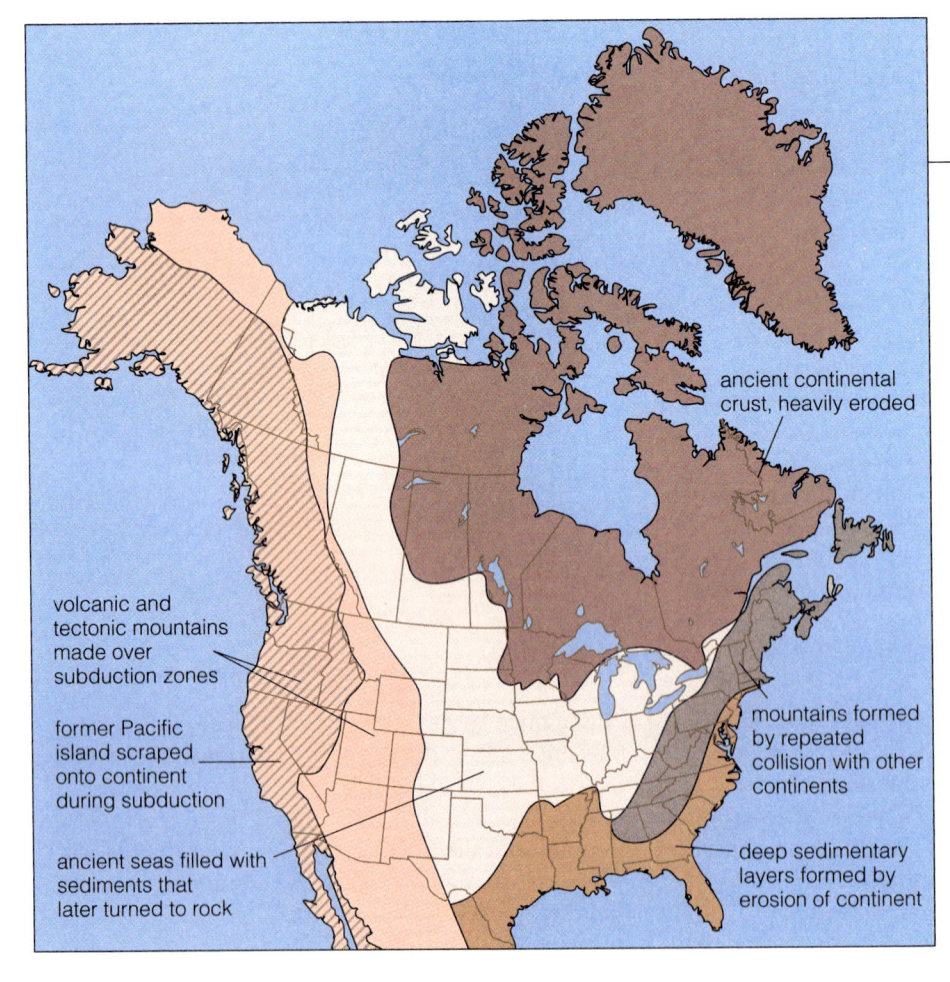

FIGURE 13.10 The major geological features of North America record the complex history of plate tectonics. Only the basic processes behind the largest features are shown here.

a Tectonic forces have torn the Arabian peninsula from Africa. (Photo from Space Shuttle.)

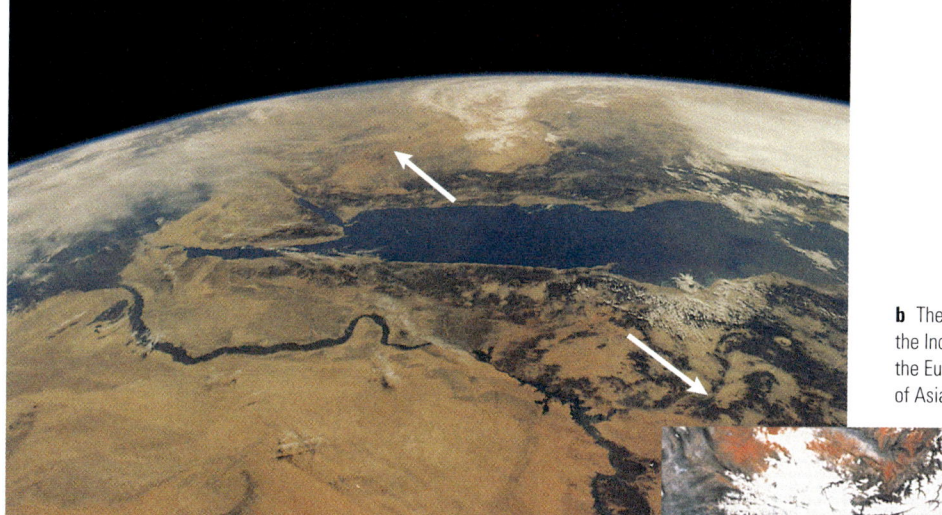

b The Himalayas are still slowly growing as the Indian plate (carrying India) pushes into the Eurasian plate (carrying most of the rest of Asia). (Satellite photo.)

FIGURE 13.11 Tectonic forces on continent.

The low-density magmas generated by subduction have relatively high viscosity, which affects the kind of volcanoes they construct. The rhyolite and andesite lavas that emerge from volcanoes along plate boundaries create steep-sided *stratovolcanoes* [Section 9.4] like Mount St. Helens in Washington and the mountains of the Andes range in South America. Granite magma usually does not erupt onto the surface but instead builds mountains through vast underground intrusions that make the rock layers above them bulge up. The upper rock layers may gradually erode away, leaving granite mountain ranges such as the Sierra Nevada in California.

Tectonic Stresses on Continents Continental crust is relatively thick, strong, and low-density, so it resists both spreading and subduction. Instead of simply separating like seafloor crust on either side of a mid-ocean ridge, continental crust thins and creates a *rift valley* when mantle convection tugs it apart. One such rift valley runs through Africa; it will eventually tear the continent apart and create a seafloor spreading center. A similar process tore the Arabian peninsula from Africa in the past, creating the Red Sea (Figure 13.11a).

When two continent-bearing plates collide, neither plate can subduct (because of the low density of the crust), and the resulting pressure creates tremendous mountain ranges. The Indian Plate is currently ramming into the Eurasian Plate, forming the Himalayas (Figure 13.11b). These mountains, already the tallest on Earth, are still growing. The Appalachian range in the eastern United States formed by the same process as North America collided twice with South America and then with western Africa over a period of hundreds of millions of years. The Appalachians were probably once as tall as the Himalayas are now, but erosion has gradually transformed them into the fairly modest mountain range we see today. Similar processes contributed to forming the Rocky Mountains in the United States and Canada.

At other places on Earth, plates slip sideways relative to each other along a **fault**, or a fracture in

a Along California's San Andreas fault (and many other places in the world), plates are sliding sideways past each other. Asterisks indicate recent earthquakes.

b The photo shows a place where a road is no longer straight because of this movement.

c Earthquakes occur when plates slip violently along a fault. This photo shows damage from the 1995 earthquake in Kobe, Japan.

FIGURE 13.12 Plate tectonics on a human level.

---

the lithosphere. This is the case with the *San Andreas fault* in California, where plate motions are carrying Los Angeles on a 20-million-year trip to San Francisco (Figure 13.12a). Plates do not slip smoothly against each other. Instead, their rough surfaces resist slippage until the tension grows too high; then everything slips along the fault in the violent motion that we call an *earthquake* (Figure 13.12b,c). The motions associated with earthquakes can raise mountains, level cities, and set the whole planet vibrating with seismic waves. In contrast to the usual motion of plates at the rate of a few centimeters per year, an earthquake can move plates by several *meters* in a few seconds. Although most earthquake faults lie along plate boundaries, a few more ancient faults are found elsewhere. As a result, devastating earthquakes occasionally occur in regions that we tend to think of as at low risk from seismic events. In fact, one of the largest earthquakes in U.S. history occurred in Missouri in 1811.

Hot Spots and Mantle Plumes Not all volcanoes occur near plate boundaries. Sometimes, a plume of hot mantle material may rise in what we call a **hot spot** within a plate. The Hawaiian Islands are the

FIGURE 13.13 The Hawaiian Islands are just the most recent of a very long string of volcanic islands made by a mantle hot spot. The black-and-white image of Loihi (lower right) was obtained by sonar, as it is still entirely under water.

result of a hot spot that has been erupting basaltic lava for tens of millions of years (Figure 13.13). The low viscosity of this lava produces broad *shield volcanoes* [Section 9.4]. Today, most of the lava erupts on the "Big Island" of Hawaii, giving much of this island a young, rocky surface. But about a million years ago, the mantle plume lay farther to the northwest—or, equivalently, the entire plate lay farther to the southeast. Back then, the eruptions built the island of Maui, parts of which are now heavily eroded and covered in lush vegetation. Before that, the mantle hot spot built the other islands of Hawaii, including Oahu (3 million years ago) and Kauai (5 million years ago); it also created Midway Island (27 million years ago), which has now almost completely eroded back to sea level. Thus, this single hot spot created a vast string of islands (and smaller underwater mountains) as the plate gradually moved over it. The process continues today: A new island named Loihi is currently forming beneath the ocean surface and will rise above sea level southeast of Hawaii in a million years or so. Hot spots can also appear beneath continental crust. The geysers and hot springs of Yellow-

CHAPTER 13 PLANET EARTH AND ITS LESSONS ON LIFE IN THE UNIVERSE 361

stone National Park result from the heating of a mantle plume. This hot spot is currently migrating to the northeast (relative to the continental crust) because of the southwestern motion of the plate.

Summary of Earth's Geology

Two fundamental differences between the geology of Earth and that of other terrestrial worlds stand out:

1. Rampant erosion driven by the action of plentiful water—a direct result of Earth's atmospheric evolution, as we will see.

2. A different style of tectonics—plate tectonics—that recycles crust, drives high-viscosity volcanism to create continental crust, and causes frequent earthquakes.

Together, these two processes have reshaped the Earth's surface beyond recognition from its original state billions of years ago. They make the Earth's surface very young, perhaps the youngest in the solar system besides that of Io [Section 11.5].

But why does Earth's geology differ from that of our neighbors in these two ways? In the next section, we'll see how our unique atmosphere leads to the strong erosion on Earth. We'll also find that the interactions between our atmosphere and geology might help explain why Earth has plate tectonics while similar-size Venus does not.

13.3 Our Unique Atmosphere

In Chapter 10, we learned that the superficial differences in the atmospheres of the terrestrial planets are fairly easy to understand. For example, the Moon and Mercury lack substantial atmospheres because of their small sizes, and a very strong greenhouse effect causes the high surface temperature on Venus. We also learned how solar heating, seasonal changes, and the Coriolis effect that comes from rotation lead to differing global circulation patterns on different worlds. But explaining how the compositions of the atmospheres came to be so different is more challenging, especially when we consider the cases of Venus, Earth, and Mars. Outgassing produced all three of these atmospheres, and the same volatiles—primarily water and carbon dioxide—were outgassed in all three cases. How, then, did Earth's atmosphere end up so different? In particular, any study of Earth's atmosphere must address four major questions:

1. Why did Earth retain most of its water—in the form of the oceans and other components of the hydrosphere—while Venus and Mars lost theirs?

2. Why does Earth have so little carbon dioxide (CO_2) in its atmosphere, when Earth should have outgassed about as much of it as Venus?

3. Why does Earth have so much more oxygen (O_2) than Venus or Mars? Where did it come from, and how does this highly reactive gas remain present in the atmosphere?

4. Why does Earth have an ultraviolet-absorbing stratosphere, while Venus and Mars do not?

To answer these questions, we must look into the history of Earth's atmosphere. The answers to all four questions turn out to be closely connected.

TIME OUT TO THINK *Recall that water ice could not condense in the region of the solar nebula where the terrestrial planets formed. How, then, did the water outgassed from volcanoes get here in the first place? (Hint: See Section 8.5.)*

Water and the Origin of the Hydrosphere

On a basic level, it's fairly easy to explain why the Earth retains so much water while Venus and Mars do not. As we learned in Chapter 10, Venus probably lost an "ocean-full" of water as solar ultraviolet photons split apart water molecules and the hydrogen atoms subsequently escaped to space. Mars probably lost some of its water in a similar way, and the rest is frozen at the polar caps and under the surface.

Let's look a little more closely at the sequence of events that led to these different histories. Picture a generic terrestrial planet, billions of years ago, with erupting volcanoes outgassing carbon dioxide, water vapor, and small amounts of nitrogen. As the outgassing continued, the water vapor might have either condensed into oceans, frozen out as snow, or remained gaseous in the atmosphere. In the case of Venus, proximity to the Sun made the planet warm enough to keep all its water gaseous in the atmosphere, even before it had a strong greenhouse effect. Because water vapor itself is a greenhouse gas, this caused significant greenhouse warming and ensured that additional water vapor would also remain gaseous and warm the planet further. This tendency of the greenhouse effect to reinforce itself is called the **runaway greenhouse effect**. At the other extreme, Mars's temperature was eventually low enough for the water vapor to freeze out of the atmosphere, resulting in thick polar caps.

On Earth, moderate temperatures allowed most of the water vapor to condense as rain, leading to the accumulation of liquid water in the oceans. A small amount of water vapor remained in the atmosphere and caused greenhouse warming, but not enough to create a runaway greenhouse effect. Additional water vapor released from volcanoes therefore also condensed, gradually forming the oceans and the other components of the hydrosphere. Thus, Venus lost its oceans because it was too hot, Mars lost its oceans because it was too cold, and Earth retained its oceans because conditions were "just right."

Where Is All the CO_2?

Measurements of outgassing by active volcanoes and estimates of past volcanic activity suggest that Earth must have outgassed nearly as much carbon dioxide throughout its history as Venus. But Venus today has an atmosphere with a CO_2 concentration of 96%, a surface pressure 90 times greater than that on Earth, and about 500°C of greenhouse warming. In contrast, the proportion of carbon dioxide in Earth's atmosphere is much less than 1%; in fact, we usually measure CO_2 concentration on Earth in units of *parts per million*. (For example, 300 parts per million means 300 ÷ 1,000,000 = 0.0003 = 0.03%.) Clearly, some atmospheric loss process must have removed nearly all of the outgassed CO_2 on Earth.

The secret of Earth's CO_2 is wrapped up in the history of our oceans. The primary process that removes atmospheric CO_2 on Earth involves liquid water. Carbon dioxide can dissolve in water, and the oceans actually contain about 60 times more carbon dioxide than the atmosphere—but this is still a very small amount compared to the total amount that must have been outgassed. Most of the carbon dioxide is locked up in rocks on the seafloor through chemical reactions. Rainfall erodes silicate rocks on the Earth's surface and carries the eroded minerals to the oceans. There the minerals react with dissolved carbon dioxide to form *carbonate* minerals [Section 10.5], which fall to the ocean floor, building up thick layers of carbonate rock such as *limestone*. This is part of the **carbonate–silicate cycle** (Figure 13.14). Because the process requires liquid water, it operates only on Earth, where it has removed about as much CO_2 from Earth's atmosphere as now remains in Venus's atmosphere. (If Venus's CO_2 could somehow be removed, the remaining atmosphere would have about as much nitrogen as Earth's.) Had Earth been slightly closer to the Sun, its warmer temperature might have evaporated the oceans, leaving the CO_2 in the atmosphere and causing a runaway greenhouse effect as on Venus. Conditions might not have been so conducive to "life as we know it."

The interplay of atmosphere and geology may also explain why Earth has plate tectonics while Venus does not. The apparent lack of plate tectonics on Venus suggests that its lithosphere is stronger and thicker than Earth's. How can this be, given that both planets probably have similar internal temperatures? The answer may lie in the history of Venus's atmosphere. As Venus lost its oceans, even the small amounts of water dissolved in the crust and mantle were baked out, removing the lubricating and softening effects of volatiles trapped in rocks. So Venus's atmospheric evolution may have thickened the lithosphere to the point where Earth-like plate tectonics cannot occur. This theory is still controversial (some planetary geologists think that Venus's lithosphere is actually thinner than Earth's), but it underscores the surprising possibility that a planet's atmosphere can affect its interior, not just the other way around.

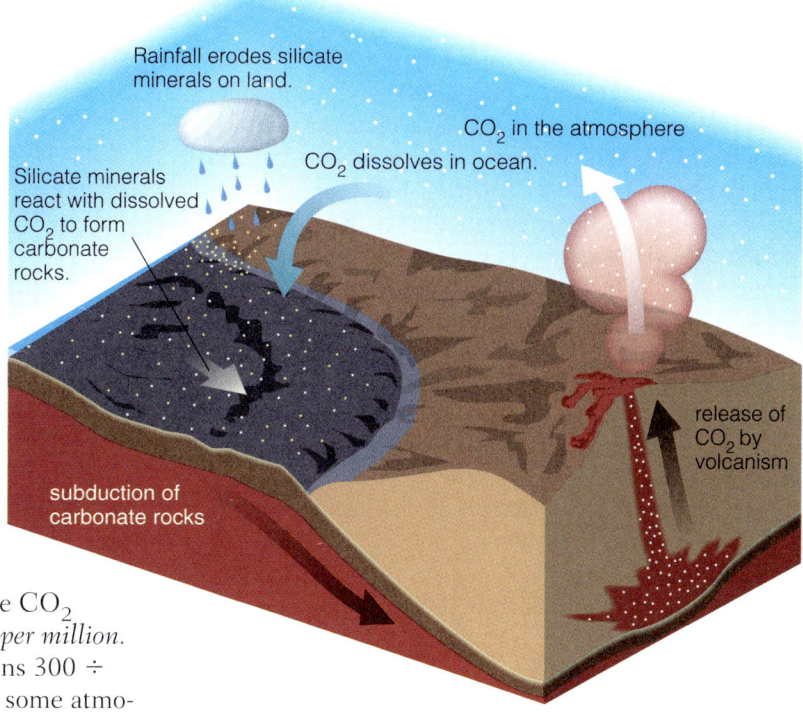

FIGURE 13.14 Recycling of CO_2 and the carbonate–silicate cycle.

The Origin of Oxygen, Ozone, and the Stratosphere

While both water and carbon dioxide are products of outgassing, molecular oxygen (O_2) is not. In fact, no geological process can explain the great abundance of oxygen (about 20%) in Earth's atmosphere. Moreover, oxygen is a highly reactive chemical that would rapidly disappear from the atmosphere if it were not continuously resupplied. Fire, rust, and the discoloration of freshly cut fruits and vegetables are everyday examples of **oxidation**—chemical reactions that remove oxygen from the atmosphere. Similar reactions between oxygen and surface materials (especially iron-bearing minerals) give rise to the reddish appearance of many of Earth's rock layers, such as those in Arizona's Grand Canyon. Thus, we must explain not only how oxygen got into the Earth's atmosphere in the first place, but also how the amount of oxygen remains relatively steady even while oxidation reactions tend to bind oxygen into rocks at a rapid rate.

The answer to the oxygen mystery is *life*. The process that supplies oxygen to the atmosphere is *photosynthesis*, which converts CO_2 to O_2. The carbon becomes incorporated into amino acids, proteins, and other components of living organisms. Today, plants and single-celled photosynthetic organisms return oxygen to the atmosphere in approximate balance with the rate at which animals and oxidation reactions consume oxygen, so the oxygen content of the atmosphere stays relatively steady. Earth originally developed its oxygen atmosphere when photosynthesis added oxygen at a rate greater than these processes could remove it from the atmosphere.

TIME OUT TO THINK *Suppose that, somehow, photosynthetic life (e.g., plants) all died out. What would happen to oxygen in our atmosphere? Could animals (such as humans) survive?*

Life and oxygen also explain the presence of Earth's ultraviolet-absorbing stratosphere. In the upper atmosphere, chemical reactions involving solar ultraviolet light transform some of the O_2 into molecules of O_3, or *ozone*. The O_3 molecule is more weakly bound than O_2, which allows it to absorb solar ultraviolet energy even better, giving rise to the warm stratosphere and preventing harmful ultraviolet radiation from reaching the surface. Mars and Venus lack photosynthetic life and therefore have too little O_2 and too little ozone to form a stratosphere.

Feedback Processes and Carbon Dioxide Balance

The Earth today is just warm enough for liquid water and "life as we know it" thanks to a moderate greenhouse effect caused by the presence of water vapor and carbon dioxide in our atmosphere. Our good fortune might seem to be based on atmospheric properties alone, but in fact our luck lies deeper—with Earth's geological processes. Much of the CO_2 is injected into the atmosphere by volcanoes near subduction zones: Carbonate rocks carried beneath the seafloor by subduction melt and release their CO_2 (see Figure 13.14). The balance between the rate at which carbonate rocks form in the oceans and the rate at which carbonate rocks melt in subduction zones largely determines the amount of CO_2 in Earth's atmosphere. If plate tectonics—especially subduction—stopped on Earth, or if it occurred at a substantially different rate, the amount of atmospheric

Common Misconceptions: Ozone—Good or Bad?

Ozone often generates confusion, because in human terms it is sometimes good and sometimes bad. In the stratosphere, ozone acts as a protective shield from the Sun's ultraviolet radiation. However, ozone is poisonous to most living creatures and therefore is a bad thing when it is found near the Earth's surface. In fact, ozone is one of the main ingredients in urban air pollution, produced as a by-product of automobiles and industry. Some people wonder whether we might be able to transport this ozone to the stratosphere and thereby alleviate the effects of ozone depletion. Unfortunately, this plan won't work. Even if we could find a way to transport this ozone, all the ozone ever produced in urban pollution would barely make a dent in the amount lost from the stratosphere.

CO_2 would undoubtedly be different. The eventual effect on Earth's temperature might prove fatal to us.

Life also plays a role in the CO_2 balance, because the photosynthetic cycle consumes carbon dioxide as it produces oxygen. (Life in the oceans plays an additional role: One mechanism by which carbonate rocks form in the ocean involves seashells that sink to the ocean floor.) Moreover, just as the carbon dioxide locked up in carbonate rocks can later be released, so can the carbon dioxide consumed by living organisms. When organic materials burn, their carbon reacts with atmospheric oxygen to produce carbon dioxide. Fossil fuels—oil, coal, and natural gas—are the remains of living organisms that died long ago. When we burn these fuels, we add carbon dioxide to the atmosphere, changing the CO_2 balance.

The full story of the Earth's carbon dioxide balance is even more complex because of **feedback relationships**—relationships in which a change in one property amplifies (*positive feedback*) or counteracts (*negative feedback*) the behavior of the rest of the system. You are probably familiar with audio feedback: If you bring a microphone too close to a loudspeaker, it picks up and amplifies small sounds from the speaker, and these amplified sounds are again picked up by the microphone and further amplified, causing a loud screech. This is an example of positive feedback. The screech usually leads to a form of negative feedback: The embarrassed person holding the microphone moves away from the loudspeaker, thereby stopping the positive audio feedback.

A planet's carbon dioxide balance, and hence its greenhouse-induced temperature, is also subject to both positive and negative feedback. The hotter a planet's surface, the more of its greenhouse gases will be in the atmosphere instead of in the oceans and crust. But this increase in greenhouse gases further increases the temperature—an example of positive feedback (which ultimately led to the runaway greenhouse effect on Venus). However, in some cases, warming temperatures can also lead to negative feedback, at least on planets with liquid water. The warmer temperature leads to more evaporation and cloud formation, and the higher albedo of clouds causes cooling as more sunlight is reflected back to space. Furthermore, higher temperatures increase the rate of carbonate rock formation, pulling more CO_2 from the atmosphere and cooling the planet. The complexity of these relationships makes predicting planetary climates difficult.

Geological evidence is mounting for an unusual stage in Earth's history that highlights the importance of feedback processes. The evidence suggests that some 600–700 million years ago glaciers resided in regions known to lie near the equator at that time. If this really occurred, the global temperature must have dipped enough for the oceans to start to freeze. Because ice has a much higher albedo than water (ice reflects about 90% of incoming sunlight, as opposed to just 5% for water), the small amount of freezing would have further cooled the planet, leading to more widespread ocean freezing. This positive feedback would have led quickly to what some geologists call "snowball Earth." How did the Earth recover from this "snowball" phase? After all, once covered with high-albedo ice, the planet must have become very cold. The answer lies in the carbonate–silicate cycle (Figure 13.14). If the oceans froze and snow fell instead of rain, the cycle was broken: CO_2 was no longer pulled from the atmosphere. It therefore built up in the atmosphere, warming the planet until the oceans melted and the carbonate–silicate cycle was restored.

TIME OUT TO THINK *How would the carbonate–silicate cycle be affected if plate tectonics did not occur? Without plate tectonics, could Earth recover from a "snowball" phase?*

Feedback processes probably also explain how the Earth has kept a fairly steady temperature despite a changing Sun: Stars like the Sun increase in brightness as they age, and billions of years ago our Sun was probably about 30% fainter than it is today [Section 14.3]. With Earth's current abundance of greenhouse gases, our planet would have been frozen over. Thus, greenhouse gases must have been more abundant early in Earth's history. Venus, on the other hand, might have experienced hospitable conditions if it possessed "only" its current inventory of greenhouse gases—but it probably also had more greenhouse gases in its early history, leading to its runaway greenhouse effect. On Mars, extra greenhouse gases once provided the warm temperatures needed for liquid water to flow, but Mars lost so much of this gas that it became colder even as the Sun grew brighter.

Summary of Earth's Atmosphere

Venus, Earth, and Mars all began on similar paths, releasing similar gases into their atmospheres by outgassing. But only Earth had conditions "just right" to maintain liquid oceans. The oceans helped remove carbon dioxide from the atmosphere, ensuring that the greenhouse effect on Earth remained just strong enough to keep conditions hospitable for life, but not so strong as to create a runaway greenhouse effect like that on Venus.

The physical connections between the Earth's geology and its atmosphere cannot fully explain the conditions on Earth today. In particular, the presence of oxygen and ozone can be explained only as products of life. Thus, to complete our understanding of the unique place of Earth in the solar system, we must turn our attention to the role of life on Earth.

13.4 Life

Fossils, the petrified remains of living organisms, tell the story of life on Earth (Figure 13.15). Most fossils formed when dead organisms fell to the bottom of a body of water, where they were gradually buried by layers of sediment and eventually compressed into

FIGURE 13.15 Dinosaur fossils.

FIGURE 13.16 The rock layers of the Grand Canyon record 2 billion years of Earth history.

rocks. In some areas, such as in the Grand Canyon, layers of sediments have been deposited one on top of the other for millions or even billions of years, burying a detailed record of life in the process (Figure 13.16). Erosion or tectonic activity later exposes the fossils. Some fossils are remarkably well preserved, and we find fossils of large animals, plants, and even microscopic, single-celled organisms.

The key to reconstructing the history of life is to determine the dates at which fossil organisms lived. The *relative* ages of fossils found in different layers can be determined easily: Deeper layers formed earlier and contain more ancient fossils. Radioactive dating confirms these relative ages and gives us fairly accurate absolute ages for fossils [Section 8.6].

For relatively recent geological history, such as the age of the dinosaurs that began about 250 million years ago and ended abruptly 65 million years ago, the numerous fossil skeletons of extinct animals prove that life on Earth has undergone dramatic changes. Life in earlier epochs is more difficult to characterize. Primitive life-forms without skeletons leave fewer fossils, erosion erases much old fossil evidence, and subduction destroys other fossil evidence deep beneath the Earth's surface. Nevertheless, geologists and biologists have developed a fairly clear picture of the history of life on Earth, at least in broad outline. As we trace this history through time, you should ask yourself three key questions:

1. How did Earth's physical conditions affect the development of life?
2. How has life influenced the physical characteristics of Earth?
3. What can we learn about the prospects for finding life elsewhere in the solar system or universe?

The Origin of Life (4.4–3.5 billion years ago)

Life arose on Earth at least 3.5 billion years ago, and possibly much earlier than that. We've found recognizable fossils in rocks as old as 3.5 billion years—microscopic fossils of single-celled bacteria and larger fossils of bacterial "colonies" called *stromatolites* (Figure 13.17). But indirect evidence suggests that life originated at least 350 million years earlier (3.85 billion years ago), even though such ancient rock samples are too contorted for fossils to remain recognizable. This evidence comes from analysis of different isotopes of carbon [Section 4.3]. Most carbon atoms are atoms of carbon-12 (6 protons and 6 neutrons), but a small proportion of carbon atoms are instead carbon-13 (6 protons and 7 neutrons). Living organisms incorporate carbon-12 slightly more easily than carbon-13, and as a result the fraction of carbon-13 is always a bit lower in fossils than in rock samples that lack fossils. In fact, all life and all fossils tested to date show the same characteristic ratio of the two carbon isotopes. Rocks as old as 3.85 billion years also show the same ratio, strongly suggesting that these rocks contain remnants of life. Beyond 3.85 billion years, we cannot yet determine whether life existed, because rocks more ancient than this no longer exist on Earth's surface. It is conceivable that life arose as early as 4.4 billion years ago.

TIME OUT TO THINK *You may have noticed that, while we've been talking about life, we haven't actually defined the term. In fact, it's surprisingly difficult to draw a clear boundary between life and nonlife. How would you define life? Explain your reasoning.*

If life did arise much before 3.85 billion years ago, it may have been in for a rough time. The transition from intolerable conditions during planetary

b Stromatolite fossils (up to 3.5 billion years old).

a Microscopic fossil bacteria.

c (below) Living stromatolites today. The biological origin of fossil stromatolites was questioned until living examples were found.

FIGURE 13.17 Ancient life.

formation to more hospitable conditions was a gradual one. The early bombardment that continued after the end of accretion gradually died away over a period of a few hundred million years. But large impacts capable of sterilizing the planet by temporarily boiling all the oceans may have occurred as late as 4.0–3.8 billion years ago. Earthquakes and volcanic eruptions were probably more frequent and more violent as well, due to greater radioactive heating in Earth's interior. Plate tectonics may not yet have begun, and oceans may have covered the entire surface. Indeed, some biologists speculate that life may have arisen several times in the Earth's early history, only to be extinguished by the hostile conditions. If so, life might have turned out very different if one less—or one more—sterilizing impact had occurred.

An even deeper question concerns not just *when* life arose, but *how*. No one knows the precise answer to this question, but we are confident of one thing: Whether life arose once or multiple times, one particular type of organism came to dominate the entire Earth. That is, every organism living today apparently developed from a single ancestor.

The idea of a common ancestor comes from several lines of evidence based on the fact that all known life-forms share uncanny chemical resemblances to one another—similarities that are not expected on simple chemical grounds and that are far too improbable and numerous to be considered coincidences. First, all known organisms use virtually the same limited set of chemical building blocks. For example, proteins are made from building blocks called **amino acids**, and all living organisms use the same set of 20 different amino acids—even though more than 70 amino acids exist. Similarly, all living organisms use the same basic molecule (called ATP) to store energy within cells, and all use molecules of DNA to transmit their genes from one generation to the next. (Some viruses use RNA, rather than DNA, as their genetic material.)

A second line of evidence for a common ancestor comes from the fact that all living organisms share nearly the same **genetic code**—the "language" that living cells use to read the instructions chemically encoded in DNA. You are probably familiar with the structure of DNA, which consists of two long strands that look somewhat like train tracks, wound together in the shape of a double helix (Figure 13.18). Each letter shown along the DNA strands represents one of four *chemical bases*, denoted by A, G, T, and C (for the first letters of their chemical names). The instructions for assembling the cell are written in the precise arrangement of these four chemical bases: Genetic "words" composed of three consecutive bases represent particular amino acids. (For example, the "word" CAG represents one amino acid, and TCA represents a completely different amino acid. Chemical reactions between the DNA and its surroundings in effect read the codes.) There is no known reason why living organisms should follow a particular genetic code, and biochemists believe that DNA could use different sequences of these bases—or possibly even different bases altogether—to encode the same genetic information. Thus, the fact that all living organisms share a common genetic code implies a common ancestry.

A third line of evidence for a common ancestor comes from detailed analysis of the sequence of chemical bases in the DNA of different organisms. In particular, biologists have mapped out the sequence in many different organisms that holds the instructions for making a molecule called rRNA (ribosomal RNA). They've discovered that this sequence includes "unused" segments that apparently do not affect the structure or function of rRNA. Biologists can trace the process of evolution by comparing these unused segments. For example, suppose the unused segments look the same in two organisms but a little bit different in a third organism. Then we can conclude that the first two organisms are more closely related to each other than to the third. By making many such comparisons, biologists have developed an evolutionary "tree of life" (Figure 13.19). Although many details in the structure of this tree remain uncertain, it clearly shows that living organisms share a common ancestry.

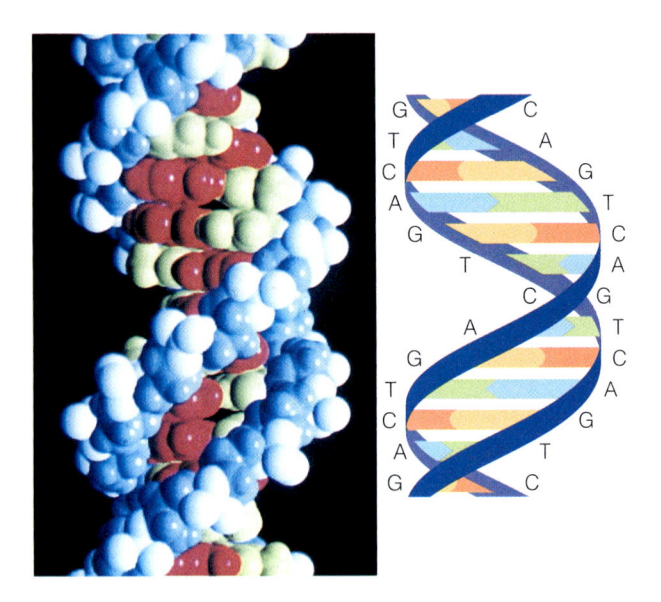

FIGURE 13.18 Left: A model of a small piece of a DNA molecule. Right: This diagram shows that a DNA molecule is made from two intertwined strands, with "ladder steps" connecting pairs of chemical bases. Note that A always connects to T, and G always connects to C. The genetic code is a language in which triplets (along a single strand) of the chemical bases make "words" that represent particular amino acids.

Comparison of DNA sequences also allows at least reasonable guesses as to which living organisms most resemble the common ancestor of all life. Surprisingly, the answer appears to be organisms living in the deep oceans around seafloor volcanic vents called *black smokers* (after the dark, mineral-rich water that flows out of them) and in hot springs in places like Yellowstone (Figure 13.20). These organisms thrive in temperatures as high as 125°C. (The high pressures at the seafloor prevent the water from

FIGURE 13.19 The tree of life, showing evolutionary relationships according to modern biology as of early 2001. Note that just two small branches represent *all* plant and animal species.

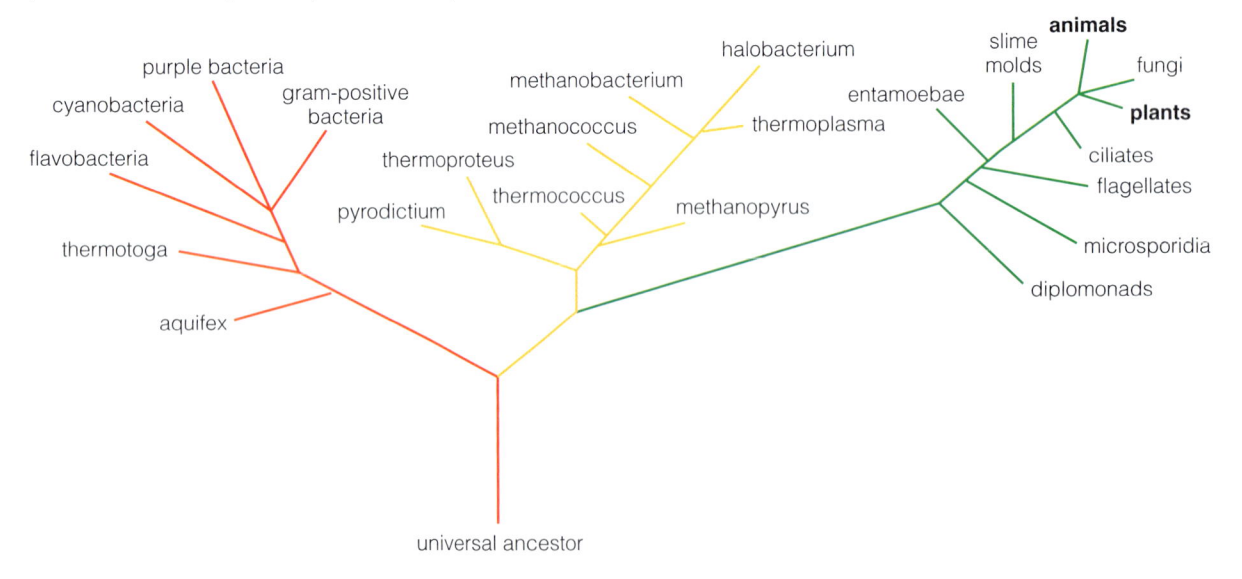

a Life around a black smoker deep beneath the ocean surface. **b** This aerial photo shows a hot spring in Yellowstone National Park that is filled with colorful bacterial life. The path in the lower left gives an indication of scale.

FIGURE 13.20 Life in hot water.

boiling until it reaches 450°C.) Unlike most life at the Earth's surface, which depends on sunlight, the ultimate energy source for these organisms is chemical reactions in water volcanically heated by the internal heat of the Earth itself.

Of course, knowing that all life shares a common ancestor still does not tell us how that ancestor first arose. The step from simple chemical building blocks to our common DNA-bearing ancestor (from nonlife to life) is huge, to say the least. No fossils record the transition, nor has it ever been duplicated in the laboratory.

The bare necessities for "life as we know it"—chemicals, energy, and water—were undoubtedly available on the early Earth. The building blocks of life, including amino acids and nucleic acids, are composed primarily of carbon, oxygen, nitrogen, and hydrogen, chemicals that were readily available. Energy to fuel chemical reactions was present on the surface in the form of lightning and ultraviolet light from the Sun and was also present in the oceans in the heated water near undersea volcanoes. We know that oceans were present, because we find ancient carbonate rocks that must have formed in water.

Laboratory experiments demonstrate that the chemical building blocks of life could have formed spontaneously and rapidly in the conditions that prevailed early in Earth's history. In these experiments, researchers replicate the chemical conditions of the early Earth by, for example, including the types of ingredients outgassed by volcanoes and "sparking" the mixture with electricity to simulate lightning or other energy sources. Within just a few days, the mixture spontaneously becomes rich in amino acids and nucleic acids. Such chemical reactions were undoubtedly an important source of organic molecules on Earth, but they may not have been the only source—some meteorites contain complex organic molecules.

With the building blocks for life plentiful, it may only have been a matter of time before chemical reactions created a molecule that could make a copy of itself. Such a *self-replicating molecule* might have spread quickly through regions with hospitable conditions. If life really began in this way, then some ancient self-replicating molecule was the ancestor of modern DNA—and of all life on Earth.

Early Evolution in the Oceans (3.5–2.0 billion years ago)

Regardless of its origin, life soon thrived in the oceans in the form of single-celled organisms, tapping a variety of energy sources including sunlight (i.e., through photosynthesis). Individual organisms that survived and reproduced passed on copies of their DNA to the next generation. However, the transmission of DNA from one generation to the next is not always perfect. **Mutations**—errors in the copying process—can change the arrangement of chemical bases in a strand of DNA, making the genetic information in a new cell slightly different from that of its parent. Mutations can be caused by many factors, including high-energy *cosmic rays* from space [Section 18.2], ultraviolet light from the Sun, particles emitted by the decay of radioactive elements in the Earth, and various toxic chemicals. Most mutations are lethal, killing the cell in which the mutation occurs. However, some mutations may make a cell better able to survive in its surroundings, and the cell then passes on this improvement to its offspring.

The process by which mutations that make an organism better able to survive get passed on to future generations is called **natural selection**. The idea of natural selection was proposed by Charles Darwin (1809–1882) as a way of explaining evolution (which simply means "change with time") in animal species. Today, we believe that natural selection is the primary mechanism by which evolution proceeds: Over time, natural selection may help individuals of a species become better able to compete for scarce resources and may also lead to the development of entirely new species from old ones. The fossil record provides strong evidence that evolution *has* occurred, while natural selection explains *how* it occurs.

Fossils show that evolution progressed remarkably slowly for most of Earth's history. For at least a billion years after life first arose, the most complex life-forms were still single-celled. Some 2 billion years ago, the land was still inhospitable because of the lack of a protective ozone layer. Continents much like those today were surrounded by oceans teeming with life, but the land itself was probably as barren as Mars is today, despite pleasant temperatures and plentiful rainfall.

Altering the Atmosphere (beginning about 2 billion years ago)

The process of photosynthesis appears to have developed quite early in the history of life. Over billions of years, the abundant single-celled organisms in the ocean pulled carbon dioxide from the atmosphere and put back oxygen. Oxygen is a highly reactive gas, and for a long time it was pulled back out of the atmosphere by reactions with surface rocks. However, by about 2 billion years ago, the oxidation of surface rocks was fairly complete, and oxygen began to accumulate in the atmosphere. Today it constitutes about 20% of the atmosphere, but this fraction may vary over periods of millions of years.

The buildup of oxygen had two far-reaching effects for life on Earth. First, it made possible the development of oxygen-breathing *animals* (Figure 13.21). No one knows precisely how or when the first oxygen-breathing organism appeared, but for that animal the world was filled with food. You can imagine how quickly these creatures must have spread around the world, and the fact that other organisms could now be eaten changed the "rules of the game" for evolution.

The second major effect of the oxygen buildup was the formation of the ozone layer, which made it safe for life to move onto the land. Although it took

FIGURE 13.21 Thanks to the buildup of oxygen in the atmosphere (and small amounts dissolved in the oceans), animals like these trilobites became possible. Trilobites were among the most complex animals alive a few hundred million years ago. The largest specimens reached 75 cm (30 in.) in length.

more than a billion years, natural selection eventually led to plants that could survive and thrive on land, and animals followed soon after, taking advantage of this new food source.

This epoch highlights the ability of life to fundamentally alter a planet. Not only did life change the Earth's atmosphere, but in doing so it made two other kinds of life possible: oxygen-breathing animals and land organisms.

An Explosion of Diversity (beginning about 0.54 billion years ago)

As recently as 540 million years ago, most life-forms were still single-celled and tiny. Note that life had already been present on Earth for at least 3 billion years by that time, demonstrating the remarkable slowness with which evolution progressed for most of Earth's history. But the fossil record reveals a dramatic diversification of life beginning about 540 million years ago. Although this change occurred over a period of about 40 million years, it was so dramatic in comparison to the events of the previous 3 billion years that it is often called the *Cambrian explosion*. (Cambrian is the name geologists give the period from about 540 million to 500 million years ago.) We can trace the origin of most of today's plant and animal species—from insects to trees to vertebrates—back to this period. Apparently, once life began to diversify, the increased competition among the many more complex species accelerated the process of natural selection and the evolution of new species.

Some species have been more successful and more adaptable than others. Dinosaurs dominated the landscape of the Earth for more than 100 million years. But their sudden demise 65 million years ago paved the way for the evolution of large mammals—including us. The earliest humans appeared on the scene only a few million years ago (after 99.9% of

Earth's history), and our few centuries of industry and technology have come after 99.99999% of Earth's history. Despite our recent arrival (in geological terms), modern humans are by some measures the most successful species ever to inhabit the Earth. Humans survive and prosper in virtually every type of land environment. With proper equipment, we can survive underwater, and a few people have lived under the sea for extended periods of time in experimental habitats. We are even developing the technology to survive away from our planet, in the inhospitable environment of space. And our population has been growing exponentially for the past few centuries, roughly doubling in just the past 40 years (Figure 13.22).

The tremendous growth of the human population suggests a chilling analogy: Exponential growth is also the mark of a cancerous tumor—which may seem very successful at surviving while it is growing but ultimately dies when it kills its host. Is it possible that our tremendous success is killing our host, the Earth? We can gain some perspective on this question by looking at a few lessons that the solar system teaches us about our relationship with the Earth.

TIME OUT TO THINK *The pace of change in human existence has accelerated since the advent of modern humans a few million years ago: Civilization arose about 10 thousand years ago, the industrial revolution began just a couple of centuries ago, and the computer era started just a few decades ago. Do you think the pace of change can continue to accelerate? What do you think the next stage will be?*

13.5 Lessons from the Solar System

The United States and other nations have carried out the exploration of the solar system for many reasons, including scientific curiosity, national pride, and technological advancement. Perhaps the most important return on this investment was one not expected: an improved understanding of our own planet and an enhanced ability to understand some very important global issues. In this section, we explore three particularly important issues about which comparative planetology offers important insights: global warming, ozone depletion, and mass extinctions.

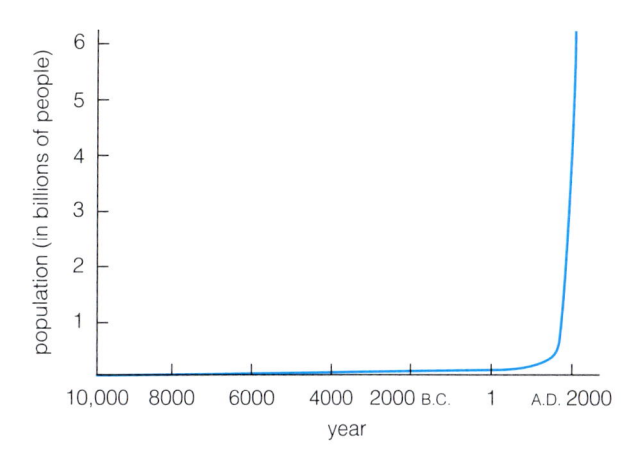

FIGURE 13.22 This graph shows human population over the past 12,000 years. Note the tremendous population growth that has occurred in just the past few centuries.

Global Warming

Venus stands as a searing example of the effects of large amounts of atmospheric carbon dioxide. Earth remains habitable only because natural processes have locked up most of its carbon dioxide on the seafloor. However, humans are now tinkering with this difference between the two planets by adding carbon dioxide and other greenhouse gases to the Earth's atmosphere. The primary way we release carbon dioxide is through the burning of fossil fuels (coal, natural gas, and oil)—the carbon-rich remains of plants buried millions of years ago. But other human actions, such as deforestation and the subsequent burning of trees, also affect the carbon dioxide balance of our atmosphere. The inescapable result is that the amount of atmospheric carbon dioxide has risen steadily since the dawn of the industrial age—and continues to rise (Figure 13.23)—nearly 20% in the past 50 years alone.

FIGURE 13.23 These data, collected for many years on Mauna Loa (Hawaii), show the increase in atmospheric carbon dioxide concentration. Yearly wiggles represent seasonal variations in the concentration, but the long-term trend is clearly upward. The concentration is measured in parts per million (ppm).

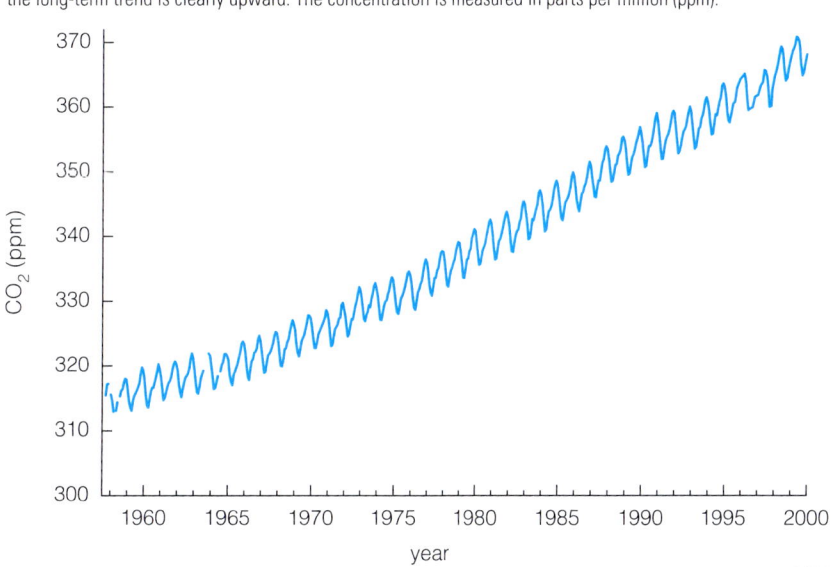

Although it may seem logical to conclude that an increase in the concentration of greenhouse gases should warm the Earth, the Earth is a very complex system. As a result, the debate over whether human activity is changing the Earth's climate is fierce. Measuring even something as seemingly simple as the average temperature of the entire planet can be surprisingly difficult. Until the advent of satellite temperature measurements, temperature data came only from ground-based weather stations and were subject to bias. For example, most weather stations are restricted to land and are located near cities (which tend to be warmer than surrounding areas). Despite measurement difficulties, however, there is general scientific consensus that the average temperature of the Earth has warmed slightly (about 0.5°C) over the past 50 years—although there is much less agreement about why. Furthermore, glaciers, ice sheets, and even the snows of Mount Kilimanjaro are in retreat, while ocean levels are slowly rising. It's likely that the cause is the human addition of greenhouse gases to the atmosphere, but it's also possible that the change is natural and would have occurred even without human activity. Some scientists even suggest that the Sun itself may be warming and thus contributing to the global temperature rise.

The question of whether human activity is inducing global warming is not merely academic: The consequences of a significant continued warming could be widespread and disastrous. The primary way scientists try to predict the effects of global warming is by making sophisticated computer models of the climate. Different models give different results, but some predict a warming of more than 5°C during the next hundred years. Past ice ages occurred when the average temperature dropped by about 5°C; a similar increase could melt polar ice and flood the Earth's densely populated coastal regions. Such warming would also increase evaporation from the oceans, tending to make storms of all types both more frequent and more destructive: Coastal regions would be hit by more hurricanes, intense thunderstorms and associated tornadoes would strike more frequently, and even winter blizzards would be more severe and more damaging. Computer models also show that global warming would not affect all regions in the same way. Some regions would warm much more than the average, while others might even cool. Rainfall patterns would also shift, and much of the world's current cropland might become unusable. The most ominous possibility is the collapse of entire ecosystems. If a regional climate changes more rapidly than local species can adapt or migrate, these species might become extinct. Given the complex interactions of the Earth's biosphere, hydrosphere, and atmosphere, the consequences of such ecosystem changes are impossible to foresee.

The uncertainty in our understanding of global warming is underscored by the fact that most current computer models predict that the Earth should already have warmed to a greater degree than we've observed. This suggests that some process may be counteracting the effects of the added greenhouse gases. One hypothesis is that increased evaporation from the oceans might lead to more clouds that prevent sunlight from reaching the surface. Another is that recent volcanic eruptions, such as the 1991 eruption of Mount Pinatubo in the Philippines, have injected particles into the stratosphere that reflect sunlight, effectively increasing the Earth's albedo. The most frightening possibility is that reflective particles are indeed being injected into the atmosphere, but their source is sulfate particles from coal-burning industries. At first, this idea might seem to suggest that the pollutants from coal burning counter the effects of the added carbon dioxide, but in reality this can only postpone and worsen the problem: Sulfate particles fall out of the atmosphere in a matter of years, while carbon dioxide may remain for half a million years.

Given the current uncertainties, it is impossible for anyone to know for sure whether global warming will even be a serious problem in the coming century, let alone to make specific predictions of its severity. Our studies of the planets show that surface temperatures depend on many factors and that positive and negative feedback processes can alter climates in surprising ways. Nevertheless, the fact that we cannot explain the current climate conditions on Earth suggests that we should be very careful about tampering with them.

TIME OUT TO THINK *The most obvious way to prevent global warming is to stop burning the fossil fuels that add greenhouse gases to the atmosphere. But much of the world economy depends on energy produced from fossil fuels. Given the uncertainties about the effects of global warming in the next hundred years, what, if any, changes do you think are justified at present? What specific proposals would you make if you were a world leader?*

Ozone Depletion

For 2 billion years, the Earth's ozone layer has shielded the surface from hazardous ultraviolet radiation, but in the past two decades humans have begun to destroy the shield. The idea that human activity might damage the ozone was first suggested in the early 1970s, but the magnitude of the problem was not recognized until the discovery of an **ozone hole** over Antarctica in the mid-1980s (Figure 13.24). The "hole" is a place where the concentration of ozone in

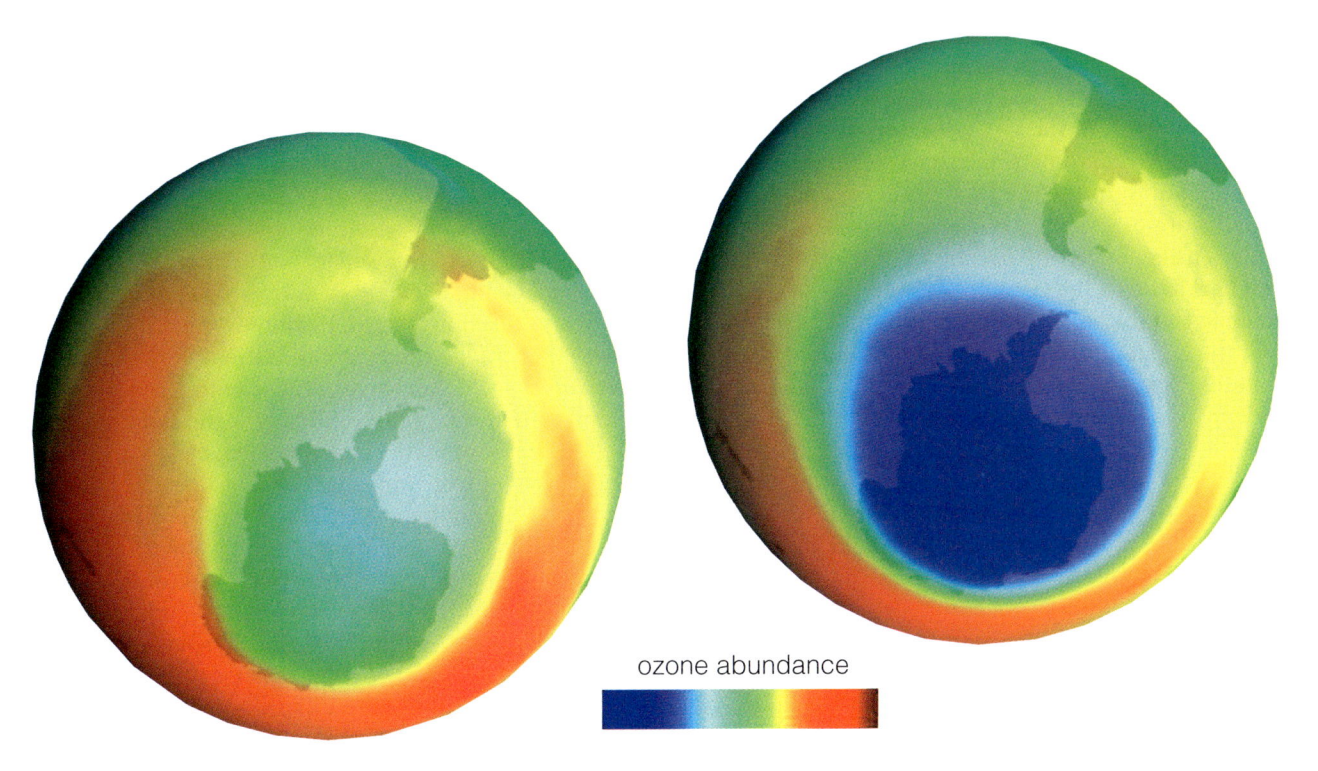

FIGURE 13.24 The extent of the ozone hole over Antarctica in 1979 (left) and 1998 (right). The ozone hole grew substantially between 1979 and 1998. Red indicates normal ozone concentration, and blue indicates severe depletion.

the stratosphere is dramatically lower than it would naturally be. The ozone hole appears for a few months during each Antarctic spring and has gradually worsened over the years. Since its discovery, satellite measurements have suggested that **ozone depletion** has caused ozone levels to fall by as much as a few percent worldwide. If ozone levels continue to decline, more solar ultraviolet radiation will reach the surface. Humans will face substantially greater risk from skin cancer, and plants and animals will suffer more genetic damage from mutations induced by ultraviolet radiation—with unknown consequences for the biosphere as a whole.

The apparent cause of the ozone depletion is human-made chemicals known as CFCs (chlorofluorocarbons). CFCs were invented in the 1930s and have been widely used in air conditioners and refrigerators, in the manufacture of packaging foam, as propellants in spray cans, in industrial solvents used in the computer industry, and for many other purposes. Indeed, CFCs once seemed like an almost ideal chemical: They are useful in many industries, cheap and easy to produce, and chemically inert (i.e., they do not burn, break down, or react with anything on the Earth's surface). Ironically, their inertness is also their downfall; because CFCs are gases and are not destroyed by chemical reactions in the lower atmosphere, they eventually rise intact into the stratosphere. There they are broken down by the Sun's ultraviolet light, and one of the by-products is chlorine—a very reactive gas that catalytically destroys ozone without being consumed in the process. On average, a single chlorine atom in the stratosphere can destroy 100,000 ozone molecules before it is consumed itself in some other chemical reaction. Chlorine's ability to attack ozone is enhanced at low temperatures, which explains why ozone depletion was first observed in the polar regions.

Mars and Venus each played an important role in our understanding of ozone and chlorine. Mars lacks an ozone layer, so ultraviolet radiation reaches the surface. As a result, the surface is sterile: Terrestrial life could not survive on the surface, and the building blocks of life would be torn apart when exposed to the Sun. But it was the study of Venus that first alerted scientists to the dangers on Earth. Chlorine is a minor but important ingredient of Venus's atmosphere, and models of its chemical cycles showed that chlorine could rapidly destroy weak molecules such as ozone. When the models were altered to apply to Earth's atmosphere, CFCs and ozone depletion were connected.

This tale of three planets may have a happy ending, thanks to the awareness it brought of the dangers posed by ozone destruction. Today, international treaties ban the production of CFCs, although previously existing CFCs may still be used and recycled. If these treaties are obeyed, continued ozone destruction probably will not be a serious problem in the future, but the ozone hole will probably take 50 years to recover. Most of the CFCs ever produced are still in intact air conditioners, refrigerators, or other products, and most of these CFCs will eventually escape into the atmosphere. Moreover, because the stratosphere lacks the weather of the troposphere [Section 10.2], CFCs (and their breakdown products) remain in the stratosphere for decades once they arrive. But the greater danger lies in the possibility of treaty failures. Some of the replacements for CFCs may be only marginally less damaging than the CFCs themselves. In addition, the replacements are generally much more expensive to produce than CFCs, and a large global black market now exists for CFCs produced in violation of the treaties. Like most global issues, the final outcome of the ozone problem remains in doubt.

TIME OUT TO THINK *You've just learned that your car air conditioner is broken and will cost $200 to fix. Then you hear about a shop that can fix it for only $100—undoubtedly by violating the laws concerning CFCs. Is the small benefit to our ozone layer worth the extra $100?*

Mass Extinctions

When geology was a young science, its practitioners thought that entire mountain ranges and huge valleys were formed suddenly by cataclysmic events. Later geologists denounced this *catastrophism* and replaced it with *uniformitarianism,* which holds that geological change occurs gradually by processes acting over very long periods. Uniformitarianism certainly holds for three of our four geological processes, but impact cratering represents the "instant geology" of catastrophism. Many geologists initially rejected the importance of impact cratering as a geological process, but they were eventually convinced by the unambiguous evidence of impacts on the Moon, by astronomical studies of comets and asteroids, and by careful analysis of the few impact craters still present on Earth. Geology today is a combination of uniformitarianism and catastrophism.

Scientific ideas about evolution are undergoing a similar transformation. Biologists once assumed that evolution always proceeded gradually, with new species arising slowly and others occasionally becoming extinct. It now appears that the vast majority of all species in Earth's history died out in sudden *mass extinctions*—for which impacts are the prime suspect. The mass extinction that killed off the dinosaurs is only the most famous of these events [Section 12.6]; the fossil record shows evidence of at least a dozen other mass extinctions, some with even greater species loss than in the dinosaur event. While some species died out from direct effects of the impact, most probably died due to the disappearance from their ecosystem of species on which they depended. Furthermore, the survival or extinction of species appears to be random during mass extinctions: Surviving species show no discernible advantage over those that perish. In a sense, mass extinctions clear the blackboard, allowing evolution to get a fresh start with a new set of species. If we re-formed the Earth from scratch, life (and even intelligent life) might well arise again, but there is no scientific basis for thinking that evolution would follow the same course.

The story of mass extinctions teaches us about the danger of large impacts, but it also teaches a potentially more important lesson: Even without an impact, the Earth may be undergoing a mass extinction right now. In the normal course of evolution, the rate of species extinction is fairly low—perhaps one species lost per century. According to some estimates, human activity is now driving species to extinction so rapidly that half of today's species may be lost by the end of the twenty-first century. It is not just that humans are hunting and fishing many species to extinction. Far more are lost due to habitat destruction and pollution. And when a few species are lost by direct human influence, entire ecosystems that depend on those species may collapse. If global warming alters the climate significantly or if ozone depletion leads to more genetic damage in plants and animals, the rate of species extinction might increase even more. In any case, the "event" of losing half the world's species in just a few hundred years certainly qualifies as a mass extinction on a geological time scale, and the biological effects could be as dramatic as those of any impact. Are we unwittingly clearing the way for a new set of dominant species?

13.6 Lessons from Earth: Life in the Solar System and Beyond

Just as comparative planetology teaches us much about the Earth, the study of the Earth can teach us much about other worlds. In particular, it teaches us about the conditions for life and may help us answer what is surely one of the deepest philosophical questions of all time: Are we alone in the universe? It is fitting to close our study of the Earth by extending our understanding of biology to the planets and beyond. The study of the possibility of life elsewhere in the universe, called *astrobiology* (or exobiology), is a logical extension of comparative planetology.

In the time between the Copernican revolution [Section 5.3] and the space age, many people expected the other planets in our solar system to be Earth-like and to harbor intelligent life. In fact, a reward was supposedly once offered for the first evidence of intelligent life on another planet *other than Mars*. Venus, a bit closer to the Sun than Earth, was often pictured as a tropical paradise. Such expectations were dashed by the bleak images of Mars returned by spacecraft and the discovery of the runaway greenhouse effect on Venus. Many scientists began to believe that only Earth has the right conditions for life, intelligent or otherwise. Recently, the pendulum has begun to swing the other way, spurred primarily by two developments. First, biologists are learning that life thrives under a much wider range of conditions than once imagined. Second, planetary scientists are developing a much better understanding of conditions on other worlds. It now seems quite likely that conditions in at least some places on other worlds might be conducive to life.

The Hardiness, Diversity, and Probability of Life

Even on Earth, biologists long assumed that many environments were uninhabitable. But recent discoveries have found life surviving in a remarkable range of conditions. The teeming life surviving at temperatures as high as 125°C near underwater volcanic vents and in hot springs is only one of many surprises. Biologists have found microorganisms living deep inside rocks in the frozen deserts of Antarctica and inside basaltic rocks buried more than a kilometer underground. Some bacteria can even survive radiation levels once thought lethal—apparently, they have evolved cellular machinery that repairs mutations as fast as they occur. The newly discovered diversity of microscopic life has forced scientists to redraw the "tree of life" (see Figure 13.19), crowding familiar plants and animals into one corner. The majority of these microorganisms need neither sunlight, oxygen, nor "food" in the form of other organisms. Instead, they tap a variety of chemical reactions for their survival.

The new view of terrestrial biology forces us to rethink the possibility of life elsewhere. First, life on Earth thrives at extremes of temperature, pressure, and atmospheric conditions that overlap conditions found on other worlds. The Antarctic valleys, for example, are as dry and cold as certain parts of Mars, and the conditions found in terrestrial hot springs may have been duplicated on a number of planets and moons at certain times in the past. Second, life harnesses energy sources readily available on other planets. The basalt-dwelling bacteria, for example, would probably survive if they were transplanted to Mars. Third, life on Earth has evolved from a common ancestor into every imaginable ecological niche (and some unimaginable ones as well). The diversification of life on Earth has basically tested the limits of our planet, and there is no reason to doubt that it would do so on other planets. Thus, if life ever had a foothold on any other planet in the past, some organisms might still survive in surprising ecological niches today even if the planet has undergone substantial changes. The only real question is whether life ever got started elsewhere in the first place.

What is the probability of life arising from nonliving ingredients? This is probably the greatest unknown in astrobiology. The fact that we exist and are asking the question does not tell us the probability; it merely tells us that it happened once. But the rapidity with which life arose on Earth may provide a clue. As we've discussed, we find fossil evidence for life dating almost all the way back to the end of the period of early bombardment in the solar system, suggesting that life arises easily and perhaps inevitably under the right conditions. In that case, we must search the solar system for those conditions, past or present.

TIME OUT TO THINK *The preceding discussion implies that the rapid appearance of life on Earth means that life is highly probable. Do you agree with this logic? What alternative conclusions could you reach?*

Life in Our Solar System

Speculation about life in the solar system usually begins with Mars, for good reason. Before it dried out billions of years ago, its early atmosphere gave the surface hospitable conditions that rivaled those on Earth, with ample running water, the necessary raw chemical ingredients for life, and a variety of familiar energy sources. Many of Earth's organisms would have thrived under early Martian conditions, and some could even survive in places in today's Martian environment. Our first attempt to search for life on Mars came with the Viking missions to Mars in the 1970s, which included two landers equipped to search for the chemical signs of life [Section 9.5]. No life was found. But the landers sampled only two locations on the planet and tested soils only very near the surface. If life once existed on Mars, it either has become extinct or is hiding in other locations.

Today, a renewed debate about Martian life is under way, thanks in part to the study of a Martian meteorite found in Antarctica in 1984. The meteorite apparently landed in Antarctica 13,000 years ago, following a 16-million-year journey through space after being blasted from Mars by an impact. The rock itself dates to 4.5 billion years ago, indicating that it solidified shortly after Mars formed and therefore was present during the time when Mars was warmer and wetter. Painstaking analysis of the meteorite

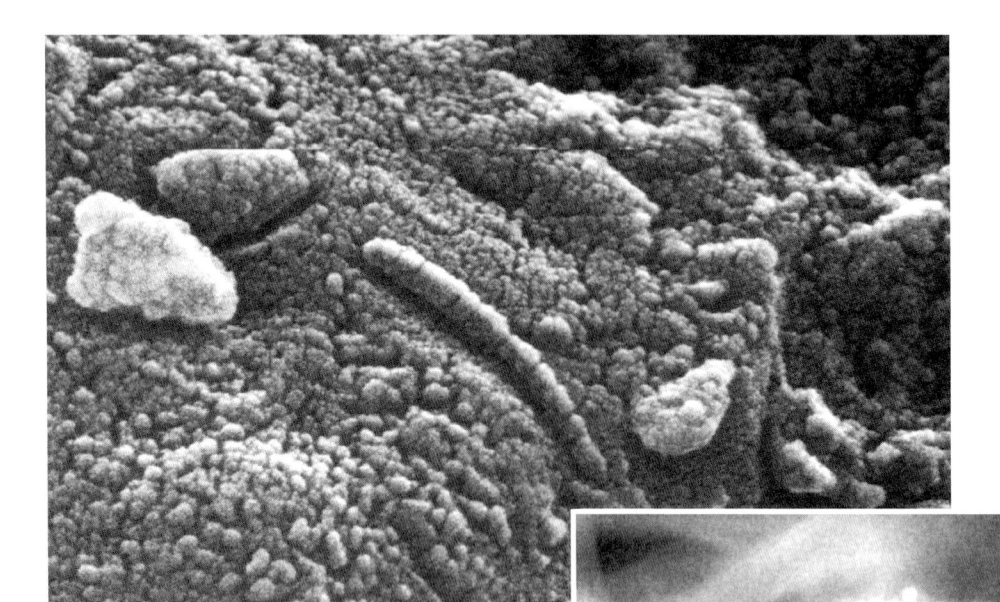

a Microscopic view of seemingly lifelike structures in a Martian meteorite.

b A comparison of microscopic chains of magnetic crystals from Earth (top) and Mars (bottom).

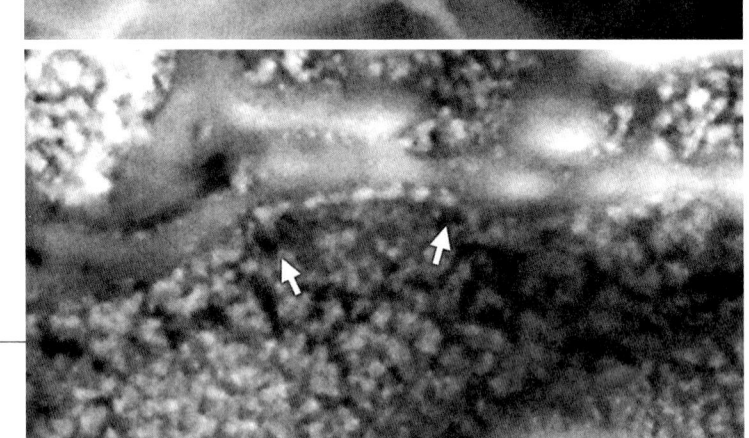

FIGURE 13.25 Evidence for life on Mars?

reveals indirect evidence of past life on Mars, including layered carbonate minerals and complex molecules (called polycyclic aromatic hydrocarbons), both of which are associated with life when they are found in Earth rocks. Even more intriguing, highly magnified images of the meteorite reveal eerily lifelike forms (Figure 13.25a). These forms bear a superficial resemblance to terrestrial bacteria, although they are about a hundred times smaller—about the same size as recently discovered terrestrial "nanobacteria" and viruses. Magnetic crystals have also been found within the rock—crystals that on Earth are only made by bacteria (Figure 13.25b). Nevertheless, many scientists dispute the conclusion that these features suggest the past existence of life on Mars, claiming that nonbiological causes can also explain the meteorite's unusual features.

Future missions to Mars will search for life using more sophisticated techniques than those used by the Viking missions. The best place to look for fossil remains of extinct life is probably in ancient valley bottoms or the sedimentary rocks of dried-up lake beds (Figure 9.27d). If any hot springs surround Mars's not-quite-dormant volcanoes, they may be a good place to look for surviving life. A thorough search for Martian life will probably require the return of rock samples to Earth or human exploration of the planet.

Martian meteorites also remind us that the planets may occasionally exchange rocks dislodged by major impacts. The harsh conditions under which some life on Earth exists suggest that living organisms might survive such impacts and even survive the journey from one planet to another. Earth's basalt-dwelling bacteria, for example, could probably survive an impact-cratering event, the ensuing millions of years in space, and a violent impact on Mars. If a meteorite from Earth once landed in hospitable conditions on Mars, it might have introduced life to Mars or wiped out life already present. In a sense, Earth, Venus, and Mars have been "sneezing" on one another for billions of years. Life could conceivably have originated on any of these three planets and been transported to the others. Some scientists even suggest that the life on Earth may have originated beyond our own solar system and been brought here by interstellar dust or meteorites.

TIME OUT TO THINK *Suppose we someday discover living organisms on Mars. How will we be able to tell whether these organisms share a common ancestor with living organisms on Earth?*

Besides Mars, the best places to search for life probably are the satellites of the jovian planets. Several of them may meet the requirements of having liquid water, appropriate chemicals, and energy sources for life. In particular, Europa may have a planet-wide ocean beneath its icy crust [Section 11.5]. The ice and rock from which Europa formed undoubtedly included the necessary chemicals, and its internal heating might lead to undersea volcanic vents. Thus, the real question about Europa may be whether it is possible that its ocean has existed for billions of years *without* developing life. NASA is considering missions to search for signs of life on Europa. The most ambitious involves landing a robotic spacecraft on the surface that will melt its way through the icy crust to reach the ocean below.

Another enticing place to look for life past or present is Saturn's moon Titan. We've already found evidence of complex chemical reactions on Titan, many involving the same elements used by life on Earth. Liquid water and energy are in shorter supply, but both might have been supplied by impacts that heated and melted the icy surface early in Titan's history. Life might have arisen in a slushy pond during the brief period that it remained liquid and might have survived after the pond froze. Unlike the case on Earth, where early impacts probably sterilized the planet, impacts on Titan may have made life possible.

Despite our new awareness of the diversity of life on Earth, we may still be underestimating the range of conditions in which life can exist. Could life arise in liquids other than water, or possibly in an atmosphere? Some people have speculated that life could develop in the clouds of Jupiter or other jovian planets. Might life be based on different elements than those used on Earth? Are there other energy sources that we have not considered? Is our definition of life too narrow? Clearly, it will be a long time before we know all the places where life might exist even within our solar system.

Life Around Other Stars

Only a few places in our solar system seem hospitable to life, but many more hospitable worlds may be orbiting some of the hundred billion other stars in the Milky Way Galaxy or stars in some of the billions of other galaxies in the universe. Might some of the stars be orbited by planets that are as hospitable as our own Earth?

Interstellar clouds throughout the galaxy contain the basic chemical ingredients of life—carbon, oxygen, nitrogen, and hydrogen, as well as other elements. All stars are born from such interstellar clouds, so it seems likely that other solar nebulae should have given rise to planetary systems similar to our own [Section 8.7]. Terrestrial planets anywhere will receive light from their parent star, and if they are large enough, like Venus or Earth, they will have plenty of internal heat. Thus, the chemicals and energy sources for life should be present on many planets throughout the universe.

Planets with liquid water on their surfaces may be somewhat rarer, particularly if we consider only planets where oceans can endure for billions of years. Even if a planet is large enough to outgas substantial quantities of water and retain its atmosphere, the stories of Venus and Mars tell us that oceans are not guaranteed. To keep oceans for billions of years, the planet must lie within a range of distances from its star, sometimes called the **habitable zone**, that has temperatures just right for liquid water. How big is the habitable zone in our own solar system? We know that Venus is too close to the Sun, so this zone must begin outside the orbit of Venus. Mars is a borderline case: If it had been large enough to retain its atmosphere and sustain a stronger greenhouse effect, Mars might still be habitable today. Overall, the habitable zone around our star probably ranges from 0.8 to 1.5 AU. This zone may be broader around brighter stars and narrower around fainter stars. Computer models of solar system formation suggest that one or more terrestrial planets will usually form within a star's habitable zone, as long as the star is not part of a binary star system. Many of the extrasolar planets discovered recently lie within their star's habitable zone. Although the planets themselves are thought to be jovian—and therefore not habitable—moons orbiting the planets could be habitable. Even moons outside the habitable zone could harbor life if other energy sources (such as tidal heating) are available.

The bottom line is that, according to our theories of solar system formation, planets with all the necessities for life should be quite common in the universe. The only major question is whether these ingredients combine to form life. The fact that life arose very early in Earth's history suggests that it may be very easy to produce life under Earth-like conditions, but we will not know for sure unless and until we find other life-bearing planets. NASA is currently developing plans for orbiting telescopes that may be able to detect ozone in the spectra of planets around other stars—and, at least in our solar system, substantial ozone implies life. In addition, radio astronomers are searching the skies in hopes of receiving a signal from some extraterrestrial civilization [Section S5.3]. Perhaps, in a decade or two, we will discover unmistakable evidence of life. On that day, if it comes, we will know that we are not alone.

TIME OUT TO THINK *Consider the following statements: (1) We are the only intelligent life in the entire universe. (2) Earth is one of many planets inhabited by intelligent life. Which do you think is true? Do you find either philosophically troubling?*

THE BIG PICTURE

Humans have observed the Sun, the Moon, and the planets for thousands of years, but only recently did we learn that these other worlds have much to teach us about our own Earth. Through our study of solar system formation and comparative planetology, we have learned to look at Earth from a very new perspective. Keep in mind the following "big picture" ideas:

- Earth has been shaped by the same geological and atmospheric processes that shaped the other terrestrial worlds. Earth is not a special case from a planetary point of view but rather a place where natural processes led to conditions conducive to life.

- Most of Earth's unique features can be traced to the fact that abundant water has remained liquid throughout our planet's history—thanks to our distance from the Sun and the size of our planet.

- Life arose early on Earth and played a crucial role in shaping our planet's history. The abundant oxygen in our atmosphere is just one of many phenomena that demonstrate how life can transform a planet.

- Humans are ideally adapted to the Earth today, but there is no guarantee that the Earth will remain as hospitable in the future. The study of our solar system teaches us how planets can change.

- The conditions necessary for "life as we know it" are probably common in the universe and may even be found in our own solar system. But so far we have no proof that life exists elsewhere.

Review Questions

1. Briefly summarize how Earth is different from the other terrestrial planets. In particular, define *hydrosphere* and *biosphere*.

2. Briefly describe the interior structure of the Earth, including its molten outer core and its lithosphere.

3. How does seafloor crust differ from continental crust? Why do we find fewer craters on the seafloor than on the continents?

4. Why is erosion more important on Earth than on Venus or Mars? Describe how erosion affects the Earth's geology.

5. Briefly describe the conveyor-like process of *plate tectonics*. What are *spreading centers* and *subduction zones*?

6. Describe how *subduction* has created the continents over billions of years. Give examples of how this process has affected Japan, the Philippines, and the western United States.

7. Briefly describe how tectonic stresses can create rift valleys, tall mountain ranges, and *faults*. Give examples of places that have been affected in each of these ways.

8. What is a *hot spot*? Describe how hot spots affect Hawaii and Yellowstone.

9. What is a *runaway greenhouse effect*? Why did it occur on Venus but not on Earth?

10. Describe how the *carbonate–silicate cycle* helps maintain the relatively small CO_2 content of our atmosphere.

11. What are *oxidation* reactions? If there were no life on Earth, would Earth's atmosphere still contain significant amounts of oxygen or ozone?

12. Give examples of positive and negative *feedback relationships*. How does feedback affect the CO_2 balance in our atmosphere? What is "snowball Earth"?

13. Approximately when did life arise on Earth? How do we know?

14. What evidence tells us that all life today shares a common ancestor? How do biologists compare DNA sequences to establish the "tree of life"? Where do plants and animals fall on this diagram?

15. What are *mutations*? How do mutations drive the process of *natural selection*?

16. Briefly describe how life gradually altered the Earth's atmosphere until it reached its current state.

17. Summarize the slow evolution of life on the early Earth. What was the Cambrian explosion?

18. What is global warming? Briefly describe some of the potential dangers of global warming and why so many uncertainties are involved in knowing whether global warming represents a real threat.

19. Briefly describe the phenomena and causes of the Antarctic *ozone hole* and worldwide *ozone depletion*.

20. Is the Earth currently undergoing a mass extinction? Explain.

21. Summarize why we now think that life can survive in a much broader range of conditions than we did just a few decades ago. Why is this important to the prospect of finding life elsewhere?

22. Describe the evidence from Martian meteorites suggesting that life may once have existed on Mars. Explain how life might have originated on one terrestrial planet and been transferred to others.

23. Why are Europa and Titan considered prospects for harboring life?

24. What is a *habitable zone*? Using this idea, briefly describe the prospect of finding life on planets around other stars.

Discussion Questions

1. *Evidence of Our Civilization.* Imagine a future archaeologist, say 10,000 years from now, trying to piece together a picture of human civilization at our time (i.e., around the year 2000). What types of evidence of our civilization are most likely to survive for 10,000 years? (Will our buildings survive? Our infrastructure, such as highways and water pipes? Information in the form of books or computer data?) What geological processes are likely to destroy evidence over the next 10,000 years? Next, discuss the evidence that will remain—and why—for an archaeologist living 100 million years from now. Be sure to consider the effects of all four geological processes, as well as continental drift.

2. *Cancer of the Earth?* In the text, we discussed how the spread of humans over the Earth resembles, at least in some ways, the spread of cancer in a human body. Cancers end up killing themselves because of the damage they do to their hosts. Do you think we are in danger of killing ourselves through our actions on the Earth? If so, what should we do to alleviate this danger? Overall, do you think the cancer analogy is valid or invalid? Defend your opinions.

3. *Contact.* Suppose we discover definitive evidence for microbial life on Mars or Europa. Would this discovery alter your view of our place in the universe? If so, how? What if we made contact with an intelligent species from another world? Do you think it is likely that either kind of life exists elsewhere in the universe? Do you think either kind will be discovered in your lifetime? Explain.

Problems

Surprising Discoveries? For **problems 1–8**, suppose we made the discoveries described. (These are *not* real discoveries.) Decide whether the discovery should be considered reasonable or surprising. Explain. (In some cases, either view can be defended.)

1. A fossil of an organism that died more than 300 million years ago, found in the crust near a mid-ocean ridge.

2. Evidence that fish once swam in a region that is now high on a mountaintop.

3. A "lost continent" on which humans had a great city just a few thousand years ago but that now resides deep underground near a subduction zone.

4. A planet in another solar system that has an Earth-like atmosphere but no life.

5. A planet in another solar system that has an ozone layer but no ordinary oxygen (O_2) in its atmosphere.

6. Evidence that the early Earth had more carbon dioxide in its atmosphere than the Earth does today.

7. Discovery of life on Mars that also uses DNA as its generic molecule and that uses a genetic code very similar to that used by life on Earth.

8. Evidence of photosynthesis occurring on a planet in another solar system that lies outside that solar system's habitable zone.

9. *Change in Formation Properties.* Consider Earth's four formation properties of size, distance, composition, and rotation rate. Choose one property, and suppose it had been different (e.g., smaller size, greater distance). Describe how this change might have affected Earth's subsequent history and the possibility of life on Earth.

10. *Growing Population.* Since about 1950, human population has grown with a doubling time of about 40 years; that is, the population doubles every 40 years. The current population is about 6 billion. If population continues to grow with a doubling time of 40 years, what will it be in 40 years? In 80 years? Do you think this will actually happen? Why or why not?

11. *Feedback Processes in the Atmosphere.*
    a. Give an everyday example (not used in this book) of a positive or negative feedback process. Explain how the process works.
    b. As the Sun gradually brightens, how can the carbonate–silicate cycle respond to reduce the warming effect? Which parts of the cycle will be affected? Is this an example of positive or negative feedback?

12. *Defining Life.* Write a definition of life and explain the basis of your definition in a few sentences. Then evaluate whether each of the following three cases meets your definition. Explain why or why not in a paragraph for each case. (i) The first self-replicating molecule on Earth lies near the boundary of life and nonlife. Does it meet your definition of life? (ii) Modern-day viruses are essentially packets of DNA encased in a microscopic shell of protein. Viruses cannot replicate themselves; instead, they reproduce by infecting living cells and "hijacking" the cell's reproduction machinery to make copies of the viral DNA and proteins. Are viruses alive? (iii) Imagine that humans someday travel to other stars and discover a planet populated by what appear to be robots programmed to mine metal, refine it, and assemble copies of themselves. Examination of fossil "robots" shows that they have improved, perhaps because cosmic rays have caused errors in their programs. Is this race of robots alive? Would your answer depend on whether another race had built the first robots?

13. *Ozone Signature.* Suppose a powerful future telescope is able to take a spectrum of a terrestrial planet around another star. The spectrum reveals the presence of significant amounts of ozone. Why would this discovery strongly suggest the presence of life on this planet? Would it tell us whether the life is microscopic or more advanced? Summarize your answers in one or two paragraphs.

14. *Explaining Ourselves to the Aliens.* Imagine that someday we make contact with intelligent aliens and even learn to communicate. Even so, many facets of life we take for granted may be utterly incomprehensible to them: music (perhaps they have no sense of hearing), money (perhaps there has never been a need), love (perhaps they don't feel this emotion), meals (perhaps they photosynthesize). Write a page attempting to explain one of these concepts, or another of your choosing. Remember that even the words in your explanation require definition; start at the lowest possible level.

Web Projects

Find useful links for Web projects on the text Web site.

1. *Life on Mars.* Find the latest information regarding the controversy over evidence for life in Martian meteorites. Write a one- to two-page report summarizing the current state of the controversy and your own opinion as to whether life once existed on Mars.

2. *Search for Life.* Learn about a proposed mission to search for life on Mars, Europa, or elsewhere in our solar system or beyond. Write a one- to two-page summary of the mission and its prospects for success.

3. *Human Threats to the Earth.* Write an in-depth research report, three to five pages in length, about current understanding and controversy regarding one of the following issues: global warming, ozone depletion, or the loss of species on Earth due to human activity. Be sure to address both the latest knowledge about the issue and proposals for alleviating any dangers associated with it. End your report by making your own recommendations about what, if anything, needs to be done to prevent damage to the Earth.

4. *Local Geology.* Write a one- to two-page report about the geology of an area you know well—perhaps the location of your campus or your hometown. Which of the four geological processes has played the most important role, in your opinion? Has plate tectonics played a direct role in shaping the area?

PART IV

A Deeper Look at Nature

Henceforth space by itself, and time by itself, are doomed to fade away into mere shadows, and only a kind of union of the two will preserve an independent reality.

Hermann Minkowski, 1908

CHAPTER S2
Space and Time

The universe consists of matter and energy moving through *space* with the passage of *time*. Up to this point in the book, we have discussed the concepts of space and time as though they are absolute and distinct—just as they appear in everyday life. But what if this appearance is deceiving?

About a century ago, Albert Einstein discovered that space and time are not what they appear to be. Instead, space and time are intertwined in a remarkable manner described by Einstein's *theory of relativity*. Because space and time are such fundamental concepts, understanding relativity is important to understanding the universe.

The theory of relativity is *not* difficult to understand, despite popular myths to the contrary. It does, however, require us to think in new ways. That is our task in this chapter and the next. By the time we are finished, you will see that Einstein brought about a revolution in human thinking with many important ramifications for understanding our place in the universe.

S2.1 Einstein's Revolution

Imagine that, with the aid of a long tape measure, you carefully measure the distance you walk from home to work to be 5.0 kilometers. You wouldn't expect any argument about this distance. For example, if a friend drives her car along the same route and measures the distance with her car's odometer, she ought to get the same measurement of 5.0 kilometers—as long as her odometer is working properly.

Likewise, you would expect agreement about the time it takes you to walk from home to work. Suppose your friend continues driving and you call her on a cellular phone just as you leave your house at 8:00 A.M. and again just as you arrive at work at 8:45 A.M. You'd certainly be surprised if she argued that your walk took an amount of time other than 45 minutes.

Distances and times appear absolute and distinct in our daily lives. We expect everyone to agree on the distance between two points, such as the locations of home and work. We also expect agreement about the time between two events, such as leaving home and arriving at work. Thus, it came as a huge surprise to everyone when in 1905 Albert Einstein showed that these expectations are not strictly correct.

With extremely precise measurements, the distance you measure between home and work will be *different* from the distance measured by a friend in a car, and you and your friend will also disagree about the time it takes you to walk to work. At ordinary speeds, the differences will be so small as to be unnoticeable. But if your friend could drive at a speed close to the speed of light, the differences would be substantial.

Disagreements about distances and times are only the beginning of the astonishing ideas contained in Einstein's **theory of relativity**. Einstein developed this theory in two parts. His **special theory of relativity**, published in 1905, showed how space and time are intertwined but did not deal with the effects of gravity. His **general theory of relativity**, published in 1915, offered a surprising new view of gravity—a view that we will use to help us understand topics such as the expansion and fate of the universe and the strange objects known as *black holes*.

In this chapter, we will focus on the new view of space and time in Einstein's special theory of relativity. In particular, we will see how this theory supports each of the following ideas:

- Nothing can travel faster than the speed of light (in a vacuum), and no material object can even reach the speed of light.

- If you carefully observe anyone or anything moving by you at a speed close to the speed of light, time will run more slowly for the moving object. That is, a person moving by you ages more slowly than you, a clock moving by you ticks more slowly than your clock, a computer moving by you runs more slowly than your similar computer, and so on.

- If you observe two events to occur simultaneously, such as flashes of light in two different places at the same time, a person moving by you at a speed close to the speed of light will not agree that the two events were simultaneous.

- If you carefully measure the size of something moving by you at a speed close to the speed of light, you will find that its length (in the direction of its motion) is shorter than it would be if the object were not moving.

- If you could measure the mass of something moving by you at a speed close to the speed of light, you would find its mass to be greater than the mass it would have if it were stationary. From this fact you can conclude, as did Einstein, that $E = mc^2$.

Although the consequences of special relativity may sound like science fiction or fantasy, their reality is supported by a vast body of observational and experimental evidence. They also follow logically from a few simple ideas. If you keep an open mind and think deeply as you read, you'll soon be confident and conversant in the ideas of relativity.

What Is Relativity?

Suppose a supersonic airplane is flying at a speed of 1,650 km/hr from Nairobi, Kenya, to Quito, Ecuador. How fast is the plane going? At first, this question sounds trivial—we have just said that the plane is going 1,650 km/hr.

But wait.... Nairobi and Quito are both nearly on the Earth's equator, and the equatorial speed of the Earth's rotation is the same 1,650 km/hr that the plane is flying [Section 1.3]. Moreover, the east-to-west flight from Nairobi to Quito is opposite the direction of the Earth's rotation (Figure S2.1). Thus, if you could observe the plane from far off in space, it would appear to stay put *while the Earth rotated beneath it*. When the flight began, you would see the plane lift straight off the ground in Nairobi. The plane would then remain stationary while the Earth's rotation carried Nairobi away from it and Quito toward it. When Quito finally reached the plane's position, the plane would drop straight down to the ground.

We have two alternative viewpoints about the plane's flight. People on Earth say that the plane is traveling westward across the surface of the Earth. Observers in space say that the plane is stationary while the Earth rotates eastward beneath it. Both viewpoints are equally valid. In fact, there are many other equally valid viewpoints about the plane's flight. Observers looking at the solar system as a whole

APPENDIXES

A Useful Numbers A-2

B Useful Formulas A-3

C A Few Mathematical Skills A-4
   - C.1 *Powers of 10* A-4
   - C.2 *Scientific Notation* A-6
   - C.3 *Working with Units* A-7
   - C.4 *The Metric System (SI)* A-10
   - C.5 *Finding a Ratio* A-11

D The Periodic Table of the Elements A-13

E Planetary Data A-15
   - Table E.1 *Physical Properties of the Sun and Planets* A-15
   - Table E.2 *Orbital Properties of the Sun and Planets* A-15
   - Table E.3 *Satellites of the Solar System* A-16

F Stellar Data A-18
   - Table F.1 *Stars Within 12 Light-Years* A-18
   - Table F.2 *Twenty Brightest Stars* A-19

G Galaxy Data A-20
   - Table G.1 *Galaxies of the Local Group* A-20
   - Table G.2 *Nearby Galaxies in the Messier Catalog* A-21
   - Table G.3 *Nearby, X-ray Bright Clusters of Galaxies* A-22

H Selected Astronomical Web Sites A-23

I The 88 Constellations A-26

J Star Charts A-29

APPENDIX A

USEFUL NUMBERS

Astronomical Distances

1 AU ≈ 1.496×10^8 km [p. 27]
1 light-year ≈ 9.46×10^{12} km [p. 8]
1 parsec (pc) ≈ 3.09×10^{13} km ≈ 3.26 light-years [p. 475]
1 kiloparsec (kpc) = 1,000 pc ≈ 3.26×10^3 light-years
1 megaparsec (Mpc) = 10^6 pc ≈ 3.26×10^6 light-years

Astronomical Times

1 solar day (average) = 24^h [p. 66]
1 sidereal day ≈ $23^h\ 56^m\ 4.09^s$ [p. 66]
1 synodic month (average) ≈ 29.53 solar days [p. 67]
1 sidereal month (average) ≈ 27.32 solar days [p. 67]
1 tropical year ≈ 365.242 solar days [p. 67]
1 sidereal year ≈ 365.256 solar days [p. 67]

Universal Constants

Speed of light [p. 8]: $c = 3 \times 10^5$ km/s = 3×10^8 m/s

Gravitational constant [p. 140]: $G = 6.67 \times 10^{-11} \dfrac{m^3}{kg \times s^2}$

Planck's constant [p. 156]: $h = 6.626 \times 10^{-34}$ joule × s

Stefan–Boltzmann constant [p. 162]: $\sigma = 5.7 \times 10^{-8} \dfrac{watt}{m^2 \times Kelvin^4}$

mass of a proton: $m_p = 1.67 \times 10^{-27}$ kg

mass of an electron: $m_e = 9.1 \times 10^{-31}$ kg

Useful Sun and Earth Reference Values

Mass of the Sun: $1 M_{Sun} \approx 2 \times 10^{30}$ kg
Radius of the Sun: $1 R_{Sun} \approx 696{,}000$ km
Luminosity of the Sun: $1 L_{Sun} \approx 3.8 \times 10^{26}$ watts
Mass of the Earth: $1 M_{Earth} \approx 5.97 \times 10^{24}$ kg
Radius (equatorial) of the Earth: $1 R_{Earth} \approx 6{,}378$ km
Acceleration of gravity on Earth: $g = 9.8$ m/s^2
Escape velocity from surface of Earth: $v_{escape} = 11$ km/s = 11,000 m/s

Energy and Power Units

Basic unit of energy [p. 114]: 1 joule = $1 \dfrac{kg \times m^2}{s^2}$

Basic unit of power [p. 114]: 1 watt = 1 joule/s

Electron-volt [p. 121]: 1 eV = 1.60×10^{-19} joule

APPENDIX B

USEFUL FORMULAS

- Universal law of gravitation for the force between objects of mass M_1 and M_2, distance d between their centers [p. 140]:

$$F = G\frac{M_1 M_2}{d^2}$$

- Newton's version of Kepler's third law; p and a are period and semimajor axis, respectively, of either orbiting mass [p. 141]:

$$p^2 = \frac{4\pi^2}{G(M_1 + M_2)} a^3$$

- Escape velocity at distance R from center of object of mass M [p. 146]:

$$v_{escape} = \sqrt{\frac{2GM}{R}}$$

- Relationship between a photon's wavelength (λ), frequency (f), and the speed of light (c) [p. 156]:

$$\lambda \times f = c$$

- Energy of a photon of wavelength λ or frequency f [p. 156]:

$$E = hf = \frac{hc}{\lambda}$$

- Stefan–Boltzmann law for thermal radiation at temperature T (in Kelvin) [p. 162]:

$$\text{emitted power per unit area} = \sigma T^4$$

- Wien's law for the peak wavelength (λ_{max}) thermal radiation at temperature T (in Kelvin) [p. 162]:

$$\lambda_{max} = \frac{2{,}900{,}000}{T} \text{ nm}$$

- Doppler shift (radial velocity is positive if the object is moving away from us and negative if it is moving toward us) [p. 167]:

$$\frac{\text{radial velocity}}{\text{speed of light}} = \frac{\text{shifted wavelength} - \text{rest wavelength}}{\text{rest wavelength}}$$

- Angular separation (α) of two points with an actual separation s, viewed from a distance d (assuming d is much larger than s) [p. 174]:

$$\alpha = \frac{s}{2\pi d} \times 360°$$

- Luminosity–distance formula [p. 473]:

$$\text{apparent brightness} = \frac{\text{luminosity}}{4\pi d^2}$$

(where d is the distance to the object)

- Parallax formula (distance d to a star with parallax angle p in arcseconds) [p. 475]:

$$d \text{ (in parsecs)} = \frac{1}{p \text{ (in arcseconds)}}$$

- The orbital velocity law [p. 542], to find the mass M_r contained within the circular orbit of radius r for an object moving at speed v:

$$M_r = \frac{r \times v^2}{G}$$

APPENDIX C

A FEW MATHEMATICAL SKILLS

This appendix reviews the following mathematical skills: powers of 10, scientific notation, working with units, the metric system, and finding a ratio. You should refer to this appendix as needed while studying the textbook, particularly if you are having difficulty with the Mathematical Insights.

C.1 Powers of 10

Powers of 10 simply indicate how many times to multiply 10 by itself. For example:

$$10^2 = 10 \times 10 = 100$$

$$10^6 = 10 \times 10 \times 10 \times 10 \times 10 \times 10 = 1{,}000{,}000$$

Negative powers are the reciprocals of the corresponding positive powers. For example:

$$10^{-2} = \frac{1}{10^2} = \frac{1}{100} = 0.01$$

$$10^{-6} = \frac{1}{10^6} = \frac{1}{1{,}000{,}000} = 0.000001$$

Table C.1 lists powers of 10 from 10^{-12} to 10^{12}. Note that powers of 10 follow two basic rules:

1. A positive exponent tells how many zeros follow the 1. For example, 10^0 is a 1 followed by no zeros, and 10^8 is a 1 followed by eight zeros.

2. A negative exponent tells how many places are to the right of the decimal point, including the 1. For example, $10^{-1} = 0.1$ has one place to the right of the decimal point; $10^{-6} = 0.000001$ has six places to the right of the decimal point.

Table C.1 Powers of 10

| \multicolumn{3}{c}{Zero and Positive Powers} | \multicolumn{3}{c}{Negative Powers} |

Power	Value	Name	Power	Value	Name
10^0	1	One	10^{-1}	0.1	Tenth
10^1	10	Ten	10^{-2}	0.01	Hundredth
10^2	100	Hundred	10^{-3}	0.001	Thousandth
10^3	1,000	Thousand	10^{-4}	0.0001	Ten thousandth
10^4	10,000	Ten thousand	10^{-5}	0.00001	Hundred thousandth
10^5	100,000	Hundred thousand	10^{-6}	0.000001	Millionth
10^6	1,000,000	Million	10^{-7}	0.0000001	Ten millionth
10^7	10,000,000	Ten million	10^{-8}	0.00000001	Hundred millionth
10^8	100,000,000	Hundred million	10^{-9}	0.000000001	Billionth
10^9	1,000,000,000	Billion	10^{-10}	0.0000000001	Ten billionth
10^{10}	10,000,000,000	Ten billion	10^{-11}	0.00000000001	Hundred billionth
10^{11}	100,000,000,000	Hundred billion	10^{-12}	0.000000000001	Trillionth
10^{12}	1,000,000,000,000	Trillion			

Multiplying and Dividing Powers of 10

Multiplying powers of 10 simply requires adding exponents, as the following examples show:

$$10^4 \times 10^7 = \underbrace{10{,}000}_{10^4} \times \underbrace{10{,}000{,}000}_{10^7} = \underbrace{100{,}000{,}000{,}000}_{10^{4+7} \,=\, 10^{11}} = 10^{11}$$

$$10^5 \times 10^{-3} = \underbrace{100{,}000}_{10^5} \times \underbrace{0.001}_{10^{-3}} = \underbrace{100}_{10^{5+(-3)} \,=\, 10^2} = 10^2$$

$$10^{-8} \times 10^{-5} = \underbrace{0.00000001}_{10^{-8}} \times \underbrace{0.00001}_{10^{-5}} = \underbrace{0.0000000000001}_{10^{-8+(-5)} \,=\, 10^{-13}} = 10^{-13}$$

Dividing powers of 10 requires subtracting exponents, as in the following examples:

$$\frac{10^5}{10^3} = \underbrace{100{,}000}_{10^5} \div \underbrace{1{,}000}_{10^3} = \underbrace{100}_{10^{5-3} \,=\, 10^2} = 10^2$$

$$\frac{10^3}{10^7} = \underbrace{1{,}000}_{10^3} \div \underbrace{10{,}000{,}000}_{10^7} = \underbrace{0.0001}_{10^{3-7} \,=\, 10^{-4}} = 10^{-4}$$

$$\frac{10^{-4}}{10^{-6}} = \underbrace{0.0001}_{10^{-4}} \div \underbrace{0.000001}_{10^{-6}} = \underbrace{100}_{10^{-4-(-6)} \,=\, 10^2} = 10^2$$

Powers of Powers of 10

We can use the multiplication and division rules to raise powers of 10 to other powers or to take roots. For example:

$$(10^4)^3 = 10^4 \times 10^4 \times 10^4 = 10^{4+4+4} = 10^{12}$$

Note that we can get the same end result by simply multiplying the two powers:

$$(10^4)^3 = 10^{4 \times 3} = 10^{12}$$

Because taking a root is the same as raising to a fractional power (e.g., the square root is the same as the 1/2 power, the cube root is the same as the 1/3 power, etc.), we can use the same procedure for roots, as in the following example:

$$\sqrt{10^4} = (10^4)^{1/2} = 10^{4 \times (1/2)} = 10^2$$

Adding and Subtracting Powers of 10

Unlike with multiplication and division, there is no shortcut for adding or subtracting powers of 10. The values must be written in longhand notation. For example:

$$10^6 + 10^2 = 1{,}000{,}000 + 100 = 1{,}000{,}100$$

$$10^8 + 10^{-3} = 100{,}000{,}000 + 0.001 = 100{,}000{,}000.001$$

$$10^7 - 10^3 = 10{,}000{,}000 - 1{,}000 = 9{,}999{,}000$$

Summary

We can summarize our findings using n and m to represent any numbers:

- To *multiply* powers of 10, *add* exponents: $10^n \times 10^m = 10^{n+m}$

- To *divide* powers of 10, *subtract* exponents: $\dfrac{10^n}{10^m} = 10^{n-m}$

- To *raise* powers of 10 to other powers, multiply exponents: $(10^n)^m = 10^{n \times m}$

C.2 Scientific Notation

When we are dealing with large or small numbers, it's generally easier to write them with powers of 10. For example, it's much easier to write the number 6,000,000,000,000 as 6×10^{12}. This format, in which a number *between* 1 and 10 is multiplied by a power of 10, is called **scientific notation**.

Converting a Number to Scientific Notation

We can convert numbers written in ordinary notation to scientific notation with a simple two-step process:

1. Move the decimal point to come after the *first* nonzero digit.

2. The number of places the decimal point moves tells you the power of 10; the power is *positive* if the decimal point moves to the left and *negative* if it moves to the right.

Examples:

$$3{,}042 \xrightarrow{\text{decimal needs to move 3 places to left}} 3.042 \times 10^3$$

$$0.00012 \xrightarrow{\text{decimal needs to move 4 places to right}} 1.2 \times 10^{-4}$$

$$226 \times 10^2 \xrightarrow{\text{decimal needs to move 2 places to left}} (2.26 \times 10^2) \times 10^2 = 2.26 \times 10^4$$

Converting a Number from Scientific Notation

We can convert numbers written in scientific notation to ordinary notation by the reverse process:

1. The power of 10 indicates how many places to move the decimal point; move it to the *right* if the power of 10 is positive and to the *left* if it is negative.

2. If moving the decimal point creates any open places, fill them with zeros.

Examples:

$$4.01 \times 10^2 \xrightarrow{\text{move decimal 2 places to right}} 401$$

$$3.6 \times 10^6 \xrightarrow{\text{move decimal 6 places to right}} 3{,}600{,}000$$

$$5.7 \times 10^{-3} \xrightarrow{\text{move decimal 3 places to left}} 0.0057$$

Multiplying or Dividing Numbers in Scientific Notation

Multiplying or dividing numbers in scientific notation simply requires operating on the powers of 10 and the other parts of the number separately.

Examples:

$$(6 \times 10^2) \times (4 \times 10^5) = (6 \times 4) \times (10^2 \times 10^5) = 24 \times 10^7 = (2.4 \times 10^1) \times 10^7 = 2.4 \times 10^8$$

$$\frac{4.2 \times 10^{-2}}{8.4 \times 10^{-5}} = \frac{4.2}{8.4} \times \frac{10^{-2}}{10^{-5}} = 0.5 \times 10^{-2-(-5)} = 0.5 \times 10^3 = (5 \times 10^{-1}) \times 10^3 = 5 \times 10^2$$

Note that, in both these examples, we first found an answer in which the number multiplied by a power of 10 was *not* between 1 and 10. We therefore followed the procedure for converting the final answer to scientific notation.

Addition and Subtraction with Scientific Notation

In general, we must write numbers in ordinary notation before adding or subtracting.

Examples:

$$(3 \times 10^6) + (5 \times 10^2) = 3{,}000{,}000 + 500 = 3{,}000{,}500 = 3.0005 \times 10^6$$

$$(4.6 \times 10^9) - (5 \times 10^8) = 4{,}600{,}000{,}000 - 500{,}000{,}000 = 4{,}100{,}000{,}000 = 4.1 \times 10^9$$

When both numbers have the *same* power of 10, we can factor out the power of 10 first.

Examples:

$$(7 \times 10^{10}) + (4 \times 10^{10}) = (7 + 4) \times 10^{10} = 11 \times 10^{10} = 1.1 \times 10^{11}$$

$$(2.3 \times 10^{-22}) - (1.6 \times 10^{-22}) = (2.3 - 1.6) \times 10^{-22} = 0.7 \times 10^{-22} = 7.0 \times 10^{-23}$$

C.3 Working with Units

Showing the units of a problem as you solve it usually makes the work much easier and also provides a useful way of checking your work. If an answer does not come out with the units you expect, you probably did something wrong. In general, working with units is very similar to working with numbers, as the following guidelines and examples show.

Five Guidelines for Working with Units

Before you begin any problem, think ahead and identify the units you expect for the final answer. Then operate on the units along with the numbers as you solve the problem. The following five guidelines may be helpful when you are working with units:

1. Mathematically, it doesn't matter whether a unit is singular (e.g., meter) or plural (e.g., meters); we can use the same abbreviation (e.g., m) for both.

2. You cannot add or subtract numbers unless they have the *same* units. For example, 5 apples + 3 apples = 8 apples, but the expression 5 apples + 3 oranges cannot be simplified further.

3. You *can* multiply units, divide units, or raise units to powers. Look for key words that tell you what to do.
   - *Per* suggests division. For example, we write a speed of 100 kilometers per hour as:

   $$100 \frac{km}{hr} \quad \text{or} \quad 100 \frac{km}{1 \, hr}$$

   - *Of* suggests multiplication. For example, if you launch a 50-kg space probe at a launch cost *of* $10,000 per kilogram, the total cost is:

   $$50 \, kg \times \frac{\$10{,}000}{kg} = \$500{,}000$$

   - *Square* suggests raising to the second power. For example, we write an area of 75 square meters as 75 m².
   - *Cube* suggests raising to the third power. For example, we write a volume of 12 cubic centimeters as 12 cm³.

4. Often the number you are given is not in the units you wish to work with. For example, you may be given that the speed of light is 300,000 km/s but need it in units of m/s for a particular problem. To convert the units, simply multiply the given number by a *conversion factor*: a fraction in which the numerator (top of the fraction) and denominator (bottom of the fraction) are equal, so that the value of the fraction is 1; the number in the denominator must have the units that you wish to change. In the case of changing the speed of light from units of km/s to m/s, you need a conversion factor for kilometers to meters. Thus, the conversion factor is:

$$\frac{1{,}000 \, m}{1 \, km}$$

Note that this conversion factor is equal to 1, since 1,000 meters and 1 kilometer are equal, and that the units to be changed (km) appear in the denominator. We can now convert the speed of light from units of km/s to m/s simply by multiplying by this conversion factor:

$$\underbrace{300{,}000 \frac{km}{s}}_{\text{speed of light in km/s}} \times \underbrace{\frac{1{,}000 \, m}{1 \, km}}_{\text{conversion from km to m}} = \underbrace{3 \times 10^8 \frac{m}{s}}_{\text{speed of light in m/s}}$$

Note that the units of km cancel, leaving the answer in units of m/s.

5. It's easier to work with units if you replace division with multiplication by the reciprocal. For example, suppose you want to know how many minutes are represented by 300 seconds. We can find the answer by dividing 300 seconds by 60 seconds per minute:

$$300 \, s \div 60 \frac{s}{min}$$

However, it is easier to see the unit cancellations if we rewrite this expression by replacing the division with multiplication by the reciprocal (this process is easy to remember as "invert and multiply"):

$$300 \, s \div 60 \frac{s}{min} = 300 \, s \times \underbrace{\frac{1 \, min}{60 \, s}}_{\text{invert and multiply}} = 5 \, min$$

We now see that the units of seconds (s) cancel in the numerator of the first term and the denominator of the second term, leaving the answer in units of minutes.

More Examples of Working with Units

Example 1. How many seconds are there in 1 day?

Solution: We can answer the question by setting up a *chain* of unit conversions in which we start with 1 *day* and end up with *seconds*. We use the facts that there are 24 hours per day (24 hr/day), 60 minutes per hour (60 min/hr), and 60 seconds per minute (60 s/min):

$$\underbrace{1 \text{ day}}_{\text{starting value}} \times \underbrace{\frac{24 \text{ hr}}{\text{day}}}_{\substack{\text{conversion} \\ \text{from} \\ \text{day to hr}}} \times \underbrace{\frac{60 \text{ min}}{\text{hr}}}_{\substack{\text{conversion} \\ \text{from} \\ \text{hr to min}}} \times \underbrace{\frac{60 \text{ s}}{\text{min}}}_{\substack{\text{conversion} \\ \text{from} \\ \text{min to s}}} = 86{,}400 \text{ s}$$

Note that all the units cancel except *seconds,* which is what we want for the answer. There are 86,400 seconds in 1 day.

Example 2. Convert a distance of 10^8 cm to km.

Solution: The easiest way to make this conversion is in two steps, since we know that there are 100 centimeters per meter (100 cm/m) and 1,000 meters per kilometer (1,000 m/km):

$$\underbrace{10^8 \text{ cm}}_{\substack{\text{starting} \\ \text{value}}} \times \underbrace{\frac{1 \text{ m}}{100 \text{ cm}}}_{\substack{\text{conversion} \\ \text{from} \\ \text{cm to m}}} \times \underbrace{\frac{1 \text{ km}}{1{,}000 \text{ m}}}_{\substack{\text{conversion} \\ \text{from} \\ \text{m to km}}} = 10^8 \text{ cm} \times \frac{1 \text{ m}}{10^2 \text{ cm}} \times \frac{1 \text{ km}}{10^3 \text{ m}} = 10^3 \text{ km}$$

Alternatively, if we recognize that the number of kilometers should be smaller than the number of centimeters (because kilometers are larger), we might decide to do this conversion by dividing as follows:

$$10^8 \text{ cm} \div \frac{100 \text{ cm}}{\text{m}} \div \frac{1{,}000 \text{ m}}{\text{km}}$$

In this case, before carrying out the calculation, we replace each division with multiplication by the reciprocal:

$$10^8 \text{ cm} \div \frac{100 \text{ cm}}{\text{m}} \div \frac{1{,}000 \text{ m}}{\text{km}} = 10^8 \text{ cm} \times \frac{1 \text{ m}}{100 \text{ cm}} \times \frac{1 \text{ km}}{1{,}000 \text{ m}}$$

$$= 10^8 \text{ cm} \times \frac{1 \text{ m}}{10^2 \text{ cm}} \times \frac{1 \text{ km}}{10^3 \text{ m}}$$

$$= 10^3 \text{ km}$$

Note that we again get the answer that 10^8 cm is the same as 10^3 km, or 1,000 km.

Example 3. Suppose you accelerate at 9.8 m/s² for 4 seconds, starting from rest. How fast will you be going?

Solution: The question asked "how fast?" so we expect to end up with a speed. Therefore, we multiply the acceleration by the amount of time you accelerated:

$$9.8 \frac{\text{m}}{\text{s}^2} \times 4 \text{ s} = (9.8 \times 4) \frac{\text{m} \times \text{s}}{\text{s}^2} = 39.2 \frac{\text{m}}{\text{s}}$$

Note that the units end up as a speed, showing that you will be traveling 39.2 m/s after 4 seconds of acceleration at 9.8 m/s².

Example 4. A reservoir is 2 km long and 3 km wide. Calculate its area, in both square kilometers and square meters.

Solution: We find its area by multiplying its length and width:

$$2 \text{ km} \times 3 \text{ km} = 6 \text{ km}^2$$

Next we need to convert this area of 6 km² to square meters, using the fact that there are 1,000 meters per kilometer (1,000 m/km). Note that we must square the term 1,000 m/km when converting from km² to m²:

$$6 \text{ km}^2 \times \left(1{,}000 \frac{\text{m}}{\text{km}}\right)^2 = 6 \text{ km}^2 \times 1{,}000^2 \frac{\text{m}^2}{\text{km}^2} = 6 \cancel{\text{km}^2} \times 1{,}000{,}000 \frac{\text{m}^2}{\cancel{\text{km}^2}}$$

$$= 6{,}000{,}000 \text{ m}^2$$

The reservoir area is 6 km², which is the same as 6 million m².

C.4 The Metric System (SI)

The modern version of the metric system, known as *Système Internationale d'Unites* (French for "International System of Units") or **SI**, was formally established in 1960. Today, it is the primary measurement system in nearly every country in the world with the exception of the United States. Even in the United States, it is the system of choice for science and international commerce.

The basic units of length, mass, and time in the SI are:

- The **meter** for length, abbreviated m
- The **kilogram** for mass, abbreviated kg
- The **second** for time, abbreviated s

Multiples of metric units are formed by powers of 10, using a prefix to indicate the power. For example, *kilo* means 10^3 (1,000), so a kilometer is 1,000 meters; a microgram is 0.000001 gram, because *micro* means 10^{-6}, or one millionth. Some of the more common prefixes are listed in Table C.2.

Metric Conversions

Table C.3 lists conversions between metric units and units used commonly in the United States. Note that the conversions between kilograms and pounds are valid only on Earth, because they depend on the strength of gravity.

Example 1. International athletic competitions generally use metric distances. Compare the length of a 100-meter race to that of a 100-yard race.

Table C.2 SI (Metric) Prefixes

Small Values			Large Values		
Prefix	Abbreviation	Value	Prefix	Abbreviation	Value
Deci	d	10^{-1}	Deca	da	10^1
Centi	c	10^{-2}	Hecto	h	10^2
Milli	m	10^{-3}	Kilo	k	10^3
Micro	μ	10^{-6}	Mega	M	10^6
Nano	n	10^{-9}	Giga	G	10^9
Pico	p	10^{-12}	Tera	T	10^{12}

Table C.3 Metric Conversions

To Metric	From Metric
1 inch = 2.540 cm	1 cm = 0.3937 inch
1 foot = 0.3048 m	1 m = 3.28 feet
1 yard = 0.9144 m	1 m = 1.094 yards
1 mile = 1.6093 km	1 km = 0.6214 mile
1 pound = 0.4536 kg	1 kg = 2.205 pounds

Solution: Table C.3 shows that 1 m = 1.094 yd, so 100 m is 109.4 yd. Note that 100 meters is almost 110 yards; a good "rule of thumb" to remember is that distances in meters are about 10% longer than the corresponding number of yards.

Example 2. How many square kilometers are in 1 square mile?

Solution: We use the square of the miles-to-kilometers conversion factor:

$$(1 \text{ mi}^2) \times \left(\frac{1.6093 \text{ km}}{1 \text{ mi}}\right)^2 = (1 \text{ mi}^2) \times \left(1.6093^2 \frac{\text{km}^2}{\text{mi}^2}\right) = 2.5898 \text{ km}^2$$

Therefore, 1 square mile is 2.5898 square kilometers.

C.5 Finding a Ratio

Suppose you want to compare two quantities, such as the average density of the Earth and the average density of Jupiter. The way we do such a comparison is by dividing, which tells us the *ratio* of the two quantities. In this case, the Earth's average density is 5.52 grams/cm^3 and Jupiter's average density is 1.33 grams/cm^3 (see Table 11.1), so the ratio is:

$$\frac{\text{average density of Earth}}{\text{average density of Jupiter}} = \frac{5.52 \text{ g/cm}^3}{1.33 \text{ g/cm}^3} = 4.15$$

Notice how the units cancel on both the top and bottom of the fraction. We can state our result in two equivalent ways:

- The ratio of the Earth's average density to Jupiter's average density is 4.15.

- The Earth's average density is 4.15 times Jupiter's average density.

Sometimes, the quantities that you want to compare may each involve an equation. In such cases, you could, of course, find the ratio by first calculating each of the two quantities individually and then dividing. However, it is much easier if you first express the ratio as a fraction, putting the equation for one quantity on top and the other on the bottom. Some of the terms in the equation may then cancel out, making any calculations much easier.

Example 1. Compare the kinetic energy of a car traveling at 100 km/hr to that of a car traveling at 50 km/hr.

Solution: We do the comparison by finding the ratio of the two kinetic energies, recalling that the formula for kinetic energy is $\frac{1}{2} mv^2$. Since we are not told the mass of the car, you might at first think that we don't have enough information to find the ratio. However, notice what happens when we put the equations for each kinetic energy into the ratio, calling the two speeds v_1 and v_2:

$$\frac{\text{K.E. car at } v_1}{\text{K.E. car at } v_2} = \frac{\frac{1}{2} m_{car} v_1^2}{\frac{1}{2} m_{car} v_2^2} = \frac{v_1^2}{v_2^2} = \left(\frac{v_1}{v_2}\right)^2$$

All the terms cancel except those with the two speeds, leaving us with a very simple formula for the ratio. Now we put in 100 km/hr for v_1 and 50 km/hr for v_2:

$$\frac{\text{K.E. car at 100 km/hr}}{\text{K.E. car at 50 km/hr}} = \left(\frac{100 \text{ km/hr}}{50 \text{ km/hr}}\right)^2 = 2^2 = 4$$

The ratio of the car's kinetic energies at 100 km/hr and 50 km/hr is 4. That is, the car has four times as much kinetic energy at 100 km/hr as it has at 50 km/hr.

Example 2. Compare the strength of gravity between the Earth and the Sun to the strength of gravity between the Earth and the Moon.

Solution: We do the comparison by taking the ratio of the Earth–Sun gravity to the Earth–Moon gravity. In this case, each quantity is found from the equation of Newton's law of gravity. (See Section 5.4.) Thus, the ratio is:

$$\frac{\text{Earth–Sun gravity}}{\text{Earth–Moon gravity}} = \frac{\cancel{G}\frac{\cancel{M_{\text{Earth}}}M_{\text{Sun}}}{(d_{\text{Earth–Sun}})^2}}{\cancel{G}\frac{\cancel{M_{\text{Earth}}}M_{\text{Moon}}}{(d_{\text{Earth–Moon}})^2}} = \frac{M_{\text{Sun}}}{(d_{\text{Earth–Sun}})^2} \times \frac{(d_{\text{Earth–Moon}})^2}{M_{\text{Moon}}}$$

Note how all but four of the terms cancel; the last step comes from replacing the division with multiplication by the reciprocal (the "invert and multiply" rule for division). We can simplify the work further by rearranging the terms so that we have the masses and distances together:

$$\frac{\text{Earth–Sun gravity}}{\text{Earth–Moon gravity}} = \frac{M_{\text{Sun}}}{M_{\text{Moon}}} \times \frac{(d_{\text{Earth–Moon}})^2}{(d_{\text{Earth–Sun}})^2}$$

Now it is just a matter of looking up the numbers (see Appendix E) and calculating:

$$\frac{\text{Earth–Sun gravity}}{\text{Earth–Moon gravity}} = \frac{1.99 \times 10^{30} \, \cancel{\text{kg}}}{7.35 \times 10^{22} \, \cancel{\text{kg}}} \times \frac{(384.4 \times 10^3 \, \cancel{\text{km}})^2}{(149.6 \times 10^6 \, \cancel{\text{km}})^2} = 179$$

In other words, the Earth–Sun gravity is 179 times stronger than the Earth–Moon gravity.

APPENDIX D

THE PERIODIC TABLE OF THE ELEMENTS

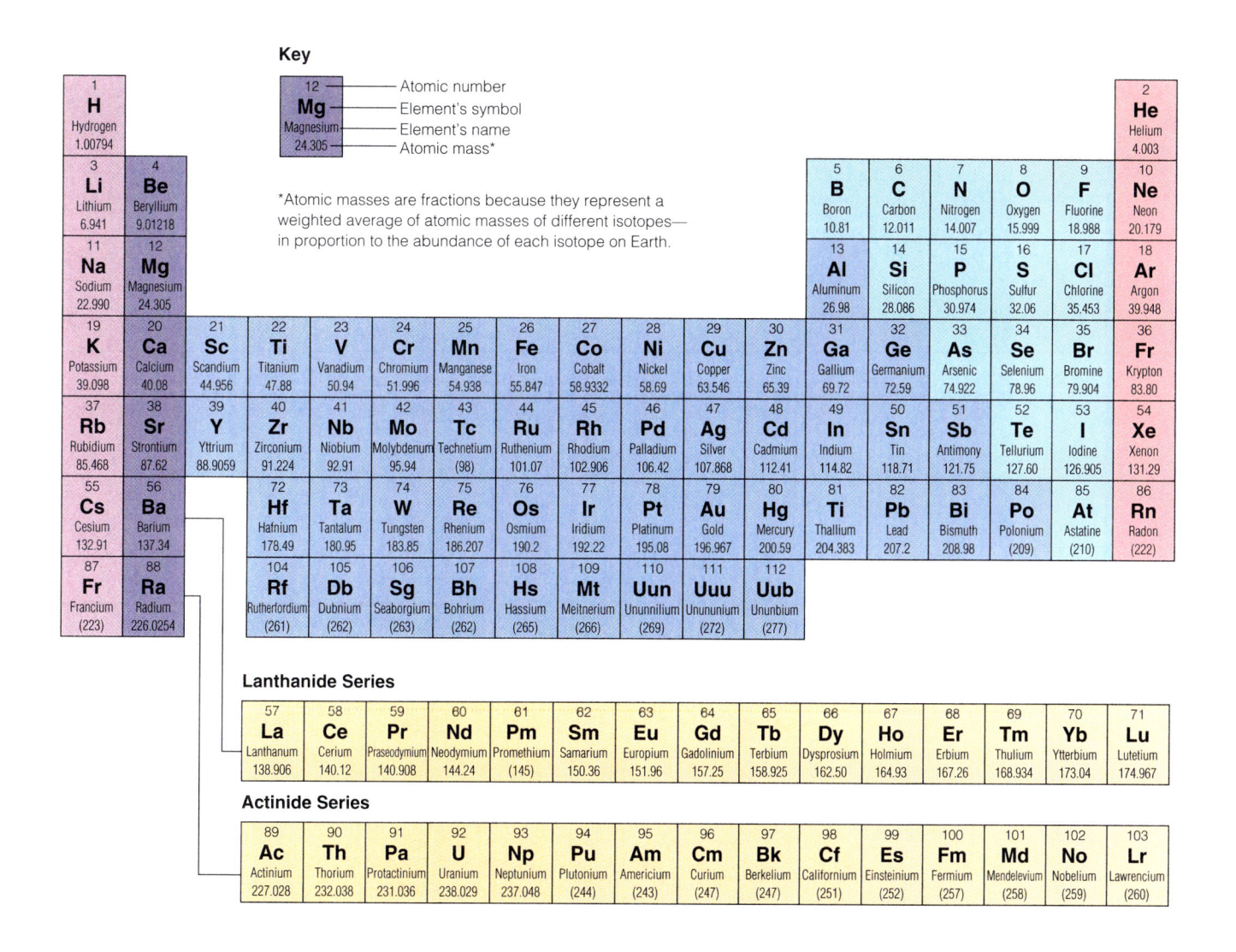

Key

12 — Atomic number
Mg — Element's symbol
Magnesium — Element's name
24.305 — Atomic mass*

*Atomic masses are fractions because they represent a weighted average of atomic masses of different isotopes—in proportion to the abundance of each isotope on Earth.

1																	2
H Hydrogen 1.00794																	**He** Helium 4.003
3 **Li** Lithium 6.941	4 **Be** Beryllium 9.01218										5 **B** Boron 10.81	6 **C** Carbon 12.011	7 **N** Nitrogen 14.007	8 **O** Oxygen 15.999	9 **F** Fluorine 18.988	10 **Ne** Neon 20.179
11 **Na** Sodium 22.990	12 **Mg** Magnesium 24.305											13 **Al** Aluminum 26.98	14 **Si** Silicon 28.086	15 **P** Phosphorus 30.974	16 **S** Sulfur 32.06	17 **Cl** Chlorine 35.453	18 **Ar** Argon 39.948
19 **K** Potassium 39.098	20 **Ca** Calcium 40.08	21 **Sc** Scandium 44.956	22 **Ti** Titanium 47.88	23 **V** Vanadium 50.94	24 **Cr** Chromium 51.996	25 **Mn** Manganese 54.938	26 **Fe** Iron 55.847	27 **Co** Cobalt 58.9332	28 **Ni** Nickel 58.69	29 **Cu** Copper 63.546	30 **Zn** Zinc 65.39	31 **Ga** Gallium 69.72	32 **Ge** Germanium 72.59	33 **As** Arsenic 74.922	34 **Se** Selenium 78.96	35 **Br** Bromine 79.904	36 **Fr** Krypton 83.80
37 **Rb** Rubidium 85.468	38 **Sr** Strontium 87.62	39 **Y** Yttrium 88.9059	40 **Zr** Zirconium 91.224	41 **Nb** Niobium 92.91	42 **Mo** Molybdenum 95.94	43 **Tc** Technetium (98)	44 **Ru** Ruthenium 101.07	45 **Rh** Rhodium 102.906	46 **Pd** Palladium 106.42	47 **Ag** Silver 107.868	48 **Cd** Cadmium 112.41	49 **In** Indium 114.82	50 **Sn** Tin 118.71	51 **Sb** Antimony 121.75	52 **Te** Tellurium 127.60	53 **I** Iodine 126.905	54 **Xe** Xenon 131.29
55 **Cs** Cesium 132.91	56 **Ba** Barium 137.34		72 **Hf** Hafnium 178.49	73 **Ta** Tantalum 180.95	74 **W** Tungsten 183.85	75 **Re** Rhenium 186.207	76 **Os** Osmium 190.2	77 **Ir** Iridium 192.22	78 **Pt** Platinum 195.08	79 **Au** Gold 196.967	80 **Hg** Mercury 200.59	81 **Ti** Thallium 204.383	82 **Pb** Lead 207.2	83 **Bi** Bismuth 208.98	84 **Po** Polonium (209)	85 **At** Astatine (210)	86 **Rn** Radon (222)
87 **Fr** Francium (223)	88 **Ra** Radium 226.0254		104 **Rf** Rutherfordium (261)	105 **Db** Dubnium (262)	106 **Sg** Seaborgium (263)	107 **Bh** Bohrium (262)	108 **Hs** Hassium (265)	109 **Mt** Meitnerium (266)	110 **Uun** Ununnilium (269)	111 **Uuu** Unununium (272)	112 **Uub** Ununbium (277)						

Lanthanide Series

57 **La** Lanthanum 138.906	58 **Ce** Cerium 140.12	59 **Pr** Praseodymium 140.908	60 **Nd** Neodymium 144.24	61 **Pm** Promethium (145)	62 **Sm** Samarium 150.36	63 **Eu** Europium 151.96	64 **Gd** Gadolinium 157.25	65 **Tb** Terbium 158.925	66 **Dy** Dysprosium 162.50	67 **Ho** Holmium 164.93	68 **Er** Erbium 167.26	69 **Tm** Thulium 168.934	70 **Yb** Ytterbium 173.04	71 **Lu** Lutetium 174.967

Actinide Series

89 **Ac** Actinium 227.028	90 **Th** Thorium 232.038	91 **Pa** Protactinium 231.036	92 **U** Uranium 238.029	93 **Np** Neptunium 237.048	94 **Pu** Plutonium (244)	95 **Am** Americium (243)	96 **Cm** Curium (247)	97 **Bk** Berkelium (247)	98 **Cf** Californium (251)	99 **Es** Einsteinium (252)	100 **Fm** Fermium (257)	101 **Md** Mendelevium (258)	102 **No** Nobelium (259)	103 **Lr** Lawrencium (260)

APPENDIX E

PLANETARY DATA

Table E.1 Physical Properties of the Sun and Planets

Name	Radius (Eq[a]) (km)	Radius (Eq) (Earth units)	Mass (kg)	Mass (Earth units)	Average Density (g/cm^3)	Surface Gravity (Earth = 1)
Sun	695,000	109	1.99×10^{30}	333,000	1.41	27.5
Mercury	2,440	0.382	3.30×10^{23}	0.055	5.43	0.38
Venus	6,051	0.949	4.87×10^{24}	0.815	5.25	0.91
Earth	6,378	1.00	5.97×10^{24}	1.00	5.52	1.00
Mars	3,397	0.533	6.42×10^{23}	0.107	3.93	0.38
Jupiter	71,492	11.19	1.90×10^{27}	317.9	1.33	2.53
Saturn	60,268	9.46	5.69×10^{26}	95.18	0.71	1.07
Uranus	25,559	3.98	8.66×10^{25}	14.54	1.24	0.91
Neptune	24,764	3.81	1.03×10^{26}	17.13	1.67	1.14
Pluto	1,160	0.181	1.31×10^{22}	0.0022	2.05	0.07

[a]Eq = equatorial.

Table E.2 Orbital Properties of the Sun and Planets

Name	Distance from Sun[a] (AU)	(10^6 km)	Orbital Period (years)	Orbital Inclination[b] (degrees)	Orbital Eccentricity	Sidereal Rotation Period (Earth days)[c]	Axis Tilt (degrees)
Sun	—	—	—	—	—	25.4	7.25
Mercury	0.387	57.9	0.2409	7.00	0.206	58.6	0.0
Venus	0.723	108.2	0.6152	3.39	0.007	−243.0	177.4
Earth	1.00	149.6	1.0	0.00	0.017	0.9973	23.45
Mars	1.524	227.9	1.881	1.85	0.093	1.026	23.98
Jupiter	5.203	778.3	11.86	1.31	0.048	0.41	3.08
Saturn	9.539	1,427	29.46	2.49	0.056	0.44	26.73
Uranus	19.19	2,870	84.01	0.77	0.046	−0.72	97.92
Neptune	30.06	4,497	164.8	1.77	0.010	0.67	29.6
Pluto	39.54	5,916	248.0	17.15	0.248	−6.39	118

[a]Semimajor axis of the orbit.
[b]With respect to the ecliptic.
[c]A negative sign indicates rotation relative to other planets.

Table E.3 Satellites of the Solar System[a]

Planet Satellite	Radius or Dimensions[b] (km)	Distance from Planet (10^3 km)	Orbital Period[c] (Earth days)	Mass[d] (kg)	Density[d] (g/cm^3)	Notes About the Satellites
Earth						**Earth**
Moon	1,738	384.4	27.322	7.349×10^{22}	3.34	*Moon:* Probably formed in giant impact.
Mars						**Mars**
Phobos	13×11×9	9.38	0.319	1.3×10^{16}	2.2	*Phobos, Deimos:* Probable captured asteroids.
Deimos	8×6×5	23.5	1.263	1.8×10^{15}	1.7	
Jupiter						**Jupiter**
Small inner moons (5 moons)	10 to 135×82×75	128–222	0.295–0.6745	—	—	*Metis, Adrastea, Amalthea, Thebe, 1999 J1:* Small moonlets within and near Jupiter's ring system.
Io	1,821	421.6	1.769	8.933×10^{22}	3.57	*Io:* Most volcanically active object in the solar system.
Europa	1,565	670.9	3.551	4.797×10^{22}	2.97	*Europa:* Possible oceans under icy crust.
Ganymede	2,634	1,070.0	7.155	1.482×10^{23}	1.94	*Ganymede:* Largest satellite in solar system; unusual ice geology.
Callisto	2,403	1,883.0	16.689	1.076×10^{23}	1.86	*Callisto:* Cratered iceball.
Irregular group 1 (5 moons)	7–22	11,400–17,100	453–829	—	—	*Leda, Himalia, Elara, Lysithea, 2000 J11:* Probable captured moons with inclined orbits.
Irregular group 2 (14 moons)	5–15	17,400–18,000	−854 to −901	—	—	*Ananke, Carme, Pasiphae, Sinope, 2000 J1–J10:* Probable captured moons in inclined backward orbits.
Saturn						**Saturn**
Small inner moons (6)	10 to 97×95×77	134–151	0.574–0.695	—	—	*Pan, Atlas, Prometheus, Pandora, Epimetheus, Janus:* Small moonlets within and near Saturn's ring system.
Mimas	199	185.52	0.942	3.70×10^{19}	1.17	*Mimas, Enceladus, Tethys:* Small and medium-size iceballs, many with interesting geology.
Enceladus	249	238.02	1.370	1.2×10^{20}	1.24	
Tethys	530	294.66	1.888	6.17×10^{20}	1.26	
Calypso	15×8×8	294.66	1.888	4×10^{15}	—	*Calypso, Telesto:* Small moonlets sharing Tethys's orbit.
Telesto	15×13×8	294.67	1.888	6×10^{15}	—	
Dione	559	377.4	2.737	1.08×10^{21}	1.44	*Dione:* Medium-size iceball, with interesting geology.
Helene	18×?×15	377.4	2.737	1.6×10^{16}	—	*Helene:* Small moonlet sharing Dione's orbit.
Rhea	764	527.04	4.518	2.31×10^{21}	1.33	*Rhea:* Medium-size iceball, with interesting geology.
Titan	2,575	1,221.85	15.945	1.3455×10^{23}	1.88	*Titan:* Dense atmosphere shrouds surface; ongoing geological activity possible.
Hyperion	180×140×112	1,481.1	21.277	2.8×10^{19}	—	*Hyperion:* Only satellite known not to rotate synchronously.

Iapetus	718	3,561.3	79.331	1.59×10^{21}	1.21
					Iapetus: Bright and dark hemispheres show greatest contrast in the solar system.
Phoebe	110	12,952	−550.4	1×10^{19}	—
					Phoebe: Very dark; material ejected from Phoebe may coat one side of Iapetus.
Irregular group 1 (4 moons)	7–22	11,400–17,100	453–829	—	—
					2000 S2, S3, S5, S6: Probable captured moons with highly inclined orbits.
Irregular group 2 (3 moons)	5–15	17,400–18,000	854–901	—	—
					2000 S4, S10, S11: Probable captured moons in inclined orbits.
Irregular group 3 (5 moons)	4–10	15,600–23,400	−723 to −1,325	—	—
					2000 S1, S7, S8, S9, S12: Probable captured moons in inclined backward orbits.
Uranus					**Uranus**
Small inner moons (11 moons)	10 to 97×95×77	134–151	0.574–0.695	—	—
					Cordelia, Ophelia, Bianca, Cressida, Desdemona, Juliet, Portia, Rosalind, Belinda, Puck, 1986 U10: Small moonlets within and near Uranus's ring system.
Miranda	236	129.8	1.413	6.6×10^{19}	1.26
Ariel	579	191.2	2.520	1.35×10^{21}	1.65
Umbriel	584.7	266.0	4.144	1.17×10^{21}	1.44
Titania	788.9	435.8	8.706	3.52×10^{21}	1.59
Oberon	761.4	582.6	13.463	3.01×10^{21}	1.50
					Miranda, Ariel, Umbriel, Titania, Oberon: Small and medium-size iceballs, with some interesting geology.
Irregular group (5 moons)	???–60	7,170–25,000	580–2,280	—	—
					Caliban, Sycorax, Stephano, Prospero, Setebos: Too recently discovered for accurate determination of their properties; several in backward orbits.
Neptune					**Neptune**
Small inner moons (5 moons)	29 to 104×?×89	48–74	0.296–0.554	—	—
					Naiad, Thalassa, Despina, Galatea, Larissa: Small moonlets within and near Neptune's ring system.
Proteus	218×208×201	117.6	1.121	—	—
Triton	1,352.6	354.59	−5.875	2.14×10^{22}	2.0
					Triton: Probable captured Kuiper belt object—largest captured object in solar system.
Nereid	170	5,588.6	360.125	3.1×10^{19}	—
					Nereid: Small, icy moon; very little known.
Pluto					**Pluto**
Charon	635	19.6	6.38718	1.56×10^{21}	1.6
					Charon: Unusually large compared to its planet; may have formed in giant impact.

[a] *Note*: Authorities differ substantially on many of the values in this table.

[b] a × b × c values for the Dimensions are the approximate lengths of the axes for irregular moons.

[c] Negative sign indicates backward orbit.

[d] Masses and densities are most accurate for those satellites visited by a spacecraft on a flyby. Masses for the smallest moons have not been measured but can be estimated from the radius and an assumed density.

APPENDIX F

STELLAR DATA

Table F.1 Stars Within 12 Light-Years

Star	Distance (ly)	Spectral Type		RA h	RA m	Dec °	Dec ′	Luminosity (L/L_{Sun})
Sun	0.000016	G2	V	—	—	—	—	1.0
Proxima Centauri	4.2	M5.5	V	14	30	−62	41	0.0006
α Centauri A	4.4	G2	V	14	40	−60	50	1.6
α Centauri B	4.4	K0	V	14	40	−60	50	0.53
Barnard's Star	6.0	M4	V	17	58	+04	42	0.005
Wolf 359	7.8	M6	V	10	56	+07	01	0.0008
Lalande 21185	8.3	M2	V	11	03	+35	58	0.03
Sirius A	8.6	A1	V	06	45	−16	42	26.0
Sirius B	8.6	DA2	—	06	45	−16	42	0.002
Luyten 726-8A	8.7	M5.5	V	01	39	−17	57	0.0009
Luyten 726-8B	8.7	M6	V	01	39	−17	57	0.0006
Ross 154	9.7	M3.5	V	18	50	−23	50	0.004
Ross 248	10.3	M5.5	V	23	42	+44	11	0.001
ε Eridani	10.5	K2	V	03	33	−09	28	0.37
Lacaille 9352	10.7	M1.5	V	23	06	−35	51	0.05
Ross 128	10.9	M4	V	11	48	+00	49	0.003
EZ Aquarii A	11.3	M5	V	22	39	−15	18	0.0006
EZ Aquarii B	11.3	M6	V	22	39	−15	18	0.0004
EZ Aquarii C	11.3	M6.5	V	22	39	−15	18	0.0003
61 Cygni A	11.4	K5	V	21	07	+38	42	0.15
61 Cygni B	11.4	K7	V	21	07	+38	42	0.09
Procyon A	11.4	F5	IV–V	07	39	+05	14	7.4
Procyon B	11.4	DA	—	07	39	+05	14	0.0005
Gliese 725 A	11.4	M3	V	18	43	+59	38	0.02
Gliese 725 B	11.4	M3.5	V	18	43	+59	38	0.01
Gliese 15 A	11.6	M1.5	V	00	18	+44	01	0.03
Gliese 15 B	11.6	M3.5	V	00	18	+44	01	0.003
DX Cancri	11.8	M6.5	V	08	30	+26	47	0.0003
ε Indi	11.8	K5	V	22	03	−56	45	0.26
τ Ceti	11.9	G8	V	01	44	−15	57	0.59
GJ 1061	11.9	M5.5	V	03	36	−44	31	0.0009

Note: These data were provided by the RECONS project, courtesy of Dr. Todd Henry. The luminosities are all total (bolometric) luminosities. The DA stellar types are white dwarfs. The coordinates are for the year 2000.

Table F.2 Twenty Brightest Stars

Star	Constellation	RA h	RA m	Dec °	Dec ′	Distance (ly)	Spectral Type		Apparent Magnitude	Luminosity (L/L_{Sun})
Sirius	Canis Major	6	45	−16	42	8.6	A1	V	−1.46	26
Canopus	Carina	6	24	−52	41	313	F0	Ib–II	−0.72	13,000
α Centauri	Centaurus	14	40	−60	50	4.4	G2	V	−0.01	1.6
							K0	V	1.3	0.53
Arcturus	Boötes	14	16	+19	11	37	K2	III	−0.06	170
Vega	Lyra	18	37	+38	47	25	A0	V	0.04	60
Capella	Auriga	5	17	+46	00	42	G0	III	0.75	70
							G8	III	0.85	77
Rigel	Orion	5	15	−08	12	772	B8	Ia	0.14	70,000
Procyon	Canis Minor	7	39	+05	14	11.4	F5	IV–V	0.37	7.4
Betelgeuse	Orion	5	55	+07	24	427	M2	Iab	0.41	38,000
Achernar	Eridanus	1	38	−57	15	144	B5	V	0.51	3,600
Hadar	Centaurus	14	04	−60	22	525	B1	III	0.63	100,000
Altair	Aquila	19	51	+08	52	17	A7	IV–V	0.77	10.5
Acrux	Crux	12	27	−63	06	321	B1	IV	1.39	22,000
							B3	V	1.9	7,500
Aldebaran	Taurus	4	36	+16	30	65	K5	III	0.86	350
Spica	Virgo	13	25	−11	09	260	B1	V	0.91	23,000
Antares	Scorpio	16	29	−26	26	604	M1	Ib	0.92	38,000
Pollux	Gemini	7	45	+28	01	34	K0	III	1.16	45
Fomalhaut	Piscis Austrinus	22	58	−29	37	25	A3	V	1.19	18
Deneb	Cygnus	20	41	+45	16	2,500	A2	Ia	1.26	170,000
β Crucis	Crux	12	48	−59	40	352	B0.5	IV	1.28	37,000

Note: Three of the stars on this list, Capella, α Centauri, and Acrux, are binary systems with members of comparable brightness. They are counted as single stars because that is how they appear to the naked eye. All the luminosities given are total (bolometric) luminosities. The coordinates are for the year 2000.

APPENDIX G

GALAXY DATA

Table G.1 Galaxies of the Local Group

Galaxy Name	Distance (millions of ly)	Type[a]	RA h	RA m	Dec °	Dec '	Luminosity (millions of L_{Sun})
Milky Way	—	Sbc	—	—	—	—	15,000
WLM	3.0	Irr	00	02	−15	30	50
NGC 55	4.8	Irr	00	15	−39	13	1,300
IC 10	2.7	dIrr	00	20	+59	18	160
NGC 147	2.4	dE	00	33	+48	30	131
And III	2.5	dE	00	35	+36	30	1.1
NGC 185	2.0	dE	00	39	+48	20	120
NGC 205	2.7	E	00	40	+41	41	370
M 32	2.6	E	00	43	+40	52	380
M 31	2.5	Sb	00	43	+41	16	21,000
And I	2.6	dE	00	46	+38	00	4.7
SMC	0.19	Irr	00	53	−72	50	230
Sculptor	0.26	dE	01	00	−33	42	2.2
LGS 3	2.6	dIrr	01	04	+21	53	1.3
IC 1613	2.3	Irr	01	05	+02	08	64
And II	1.7	dE	01	16	+33	26	2.4
M 33	2.7	Sc	01	34	+30	40	2,800
Phoenix	1.5	dIrr	01	51	−44	27	0.9
Fornax	0.45	dE	02	40	−34	27	15.5
EGB0427+63	4.3	dIrr	04	32	+63	36	9.1
LMC	0.16	Irr	05	24	−69	45	1,300
Carina	0.33	dE	06	42	−50	58	0.4
Leo A	2.2	dIrr	09	59	+30	45	3.0
Sextans B	4.4	dIrr	10	00	+05	20	41
NGC 3109	4.1	Irr	10	03	−26	09	160
Antlia	4.0	dIrr	10	04	−27	19	1.7
Leo I	0.82	dE	10	08	+12	18	4.8
Sextans A	4.7	dIrr	10	11	−04	42	56
Sextans	0.28	dE	10	13	−01	37	0.5
Leo II	0.67	dE	11	13	+22	09	0.6
GR 8	5.2	dIrr	12	59	+14	13	3.4
Ursa Minor	0.22	dE	15	09	+67	13	0.3
Draco	2.7	dE	17	20	+57	55	0.3
Sagittarius	0.08	dE	18	55	−30	29	18
SagDIG	3.5	dIrr	19	30	−17	41	6.8
NGC 6822	1.6	Irr	19	45	−14	48	94
DDO 210	2.6	dIrr	20	47	−12	51	0.8
IC 5152	5.2	dIrr	22	03	−51	18	70
Tucana	2.9	dE	22	42	−64	25	0.5
UKS2323-326	4.3	dE	23	26	−32	23	5.2
Pegasus	3.1	dIrr	23	29	+14	45	12

[a]Types beginning with S are spiral galaxies classified according to Hubble's system (see Chapter 19). Type E galaxies are elliptical or spheroidal. Type Irr galaxies are irregular. The prefix d denotes a dwarf galaxy.

Table G.2 Nearby Galaxies in the Messier Catalog[a,b]

Galaxy Name (M / NGC)[c]	RA h	RA m	Dec °	Dec '	RV_{hel}[d]	RV_{gal}[e]	Type[f]	Nickname
M 31 / NGC 224	00	43	+41	16	−300 ± 4	−122	Spiral	Andromeda
M 32 / NGC 221	00	43	+40	52	−145 ± 2	32	Elliptical	
M 33 / NGC 598	01	34	+30	40	−179 ± 3	−44	Spiral	Triangulum
M 49 / NGC 4472	12	30	+08	00	997 ± 7	929	Elliptical/Lenticular/Seyfert	
M 51 / NGC 5194	13	30	+47	12	463 ± 3	550	Spiral/Interacting	Whirlpool
M 58 / NGC 4579	12	38	+11	49	1,519 ± 6	1,468	Spiral/Seyfert	
M 59 / NGC 4621	12	42	+11	39	410 ± 6	361	Elliptical	
M 60 / NGC 4649	12	44	+11	33	1,117 ± 6	1,068	Elliptical	
M 61 / NGC 4303	12	22	+04	28	1,566 ± 2	1,483	Spiral/Seyfert	
M 63 / NGC 5055	13	16	+42	02	504 ± 4	570	Spiral	Sunflower
M 64 / NGC 4826	12	57	+21	41	408 ± 4	400	Spiral/Seyfert	Black Eye
M 65 / NGC 3623	11	19	+13	06	807 ± 3	723	Spiral	
M 66 / NGC 3627	11	20	+12	59	727 ± 3	643	Spiral/Seyfert	
M 74 / NGC 628	01	37	+15	47	657 ± 1	754	Spiral	
M 77 / NGC 1068	02	43	−00	01	1,137 ± 3	1,146	Spiral/Seyfert	
M 81 / NGC 3031	09	56	+69	04	−34 ± 4	73	Spiral/Seyfert	
M 82 / NGC 3034	09	56	+69	41	203 ± 4	312	Irregular/Starburst	
M 83 / NGC 5236	13	37	−29	52	516 ± 4	385	Spiral/Starburst	
M 84 / NGC 4374	12	25	+12	53	1,060 ± 6	1,005	Elliptical	
M 85 / NGC 4382	12	25	+18	11	729 ± 2	692	Spiral	
M 86 / NGC 4406	12	26	+12	57	−244 ± 5	−298	Elliptical/Lenticular	
M 87 / NGC 4486	12	30	+12	23	1,307 ± 7	1,254	Elliptical/Central Dominant/Seyfert	Virgo A
M 88 / NGC 4501	12	32	+14	25	2,281 ± 3	2,235	Spiral/Seyfert	
M 89 / NGC 4552	12	36	+12	33	340 ± 4	290	Elliptical	
M 90 / NGC 4569	12	37	+13	10	−235 ± 4	−282	Spiral/Seyfert	
M 91 / NGC 4548	12	35	+14	30	486 ± 4	442	Spiral/Seyfert	
M 94 / NGC 4736	12	51	+41	07	308 ± 1	360	Spiral	
M 95 / NGC 3351	10	44	+11	42	778 ± 4	677	Spiral/Starburst	
M 96 / NGC 3368	10	47	+11	49	897 ± 4	797	Spiral/Seyfert	
M 98 / NGC 4192	12	14	+14	54	−142 ± 4	−195	Spiral/Seyfert	
M 99 / NGC 4254	12	19	+14	25	2,407 ± 3	2,354	Spiral	
M 100 / NGC 4321	12	23	+15	49	1,571 ± 1	1,525	Spiral	
M 101 / NGC 5457	14	03	+54	21	241 ± 2	360	Spiral	
M 104 / NGC 4594	12	40	−11	37	1,024 ± 5	904	Spiral/Seyfert	Sombrero
M 105 / NGC 3379	10	48	+12	35	911 ± 2	814	Elliptical	
M 106 / NGC 4258	12	19	+47	18	448 ± 3	507	Spiral/Seyfert	
M 108 / NGC 3556	11	09	+55	57	695 ± 3	765	Spiral	
M 109 / NGC 3992	11	55	+53	39	1,048 ± 4	1,121	Spiral	
M 110 / NGC 205	00	38	+41	25	−241 ± 3	−61	Elliptical	

[a] Galaxies identified in the catalog published by Charles Messier in 1781; these galaxies are relatively easy to observe with small telescopes.

[b] Data obtained from NED: NASA/IPAC Extragalactic Database (http://ned.ipac.caltech.edu). The original Messier list of galaxies was obtained from SED, and the list data were updated to 2001 and M 102 was dropped.

[c] The galaxies are identified by the Messier number (M followed by a number) and by their NGC numbers, which come from the *New General Catalog* published in 1888.

[d] Radial velocity in km/s, with respect to the Sun (heliocentric). Positive values mean motion away from the Sun, and negative values are toward the Sun.

[e] Radial velocity in km/s, with respect to the Milky Way Galaxy, calculated from the RV_{hel} values with a correction for the Sun's motion around the galactic center.

[f] Galaxies are first listed by their primary type (spiral, elliptical, or irregular) and then by any other special categories that apply (see Chapters 19 and 20).

Table G.3 Nearby, X-ray Bright Clusters of Galaxies

Cluster Name	Redshift	Distance[a] (billions of ly)	Temperature of Intracluster Medium (millions of K)	Average Orbital Velocity of Galaxies[b] (km/sec)	Cluster Mass[c] ($10^{15} M_{Sun}$)
Abell 2142	0.0907	1.20	101. ± 2	1,132 ± 110	1.8
Abell 2029	0.0766	1.07	100. ± 3	1,164 ± 98	1.8
Abell 401	0.0737	1.03	95.2 ± 5	1,152 ± 86	1.6
Coma	0.0233	0.34	95.1 ± 1	821 ± 49	1.6
Abell 754	0.0539	0.77	93.3 ± 3	662 ± 77	1.6
Abell 2256	0.0589	0.83	87.0 ± 2	1,348 ± 86	1.4
Abell 399	0.0718	1.01	81.7 ± 7	1,116 ± 89	1.3
Abell 3571	0.0395	0.57	81.1 ± 3	1,045 ± 109	1.3
Abell 478	0.0882	1.22	78.9 ± 2	904 ± 281	1.2
Abell 3667	0.0566	0.80	78.5 ± 6	971 ± 62	1.2
Abell 3266	0.0599	0.85	78.2 ± 5	1,107 ± 82	1.2
Abell 1651a	0.0846	1.17	73.1 ± 6	685 ± 129	1.2
Abell 85	0.0560	0.80	70.9 ± 2	969 ± 95	1.2
Abell 119	0.0438	0.63	65.6 ± 5	679 ± 106	0.94
Abell 3558	0.0480	0.69	65.3 ± 2	977 ± 39	0.94
Abell 1795	0.0632	0.89	62.9 ± 2	834 ± 85	0.88
Abell 2199	0.0314	0.46	52.7 ± 1	801 ± 92	0.68
Abell 2147	0.0353	0.51	51.1 ± 4	821 ± 68	0.65
Abell 3562	0.0478	0.68	45.7 ± 8	736 ± 49	0.55
Abell 496	0.0325	0.47	45.3 ± 1	687 ± 89	0.54
Centaurus	0.0103	0.15	42.2 ± 1	863 ± 34	0.49
Abell 1367	0.0213	0.31	41.3 ± 2	822 ± 69	0.47
Hydra	0.0126	0.19	38.0 ± 1	610 ± 52	0.42
C0336	0.0349	0.50	37.4 ± 1	650 ± 170	0.41
Virgo	0.0038	0.06	25.7 ± 0.5	632 ± 41	0.23

Note: This table lists the 25 brightest clusters of galaxies in the X-ray sky from a catalog by J. P. Henry (2000).

[a] Cluster distances were computed using a value for Hubble's constant of 65 km/sec/Mpc.

[b] The average orbital velocities of galaxies given in this column are the velocity dispersions of the clusters' galaxies.

[c] This column gives each cluster's mass within the largest radius at which the intracluster medium can be in gravitational equilibrium. Because our estimates of that radius depend on Hubble's constant, these masses are inversely proportional to Hubble's constant, which we have assumed to be 65 km/s/Mpc.

APPENDIX H

SELECTED ASTRONOMICAL WEB SITES

The Web contains a vast amount of astronomical information. For all your astronomical Web surfing, the best starting point is the Web site for this textbook:

The Astronomy Place
www.astronomyplace.com

The following are some other sites that may be of particular use. In case any of the links change, you can always find live links to these sites, and many more, on The Astronomy Place.

Key Mission Sites

The following table lists the Web pages for major current astronomy missions.

Site	Description	Web Address
NASA's Office of Space Science Missions Page	Direct links to all past, present, and planned NASA space science missions	http://spacescience.nasa.gov/missions
Cassini/Huygens	Mission scheduled to arrive at Saturn in 2004	http://www.jpl.nasa.gov/cassini
Chandra X-Ray Observatory	Latest discoveries, educational activities, and other information from the Chandra X-Ray Observatory	http://chandra.harvard.edu
Far Ultraviolet Spectroscopic Explorer (FUSE)	Ultraviolet observatory in space	http://fuse.pha.jhu.edu
Galileo	Mission orbiting Jupiter	http://www.jpl.nasa.gov/galileo
Hubble Space Telescope	Latest discoveries, educational activities, and other information from the Hubble Space Telescope	http://hubble.stsci.edu
Mars Exploration Program	Information on current and planned Mars missions	http://mars.jpl.nasa.gov
Microwave Anisotropy Probe (MAP)	Mission to study the cosmic microwave background	http://map.gsfc.nasa.gov
Space Infrared Telescope Facility (SIRTF)	Infrared observatory scheduled for launch in 2002	http://sirtf.jpl.nasa.gov
Stratospheric Observatory for Infrared Astronomy (SOFIA)	Airborne observatory scheduled to begin flights in 2002	http://sofia.arc.nasa.gov

Key Observatory Sites

The following table lists the Web pages leading to major ground-based observatories.

Site	Description	Web Address
World's Largest Optical Telescopes	Direct links to most of the world's major optical observatories	http://www.seds.org/billa/bigeyes.html
Arecibo Observatory (Puerto Rico)	World's largest single-dish radio telescope	http://www.naic.edu
Cerro Tololo Inter-American Observatory	Links to major observatories on site in Cerro Tololo, Chile	http://www.ctio.noao.edu
European Southern Observatory	Links to European telescope projects in Chile, including the Very Large Telescope	http://www.eso.org
Mauna Kea Observatories	Links to major observatories in Hawaii, including Keck, Gemini, Subaru, CFHT, and others	http://www.ifa.hawaii.edu/mko
Mt. Palomar Observatory	Powerful telescope near San Diego	http://www.astro.caltech.edu/palomarpublic
National Optical Astronomy Observatory	Home page for United States national observatories in Arizona, Hawaii, and Chile	http://www.noao.edu
National Radio Astronomy Observatory	Home page for United States national radio observatories, including the Very Large Array (VLA)	http://www.nrao.edu

More Astronomical Web Sites

The following Web sites are some of the authors' favorites among many other non-commercial resources for astronomy.

Site	Description	Web Address
The Astronomy Place	**Don't forget to start here for all your astronomical Web surfing.**	**http://www.astronomyplace.com**
American Association of Variable Star Observers (AAVSO)	One of the largest organizations of amateur astronomers in the world. Check this site if you are interested in serious amateur astronomy.	http://www.aavso.org
Astronomical Society of the Pacific	An organization for both professional astronomers and the general public, devoted largely to astronomy education.	http://www.aspsky.org
Astronomy Picture of the Day	An archive of beautiful pictures, updated daily.	http://antwrp.gsfc.nasa.gov/apod
AstroWeb	Listing of major resources for astronomy on the Web.	http://www.cv.nrao.edu/fits/www/astronomy.html
Canadian Space Agency	Home page for Canada's space program.	http://www.space.gc.ca
European Space Agency (ESA)	Home page for this international agency.	http://www.esa.int
The Extrasolar Planets Encyclopedia	Information about the search for and discoveries of extrasolar planets.	http://cfa-www.harvard.edu/planets
NASA Home Page	Learn almost anything you want about NASA.	http://www.nasa.gov
NASA Science News	Read the latest news from NASA; has option to subscribe to e-mail notices of news releases.	http://science.nasa.gov
The Nine Planets (University of Arizona)	A multimedia tour of the solar system.	http://seds.lpl.arizona.edu/nineplanets/nineplanets/nineplanets.html
The Planetary Society	Has more than 100,000 members who are interested in planetary exploration and the search for life in the universe.	http://planetary.org
The SETI Institute	Devoted to the search for other civilizations.	http://www.seti.org
Voyage Scale Model Solar System	Take a virtual tour of the Voyage Scale Model Solar System.	http://www.voyageonline.org

APPENDIX I

THE 88 CONSTELLATIONS

Constellation Names (English equivalent in parentheses)

Andromeda (The Chained Princess)
Antlia (The Air Pump)
Apus (The Bird of Paradise)
Aquarius (The Water Bearer)
Aquila (The Eagle)
Ara (The Altar)
Aries (The Ram)
Auriga (The Charioteer)
Boötes (The Herdsman)
Caelum (The Chisel)
Camelopardalis (The Giraffe)
Cancer (The Crab)
Canes Venatici (The Hunting Dogs)
Canis Major (The Great Dog)
Canis Minor (The Little Dog)
Capricornus (The Sea Goat)
Carina (The Keel)
Cassiopeia (The Queen)
Centaurus (The Centaur)
Cepheus (The King)
Cetus (The Whale)
Chamaeleon (The Chameleon)
Circinus (The Drawing Compass)
Columba (The Dove)
Coma Berenices (Berenice's Hair)
Corona Australis (The Southern Crown)
Corona Borealis (The Northern Crown)
Corvus (The Crow)
Crater (The Cup)
Crux (The Southern Cross)
Cygnus (The Swan)
Delphinus (The Dolphin)
Dorado (The Goldfish)
Draco (The Dragon)
Equuleus (The Little Horse)
Eridanus (The River)
Fornax (The Furnace)
Gemini (The Twins)
Grus (The Crane)

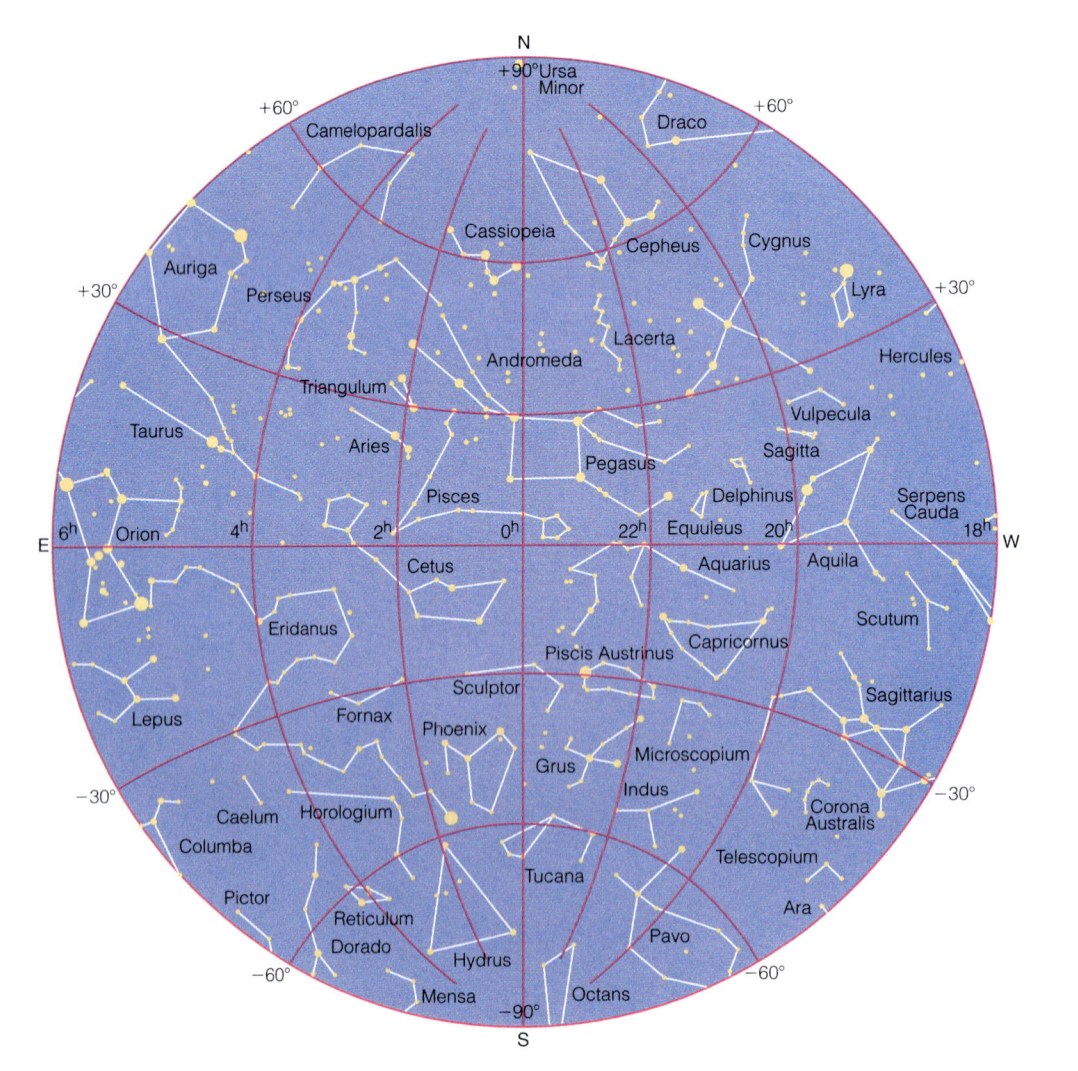

Hercules
Horologium (The Clock)
Hydra (The Sea Serpent)
Hydrus (The Water Snake)
Indus (The Indian)
Lacerta (The Lizard)
Leo (The Lion)
Leo Minor (The Little Lion)
Lepus (The Hare)
Libra (The Scales)
Lupus (The Wolf)
Lynx (The Lynx)
Lyra (The Lyre)
Mensa (The Table)
Microscopium (The Microscope)
Monoceros (The Unicorn)
Musca (The Fly)
Norma (The Level)
Octans (The Octant)
Ophiuchus (The Serpent Bearer)
Orion (The Hunter)
Pavo (The Peacock)
Pegasus (The Winged Horse)
Perseus (The Hero)
Phoenix (The Phoenix)
Pictor (The Painter's Easel)
Pisces (The Fish)
Piscis Austrinus (The Southern Fish)
Puppis (The Stern)
Pyxis (The Compass)
Reticulum (The Reticle)
Sagitta (The Arrow)
Sagittarius (The Archer)
Scorpius (The Scorpion)
Sculptor (The Sculptor)
Scutum (The Shield)
Serpens (The Serpent)
Sextans (The Sextant)
Taurus (The Bull)
Telescopium (The Telescope)
Triangulum (The Triangle)
Triangulum Australe (Southern Triangle)
Tucana (The Toucan)
Ursa Major (The Great Bear)
Ursa Minor (The Little Bear)
Vela (The Sail)
Virgo (The Virgin)
Volans (The Flying Fish)
Vulpecula (The Fox)

Constellation Locations

Each of the charts on these pages shows half of the celestial sphere in projection, so you can use them to learn the approximate locations of the constellations. The grid lines are marked by right ascension and declination [Section S1.4].

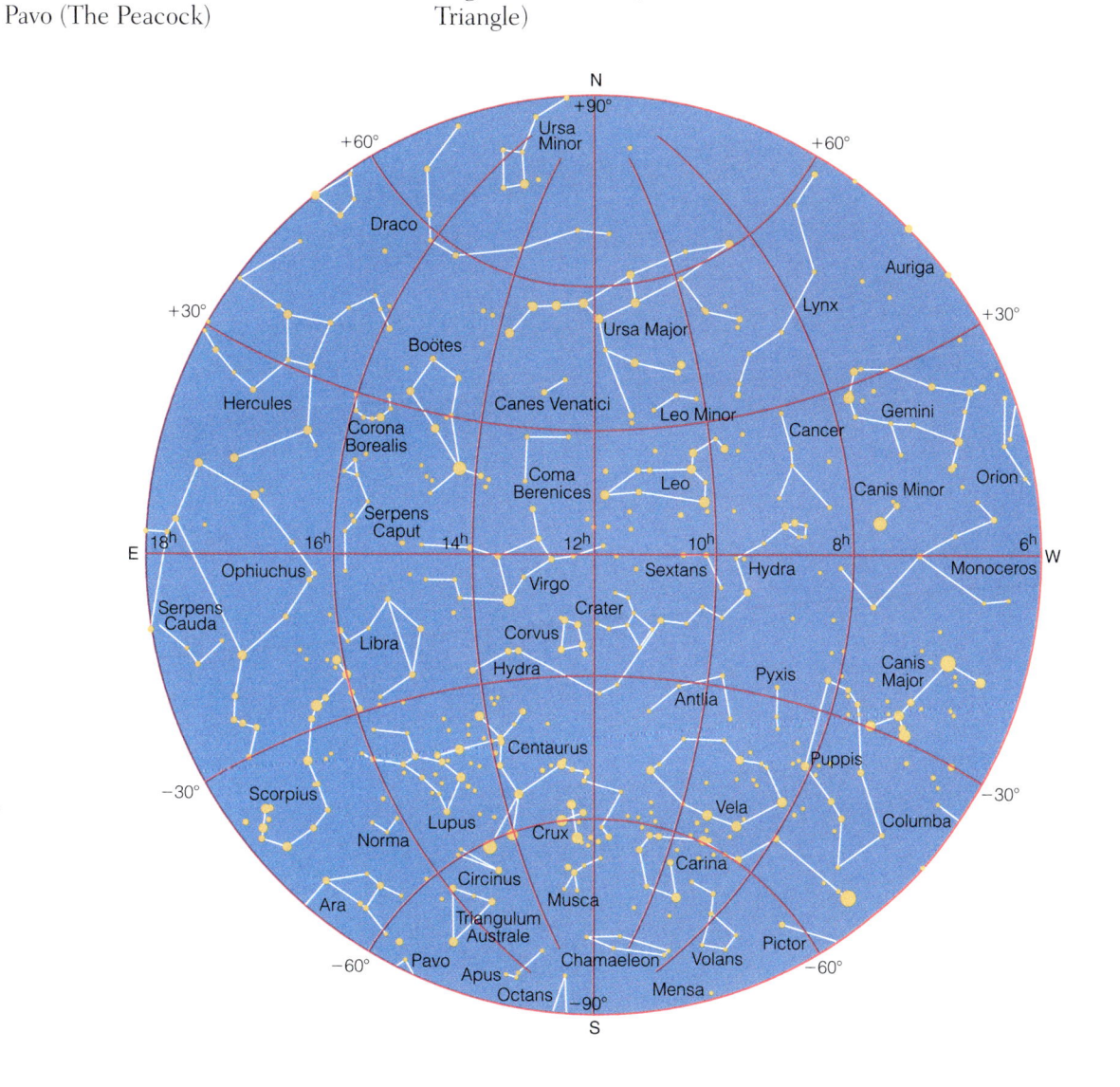

APPENDIX J

STAR CHARTS

How to use the star charts:

Check the times and dates under each chart to find the best one for you. Take it outdoors within an hour or so of the time listed for your date. Bring a dim flashlight to help you read it.

On each chart, the round outside edge represents the horizon all around you. Compass directions around the horizon are marked in yellow. Turn the chart around so the edge marked with the direction you're facing (for example, north, southeast) is down. The stars above this horizon now match the stars you are facing. Ignore the rest until you turn to look in a different direction.

The center of the chart represents the sky overhead, so a star plotted on the chart halfway from the edge to the center can be found in the sky halfway from the horizon to straight up.

The charts are drawn for 40°N latitude (for example, Denver, New York, Madrid). If you live far south of there, stars in the southern part of your sky will appear higher than on the chart and stars in the north will be lower. If you live far north of there, the reverse is true.

Jan–March

© Sky Publishing Corp.

Use this chart January, February, and March.

Early January — 1 A.M. Early February — 11 P.M. Early March — 9 P.M.
Late January — Midnight Late February — 10 P.M. Late March — Dusk

A-30 APPENDIX J

Apr–June
© Sky Publishing Corp.

Use this chart April, May, and June.

Early April — 3 A.M.*   Early May — 1 A.M.*   Early June — 11 P.M.*
Late April — 2 A.M.*   Late May — Midnight*   Late June — Dusk

*Daylight Saving Time

©1999 Sky & Telescope

STAR CHARTS   A-31

July–Sept.
© Sky Publishing Corp.

Use this chart July, August, and September.

Early July — 1 A.M.*
Late July — Midnight*

Early August — 11 P.M.*
Late August — 10 P.M.*

Early September — 9 P.M.*
Late September — Dusk

*Daylight Saving Time

A-32　APPENDIX J

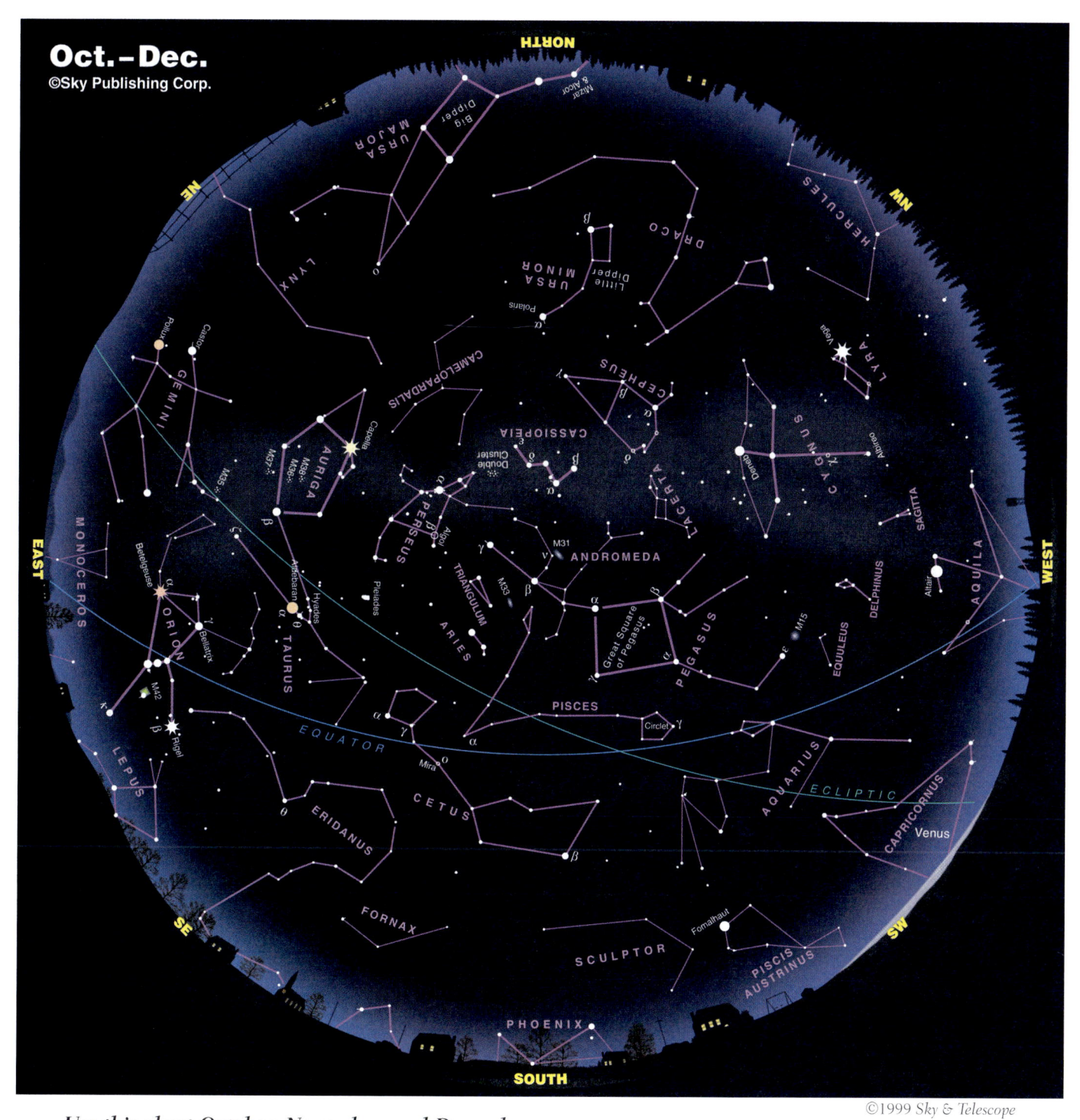

Oct.–Dec.
©Sky Publishing Corp.

Use this chart October, November, and December.

Early October — 1 A.M.* Early November — 10 P.M. Early December — 8 P.M.
Late October — Midnight* Late November — 9 P.M. Late December — 7 P.M.

*Daylight Saving Time

©1999 Sky & Telescope

GLOSSARY

21-cm line A spectral line from atomic hydrogen with wavelength 21 cm (in the radio portion of the spectrum).

absolute magnitude A measure of an object's luminosity; defined to be the apparent magnitude the object would have if it were located exactly 10 parsecs away.

absolute zero The coldest possible temperature, which is 0 K.

absorption (of light) The process by which matter absorbs radiative energy.

absorption-line spectrum A spectrum that contains absorption lines.

accelerating universe The possible fate of our universe in which a repulsive force (see cosmological constant) causes the expansion of the universe to accelerate with time. Its galaxies will recede from one another increasingly faster, and it will become cold and dark more quickly than a coasting universe.

acceleration The rate at which an object's velocity changes. Its standard units are m/s^2.

acceleration of gravity The acceleration of a falling object. On Earth, the acceleration of gravity, designated by g, is $9.8\ m/s^2$.

accretion The process by which small objects gather together to make larger objects.

accretion disk A rapidly rotating disk of material that gradually falls inward as it orbits a starlike object (e.g., white dwarf, neutron star, or black hole).

active galactic nuclei The unusually luminous centers of some galaxies, thought to be powered by accretion onto supermassive black holes. Quasars are the brightest type of active galactic nuclei; radio galaxies also contain active galactic nuclei.

adaptive optics A technique in which telescope mirrors flex rapidly to compensate for the bending of starlight caused by atmospheric turbulence.

albedo Describes the fraction of sunlight reflected by a surface; albedo = 0 means no reflection at all (a perfectly black surface); albedo = 1 means all light is reflected (a perfectly white surface).

altitude (above horizon) The angular distance between the horizon and an object in the sky.

amino acids The building blocks of proteins.

analemma The figure-8 path traced by the Sun over the course of a year when viewed at the same place and the same time each day; represents the discrepancies between apparent and mean solar time.

Andromeda Galaxy (M 31; the Great Galaxy in Andromeda) The nearest large spiral galaxy to the Milky Way.

angular momentum Momentum attributable to rotation or revolution. The angular momentum of an object moving in a circle of radius r is the product $m \times v \times r$.

angular resolution (of a telescope) The smallest angular separation that two point-like objects can have and still be seen as distinct points of light (rather than as a single point of light).

angular size (or **angular distance**) A measure of the angle formed by extending imaginary lines outward from our eyes to span an object (or between two objects).

annihilation See matter–antimatter annihilation

annular solar eclipse A solar eclipse during which the Moon is directly in front of the Sun but its angular size is not large enough to fully block the Sun; thus, a ring (or *annulus*) of sunlight is still visible around the Moon's disk.

Antarctic Circle The circle on the Earth with latitude 66.5°S.

antimatter Refers to any particle with the same mass as a particle of ordinary matter but whose other basic properties, such as electrical charge, are precisely opposite.

aphelion The point at which an object orbiting the Sun is farthest from the Sun.

apogee The point at which an object orbiting the Earth is farthest from the Earth.

apparent brightness The amount of light reaching us *per unit area* from a luminous object; often measured in units of $watts/m^2$.

apparent magnitude A measure of the apparent brightness of an object in the sky, based on the ancient system developed by Hipparchus.

apparent retrograde motion Refers to the apparent motion of a planet, as viewed from Earth, during the period of a few weeks or months when it moves westward relative to the stars in our sky.

apparent solar time Time measured by the actual position of the Sun in your local sky; defined so that noon is when the Sun is *on* the meridian.

arcminutes (or **minutes of arc**) One arcminute is 1/60 of 1°.

arcseconds (or **seconds of arc**) One arcsecond is 1/60 of an arcminute, or 1/3,600 of 1°.

Arctic Circle The circle on the Earth with latitude 66.5°N.

asteroid A relatively small and rocky object that orbits a star; asteroids are sometimes called *minor planets* because they are similar to planets but smaller.

asteroid belt The region of our solar system between the orbits of Mars and Jupiter in which asteroids are heavily concentrated.

astrobiology The study of life on Earth and beyond; emphasizes research into questions of the origin of life, the conditions under which life can survive, and the search for life beyond Earth.

astronomical unit (AU) The average distance (semimajor axis) of the Earth from the Sun, which is about 150 million km.

atmospheric pressure The surface pressure resulting from the overlying weight of an atmosphere.

atomic mass The combined number of protons and neutrons in an atom.

atomic number The number of protons in an atom.

atoms Consist of a nucleus made from protons and neutrons surrounded by a cloud of electrons.

aurora Dancing lights in the sky caused by charged particles entering our atmosphere; called the *aurora borealis* in the Northern Hemisphere and the *aurora australis* in the Southern Hemisphere.

azimuth (usually called *direction* in this book) Direction around the horizon from due north, measured clockwise in degrees. E.g., the azimuth of due north is 0°, due east is 90°, due south is 180°, and due west is 270°.

bar The standard unit of pressure, approximately equal to the Earth's atmospheric pressure at sea level.

baryonic matter Refers to ordinary matter made from atoms (because the nuclei of atoms contain protons and neutrons, which are both baryons).

baryons Particles, including protons and neutrons, that are made from three quarks.

basalt A type of volcanic rock that makes a low-viscosity lava when molten.

belts (on a jovian planet) Dark bands of sinking air that encircle a jovian planet at a particular set of latitudes.

Big Bang The event that gave birth to the universe.

Big Crunch If gravity ever reverses the universal expansion, the universe will someday begin to collapse and presumably end in a Big Crunch.

binary star system A star system that contains two stars.

biosphere Refers to the "layer" of life on Earth.

black hole A bottomless pit in spacetime. Nothing can escape from within a black hole, and we can never again detect or observe an object that falls into a black hole.

black smokers Structures around seafloor volcanic vents that support a wide variety of life.

blueshift A Doppler shift in which spectral features are shifted to shorter wavelengths, caused when an object is moving toward the observer.

bosons Particles, such as photons, to which the exclusion principle does not apply.

bound orbits Orbits on which an object travels repeatedly around another object; bound orbits are elliptical in shape.

brown dwarf An object too small to become an ordinary star because electron degeneracy pressure halts its gravitational collapse before fusion becomes self-sustaining; brown dwarfs have mass less than $0.08 M_{Sun}$.

bubble (interstellar) The surface of a bubble is an expanding shell of hot, ionized gas driven by stellar winds or supernovae; inside the bubble, the gas is very hot and has very low density.

bulge (of a spiral galaxy) The central portion of a spiral galaxy that is roughly spherical (or football shaped) and bulges above and below the plane of the galactic disk.

Cambrian explosion The dramatic diversification of life on Earth that occurred between about 540 and 500 million years ago.

carbon stars Stars whose atmospheres are especially carbon-rich, thought to be near the ends of their lives; carbon stars are the primary sources of carbon in the universe.

carbonate rock A carbon-rich rock, such as limestone, that forms underwater from chemical reactions between sediments and carbon dioxide. On Earth, most of the outgassed carbon dioxide currently resides in carbonate rocks.

carbonate–silicate cycle The process that cycles carbon dioxide between the Earth's atmosphere and surface rocks.

CCD (charge coupled device) A type of electronic light detector that has largely replaced photographic film in astronomical research.

celestial coordinates The coordinates of right ascension and declination that fix an object's position on the celestial sphere.

celestial equator (CE) The extension of the Earth's equator onto the celestial sphere.

celestial navigation Navigation on the surface of the Earth accomplished by observations of the Sun and stars.

celestial sphere The imaginary sphere on which objects in the sky appear to reside when observed from Earth.

Celsius (temperature scale) The temperature scale commonly used in daily activity internationally. Defined so that, on Earth's surface, water freezes at 0°C and boils at 100°C.

central dominant galaxy A giant elliptical galaxy found at the center of a dense cluster of galaxies, apparently formed by the merger of several individual galaxies.

Cepheid *See Cepheid variable*

Cepheid variable A particularly luminous type of pulsating variable star that follows a period–luminosity relation and hence is very useful for measuring cosmic distances.

charged particle belts Zones in which ions and electrons accumulate and encircle a planet.

chemical enrichment The process by which the abundance of heavy elements (heavier than helium) in the interstellar medium gradually increases over time as these elements are produced by stars and released into space.

chromosphere The layer of the Sun's atmosphere below the corona; most of the Sun's ultraviolet light is emitted from this region, in which the temperature is about 10,000 K.

circulation cells (also called *Hadley cells*) Large-scale cells (similar to convection cells) in a planet's atmosphere that transport heat between the equator and the poles.

circumpolar star A star that always remains above the horizon for a particular latitude.

climate Describes the long-term average of weather.

close binary A binary star system in which the two stars are very close together.

closed universe The universe is closed if its average density is greater than the critical density, in which case spacetime must curve back on itself to the point where its overall shape is analogous to that of the surface of a sphere. In the absence of a repulsive force (*see* cosmological constant), a closed universe would someday stop expanding and begin to contract.

cluster of galaxies A collection of a few dozen or more galaxies bound together by gravity; smaller collections of galaxies are simply called *groups*.

cluster of stars A group of anywhere from several hundred to a million or so stars; star clusters come in two types—open clusters and globular clusters.

CNO cycle The cycle of reactions by which intermediate- and high-mass stars fuse hydrogen into helium.

coasting universe The possible fate of our universe in which the mass density of the universe is *smaller* than the critical density, so that the collective gravity of all matter cannot halt the expansion. In the absence of a repulsive force (*see* cosmological constant), such a universe would keep expanding forever with little change in its rate of expansion.

coma (of a comet) The dusty atmosphere of a comet created by sublimation of ices in the nucleus when the comet is near the Sun.

comet A relatively small, icy object that orbits a star.

comparative planetology The study of the solar system by examining and understanding the similarities and differences among worlds.

compound (chemical) A substance made from molecules consisting of two or more atoms with different atomic number.

condensates Solid or liquid particles that condense from a cloud of gas.

condensation The formation of solid or liquid particles from a cloud of gas.

conduction (of energy) The process by which thermal energy is transferred by direct contact from warm material to cooler material.

conjunction (of a planet with the Sun) When a planet and the Sun line up in the sky.

conservation of angular momentum (law of) The principle that, in the absence of net torque (twisting force), the total angular momentum of a system remains constant.

conservation of energy (law of) The principle that energy (including mass-energy) can be neither created nor destroyed, but can only change from one form to another.

conservation of momentum (law of) The principle that, in the absence of net force, the total momentum of a system remains constant.

constellation A region of the sky; 88 official constellations cover the celestial sphere.

convection The energy transport process in which warm material expands and rises, while cooler material contracts and falls.

convection cell An individual small region of convecting material.

convection zone (of a star) A region in which energy is transported outward by convection.

core (of a planet) The dense central region of a planet that has undergone differentiation.

core (of a star) The central region of a star, in which nuclear fusion can occur.

Coriolis effect Causes air or objects moving on a rotating planet to deviate from straight-line trajectories.

corona (solar) The tenuous uppermost layer of the Sun's atmosphere; most of the Sun's X rays are emitted from this region, in which the temperature is about 1 million K.

coronal holes Regions of the corona that barely show up in X-ray images because they are nearly devoid of hot coronal gas.

cosmic microwave background The remnant radiation from the Big Bang, which we detect using radio telescopes sensitive to microwaves (which are short-wavelength radio waves).

cosmic rays Particles such as electrons, protons, and atomic nuclei that zip through interstellar space at close to the speed of light.

cosmological constant The name given to a term in Einstein's equations of general relativity. If it is not zero, then it represents a repulsive force or a type of energy (sometimes called *dark energy* or *quintessence*) that might cause the expansion of the universe to accelerate with time.

cosmological horizon The boundary of our observable universe, which is where the lookback time is equal to the age of the universe. Beyond this boundary in spacetime, we cannot see anything at all.

cosmological redshift Refers to the redshifts we see from distant galaxies, caused by the fact that expansion of the universe stretches all the photons within it to longer, redder wavelengths.

critical density The precise average density for the entire universe that marks the dividing line between a recollapsing universe and one that will expand forever.

critical universe The possible fate of our universe in which the mass density of the universe *equals* the critical density. The universe will never collapse, but in the absence of a repulsive force it will expand more and more slowly as time progresses. *See* cosmological constant

crust (of a planet) The low-density surface layer of a planet that has undergone differentiation.

cycles per second Units of frequency for a wave; describes the number of peaks (or troughs) of a wave that pass by a given point each second. Equivalent to *hertz*.

dark energy Name sometimes given to energy that could be causing the expansion of the universe to accelerate. *See* cosmological constant.

dark matter Matter that we infer to exist from its gravitational effects but from which we have not detected any light; dark matter apparently dominates the total mass of the universe.

daylight saving time Standard time plus 1 hour, so that the Sun appears on the meridian around 1 P.M. rather than around noon.

declination (dec) Analogous to latitude, but on the celestial sphere; it is the angular north-south distance between the celestial equator and a location on the celestial sphere.

degeneracy pressure A type of pressure unrelated to an object's temperature, which arises when electrons (electron degeneracy pressure) or neutrons (neutron degeneracy pressure) are packed so tightly that the exclusion and uncertainty principles come into play.

degenerate object An object in which degeneracy pressure is the primary pressure pushing back against gravity, such as a brown dwarf, white dwarf, or neutron star.

deuterium A form of hydrogen in which the nucleus contains a proton and a neutron, rather than only a proton (as is the case for most hydrogen nuclei).

differential rotation Describes the rotation of an object in which the equator rotates at a different rate than the poles.

differentiation The process in which gravity separates materials according to density, with high-density materials sinking and low-density materials rising.

diffraction grating A finely etched surface that can split light into a spectrum.

diffraction limit The angular resolution that a telescope could achieve if it were limited only by the interference of light waves; it is smaller (i.e., better angular resolution) for larger telescopes.

dimension (mathematical) Describes the number of independent directions in which movement is possible; e.g., the surface of the Earth is two-dimensional because only two independent directions of motion are possible (north-south and east-west).

direction (in local sky) One of the two coordinates (the other is altitude) needed to pinpoint an object in the local sky. It is the direction, such as north, south, east, or west, in which you must face to see the object. *See also* azimuth

disk component (of a galaxy) The portion of a spiral galaxy that looks like a disk and contains an interstellar medium with cool gas and dust; stars of many ages are found in the disk component.

Doppler effect (shift) The effect that shifts the wavelengths of spectral features in objects that are moving toward or away from the observer.

down quark One of the two quark types (the other is the up quark) found in ordinary protons and neutrons. Has a charge of $-\frac{1}{3}$.

dust (or **dust grains**) Tiny solid flecks of material; in astronomy, we often discuss interplanetary dust (found within a star system) or interstellar dust (found between the stars in a galaxy). *See also* interstellar dust grains

dust tail (of a comet) One of two tails seen when a comet passes near the Sun (the other is the plasma tail); composed of small solid particles pushed away from the Sun by the radiation pressure of sunlight.

dwarf elliptical galaxy A small elliptical galaxy with less than about a billion stars.

Earth-orbiters (spacecraft) Spacecraft designed to study the Earth or the universe from Earth orbit.

eccentricity A measure of how much an ellipse deviates from a perfect circle; defined as the center-to-focus distance divided by the length of the semimajor axis.

eclipse Occurs when one astronomical object casts a shadow on another or crosses our line of sight to the other object.

eclipse seasons Periods during which lunar and solar eclipses can occur because the nodes of the Moon's orbit are nearly aligned with the Earth and Sun.

eclipsing binary A binary star system in which the two stars happen to be orbiting in the plane of our line of sight, so that each star will periodically eclipse the other.

ecliptic The Sun's apparent annual path among the constellations.

ecliptic plane The plane of the Earth's orbit around the Sun.

ejecta (from an impact) Debris ejected by the blast of an impact.

electromagnetic field An abstract concept used to describe how a charged particle would affect other charged particles at a distance.

electromagnetic force One of the four fundamental forces; it is the force that dominates atomic and molecular interactions.

electromagnetic spectrum The complete spectrum of light, including radio waves, infrared, visible light, ultraviolet light, X rays, and gamma rays.

electromagnetic wave A synonym for light, which consists of waves of electric and magnetic fields.

electron degeneracy pressure Degeneracy pressure exerted by electrons, as in brown dwarfs and white dwarfs.

electrons Fundamental particles with negative electric charge; the distribution of electrons in an atom gives the atom its size.

electron-volt (eV) A unit of energy equivalent to 1.60×10^{-19} joule.

electroweak era The era of the universe during which only three forces operated (gravity, strong force, and electroweak force), lasting from 10^{-38} second to 10^{-10} second after the Big Bang.

electroweak force The force that exists at high energies when the electromagnetic force and the weak force exist as a single force.

element (chemical) A substance made from individual atoms of a particular atomic number.

ellipse A type of oval that happens to be the shape of bound orbits. An ellipse can be drawn by moving a pencil along a string whose ends are tied to two tacks; the locations of the tacks are the foci (singular, focus) of the ellipse.

elliptical galaxies Galaxies that appear rounded in shape, often longer in one direction, like a football. They have no disks and contain very little cool gas and dust compared to spiral galaxies, though they often contain very hot, ionized gas.

elongation (greatest) For Mercury or Venus, the point at which it appears farthest from the Sun in our sky.

emission (of light) The process by which matter emits energy in the form of light.

emission-line spectrum A spectrum that contains emission lines.

energy Broadly speaking, energy is what can make matter move. The three basic types of energy are kinetic, potential, and radiative.

equation of time Describes the discrepancies between apparent and mean solar time.

equivalence principle The fundamental starting point for general relativity, which states that the effects of gravity are exactly equivalent to the effects of acceleration.

era of atoms The era of the universe lasting from about 500,000 years to about 1 billion years after the Big Bang, during which it was cool enough for neutral atoms to form.

era of galaxies The present era of the universe, which began with the formation of galaxies when the universe was about 1 billion years old.

era of nuclei The era of the universe lasting from about 3 minutes to about 500,000 years after the Big Bang, during which matter in the universe was fully ionized and opaque to light. The cosmic background radiation was released at the end of this era.

era of nucleosynthesis The era of the universe lasting from about 0.001 second to about 3 minutes after the Big Bang, by the end of which virtually all of the neutrons and about one-seventh of the protons in the universe had fused into helium.

erosion The wearing down or building up of geological features by wind, water, ice, and other phenomena of planetary weather.

eruption The process of releasing hot lava on the planet's surface.

escape velocity The speed necessary for an object to completely escape the gravity of a large body such as a moon, planet, or star.

evaporation The process by which atoms or molecules escape into the gas phase from a liquid.

event Any particular point along a worldline represents a particular event; all observers will agree on the reality of an event but may disagree about its time and location.

event horizon The boundary that marks the "point of no return" between a black hole and the outside universe; events that occur within the event horizon can have no influence on our observable universe.

exchange particle According to the standard model of physics, each of the four fundamental forces is transmitted by the transfer of particular types of exchange particles.

excited state (of an atom) Any arrangement of electrons in an atom that has more energy than the ground state.

exclusion principle The law of quantum mechanics that states that two fermions cannot occupy the same quantum state at the same time.

exosphere The hot, outer layer of an atmosphere, where the atmosphere "fades away" to space.

exposure time The amount of time for which light is collected to make a single image.

extrasolar planet A planet orbiting a star other than our Sun.

Fahrenheit (temperature scale) The temperature scale commonly used in daily activity in the United States. Defined so that, on Earth's surface, water freezes at 32°F and boils at 212°F.

fall equinox (autumnal equinox) Refers both to the point in Virgo on the celestial sphere where the ecliptic crosses the celestial equator and to the moment in time when the Sun appears at that point each year (around September 21).

false-color image An image displayed in colors that are *not* the true, visible-light colors of an object.

fault (geological) A place where rocks slip sideways relative to one another.

feedback relationships Processes in which one property amplifies (positive feedback) or counteracts (negative feedback) the behavior of properties.

fermions Particles, such as electrons, neutrons, and protons, that obey the exclusion principle.

field An abstract concept used to describe how a particle would interact with a force. For example, the idea of a *gravitational field* describes how a particle would react to the local strength of gravity, and the idea of an *electromagnetic field* describes how a charged particle would respond to forces from other charged particles.

filter (for light) A material that transmits only particular wavelengths of light.

fireball A particularly bright meteor.

flare star A small, spectral type M star that displays particularly strong flares on its surface.

flat (or Euclidean) geometry Refers to any case in which the rules of geometry for a flat plane hold, such as that the shortest distance between two points is a straight line.

flat universe A universe in which the overall geometry of spacetime is flat (Euclidean), as would be the case if the density of the universe is equal to the critical density.

flybys (spacecraft) Spacecraft that fly past a target object (such as a planet), usually just once, as opposed to entering a bound orbit of the object.

focal plane The place where an image created by a lens or mirror is in focus.

focus (of a lens or mirror) The point at which rays of light that were initially parallel (such as light from a distant star) converge.

force Anything that can cause a change in momentum.

formation properties (of planets) In this book, for the purpose of understanding geological processes, planets are defined to be born with four formation properties: size (mass and radius), distance from the Sun, composition, and rotation rate.

frame of reference (in relativity) Two (or more) objects share the same frame of reference if they are *not* moving relative to each other.

free-fall Refers to conditions in which an object is falling without resistance; objects are weightless when in free-fall.

free-float frame A frame of reference in which all objects are weightless and hence float freely.

frequency Describes the rate at which peaks of a wave pass by a point; measured in units of 1/s, often called *cycles per second* or *hertz*.

frost line The boundary in the solar nebula beyond which ices could condense; only metals and rocks could condense within the frost line.

fundamental forces There are four known fundamental forces in nature: gravity, the electromagnetic force, the strong force, and the weak force.

fundamental particles Subatomic particles that cannot be divided into anything smaller.

galactic disk (of a spiral galaxy) *See* disk component

galactic fountain Refers to a model for the cycling of gas in the Milky Way Galaxy in which fountains of hot, ionized gas rise from the disk into the halo and then cool and form clouds as they sink back into the disk.

galactic wind A wind of low-density but extremely hot gas flowing out from a starburst galaxy, created by the combined energy of many supernovae.

galaxy A huge collection of anywhere from a few hundred million to more than a trillion stars, all bound together by gravity.

galaxy cluster *See* cluster of galaxies

galaxy evolution The formation and development of galaxies.

gamma-ray burst A sudden burst of gamma rays from deep space; such bursts apparently come from distant galaxies, but their precise mechanism is unknown.

gamma rays Light with very short wavelengths (and hence high frequencies)—shorter than those of X rays.

gas phase The phase of matter in which atoms or molecules can move essentially independently of one another.

gas pressure Describes the force (per unit area) pushing on any object due to surrounding gas. *See also* pressure

genetic code The "language" that living cells use to read the instructions chemically encoded in DNA.

geocentric universe (ancient belief in) The idea that the Earth is the center of the entire universe.

geological controlling factors In this book, for the purpose of understanding geological processes, geology is considered to be influenced primarily by four geological controlling factors: surface gravity, internal temperature, surface temperature, and the presence (and extent) of an atmosphere.

geological processes The four basic geological processes are impact cratering, volcanism, tectonics, and erosion.

geology The study of surface features (on a moon, planet, or asteroid) and the processes that create them.

giants (luminosity class III) Stars that appear just below the supergiants on the H–R diagram because they are somewhat smaller in radius and lower in luminosity.

global positioning system (GPS) A system of navigation by satellites orbiting the Earth.

global wind patterns (or **global circulation**) Wind patterns that remain fixed on a global scale, determined by the combination of surface heating and the planet's rotation.

globular cluster A spherically shaped cluster of up to a million or more stars; globular clusters are found primarily in the halos of galaxies and contain only very old stars.

gluons The exchange particles for the strong force.

grand unified theory (GUT) A theory that unifies three of the four fundamental forces—the strong force, the weak force, and the electromagnetic force (but not gravity)—in a single model.

granulation (on the Sun) The bubbling pattern visible in the photosphere, produced by the underlying convection.

gravitation (law of) *See* universal law of gravitation

gravitational constant The experimentally measured constant G that appears in the law of universal gravitation;
$$G = 6.67 \times 10^{-11} \frac{m^3}{kg \times s^2}.$$

gravitational contraction The process in which gravity causes an object to contract, thereby converting gravitational potential energy into thermal energy.

gravitational encounter Occurs when two (or more) objects pass near enough so that each can feel the effects of the other's gravity and can therefore exchange energy.

gravitational equilibrium Describes a state of balance in which the force of gravity pulling inward is precisely counteracted by pressure pushing outward.

gravitational lensing The magnification or distortion (into arcs, rings, or multiple images) of an image caused by light bending through a gravitational field, as predicted by Einstein's general theory of relativity.

gravitational redshift A redshift caused by the fact that time runs slow in gravitational fields.

gravitational time dilation The slowing of time that occurs in a gravitational field, as predicted by Einstein's general theory of relativity.

gravitational waves Predicted by Einstein's general theory of relativity, these waves travel at the speed of light and transmit distortions of space through the universe. Although not yet observed directly, we have strong indirect evidence that they exist.

gravitationally bound system Any system of objects, such as a star system or a galaxy, that is held together by gravity.

gravitons The exchange particles for the force of gravity.

gravity One of the four fundamental forces; it is the force that dominates on large scales.

grazing incidence (in telescopes) Reflections in which light grazes a mirror surface and is deflected at a small angle; commonly used to focus high-energy ultraviolet light and X rays.

great circle A circle on the surface of a sphere whose center is at the center of the sphere.

Great Red Spot A large, high-pressure storm on Jupiter.

greenhouse effect The process by which greenhouse gases in an atmosphere make a planet's surface temperature warmer than it would be in the absence of an atmosphere.

greenhouse gases Gases, such as carbon dioxide, water vapor, and methane, that are particularly good absorbers of infrared light but are transparent to visible light.

Gregorian calendar Our modern calendar, introduced by Pope Gregory in 1582.

ground state (of an atom) The lowest possible energy state of the electrons in an atom.

group (of galaxies) A few to a few dozen galaxies bound together by gravity. *See also* cluster of galaxies

GUT era The era of the universe during which only two forces operated (gravity and the grand-unified-theory or GUT force), lasting from 10^{-43} second to 10^{-38} second after the Big Bang.

GUT force The proposed force that exists at very high energies when the strong force, the weak force, and the electromagnetic force (but not gravity) all act as one.

habitable zone The region around a star in which planets could potentially have surface temperatures at which liquid water could exist.

half-life The time it takes for half of the nuclei in a given quantity of a radioactive substance to decay.

halo (of a galaxy) The spherical region surrounding the disk of a spiral galaxy.

Hawking radiation Radiation predicted to arise from the evaporation of black holes.

heavy elements In astronomy, *heavy elements* generally refers to all elements *except* hydrogen and helium.

helium-capture reactions Fusion reactions that fuse a helium nucleus into some other nucleus; such reactions can fuse carbon into oxygen, oxygen into neon, neon into magnesium, and so on.

helium flash The event that marks the sudden onset of helium fusion in the previously inert helium core of a low-mass star.

helium fusion The fusion of three helium nuclei into one carbon nucleus; also called the *triple-alpha reaction*.

hertz (Hz) The standard unit of frequency for light waves; equivalent to units of 1/s.

Hertzsprung–Russell (H–R) diagram A graph plotting individual stars as points, with stellar luminosity on the vertical axis and spectral type (or surface temperature) on the horizontal axis.

high-mass stars Stars born with masses above about $8M_{Sun}$; these stars will end their lives by exploding as supernovae.

horizon A boundary that divides what we can see from what we cannot see.

horizontal branch The horizontal line of stars that represents helium-burning stars on an H–R diagram for a cluster of stars.

horoscope A predictive chart made by an astrologer; in scientific studies, horoscopes have never been found to have any validity as predictive tools.

hot spot (geological) A place within a plate of the lithosphere where a localized plume of hot mantle material rises.

hour angle (HA) The angle or time (measured in hours) since an object was last on the meridian in the local sky. Defined to be 0 hours for objects that *are* on the meridian.

Hubble's constant A number that expresses the current rate of expansion of the universe; designated H_0, it is usually stated in units of km/s/Mpc. The reciprocal of Hubble's constant is the age the universe would have *if* the expansion rate had never changed.

Hubble's law Mathematically expresses the idea that more distant galaxies move away from us faster; its formula is $v = H_0 \times d$, where v is a galaxy's speed away from us, d is its distance, and H_0 is Hubble's constant.

hydrogen compounds Compounds that contain hydrogen and were common in the solar nebula, such as water (H_2O), ammonia (NH_3), and methane (CH_4).

hydrogen-shell burning Hydrogen fusion that occurs in a shell surrounding a stellar core.

hydrosphere Refers to the "layer" of water on the Earth consisting of oceans, lakes, rivers, ice caps, and other liquid water and ice.

hyperbola The precise mathematical shape of one type of unbound orbit (the other is a parabola) allowed under the force of gravity; at great distances from the attracting object, a hyperbolic path looks like a straight line.

hypernova A term sometimes used to describe a supernova (explosion) of a star so massive that it leaves a black hole behind.

hyperspace Any space with more than three dimensions.

hypothesis A tentative model proposed to explain some set of observed facts, but which has not yet been rigorously tested and confirmed.

ices (in solar system theory) Materials that are solid only at low temperatures, such as the hydrogen compounds water, ammonia, and methane.

image A picture of an object made by focusing light.

imaging (in astronomical research) The process of obtaining pictures of astronomical objects.

impact The collision of a small body (such as an asteroid or comet) with a larger object (such as a planet or moon).

impact basin A very large impact crater often filled by a lava flow.

impact crater A bowl-shaped depression left by the impact of an object that strikes a planetary surface (as opposed to burning up in the atmosphere).

impact cratering The excavation of bowl-shaped depressions (*impact craters*) by asteroids or comets striking a planet's surface.

impactor The object responsible for an impact.

inflation (of the universe) A sudden and dramatic expansion of the universe thought to have occurred at the end of the GUT era.

infrared light Light with wavelengths that fall in the portion of the electromagnetic spectrum between radio waves and visible light.

inner solar system Generally considered to encompass the region of our solar system out to about the orbit of Mars.

intensity (of light) A measure of the amount of energy coming from light of specific wavelength in the spectrum of an object.

interferometry A telescopic technique in which two or more telescopes are used in tandem to produce much better angular resolution than the telescopes could achieve individually.

intermediate-mass stars Stars born with masses between about $2-8M_{Sun}$; these stars end their lives by ejecting a planetary nebula and becoming a white dwarf.

interstellar cloud A cloud of gas and dust between the stars.

interstellar dust grains Tiny solid flecks of carbon and silicon minerals found in cool interstellar clouds; they resemble particles of smoke and form in the winds of red giant stars.

interstellar medium Refers to gas and dust that fills the space between stars in a galaxy.

intracluster medium Hot, X-ray-emitting gas found between the galaxies within a cluster of galaxies.

inverse square law Any quantity that decreases with the square of the distance between two objects is said to follow an inverse square law.

inversion (atmospheric) A local weather condition in which air is colder near the surface than higher up in the troposphere— the opposite of the usual condition, in which the troposphere is warmer at the bottom.

Io torus A donut-shaped charged-particle belt around Jupiter that approximately traces Io's orbit.

ionization The process of stripping an electron from an atom.

ionization nebula A colorful, wispy cloud of gas that glows because neighboring hot stars irradiate it with ultraviolet photons that can ionize hydrogen atoms.

ionosphere A portion of the thermosphere in which ions are particularly common (due to ionization by X rays from the Sun).

ions Atoms with a positive or negative electrical charge.

irregular galaxies Galaxies that look neither spiral nor elliptical.

isotopes Each different isotope of an element has the *same* number of protons but a *different* number of neutrons.

jets High-speed streams of gas ejected from an object into space.

joule The international unit of energy, equivalent to about 1/4,000 of a Calorie.

jovian nebulae The clouds of gas that swirled around the jovian planets, from which the moons formed.

jovian planets Giant gaseous planets similar in overall composition to Jupiter.

Julian calendar The calendar introduced in 46 B.C. by Julius Caesar and used until it was replaced by the Gregorian calendar.

Kelvin (temperature scale) The most commonly used temperature scale in science, defined such that absolute zero is 0 K and water freezes at 273.15 K.

Kepler's first law States that the orbit of each planet about the Sun is an ellipse with the Sun at one focus.

Kepler's laws of planetary motion Three laws discovered by Kepler that describe the motion of the planets around the Sun.

Kepler's second law States that, as a planet moves around its orbit, it sweeps out equal areas in equal times. This tells us that a planet moves faster when it is closer to the Sun (near perihelion) than when it is farther from the Sun (near aphelion) in its orbit.

Kepler's third law States that the square of a planet's orbital period is proportional to the cube of its average distance from the Sun (semimajor axis), which tells us that more distant planets move more slowly in their orbits. In its original form, written $p^2 = a^3$. *See also* Newton's version of Kepler's third law

kinetic energy Energy of motion, given by the formula $\frac{1}{2}mv^2$.

Kuiper belt The comet-rich region of our solar system that spans distances of about 30–100 AU from the Sun; Kuiper belt comets have orbits that lie fairly close to the plane of planetary orbits and travel around the Sun in the same direction as the planets.

Large Magellanic Cloud One of two small, irregular galaxies (the other is the Small Magellanic Cloud) located about 150,000 light-years away; it probably orbits the Milky Way Galaxy.

large-scale structure (of the universe) Generally refers to structure of the universe on size scales larger than that of clusters of galaxies.

latitude The angular north-south distance between the Earth's equator and a location on the Earth's surface.

leap year A calendar year with 366 rather than 365 days; our current calendar (the Gregorian calendar) has a leap year every 4 years (by adding February 29) except in century years that are not divisible by 400.

length contraction Refers to the effect in which you observe lengths to be shortened in reference frames moving relative to you.

lenticular galaxies Galaxies that look lens-shaped when seen edge-on, resembling spiral galaxies without arms. They tend to have less cool gas than normal spiral galaxies but more gas than elliptical galaxies.

leptons Fermions *not* made from quarks, such as electrons and neutrinos.

life track A track drawn on an H–R diagram to represent the changes in a star's surface temperature and luminosity during its life; also called an *evolutionary track*.

light-collecting area (of a telescope) The area of the primary mirror or lens that collects light in a telescope.

light curve A graph of an object's intensity against time.

light gases (in solar system theory) Refers to hydrogen and helium, which never condense under solar nebula conditions.

light pollution Human-made light that hinders astronomical observations.

light-year The distance that light can travel in 1 year, which is 9.46 trillion km.

liquid phase The phase of matter in which atoms or molecules are held together but move relatively freely.

lithosphere The relatively rigid outer layer of a planet; generally encompasses the crust and the uppermost portion of the mantle.

Local Group The group of over 30 galaxies to which the Milky Way Galaxy belongs.

local sidereal time (LST) Sidereal time for a particular location, defined according to the position of the spring equinox in the local sky. More formally, the local sidereal time at any moment is defined to be the hour angle of the spring equinox.

local sky The sky as viewed from a particular location on Earth (or another solid object). Objects in the local sky are pinpointed by the coordinates of *altitude* and *direction* (or azimuth).

Local Supercluster The supercluster of galaxies to which the Local Group belongs.

longitude The angular east-west distance between the prime meridian (which passes through Greenwich) and a location on the Earth's surface.

lookback time Refers to the amount of time since the light we see from a distant object was emitted. I.e., if an object has a

lookback time of 400 million years, we are seeing it as it looked 400 million years ago.

low-mass stars Stars born with masses less than about $2M_{Sun}$; these stars end their lives by ejecting a planetary nebula and becoming a white dwarf.

luminosity The total power output of an object, usually measured in watts or in units of solar luminosities (L_{Sun} = 3.8 × 10^{26} watts).

luminosity class Describes the region of the H–R diagram in which a star falls. Luminosity class I represents supergiants, III represents giants, and V represents main-sequence stars; luminosity classes II and IV are intermediate to the others.

luminosity–distance formula The formula that relates apparent brightness, luminosity, and distance:

$$\text{apparent brightness} = \frac{\text{luminosity}}{4\pi \times (\text{distance})^2}$$

lunar eclipse Occurs when the Moon passes through the Earth's shadow, which can occur only at full moon; may be total, partial, or penumbral.

lunar maria The regions of the Moon that look smooth from Earth and actually are impact basins.

lunar month *See* synodic month

lunar phase Describes the appearance of the Moon as seen from Earth.

MACHOs Stands for *massive compact halo objects* and represents one possible form of dark matter in which the dark objects are relatively large, like planets or brown dwarfs.

magma Underground molten rock.

magnetic braking The process by which a star's rotation slows as its magnetic field transfers its angular momentum to the surrounding nebula.

magnetic field Describes the region surrounding a magnet in which it can affect other magnets or charged particles in its vicinity.

magnetic-field lines Lines that represent how the needles on a series of compasses would point if they were laid out in a magnetic field.

magnetosphere The region surrounding a planet in which charged particles are trapped by the planet's magnetic field.

magnitude system A system of describing stellar brightness by using numbers, called *magnitudes*, based on an ancient Greek way of describing the brightnesses of stars in the sky. This system uses *apparent magnitude* to describe a star's apparent brightness and *absolute magnitude* to describe a star's luminosity.

main sequence (luminosity class V) The prominent line of points running from the upper left to the lower right on an H–R diagram; main-sequence stars shine by fusing hydrogen in their cores.

main-sequence fitting A method for measuring the distance to a cluster of stars by comparing the apparent brightness of the cluster's main sequence with the standard main sequence.

main-sequence lifetime The length of time for which a star of a particular mass can shine by fusing hydrogen into helium in its core.

main-sequence turnoff A method for measuring the age of a cluster of stars from the point on its H–R diagram where its stars turn off from the main sequence; the age of the cluster is equal to the main-sequence lifetime of stars at the main-sequence turn-off point.

mantle (of a planet) The rocky layer that lies between a planet's core and crust.

Martian meteorite This term is used to describe meteorites found on Earth that are thought to have originated on Mars.

mass A measure of the amount of matter in an object.

mass-energy The potential energy of mass, which has an amount $E = mc^2$.

mass exchange (in close binary star systems) The process in which tidal forces cause matter to spill from one star to a companion star in a close binary system.

mass extinction An event in which a large fraction of the species living on Earth go extinct, such as the event in which the dinosaurs died out about 65 million years ago.

mass increase (in relativity) Refers to the effect in which an object moving past you seems to have a mass greater than its rest mass.

mass-to-light ratio The mass of an object divided by its luminosity, usually stated in units of solar masses per solar luminosity. Objects with high mass-to-light ratios must contain substantial quantities of dark matter.

massive-star supernova A supernova that occurs when a massive star dies, initiated by the catastrophic collapse of its iron core; often called a Type II supernova.

matter–antimatter annihilation Occurs when a particle of matter and a particle of antimatter meet and convert all of their mass-energy to photons.

mean solar time Time measured by the average position of the Sun in your local sky over the course of the year.

meridian A half-circle extending from your horizon (altitude 0°) due south, through your zenith, to your horizon due north.

metals (in solar system theory) Elements, such as nickel, iron, and aluminum, that condense at fairly high temperatures.

meteor A flash of light caused when a particle from space burns up in our atmosphere.

meteor shower A period during which many more meteors than usual can be seen.

meteorite A rock from space that lands on Earth.

Metonic cycle The 19-year period, discovered by the Babylonian astronomer Meton, over which the lunar phases occur on the same dates.

Milky Way Used both as the name of our galaxy and to refer to the band of light we see in the sky when we look into the plane of the Milky Way Galaxy.

millisecond pulsars Pulsars with rotation periods of a few thousandths of a second.

model (scientific) A representation of some aspect of nature that can be used to explain and predict real phenomena without invoking myth, magic, or the supernatural.

molecular bands The tightly bunched lines in an object's spectrum that are produced by molecules.

molecular clouds Cool, dense interstellar clouds in which the low temperatures allow hydrogen atoms to pair up into hydrogen molecules (H_2).

molecular dissociation The process by which a molecule splits into its component atoms.

molecule Technically the smallest unit of a chemical element or compound; in this text, the term refers only to combinations of two or more atoms held together by chemical bonds.

momentum The product of an object's mass and velocity.

moon An object that orbits a planet.

mutations Errors in the copying process when a living cell replicates itself.

natural selection The process by which mutations that make an organism better able to survive get passed on to future generations.

nebula A cloud of gas in space, usually one that is glowing.

nebular capture The process by which icy planetesimals capture hydrogen and helium gas to form jovian planets.

nebular theory The detailed theory that describes how our solar system formed from a cloud of interstellar gas and dust.

net force The overall force to which an object responds; the net force is equal to the rate of change in the object's momentum, or equivalently to the object's mass × acceleration.

neutrino A type of fundamental particle that has extremely low mass and responds only to the weak force; neutrinos are leptons and come in three types—electron neutrinos, mu neutrinos, and tau neutrinos.

neutron degeneracy pressure Degeneracy pressure exerted by neutrons, as in neutron stars.

neutron star The compact corpse of a high-mass star left over after a supernova; typically contains a mass comparable to the mass of the Sun in a volume just a few kilometers in radius.

neutrons Particles with no electrical charge found in atomic nuclei, built from three quarks.

newton The standard unit of force in the metric system:

$$1 \text{ newton} = 1 \frac{\text{kg} \times \text{m}}{\text{s}^2}$$

Newton's first law of motion States that, in the absence of a net force, an object moves with constant velocity.

Newton's laws of motion Three basic laws that describe how objects respond to forces.

Newton's second law of motion States how a net force affects an object's motion. Specifically: force = rate of change in momentum, or force = mass × acceleration.

Newton's third law of motion States that, for any force, there is always an equal and opposite reaction force.

Newton's universal law of gravitation *See* universal law of gravitation

Newton's version of Kepler's third law This generalization of Kepler's third law can be used to calculate the masses of orbiting objects from measurements of orbital period and distance. Usually written as:
$$p^2 = \frac{4\pi^2}{G(M_1 + M_2)} a^3$$

nodes (of Moon's orbit) The two points in the Moon's orbit where it crosses the ecliptic plane.

nonbaryonic matter Refers to exotic matter that is not part of the normal composition of atoms, such as neutrinos or the hypothetical WIMPs.

nonscience As defined in this book, nonscience is any way of searching for knowledge that makes no claim to follow the scientific method, such as seeking knowledge through intuition, tradition, or faith.

north celestial pole (NCP) The point on the celestial sphere directly above the Earth's North Pole.

nova The dramatic brightening of a star that lasts for a few weeks and then subsides; occurs when a burst of hydrogen fusion ignites in a shell on the surface of an accreting white dwarf in a binary star system.

nuclear fission The process in which a larger nucleus splits into two (or more) smaller particles.

nuclear fusion The process in which two (or more) smaller nuclei slam together and make one larger nucleus.

nucleus (of an atom) The compact center of an atom made from protons and neutrons.

nucleus (of a comet) The solid portion of a comet, and the only portion that exists when the comet is far from the Sun.

observable universe The portion of the entire universe that, at least in principle, can be seen from Earth.

Olbers' paradox Asks the question of how the night sky can be dark if the universe is infinite and full of stars.

Oort cloud A huge, spherical region centered on the Sun, extending perhaps halfway to the nearest stars, in which trillions of comets orbit the Sun with random inclinations, orbital directions, and eccentricities.

opacity A measure of how much light a material absorbs compared to how much it transmits; materials with higher opacity absorb more light.

opaque (material) Describes a material that absorbs light.

open cluster A cluster of up to several thousand stars; open clusters are found only in the disks of galaxies and often contain young stars.

open universe The universe is open if its average density is less than the critical density, in which case spacetime has an overall shape analogous to the surface of a saddle.

opposition The point at which a planet appears opposite the Sun in our sky.

optical quality Describes the ability of a lens, mirror, or telescope to obtain clear and properly focused images.

orbital resonance Describes any situation in which one object's orbital period is a simple ratio of another object's period, such as 1/2, 1/4, or 5/3. In such cases, the two objects periodically line up with each other, and the extra gravitational attractions at these times can affect the objects' orbits.

orbiters (of other worlds) Spacecraft that go into orbit of another world for long-term study.

outer solar system Generally considered to encompass the region of our solar system beginning at about the orbit of Jupiter.

outgassing The process of releasing gases from a planetary interior, usually through volcanic eruptions.

oxidation Refers to chemical reactions, often with the surface of a planet, that remove oxygen from the atmosphere.

ozone The molecule O_3, which is a particularly good absorber of ultraviolet light.

ozone depletion Refers to the declining levels of atmospheric ozone found worldwide on Earth, especially in Antarctica, in recent years.

ozone hole A place where the concentration of ozone in the stratosphere is dramatically lower than is the norm.

pair production The process in which a concentration of energy spontaneously turns into a particle and its antiparticle.

parabola The precise mathematical shape of a special type of unbound orbit allowed under the force of gravity; if an object in a parabolic orbit loses only a tiny amount of energy, it will become bound.

paradigm (in science) Refers to general patterns of thought that tend to shape scientific beliefs during a particular time period.

paradox A situation that, at least at first, seems to violate common sense or contradict itself. Resolving paradoxes often leads to deeper understanding.

parallax The apparent shifting of an object against the background, due to viewing it from different positions. *See also* stellar parallax

parallax angle Half of a star's annual back-and-forth shift due to stellar parallax; related to the star's distance according to the formula

$$\text{distance in parsecs} = \frac{1}{p}$$

where p is the parallax angle in arcseconds.

parsec (pc) Approximately equal to 3.26 light-years; it is the distance to an object with a parallax angle of 1 arcsecond.

partial lunar eclipse A lunar eclipse in which the Moon becomes only partially covered by the Earth's umbral shadow.

partial solar eclipse A solar eclipse during which the Sun becomes only partially blocked by the disk of the Moon.

particle accelerator A machine designed to accelerate subatomic particles to high speeds in order to create new particles or to test fundamental theories of physics.

particle era The era of the universe lasting from 10^{-10} second to 0.001 second after the Big Bang, during which subatomic particles were continually created and destroyed and ending when matter annihilated antimatter.

peculiar velocity (of a galaxy) The component of a galaxy's velocity relative to the Milky Way that deviates from the velocity expected by Hubble's law.

penumbra The lighter, outlying regions of a shadow.

penumbral (lunar) eclipse A lunar eclipse in which the Moon passes only within the Earth's penumbral shadow and does not fall within the umbra.

perigee The point at which an object orbiting the Earth is nearest to the Earth.

perihelion The point at which an object orbiting the Sun is closest to the Sun.

period–luminosity relation The relation that describes how the luminosity of a Cepheid variable star is related to the period between peaks in its brightness; the longer the period, the more luminous the star.

phase (of matter) Describes the way in which atoms or molecules are held together; the common phases are solid, liquid, and gas.

photon An individual particle of light, characterized by a wavelength and a frequency.

photosphere The visible surface of the Sun, where the temperature averages just under 6,000 K.

pixel An individual "picture element" on a CCD.

Planck era The era of the universe prior to the Planck time.

Planck time The time when the universe was 10^{-43} second old, before which random energy fluctuations were so large that our current theories are powerless to describe what might have been happening.

Planck's constant A universal constant, abbreviated h, with value $h = 6.626 \times 10^{-34}$ joule × s.

planet An object that orbits a star and that, while much smaller than a star, is relatively large in size; there is no "official" minimum

GLOSSARY G-8

size for a planet, but the nine planets in our solar system all are at least 2,000 km in diameter.

planetary nebula The glowing cloud of gas ejected from a low-mass star at the end of its life.

planetesimals The building blocks of planets, formed by accretion in the solar nebula.

plasma A gas consisting of ions and electrons.

plasma tail (of a comet) One of two tails seen when a comet passes near the Sun (the other is the dust tail); composed of ionized gas blown away from the Sun by the solar wind.

plate tectonics The geological process in which plates are moved around by stresses in a planet's mantle.

plates (on a planet) Pieces of a lithosphere that apparently float upon the denser mantle below.

potential energy Energy stored for later conversion into kinetic energy; includes gravitational potential energy, electrical potential energy, and chemical potential energy.

power The rate of energy usage, usually measured in watts (1 watt = 1 joule/s).

precession The gradual wobble of the axis of a rotating object around a vertical line.

pressure Describes the force (per unit area) pushing on an object. In astronomy, we are generally interested in pressure applied by surrounding gas (or plasma). Ordinarily, such pressure is related to the temperature of the gas (see thermal pressure). In objects such as white dwarfs and neutron stars, pressure may arise from a quantum effect (see degeneracy pressure). Light can also exert pressure. (See radiation pressure.)

primary mirror The large, light-collecting mirror of a reflecting telescope.

prime focus (of a reflecting telescope) The first point at which light focuses after bouncing off the primary mirror; located in front of the primary mirror.

prime meridian The meridian of longitude that passes through Greenwich, England, defined to be longitude 0°.

primitive meteorites Meteorites that formed at the same time as the solar system itself, about 4.6 billion years ago.

processed meteorites Meteorites that apparently once were part of a larger object that "processed" the original material of the solar nebula into another form.

protogalactic cloud A huge, collapsing cloud of intergalactic gas from which an individual galaxy formed.

proton–proton chain The chain of reactions by which low-mass stars (including the Sun) fuse hydrogen into helium.

protons Particles found in atomic nuclei with positive electrical charge, built from three quarks.

protoplanetary disk A disk of material surrounding a young star (or protostar) that may eventually form planets.

protostar A forming star that has not yet reached the point where sustained fusion can occur in its core.

protostellar disk A disk of material surrounding a protostar; essentially the same as a protoplanetary disk, but may not necessarily lead to planet formation.

protostellar wind The relatively strong wind from a protostar.

protosun The central object in the forming solar system that eventually became the Sun.

pseudoscience Something that purports to be science or may appear to be scientific but that does not adhere to the testing and verification requirements of the scientific method.

pulsar A neutron star from which we see rapid pulses of radiation as it rotates.

pulsating variable stars Stars that alternately grow brighter and dimmer as their outer layers expand and contract in size.

quantum mechanics The branch of physics that deals with the very small, including molecules, atoms, and fundamental particles.

quantum state Refers to the complete description of the state of a subatomic particle, including its location, momentum, orbital angular momentum, and spin, to the extent allowed by the uncertainty principle.

quantum tunneling The process in which, thanks to the uncertainty principle, an electron or other subatomic particle appears on the other side of a barrier that it does not have the energy to overcome in a normal way.

quarks The building blocks of protons and neutrons, quarks are one of the two basic types of fermions (leptons are the other).

quasar The brightest type of active galactic nucleus.

radar ranging A method of measuring distances within the solar system by bouncing radio waves off planets.

radial motion The component of an object's motion directed toward or away from us.

radiation pressure Pressure exerted by photons of light.

radiation zone (of a star) A region of the interior in which energy is transported primarily by radiative diffusion.

radiative diffusion The process by which photons gradually migrate from a hot region (such as the solar core) to a cooler region (such as the solar surface).

radiative energy Energy carried by light; the energy of a photon is Planck's constant times its frequency, or $h \times f$.

radio galaxy A galaxy that emits unusually large quantities of radio waves; thought to contain an active galactic nucleus powered by a supermassive black hole.

radio waves Light with very long wavelengths (and hence low frequencies)—longer than those of infrared light.

radioactive dating The process of determining the age of a rock (i.e., the time since it solidified) by comparing the present amount of a radioactive substance to the amount of its decay product.

radioactive element (or **radioactive isotope**) A substance whose nucleus tends to fall apart spontaneously.

recession velocity (of a galaxy) The speed at which a distant galaxy is moving away from us due to the expansion of the universe.

recollapsing universe The possible fate of our universe in which the collective gravity of all its matter eventually halts and reverses the expansion. The galaxies will come crashing back together, and the universe will end in a fiery Big Crunch.

red giant A giant star that is red in color.

red-giant winds The relatively dense but slow winds from red giant stars.

redshift (Doppler) A Doppler shift in which spectral features are shifted to longer wavelengths, caused when an object is moving away from the observer.

reflecting telescope A telescope that uses mirrors to focus light.

reflection (of light) The process by which matter changes the direction of light.

refracting telescope A telescope that uses lenses to focus light.

rest wavelength The wavelength of a spectral feature in the absence of any Doppler shift or gravitational redshift.

retrograde motion Motion that is backward compared to the norm; e.g., we see Mars in apparent retrograde motion during the periods of time when it moves westward, rather than the more common eastward, relative to the stars.

revolution The orbital motion of one object around another.

right ascension (RA) Analogous to longitude, but on the celestial sphere; it is the angular east-west distance between the vernal equinox and a location on the celestial sphere.

rings (planetary) Consist of numerous small particles orbiting a planet within its Roche zone.

Roche tidal zone The region within two to three planetary radii (of any planet) in which the tidal forces tugging an object apart become comparable to the gravitational forces holding it together; planetary rings are always found within the Roche tidal zone.

rocks (in solar system theory) Material common on the surface of the Earth, such as silicon-based minerals, that are solid at temperatures and pressures found on Earth but typically melt or vaporize at temperatures of 500–1,300 K.

rotation The spinning of an object around its axis.

rotation curve A graph that plots rotational (or orbital) velocity against distance from the center for any object or set of objects.

runaway greenhouse effect A positive feedback cycle in which heating caused by the greenhouse effect causes more greenhouse gases to enter the atmosphere, which further enhances the greenhouse effect.

saddle-shaped (or **hyperbolic**) **geometry** Refers to any case in which the rules of geometry for a saddle-shaped surface hold, such as that two lines that begin parallel eventually diverge.

Sagittarius Dwarf A small, dwarf elliptical galaxy that is currently passing through the disk of the Milky Way Galaxy.

saros cycle The period over which the basic pattern of eclipses repeats, which is about 18 years $11\frac{1}{3}$ days.

satellite Any object orbiting another object.

scattered light Light that is reflected into random directions.

Schwarzschild radius A measure of the size of the event horizon of a black hole.

science The search for knowledge that can be used to explain or predict natural phenomena in a way that can be confirmed by rigorous observations or experiments.

scientific method An organized approach to explaining observed facts through science.

scientific theory A model of some aspect of nature that has been rigorously tested and has passed all tests to date.

secondary mirror A small mirror in a reflecting telescope, used to reflect light gathered by the primary mirror toward an eyepiece or instrument.

sedimentary rock A rock that formed from sediments created and deposited by erosional processes.

seismic waves Earthquake-induced vibrations that propagate through a planet.

semimajor axis Half the distance across the long axis of an ellipse; in this text, it is usually referred to as the *average* distance of an orbiting object, abbreviated *a* in the formula for Kepler's third law.

SETI (search for extraterrestrial intelligence) The name given to observing projects designed to search for signs of intelligent life beyond Earth.

shield volcano A shallow-sloped volcano made from the flow of low-viscosity basaltic lava.

shock wave A wave of pressure generated by gas moving faster than the speed of sound.

sidereal day The time of 23 hours 56 minutes 4.09 seconds between successive appearances of any particular star on the meridian; essentially the true rotation period of the Earth.

sidereal month About $27\frac{1}{4}$ days, the time required for the Moon to orbit the Earth once (as measured against the stars).

sidereal period (of a planet) A planet's actual orbital period around the Sun.

sidereal time Time measured according to the position of stars in the sky rather than the position of the Sun in the sky. *See also* local sidereal time

sidereal year The time required for the Earth to complete exactly one orbit as measured against the stars; about 20 minutes longer than the tropical year on which our calendar is based.

silicate rock A silicon-rich rock.

singularity The place at the center of a black hole where, in principle, gravity crushes all matter to an infinitely tiny and dense point.

Small Magellanic Cloud One of two small, irregular galaxies (the other is the Large Magellanic Cloud) located about 150,000 light-years away; it probably orbits the Milky Way Galaxy.

snowball Earth Name given to a hypothesis suggesting that, some 600–700 million years ago, the Earth experienced a period in which it became cold enough for glaciers to exist worldwide, even in equatorial regions.

solar activity Refers to short-lived phenomena on the Sun, including the emergence and disappearance of individual sunspots, prominences, and flares; sometimes called *solar weather*.

solar circle The Sun's orbital path around the galaxy, which has a radius of about 28,000 light-years.

solar day Twenty-four hours, which is the average time between appearances of the Sun on the meridian.

solar eclipse Occurs when the Moon's shadow falls on the Earth, which can occur only at new moon; may be total, partial, or annular.

solar flares Huge and sudden releases of energy on the solar surface, probably caused when energy stored in magnetic fields is suddenly released.

solar luminosity The luminosity of the Sun, which is approximately 4×10^{26} watts.

solar maximum The time during each sunspot cycle at which the number of sunspots is the greatest.

solar minimum The time during each sunspot cycle at which the number of sunspots is the smallest.

solar nebula The piece of interstellar cloud from which our own solar system formed.

solar neutrino problem Refers to the disagreement between the predicted and observed number of neutrinos coming from the Sun.

solar prominences Vaulted loops of hot gas that rise above the Sun's surface and follow magnetic-field lines.

solar system (or **star system**) Consists of a star (sometimes more than one star) and all the objects that orbit it.

solar wind A stream of charged particles ejected from the Sun.

solid phase The phase of matter in which atoms or molecules are held rigidly in place.

sound wave A wave of alternately rising and falling pressure.

south celestial pole (SCP) The point on the celestial sphere directly above the Earth's South Pole.

spacetime The inseparable, four-dimensional combination of space and time.

spacetime diagram A graph that plots a spatial dimension on one axis and time on another axis.

spectral lines Bright or dark lines that appear in an object's spectrum, which we can see when we pass the object's light through a prismlike device that spreads out the light like a rainbow.

spectral resolution Describes the degree of detail that can be seen in a spectrum; the higher the spectral resolution, the more detail we can see.

spectral type A way of classifying a star by the lines that appear in its spectrum; it is related to surface temperature. The basic spectral types are designated by a letter (OBAFGKM, with O for the hottest stars and M for the coolest) and are subdivided with numbers from 0 through 9.

spectroscopic binary A binary star system whose binary nature is revealed because we detect the spectral lines of one or both stars alternately becoming blueshifted and redshifted as the stars orbit each other.

spectroscopy (in astronomical research) The process of obtaining spectra from astronomical objects

spectrum (of light) *See* electromagnetic spectrum

speed The rate at which an object moves. Its units are distance divided by time, such as m/s or km/hr.

speed of light The speed at which light travels, which is about 300,000 km/s.

spherical geometry Refers to any case in which the rules of geometry for the surface of a sphere hold, such as that lines that begin parallel eventually meet.

spheroidal component (of a galaxy) The portion of any galaxy that is spherical (or football-like) in shape and contains very little cool gas; generally contains only very old stars. Elliptical galaxies have only a spheroidal component, while spiral galaxies also have a disk component.

spin (quantum) *See* spin angular momentum

spin angular momentum Often simply called *spin*, it refers to the inherent angular momentum of a fundamental particle.

spiral arms The bright, prominent arms, usually in a spiral pattern, found in most spiral galaxies.

spiral density waves Gravitationally driven waves of enhanced density that move through a spiral galaxy and are responsible for maintaining its spiral arms.

spiral galaxies Galaxies that look like flat, white disks with yellowish bulges at their centers. The disks are filled with cool gas and dust, interspersed with hotter ionized gas, and usually display beautiful spiral arms.

spreading centers (geological) Places where hot mantle material rises upward between plates and then spreads sideways creating new seafloor crust.

spring equinox (vernal equinox) Refers both to the point in Pisces on the celestial sphere where the ecliptic crosses the celestial equator and to the moment in time when the Sun appears at that point each year (around March 21).

standard candle An object for which we have some means of knowing its true luminosity, so that we can use its apparent brightness to determine its distance with the luminosity–distance formula.

standard model (of physics) The current theoretical model that describes the fundamental particles and forces in nature.

standard time Time measured according to the internationally recognized time zones.

star A large, glowing ball of gas that generates energy through nuclear fusion in its core. The term *star* is sometimes applied to objects that are in the process of becoming true stars (e.g., protostars) and to the remains of stars that have died (e.g., neutron stars).

star cluster *See* cluster of stars

starburst galaxy A galaxy in which stars are forming at an unusually high rate.

state (quantum) *See* quantum state

steady state theory A now-discredited theory that held that the universe had no beginning and looks about the same at all times.

Stefan–Boltzmann constant constant that appears in the laws of thermal radiation, with value

$$\sigma = 5.7 \times 10^{-8} \frac{\text{watt}}{\text{m}^2 \times \text{Kelvin}^4}.$$

stellar evolution The formation and development of stars.

stellar parallax The apparent shift in the position of a nearby star (relative to distant objects) that occurs as we view the star from different positions in the Earth's orbit of the Sun each year.

stellar wind A stream of charged particles ejected from the surface of a star.

stratosphere An intermediate-altitude layer of the atmosphere that is warmed by the absorption of ultraviolet light from the Sun.

stratovolcano A steep-sided volcano made from viscous lavas that can't flow very far before solidifying.

stromatolites Large bacterial "colonies."

strong force One of the four fundamental forces; it is the force that holds atomic nuclei together.

subduction (of tectonic plates) The process in which one plate slides under another.

subduction zones Places where one plate slides under another.

subgiant A star that is between being a main-sequence star and being a giant; subgiants have inert helium cores and hydrogen-burning shells.

sublimation The process by which atoms or molecules escape into the gas phase from a solid.

summer solstice Refers both to the point on the celestial sphere where the ecliptic is farthest north of the celestial equator and to the moment in time when the Sun appears at that point each year (around June 21).

sunspot cycle The period of about 11 years over which the number of sunspots on the Sun rises and falls.

sunspots Blotches on the surface of the Sun that appear darker than surrounding regions.

superbubble Essentially a giant interstellar bubble, formed when the shock waves of many individual bubbles merge to form a single, giant shock wave.

supercluster Superclusters consist of many clusters of galaxies, groups of galaxies, and individual galaxies and are the largest known structures in the universe.

supergiants (luminosity class I) The very large and very bright stars that appear at the top of an H–R diagram.

supermassive black hole Giant black hole, with a mass millions to billions of times that of our Sun, thought to reside in the centers of many galaxies and to power active galactic nuclei.

supernova The explosion of a star.

Supernova 1987A A supernova witnessed on Earth in 1987; it was the nearest supernova seen in nearly 400 years and helped astronomers refine theories of supernovae.

supernova remnant A glowing, expanding cloud of debris from a supernova explosion.

synchronous rotation Describes the rotation of an object that always shows the same face to an object that it is orbiting because its rotation period and orbital period are equal.

synodic month (or **lunar month**) The time required for a complete cycle of lunar phases, which averages about $29\frac{1}{2}$ days.

synodic period (of a planet) The time between successive alignments of a planet and the Sun in our sky; measured from opposition to opposition for a planet beyond Earth's orbit, or from superior conjunction to superior conjunction for Mercury and Venus.

tangential motion The component of an object's motion directed across our line of sight.

tectonics The disruption of a planet's surface by internal stresses.

temperature A measure of the average kinetic energy of particles in a substance.

terrestrial planets Rocky planets similar in overall composition to Earth.

theories of relativity (*special* and *general*) Einstein's theories that describe the nature of space, time, and gravity.

thermal emitter An object that produces a thermal radiation spectrum; sometimes called a "blackbody."

thermal energy Represents the collective kinetic energy, as measured by temperature, of the many individual particles moving within a substance.

thermal escape The process in which atoms or molecules in a planet's exosphere move fast enough to escape into space.

thermal pressure The ordinary pressure in a gas arising from motions of particles that can be attributed to the object's temperature.

thermal pulses The predicted upward spikes in the rate of helium fusion, occurring every few thousand years, that occur near the end of a low-mass star's life.

thermal radiation The spectrum of radiation produced by an opaque object that depends only on the object's temperature; sometimes called "blackbody radiation."

thermosphere A high, hot X-ray-absorbing layer of an atmosphere, just below the exosphere.

tidal force A force that is caused when the gravity pulling on one side of an object is larger than that on the other side, causing the object to stretch.

tidal friction Friction within an object that is caused by a tidal force.

tidal heating A source of internal heating created by tidal friction. It is particularly important for satellites with eccentric orbits such as Io and Europa.

time dilation Refers to the effect in which you observe time running slower in reference frames moving relative to you.

timing (in astronomical research) The process of tracking how the light intensity from an astronomical object varies with time.

torque A twisting force that can cause a change in an object's angular momentum.

total apparent brightness *See* apparent brightness. We sometimes say "total apparent brightness" to distinguish it from wavelength-specific measures such as the apparent brightness measured in visible light.

total luminosity *See* luminosity. We sometimes say "total luminosity" to distinguish it from wavelength-specific measures such as the luminosity emitted in visible light or the X-ray luminosity.

total lunar eclipse A lunar eclipse in which the Moon becomes fully covered by the Earth's umbral shadow.

total solar eclipse A solar eclipse during which the Sun becomes fully blocked by the disk of the Moon.

totality (eclipse) The portion of either a total lunar eclipse during which the Moon is fully within the Earth's umbral shadow or a total solar eclipse during which the Sun's disk is fully blocked by the Moon.

transmission (of light) The process in which light passes through matter without being absorbed.

transparent (material) Describes a material that transmits light.

triple-alpha reaction *See* helium fusion

Trojan asteroids Asteroids found within two stable zones that share Jupiter's orbit but lie 60° ahead of and behind Jupiter.

tropic of Cancer The circle on the Earth with latitude 23.5°N. It is the northernmost latitude at which the Sun ever passes directly overhead (at noon on the summer solstice).

tropic of Capricorn The circle on the Earth with latitude 23.5°S. It is the southernmost latitude at which the Sun ever passes directly overhead (at noon on the winter solstice).

tropical year The time from one spring equinox to the next, on which our calendar is based.

troposphere The lowest atmospheric layer, in which convection and weather occur.

Tully–Fisher relation A relationship among spiral galaxies showing that the faster a spiral galaxy's rotation speed, the more luminous it is; it is important because it allows us to determine the distance to a spiral galaxy once we measure its rotation rate and apply the luminosity–distance formula.

turbulence Rapid and random motion.

ultraviolet light Light with wavelengths that fall in the portion of the electromagnetic spectrum between visible light and X rays.

umbra The dark central region of a shadow.

unbound orbits Orbits on which an object comes in toward a large body only once, never to return; unbound orbits may be parabolic or hyperbolic in shape.

uncertainty principle The law of quantum mechanics that states that we can never know both a particle's position and its momentum, or both its energy and the time it has the energy, with absolute precision.

universal law of gravitation The law expressing the force of gravity (F_g) between two objects, given by the formula
$$F_g = G \frac{M_1 M_2}{d^2}$$
$$\left(G = 6.67 \times 10^{-11} \frac{\text{m}^3}{\text{kg} \times \text{s}^2} \right).$$

universal time (UT) Standard time in Greenwich (or anywhere on the prime meridian).

universe The sum total of all matter and energy.

up quark One of the two quark types (the other is the down quark) found in ordinary protons and neutrons. Has a charge of $+\frac{2}{3}$.

velocity The combination of speed and direction of motion; it can be stated as a speed in a particular direction, such as 100 km/hr due north.

virtual particles Particles that "pop" in and out of existence so rapidly that, according to the uncertainty principle, they cannot be directly detected.

viscosity Describes the "thickness" of a liquid in terms of how rapidly it flows; low-viscosity liquids flow quickly (e.g., water), while high-viscosity liquids flow slowly (e.g., molasses).

visible light The light our eyes can see, ranging in wavelength from about 400 to 700 nm.

visual binary A binary star system in which we can resolve both stars through a telescope.

voids Huge volumes of space between superclusters that appear to contain very little matter.

volatiles Refers to substances, such as water, carbon dioxide, and methane, that are usually found as gases, liquids, or surface ices on the terrestrial worlds.

volcanism The eruption of molten rock, or lava, from a planet's interior onto its surface.

wavelength The distance between adjacent peaks (or troughs) of a wave.

weak bosons The exchange particles for the weak force.

weak force One of the four fundamental forces; it is the force that mediates nuclear reactions; also the only force besides gravity felt by weakly interacting particles.

weakly interacting particles Particles, such as neutrinos and WIMPs, that respond only to the weak force and gravity; that is, they do not feel the strong force or the electromagnetic force.

weather Describes the ever-varying combination of winds, clouds, temperature, and pressure in a planet's troposphere.

weight The net force that an object applies to its surroundings; in the case of a stationary body on the surface of the Earth, weight = mass × acceleration of gravity.

weightless A weight of zero, as occurs during free-fall.

white-dwarf limit (also called the *Chandrasekhar limit*) The maximum possible mass for a white dwarf, which is about $1.4 M_{\text{Sun}}$.

white dwarf supernova A supernova that occurs when an accreting white dwarf reaches the white-dwarf limit, ignites runaway carbon fusion, and explodes like a bomb; often called a *Type Ia supernova*.

white dwarfs The hot, compact corpses of low-mass stars, typically with a mass similar to the Sun compressed to a volume the size of the Earth.

WIMPs Stands for *weakly interacting massive particles* and represents a possible form of dark matter consisting of subatomic particles that are dark because they do not respond to the electromagnetic force.

winter solstice Refers both to the point on the celestial sphere where the ecliptic is farthest south of the celestial equator and to the moment in time when the Sun appears at that point each year (around December 21).

worldline A line that represents an object on a spacetime diagram.

wormholes The name given to hypothetical tunnels through hyperspace that might connect two distant places in our universe.

X rays Light with wavelengths that fall in the portion of the electromagnetic spectrum between ultraviolet light and gamma rays.

X-ray binary A binary star system that emits substantial amounts of X rays, thought to be from an accretion disk around a neutron star or black hole.

X-ray burster An object that emits a burst of X rays every few hours to every few days; each burst lasts a few seconds and is thought to be caused by helium fusion on the surface of an accreting neutron star in a binary system.

zenith The point directly overhead, which has an altitude of 90°.

zodiac The constellations on the celestial sphere through which the ecliptic passes.

zones (on a jovian planet) Bright bands of rising air that encircle a jovian planet at a particular set of latitudes.

ACKNOWLEDGMENTS

ILLUSTRATION BY JOE BERGERON: Figures 1.1, 1.3, 1.15, 1.16, 1.17, 2.6, 2.21, 8.6, 8.7, 8.14, 8.15, 8.16, 11.20, 12.20g,

Part Opener I and Chapter Openers 1–2, S1: ©Roger Ressmeyer/CORBIS

Part Opener II and Chapter Openers 3–7: "An Expanding Bubble in Space": NASA, Donald Walter, South Carolina State Univ., Paul Scowen and Brian Moore, Arizona State Univ.

Part Opener III and Chapter Openers 8–13: "Valles Marineris—Point Perspective": Jody Swann, Tammy Becker, and Alfred McEwen of U.S. Geological Survey in Flagstaff, Arizona

All Part Opener backgrounds, Visual Language Library Antique Celestial Charts and Illustrations, 1680–1880, Volume 3

CHAPTER 1 **1.2** NASA **1.4** Jerry Lodriguss **1.5** Created by Vincent Ciulla Design. From the Voyage scale model solar system, developed by Challenger Center for Space Science Education, the Smithsonian Institution, and NASA **1.6** Stan Maddock **Page 12** (**Sun**-left), Big Bear Solar Observatory/New Jersey Institute of Technology and NASA Marshall Space Flight Center; (**Sun**-right) courtesy of SOHO. SOHO is a project of international cooperation between ESA and NASA. **Page 13** (**Mercury**-left) From the Voyage scale model solar system, developed by Challenger Center for Space Science Education, the Smithsonian Institution, and NASA. Image created by ARC Science Simulations, ©2001; (**Mercury**-right) NASA/USGS **Page 14** (**Venus**-left) NASA, courtesy of NSSDC; (**Venus**-right) From the Voyage scale model solar system, developed by Challenger Center for Space Science Education, the Smithsonian Institution, and NASA. ©2001 David P. Anderson, Southern Methodist Univ. **Page 15** (**Earth**-left) From the Voyage scale model solar system, developed by Challenger Center for Space Science Education, the Smithsonian Institution, and NASA. Image created by ARC Science Simulations, ©2001; (**Earth**-right) NASA **Page 16** (**Mars**-top) NASA/USGS, courtesy of NSSDC; (**Mars**-bottom) NASA **Page 17** (**Asteroid**) NEAR Project, John Hopkins Univ./APL, NASA **Page 18** (**Jupiter**) From the Voyage scale model solar system, developed by Challenger Center for Space Science Education, the Smithsonian Institution, and NASA. Image created by ARC Science Simulations, ©2001 **Page 19** (**Saturn**) From the Voyage scale model solar system, developed by Challenger Center for Space Science Education, the Smithsonian Institution, and NASA. Image created by ARC Science Simulations, ©2001 **Page 20** (**Uranus**) From the Voyage scale model solar system, developed by Challenger Center for Space Science Education, the Smithsonian Institution, and NASA. Image created by ARC Science Simulations, ©2001 **Page 21** (**Neptune**) From the Voyage scale model solar system, developed by Challenger Center for Space Science Education, the Smithsonian Institution, and NASA. Image created by ARC Science Simulations, ©2001 **Page 22** (**Pluto and Charon**) Dr. R. Albrecht (ESA/ESO) and NASA **Page 23** (**Comet Hale–Bopp**) Niescja Turner and Carter Emmart **1.7** Akira Fujii **1.8** Megan Donahue (STScI) **1.9** (**clockwise, starting far-top right**) ©Photo by Corel; ©Christian Jegou/Publiphoto/Photo Researchers, Inc.; ©David Gifford/Science Photo Library/Photo Researchers, Inc.; ©Blakeley Kim; ©Photo by Corel; NASA; Quade Paul/fiVth.com; ©John Eastcott and Yva Momatiuk/Photo Researchers, Inc. **1.19** Quade Paul/fiVth.com, PhotoDisc, and Hubble Space Telescope

CHAPTER 2 **2.1** Andrea Dupree (Harvard-Smithsonian CFA), Ronald Gilliland (STScI), ESA, and NASA **2.5** Gordon Garradd **2.10** (**c**) Richard Tauber Photography, San Francisco **2.11** (**b**) ©Dennis diCicco **2.13** ©David Nunuk **2.16** ©Dennis diCicco **2.17** ©Husmo-foto **2.19** (**Moons**) Akira Fujii; (**Earth**) NASA **2.23** (**top**, **bottom**) Akira Fujii; (**center**) Dennis diCicco **2.24**, **2.25** Akira Fujii **2.26** Adapted from eclipse map by Fred Espenak, NASA/GSF: (http://sunearth.gsfc.nasa.gov/eclipse/eclipse.html) **2.27** ©Frank Zullo

CHAPTER S1 **S1.5** Image from TRACE (Transition Region and Coronal Explorer), a mission of the Stanford-Lockheed Institute for Space Research (a joint program of the Lockheed-Martin Advanced Technology Center's Solar and Astrophysics laboratory and Stanford's Solar Observatories Group), and part of the NASA Small Explorer program **S1.6** ©Bernd Wittich/Visuals Unlimited **S1.8** NASA/MSFC **S1.25** (**a**, **b**) ©Bettmann/CORBIS; (**c**) The Granger Collection, NY; (**d**) ©Science VU/Visuals Unlimited

CHAPTER 3 **3.2** ©Michael Yamashita/CORBIS **3.3** (**a**) ©N. Pecnik/Visuals Unlimited **3.4** ©Kenneth Garrett **3.5** (**a**) ©Wm. E. Woolam/Southwest Parks; (**b**) ©2001 Stone/Richard A. Cooke, III **3.6** ©1987 by Margaret R. Curtis **3.7** ©Richard A. Cooke, III **3.8** ©Loren McIntyre/Woodfin Camp & Assoc. **3.9** (**left**) ©John A. Eddy/Visuals Unlimited; (**right**) ©Jeff Henry/Peter Arnold, Inc. **3.10** ©Oliver Strewe/Wave Productions Pty, Ltd. **3.11** Werner Forman Archive/Art Resource, NY **3.14** Courtesy of Carl Sagan Productions, Inc. From Cosmos (Random House), © 1980 Carl Sagan **3.15** ©Bettman/CORBIS **Page 107** The Granger Collection, NY

CHAPTER 4 **4.1** ©2001 Stone/Dimitri Iundt **4.5** ©Science VU/Visuals Unlimited **Page 117** ©Bettmann/CORBIS

C-1

CHAPTER 5   Page 134 (**top**) Giraudon/Art Resource, NY; (**bottom**) Archive Photos Page 135 (**left**) The Granger Collection, NY; (**right**) ©Erich Lessing/Art Resource, NY **5.13** ©Jerry Lodriguss   Page 138 (**right**) ©Bettmann/CORBIS   Page 139 ©Bettmann/CORBIS   **5.17** NASA   **5.19** (**both**) ©Bill Bachmann/Gnass Photo Images

CHAPTER 6   **6.1** ©Runk/Schoenberger from Grant Heilman Photography

CHAPTER 7   **7.4** (**a**) Emiko-Rose Koike/fiVth.com   **7.6** ©Craig M. Moore   **7.7** (**b**) Yerkes Observatory   **7.8** (**b**) Palomar Observatory/Caltech   **7.10** (**a**) ©Richard Wainscoat; (**b**) Russ Underwood (W. M. Keck Observatory)   **7.12** Mark Voit (STScI)   **7.13** ©Anglo-Australian Observatory, photograph by D. Malin   **7.15** NASA/CXC/SAO   **7.17** ©Richard Wainscoat   **7.18** NASA/Ames Research Center   **7.19** (**a, b**) Canada-France-Hawaii Telescope Corporation, Hawaii   **7.20** (**a**) NASA; (**b**) Don Foley/National Geographic Image Collection   **7.21** Jodi Schoemer   **7.22** (**c**) Eastman-Kodak   **7.23** NASA   **7.24** David Parker, 1997/Science Library. *The Arecibo Observatory is part of the National Astronomy and Ionosphere Center, which is operated by Cornell Univ. under a cooperative agreement with the National Science Foundation*   **7.26** ©Joel Gordon Photography   **7.28** (**right**) NASA/JPL   **7.30** NASA/JPL/Caltech

CHAPTER 8   **8.2** (**left**) Johns Hopkins Univ./APL/NASA   **8.3** (**left**) ©1997 H. Mikuz and B. Kambic (Crni Vrh Observatory, Slovenia)   **8.4** C. R. O'Dell (Rice Univ.) and NASA   **8.8** (**a-left**) NASA/STScI, courtesy of Alfred Schultz and Helen Hart; (**a-right**) JPL/Caltech, Franklin & Marshall College, and NASA at W. M. Keck Observatory; (**b-left**) M. J. McCaughrean (MPIA), C. R. O'Dell (Rice Univ.), and NASA; (**b-right**) J. Bally, D. Devine, and R. Sutherland   **8.12** Quade Paul/fiVth.com   **8.13** Meteorite specimen courtesy of Robert Haag Meteorites   **8.17** (**a, b**) NASA/JPL, courtesy NSSDC   **8.18** ©William K. Hartmann   **8.22** Data adapted from STARE project: (www.hao.ucar.edu/public/research/stare/hd209458.html)

CHAPTER 9   **9.1** (**Moon**) Akira Fujii; (**others**) NASA   **9.2** (**a**) ©Don Davis; (**b**) NASA, courtesy NSSDC; (**c**) ©Barrie Rokeach; (**d**) Artis Planetarium/The Netherlands; (**e**) NASA/JPL   **9.3** (**both**) ©Roger Ressmeyer/CORBIS   **9.9** (**a**) ©Jules Bucher/Photo Researchers, Inc.   **9.11** Don Gault, Peter Schmultz, and NASA Ames Research Center   **9.12** (**a**) NASA, from the Apollo 16 crew's mapping camera; (**b**) NASA/USGS   **9.13** (**a**) NASA/JPL (Viking Orbiter); (**b**) NASA/USGS; (**c**) NASA/JPL/Malin Space Science Systems   **9.14** NASA   **9.16** (**b**) ©2001 Stone/Paul Chesley   **9.17** (**top**) NASA, courtesy of LPL; (**center**) NASA/JPL; (**bottom**) ©Forest Buchanan/Visuals Unlimited   **9.19** (**left**) NASA/USGS; (**right**) NASA/JPL (Magellan Mission)   **9.22** (**all**) NASA   **9.23** (**a, c**) NASA/JPL; (**b**) USGS (Mariner 10 Mission)   **9.24** (**a**) NASA, courtesy of Mark Robinson   **9.25** Lowell Observatory Photographs   **9.26** (**all**) NASA/USGS   **9.27** (**a–c**) NASA/JPL; (**d, f**) NASA/JPL/Maline Space Science Systems; (**e**) Dr. David E. Smith, NASA, and MOLA Science Team   **9.28** (**all**) NASA/JPL

CHAPTER 10   **10.1** (**Mercury**-globe) NASA/courtesy NSSDC, (**Mercury**-surface) ©Don Davis; (**Venus**-globe) NASA/JPL, (**Venus**-surface) ©Don Davis; (**Earth**-globe) NASA, (**Earth**-surface) ©Barrie Rokeach; (**Moon**-globe) Akira Fujii, (**Moon**-surface) Artis Planetarium/The Netherlands; (**Mars**-globe) NASA/courtesy of Calvin Hamilton, (**Mars**-surface) NASA/JPL/Caltech   **10.2** (**a, b**) NASA   **10.10** (**b**) ©Hinrich Baesemann (www.polarfoto.com); (**c**) Dr. L. A. Frank, Univ. of Iowa; (**d**) NASA   **10.12** ©Tom Van Sant/The Geosphere Project/Corbis Stock Market   **10.14** (**a**) ©Tom Van Sant/The Geosphere Project/Corbis Stock Market; (**b**) GOES-10 satellite image. Colorization by ARC Science Simulations, ©1998   **10.15** ©Tom Van Sant/The Geosphere Project/Corbis Stock Market   **10.16** (**a, c**) NASA/JPL; (**b**) Malin Space Science Systems and Caltech   **10.18** NASA, courtesy NSSDC   **10.19** NASA/JPL and Malin Space Science Systems   **10.21** ©D. Cavagnaro/Visuals Unlimited   **10.24** (**a**) D. Potter, T. Morgan, and R. Killen; (**b**) M. Mendillo, J. Baumgartner, and J. Wilson of Boston Univ.

CHAPTER 11   **11.1** NASA   **11.3** (**b**) NASA/JPL   **11.6** (**a**) ©Michael Carroll   **11.7** (**b**) STScI and R. Beebe and A. Simon (NMSU); (**c**) NASA Infrared Telescope Facility and Glenn Orton   **11.8** (**a**) NASA/JPL   **11.11** (**Uranus**) E. Karkoscha (Univ. of Arizona and NASA); (**others**) NASA/JPL   **11.12** (**b**) J. Clarke (Univ. of Michigan) and NASA   **11.13** John Spencer/Lowell Observatory   **11.14** Data provided by Fran Bagenal   **11.15** Courtesy Tim Parker/JPL   **11.16** NASA/JPL   **11.17** (**a**) PIRL/Lunar & Planetary Laboratory (Univ. of Arizona) and NASA; (**b**) Univ. of Arizona and NASA; (**c-left**) NASA/JPL; (**c-right**) Univ. of Arizona and NASA; (**d**) PIRL/Lunar & Planetary Laboratory (Univ. of Arizona) and NASA; (**e**) NASA/USGS; (**f**) NASA/JPL/Ames Research Center; (**g**) Univ. of Arizona and NASA   **11.19** (**a**) DLR/NASA/JPL; (**b**) Arizona State Univ. and NASA; (**c**) NASA/JPL/Arizona State Univ.   **11.21** (**a, c**) NASA/JPL; (**b**) NASA/JPL and Brown Univ.   **11.22** (**a**) NASA/JPL; (**b**) NASA/JPL and Arizona State Univ.   **11.23** (**a**) NASA/JPL/Caltech; (**b**) ©Don Davis; (**c**) Seran Gibbard, Bruce Macintosh, Don Gavel, Claire Max (Lawrence Livermore Natonal Laboratory)   **11.24, 11.25** NASA/JPL   **11.26** (**a**) NASA/USGS; (**b**) NASA/JPL   **11.27** (**a**) S. Larson (Univ. of Arizona/LPL); (**b**) NASA/JPL; (**c**) ©William K. Hartmann   **11.28** NASA/JPL/Caltech   **11.29** NASA/JPL, courtesy of M. Showalter (Plantary Data Systems Ring Node)   **11.30** (**a, b**) NASA/JPL   **11.31** (**b**) NASA/JPL, courtesy of M. Showalter (Plantary Data Systems Ring Node)   **11.32** (**Uranus**) Erich Karkoschka (Univ. of Arizona/LPL) and NASA; (**others**) NASA/JPL

CHAPTER 12   **12.1, 12.2** Data provided by Dave Tholen (Univ. of Hawaii)   **12.3** Discovery photograph made January 7, 1976 by Eleanor F. Helin/JPL (Helin then associated with Caltech/Palomar Observatory)   **12.4** (**a**) NASA/USGS; (**b**) NASA/JPL; (**c, d**) Johns Hopkins Univ./APL/NASA   **12.6** Laird M. Close (Univ. of Arizona) and Wm. J. Merline (Southwest Research Institute, Boulder, Colorado)   **12.7** (**b**) Stephen Ostro et al (JPL, Arecibo Radio Telescope, NSF, and NASA)   **12.8** (**a**) Geological Survey of Canada; (**b**) courtesy of Calvin Hamilton   **12.9** (**a**) Walt Radomski (nyrockman.com); (**b**) PEANUTS reprinted by permission of United Feature Syndicate, Inc.   **12.10** Meteorite specimens courtesy of Robert Haag Meteorites   **12.11** (**a**) ©Peter Ceravolo; (**b**) Tony and Daphne Hallas   **12.12** LASCO, SOHO Consortium, NRL, ESA, and NASA   **12.13** (**left**) Astuo Kuboniwa, March 9, 1997, 19:25:00–19:45:out, BISTAR Astronomical Observatory, Japan, Telescope: d = 125mm, f = 500mm refractor, Film: Ektracrome E100S; (**right**) Halley Multicolour Camera Team, Giotto, ESA, ©MPAE   **12.17** (**a**) U.S. Naval Observatory; (**b**) Dr. R. Albrecht, ESA/ESO, and NASA   **12.18** Eliot Young/Southwest Research Institute   **12.19** (**a**) Hal Weaver and T. E. Smith (STScI) and NASA; (**b**) Courtesy of Paul Schenk (Lunar and Planetary Institute)   **12.20** (**a**) NASA/JPL/Caltech; (**b**) HST Jupiter Imaging Science Team; (**c**) MSSO, ANU/Science Library/Photo Researchers, Inc.; (**d**) Courtesy of Richard Wainscoat et al (Univ. of Hawaii); (**e**) H. Hammel (MIT) and NASA; (**f**) HST Comet Team and NASA   **12.21** (**c**) James W. Young/TMO/JPL/NASA   **12.22** Brad Snowder   **12.23** Kirk Johnson/Denver Museum of Natural History   **12.24** Image courtesy of Dr. Virgil Sharpton, Univ. of Alaska-Fairbanks   **12.25** (**all**) Quade Paul/fiVth.com   **12.26** TASS/Sovfoto

CHAPTER 13   **13.1** NASA   **13.2** (**Venus**) NASA/Magellan data, courtesy Peter Ford (MIT); (**Earth**) NOAA; (**Mars**) NASA/Viking data, courtesy Mike Mellon (Univ. of Colorado)   **13.4** ©Tom Van Sant/The Geosphere Project/Corbis Stock Market   **13.5** (**a**) ©Gene Ahrens/Bruce Coleman, Inc.; (**b**) ©J. Messerschmidt/Bruce Coleman, Inc.; (**c**) Biological Photo Service; (**d**) ©C. C. Lockwood/DDB Stock Photo   **13.7** Digital image by Dr. Peter W. Sloss (NOAA/NESDIS/NGDCO)   **13.11** (**a**) NASA; (**b**) Earth Satellite Corp./Science Photo Library/Photo Researchers, Inc.   **13.12** (**b**) ©Roger Ressmeyer/CORBIS; (**c**) Mike Yamashita/Woodfin Camp & Assoc.   **13.13** (**clockwise from top**) ©Philip Rosenberg; ©Philip Rosenberg; Univ. of Hawaii; R. Shallenberger/Midway Atol National Wildlife Refuge   **13.15** ©James L. Amos/Photo Researchers, Inc.   **13.16** ©Jeff Greenberg/Visuals Unlimited   **13.17** (**a, c**) Biological Photo Services; (**b**) ©Kevin Collins/Visuals Unlimited   **13.18** ©Nih R. Feldman/Visuals Unlimited   **13.20** (**a**) Woods Hole Oceanographic Institute; (**b**) ©Barrie Rokeach   **13.21** ©Ken Lucas/Visuals Unlimited   **13.23** Data obtained from C. D. Keeling and T. P. Whorf, Scripps Institution of Oceanography, Univ. of California, La Jolla: (http://cdiac.esd.ornl.gov/ftp/maunaloa-co2/maunaloa-co2)   **13.24** NASA/GSFC (Science Visualization Studio)   **13.25** (**a**) NASA; (**b**-both) NASA/Ames Research Center

ACKNOWLEDGMENTS   C-2

INDEX

Page references in italics refer to figures. Page references preceded by a "t" refer to tables. Page references followed by an "n" refer to footnotes. Page references in boldface refer to main discussions.

Aberration of starlight, 62
Absolute zero, 113, *113*
Absorption, light, 154, **160–162**, *164*, 263–266, *264*
Absorption line spectra, 161, *164*, *165*
Acceleration
 around curves, 130, *130*
 definition, 126
 gravitational, 126, *127*, **147–148**, A-2
 weight and, 127
Accretion, **208–210**, *209*, 217, 232, 247
Accretion disks, 204
Acetylene, 294
Achondrites, 332n
Acrux, A-19
Adams, John Couch, 21
Adaptive optics, 184, *184*
Adrastea (moon of Jupiter), A-16
Aerosols, 277
Air pollution, 268, 364
Albedo
 asteroids, 328, *328*
 climate change, 277–278
 glaciers, 365
 overview, **263–264**, t 264, 266
 positive feedback, 365
Aldebaran, A-19
Aldrin, Buzz, *15*
Aleutian Islands, 358
Alexander the Great, 100
Alexandria, ancient science in, 100–101, *102*
Alfonsine Tables, 133n
Allende meteorite, 216
Almagest, 103
Al-Mamun, 103
Alpha Centauri A, 24, *24*, A-18, A-19
Alpha Centauri B, A-18
Altair, A-19
Altitude, 44, *44*, 75, **78–84**, *80*, *81*
Aluminum, condensation of, 207, *207*
Alvarez, Luis and Walter, 345
Amalthea (moon of Jupiter), A-16
Amino acids, 367–368
Ammonia, 207, *207*, 294–295, 297, 298, 311
Ammonium hydrosulfide, 295, 297, 298
Analemma, 71, *71*
Ananke (moon of Jupiter), A-16

Anaxagoras, 330
Anaximander, 44
Andesite, 358–359
Andes mountains, 359
Andromeda Galaxy, *9*, 32, 36, *42*, *42*
Angle of incidence, 155
Angle of reflection, 155
Angular distance, 44, *45*
Angular momentum, *132*, **132**, *133*, 140–141. *See also* Conservation of angular momentum
Angular resolution, 173, **178–179**, 179n, *180*, 188
Angular separation, 173, **174**, A-3
Angular sizes, 44, *45*, 52, 288
Annular eclipse, 56, *56*
Antarctica, 331, 372–373, *373*, 375
Antarctic Circle, 50, 82, 84
Antares, 163, A-19
Ante meridian (a.m.), 69
Aperture, 173
Aphelion, 136, *137*
Apollo missions, *15*, 239, *239*, 247, *247*
Apollonius, 102
Appalachian Mountains, 244, *244*, 359
Apparent brightness (luminosity), 221, A-2, A-3, A-18, A-20
Apparent retrograde motion, 59, **59–61**, *60*, *61*, 102, *103*
Apparent solar time, 69–71, *71*
Arabian peninsula, 359, *359*
Arches National Park (Utah), *47*
Arcminutes, 44
Arcseconds, 44
Arctic Circle, 50, 82, 84, *84*
Arcturus, A-19
Area, formulas for, 178, 266
Arecibo telescope (Puerto Rico), 187, *188*, A-24
Argon gas, 216–217
Ariel (moon of Uranus), 314, A-17
Aristarchus, 60–61
Aristotle, 44, 101, 137, **138**
Armstrong, Neil, 15
Artificial stars, 184
Ashen light, 53
Asteroid belt, *17*, **17**, 201, 212, 324, *324*
Asteroids, **324–330**. *See also* Planetesimals
 atmosphere formation and, 278
 captured moons from, 214, *214*
 definition, 4, 324
 discovery of, 324
 Earth-approaching, 326, 346–347

 formation, *201*, 201–202, 212
 frequency of impacts, 347–348, *348*
 future impacts, 346–348
 orbits, *201*, 325
 origin and evolution, 325–326
 properties, t 200
 size determination, 328, *328*
 studies of, 326–330, *327*, *328*, *329*
 Trojan, 201, *201*, 325, 326
Asthenosphere, 230
Astrobiology, 374–378
Astrolabes, 85–86, *86*
Astrology, 30, 106–108
Astronomical distances, A-2
Astronomical units (AU), 27, A-2
Astronomy
 ancient models of the universe, 101–103, *102*, *103*
 ancient observations, 92–100
 astrology and, 106–108
 Copernicus, 34, 61, 69, 103–104, 133–134, *134*
 history of modern, 100–103
 Islamic role, 103
 modern, 106
 overview, 34–35
 scientific method, 103–106
 Tycho Brahe, 134, **134**, 333
 Web sites, A-23 to A-25
Astrophysics, 139
Atlas (moon of Saturn), A-16
Atmosphere. *See also* Weather
 comparative structures, 269, 269–270
 definition, 262
 Earth, t 261, *261*, 263, 269, 269–270, 362–365
 effects of chlorine, 373
 effects of no atmosphere, 263–264
 erosion and, 245–246, *246*
 gas pressure, t 261, 262, 274
 generic structures, 264–267, *265*
 geological controlling factors, 236, *240*, *243*
 jovian moons, 311, *312*, 313, 315
 jovian planets, 260, 293–300
 limit of, 262, 269
 loss of, 279, 279–281
 magnetospheres, 235, 270, *271*
 with no greenhouse effect, 263–264, t 264, 266
 origin, 213, 278, **278–282**, *279*
 ozone depletion, 372–374
 Pluto, 339
 turbulence, 183–184

 Venus, 260, t 261, 269, 269–270, 280, 283–284
 wavelength penetration, *185*, **265**, *265*, 267
Atmospheric cratering, 279, 281, 313
Atomic mass, 118
Atomic number, 118
Atomic weight, 118
Atoms, 117–120, *118*, *119*
ATP molecule, 367
Auriga, *206*
Auroras, **270**, *271*, 300, *300*, 302
Autumnal equinox, *29*, 29, 74, 81–84
Axis tilts
 climate change, 277
 Earth, 28, 28–29, *29*
 from giant impacts, 214–215
 planets, A-15
 seasons, 29, *29*, 48
 solar days, 71
Azimuth, 44n, 75

Bacon, Francis, 151
Bacteria, 375–376, *376*
Balmer transitions, 160n
Barnard's Star, A-18
Bars (units), 262
Basalt, 241, 333
Belinda (moon of Uranus), A-17
Belts, falling air, 295–296, *296*
Bessel, Friedrich, 62
Beta Pictoris, *206*
Betelgeuse, *40*, 163, A-19
Bianca (moon of Uranus), A-17
Big Bang, 5
Big Bang theory, 6–7
Big Dipper, 44, *45*, 48
Big Horn Medicine Wheel (Wyoming), 98, *98*
Binary stars, 218
Biosphere, 352
Blackbody radiation. *See* Thermal radiation
Black holes, supermassive, 181
Black smokers, 368, *369*
Blueshifts, **165–167**, *166*, *168*, 219
Bode, Johann, 20
Boltzmann's constant, 280
Bombardment, **212–213**, 278, 279, 279–281, 283
Bradley, James, 62
Brightness (luminosity) *See also* Apparent brightness, 221, A-2, A-3, A-18, A-20
British thermal units (BTUs), 112
Brown ovals, 297
Bruno, Giordano, 151
Bulges, tidal, *142*, **142–143**, 145, *145*
Butterfly effect, 272

I-1

Calculus, 140
Calendars, **72–73**, *94*, 94–96, *95*
Caliban (moon of Uranus), A-17
Callanish sacred stone circle (Scotland), 96, *97*
Callisto (moon of Jupiter), 305, 310–311, *311*, *341*, A-16
Calories, 112
Caloris Basin (Mercury), *248*, 248–249
Calypso (moon of Saturn), A-16
Cambrian explosion, 370–371
Cameras, 173–175, *174*
Canopus, A-19
Cantaloupe terrain, 314
Capella, A-19
Carbon
  carbon-13, 366
  in comets, 338
  isotopes, 119, *119*
  in meteorites, 332, 332–333
  in Pluto, 340
Carbonate rocks, 281, 363–365
Carbonate-silicate cycle, 363, 365
Carbon dioxide
  absorption lines, *165*
  on Earth, 363, *363*
  feedback processes, 364–365
  as greenhouse gas, 269
  increase in atmospheric, *371*
  on Mars, 277, 279, 281, 283
  on Venus, 283, 363
Carl Sagan Memorial Station, 252
Carme (moon of Jupiter), A-16
Cassegrain focus, 177, *177*
Cassen, Pat, 305
Cassini division, 318, *318*
Cassini spacecraft, 19, t 190, 191, *191*, 312, A-23
Catastrophism, 374
CCDs (charge-coupled devices), 175, *175*
Celestial coordinates, **74–77**, *75*, *77*, t 78
Celestial equator, 41, *41*, 43
Celestial navigation. *See* Navigation, celestial
Celestial poles
  celestial coordinates, 80
  circumpolar stars, 45, *47*, *48*
  concept, 41, *41*, 43
Celestial sphere
  concept, 41, **41**
  map of, *73*, 73–74
  path of the Sun, *81*, **81–84**, *82*, *83*, *84*
  proposed by Anaximander, 44
  star tracks, *78*, 78–81, *79*, *80*, *81*
Celsius scale, 113, *113*
Centaurus, 24, A-22
Center of mass, 140
Centripetal force, 130, *130*
Ceres, 17, 324, 329
Cerro Tololo Inter-American Observatory, A-24
Ceti, A-18
CFCs (chlorofluorocarbons), 268, 373–374
Chandra X-Ray Observatory, *181*, t 186, A-23
Chaotic systems, 272
Charged particle belts, 270, *300*, **300–301**, *301*
Charge, particle, 118

Charon, *144*, 145, 150, 338–339, *339*, A-17
Chemical bases, 367–368, *368*
Chemical energy density, 115
Chemical potential energy, 112
Chemical reactions, atmospheric loss by, 279
Chichén Itzá observatory, *97*, 97
Chicxulub crater (Mexico), 345, *346*
Chinese astronomy, ancient, 99, *99*
Chiron, 338
Chlorine, 373–374
Chlorofluorocarbons (CFCs), 268, 373–374
Chondrites, 332n
Chondrules, 332n
Chromatic aberration, 177
Circles, 136, *136*
Circulation cells, **272–274**, *273*, *275*, *295*, *296*
Circumpolar, 45–46, *47*, *48*
Climate change, 270, 277–278, 283, **371–372**
Climate, definition, 270–271. *See also* Weather
Clocks, 70, *70*, **87**, 93–94. *See also* Timekeeping
Clouds
  formation, 276–277
  greenhouse effect, 268
  interstellar, *203*, **203–205**, *205*
  jovian planets, 295, 298, *299*
  Oort, 202, *202*, 212, 324, **337**, *337*
  terrestrial planets, t 261
  Venus, *276*, 276–277
Clusters, galaxy, 4–5, A-22
Clusters, star, 4–5
Coal-burning emissions, 372
Colors
  false-color images, 182
  jovian planets, 297–298, *298*
  light, *154*
  Pluto, 340
  sky, 267
  thermal radiation, 163
Columbus, Christopher, 44
Coma Cluster, A-22
Coma, comet, *335*, 335
Comets. *See also* Planetesimals
  anatomy, *335*, 335
  atmosphere formation and, 278
  definition, 4, 324
  impacts by, *341*, 341–342, *342*, *343*, 346
  Kuiper belt, 202, *202*, 212, 324, *337*, **337–338**
  life span, 336
  mass extinctions, 345–346, 347
  Oort cloud, 202, *202*, 212, 324, **337**, *337*
  orbits, 212, 336
  origin, 201–202, 212, 337–338
  Pluto, 22
  properties, t 200
  Shoemaker-Levy 9, 291
  tails, *335*, 342–344
Compact fluorescent light bulbs, 170
Comparative planetology, 198
Compass directions, 84
Compounds, 117
Compression, 244

Compton Gamma Ray Observatory, 187, *188*
Computer simulations, 206
Comte, Auguste, 169
Condensates, 207
Condensation
  from atmospheres, 275, *279*, 281
  jovian moons, 313
  planets, **206–208**, *207*, *208*
Condensation temperatures, 207–208
Conduction, planetary cooling by, *233*, 233
Cones, in eyes, 155, 172
Conjunction, 68
Conscience, origin of term, 104
Conservation of angular momentum
  changes in daylength, 144
  definition, *132*, **132–133**
  Kepler's laws of planetary motion, 140–141
  in sink drains, 274
  solar nebulae, 204, 211
Conservation of energy, 116–117
Conservation of momentum, *131*, 131–132. *See also* Conservation of angular momentum
Constellations, 40, 40–41, 48, A-26 to A-27, A-30 to A-33. *See also* Skywatching; Stars
Continental crust, 353–354, 358–360
Continental drift, 230, **354**, *355*, 356
Continents, origin of, 358–359
Contour spacecraft, t 190
Contraction, 249, 292
Convection
  atmospheric, 268, *276*, 276, 295
  planetary cooling, *233*, 233
  tectonics, 243–244
Convection cells, 233, *354*, 354–356
Conversion factors, A-2, A-8, A-9 to A-10
Copernicus, 34, 61, 69, 103–104, 133–134, *134*
Cordelia (moon of Uranus), A-17, *317*
Cores
  Earth, 353
  jovian planets, *289*, 289, 291, 293, *293*
  magnetic fields and, *234*, **234–235**, *235*
  terrestrial planets, 229–230, *230*
Coriolis effect, 62, *273*, **273–276**, 295–296, *296*, *297*
Coriolis, Gustave, 62
Coronas, 56, 253, *254*
Cosmic calendar, 25–27, *26*
Cosmic distances, A-2, A-3
Cosmic rays, 369–370
Cosmos. *See* Universe
Coudé focus, 177, *177*
Creation of the universe, 6–7
Cressida (moon of Uranus), A-17
Cross-staff, 86, *86*
Crucis, A-19
Crust, 229
Crustal compression, 244
Crusts, 229
Cycles per second, 156

Cygnus A, A-18
Cygnus B, A-18

Dactyl, 327, *327*, 328
Dark matter, 32
Darth Crater (Mimas), 314
Darwin, Charles, 370
Dating, radioactive, **215–217**, 238, 366
Daughter products, 215, 218
Daylength changes, 144
Daylight saving time, 70–72
Day names, t 93
Decay, radioactive, 215–216
Declination (dec), 75, *75*, *76*, **77–84**, *78*, *79*, *80*
Deductive arguments, 104
Deforestation, 371
Deimos (moon of Mars), *214*, 214, 327, A-16
Delta Aquarids meteor shower, t 344
Democritus, 117, *117*, 123
Deneb, A-19
Density
  asteroids, 328
  chemical energy, 115
  comets, 335
  differentiation, 229
  jovian planets, 210, t 289, 291
  meteorites, 332
  Moon of Earth, 215
  moons, A-16 to A-17
  overview, 115
  planets, t 200, *208*, 208–209, 235, A-15
  Pluto, 339
  population, 115
  rocks, 115, 229
  water, 115
Desdemona (moon of Uranus), A-17
Despina (moon of Neptune), A-17
Deuterium, 119, 284, 284n
Differential rotation, 290
Differentiation, **229**, 232, *232*, 293, 333
Diffraction gratings, 154, *159*, *181*
Diffraction limit, 179–180
Dinosaurs, 27, 345–346, 365, 366, 370
Dione (moon of Saturn), 313, *313*, A-17
Direction, 44, *44*, 75
Disk, galactic, 33
Distance measurements, A-3
Distances, astronomical, A-2
DNA, 367–369, *368*
Doppler shifts
  blueshifts, **165–167**, *166*, 219
  calculations, 167, A-3
  extrasolar planet detection, 218–219, *219*, 221
  overview, **165–167**
  redshifts, **165–167**, *166*, *168*, 219, 525–526, A-22
  rotations, 167, *168*
  sound waves, 165, *166*
Dust storms, 275–276
Dust tails, *335*, 335–336
DX Cancri, A-18

Earth
  atmosphere, t 261, *261*, *263*, 269, **269–270**
  cosmic address, 2–3, *4*

differences from other worlds, 352
early bombardment, 212–213, *213*
escape velocity, 146–147, A-2
as flat, 44
geological processes, 354, 358
greenhouse effect, 269–270
interior layering, **229–232**, *230*, 353–354, *354*
magnetic field, 235, *235*
mass, A-2
measured by Eratosthenes, 101, *101*
no-greenhouse temperature, t 264, 266
origin of water on, 362–363
photographs, 2, 15, 288, 352
plate tectonics (See Plate tectonics)
properties, t 200, A-15
radius, A-2
rotation and revolution, 27–30, *28*, *29*, *30*
seasons, 28–29, *29*
snowball, 365
surface geology, 226, *226*, *228*, 352–353, *353*
surface temperature, 232, t 264
as terrestrial planet, 199
view from the Moon, 53
walking tour, 15
Earth-orbiting spacecraft, 189
Earthquakes, 231, *231*, 360, *360*
Earthshine, 53
Easter, 96
Eccentricity, 136, *136*, *137*, A-15
Eclipses
definition, 53–54
lunar, 54–56, *55*
prediction, 56–58, *58*, t 59
solar, 54–56, *56*, *57*, *58*
Eclipse seasons, 57–58
Ecliptic, 41, *41*, 48, 73, *73*–74
Ecliptic plane, 28, *28*, 54, *54*
Einstein, Albert, 111
Einstein's equation, 115–116
Ejecta, 238, *239*, *240*, 345
Elara (moon of Jupiter), A-16
Electrical charge, 118
Electrical potential energy, 112, 120
Electric fields, 156
Electromagnetic radiation, 158
Electromagnetic spectrum, 154, 158, *158*, 185
Electromagnetic waves, 157, *157*
Electromagnets, 234–235, *235*
Electrons
in atoms, 118, *118*
energy states, 120–122, *121*
mass and charge, A-2
transitions, 160, 160n, *160*
wave-particle duality, 157
Electron-volts (eV), 121, A-2
Elements
in Big Bang theory, 8
condensation temperatures, 207, *208*
definition, 117
Democritus on, 117
periodic table, A-13
recycling of, 5–8, 203–204, *204*
spectral fingerprints, *160*, 160–162, *161*
Ellipses, 136, *136*

Elliptical orbits, 136, *136*, *137*, 140–141
Elongation, 68
Emerson, Ralph Waldo, 225
Emission, light, 154, *160*, **160–162**, *161*, *164*
Emission line spectra, 160, *160*, *161*, *164*
Enceladus (moon of Saturn), 313, *313*, A-16
Energy
chemical energy density, 115
chemical potential, 112
conservation of, 116–117
definition, 112
electrical potential, 112, 120
expenditures, t 112
gravitational potential, 112, **115**, 232, *232*
kinetic, 112, **113–115**, 280n
mass-energy, 115–116
orbital, 145–147
potential, 112, **115–116**
radiative, 112, 154, 157
from the Sun, 232
thermal, 114, *114*, *232*, 232–235, *233*, 292
units, 112, A-2
Epimetheus (moon of Saturn), A-16
Epsilon Eridani, 218
Equation of time, 71, *71*
Equatorial bulges, 289, *290*
Equator, star tracks, 79, 83, *83*
Equinoxes, 28, *29*, 72–74, 76–77, 81–84, *82*
Eratosthenes, *101*
Eridani, A-18
Eros, *17*, 327, *327*, 346
Erosion
Earth, 354, *355*
jovian moons, 304
Mars, 251–253, *252*, 282–283
processes, 235, **244–246**
Venus, 254
Eruption, planetary cooling and, *233*, 233
Escape velocity
calculations, 146, A-3
Earth, A-2
orbits, 128, *129*
spacecraft, 146–147
thermal escape, *280*, 280–281
Eta Aquarids meteor shower, t 344
Ethane, 294, 311–312
Ether, 149
Eugenia, 328
Europa (moon of Jupiter), 18, 305–307, *308*, *309*, 377, A-16
European Southern Observatory, A-24
Evaporation, 119, *276*, 276–278, *279*, 372
Evolution, *368*, **368–371**, 375. *See also* Life on Earth
Excited state, 121, *121*
Exospheres, 265, 267, 269
Expansion of the universe, 5–6, **32–34**, *33*, t 34
Exponential decay, 216, *216*
Exponential growth, 371
Exponents, A-4 to A-6
Exposure time, 173, 175
Extension, 244
Extinctions, 345–346, *347*, *348*, 374

Extrasolar planets, **218–221**, *219*, *220*, *221*, 377–378
Extraterrestrial civilizations, 375–378
Eyes, *172*, 172–173
EZ Aquarii A/B/C, A-18

Fahrenheit scale, 113, *113*
Fall equinox, 29, *29*, 74, 81–84
False-color images, 182
Far Ultraviolet Spectroscopic Explorer (FUSE), t 186, A-23
Faults, 359
Feedback relationships, 364–365
Film, 173–175, *174*
Fireballs, 330, *342*. *See also* Meteorites
First Law of Motion (Newton), 129, 137
First Law of Planetary Motion (Kepler), 136, *137*
Fission, nuclear, 115–116, 216, *216*
Flat Earth Society, 63
Flybys, 189–190
Focal plane, 172
Foci (singular, focus), 136, *136*, *172*, 172, 177
Fomalhaut, A-19
Forces
definition, 126–127
gravity (See Gravity)
net, 127
Newton's laws of motion, 129–131, 137
strong, 118
tectonic, 243–244, *244*
tidal forces, 143–145
torque, 132, *132*, 133n
units, 130
Formation properties, 236, *236*, 240
Formulas, A-3
Fossil fuel combustion, 364, 371
Fossils, 365, 365–366, *367*, 370
Foucault, Jean, 62
Foucault pendulums, 62
Frederick II (King of Denmark), 134
Free-fall, 126, 127–128, 147–148
Frequency, 156–157, A-3
Frost line, 208, *208–209*, 324
Full moon, 51–52
FUSE (Far Ultraviolet Spectroscopic Explorer), t 186, A-23
Fusion, nuclear, 8, 12, 290

Galatea (moon of Neptune), A-17
Galaxies, 4–8, 203–204, *204*, A-20, A-21. *See also* Milky Way Galaxy
Galaxy clusters, 4–5, A-22
Galilean moons, 18
Galileo
acceleration of gravity, 126
astrology, 107
jovian planets, 288, 304–305
opposition by the Church, 34, 62
Sun-centered solar system, 61, 137–139, *138*, *139*
telescope building, 175
Galileo spacecraft
asteroids, 327, *327*
Jupiter probe, 291, *294*, 294–295
mission data, t 190

Shoemaker-Levy 9 comet impact, *341*, *342*
Web site, A-23
Galle, Johann, 21
Galle, Philip, 86
Gamma rays, 158, *158*, 185, 187
Ganymede (moon of Jupiter), 305, *308*, 310, *310*, A-16
Gap moons, 317, 317n, 319
Gaps, 317–318, *318*, 325, 326
Gases
absorption, transmission, and scattering by, *160*, 160–162, *164*, 264–267, *265*
atmospheric pressure, t 261, 262, 274
definition, 117, 119
escape from atmosphere, *279*, 280–281
greenhouse, 265, 267, 365, 372
velocity, *280*
Gaspra, 327, *327*, 329, *329*
Gas pressures, 262
Geminids meteor shower, t 344
Gemini telescopes, t 178
Genetic codes, 367–368, *368*
Geological controlling factors, 236, **236–237**, 240, 243, 245, 246
Geology, definition, 226. *See also* Planetary geology, comparative
Georgium Sidus, 20
Geostationary satellites, 142
Geosynchronous satellites, 142, 146
Giant impacts, 214
Gibbous moon, 52
Gilbert, William, 140, 151
Giotto, 335
Giotto spacecraft, 335
GJ 1061, A-18
Gliese 15 A/B, A-18
Gliese 725 A/B, A-18
Global positioning system (GPS), 87
Global warming, 270, **371–372**. *See also* Climate change
Global wind patterns, 272–276, 296
Gnomons, 70
Grand Canyon, 355, 364, 366, *366*
Granite, 229, 358–359
Gravitational constant, 140, A-2
Gravitational contraction, 292
Gravitational encounters, 146, *146*
Gravitational equilibrium, 262
Gravitational fields, 156
Gravitational potential energy, 112, **115**, 232, *232*
Gravity
acceleration of, 126, *127*, **147–148**, A-2
direction of motion, 129n
discovery of Neptune, 21
extrasolar planet detection, 218–219, *219*
layered interiors of planets, 229–232
nebular capture, 210
planet density misconception, 208, *208*
planet surface, A-15

INDEX I-3

spherical shape of planets, 227, 229
torque, 133n
universal law of gravitation, 140, *140*, 142
weightlessness and, 128
Grazing incidence mirrors, 187, *187*
Great Dark Spot (Neptune), 299
Great Galaxy (M 31), *9*, 9–10, 42, *42*
Great Red Spot (Jupiter), 18, 297, *297*, 299
Greenhouse effect
climate change, 278
infrared radiation, **267–268**, *268*
on Mars, 283
misconceptions, 270
origin of term, 267n
runaway, 362, 365
surface temperature and, 263–264, 266
on terrestrial planets, 269
on Titan, 311
on Venus, 14, 283–284
Greenhouse gases, **265**, 267, 365, 372
Greenwich mean time (GMT), 72, *72*
Greenwich meridian, 45, *46*, 76
Gregorian calendar, 72–73
Grooved terrain, 310, *310*, 313
Ground ice, 277
Ground state, 120–122, *121*
Guinevere Plains (Venus), *244*

Habitable zones, 377
Hadar, A-19
Hadley cells, 272–274, *273*, *275*
Hale-Bopp comet, *23*, *42*, 333, *334*, 337
Hale telescope, 178
Half-lives, 216, *216*
Halley, Edmund, 333
Halley's comet, 333, 335
Halo, galactic, 31, 33
Harrison, John, 87
Hawaiian Islands, *241*, 241–242, 244, **360–361**, *361*
HD209548, 219, 220, *221*, 224
Heat, temperature and, 114–115
Helene (moon of Saturn), A-17
Helium
condensation of, 207, *207*
emission line spectrum, *161*
helium-4, 119
in jovian planets, 289, 293
loss from atmosphere, 280
Heracleides, 60
Herschel crater (Mimas), 314
Herschel, William, 20
Hertz (Hz), 156, *158*
Hesiod, 97
High Energy Transient Explorer (HETE-2), t 186
Himalaya Mountains, 359, *359*
Himalia (moon of Jupiter), A-16
Hipparchus, 102
Hobby-Eberly telescope, t 178
Horizon, 44, *44*
Horoscopes, 107
Hot spots, 360–362
Hour angle (HA), 76
Hourglasses, 94n
House of Wisdom (Baghdad), 103

Hubble Space Telescope
angular resolution, 178
appearance, *185*
diffraction limits, 180
field of view, *194*
launch year, t 186
meteorology of jovian planets, 298, *298*
misconceptions of advantages, 186
Shoemaker-Levy 9 impact, *342*, 342
Web site, A-23
Humans, 370–371. *See also* Life on Earth
Hurricanes, 297
Huygens, Christiaan, 1
Huygens probe, 312
Hyakutake comet, 333, *334*, 337
Hydrogen
condensation of, 207, *207*
deuterium, 119, 284, 284n
emission line spectrum, *160*
energy levels, 121, *121*, *160*
in jovian planets, 289, 291, 293
loss from atmosphere, 280–281
metallic, 291, 293
molecular, 162, *162*
tritium, 119
Hydrogen compounds, **199**, 207, 208, *208*, 213, 294
Hydrosphere, 352
Hydrostatic equilibrium, 262
Hydrothermal vents, 368, 375
Hypatia, 101
Hyperbolic orbits, 141, *141*
Hyperion (moon of Saturn), A-17
Hypothesis, in scientific method, 105

Iapetus (moon of Saturn), 313, *313*, A-17
Ice ages, 277, 365
Ice geology, 304, 313
Ices
jovian moons, 304, 306–307, 308, *309*
Mars, 251, 277
planetary rings, 316
solar system formation, 207–209
Iceteroids. *See* Kuiper belt
Ida, 327, *327*, 328, 329
Images, *172*
Imaging, 180, *181*
Impact basins, 238, *238*
Impact cratering
from comets, *341*, 341–342, *342*, *343*
Earth, 330, 354
icy moons, 304
Mars, 239
mass extinctions, 345–346, 347
Moon, 238, 239, 247, *247*
process of, 237, **237–240**, *238*, *239*, *240*, *243*
Venus, 253, *254*
Impact craters, 212–213, 330
Impactors, 212–213
Impacts, 212, 214, *214*. *See also* Bombardment; Planetesimals
Inclination, A-15
Indi, A-18
Inductive arguments, 104

Infrared radiation. *See also* Thermal radiation
effect on atmospheric gases, 265, *265*
greenhouse effect, 267–268, *268*
from nebular collapse, 206
penetration through the atmosphere, *185*
telescopes, 184, 186
from terrestrial planets, 233
wavelength of, 158, *158*
Intensity, 159, *159*
Interference, 179, *179*
Interferometry, 188–189
Interiors of planets
cooling of, 233, *233*
heating of, 232, *232*–233
jovian planets, 289, *289*, 291–293, *293*
layering, **229–232**, *230*
surface-area-to-volume ratio, 234
terrestrial planets, 227–232
International Astronomical Union (IAU), 40
Internet, Web sites on, A-23 to A-25
Interstellar clouds, *203*, **203–205**, *205*, 377
Interstellar dust, 198, 336
Inverse square law, 140, 266
Inversions, 268
Iodine-129, 218
Io (moon of Jupiter)
colors, 297, 306, *307*
Jupiter's magnetosphere and, *300*, *301*, 301
orbit, 151, 308
properties, A-16
tidal heating, 232, 305
torus, 301, 305
volcanism, 18, 305, *306*, *307*
Ionization, 120–121, *121*
Ionospheres, 265, 267–269
Ions, 120
Io torus, 301, 305
Iridium, 331, 345
Iron, 207, *207*, 332, *332*
Ishango bone, 93
Islamic science, 103
Isotopes, 119, *119*, 216, *216*

Jacob's staff, 86, *86*
Janus (moon of Saturn), A-16
Jefferson, Thomas, 330
Joules (units), 112
Jovian nebulae, *210*, 210, 291
Jovian planets. *See also names of individual planets*
bulk properties, t 289
classification, **198–201**, t 201
interior structure, 289, *289*, 291–293, *293*
magnetospheres, *300*, 300–302, *301*, *302*
moon data, A-16 to A-17
nebular capture, 210, *210*
origin of names, 287
rings, 316, **316–320**, *317*, *318*, *319*, 320
spherical nature of, 227, 290, *290*
temperature, 294–295, *295*
Julian calendar, 72–73
Juliet (moon of Uranus), A-17
Jupiter. *See also* Jovian planets
apparent retrograde motion, 59

atmosphere, **293–297**, *294*, *295*, *296*, *297*
captured moons, 214
colors, 297–298
effects on asteroid belt, 212, 325, 326
interior structure, 289, *289*, 291–293, *293*
as jovian planet, 199
life on, 377
magnetosphere, *300*, 300–302, *301*, *302*
moons, 139, 145, 302, 303, **304–311**, A-16
orbital period, *137*
overview, 18, *18*
photographs, *18*, 288, *296*
properties, t 200, t 289, A-15
rings, 320
Shoemaker-Levy 9 impact, *341*, 341–342
sidereal period, 68
temperature, 298
viewed from Earth, 59, *59*

Kauai, 361, *361*
Keck telescopes (Mauna Kea), 178, t 178, *179*, 189
Kelvin scale, 113, *113*
Kepler, Johannes, 61, 105, 107, **134–136**, *135*
Kepler's laws of planetary motion, **135–136**, *136*, *137*, 140–142, *141*
Kepler's third law, Newton's version of, *141*, **141–142**, 288, A-3
Kilauea (Hawaii), *241*
Kilograms (units), 130
Kiloparsecs (units), A-2
Kilowatt-hours, 112
Kinetic energy, 112, **113–115**, 280n
Kirchhoff's laws, 163–164
Kirkwood gaps, 325
Kleopatra, 329, *329*
Kobe earthquake (Japan), 360, *360*
Kuiper belt. *See also* Comets
location, *23*, 202, *202*
orbits, 324
origin, 212, 337, **337–338**
Kuiper belt objects, 339–340

Lacaille 9352, A-18
Lagrange, Joseph, 326
Lagrange points, 326n
Lalande 21185, A-18
Land of the midnight sun, 50
Large Binocular Telescope, t 178
Larissa (moon of Neptune), A-17
Latitude
altitude of celestial pole, 47, *48*, **80**, *80*, 84–85
celestial navigation, 84–85, *85*
declination and, 75
definition, 45, *46*, 75
skywatching and, 44–47
star tracks, 78, **78–81**, *79*, *80*, *81*
from Sun's altitude, 85, *85*
Lava, *241*, 241
Lava plains, *242*, 248
Law of conservation of angular momentum. *See* Conservation of angular momentum
Law of conservation of energy, 116–117

I-4 INDEX

Laws of thermal radiation, 162
Leap year, 72–73
Leda (moon of Jupiter), A-16
Lemaître, G., 197
Lenses, 172, *172*
Leonardo da Vinci, 151
Leonids meteor shower, 344, t 344, *344*
Leverrier, Urbain, 21
Levy, David, 341
Library of Alexandria, 101, *102*
Libration, 52n
Life on Earth
  atmosphere alterations, 370
  Cambrian explosion, 370–371
  carbon dioxide regulation, 364
  cosmic calendar, 26, *27*
  early evolution in the oceans, 369–370
  evidence for common ancestry, 367–368
  fossils, 365, 365–366, *367*
  hardiness, diversity, and probability, 375
  mass extinctions, 345–346, *347*, 348, 374
  origin, 366–369, 375
  oxygen-breathing animals, 370
  oxygen regulation, 364
  tree of life, 368, 368–369, 375
Life outside Earth, **375–378**
Light. *See also* Luminosity
  absorption, 154, **160–162**, *164*, 263–266, *264*
  diffraction limit, 179–180
  emission, 154, *160*, **160–162**, *161*, *164*
  forms of, *158*, 158–159
  intensity, 159
  interference, 179, *179*
  radiative energy, 112, 154
  reflection, 154, *155*, **164–165**, 263–266, *264*
  scattering, 154, 265, **267**, *267*, *318*, *319*
  size determination by, 328, *328*
  spectral formation, 163–164
  spectrum, 154, 159, *159*
  speed of, 8
  transmission, 154
  visible, *158*, 185, 267
  wave-particle duality, **155–156**
Light bulbs, *164*, 170
Light-collecting area, 177–178
Light gases, 207, *207*. *See also* Helium; Hydrogen
Light pollution, 182
Light-years, 8–9, A-2
Limestone, 363
Lippershey, Hans, 137
Liquids, 117, *119*
Lithospheres, **230**, *230*, 243–244, 353, 363
Local Group, 2–3, *5*, t 34, A-20
Local sidereal time (LST), 76
Local sky, 44
Local solar neighborhood, 31, *31*, t 34
Local solar time, 76
Local Supercluster, 2–3, *5*
Logic, scientific method and, 104
Loihi (Hawaii), 361, *361*
Longitude
  celestial navigation, 85
  clock development and, 87
  definition, 45, *46*, 75
  right ascension and, 75–76
  zero, 45, *46*, 76

Lowell, Percival, 249, *250*, 253
Luminosity, 221, A-2, A-3, A-18, A-20
Luminosity-distance formula, A-3
Lunar eclipses, 54–56, *55*, t 59
Lunar highlands, 238, 238, 247
Lunar maria, 238, *238*, 241, 247
Lunar phases, 50–53, *51*, *52*, *53*
Lunar Prospector spacecraft, 248, 281
Luyten 726-8A/B, A-18
Lyman series of transitions, 160n
Lyrids meteor shower, t 344
Lysithea (moon of Jupiter), A-16

M 31, 9, 9–10, 42, *42*
Magellan spacecraft, 14, t 190, 253, *254*
Magellan telescopes, t 178
Magma, 241, *241*
Magma chambers, 241
Magma oceans, 247
Magnetic braking, *211*, 211–212
Magnetic fields, *234*, **234–235**, *235*, 301
Magnetic-field strength, 230–231
Magnetic north, 84
Magnetic poles, 235
Magnetospheres
  Earth, 235, 270, *271*
  jovian planets, 290, 300–302, *301*, *302*
  Mars, 283
Magnification, 177
Manicouagan Lakes crater (Canada), *330*
Mantle plumes, 360–362
Mantles, 229–230, *230*, 243–244, 254, 353
MAP (Microwave Anisotropy Probe), t 186, A-23
Mare Imbrium (Moon), *247*
Mare Orientale (Moon), 238
Mariner 4 spacecraft, 250
Mariner 9 spacecraft, 250
Mariner 10 spacecraft, 13, 248
Mars
  atmosphere, t 261, *261*, *269*, 269–270, 279–283
  canals on, 249–250, *250*
  Coriolis effect, 274–276
  greenhouse gases, 365
  interior structures, 230, 235
  interior temperatures, 234
  lack of ozone layer, 373
  meteorites from, 283, 333, 375–377, *376*
  mission Web sites, A-23
  moons, 214, *214*, A-16
  overview, 16, *16*
  properties, t 200, A-15
  search for life on, 375
  seasons, 29, 271–272, *273*
  spectrum of, *159*, 165
  surface geology, 226, *227*, *228*, *239*, **249–253**, 353
  surface temperature, t 264, 275
  as terrestrial planet, 199
  viewed from Earth, 59
  volcanoes, *242*, 250–251, *251*
  water on, 251–253, 277, 362–363
Mars Exploration Rovers, t 190
Mars Express, t 190
Mars Global Surveyor spacecraft, t 190, 190–191, 250, 252, 253, 283
Mars Odyssey 2001, t 190

Mars Pathfinder spacecraft, *16*, 190, t 190, *191*, 250–253, *252*
Mass density. *See* Density
Mass-energy, 115–116
Masses
  asteroids, 328, *328*
  atomic, 118
  definition, 127
  Earth, A-2, A-15
  electrons, A-2
  galaxy clusters, A-22
  jovian planets, t 289, 290–292, *292*, A-15
  mass-energy, 115–116
  moons, A-16 to A-17
  from orbital velocity law, 141–142
  Pluto, 338n, A-15
  protons, A-2
  Sun, 141–142, A-2, A-15
  units, 130
  weight and, *128*
Mass extinctions, 345–346, *347*, 348, 374
Mathematical Insights
  acceleration of gravity, 147
  angular separation, 174
  density, 115
  diffraction limit, 180
  Doppler shifts, 167
  escape velocity, 146
  laws of thermal radiation, 162
  light-years, 8
  mass-energy, 116
  metric system, A-10
  "no greenhouse" temperatures, 266
  powers of ten, A-4 to A-6
  radioactive decay, 216
  ratios, A-11 to A-12
  scientific notation, A-6 to A-7
  surface-area-to-volume ratio, 234
  thermal escape from an atmosphere, 280–281
  time by the stars, 76
  units of force, mass, and weight, 130
  using Newton's version of Kepler's third law, 142
  wavelength, frequency, and energy, 157
Mathilde, 327, 327–329
Matter. *See also* Particles
  atomic structure, 118–119
  definition, 112
  energy in atoms, 120–122, *121*
  ionization, 120–121, *121*
  phases of, 117, 119–120, *120*
  properties of, 117
Maui, 361
Mauna Kea observatories, 178, t 178, *179*, 189, A-24
Mayan calendar, 96
Mean solar time, 70–71, *71*
Media misrepresentations, 159
Medicine wheels, 98, *98*
Megaparsecs (units), A-2
Melting, 229
Mercury
  atmosphere, 260, t 261, **269–270**, 279–280, 282, *282*
  contraction, 249
  giant impacts, 215
  interior structures, 230, 235
  orbit, 139, 145
  overview, 13, *13*

  properties, t 200, A-15
  surface geology, 226, *226*, *228*, *248*, **248–249**, *249*
  surface temperature, 13, t 264, 266
  as terrestrial planet, 199
  tidal bulges, 145
  viewed from Earth, 69, *69*
  water on, 281
Meridian, 44, *44*
Mesosphere, 267n
Messier Catalog, A-21
Metals, 207, *207*
Meteor Crater (Arizona), 345, *345*
Meteorites. *See also* Asteroids
  from asteroids, 333
  asteroid study from, 330–331
  composition, 331–332, *332*
  condensation in, *209*, 209–210
  definition, 17, 324
  Martian, 283, 375–377, *376*
  mass extinctions, 345–346, *347*
  origin, 332–333
  primitive, 331–333
  processed, 332n, 332–333
  radioactive dating, 217
  risks from, *331*
Meteorologists, 270
Meteors, 270, 324. *See also* Meteorites
Meteor showers, 342–344, *344*
Methane
  condensation, 207, *207*
  on jovian moons, 311–312
  on jovian planets, 294, 298, *298*, *299*
  on Pluto, 340
Metis (moon of Jupiter), A-16
Metonic cycle, 96
Metric system, A-10
Micrometeorites, 239, 248, 279–280
Microwave Anisotropy Probe (MAP), t 186, A-23
Microwave ovens, 169
Microwaves, 169
Mid-ocean ridges, 354, *355*, 356
Midway Island, 361, *361*
Milky Way Galaxy
  age of, 27
  Andromeda Galaxy and, 32, *37*
  location, 2–3, *4*
  luminosity, A-20
  motion within, 30–32, *31*, *32*
  overview, 25
  rotation, 32, t 34
  viewing, 42, *42*, *43*
Mimas (moon of Saturn), 313, 313–314, 318, *318*, A-16
Miranda (moon of Uranus), 314, *314*, A-17
Misconceptions
  cause of seasons, 29
  compass directions, 84
  confusing solar system and galaxy, 25
  dark side of the moon, 53
  direction of toilet flush, 274
  eggs on the equinox, 105
  flat Earth, 44
  greenhouse effect is bad, 270
  hearing radio waves and seeing X-rays, 159
  higher altitudes are always colder, 269

high noon, 50
illusion of solidity, 118
is radiation dangerous?, 158
light-years, 9
magnification and telescopes, 177
moon in the daytime, 51
moon on the horizon, 52
no gravity in space, 128
North Star (Polaris), 47
one phase at a time, 119
origin of tides, 143
ozone - good or bad?, 364
pressure and temperature, 233
solar gravity and planet density, 208
space telescopes closer to the stars, 186
Stars in the daytime, 41
sun signs, 30
toilet flushing in the Southern Hemisphere, 274
twinkling stars, 183
what makes a rocket launch, 131
MMT telescopes, t 178
Models of the universe, 101–103, *102*, *103*
Molecular bands, 161–162, *162*
Molecular dissociation, 120, *120*, 265, *265*
Molecules, 117, 161–162, *162*
Momentum. *See also* Conservation of angular momentum
 angular, *132*, **132**, *133*, 140–141
 conservation of, *131*, 131–132
 definition, 126–127
Mont-Saint-Michel, France, 143, *143*
Moon (moon of Earth)
 angular size, 44, *45*
 atmosphere, t 261, *261*, 269–270, 279–282, *282*
 dark side misconception, 53
 eclipses, 54–56, *55*, *56*, t 59
 escape velocity, 146
 footprints on, 239, *239*
 formation, *214*, 214–215, 247
 full, 51–52
 future observatory on, 192
 human behavior and, 57
 interior structures, *230*, 235
 libration, 52n
 lunar calendar, 96
 lunar phases, 50–53, *51*, *52*, 53
 meteorites from, 333
 Metonic cycle, 96
 orbital period, 67, *67*, 144
 properties, A-16
 radioactive dating, 217, 238
 surface geology, 226, *226*, *228*, *238*, 238, 247
 surface temperature, t 264, 281
 tidal forces from Earth, 145
 tidal forces on Earth, 142–145
 view of Earth from, 53
 volcanoes, 247, 247–248
 water on, 281
 weather prediction, 92, *92–93*
Moons
 of asteroids, 328
 captured, 213–214, *214*, 302
 definition, 4

Galilean, 18
gap, 317, 317n, 319
icy, 304
of jovian planets, 302–315, *303*, 377
shepherd, 317, *317*
Morabito, Linda, 305
Motion
 apparent retrograde, 59, **59–61**, *60*, *61*, 102, *103*
 basic concepts, 126–128
 Copernicus, 133–134
 Kepler's laws of planetary, **135–136**, *136*, *137*, 140–142, *141*
 Newton's laws of, 129–133, 137
Mountains
 cloud formation, 277, *277*
 erosion, 245
 high-altitude temperatures, 269
 tectonics and, 244, *244*, 358, 359
Mount Palomar observatory, 176, 178, A-24
Mount Pinatubo (Philippines), 372
Mount St. Helens (Washington), 242, 359
Mount Wilson telescope, 182
Movie Madness, 159
Multi ring basins, 238, *238*
Mutations, 369–370

Naiad (moon of Neptune), A-17
NASA, 221, A-25
Nasmyth focus, 177, *177*
National Optical Astronomy Observatory, A-24
National Radio Astronomy Observatory, A-24
Natural selection, 370
Navigation, celestial
 celestial coordinates, **74–77**, *75*, *77*, t 78
 celestial sphere, 75, *75*
 global positioning system, 87
 latitude and, 84–85, *85*
 path of the Sun, *81*, **81–84**, *82*, *83*, *84*
 Polaris, 47, *48*
 Polynesian, 98–99, *99*
 practice of, 85–87, *86*, 99
 star tracks, **78–81**, *79*, *80*, *81*
Neap tides, 144, *144*
Near-Earth Asteroid Rendezvous (NEAR), *17*, t 190, 327, 329
Nebulae
 definition, 203
 jovian, 210, *210*, 291
 solar, **204–208**, *205*, *206*, *207*, 278
Nebular capture, 210, *210*
Nebular theory, **203–206**, *205*, *206*, 215
Negative feedback, 364
Neon, emission line spectra, *161*
Neptune. *See also* Jovian planets
 colors, 298, *299*
 interior, 292–293, *293*
 as jovian planet, 199
 magnetosphere, 301–302, *302*
 moons, 21, 214, *303*, 314–315, *315*, A-17
 orbital period, 137
 overview, 21, *21*
 photographs, 288

Pluto's orbit and, 338, *338*
properties, t 200, t 289, A-15
rings, 320
temperature, 298
Nereid (moon of Neptune), 314, A-17
Net force, 127
Neutrons, 118, *118*
Newtonian focus, 177, *177*
Newton, Sir Isaac
 jovian planets, 288n
 Kepler's laws and, 140–142
 Kepler's third law, *141*, **141–142**, 288, A-3
 photograph, *139*
 universal law of gravitation, **139–142**, *140*, 333
Newton's laws of motion, 129–131, 137
Newtons (units), 130
Next Generation Space Telescope (NGST), t 186, 192
Nicholas of Cusa, 61, 151
Nickel, 207, *207*
Nitrogen, molecular, 262
Nodes, of the Moon's orbit, 54
Nonscience, 105, 107
North America, formation of, 358
North celestial pole (NCP), 41, *41*, 45, *47*, *48*, 80
Northern Hemisphere, skywatching in, 45–46, *46*
Northern lights (aurora borealis), **270**, *271*, 300, *300*, 302
North Pole, star tracks and, 78, 78–79, *79*, 82, 82–83
North Star. *See* Polaris
Novae, 134
Nova Reperta, 86
Nozomi spacecraft, t 190
Nuclear fission, 115–116, 216, *216*
Nuclear fusion, 8, 12, 290
Nuclear power generation, 115–116, 216, *216*
Nucleus, atomic, 118, *118*
Nucleus, comet, 335, *335*, 340

Oberon (moon of Uranus), 314, A-17
Observable universe, 9
Observation, in scientific method, 105
Observatories, t 186, 192, A-24. *See also names of individual observatories*
Occultation, stellar, 318
Oceans, 36, 363, 367–368
Olympus Mons (Mars), 242, 250, *251*, 277
Omniscience, origin of term, 104
Oort cloud. *See also* Comets; Planetesimals
 location, 23, 202, *202*
 orbits, 324
 origin, 212, **337**, *337*
Opacity, 155
Opaque materials, 155, *164*
Ophelia (moon of Uranus), 317, A-17
Ophiuchus, 48, 74
Opposition, 68
Orbital energy, 145–147
Orbital periods, 67, *67*, **136**, *137*, 144
Orbital resonances
 asteroid belt, 325
 moons of Saturn, 318, *318*
 moons of Jupiter, 305, 308

Pluto and Neptune, 338, *338*
Orbital velocity, 129, 136, A-3, A-22
Orbital velocity law, A-3
Orbits
 asteroids, *201*, 212, 326, *327*
 bound *vs.* unbound, 140, *140*, 146
 captured moons, 213–214, 302
 changes in, *146*, 146–147
 comets, 212, 336, 337
 Earth, **27–30**, *28*, *29*
 elliptical, *136*, 136, *137*, 140–141
 geosynchronous and geostationary, 142
 hyperbolic, 141, *141*
 jovian planet satellites, 210, 302–305
 Kepler's laws of motion, 135–136, *136*, *137*
 Moon, 67, *67*
 parabolic, 141, *141*
 planetary, 199, 202, 205, A-15
 planetary moons, A-16 to A-17
 Pluto, 338, *338*
 rings, 317, 319
 seasons and, 29, 271–272, *273*
 solar nebula, 205
 Sun, A-15
Orion, 40, *40*, 48, 93
Orionids meteor shower, t 344
Orion Nebula
 distance from Earth, 8
 emission line spectrum, *161*
 photographs, *203*
 protoplanetary disks, *206*
Outgassing, 278, *278*
Oxidation, 363, 370
Oxygen, 120, 363–364, 370
Ozone, 265, 268, 364, **372–374**, *373*, 378
Ozone depletion, 372–374
Ozone hole, 372–374, *373*

Pallas, 329
Pandora (moon of Saturn), A-16
Pangaea, 357
Pan (moon of Saturn), A-16
Parabolic orbits, 141, *141*
Paradigms, 106
Parallax formula, A-3
Parallax, stellar, 61–62, *62*, 137–138, A-3
Parsecs (units), A-2
Partial lunar eclipse, 55, *55*
Partial solar eclipse, 56, *56*
Particle radiation, 158
Particles, 117–118, 155–156, 204
Parts per million, 363
Pasiphae (moon of Jupiter), A-16
Pawnee lodges, 98
Peak thermal velocity, *280*, 280–281
Peale, Stan, 305
51 Pegasi, 219, *220*, 223
Penumbra, 55, *55–56*
Penumbral lunar eclipse, 55, *55*
Perihelion, 136, *137*
Periodic table of the elements, A-13
Perseids meteor shower, 344, t 344
Phaethon, 344
Phases, of matter, 117, 119–120, *120*
Phobos (moon of Mars), 214, *214*, 327, A-16

Phoebe (moon of Saturn), A-17
Phosphine, 297
Phosphorus compounds, 297
Photography, 173–175, *174*
Photometry, 181
Photons, **156**, 162, 280–281
Photosynthesis, 112, 364, 370
Pioneer 10/11 spacecraft, 288
Pioneer Venus Orbiter spacecraft, *14*, 276
Pixels, 175
Planck's constant, 157, A-2
Planetary geology, comparative, 225–258. *See also* Plate tectonics
   continental vs. seafloor crusts, 353, *353*
   geological relationships, *236*, **236–237**, *240, 243, 245, 246*
   impact cratering, *237*, **237–240**, *238, 239, 240*
   interiors, *232*, **232–235**, *233*
   magnetic fields, *234*, **234–235**, *235*
   North America, 358
   rock strength, 227–229
   terrestrial-world layering, 229–232
   uniformitarianism vs. catastrophism, 374
   volcanism, 235–236, *241*, **241–242**, *242, 243*
Planetesimals. *See also* Asteroids; Comets
   captured moons, 213–214, *214*
   early bombardment by, **212–213**, *213, 278*
   giant impacts, *214*, 214–215
   impact cratering, *237*, **237–240**, *238, 239, 240*
   origin of asteroids and comets, 212
   planetary atmospheres from, *278*, 278–281
   in planet formation, 208–210, *209, 210*
Planetology, 198
Planet-quakes, 244. *See also* Earthquakes
Planets
   apparent retrograde motion, *59*, **59–61**, *60, 61, 102, 103*
   captured moons, 213–214, *214*
   categories, 198–201
   comparative planetology, 198
   condensation, 206–208, *207*
   definition, 4
   density, 208, *208*
   early bombardment of, 212–213, *213, 255*
   extrasolar, 24, **218–221**, *219, 220, 221, 377–378*
   formation, 206–212
   geological tours, 246–255
   interior layering, 229–232
   interiors (*See* Interiors of planets)
   Kepler's laws of planetary motion, **135–136**, *136, 137, 140–142*
   life on, 375–377 (*See also* Life on Earth)
   nebular capture, 210, *210*
   origin of term, 59
   physical data, A-15

planetesimals, 208–210, *209, 210*
   shape of, 227, 229
   solar wind, *211*, **211–212**, *283, 301*
   surfaces (*See* Surfaces of planets)
   synodic vs. sidereal periods, 68, 68–69
Planet X, 23
Planispheres, 85
Plants, oxygen regulation, 364
Plasma, *120*, **120**
Plasma tails, 335, *335–336*
Plates, tectonic, 354
Plate tectonics. *See also* Tectonics
   carbonate-silicate cycle, 365
   continental drift, 230, **354**, *355, 356, 357, 357*
   definition, 243, 356
   discovery of, 354–356
   hot spots and mantle plumes, 360–362
   origin of continents, 358–360, *359*
   processes, 356–357, *357*
   rate of movement, 357
   stresses on continents, 359–360
   uniqueness of Earth, 352–353
   Venus, 254
Plato, 65, 102, 138
Pleiades Cluster, 97
Plutinos, 340. *See also* Kuiper belt
Pluto
   atmosphere, 339
   future missions to, 340
   as Kuiper belt object, 339–340
   moon of, 338–339, *339*, A-17
   orbit, 338, *338*
   overview, 22, *22*, 338–340
   properties, t 200, 201, 338, A-15
   synchronous rotation, 145, *145*
Polaris
   latitude from, *47, 48*, **80**, *80, 84–85*
   misconceptions about, 47
   precession of Earth and, *30, 30*
Pollution, light, 182
Pollux, A-19
Polycyclic aromatic hydrocarbons, 376
Polynesian navigators, 98–99, *99*
Pope, Alexander, 140
Population density, 115
Population growth, 371, *371*, 379
Portia (moon of Uranus), A-17
Positive feedback, 364–365
Post meridian (p.m.), 69–70
Potassium, radioactive decay, *216*, 216–217
Potential energy, 112, **115–116**
Power, **154**, 162, 266, A-2
Powers of ten, A-4 to A-6
Precession, 30, *30*
Precipitation (weather), *276*, 276–277
Pressure
   atmospheric, t 261, 297
   effect on rocks, 229, 233
   jovian planets, 291
   radiation, 335
   temperature and, 233
Primary mirrors, 175, *175*
Prime focus, 175
Prime meridian, 45, *46*, 76

Principia, 140
Prisms, 154, *154, 159*
Probes, 189
Procyon, *40*, A-19
Procyon A/B, A-18
Prometheus (moon of Saturn), A-16
Propane, 294
Prospero (moon of Uranus), A-17
Proteus (moon of Neptune), 314, A-17
Protons, 118, *118*, A-2
Protoplanetary disks, **204–205**, *206, 218*
Protosuns, 204
Proxima Centauri, 24, A-18
Pseudoscience, 105
Ptolemy, 30, 86, **101–103**, *107, 107*
Puck (moon of Uranus), A-17
Pupils, *172*, 172–173
P waves, 231, *231*
Pythagoras, 44

1992QB1, 337
Quadrantids meteor shower, t 344
Quantum mechanics, 122
Quintessence, 149

Radar mapping, 253
Radial velocity, A-21
Radiants, 350
Radiation
   dangers of, 158
   electromagnetic, 158
   infrared (*See* Infrared radiation)
   particle, 158
   pressure, 335
   thermal (*See* Thermal radiation)
   ultraviolet (*See* Ultraviolet (UV) radiation)
Radiation pressure, 335
Radiative energy, 112, 157
Radioactive dating, **215–217**, *238, 366*
Radioactive decay, 215–216
Radioactivity, interior heating of planets, 232–233, *233*
Radio broadcasts, 157, 159, 269
Radio signals, asteroid shapes and, 329
Radio telescopes, **187–189**, *188*
Radio waves, 158, *158, 185*
Rainbows, 163–164
Ratios, A-11 to A-12
Redshifts, **165–167**, *166, 168, 219, 525–526*, A-22
Reflecting telescopes, 175, *176, 181*
Reflection, 154, 155, **164–165**, *263–266, 264*
Refracting telescopes, 175, *176*
Relative dating, 366
Rest wavelengths, *166, 167*
Retinas, *172, 172*
Retrograde motion, *59*, **59–61**, *60, 61, 102, 103*
Revolution, Earth, **27–30**, *28, 29, 30*, t 34
Reynolds, Ray, 305
Rhea (moon of Saturn), 313, *313*, A-17
Rheticus, 151
Rhyolite, 358–359
Ribosomal RNA, 368
Rift valleys, 359

Rigel, A-19
Right ascension (RA), *75*, **75–77**, *77, 79*, A-18, A-20
Rings, planetary, *316*, **316–320**, *317, 318, 319, 320*
Rio Grande Valley (New Mexico), 244
RNA, 367–368, *368*
Roche tidal zone, 317, 319, 341
Rocks
   age of oldest, 215, 217
   carbonate, 281, 363–365
   condensation, 207, *207*
   density, 115
   initial oxidation of, 370
   insulation by, 233–234
   Moon, 217, 226, 238, 247
   radioactive dating, 215–217, *216*
   sedimentary, 245, 253
   silicate, 363
   solidity of, 227–229
   strength, 227–230
Rods, in eyes, 155, 172
Rosalind (moon of Uranus), A-17
Ross 128, A-18
Ross 154, A-18
Ross 248, A-18
Rossi X-Ray Timing Experiment (RXTE), t 186
Rotation
   backward, 339
   differential, 290
   Doppler shifts from, 167, *168*
   Earth, *27, 28*, t 34, 66–67, 144
   galaxies, 32
   global wind patterns, 273–276
   jovian planets, 290
   molecules, *162*
   Moon, 53, 144–145
   planets, *199*, A-15
   relationships with geological controlling factors, *236, 240*
   solar nebula, 204
   Sun, 211
   surface temperature and, 264
   synchronous, **144–145**, 304, 339
   water, 274
Runaway greenhouse effect, 362, 365

Sacred rounds, 96
Sagan, Carl, 9
Sagittarius Dwarf Elliptical galaxy, A-20
San Andreas fault (California), 360, *360*
Saros cycles, 58, *58*, 96
Satellites, 4, 141, *146*. *See also* Moons
Saturn. *See also* Jovian planets
   colors, 297–298
   internal structure, 291–293, *293*
   as jovian planet, 199
   magnetosphere, 301, *302*
   moons, 303, **311–314**, *318, 377*, A-16 to A-17
   orbital period, 137
   overview, 19, *19*
   photographs, 288
   properties, t 200, t 289, A-15
   rings, 19, *316*, **316–318**, *317, 318, 320*
   shape, 290, *290*
   temperature, 298

viewed from Earth, 59
weather, 299
Scattering of light, 154, 265, **267**, 267, 318, 319
Schiaparelli, Giovanni, 249–250
Science, origin of term, 104
Scientific method, 103–106
Scientific models, 101–103
Scientific notation, A-6 to A-7
Scientific prediction, 272
Scientific theories, 105
Seafloor crust, 353–354, 356
Search for extraterrestrial intelligence (SETI), A-25
Seasons
    changes in skywatching, 48–50, 49, 50
    Earth, 28–29, 29
    jovian planets, 20, 299
    Mars, 29, **271–272**, 273
Secondary mirrors, 175, 177
Second Law of Motion (Newton), 129–130
Second Law of Planetary Motion (Kepler), 136, 137
Sedimentary rocks, 245, 253
Seismic waves, 226, 231, **231**
Seismographs, 231
Self-replicating molecules, 369
Semimajor axis, 136, 136, 137
Setebos (moon of Uranus), A-17
Sextants, 86, 86
Shen (Orion), 40, 40, 48, 93
Shepherd moons, 317, 317
Shield volcanoes, **241**, 242, 248, 254, 307, 361
Shoemaker, Gene and Carolyn, 341
Shoemaker-Levy 9, 291, 341, 341–342
Shooting stars (meteors), 270, 324. See also Meteorites
Sidereal days, 66, 66–67, 67, A-2
Sidereal months, 67, 67
Sidereal periods, 68, 68–69
Sidereal time, 76–77
Sidereal years, 67–68, A-2
Sierra Nevada, 359
Silicates, 229
SIM (Space Interferometry Mission), t 186
Sinope (moon of Jupiter), A-16
Sirius, 8, 40, A-18, A-19
SIRTF (Space Infrared Telescope Facility), 186, t 186, A-23
SI units, A-10
Skanda (Orion), 40, 40, 48, 93
Skywatching
    constellation locations, A-26 to A-27
    dome of the sky, 44, 44
    equinoxes and solstices, 74
    Milky Way Galaxy, 42, 42, 43
    Moon, 50–58
    seasonal changes, 48–50, 49, 50
    star charts, A-30 to A-33
    variations with latitude, 44–47, 46, 47
    Web sites, A-24 to A-25
Snowball Earth, 365
Socrates, 65
Sodium, 161, 281
SOFIA airborne observatory, 184, 184, A-23
SOHO spacecraft, 333, 334
Sojourner rover, 16, 253

Solar days, 66, **66–67**, 67, A-2
Solar eclipses, 54–56, 56, 57, 58
Solar luminosity, A-2
Solar nebulae
    collapse of, 204–206, 205, 206
    condensation, 206–208, 207
    planetary atmospheres, 278
Solar system, 197–224
    age of, 215–218
    captured moons, 213–214, 214
    classification of planets, 198–201
    definition, 4
    extrasolar planets, lessons from, 219–221
    four major characteristics, 198–203, t 203
    inner *vs.* outer, 17
    layout of, 198, 199
    life in, 375–377 (See also Life on Earth)
    moon data, A-16 to A-17
    nebular theory, 203–206, 204, 205
    scale model, 10, t 11, A-25
    sizes and distances, 10, t 11, 11
Solar time, 76
Solar wind
    atmospheric loss, 280–281
    jovian planets, 301
    magnetospheres and, 270, 271
    Mars, 283
    overview, 211, **211–212**
Solids, 117, 119
Solstices, 29, 29, 74, 81–84, 82, 94
Sound waves, 165, 166
South celestial pole (SCP), 41, 41, 46, 48
Southern Hemisphere, skywatching in, 45–46, 46
Southern lights (aurora australis), **270**, 271, 300, 300, 302
South Pole, star tracks, 82–83
Space colonies, at Lagrange points, 326n
Spacecraft, 146–147, 189–191, t 190, A-23. See also names of *individual spacecraft*
Space Infrared Telescope Facility (SIRTF), 186, t 186, A-23
Space Interferometry Mission (SIM), t 186
Space Shuttle, 263
Space Station, 146–147, 269
Spectra, 154, 159, 159, 164
Spectral formation, 163–164
Spectral resolution, 182, 182
Spectral types, A-18
Spectroscopy, 180, 181, 182
Speed. See also Velocity
    definition, 126
    of light, 8
    Local Group, t 34
    of universal expansion, t 34
    wave, 156
Spica, A-19
Spinning (angular momentum), 132, **132**, 133, 140–141
Spokes, 317, 317
Spreading centers, 356
Spring equinox, 28, 29, 72–74, 76–77, 81–84
Spring tides, 144, 144
Stadia (units), 101
Standard time, 70
Star charts, A-30 to A-33

Star clocks, 94
Star clusters, 4–5
Stardust mission, 336
Star-forming clouds, 204
Star-gas-star cycle, t 52
Stargazing. See Skywatching
Stars
    aberration of starlight, 62
    artificial, 184
    brightest twenty, A-19
    celestial coordinates, 77
    circumpolar, 45–46, 47, 48
    constellations, 40, 40–41
    in the daytime, 41
    definition, 4
    life around, 377–378
    within 12 light-years, A-18
    number in observable universe, 25
    properties, A-18 to A-19
    rotation, 211–212
    twinkling, 183
    velocity, 218
Star systems, definition, 4
Star tracks, 78, 78–81, 79, 80, 81
Stefan-Boltzmann law, 162, A-2, A-3
Stellar occultation, 318
Stellar parallax, 61–62, 62, 137–138, A-3
Stephano (moon of Uranus), A-17
Stonehenge (England), 94, 94
Storms, 276, 297, 372. See also Weather
Stratospheres, 265, 267–269, 295
Stratospheric Observatory for Infrared Astronomy (SOFIA), 184, 184, A-23
Stratovolcanoes, 242, 242, 359
Stromatolites, 366, 367
Strong force, 118
Subaru telescope, t 178
Subduction, 356, 358–359
Subduction zones, 357, 357
Sublimation
    comets, 335
    definition, 119
    jovian moons, 306, 315
    planetary atmospheres, 275, 276, 278, 279
Sulfur compounds, 297, 372
Sulfuric acid rain, 277
Summer solstice, 28, 29, 74, 81–84, 94
Sun
    basic properties, A-18
    celestial coordinates, 77, t 78, 81, 81–82
    color, 163
    determining time of year, 94, 94–95, 95
    eclipses, 54–56, 56, 57, 58
    formation, 204
    increasing brightness, 277
    light absorption and reflection, 263–264, 264, 265, 266
    luminosity, A-2
    mass, 141–142, A-2
    overview, 12, 12
    path through the sky, 81, **81–84**, 82, 83, 84
    physical data, A-15
    rotation, 211
    seen from Earth, 48–50, 49, 50
    sunspots, 211
    tidal force on Earth, 144
    timekeeping and, 69–70
Sun Dagger, 95, 95

Sundials, 70, 70
Sunset colors, 267
Sunspots, 211
Superclusters, 4–5
Supermassive black holes, 181
Supernovae, 134
Surface area, 266
Surface-area-to-volume ratio, 234
Surfaces of planets
    erosion, 235, **244–246**, 251–254, 282–283, 354
    geological relationships, 236, **236–237**, 240, 243, 245, 246
    impact cratering, 235, 237, **237–240**, 238, 239, 240
    Mars, 226, 226, 228, 239, 239, **249–253**
    Mercury, 226, 226, 228, 248, **248–249**, 249
    tectonics, 235
    temperature, 232, 236, 246, **263–264**, t 264
    volcanism, 235, 241, 241–242, 242
S waves, 231, 231
Swift-Tuttle comet, 344
Sycorax (moon of Uranus), A-17
Synchronous rotation, **144–145**, 304–305
Synodic months, 67, 67, A-2
Synodic periods, 68, 68–69

Tails, comet, 335, 342–344
Taurids meteor shower, t 344
Tectonics. See also Plate tectonics
    Earth, 354
    forces driving, 243
    jovian moons, 304–305, 310
    Mercury, 249, 249
    origin of term, 243
    planetary size and, 255
    planet surfaces, 235
    Venus, 253–254, 254
    volcanism, 243
Telescopes. See also names of *individual telescopes*
    adaptive optics, 328
    angular resolution, 173, **178–180**, 179n, 188
    atmospheric effects, 182–184, 184, 185
    basic design, 175–177, 176, 177
    Galileo and, 137
    gamma ray, 187
    infrared, 184, 186
    invention of, 137
    largest optical, t 178
    light-collecting area, 177–178
    magnification and, 177
    observatories in space, t 186
    orbital, t 186, 192, 221
    radio, **187–189**, 188
    reflecting, 175, 176, 181
    refracting, 175, 176
    spectral resolution, 182, 182
    ultraviolet, 186
    uses, 180–182
    wavelength and, 185, 185–189
    X-ray, 181, t 186, **186–187**, 187
Telesto (moon of Saturn), A-16
Television, 159
Temperature
    axis tilts, 277
    condensation, 207, 207–208, 208
    current trends on Earth, 372

definition, 114
effect on rocks, 229
galaxy clusters, A-22
greenhouse effect, 365
heat and, *114*, 114–115
high altitudes, 269
jovian planets, 291, 294–295, *295*, *296*, 298
lithosphere, 230
Mercury, 13
with no greenhouse effect, 263–264, t 264, 266
phase changes, 120, *120*
planets, t 200, 232, *236*, *243*
pressure and, 233
scales, 113, *113*
solar nebula collapse, 204
spectral lines, 161
surface, 232, *236*, *246*, 263–264, t 264
tectonics and, 244, *245*
thermal escape, 280–281
thermal radiation, 162–163, *165*
Triton, 314
Templo Mayor (Mexico), 94–95, *95*
Terrestrial Planet Finder (TPF), t 186
Terrestrial planets. *See also* Earth; Mars; Mercury; Venus
classification, 198–199, t 201
geological tour, 246–255
interiors, 227–232
layering, 229–232
magnetic fields, *234*, **234–235**, *235*
surface shaping, 235–246, 255
Tethys (moon of Saturn), 313, *313*, A-16
Tetrabiblios, 30, 107
Thalassa (moon of Neptune), A-17
Thales, 56
Tharsis Bulge (Mars), 250, *251*
Thebe (moon of Jupiter), A-16
Thermal energy, 114, *114*, 232, **232–235**, 233, 292
Thermal escape, 279, 280–281
Thermal radiation. *See also* Infrared radiation
from astronauts, 114–115
laws of, 162
from nebular collapse, 206
with no greenhouse effect, 266
overview, 162–163, *163*
planetary cooling, 233
Thermospheres, 265, 267–269, 295
Third Law of Motion (Newton), 130–131
Third Law of Planetary Motion (Kepler), 136, *137*, *141*, 141–142
Tidal bulges, *142*, **142–143**, *145*, 145
Tidal flexing, 306, *308*
Tidal forces, **143–145**
Tidal friction, 144–145
Tidal heating, 232, **305–307**, *308*
Time. *See also* Timekeeping
apparent solar time, 69–71, *71*
calendars, 72–73, *94*, 94–96, *95*
cosmic calendar, 25–27, *26*
daylight, 70–72
daylight saving, 70–72
equation of, 71, *71*

light from distant galaxies and, 8–10
mean solar, 70–71, *71*
solar *vs.* sidereal days, *66*, 66–67, *67*
standard, 70
synodic *vs.* sidereal months, 67, *67*
synodic *vs.* sidereal periods, 68, 68–69
tropical *vs.* sidereal years, 67–68
unit conversions, A-2
universal, 72, *72*
Timekeeping
ancient observations, 93–94
apparent solar time, 69–71, *71*
calendars, 72–73, *94*, 94–96, *95*
clocks, 70, *70*, 87, 93–94
hours, 93–94
mean solar time, 70–71, *71*
sidereal time, 76–77
standard, daylight, and universal time, 70–72, *72*
time of year, *94*, 94–95, *95*
Time zones, 70, *72*
Timing, by telescopes, 180–181
Titania (moon of Uranus), 314, A-17
Titan (moon of Saturn), 19, **311–313**, *312*, 377, A-17
Tombaugh, Clyde, 22, 338
Torque, 132, *132*, 133n
Totality, 56, *58*
Total lunar eclipse, 55, *55*
Total solar eclipse, 56, *56*
Toutatis, 350
TRACE satellite, *69*
Transit, 68
Transmission, 154
Transparent materials, 155
Trenches, oceanic, 354, *356*, 357
Trilobites, 370
Tritium, 119
Triton (moon of Neptune), 21, *21*, 214, **314–315**, 340, A-17
Trojan asteroids, 201, *201*, 325, 326
Tropical years, 67–68, A-2
Tropic of Cancer, 82, *83*, 83–84
Tropic of Capricorn, 82, 83–84
Tropics, 83
Tropospheres, 265, 267–268, 295
Truth, Sojourner, 253
Tunguska event (Siberia), 346, *348*
Turbulence, 183–184
Twain, Mark, 39
Tycho Brahe, 134, **134**, 137–138, 333

Ultraviolet (UV) radiation
loss of water from planets, 283
molecular dissociation, 265, *265*
penetration through the atmosphere, *185*
stratosphere, 268
telescopes, 186
wavelengths of, 158, *158*
Umbra, 55, 55–56, *56*
Umbriel (moon of Uranus), 314, A-17
Uniformitarianism, 374
Units
astronomical, 27, A-2
conversions, A-2, A-9 to A-10

distance, A-2
energy, 112, A-2
force, mass, and weight, 130
guidelines, A-7 to A-10
power, 154, A-2
prefixes, A-10
SI (metric), A-10
Universal law of gravitation, **139–140**, *140*, 142, A-3
Universal time, 72, *72*
Universe
cosmic addresses, 2–3, *4*
definition, 4–5
Earth-centered, 4, 60, 137–139, *139*
expansion of, 5–6, **32–34**, *33*, t 34
observable, 9
Ptolemaic model, 101–103, *103*
Upsilon Andromedae, 218, *220*
Uranium, 217
Uranus. *See also* Jovian planets
colors, 298, *299*
interior, 292–293, *293*
as jovian planet, 199
magnetosphere, 301–302, *302*
moons, 303, 314, *314*, A-17
orbital period, *137*
overview, 20, *20*
photographs, 288
properties, t 200, t 289, A-15
rings, 318, *319*, 320
rotation and tilt, 291
temperature, 298
weather, 299
Ursids meteor shower, t 344
UV radiation. *See* Ultraviolet (UV) radiation

Valles Marineris (Mars), 16, 250–251, *251*, 252
Van Allen belts, 270
Vega
celestial coordinates, 76, *77*, 85
Doppler shifts, 167
future North Star, 30, *30*
properties, A-19
Velocity. *See also* Speed
constant, 126, 129
definition, 126
escape (*See* Escape velocity)
gas, 280
orbital, *129*, 136, A-3, A-22
radial, A-21
star, 218
Venus
atmosphere, 260, t 261, 269, **269–270**, 280, 283–284
clouds, 276, 276–277
Coriolis effect, 274
Galileo's observation of phases, 139, *139*
giant impacts, 215
greenhouse effect, 14, 362–363, 371
interior structures, 230, 235
lack of plate tectonics, 363
overview, 14, *14*
ozone depletion, 373
properties, t 200, A-15
surface geology, 226, *226*, 228, **253–254**, *254*, 353
surface temperature, t 264
as terrestrial planet, 199
viewed from Earth, 59, 68–69
water on, 283–284, 284n, 362
wind patterns, 273

Vernal (spring) equinox, 28, *29*, 72–74, 76–77, 81–84
Very Large Array (VLA), 188, *188*
Very Large Telescope (Chile), t 178
Vespucci, Amerigo, 86
Vesta, 329, 333
Viking spacecraft, t 190, 250, 252, 375–376
Virgo Cluster, A-22
Viruses, 380
Viscosity, molten rocks, 229, 241, *242*
Visible light, 158, *185*, 267
Vision, 155, 172, **172–173**
Volatiles, atmospheric, 262
Volatiles, rock, 245, 247–248
Volcanic (hydrothermal) vents, 368, 375
Volcanic plains, 241
Volcanoes
albedo effects, 277–278, 372
asteroids, 333
atmosphere formation, 278, 281
Earth, 354
jovian moons, 304–305, *306*, 313
Mars, 242, 250–251, 283
Moon, 247, 247–248
planetary cooling, 233
planetary size and, 255
plate tectonics and, 358
shield, *241*, *242*, 248, 254, 307, 361
stratovolcanoes, 242, *242*
surface shaping, 235–236, *241*, **241–242**, *242*, *243*, 246
tectonics, *243*, *245*
Venus, 253
Volcanoes National Park (Hawaii), 279
Voyager 1 spacecraft, 19, t 190
Voyager 2 spacecraft
mission data, 24, t 190
photographs from, *20*, 288
rings, 317–318
trajectory, *189*
weather detection, 299–300
Voyage scale model solar system, 10, A-25

Waning moon, *52*
Water
circulation by hemisphere, 274
condensation, *207*, 207–208
density, 115
on Earth, 352, 362–363
erosion, 354, *354*
as greenhouse gas, 268
hydrologic cycle, *276*, 276–277
hydrosphere, 362–363
on Jupiter, 294, 298
on Mars, **251–253**, 277, 282–283
on the Moon, 281
origin of, 213, 278–279
phases of, 119–120, *120*
on Venus, **283–284**, 284n, 362
Water clocks, 94
Watts (units), 154
Wavelength
atmospheric penetration, *185*, **265**, 265, 267

Wavelength (*continued*)
  frequency, energy, and, 156, **157**, *157*, A-3
  rest, *166*, 167
Waves
  electromagnetic, 157, *157*
  properties, 156, *156*
  seismic, 226, *231*, **231**
  sound, 165, *166*
  wave-particle duality, 155–156
  wave speed, 156
Waxing moon, 52
Weather
  climate, 270–271
  clouds and precipitation, 276–277
  definition, 270
  global wind patterns, 272–276
  inversions, 268
  jovian planets, 295–299, *297*
  prediction, 92, *92–93*, 272
  seasonal patterns, 271–272, *273*
  storms, 274, *274*, 276

Web sites, astronomical, A-23 to A-25
Wegener, Alfred, 354, 356
Weight, 118, 127, 127n, *128*, 130. *See also* Masses
Weightlessness, 127–128
Wien's law, 162, A-3
Winds, 272–276, 296–297, 299
Winter solstice, 29, *29*, 74, 82–84
Winter Triangle, *40*
Wolf 359, A-18
2000WR106, 338

Xenon-129, 218
X-ray binaries, 181
X-Ray Multi-Mirror Mission (XMM), t 186
X rays
  atmospheric penetration, **265**, *265*, 268–269
  imaging, *181*, 181–182
  penetration through the atmosphere, 185
  on photographic film, 159
  telescopes, *181*, t 186, **186–187**, *187*
  wavelengths of, 158, *158*
X-ray vision, 159

Yellowstone National Park, 361–362, 368, *369*
Yerkes Observatory (Chicago), *176*
Yogi (rock), *16*
Yosemite Valley (California), 355
Yucatán crater, 345, *346*

Zenith, 44, *44*
Zodiac, 48, *49*
Zones, rising air, 295–296, *296*